Valorization of Solid Wastes to Biofuels and Chemical Products for Sustainable World

Abha Kumari • Monika Prakash Rai
Ashokkumar Veeramuthu • Abha Mishra
Editors

Valorization of Solid Wastes to Biofuels and Chemical Products for Sustainable World

Springer

Editors
Abha Kumari
Amity Institute of Biotechnology
Amity University
Noida, Uttar Pradesh, India

Ashokkumar Veeramuthu
Center for Waste Management
and Renewable Energy
Saveetha Institute of Medical and Technical Sciences
Saveetha University
Chennai, Tamil Nadu, India

Monika Prakash Rai
Department of Biotechnology
Motilal Nehru National Institute of Technology
Prayagraj, Uttar Pradesh, India

Abha Mishra
School of Biochemical Engineering
Indian Institute of Technology (BHU)
Varanasi, Uttar Pradesh, India

ISBN 978-981-96-8593-6 ISBN 978-981-96-8594-3 (eBook)
https://doi.org/10.1007/978-981-96-8594-3

This Springer imprint is published by the registered company Springer Nature Singapore Pte Ltd.
The registered company address is: 152 Beach Road, #21-01/04 Gateway East, Singapore 189721, Singapore

Preface

The escalating global waste crisis, fuelled by rapid industrialization, urbanization, and agricultural expansion, generates vast quantities of solid waste, posing severe environmental, economic, and health challenges. Traditional disposal methods like landfilling and incineration exacerbate pollution, land degradation, and greenhouse gas emissions, threatening ecosystems and human well-being. Yet, within this challenge lies an opportunity: transforming municipal, agricultural, and industrial solid waste into valuable biofuels like bioethanol and biogas, and high-value chemicals such as bioplastics and organic acids. This book, *Valorization of Solid Wastes to Biofuels and Chemical Products for a Sustainable World*, presents a transformative vision for sustainable waste management, leveraging advanced thermochemical and biochemical technologies to convert waste into renewable resources. By reducing carbon footprints, promoting green innovation, and fostering resource circularity, this approach aligns with global sustainability goals, offering a pathway to a cleaner, more resilient future.

This book explores the potential of municipal solid waste (MSW), including food, fruit, vegetable, paper, packaging, and plastic waste, as feedstocks for biofuels and chemicals through processes like anaerobic digestion, pyrolysis, and enzymatic hydrolysis. It examines agricultural solid waste (ASW), such as wheat straw, rice husk, corn stover, sawdust, forestry waste, millet, and pulse residues, highlighting their conversion into bioethanol, bio-oil, and platform chemicals like furfural via fermentation and torrefaction. Industrial solid waste (ISW) from sectors like petrochemicals, textiles, pharmaceuticals, sugar, dairy, leather, and pulp/paper is also addressed, with technologies like gasification and microbial conversion transforming wastes like bagasse and lignin into syngas and biopolymers. The environmental sustainability of these processes is rigorously evaluated through life cycle assessment (LCA), quantifying impacts like carbon emissions and energy use and comparing valorization to traditional disposal methods. This book concludes by illustrating how waste valorization advances a circular bioeconomy, enhancing resource efficiency and reducing fossil fuel dependency through policy incentives, digital tools like AI-driven waste sorting, and global collaboration.

This work is a collective effort by experts from institutions worldwide, including Amity University, Delhi University, IIT BHU, IARI, Nottingham Trent University, UK, North Carolina State University, Chungnam National University, and others in regions like India, Brazil, the USA, South Korea, and Malaysia, bringing expertise in waste-to-resource technologies, microbial processes, and environmental assessments to ensure a comprehensive analysis. This book is organized into five interconnected sections that collectively advance sustainable waste management through valorization. The first section explores the transformation of municipal solid waste, such as food, plastic, and packaging, into biofuels and high-value chemicals using green and sustainable technologies, addressing urban waste challenges. Building on this, the second section focuses on agricultural solid waste, converting residues like wheat straw and rice husk into biogas, bioethanol, and furfural through biochemical processes, emphasizing rural resource potential. The third section examines industrial solid waste from sectors like petrochemicals, pharma, and pulp/paper, producing syngas and biopolymers via gasification to mitigate industrial impacts. The fourth section employs life cycle assessment to quantify the environmental benefits of these processes against traditional disposal methods, ensuring sustainability. Finally, the fifth section integrates these efforts into a circular bioeconomy framework, promoting resource efficiency and advocating policy incentives like carbon credits to drive global adoption.

Spanning approximately 600 pages, this book integrates case studies, infographics, and a glossary to make complex concepts accessible to researchers, industry professionals, and policymakers. It offers actionable insights for developing scalable biorefineries, implementing sustainable policies, and advancing a circular bioeconomy. By bridging science, industry, and policy, this book aims to inspire transformative solutions for a sustainable, resource-efficient world, addressing the urgent need to convert waste into wealth for a sustainable world and future generations.

Noida, Uttar Pradesh, India — Abha Kumari
Prayagraj, Uttar Pradesh, India — Monika Prakash Rai
Chennai, Tamil Nadu, India — Ashokkumar Veeramuthu
Varanasi, Uttar Pradesh, India — Abha Mishra

Acknowledgements

First and foremost, I am indebted to the Almighty God, whose unwavering grace and blessings, as well as His ever-present presence, have inspired me to complete this book. I take this opportunity to acknowledge and extend my sincere gratitude to all who made this book a possibility. The encouraging support of numerous people contributed in many ways to this book and made it an unforgettable experience for me. It has been a long journey for editors, authors, and publishers from the basic outline to whole chapters and ultimately to a book. I take this opportunity to acknowledge and extend my sincere gratitude to all authors who made this book possible. I appreciate the authors who continued to support and accomplish their task efficiently and contribute to the chapters. My sincere thanks, gratitude, and appreciation to Dr. Monika Prakash Rai, Dr. Ashokkumar Veeramuthu, and Dr. Abha Mishra for their invaluable assistance and support in editing, invaluable guidance, encouragement and expertise.

We thank the team at Springer Publishers for their support; without their support, it would not have been possible to publish this book.

I thank and express deep gratitude to the Indian Institute of Science, Bangalore, and eminent professors with whom I worked for imparting extensive knowledge in advanced science and engineering and instilling in-depth understanding in me. I would also like to acknowledge the Department of Biotechnology, Government of India, for funding projects on waste. I take this opportunity to acknowledge and express my deep gratitude to the Founder President, the Chancellor, and the Vice Chancellor of the Amity Group of Institutions for their continuous encouragement and support.

My special thanks to Dr. Manju for assisting with editing the chapters, as well as for her persuasion, discussion, continuous encouragement, and constant support. My special thanks to my team, Dr. Meenu, Surbhi, Anshika, Prisha, and Nishika, for their assistance in editing, their continuous support, and their sincere endeavour.

I dedicate this book to my late father, Prof. Devendra Prasad, and my late mother, Mrs. Usha Prasad, who planted the seed of knowledge, humility, honesty, and values in me and nurtured it, taught me to be an independent and determined person, and poured hearts and souls into us. They inspired me in shaping this book. My special thanks to my brothers and sisters for their affection, unwavering source of strength, inspiration, support, and constant encouragement. I thank my brother-in-laws for their support and constant encouragement. I thank my nephew Keshav and nieces Stuti, Chitranshi, and Smriti for bringing joy with their smiles and laughter. My special thanks to my friends for their endearment, constant encouragement, and support.

This herculean task could not have cascaded itself into such a refined work without the help of my friends. I am indebted to my friends Alka, Shivani, Bharat, Manoj, Arvind, Sarwa, Chandraprabha, Kumud, Aruna, and Sarika, for their endearment, constant encouragement, and support throughout.

Abha Kumari

I feel grateful to Prof. R. S. Verma, Director, Motilal Nehru National Institute of Technology (MNNIT) Allahabad for providing me the required facilities and support for the conduct of the assignment. I thank all the co-editors of this book for the insightful discussions, suggestions, and successful teamwork. I want to express my gratitude to my family, friends, and colleagues for providing emotional support and encouragement during the writing process.

Monika Prakash Rai

I sincerely thank the Springer Nature team for the opportunity to work with them in editing this book. I also express my deep gratitude to the Centre for Waste Management and Renewable Energy, SIMATS, for their constant support in my research and in bringing out this book. Special thanks to my research scholars, Ms. Agila, and my colleague, Dr. Flora for their assistance in editing the chapters.

Ashokumar Veeramuthu

I express my deepest gratitude to the Baba Kashi Vishwanath and to our founder Pt. Madan Mohan Malviya Ji, whose blessings have bestowed upon me the gift of education, knowledge, and wisdom. I am forever indebted to my parents, my father Prof. Krishna Dutt Mishra (Plasma Physicist) and mother Ms. Gayatri Mishra who instilled in me the values of honesty, hard work, and sincerity and taught me the dignity of labour in any task undertaken. My heartfelt thanks to my daughter, Alnkrita Singh, whose presence has opened a new dimension in my life. Her constant support, insight, and friendship have been a guiding light through both personal and professional challenges. I am equally grateful to my husband, Prof. Amit Singh, who has stood by me as a strong and unwavering pillar of strength. To all my students, past and present, I extend my sincere appreciation. Their inquisitive minds and collaborative spirit have enriched my journey, often leading me to explore new perspectives and learn alongside them. I owe a profound sense of gratitude to every individual who, in some way, has contributed to shaping me both personally and professionally into who I am today.

Abha Mishra

Contents

The original version of the book has been revised. A correction to this book can be found at https://doi.org/10.1007/978-981-96-8594-3_35

Editors and Contributors

About the Editors

Abha Kumari is currently working as an Associate Professor at the Centre for Biotechnology and Biochemical Engineering, Amity Institute of Biotechnology, Amity University, Noida, India. She has earlier served as an Associate Professor in the Department of Biotechnology at Delhi Technological University, Delhi, India, Visiting Researcher at the Tohoku Institute of Technology, Sendai, Japan, DBT Postdoctoral Fellow at National Chemical Laboratory, Pune, and Postdoctoral Fellow at Lakehead University, Thunder Bay, Ontario, Canada. She was awarded a Pool Officer fellowship from the Council of Scientific Industrial Research, India. Dr. Abha earned her Ph.D. and M.Sc. (Engg.) from the Indian Institute of Science (IISc), Bangalore, India. Her research interest is in the field of biochemical engineering and biotechnology with a specialization in waste-to-value, municipal solid waste-to-biogas and carotenoid production, their application and biorefinery, bio-enzyme from waste, waste management, bioremediation, industrial effluent treatment using photochemical techniques, and metals recovery from waste. She has more than 20 years of research and teaching experience in the field of bioprocess technology, biochemical engineering, downstream processing, enzyme technology, and principles of chemical engineering. She has published research articles in peer-reviewed international journals and authored numerous book chapters. She has worked as a Principal Investigator on two government-funded projects and as a Co-investigator on one government-funded project. She is a life member of The Biotech Research Society, India (BRSI) India, Indian Institute of Metals (IIM) India, and the Indian Science Congress Association.

Monika Prakash Rai is working as Associate Professor at Department of Biotechnology, Motilal Nehru National Institute of Technology Allahabad (MNNIT Allahabad), Prayagraj, India. She has earlier served as Professor in the Amity Institute of Biotechnology at Amity University Uttar Pradesh, Noida, India (2008-2023), Visiting Professor at University of Nottingham, UK (2011), and Assistant Professor at IILM-AHL, Greater Noida, India (2006–2008). She did her Ph.D. from School of Biochemical Engineering, Indian Institute of Technology (BHU) Varanasi, India, in collaboration with the Department of Biochemistry and Pharmacy, Johannes Gutenberg University, Mainz, Germany, in the year 2005. Her research interest are in bioprocess, biofuels, bioremediation, waste management, nanobiotechnology, and industrial biotechnology with specialization in algal cultivation, phycoremediation, and algal-based bioproducts for industrial applications and biorefinery. She has received Distinguished Indian Visiting Fellowship and Travel grant from University of Nottingham, UK, for collaborative research in the year 2011 and 2019. She has more than 18 years of teaching experience in Biochemistry, Enzyme Technology, Bioprocess and Biotechnology, Downstream Processing, and Green Technology. She has published more than 50 research articles in the peer-reviewed international journals and authored more than 20 books and book chapters. She has worked as Principal Investigator on 4 Government-funded projects and filed more than 10 patents in Indian Patent office.

Ashokkumar Veeramuthu earned his Ph.D. from the University of Madras, Chennai, India, in 2013. During this period, he was honored with an Indo-Australian fellowship, which led to his role as a visiting scientist at Adelaide University, Australia. His academic journey also includes obtaining M. Phil and M.Sc. degrees in Plant Biotechnology from Indian universities. Currently serving as Professor and Head at the Center for Waste Management and Renewable Energy within Saveetha Institute of Medical and Technical Sciences (SIMATS), Saveetha University, Chennai, India, Dr. Ashokkumar is a Visiting Professor at Chulalongkorn University, Thailand, and the National University of Singapore (NUS), Singapore. Previously, he worked as a Postdoctoral Scientist in Microalgal Biorefinery at the Department of Mechanical Engineering, University Technology Malaysia (UTM), Malaysia, and as a Postdoctoral Scientist at the Department of Aeronautics and Astronautics, National Cheng Kung University, Taiwan. His international exposure includes being a visiting scientist at European Universities under the Marie Skłodowska-Curie Actions Fellowship scheme. With over 15 years of research experience in energy and environment, he has contributed significantly to both academic and industrial fields. His expertise covers a broad spectrum, including value-added bioproducts, biomass thermochemical techniques, torrefaction, biochar synthesis, bio-jet fuels, and metallic iron production. Dr. Ashok also specializes in biofuels, biorefinery, wastewater treatment using microalgae, and CO2 sequestration. He has delivered over 90 invited talks at various institutions and conferences. His impressive publication record includes more than 115 peer-reviewed international journal articles and 7 book chapters.

Abha Mishra is a Professor of School of Biochemical Engineering, Indian Institute of Technology (BHU), Varanasi, India. She has received her M. Tech and Ph.D. degree in Biochemical Engineering, from IIT (BHU). She has 22 years of teaching and research experience in the field of biochemistry, structural biology, bioinformatics, fermentation technology, molecular biology and genetic engineering, and intellectual property rights in biotechnology. Her research interest is in the area of bioprocess technology, drug discovery, and drug delivery techniques. She is fellow of Institution of Engineers. She has supervised 8 Ph.D. and Co-supervised 3 Ph.D. students in Biochemical Engineering. Currently, 5 Ph.D. students under supervision and 06 co-supervision students are working with her. She has supervised 52 master's degree students. She has more than 97 papers as corresponding author in various internationally acclaimed peer-reviewed journals. She has authored more than 12 book chapters. She is life member of Biological Engineering Society, Biotechnology Research Society of India, and Indian Science Congress, AICTE society.

Contributors

Kavya Adlakha Amity Institute of Biotechnology, Amity University, Noida, Uttar Pradesh, India

Nikki Agrawal Atal Bihari Vajpayee Vishwavidyalaya, Bilaspur, Chhattisgarh, India

Vajiha Aafrin Basheer Department of Biotechnology, Vel Tech Rangarajan Dr. Sagunthala R&D Institute of Science and Technology, Chennai, Tamil Nadu, India

Centre for Food Technology, Anna University, Chennai, Tamil Nadu, India

Jagadeesh Chandra Bose University Institute of Biotechnology, Chandigarh University, Mohali, Punjab, India

Rajasekhar Reddy Busigari Department of Fuel, Mineral and Metallurgical Engineering, Indian Institute of Technology (Indian School of Mines), Dhanbad, Jharkhand, India

V. P. Chandramughi Center for Waste Management and Renewable Energy, Saveetha Dental College and Hospitals, Saveetha Institute of Medical and Technical Sciences, Saveetha University, Chennai, Tamil Nadu, India

D. P. Chaudhary ICAR-Indian Institute of Maize Research, Ludhiana, Punjab, India

Vaishali Chaudhary Department of Biotechnology, Motilal Nehru National Institute of Technology Allahabad, Prayagraj, Uttar Pradesh, India

Savita Chaurasia Biochemistry and Molecular Biology Program, Department of Chemistry, Bellarmine University, Louisville, KY, USA

Anshika Dedha Amity Institute of Biotechnology, Amity University, Noida, Uttar Pradesh, India

Anurag Deepak Department of Mechanical Engineering, Motilal Nehru National Institute of Technology Allahabad, Prayagraj, Uttar Pradesh, India

Ravi K. Deshwal Institute of Bioscience and Technology Sri Ramswaroop Memorial University, Lucknow, Uttar Pradesh, India

Neeraj Dwivedi Council of Industrial Innovation and Research, Noida, Uttar Pradesh, India

Padmanabh Dwivedi Department of Plant Physiology, Institute of Agricultural Sciences, Banaras Hindu University, Varanasi, Uttar Pradesh, India

Surya Prakash Dwivedi Council of Industrial Innovation and Research, Noida, Uttar Pradesh, India

Agila Elango Department of Biochemistry, Saveetha Dental College and Hospitals, Saveetha Institute of Medical & Technical Sciences, Saveetha University, Chennai, Tamil Nadu, India

G. Flora Department of Botany, St. Mary's College (Autonomous), Thoothukudi, Tamil Nadu, India

Sanghdeep Gautam Department of Biosciences, IMS Ghaziabad, University Courses Campus, Ghaziabad, Uttar Pradesh, India

Sushant Gawali Amity Institute of Biotechnology, Amity University, Mumbai, Maharashtra, India

Department of Chemical Engineering, State University of New York College of Environmental Science and Forestry, Syracuse, NY, USA

Ronalds Gonzalez Conversion Economics and Sustainability, Department of Forest Biomaterials, North Carolina State University, Raleigh, NC, USA

Abhishek Guldhe Amity Institute of Biotechnology, Amity University, Mumbai, Maharashtra, India

Mamta Gupta ICAR-Indian Institute of Maize Research, Ludhiana, Punjab, India

Manju M. Gupta Department of Botany, Sri Aurobindo College, University of Delhi, New Delhi, India

Amity Institute of Biotechnology, Amity University, Noida, Uttar Pradesh, India

Prisha Gupta Amity Institute of Biotechnology, Amity University, Noida, Uttar Pradesh, India

Priya Gupta Amity Institute of Biotechnology, Amity University Chhattisgarh, Raipur, Chhattisgarh, India

Zeeshan Hussain Hashmi Amity Law School, Amity University, Jaipur, Rajasthan, India

Bhawna Jain Amity Institute of Biotechnology, Amity University, Noida, Uttar Pradesh, India

Vivek Jaiswal Department of Textile and Fibre Engineering, Indian Institute of Technology Delhi, New Delhi, India

Hasan Jameel Conversion Economics and Sustainability, Department of Forest Biomaterials, North Carolina State University, Raleigh, NC, USA

H. S. Jat ICAR-Indian Institute of Maize Research, Ludhiana, Punjab, India

Naveen Kumar Karri Department of Chemical Engineering, Indian Institute of Petroleum and Energy, Visakhapatnam, Andhra Pradesh, India

Rajeev Kaushik Division of Microbiology, ICAR-Indian Agricultural Research Institute, New Delhi, India

Anwesha Khanra Department of Biotechnology, Parul Institute of Technology, Parul University, Vadodara, Gujarat, India

Mohammed Ali Khasawneh Department of Civil Engineering, Prince Mohammad Bin Fahd University, Al Khobar, Kingdom of Saudi Arabia

Minsik Kim Department of Biological Engineering, Inha University, Incheon, Republic of Korea

Abha Kumari Department of Botany, Sri Aurobindo College, University of Delhi, New Delhi, India

Amity Institute of Biotechnology, Amity University, Noida, Uttar Pradesh, India

Manish Kumar Department of Plastic Technology, Harcourt Butler Technical University, Kanpur, Uttar Pradesh, India

Manoj Kumar Department of Mechanical Engineering, ABES Engineering College, Ghaziabad, Uttar Pradesh, India

Prasann Kumar Department of Agronomy, School of Agriculture, Lovely Professional University, Phagwara, Punjab, India

Sumit Kumar Amity Institute of Biotechnology, Amity University, Noida, Uttar Pradesh, India

Sunil Kumar ICAR-Indian Institute of Maize Research, Ludhiana, Punjab, India

A. Lakshmi Department of Chemistry, St. Mary's College (Autonomous), Thoothukudi, Tamil Nadu, India

Abhishek Kumar Lal Department of Leather Technology, School of Chemical Technology, HBTU, Kanpur, Uttar Pradesh, India

Sarika Loni Amity Institute of Biotechnology, Amity University, Mumbai, Maharashtra, India

Organic Recycling Systems Ltd., Mumbai, India

Ronald Marquez Conversion Economics and Sustainability, Department of Forest Biomaterials, North Carolina State University, Raleigh, NC, USA

Yukta Menon Department of Biotechnology, Jaypee Institute of Information Technology, Noida, Uttar Pradesh, India

Abha Mishra Biomolecular Engineering Laboratory, School of Biochemical Engineering, Indian Institute of Technology (BHU), Varanasi, Uttar Pradesh, India

Swechchha Mishra Department of Biotechnology, Kanpur Institute of Technology, Kanpur, Uttar Pradesh, India

Nayeemuddin Mohammed Department of Civil Engineering, Prince Mohammad Bin Fahd University, Al Khobar, Kingdom of Saudi Arabia

Ranjan Kumar Mohapatra Department of Environmental and IT Convergence Engineering, Chungnam National University, Daejeon, South Korea

Simran Neeraj Department of Biotechnology, Jaypee Institute of Information Technology, Noida, Uttar Pradesh, India

Anshul Nigam Department of Biotechnology, Kanpur Institute of Technology, Kanpur, Uttar Pradesh, India

Azhagu Saravana Babu Packirisamy Department of Biotechnology, Vel Tech Rangarajan Dr. Sagunthala R&D Institute of Science and Technology, Chennai, Tamil Nadu, India

Dibyarupa Pal Department of Biosciences, JIS University, Agarpara, Kolkata, West Bengal, India

Venkata Chandra Sekhar Palla Material Resource Efficiency Division (MRED), CSIR-Indian Institute of Petroleum (IIP), Dehradun, Uttarakhand, India

Sudheer D. V. N. Pamidimarri Discipline of Industrial Biotechnology, Gujarat Biotechnology University, Gandhinagar, Gujarat, India

Siddharth Pandey Department of Chemical Engineering, Parul Institute of Technology, Parul University, Vadodara, Gujarat, India

Disha Patra Amity Institute of Biotechnology, Amity University, Noida, Uttar Pradesh, India

Aryan Prajapati Amity Institute of Biotechnology, Amity University, Noida, Uttar Pradesh, India

Dev Pramanick Department of Biotechnology, Kanpur Institute of Technology, Kanpur, Uttar Pradesh, India

V. Vishnu Priya Department of Biochemistry, Saveetha Dental College and Hospitals, Saveetha Institute of Medical & Technical Sciences, Saveetha University, Chennai, Tamil Nadu, India

Monika Prakash Rai Department of Biotechnology, Motilal Nehru National Institute of Technology Allahabad, Prayagraj, Uttar Pradesh, India

Meenal Rastogi CSIR-Institute of Microbial Technology, Chandigarh, Punjab, India

Shruti S. Raut Biomolecular Engineering Laboratory, School of Biochemical Engineering, Indian Institute of Technology (BHU), Varanasi, Uttar Pradesh, India

Sai Nandhini Ravi Department of Biotechnology, Vel Tech Rangarajan Dr. Sagunthala R&D Institute of Science and Technology, Chennai, Tamil Nadu, India

Ayesha Sadaf Interdisciplinary Biomedical Research Centre, School of Science and Technology, Nottingham Trent University, Nottingham, UK

Sagish Sagia Division of Microbiology, ICAR-Indian Agricultural Research Institute, New Delhi, India

Biswajit Saha Amity Institute of Biotechnology, Amity University, Noida, Uttar Pradesh, India

Rishika Saha Amity Institute of Biotechnology, Amity University, Noida, Uttar Pradesh, India

Vishal Kumar Sandhwar Department of Chemical Engineering, Parul Institute of Technology, Parul University, Vadodara, Gujarat, India

Sudipto Sarkar Department of Agricultural Engineering, Assam University, Silchar, Assam, India

Diksha Saxena Department of Chemical Engineering, Parul Institute of Technology, Parul University, Vadodara, Gujarat, India

Shivendu Saxena Department of Chemical Engineering, Parul Institute of Technology, Parul University, Vadodara, Gujarat, India

Silambarasan Tamil Selvan Helix Research Center, Department of General Medicine, School of Allied Health Sciences, Vinayaka Missions Research Foundation (Deemed University), Chennai, Tamil Nadu, India

Feroz Shaik Department of Mechanical Engineering, Prince Mohammad Bin Fahd University, Al Khobar, Kingdom of Saudi Arabia

Anirudh Sharma Department of Biotechnology, Jaypee Institute of Information Technology, Noida, Uttar Pradesh, India

Arpit Sharma Biomolecular Engineering Laboratory, School of Biochemical Engineering, Indian Institute of Technology (BHU), Varanasi, Uttar Pradesh, India

Department of Pharmacology, Institute of Medical Science-Banaras Hindu University (IMS-BHU), Varanasi, Uttar Pradesh, India

Nishika Sharma Amity Institute of Biotechnology, Amity University, Noida, Uttar Pradesh, India

Pushpendra Sharma Division of Microbiology, ICAR-Indian Agricultural Research Institute, New Delhi, India

Surbhi Sharma Amity Institute of Biotechnology, Amity University, Noida, Uttar Pradesh, India

Smriti Shrivastava Amity Institute of Biotechnology, Amity University, Noida, Uttar Pradesh, India

Priya Shukla School of Biochemical Engineering, Indian Institute of Technology (Banaras Hindu University), Varanasi, Uttar Pradesh, India

Aditya Singh Department of Biotechnology, Kanpur Institute of Technology, Kanpur, Uttar Pradesh, India

Akhilesh Kumar Singh Department of Biotechnology, School of Life Sciences, Mahatma Gandhi Central University, Motihari, Bihar, India

Alla Singh ICAR-Indian Institute of Maize Research, Ludhiana, Punjab, India

Chandra Jeet Singh Department of Textile and Fibre Engineering, Indian Institute of Technology Delhi, New Delhi, India

Gurpinder Singh ICAR-Indian Institute of Maize Research, Ludhiana, Punjab, India

Ravi Kant Singh Amity Institute of Biotechnology, Amity University, Noida, Uttar Pradesh, India

Rishi Kumar Singh Department of Biosciences, IMS Ghaziabad, University Courses Campus, Ghaziabad, Uttar Pradesh, India

Sangeeta Devendra Kumar Singh Amity Institute of Biotechnology, Amity University Chhattisgarh, Raipur, Chhattisgarh, India

Shweta Singh Council of Industrial Innovation and Research, Noida, Uttar Pradesh, India

Sujeet Pratap Singh Amity Institute of Biotechnology, Amity University Lucknow Campus, Lucknow, Uttar Pradesh, India

Sujit Pratap Singh Amity Institute of Biotechnology, Amity University, Lucknow, Uttar Pradesh, India

Nalini Soni Amity Institute of Biotechnology, Amity University Chhattisgarh, Raipur, Chhattisgarh, India

Irudaya Antonat Sophia Department of Chemistry, St. Mary's College (Autonomous), Thoothukudi, Tamil Nadu, India

Vithyasree Srikanth Department of Biotechnology, Vel Tech Rangarajan Dr. Sagunthala R&D Institute of Science and Technology, Chennai, Tamil Nadu, India

Mayuri Srivastava School of Interdisciplinary Research, Indian Institute of Technology Delhi, New Delhi, India

Pradeep Srivastava School of Biochemical Engineering, Indian Institute of Technology (Banaras Hindu University), Varanasi, Uttar Pradesh, India

Tasneem Sultana Artificial Intelligence and Machine Learning, Godutai Engineering College for Women, Sharnbasva University, Kalaburagi, India

Hemanth Kumar Tanneru Department of Chemical Engineering, Indian Institute of Petroleum and Energy, Visakhapatnam, Andhra Pradesh, India

Manju D. Tanwar Organic Recycling Systems Ltd., Mumbai, India

Alok Tiwari Department of Chemical Engineering, Parul Institute of Technology, Parul University, Vadodara, Gujarat, India

Fernando Urdaneta Conversion Economics and Sustainability, Department of Forest Biomaterials, North Carolina State University, Raleigh, NC, USA

Manasvi Vashisht Amity Institute of Biotechnology, Amity University, Noida, Uttar Pradesh, India

Shrestha Vashista Amity Institute of Biotechnology, Amity University, Noida, Uttar Pradesh, India

Ashokkumar Veeramuthu Center for Waste Management and Renewable Energy, Saveetha Dental College and Hospitals, Saveetha Institute of Medical and Technical Sciences, Saveetha University, Chennai, Tamil Nadu, India

Balasubramanian Velramar Amity Institute of Biotechnology, Amity University Chhattisgarh, Raipur, Chhattisgarh, India

Ramon E. Vera Conversion Economics and Sustainability, Department of Forest Biomaterials, North Carolina State University, Raleigh, NC, USA

Alisha Verma Symbiosis Law School, Pune, Maharashtra, India

Ananya Verma Amity Institute of Biotechnology, Amity University, Mumbai, Maharashtra, India

Keren A. Vivas Conversion Economics and Sustainability, Department of Forest Biomaterials, North Carolina State University, Raleigh, NC, USA

Purnima Yadav Amity Institute of Biotechnology, Amity University, Noida, Uttar Pradesh, India

Renu Yadav NorthCap University, Gurugram, Haryana, India

Zia Zehra Zaidi GNIOT Institute of Management Studies (GIMS), Greater Noida, Uttar Pradesh, India

Part I

Municipal Solid Waste to Biofuels and Chemical Products

Municipal Solid Waste as a Feedstock for Biofuels and Chemicals Production

1

Surbhi Sharma, Anshika Dedha, Manju M. Gupta, Purnima Yadav, and Abha Kumari

Abstract

The demand for sustainable waste management systems solutions has increased due to the expanding population and industrialization. This chapter explores the production of biofuels and chemicals from various types of municipal solid waste (MSW), focusing on food, plastic, pulp and paper, and poultry waste. Anaerobic digestion and gasification are two methods that can be used to turn food waste, which makes up a large amount of household waste, into biofuels like bioethanol and biodiesel. Plastic waste, which is abundant in today's world, can be converted by thermal and catalytic processes into valuable chemicals and fuels. The pulp and paper industry, which uses an extensive amount of water, can recycle its solid waste and wastewater through fermentation and enzymatic hydrolysis to produce biofuels and chemicals. The poultry industry generates a significant amount of waste, which can be converted into valuable fuels through processes such as anaerobic digestion, pyrolysis, and gasification, thereby enhancing waste management and sustainability. This thorough analysis emphasizes the potential of converting MSW into useful resources, promoting a circular economy, and reducing environmental impact.

Keywords

Lignocellulosic biomass · Pretreatment technologies · Sustainable waste management · Municipal solid waste (MSW) · Biofuels and biochemicals

S. Sharma · A. Dedha · P. Yadav · A. Kumari (✉)
Amity Institute of Biotechnology, Amity University, Noida, Uttar Pradesh, India
e-mail: akumari@amity.edu

M. M. Gupta
Department of Botany, University of Delhi, Sri Aurobindo College, Delhi, India

A. Kumari et al. (eds.), *Valorization of Solid Wastes to Biofuels and Chemical Products for Sustainable World*, https://doi.org/10.1007/978-981-96-8594-3_1

1.1 Introduction

Municipal solid waste (MSW) refers to the materials that are disposed of in urban areas, predominantly consisting of residential waste, but can also occasionally include commercial waste. Paper, food waste, yard and wood trimmings, cotton, leather, and other biomass-producing materials make up a significant portion of MSW. Furthermore, MSW contains materials derived from fossil fuels, such as textiles, rubber, and plastics. Waste-to-energy (WTE) processes recover energy from waste by either creating combustible fuels in the form of hydrogen, methane, or other synthetic fuels or by directly burning the waste (Kumari et al. 2022a). The two main WTE technologies that have been effectively applied globally are incineration and gasification.

The conversion of MSW to biofuels and chemicals is becoming significantly important at present. More efficient waste management techniques are required due to the rise in MSW brought about by the global population expansion and fast urbanization. Significant amounts of pollutants and greenhouse gases are released during traditional waste disposal processes like incineration and landfilling (Ali et al. 2020; Gupta et al. 2024). MSW can be converted into chemicals and biofuels to reduce its impact on the environment and burden on landfills. Additionally, in contrast to finite fossil fuels, biofuels made from MSW are a renewable energy source that helps mitigate climate change (Nanda and Berruti, 2021). The conversion process creates economic opportunities while transforming waste into a valuable resource. Another important benefit is increased energy security since domestic biofuel production reduces dependency on imported fossil fuels, which is important for nations without natural fossil fuel reserves (Kanwal et al. 2022). In keeping with the tenets of the circular economy, this strategy considers waste as a resource that can be recycled to encourage sustainability.

In Asian nations such as Bangladesh, India, and Nepal, managing solid waste is a major problem. India produces approximately 68.8 million tons of MSW annually, with 366 cities (housing 70% of the urban population) generating 47.2 million tons of this waste. If present trends continue, these cities are predicted to produce 160.96 million tons of MSW by 2041, a fivefold increase in just 40 years (Singh et al. 2024). In India, MSW generation has increased sharply, from 31 million tons in 2001 to 62 million tons in 2015, as reported by the Ministry of Housing and Urban Affairs (MoHUA) in 2016. This figure is projected to reach 165 million tons by 2030. Despite this rapid increase, waste management remains a significant challenge, with only 20% (27,000 tons/day) of the collected MSW is processed, while the remaining 80% (0.11 million tons/day) is disposed of untreated in landfills and water bodies, leading to severe environmental contamination (Ram and Kumar 2021; Siddiqua et al. 2022). This concerning forecast emphasizes the urgent need for proactive waste management strategies that emphasize eco-friendly methods and innovative solutions to mitigate the long-term negative effects of exponential waste generation on society and the environment.

This chapter explores the various kinds of MSW utilized to produce biofuels and chemicals while also analyzing the MSW composition and generation. By transforming materials such as food waste, paper, pulp, plastics, and poultry waste into different biofuels and chemicals such as ethanol, biodiesel, biogas, and industrial chemicals, MSW plays a significant role in promoting environmental sustainability and a circular economy. This approach reduces waste volume, produces chemical feedstock and renewable energy, and improves energy security by reducing dependency on fossil fuels.

1.2 Generation and Composition of MSW Around the World

The generation and composition of MSW vary globally due to differences in population density, urbanization, economic growth, cultural habits, waste treatment infrastructure, and legal frameworks. The global urban population rose from 751 million in 1950 to 4.2 billion in 2018, with cities now housing 55% of the world's population (Kadam and Thakur 2020). This trend is projected to continue, reaching 68% by 2050 (Gu et al. 2021; Sharma and Jain 2020). Developing countries such as India, Nigeria, and China are expected to add hundreds of millions of urban residents by mid-century, significantly increasing waste production (UN DESA 2018; Sharma and Jain 2020). Currently, the world generates over 2 billion tons of MSW annually, and this figure is projected to increase by 70%, reaching 3.88 million tons by 2050 (World Bank Group 2018). In 2020, the world produced an average of 0.79 kg of waste per person per day, with variations based on income level and geographic region. Income disparities significantly impact waste generation rates: high-income countries produce an average of 1.60 kg/person/day, upper-middle-income countries 0.91 kg, lower-middle-income countries 0.47 kg, and low-income countries 0.41 kg. Currently, the East Asia and Pacific region leads in waste production, followed by Europe and Central Asia. However, by 2050, South Asia and Sub-Saharan Africa are expected to surpass these regions in waste generation (World Bank Group 2018). Figures 1.1 and 1.2 illustrate the projected global MSW trends, highlighting the impact of economic disparities on waste generation patterns over time.

MSW is composed of both non-biodegradable and biodegradable components. However, its composition and classification may vary significantly across municipalities globally. For example, in high-income countries, MSW typically contains a

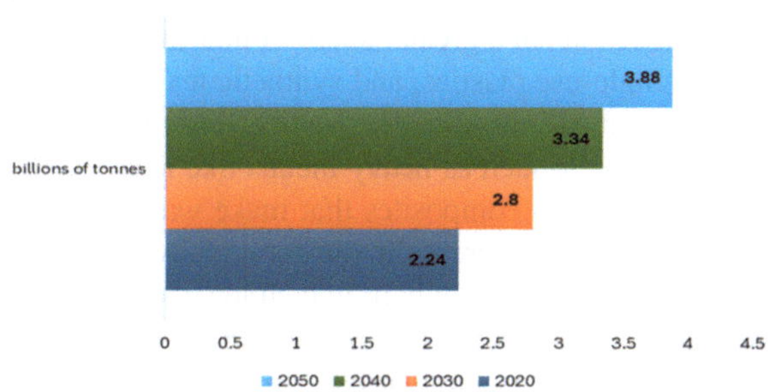

Fig. 1.1 Projected generation of waste on a global scale (World Bank Group 2018)

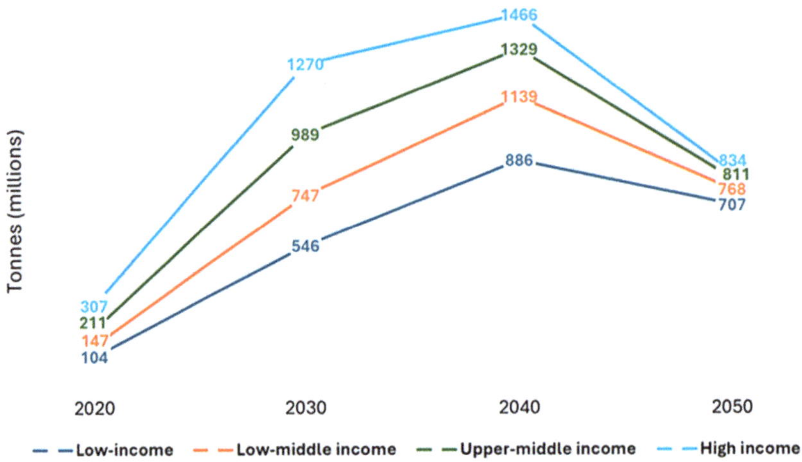

Fig. 1.2 Projected trends of waste generation based on global income disparities (World Bank Group 2018)

higher proportion of packaging materials, plastics, and electronic waste, while in low-income countries, organic waste such as food scraps, yard trimmings, and agricultural residues dominate. Additionally, some municipalities classify waste such as recyclables, compostables, and hazardous waste, whereas others use classifications, such as wet and dry waste.

Around the world, MSW is typically composed of around 44% organic matter (food and green waste), 12% plastic, 4% metal, 2% wood, 5% glass, 2% rubber and leather, 17% cardboard and paper, and 14% other materials (US-EPA 2016; World Bank Group 2018). Usually, MSW in prosperous countries comprises 25–35% organic waste, 25–35% paper and cardboard, 10–15% plastics, 5–10% glass, 5–10% metals, and lower amounts of textiles and miscellaneous waste. Over 50% of the MSW produced in middle- and low-income countries comprises of organic waste (Cayumil et al. 2021). A combination is seen in emerging economies, with 40–55% organic waste and rising levels of paper, plastics, and metals. The majority of waste produced by agricultural activities in rural areas (60–75%) is organic waste; packaging waste is negligible. Due to expanding consumerism and the spread of technology, there is a global trend toward an increase in plastic and electronic waste, which has a substantial influence on the composition of urban waste. The increase in plastic waste results in a greater share of non-biodegradable materials, including packaging, single-use plastics, and synthetic textiles, which are difficult to manage and recycle. In the same vein, the rise in electronic waste (e-waste) brings forth dangerous substances, such as heavy metals like lead, cadmium, and mercury, as well as plastic and metal composites that make waste sorting and disposal more difficult. This has led to an increase in the complexity of urban waste streams, as the proportion of materials that are hard to degrade and need specialized recycling methods is on the rise (Mihai et al. 2021). The overall proportion of components in MSW generated worldwide is summarized in Fig. 1.3.

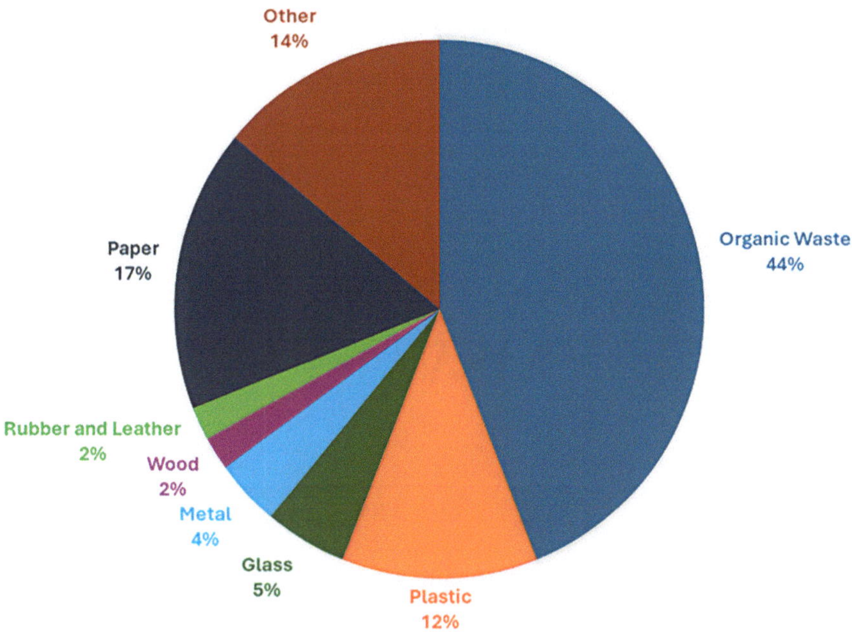

Fig. 1.3 Overall composition of municipal solid waste (MSW) (World Bank Group 2018)

1.3 Pretreatment of MSW for Their Conversion to Biofuels and Chemicals

MSW is conventionally turned into many valuable products, including biofuels (such as biodiesel, bioethanol, and biogas), platform chemicals (such as syngas, hydrogen, and methane), and fertilizers. This transformation is achieved through a range of technologies, including thermochemical processes (such as gasification, incineration, and pyrolysis) and physicochemical processes (transesterification and hydrolysis) (Nandhini et al. 2022). However, the efficiency of these technologies can be hindered by several obstacles. One major challenge is the presence of refractory lignocellulosic materials in MSW, which are difficult to ferment. Additionally, harmful inhibitory substances within MSW can further hinder the bioconversion process. Therefore, a suitable pre-treatment process is essential to break down these resistant materials and reduce inhibitors, ultimately improving bioconversion efficiency (Cesaro and Belgiorno 2014; Ebrahimian et al. 2023). The initial step of pretreatment is crucial to maximizing the output and efficiency of the conversion processes into chemicals and biofuels. The goal of this method is to improve the waste's conversion potential by altering its chemical, biological, and physical characteristics. There are several pre-treatment techniques used, each having unique advantages and disadvantages, including physical, chemical, physicochemical, biological, and mixed pretreatment techniques (Maurya et al. 2015). Figure 1.4 summarizes the various pre-treatment techniques employed for the conversion of MSW to biofuels and chemicals.

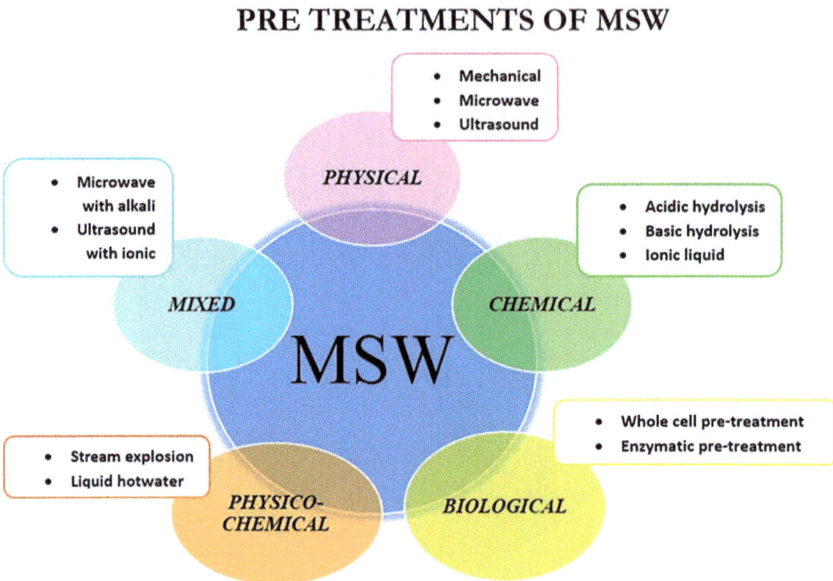

Fig. 1.4 Different pre-treatment techniques employed for the conversion of MSW to biofuels and chemicals

The purpose of physical pretreatment is to decrease the degree of polymerization and crystallinity of the lignocellulosic materials while increasing the surface area available to microorganisms. The three most often used methods are extrusion, milling, and microwave irradiation (Ebrahimian et al. 2023). By reducing the waste's particle size through shredding or grinding, more surface area is made accessible for microbial or chemical activity. Particle size reduction increases the effectiveness of anaerobic digestion and hydrolysis, two downstream processes. In addition, non-biodegradable elements like metals and plastics are sorted out of the waste using separation techniques including air classification, magnetic separation, and eddy current separation (Zhang et al. 2019; Kumari et al. 2022b). Chemical pretreatment entails using a variety of chemicals to break down complex polymers in the waste materials. Strong acids such as sulfuric acid are used in acid hydrolysis to break down cellulose and hemicellulose into sugars that can be fermented. Although this method is quite efficient, the acidic wastes must be handled and disposed of carefully. By solubilizing lignin and hemicellulose with bases like sodium hydroxide, alkaline pre-treatment increases cellulose's accessibility for enzymatic hydrolysis. Another approach is oxidative degradation, which employs agents, such as hydrogen peroxide, to break down lignin and increase the biomass digestibility. Although chemical pre-treatments can be costly and produce hazardous waste that needs to be controlled, they are effective (Chaturvedi and Verma 2013; Sołowski et al. 2020).

The term "physicochemical pretreatment" refers to a combination of chemical and physical techniques (Abraham et al. 2020). Usually, this method starts with mechanical size reduction, like shredding or grinding, to make the waste materials' surface area larger. Subsequently, complicated organic polymers are broken down by chemical treatments. This method commonly includes techniques such as hot water pretreatment and steam explosion (Ebrahimian et al. 2023). Additionally, the process of biological pretreatment incorporates the use of microorganisms or enzymes to facilitate enzymatic hydrolysis and break down hemicellulose and lignin. Although the operational costs are cheap and inhibitors are rarely generated, this pretreatment method takes a lot of time, thus, it is not the best option for commercialization (Andlar et al. 2018; Ebrahimian et al. 2023). To get better outcomes, pre-treatment techniques are frequently combined. Effective pre-treatment is necessary to obtain the maximum yield of biofuels and chemicals. Every pre-treatment technique has benefits and drawbacks, and the best approach will frequently rely on the particulars of the waste and the final products that are intended. These pre-treatment methods are essential to the sustainable conversion of waste into useful biofuels and chemicals because they increase the accessibility of the organic components in MSW.

1.4 Production of Biofuels and Chemicals from Different MSW Feedstocks

1.4.1 Food Waste to Biofuel

Food waste represents a major challenge in global waste management and resource utilization efforts. The Food and Agriculture Organization (FAO) emphasizes reducing food waste to achieve sustainable development goals (SDGs) aimed at ensuring food security amidst a rapidly growing population. A substantial portion, between 44% and 47% of all household waste, is attributed to fruit and vegetable waste (De Laurentiis et al. 2018). According to De Laurentiis et al. (2018), a significant amount of this waste (23–28%) comes from inedible sections of fresh fruits and vegetables. If there are no changes in consumption habits, these wastes are deemed unavoidable and will still be generated despite preventive actions. As a result, it is crucial to take into account methods of processing and treatment to effectively manage and utilize these resources, including their potential for value-adding goods like chemicals and biofuels. Several technologies convert food waste, including fruits, vegetables, nuts, and cereals, into biofuels. Gasification is an efficient technique that turns waste into syngas, which is a blend of hydrogen, carbon monoxide, carbon dioxide, and sometimes methane. Syngas can be used to generate bioethanol, but this is usually done via catalytic or microbial methods like the Fischer-Tropsch process, rather than through direct fermentation (Ramachandriya et al. 2016). Additionally, food waste can be processed into solid biofuels like wood pellets, utilizing materials like mango trees for renewable electrochemical fuel cells (Paul and Kumar 2016). Similarly, wood from plum, apple, and pear trees contributes as solid biofuel, generating an energy potential of more than 100,000 MJ/hectare (Winzer et al. 2017). The oil palm frond and woody biomass of palm trees are also valuable bioenergy feedstock (Matali et al. 2016).

Liquid biofuel, such as ethanol, is derived through the fermentation of plant-based materials. For example, the United States produces bioethanol from 42% (114 million tons) of its harvested maize grains annually, offsetting a significant portion (10%) of petrol consumption (Energy Information Administration and U.S. Energy Information Agency 2013). Biodiesel, another significant liquid biofuel, can be synthesized from various sources, offering a renewable alternative to petro-diesel. Pyrolysis, involving high temperature (300–900 °C) heating of plant biomass without air, produces pyrolysis bio-oil (the brown vapor condensate), biochar (the black solid residue), and syngas (the non-condensable vapor) (Oyebanji et al. 2018; Dalal et al. 2022). Pyrolysis bio-oil can be produced from almost any organic material, including wood, crop straw, sugar cane bagasse, switchgrass, peanut hulls, and chicken litter (Fahmy et al. 2020). Anaerobic digestion is employed to produce biogas from organic wastes, presenting a way to recover renewable bioenergy. With continued technological advancements, biogas has the potential to replace around 25% of global natural gas consumption (Shehu et al. 2019). Beyond waste management, anaerobic digestion generates biogas, suitable for heating, producing electricity, and conversion to biomethane for vehicle fuel.

Food waste can be processed in a variety of ways to produce useful compounds such as chemicals, in addition to biofuels. These substances can be used as raw materials in a wide range of industrial processes. For instance, the fermentation of carbohydrate-rich food waste generates lactic acid, a precursor to biodegradable polymers like polylactic acid (PLA). Tang et al. (2017) investigated the effects of anaerobic sludge, fresh food waste, and methanogenic sludge on the yield of lactic acid (LA). According to their findings, reactors that were inoculated with fresh food waste produced much more LA, with a yield of 28.4 g/L. The fermentation of food waste also produces acetic acid, utilized in chemical synthesis, vinegar production, and various other industrial applications. Succinic acid, derived from sugars fermentation in food waste (Stylianou et al. 2020), is utilized in the chemical industry as an intermediary and in the manufacturing of solvents and biodegradable polymers. Table 1.1 depicts different studies done on the conversion of various types of MSW to biofuels and chemicals.

1.4.2 Plastic Waste to Biofuel and Chemicals

Plastic waste poses a great threat to the environment due to its extensive use in packaging, engineering, healthcare, construction, and other sectors (Ugoeze et al. 2021). To address plastic pollution effectively, innovative approaches for the management and utilization of plastic waste to create chemicals and biofuels are necessary. Pyrolysis and gasification are promising techniques that are employed to break down polymeric chains into smaller, more valuable molecules, promoting resource recovery and sustainability (Inayat et al. 2019).

Table 1.1 Studies conducted on the production of different types of biofuels and chemicals from a diverse range of municipal solid waste using various processes

S. No.	Type of MSW	Biofuel generated	Process used for biofuel generation	Experimental conditions	References
(a) Food waste					
(1)	Banana peels	Bioethanol	Acid hydrolysis and fermentation	0.1M H_2SO_4, 90 °C, 20 min, and 9 g immobilized yeast cell loading, pH 5, 30 °C for 24 h	Abdulla et al. (2022)
(2)	Apple pomace	Biodiesel	Supercritical CO_2 extraction and methanolysis	350 °C, 20 MPa, and oil-to-methanol molar ratio 1:42	Tasić et al. (2022)
(3)	Potato peels	Bioethanol and biogas	Enzymatic hydrolysis/ alcoholic fermentation, and anaerobic digestion	NaOH 1% (w/v), enzymatic hydrolysis with amylolytic (50 μL/g starch) and cellulolytic (300 μL/g cellulose) enzymes. 2% w/w S. cerevisiae, 30 °C, 24 h followed by anaerobic digestion	Barampouti et al. (2023)
(4)	Cabbage	Biochar	Pyrolysis	360 °C, feed particle size 0.90 mm	Pradhan et al. (2020)

(continued)

Table 1.1 (continued)

S. No.	Type of MSW	Biofuel generated	Process used for biofuel generation	Experimental conditions	References
(5)	Corncob	Bioethanol	Acid hydrolysis/ fermentation	Iron oxide nanoparticles as nanocatalysts, 72 h	Rekha and Saravanathamizhan (2021)
(b) Paper and pulp waste					
(6)	Corrugated cardboard (CC)	Bio-oil and biochar	Pyrolysis	350 °C and 450 °C	Sotoudehnia et al. (2020)
(7)	Sanitary napkins	Pyro-oil and pyro-char	Pyrolysis	550–600 °C	Hameed et al. (2024)
(8)	Newspaper and office paper	Bioethanol	Acid hydrolysis and fermentation	5% conc. H_2SO_4, 121 °C, 30 min	Thakare and Trivedi (2021)
(c) Plastic waste					
(9)	Polyolefin	Bio-oil	Co-pyrolysis with lignocellulosic biomass waste	410 °C to 500 °C	Salvilla et al. (2020)
(10)	High-density polyethylene (HDPE)	Bio-oil	Co-pyrolysis with brown salwood	Nitrogen rates of 40 to 160 ml min^{-1}, an HDPE-biomass ratio of 0.1 to 0.9, at 600 °C	Charusiri et al. (2022)
(11)	Polyvinyl chloride (PVC)	Bioethanol and syngas	Dechlorination of PVC to HCl, followed by acid treatment with oak sawdust, along with CO_2-assisted pyrolysis of residues to recover additional energy	5% HCl, 130 °C, 2 h	Park et al. (2024)
(12)	Polypropylene	Polypropylene pyrolysis oil (PPO) (transport fuel)	Pyrolysis and hydroprocessing of PPO	Ni-Au/MOR bimetal catalyst at a hydrogen pressure of 70 bar and 350 °C	Mangesh et al. (2020)

(13)	Polystyrene	Liquid fuel	Liquification with ethanol as a solvent	350 °C, with an ethanol-to-polystyrene ratio 0.5:1, 60 min	Ahmad et al. (2020)
(d) Poultry waste					
(14)	Chicken manure and cow dung	Biogas	Anaerobic digestion	Cow dung: chicken	Malolan et al. (2021)
(15)	Camel manure	Bio-gasoline	Hydrothermal liquification	250–400 °C, 15–60 min, 4–25 MPa	Alherbawi et al. (2021)
(16)	Chicken litter	Bio-oil	Catalytic pyrolysis	500 °C, heating rate 50 °C/min	Syazaidah et al. (2021)
(17)	Swine manure	Biodiesel, biochar, and syngas	Transesterification and CO_2-assisted pyrolysis	400 °C (biogas) and 220 °C, Ni/SiO_2	Cho et al. (2021)
(e) Fats and oils					
(18)	Frying oil	Biodiesel	Methanolysis	Biochar-based catalyst by carbonizing banana peels and alkali carbonates	Soares Dias et al. (2024)
(19)	Clay oil	High-grade biofuels	Catalytic pyrolysis	5% seaweed char, 800 °C, feed/catalyst ratio of 10:1	Yuan et al. (2021)
(20)	Shark liver oil	Biodiesel	Acid-catalyzed transesterification	10.3M methanol/oil ratio, 6.5 h, 60 °C, and 5.9 wt % H_2SO_4 catalyst	Al Hatrooshi et al. (2020)

Phetyim and Pivsa-Art (2018) investigated the process of turning spent lubricating oil and waste plastic into diesel. They used a variety of plastic kinds and oil ratios in their investigation, including high-density polyethylene (HDPE), polypropylene (PP), and polystyrene (PS). With the exception of a lower ignition point, they discovered that diesel generated at a ratio of 50:30:20:0 (oil:HDPE: PP) was nearly identical to the specifications specified by Thailand's Ministry of Energy. The pyrolysis reactor showed its best characteristics while it was running between 400 °C and 425 °C. In a related study, Brindhadevi et al. (2022) used manufactured catalysts in a solid reactor to demonstrate the catalytic breakdown of low-density polyethylene (LDPE). Gas, coke, and hydrocarbon-rich liquid fuel were produced during the process in addition to petrol. A TiO_2 catalyst produced the maximum yield of liquid fuel, but with time, its efficiency declined. Petrol content rose from 45.6% to 85.4% with the use of a TiO_2/AlSBA-15 composite catalyst, which also had 89.1% liquid fuel efficiency and a 98.4% conversion rate. With a calorific value of 47.8 MJ/kg, the liquid fuel produced was more than twice as high as regular petroleum.

Connecting these thermal processes, pyrolysis and gasification play important role in resource recovery by converting plastic waste into energy-rich products: pyrolysis yields pyrolysis oil, gas, and char, while gasification produces syngas, a mixture of hydrogen (H_2), carbon monoxide (CO), carbon dioxide (CO_2), and methane (CH_4) (Inayat et al. 2019). The oil derived from pyrolysis can be used as a feedstock for the synthesis of chemicals or fuels. In contrast, syngas from gasification can be utilized in catalytic synthesis to produce hydrogen, methanol, and other valuable chemicals. Chemical recycling methods, such as depolymerization, disassemble polymers into their component monomers or other useful compounds. For instance, polyethylene terephthalate (PET) can be depolymerized to produce ethylene glycol and terephthalic acid, crucial for manufacturing new PET (Singh and Demirsöz, 2022). These procedures not only provide a viable way to manage plastic waste, but they also support the circular economy by turning waste products into useful resources for industrial applications.

1.4.3 Pulp and Paper to Biofuel and Chemicals

This is a significant sector with about 5000 mills generating about 400 million tons of paper annually worldwide (Bajpai 2015). Nevertheless, the production of paper is highly water-intensive, requiring between 300 and 2600 m^3 of water per ton of paper (Van Oel and Hoekstra 2012). This explores the importance of recycling wastewater from these mills to address sustainability challenges. Chemical Oxygen Demand (COD), which usually ranges from 1 to 80 g/L, is present in quite significant amounts in the enormous amount of paper mill effluent (PME) (Lokman et al. 2021). Efficient biological processes can convert this wastewater into bioenergy.

Utilizing crude cellulase derived from *Trichoderma reesei*, Lakshmidevi and Muthukumar (2010) demonstrated the enzymatic hydrolysis of PME. From the resulting hydrolysate, Enterobacter aerogenes was then used to produce biohydrogen. Similarly, Zwain et al. (2013) studied the biomethane production and treatment of recycled PME through a modified anaerobic baffled reactor, achieving high methane output and COD removal efficiency during the startup period. Additionally, Buzała et al. (2015) discovered that waste fines from paper production lines and pine kraft cellulosic pulps are promising feedstock for biofuel synthesis, yielding significant amounts of glucose and reducing sugars through the enzymatic hydrolysis. Around 75–80% of glucose and total reducing sugars were produced by the enzymatic hydrolysis of the pine wood processing fines in a paper mill. This amounted to around 34% and 36% on a dry weight basis for pine wood. It is worth noting that removing lignin from pine wood did not always result in the maximum yields, suggesting that producing biofuel from this waste stream is feasible. All things considered, the industry's conversion of waste pulp and paper into biofuels offers a viable path towards resource management and energy generation that is sustainable.

An innovative method of recycling waste materials and reducing the pulp and paper industry's environmental impact is the conversion of waste pulp and paper into chemicals. The lignocellulosic components of pulp and paper waste, such as cellulose, hemicellulose, and lignin, can be transformed into valuable raw materials for chemical manufacturing (Przybysz Buzala et al. 2016). Enzymatic hydrolysis converts the cellulose into glucose, while hemicellulose is broken down into a mixture of sugars, including xylose, arabinose, and other pentoses, depending on the feedstock. These simple sugars can be further processed into various compounds (Buzała et al. 2015). Lactic acid, succinic acid, and ethanol are examples of bio-based compounds that can be produced through the fermentation of glucose and are used in the food, pharmaceutical, and biofuels industries. Similarly, xylose can be transformed into derivatives that are used to make polymers, resins, and medications, such as xylitol, furfural, or furan (Narisetty et al. 2022). An alternative method is to use the complex aromatic polymer lignin, which is present in waste paper and pulp, as a precursor to produce high-value compounds such as phenols, guaiacols, and syringols via thermochemical or catalytic depolymerization of lignin (Nguyen et al. 2021). These compounds are useful precursors for the synthesis of polymers, resins, and specialty chemicals.

1.4.4 Poultry Waste to Biofuel and Chemicals

The production of poultry contributes significantly to the worldwide agricultural industry and is an essential source of protein for human consumption (Mottet and Tempio 2017). However, the business also produces large amounts of waste in the form of poultry abattoir sludge (PAS) and nutrient-rich effluent in addition to producing chicken meat. For the poultry industry to promote sustainable practices and reduce environmental contamination, waste management must be done effectively (Onwosi et al. 2020). Growing concerns about environmental sustainability and energy security have sparked interest in exploring poultry waste, particularly in the manufacture of chemicals and biofuels. Organic materials included in poultry manure can be processed by anaerobic digestion, fermentation, and other methods to produce valuable biofuels such as biogas and biodiesel. The value proposition of using poultry waste is further enhanced by the fact that the organic compounds found in it may be used as feedstock to produce a variety of chemicals, such as organic acids, solvents, and specialty chemicals.

To overcome the challenge of utilizing poultry waste, Ozdemir et al. (2022) investigated the production of bio-pellets for boiler heating applications utilizing the biomass of the energy crop *Arundo donax* (AD), which is cultivated using treated abattoir effluent and solar-dried PAS. In the study, ground AD straw and sun-dried PAS were extruded into pellets with different amounts of PAS and AD. The resultant pellets demonstrated energy densities of 10.86–22.71 GJ m^{-3}, bulk densities of 646–779 kg m^{-3}, and calorific values ranging from 16.81 to 28.78 MJ kg^{-1}. Combustion analysis revealed that the red color intensity and flame richness of the PAS + AD pellets were higher than those of the PAS or AD pellets alone. Based on these results, the study concluded that chicken waste has the potential to be a sustainable source of biofuel, with renewable PAS + AD pellets being able to replace up to 58% of the daily use of fossil fuel coal for hot water generation. This study shows that it is feasible to produce bioenergy from poultry waste streams, which can help to value waste and reduce the need for fossil fuels.

1.5 Conclusion

Investigating the possibility of converting MSW into biofuels and chemicals offers significant opportunities to improve sustainability and address environmental issues. Food waste, plastic waste, pulp and paper waste, and poultry waste are just a few of the waste streams that show tremendous promise for being converted into useful resources by means of innovative technologies and procedures. Food waste, a major component of household waste, can be efficiently transformed into biofuels such as bioethanol and biodiesel by gasification, anaerobic digestion, and fermentation, reducing its environmental impact and assisting in the generation of renewable energy Similarly, thermal and catalytic processes can be used to convert plastic waste, a serious environmental challenge, into useful chemicals such as organic acids, solvents, resins, and syngas, along with fuels like biodiesel, biogas, thereby reducing pollution and fostering resource recovery. The pulp and paper industry, a significant consumer of water and producer of a lot of effluent, can recycle its solid waste and wastewater by fermenting and hydrolyzing it to generate biofuels and chemicals. Additionally, the poultry industry generates significant amounts of waste such as sludge from poultry abattoirs, manure rich in nutrients, and feathers. This waste can be treated via anaerobic digestion to yield biogas, fermentation to create bioethanol, and bio-pellet production for use as solid biofuel. These approaches offer sustainable biofuel options, decrease reliance on fossil fuels, and improve waste management practices. However, issues like excessive costs and technological constraints need to be resolved if these technologies are to reach their full potential. To improve the effectiveness and scalability of these procedures, targeted research and development initiatives are needed. The future scope includes cost-effective process development, conversion technique optimization, and expanding the applications of these technologies across a wider range of industries. In addition, to surmount these obstacles and encourage broad adoption, encouraging legislative frameworks, monetary incentives, and market processes are mandatory.

References

Abdulla R, Johnny Q, Jawan R, Sani SA (2022) Agrowastes of banana peels as an eco-friendly feedstock for the production of biofuels using immobilized yeast cells. IOP Conf Ser Earth Environ Sci 1103(1):012022

Abraham A, Mathew AK, Park H, Choi O, Sindhu R, Parameswaran B, Sang BI (2020) Pretreatment strategies for enhanced biogas production from lignocellulosic biomass. Bioresour Technol 301:122725. https://doi.org/10.1016/j.biortech.2019.122725

Ahmad N, Ahmad N, Maafa IM, Ahmed U, Akhter P, Shehzad N, Hussain M (2020) Thermal conversion of polystyrene plastic waste to liquid fuel via ethanolysis. Fuel 279:118498. https://doi.org/10.1016/j.fuel.2020.118498

Al Hatrooshi AS, Eze VC, Harvey AP (2020) Production of biodiesel from waste shark liver oil for biofuel applications. Renew Energy 145:99–105. https://doi.org/10.1016/j.renene.2019.06.002

Alherbawi M, Parthasarathy P, Al-Ansari T, Mackey HR, McKay G (2021) Potential of drop-in biofuel production from camel manure by hydrothermal liquefaction and biocrude upgrading: a Qatar case study. Energy 232:121027. https://doi.org/10.1016/j.energy.2021.121027

Ali J, Rasheed T, Afreen M, Anwar MT, Nawaz Z, Anwar H, Rizwan K (2020) Modalities for conversion of waste to energy—challenges and perspectives. Sci Total Environ 727:138610. https://doi.org/10.1016/j.scitotenv.2020.138610

Andlar M, Rezić T, Marđetko N, Kracher D, Ludwig R, Šantek B (2018) Lignocellulose degradation: an overview of fungi and fungal enzymes involved in lignocellulose degradation. Eng Life Sci 18(11):768–778. https://doi.org/10.1002/elsc.201800039

Bajpai P (2015) Management of pulp and paper mill waste, vol 431. Springer, Cham

Barampouti EM, Christofi A, Malamis D, Mai S (2023) A sustainable approach to valorize potato peel waste towards biofuel production. Biomass Convers Biorefinery 13(9):8197–8208. https://doi.org/10.1007/s13399-021-01811-4

Brindhadevi K, Hiep BT, Khouj M, Garalleh HA (2022) A study on biofuel produced from cracking of low density poly ethylenes using TiO2/AlSBA-15 nanocatalysts. Fuel 323:124299. https://doi.org/10.1016/j.fuel.2022.124299

Buzała K, Przybysz P, Rosicka-Kaczmarek J, Kalinowska H (2015) Production of glucose-rich enzymatic hydrolysates from cellulosic pulps. Cellulose 22:663–674. https://doi.org/10.1007/s10570-014-0522-x

Cayumil R, Khanna R, Konyukhov Y, Burmistrov I, Kargin JB, Mukherjee PS (2021) An overview on solid waste generation and management: current status in Chile. Sustain For 13(21):11644. https://doi.org/10.3390/su132111644

Cesaro A, Belgiorno V (2014) Pretreatment methods to improve anaerobic biodegradability of organic municipal solid waste fractions. Chem Eng J 240:24–37. https://doi.org/10.1016/j.cej.2013.11.055

Charusiri W, Phowan N, Vitidsant T (2022) Pyrolysis of lignocellulosic biomass with high-density polyethylene to produce chemicals and bio-oil with high liquid yields. Sustain Chem Pharm 25:100567. https://doi.org/10.1016/j.scp.2021.100567

Chaturvedi V, Verma P (2013) An overview of key pretreatment processes employed for bioconversion of lignocellulosic biomass into biofuels and value added products. Biotech 3:415–431

Cho SH, Jung S, Park YK, Lin KYA, Chen WH, Tsang YF, Kwon EE (2021) Biofuel production as an example of virtuous valorization of swine manure. ACS Sustain Chem Eng 9(41):13761–13772

Dalal R, Wathore R, Labhasetwar N (2022) Sustainable production of biochar, bio-gas and bio-oil from lignocellulosic biomass and biomass waste. In: Zero waste biorefinery. Springer, Singapore, pp 177–205

De Laurentiis V, Corrado S, Sala S (2018) Quantifying household waste of fresh fruit and vegetables in the EU. Waste Manag 77:238–251. https://doi.org/10.1016/j.wasman.2018.04.001

Ebrahimian F, Denayer JF, Mohammadi A, Khoshnevisan B, Karimi K (2023) A critical review on pretreatment and detoxification techniques required for biofuel production from the organic fraction of municipal solid waste. Bioresour Technol 368:128316. https://doi.org/10.1016/j.biortech.2022.128316

Fahmy TY, Fahmy Y, Mobarak F, El-Sakhawy M, Abou-Zeid RE (2020) Biomass pyrolysis: past, present, and future. Environ Dev Sustain 22:17–32. https://doi.org/10.1007/s10668-018-0200-5

Gu D, Andreev K, Dupre ME (2021) Major trends in population growth around the world. China CDC Wkly 3(28):604

Gupta MM, Dutta R, Kumari A, Bala K (2024) State-of-the-art of microbial whole-cell catalysts for biofuel production. In: Whole-cell biocatalysis. Elsevier, Amsterdam, pp 355–375

Hameed MSA, Sreedharan SP, Thampan D, Hussain R, Sivagami K (2024) Pyrolysis-based resource recovery from sanitary napkins: a sustainable waste management approach (No. 11667). EasyChair, Stockport

Inayat M, Sulaiman SA, Kurnia JC, Shahbaz M (2019) Effect of various blended fuels on syngas quality and performance in catalytic co-gasification: a review. Renew Sust Energ Rev 105:252–267. https://doi.org/10.1016/j.rser.2019.01.059

Kadam AR, Thakur S (2020) Pattern of urbanization and urban population growth in Mumbai District, Maharashtra, India. Akwapoly J Commun Sci Res 5(1):1

Kanwal S, Mehran MT, Hassan M, Anwar M, Naqvi SR, Khoja AH (2022) An integrated future approach for the energy security of Pakistan: replacement of fossil fuels with syngas for better environment and socio-economic development. Renew Sust Energ Rev 156:111978. https://doi.org/10.1016/j.rser.2021.111978

Kumari A, Dutta R, Gupta MM, Rai MP, Srivastava S, Chandra S, Nimkar V (2022a) Wastewater and solid waste as feedstock for energy production. In: Novel feedstocks for biofuels production. Springer, Singapore, pp 219–270. https://doi.org/10.1007/978-981-19-3582-4_8

Kumari A, Kundu PK, Gupta MM, Bala K, Chandra S, Dutta R, Das A (2022b) Role of microorganisms in production of biofuels. In: Novel feedstocks for biofuels production. Springer, Singapore, pp 65–116

Lakshmidevi R, Muthukumar K (2010) Enzymatic saccharification and fermentation of paper and pulp industry effluent for biohydrogen production. Int J Hydrog Energy 35(8):3389–3400. https://doi.org/10.1016/j.ijhydene.2009.12.165

Lokman NA, Ithnin AM, Yahya WJ, Yuzir MA (2021) A brief review on biochemical oxygen demand (BOD) treatment methods for palm oil mill effluents (POME). Environ Technol Innov 21:101258. https://doi.org/10.1016/j.eti.2020.101258

Malolan R, Jayaraman RS, Adithya S, Arun J, Gopinath KP, SundarRajan P, Govarthanan M (2021) Anaerobic digestate water for Chlorella pyrenoidosa cultivation and employed as co-substrate with cow dung and chicken manure for methane and hydrogen production: a closed loop approach. Chemosphere 266:128963

Mangesh VL, Tamizhdurai P, Krishnan PS, Narayanan S, Umasankar S, Padmanabhan S, Shanthi K (2020) Green energy: hydroprocessing waste polypropylene to produce transport fuel. J Clean Prod 276:124200. https://doi.org/10.1016/j.jclepro.2020.124200

Matali S, Rahman NA, Idris SS, Yaacob N, Alias AB (2016) Lignocellulosic biomass solid fuel properties enhancement via torrefaction. Proc Eng 148:671–678. https://doi.org/10.1016/j.proeng.2016.06.550

Maurya DP, Singla A, Negi S (2015) An overview of key pretreatment processes for biological conversion of lignocellulosic biomass to bioethanol. Biotech 5:597–609

Mihai FC, Gündoğdu S, Markley LA, Olivelli A, Khan FR, Gwinnett C, Molinos-Senante M (2021) Plastic pollution, waste management issues, and circular economy opportunities in rural communities. Sustain For 14(1):20. https://doi.org/10.3390/su14010020

Mottet A, Tempio G (2017) Global poultry production: current state and future outlook and challenges. Worlds Poult Sci J 73(2):245–256
Nanda S, Berruti F (2021) Municipal solid waste management and landfilling technologies: a review. Environ Chem Lett 19(2):1433–1456. https://doi.org/10.1007/s10311-020-01100-y
Nandhini R, Berslin D, Sivaprakash B, Rajamohan N, Vo DVN (2022) Thermochemical conversion of municipal solid waste into energy and hydrogen: a review. Environ Chem Lett 20(3):1645–1669. https://doi.org/10.1007/s10311-022-01410-3
Narisetty V, Cox R, Bommareddy R, Agrawal D, Ahmad E, Pant KK, Kumar V (2022) Valorisation of xylose to renewable fuels and chemicals, an essential step in augmenting the commercial viability of lignocellulosic biorefineries. Sustainable Energy Fuels 6(1):29–65
Nguyen LT, Phan DP, Sarwar A, Tran MH, Lee OK, Lee EY (2021) Valorization of industrial lignin to value-added chemicals by chemical depolymerization and biological conversion. Ind Crop Prod 161:113219. https://doi.org/10.1016/j.indcrop.2020.113219
Onwosi CO, Igbokwe VC, Odimba JN, Nwagu TN (2020) Anaerobic bioconversion of poultry industry-derived wastes for the production of biofuels and other value-added products. In: Biovalorisation of wastes to renewable chemicals and biofuels. Elsevier, Amsterdam, pp 113–131
Oyebanji JA, Okekunle PO, Lasode OA, Oyedepo SO (2018) Chemical composition of bio-oils produced by fast pyrolysis of two energy biomass. Biofuels 9(4):479–487. https://doi.org/10.1080/17597269.2017.1284473
Ozdemir S, Şimşek A, Ozdemir S, Dede C (2022) Investigation of poultry slaughterhouse waste stream to produce bio-fuel for internal utilization. Renew Energy 190:274–282. https://doi.org/10.1016/j.renene.2022.03.113
Park J, Yim JH, Cho SH, Jung S, Tsang YF, Chen WH, Kwon EE (2024) A virtuous cycle for thermal treatment of polyvinyl chloride and fermentation of lignocellulosic biomass. Appl Energy 362:123011. https://doi.org/10.1016/j.apenergy.2024.123011
Paul S, Kumar U (2016) Study of enzymatic saccharification and fermentation of hydrolyzed mango wood to produce biofuel for electrochemical renewable energy in fuel cell. SCIREA J Energy 1:39–59
Phetyim N, Pivsa-Art S (2018) Prototype co-pyrolysis of used lubricant oil and mixed plastic waste to produce a diesel-like fuel. Energies, 11(11):2973. https://doi.org/10.3390/en11112973
Pradhan S, Shahbaz M, Abdelaal A, Al-Ansari T, Mackey HR, McKay G (2020) Optimization of process and properties of biochar from cabbage waste by response surface methodology. Biomass Convers Biorefinery. https://doi.org/10.1007/s13399-020-01101-5
Przybysz Buzała K, Przybysz P, Kalinowska H, Przybysz K, Kucner M, Dubowik M (2016) Evaluation of pine Kraft cellulosic pulps and fines from papermaking as potential feedstocks for biofuel production. Cellulose 23:649–659. https://doi.org/10.1007/s10570-015-0808-7
Ram C, Kumar A (2021) Municipal solid waste management: recent practices. In: Nanobiotechnology for green environment. CRC Press, Boca Raton, pp 37–65
Ramachandriya KD, Kundiyana DK, Sharma AM, Kumar A, Atiyeh HK, Huhnke RL, Wilkins MR (2016) Critical factors affecting the integration of biomass gasification and syngas fermentation technology. AIMS Bioeng 3:188–210
Rekha B, Saravanathamizhan R (2021) Catalytic conversion of corncob biomass into bioethanol. Int J Energy Res 45(3):4508–4518
Salvilla JNV, Ofrasio BIG, Rollon AP, Manegdeg FG, Abarca RRM, de Luna MDG (2020) Synergistic co-pyrolysıs of polyolefin plastics with wood and agricultural wastes for biofuel production. Appl Energy 279:115668. https://doi.org/10.1016/j.apenergy.2020.115668
Sharma KD, Jain S (2020) Municipal solid waste generation, composition, and management: the global scenario. Soc Responsibility J 16(6):917–948
Shehu I, Akanbi TO, Wyatt V, Aryee AN (2019) Fruit, nut, cereal, and vegetable waste valorization to produce biofuel. Byproducts from agriculture and fisheries: adding value for food, feed, pharma, and fuels, 665–684. https://doi.org/10.1002/9781119383956.ch30
Siddiqua A, Hahladakis JN, Al-Attiya WAK (2022) An overview of the environmental pollution and health effects associated with waste landfilling and open dumping. Environ Sci Pollut Res Int 29(39):58514–58536. https://doi.org/10.1007/s11356-022-21578-z

Singh N, Demirsöz R (2022) Recycling of traditional plastics: PP, PS, PVC, PET, HDPE, and LDPE, and their blends and composites. In: Nanomaterials in manufacturing processes. CRC Press, Boca Raton, pp 235–258

Singh M, Singh M, Singh SK (2024) Tackling municipal solid waste crisis in India: insights into cutting-edge technologies and risk assessment. Sci Total Environ 917:170453

Soares Dias AP, Pedra I, Salvador É, Rijo B, Costa Pereira MF, Serralha F, Nogueira I (2024) Biodiesel production over banana peel biochar as a sustainable catalyst. Catalysts 14(4):266

Sołowski G, Konkol I, Cenian A (2020) Production of hydrogen and methane from lignocellulose waste by fermentation. A review of chemical pretreatment for enhancing the efficiency of the digestion process. J Clean Prod 267:121721. https://doi.org/10.1016/j.jclepro.2020.121721

Sotoudehnia F, Rabiu AB, Alayat A, McDonald AG (2020) Characterization of bio-oil and biochar from pyrolysis of waste corrugated cardboard. J Anal Appl Pyrolysis 145:104722. https://doi.org/10.1016/j.jaap.2019.104722

Stylianou E, Pateraki C, Ladakis D, Cruz-Fernández M, Latorre-Sánchez M, Coll C, Koutinas A (2020) Evaluation of organic fractions of municipal solid waste as renewable feedstock for succinic acid production. Biotechnol Biofuels 13:1–16. https://doi.org/10.1186/s13068-020-01708-w

Syazaidah I, Bakar MSA, Reza MS, Azad AK (2021) Ex-situ catalytic pyrolysis of chicken litter for bio-oil production: experiment and characterization. J Environ Manag 297:113407. https://doi.org/10.1016/j.jenvman.2021.113407

Tang J, Wang XC, Hu Y, Zhang Y, Li Y (2017) Effect of pH on lactic acid production from acidogenic fermentation of food waste with different types of inocula. Bioresour Technol 224:544–552. https://doi.org/10.1016/j.biortech.2016.11.111

Tasić MB, Stanković MS, Kostić MD, Stamenković OS, Veljković VB (2022) Circular economy in apple processing industry: biodiesel production from waste apple seeds. Chem Ind Chem Eng Q 28(3):237–245. https://doi.org/10.2298/CICEQ210819035T

Thakare K, Trivedi M (2021) Bioethanol production from waste paper. Int J Sci Res Biol Sci 8(2):49–51

Ugoeze KC, Amogu EO, Oluigbo KE, Nwachukwu N (2021) Environmental and public health impacts of plastic wastes due to healthcare and food products packages: a review. J Environ Sci Publ Health 5(1):1–31

UN DESA (2018) 2018 Revision of world urbanization prospects by united nations, department of economic and social affairs (UN DESA). Available https://esa.un.org/unpd/wup/Publications/Files/WUP2018-PressRelease.pdf

US-EPA (2016) Wastes. Non-Hazardous Waste. Municipal Solid Waste. Energy Recovery from Waste, https://archive.epa.gov/epawaste/nonhaz/municipal/web/html/index-11.html; [accessed 24 February, 2024].

Van Oel PR, Hoekstra AY (2012) Towards quantification of the water footprint of paper: a first estimate of its consumptive component. Water Resour Manag 26:733–749. https://doi.org/10.1007/s11269-011-9942-7

Winzer F, Kraska T, Elsenberger C, Kötter T, Pude R (2017) Biomass from fruit trees for combined energy and food production. Biomass Bioenergy 107:279–286. https://doi.org/10.1016/j.biombioe.2017.10.027

World Bank Group (2018) What a waste 2.0: a global snapshot of solid waste management to 2050-the urban development series. World Bank, Washington. Available at https://openknowledge.worldbank.org/handle/10986/30317

Yuan C, Abomohra AEF, Wang S, Liu Q, Zhao S, Cao B, Hu Y (2021) High-grade biofuel production from catalytic pyrolysis of waste clay oil using modified activated seaweed carbon-based catalyst. J Clean Prod 313:127928. https://doi.org/10.1016/j.jclepro.2021.127928

Zhang Y, Kusch-Brandt S, Gu S, Heaven S (2019) Particle size distribution in municipal solid waste pre-treated for bioprocessing. Resources 8(4):166. https://doi.org/10.3390/resources8040166

Zwain HM, Hassan SR, Zaman NQ, Aziz HA, Dahlan I (2013) The start-up performance of modified anaerobic baffled reactor (MABR) for the treatment of recycled paper mill wastewater. J Environ Chem Eng 1(1–2):61–64. https://doi.org/10.1016/j.jece.2013.03.007

Valorization of Fruit Waste to Biofuels and Chemicals

2

Surya Prakash Dwivedi, Neeraj Dwivedi, and Shweta Singh

Abstract

The valorization of food waste into chemicals and biofuels is an important aspect of sustainable waste management and resource utilization. Food waste is a significant global issue, and finding ways to convert it into valuable products can help reduce environmental impacts, conserve resources, and contribute to a more circular economy. In this chapter, we discuss several process optimization methods that may be considered for the valorization of food waste into biofuels, leading to the sustainable development of waste utilization techniques. The optimization techniques include anaerobic digestion, fermentation, bioconversion, hydrothermal processing, and nutrient recovery, specifically focusing on nitrogen and phosphorus for a self-sustainable system for waste utilization. The exploration of some food waste components, such as starch and cellulose, is to be used to produce bioplastics using the fermentation process. Biodegradable plastics made from food waste can be environmentally friendly alternatives to traditional plastics. The impact assessment of optimization techniques was also evaluated for the regulatory parameters and environmental impact. The optimized techniques can play a crucial role in the efficient valorization of food and organic waste. It is also ensured that the processes are environmentally sustainable and compliant with local regulations. This chapter concluded that food waste valorization is a multidisciplinary field that combines elements of microbial technology and sustainability. The findings aim to develop efficient and cost-effective methods for transforming food waste into valuable products, thereby promoting sustainability and advancing the principles of a circular economy. A circular economy is an economic model that minimizes waste and maximizes resource efficiency by continuously repurposing materials through reuse, recy-

S. P. Dwivedi (✉) · N. Dwivedi · S. Singh
Council of Industrial Innovation and Research, Noida, Uttar Pradesh, India

A. Kumari et al. (eds.), *Valorization of Solid Wastes to Biofuels and Chemical Products for Sustainable World*, https://doi.org/10.1007/978-981-96-8594-3_2

cling, and regeneration. Food waste valorization aligns with this concept by converting organic waste into biofuels, bioplastics, and nutrient-rich byproducts, reducing environmental impact while enhancing resource recovery and sustainability.

Keywords

Anaerobic digestion · Biofuels · Valorization · Circular economy · Food waste management

2.1 Introduction

Fruit processing generates substantial solid waste, accounting for 20–50% of processed fruit weight, and improper disposal methods like landfilling and incineration pose significant environmental risks, including methane emissions and water contamination (Smith et al. 2022; Chen et al. 2021). Current repurposing methods such as animal feed, compost, and biogas face challenges in wastewater management, particularly due to high chemical oxygen demand (COD), which can lead to water contamination and hinder microbial degradation Industries are increasingly exploring clean energy production from fruit processing waste (FPW), which is rich in fermentable sugars and carbohydrates, suitable for bioenergy and biofuels. Integrated biorefineries are advanced processing facilities that utilize various biological, chemical, and thermal technologies to efficiently convert food processing waste (FPW) into bioenergy, biochemicals, and biomaterials. These systems operate on the principle of maximizing resource utilization by integrating multiple conversion processes, such as fermentation, anaerobic digestion, and thermochemical treatments, to extract valuable compounds while minimizing waste. By adopting this approach, integrated biorefineries enhance the economic viability of food waste valorization, contributing to a more sustainable and circular economy (Chen et al. 2021). However, low carbohydrate content and process design remain obstacles. Valorizing fruit waste for biofuels and chemicals offers environmental and economic benefits, supporting renewable energy production and a circular economy. Fruit waste from diverse sources, including food processing industries, markets, and households, presents a pressing waste management issue, with conventional methods contributing to environmental pollution and resource wastage. Sustainable alternatives utilize advanced technologies to valorize fruit waste, which is abundant in organic materials such as carbohydrates, lipids, proteins, and bioactive compounds (Gupta et al. 2020). Understanding the complex composition of fruit waste is crucial for designing efficient valorization processes, with tailored strategies required for different fruits. Conversion into biofuels and chemicals aligns with bioeconomy principles, offering sustainable alternatives to fossil-based products. Advances in biotechnology, catalysis, and process engineering have improved the efficiency and feasibility of valorization processes, such as enzymatic hydrolysis, fermentation, and pyrolysis. Economic viability, market trends, potential revenue streams, and environmental impacts like carbon footprints and life cycle analyses are critical factors in the widespread adoption and sustainability of fruit waste valorization.

This chapter delves into the intricacies of environmental impacts and sustainability benchmarks associated with fruit waste valorization. The primary objective of this chapter is to explore the various techniques and strategies for the conversion of fruit waste into biofuels and chemicals. The chapter discusses the challenges associated with fruit waste valorization. It also highlights the importance of sustainable practices in achieving efficient and economically viable solutions.

2.2 Fruit Waste Composition and Characteristics

Understanding the variations in composition, acidity, and moisture content among different types of fruit waste is essential for tailoring optimization processes (Li et al. 2019). This knowledge allows researchers and industry professionals to design and implement efficient and specific valorization techniques, maximizing the yield of biofuels and chemicals from diverse fruit feedstocks.

2.2.1 Sources of Food Waste

In a comprehensive examination of food waste sources, we can categorize them into four main groups: food production and harvesting, food processing and storage, domestic waste, and retail waste. Pre-harvest losses in crop production are influenced by factors such as insect infestations and adverse weather conditions from planting to harvesting. The utilization of heavy machinery during crop harvesting can result in waste, as these machines may struggle to differentiate between ripe and immature crops, sometimes collecting only a fraction of the entire yield (Kantor et al. 1997).

Processing-induced food waste occurs due to losses in nutritional value, caloric content, and edibility caused by factors such as extreme temperature fluctuations, high humidity, or the presence of undesirable microbes (FAO 2012). Storage-related food waste is exacerbated by heat and high humidity, providing optimal conditions for pest proliferation (FAO 2012). Retailers, enforcing strict appearance criteria for food items, contribute significantly to food waste. Damaged or bruised fruits and vegetables are often excluded from display, adding to the overall waste.

2.2.2 Types of Fruit Waste

Understanding the intricate chemical composition of fruit waste serves as the cornerstone for crafting optimization processes (Wang et al. 2018). Designing tailored valorization strategies based on the distinct components present enables highly

Table 2.1 Chemical composition and characteristics of various fruit wastes

Category	Composition	Acidity	Moisture content
Citrus waste	Waste generated from citrus fruits (oranges, lemons, and grapefruits) includes peels, pulp, and seeds; rich in pectin, limonene, and bioflavonoids	Marked by high acidity due to citric acid; influences processing methods	Variable moisture content impacts drying efficiency
Apple pomace	Apple waste, comprising peels, cores, and seeds; significant presence of pectin, polyphenols, and dietary fibers	Moderately acidic with notable malic acid content	Variable moisture content influences drying and extraction processes
Banana residues	Banana waste, including peels and stems; abundant in starch, cellulose, and hemicellulose	Relatively low acidity impacting fermentation or enzymatic processes	Typically, high moisture content, affecting drying and storage
Berry by-products	Waste from berries (strawberries, blueberries, raspberries); rich in antioxidants like anthocyanins and polyphenols	Varying acidity levels influence extraction and processing methods	Significant moisture content impacts drying and preservation
Stone fruit residues	Waste from stone fruits (peaches, plums, cherries); contains pits, skins, and pulp with high sugar and phenolic content	Acidity varies, impacting processing methods like enzymatic hydrolysis or fermentation	Moisture content may vary, influencing the efficiency of drying

efficient and targeted valorization, unlocking the full potential for biofuel and chemical production from a diverse array of fruit feedstocks.

Table 2.1 presents a comprehensive overview of the chemical composition and characteristics of diverse fruit wastes, including citrus waste, apple pomace, banana residues, berry by-products, and stone fruit residues. Each category delineates specific composition, acidity levels, and moisture content, offering insights for tailored valorization processes. From citrus waste's rich pectin content to berry by-products' antioxidant richness, the table serves as a detailed guide for optimizing biofuel and chemical production from various fruit feedstocks.

2.2.3 Composition of Various Fruit Wastes

Fruit waste composition varies by type, processing method, and waste stream, including peels, seeds, pulp, and residues, each rich in specific nutrients. Peels are high in fiber, vitamins, antioxidants, and protective waxes, while seeds contain oils, proteins, and antioxidants. Pulp consists of sugars, organic acids, vitamins, and flavor compounds. Spoiled fruit shows cell breakdown, and excess production waste varies. Citrus peels have essential oils and flavonoids; apple seeds contain amygdalin; and banana peels are high in fiber. Key components include cellulose, hemicellulose, pectin, and lignin, crucial for bioenergy and biochemical production.

Understanding these components helps optimize processes and extract valuable compounds, supporting sustainability. Fruit waste's diverse chemical composition, including sugars, pectin, cellulose, hemicellulose, lignin, lipids, proteins, polyphenols, and organic acids, is essential for biofuel, bio-based chemicals, biodiesel, fertilizers, pharmaceuticals, and cosmetics production. Essential vitamins, minerals, and water content affect drying processes and compound concentration, necessitating tailored valorization approaches for sustainable and efficient utilization.

Table 2.2 details the chemical composition of fruit waste, highlighting components crucial for various valorization processes. Sugars such as glucose, fructose, and sucrose are essential for biofuel production. Polysaccharides such as pectin from citrus peels can be converted into galacturonic acid for bio-based chemicals. Cellulose and hemicellulose are broken down into fermentable sugars through enzymatic hydrolysis, which can then be used for biofuel production, while lignin is useful for specialty chemicals and thermochemical applications. Lipids, proteins, polyphenols, organic acids, vitamins, and minerals offer unique valorization opportunities. The water content affects drying processes and compound concentration, underscoring the need for tailored and sustainable valorization approaches.

2.2.4 Quantitative Production of Fruit Waste

According to the Indian Agricultural Research Data Book 2024, India has seen significant growth in fruit and vegetable production over the past decades. In the early 2000s, production was approximately 150 million tons, generating about 50 million tons of waste. Recent data indicates that production has increased to around 246.55 million tons, with India contributing 14% of the world's vegetable production. However, despite this growth, the country continues to face substantial post-harvest losses. Estimates suggest that approximately 18% of India's fruit and vegetable production, valued at INR 133 billion, is wasted annually.

These losses occur at various stages, including harvesting, transportation, storage, and distribution, and encompass both edible and inedible waste. Efforts are ongoing to reduce these losses through improved post-harvest management, better storage facilities, and enhanced supply chain logistics. The overall loss in these commodities is estimated to be around 20–30% of total production, resulting in an annual economic loss of ₹30,000 crore. The 50 million tons of waste represent the overall waste generated from total production, including losses occurring at various stages such as harvesting, transportation, storage, and distribution. The 20–30% overall loss also pertains to total production, not just processing. This loss includes both edible and inedible waste. The edible waste being fruits and vegetables that are discarded due to spoilage or quality issues, and the inedible waste including peels, seeds, and other non-consumable parts. The Food and Agriculture Organization (FAO) reported in 2003 that fruit waste alone amounted to 3.36 million tons out of a total fruit production of 16.8 million tons, with banana waste contributing significantly at 6.4 million tons. India's citrus fruit production, including mandarins, lime, lemon, and sweet orange, amounts to three million tons. Citrus waste is identified

Table 2.2 Comprehensive overview of fruit waste components and their applications for sustainable valorization

Component	Description	Application/impact
Sugars	Glucose and fructose: Common fermentable sugars in fruit waste Sucrose: Some fruits contain sucrose, requiring enzymatic hydrolysis for efficient conversion	Influences fermentation processes for biofuel production
Pectin	Complex polysaccharide found in fruit peels, especially in citrus fruits. Hydrolysis yields galacturonic acid, further processed into bio-based chemicals	Potential for bio-based chemical production
Cellulose and hemicellulose	Cellulose: Major component of plant cell walls, provides fermentable sugars through enzymatic hydrolysis Hemicellulose: Comprises various sugars, supplementing fermentable sugars for biofuel production	Source of fermentable sugars for biofuel production
Lignin	Aromatic compounds in lignin can be valorized into specialty chemicals or used for energy generation through thermochemical processes	Potential applications in specialty chemicals and energy generation
Lipids	Fatty acids in fruit seeds or peels can be converted into biodiesel through transesterification. Fatty acid composition influences biodiesel quality	Potential for biodiesel production, quality influenced by fatty acid composition
Proteins	Fruit waste proteins can be hydrolyzed into amino acids, finding applications in bio-based chemicals and fertilizers	Amino acids derived from proteins can be used in various industries
Polyphenols	Rich in antioxidants, polyphenols in fruit waste have applications in pharmaceuticals, cosmetics, and as natural additives	Potential applications in pharmaceuticals, cosmetics, and natural additives
Organic acids	Citric acid: Abundant in citrus fruits, used in food and pharmaceutical industries Malic acid: Found in apples, utilized in bio-based chemicals	Application in food and pharmaceutical industries, bio-based chemical production
Vitamins and nutrients	Vitamin C (ascorbic acid): Extracted for use in food and cosmetics. Minerals: Essential minerals in fruit waste can be recovered as supplements or soil amendments	Applications in food, cosmetics, and the recovery of essential minerals
Water content	Water content affects drying processes and the overall concentration of valuable compounds	Efficiency of drying processes and the concentration of valuable compounds are influenced by water content

as a valuable source of oil, pectin, and various by-products. The failure to economically salvage and reuse valuable materials like pectin, oils, and fibers from citrus waste contributes to unnecessary waste and resource depletion. Quantitative data on the percentage of processing waste generated from various fruits and vegetables were reported by Gera and Kramer (1969), according to which the waste percentages vary significantly among different commodities. For instance, apples and pears have a similar waste range, with apple processing generating between 12% and 47% waste by weight, while pear waste ranges from 12% to 46%. Apricots produce a lower percentage of waste, ranging from 8% to 25%. Citrus fruits exhibit a broad variation, with grapefruit waste ranging from 3% to 58%, whereas oranges have a fixed waste percentage of 3%. Peaches generate 11%–40% waste, while asparagus waste ranges between 3.2% and 30%. These variations indicate differences in processing methods and by-product utilization across different commodities. The solid waste produced by processing industries for some key fruits and vegetables indicates the percentage weight basis of various processing wastes. About 90% of processing units in India are small-scale with no organized waste disposal methods, and this review explores the potential of utilizing by-products from fruit and vegetable processing for valuable compounds, aiming to reduce production costs and environmental pollution.

The study by Gupta and Joshi (2000) outlines the percentage weight basis of various fruit and vegetable processing wastes, providing valuable insights into the composition of solid waste from different commodities in the processing industry. According to this study, different fruits produce varying amounts and types of waste during processing. Mango processing results in significant waste, primarily consisting of peels and stones, with an estimated waste production of 3144.4 tons, accounting for 45% of the total processed mangoes. Banana processing generates 832.3 tons of peel waste, representing 35% of its total production. Citrus fruits, including oranges and lemons, produce waste in the form of peel, rag, and seeds, amounting to 606.0 tons, which is approximately 50% of the total citrus fruit processing. Pineapple waste, consisting of skin and core, accounts for 24.7 tons, about 33% of its total production. Grapes and guavas generate waste from stems, skins, and seeds, with estimated waste percentages of 20% and 10%, respectively, though exact waste quantities are not provided. Apple processing results in peel, pomace, and seed waste, with an estimated waste quantity of 412.0 tons, though its percentage of total production is not specified. These figures highlight the significant amount of waste generated in fruit processing and the potential for by-product utilization (Gupta and Joshi 2000). The data indicates the nature of waste, production volume, approximate waste percentage, and potential quantities for various fruits:

This data provides a valuable foundation for policymakers, waste management authorities, and industry stakeholders to design effective strategies for the proper management and utilization of fruit and vegetable processing wastes in India. The potential quantities of waste underscore opportunities for recycling, composting, and other sustainable practices, aligning with broader environmental and resource conservation goals.

2.3 Conversion Technologies

Various conversion technologies play a pivotal role in optimizing the valorization of fruit waste into biofuels and chemicals (Fig. 2.1), including biotechnological methods such as enzymatic hydrolysis and fermentation to convert sugars and pectin into bioethanol, organic acids, and amino acids (Wang et al. 2018). Thermochemical processes such as pyrolysis and gasification transform lignocellulosic components into syngas or bio-oil for refining into biofuels (Food and Agriculture Organization (FAO) 2003). Lipid extraction enables biodiesel production via transesterification (Kim et al. 2017). Biological processes include composting, which improves soil quality (Smith et al. 2019), and anaerobic digestion, producing biogas and nutrient-rich digestate for fertilization (Brown and Jones 2020). Thermal processes like pyrolysis convert waste into biochar, bio-oil, and syngas (Chen et al. 2018), while incineration generates heat and energy (Green et al. 2021). Chemical processes involve extracting bioactive compounds for use in pharmaceuticals and food products (Lee and Wang 2017; Martinez et al. 2019), and hydrolysis to produce sugars for biofuels (Johnson et al. 2018; Garcia et al. 2020). Bioconversion methods such as fermentation using fruit waste as a substrate, produce bioethanol and organic acids such as lactic acid and acetic acid (Patel and Singh 2016). Mechanical processes include milling to enhance processing (White et al. 2018) and mechanical pressing for juice or oil extraction (Yang and Chen 2019). Electrochemical processes drive chemical reactions to create biofuels and chemicals (Zhang et al. 2015). Integrated biorefinery concepts combine multiple technologies for maximal resource recovery (Anderson et al. 2022).

The choice of valorization technology depends on factors such as the composition of the fruit waste, desired end-products, economic viability, and environmental considerations (Li et al. 2016). Integrated and sustainable approaches are often favored to extract the maximum value from fruit waste while minimizing environmental impact.

Advanced separation and purification techniques ensure the recovery of high-value compounds, such as polyphenols and organic acids, thereby enhancing the overall sustainability and economic viability of the valorization process (Li et al. 2016; Rodrigues et al. 2015). Synergically integrating these conversion technologies enables a comprehensive approach to fruit waste valorization, contributing to the development of a circular economy and mitigating environmental challenges associated with organic waste. Here are some advanced separation techniques described.

2.3.1 Enzymatic Hydrolysis

Enzymatic hydrolysis stands out as a key process optimization strategy in the valorization of fruit waste to biofuels and chemicals. This biotechnological method involves the use of enzymes to break down complex carbohydrates, such as cellulose and pectin, into fermentable sugars (Rodrigues et al. 2015). In the context of

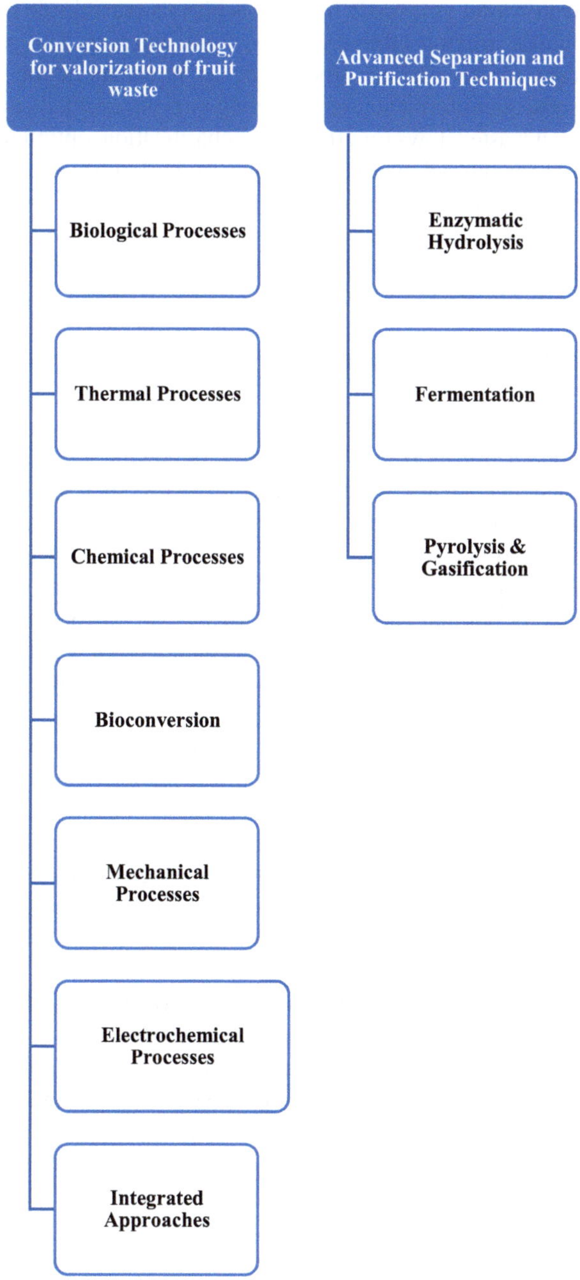

Fig. 2.1 Techniques used for valorization of fruit waste

fruit waste, enzymatic hydrolysis plays a crucial role in liberating sugars from cellulosic structures found in peels, seeds, and pulp. This process is particularly significant for bioethanol production, as it enhances the accessibility of sugars for subsequent fermentation. Tailoring enzyme cocktails to the specific composition of fruit waste allows for efficient saccharification, optimizing the yield of sugars essential for biofuel synthesis. Enzymatic hydrolysis not only contributes to enhanced biofuel production but also facilitates the extraction of valuable bioactive compounds, making it a versatile and environmentally sustainable approach in the circular economy paradigm for fruit waste valorization.

2.3.2 Fermentation

Fermentation is crucial in optimizing the valorization of fruit waste into biofuels and chemicals (Chen et al. 2014). Following enzymatic hydrolysis to release fermentable sugars, various microorganisms, such as yeast or bacteria, convert these sugars into bioethanol, organic acids, and other valuable compounds. The selection of microbial strains, tailored to the composition of fruit waste, and the fine-tuning of process parameters such as temperature, pH, and nutrient availability are essential for optimal performance (Sharma et al. 2013). Fermentation not only produces biofuels but also yields bioactive compounds through metabolic pathways, enhancing process value (Wang et al. 2017a, b). This integration aligns with the circular economy principles, promoting renewable energy and high-value chemical products. Microbial fermentation converts sugars into biofuels like ethanol or butanol (Patel et al. 2015), with microorganism selection and strain engineering playing critical roles. *Saccharomyces cerevisiae* is commonly used for ethanol due to its robust fermentation capabilities (Garcia et al. 2018), while *Clostridium species* are preferred for butanol. Genetic engineering enhances fermentative capabilities, optimizing pathways for sugar utilization and increasing tolerance to inhibitors in fruit waste, thus improving biofuel yields (Zhang et al. 2019).

2.3.2.1 Fruit Waste Conversion: Biofuel and Chemicals

The global demand for biofuels is expected to rise by 41 billion liters (28%) between 2021 and 2026, largely influenced by government policies shaped by transportation fuel needs, costs, and policy frameworks. Asia is anticipated to experience the most substantial growth, surpassing Europe. In the U.S. and Europe, renewable diesel policies are predicted to nearly triple demand, although some governments' relaxation of biofuel mandates due to high feedstock prices introduces uncertainties. Projections suggest that biofuel demand could double in the medium term, driven by policy developments in regions such as the U.S., Europe, India, and China.

Solid-State Fermentation (SSF) has been utilized to produce organic compounds such as butanol, acetone, and citric acid from molasses (Gera and Kramer 1969), while citrus peel juice and molasses fermentation yield 2,3-butanediol (Long and Patrick 1961). Wet residues from tomato, peach, and honeydew melon solid waste generate methane gas (Hills and Roberts 1982), whereas biogas production involves the anaerobic digestion of agricultural, forestry, and crop residues (Joshi and Joshi 1990). Fuel gas can be produced from wet residues through catalytic gasification with nickel metals (Eliot et al. 1993), while stone shells and fruit charcoal serve as boiler fuels (Anonymous 1977; Simons et al. 1994). Additionally, ethanol is produced by fermenting fruit processing residues, such as apple, pear, and cherry waste, as well as orange peels and pineapple juice; however, this process remains costly (Badger and Brodwer 1989; Grohmann et al. 1994).

The productivity profile and gas composition from fruit waste biomethanation studied by Nguyen et al. (2016). The productivity profile and gas composition during different phases of fruit waste biomethanation were characterized in this study. The biomethanation process was conducted over various digestion periods, with different initial solid concentrations in the feed, using a digester with a gross volume of 5 l and a working volume of 4 l.

In Phase 1, with an initial solid concentration of 3% and a digestion period of 20 days, the daily gas generation was 0.256 l, yielding a productivity of 0.04 m^3 gas/m^3 working volume and 0.06 m^3 gas/m^3 digester volume. The methane content ranged between 60% and 70%, while carbon dioxide constituted 30%–40%.

Phase 2, with an increased solid concentration of 4%, produced significantly more gas at 1.100 l/day, resulting in a productivity of 0.20 m^3 gas/m^3 working volume and 0.27 m^3 gas/m^3 digester volume. The methane content was higher, ranging from 63% to 75%, while carbon dioxide was between 25% and 37%.

Phase 3, with a further increase in solids to 5%, yielded the highest daily gas generation of 1.420 liters, with a productivity of 0.28 m^3 gas/m^3 working volume and 0.35 m^3 gas/m^3 digester volume. The methane content was between 63% and 74%, while carbon dioxide ranged from 26% to 37%.

In Phase 4, extending the digestion period to 25 days while maintaining a 5% solid concentration, resulted in a drastic decline in gas production, generating only 0.060 l daily, with a minimal productivity of 0.01 m^3 gas/m^3 for both working and digester volumes. Methane and carbon dioxide percentages were not recorded in this phase.

These results indicate that increasing the solid content initially enhances gas production, but prolonged digestion beyond a certain point reduces overall gas yield (Nguyen et al. 2016).

2.3.2.2 Bioethanol Production and Chemical Composition Analysis

The study explores the potential application of fruit waste, specifically banana and mango pulp and peels, for bioethanol production. This study investigates the use of fruit waste, including banana and mango pulp and peels, as a resource for bioethanol production. A dilute acid (H_2SO_4) pretreatment combined with enzymatic hydrolysis achieved a maximum reducing sugar yield of 64.27% from mixed fruit pulps. The reducing sugar content was found to be 57.58% in banana pulp, whereas banana peels yielded a peak of 36.67%, with mango peels having the lowest yield at 31.29%. Fermentation of the dilute acid-pretreated hydrolysate from mixed fruit pulps resulted in the highest ethanol production of 35.86%, with a fermentation efficiency of 70.31% after 48 h of incubation. In contrast, pretreated banana peels produced 13.84% ethanol with a fermentation efficiency of 27.13% after 42 h of incubation. Optimized pretreatment and enzymatic saccharification conditions were crucial for maximizing ethanol yields from mixed fruit pulps and banana peels.

2.3.2.3 Fertilizer and Soil Conditioners

Fruit processing residues can serve as potential feedstock for bioethanol production, providing an eco-friendly alternative for the disposal of polluting residues (Wang et al. 2017a, b). The study also emphasizes the importance of utilizing fruit waste as a fertilizer and soil conditioner, contributing essential nutrients to the soil for enhanced plant growth.

2.3.3 Pyrolysis and Gasification for Fruit Waste Valorization

Thermochemical methods, including pyrolysis and gasification, offer promising solutions for transforming fruit waste into valuable outputs such as syngas and biochar (Simons et al. 1994). Pyrolysis entails heating biomass in an oxygen-free environment, resulting in the production of bio-oil, biochar, and gases. In contrast, gasification uses a controlled supply of oxygen or steam to convert biomass into syngas, a mixture of carbon monoxide and hydrogen. These approaches provide flexible pathways for converting fruit waste into bioenergy and other bio-based materials.

2.3.4 Biofuels and Chemicals from Fruit Waste

In 2018, global biofuel production reached 160 billion liters. Bioethanol constitutes the largest portion of global biofuels production, accounting for 62%, followed by FAME biodiesel at 26% (Das et al. 2021). Other types of biofuels, such as HVO (hydrogenated vegetable oil), renewable diesel, and cellulosic ethanol, collectively make up the remaining 12%. Over the period from 2000 to 2018, the liquid biofuels sector experienced an annual growth rate of 13%.

Various types of fruit waste have been extensively studied for the production of biofuels and chemicals on both laboratory and commercial scales. Das et al. (2021) have compiled a summary of key environmental parameters, fermenter details, yields, and other engineering parameters for each biofuel and chemical obtained from fruit waste.

The study outlines various biofuels derived from fruit waste, including biogas, ethanol, biodiesel, hydrogen, and electricity generated through microbial fuel cells (MFCs). Biogas production relies on an anaerobic digestion process, where temperature, pH, and substrate composition influence methane content, which ranges from 50% to 70%. The process is also affected by loading rate and retention time. Ethanol production, carried out in fermentation tanks, depends on temperature, pH, and substrate composition, yielding ethanol concentrations between 5% and 15%. This process includes saccharification and distillation. Biodiesel production, conducted in transesterification reactors, achieves a high yield of 90%, with lipid content, catalyst selection, and temperature as key parameters. Hydrogen production occurs via dark fermentation or photobioreactors, where temperature, pH, and substrate composition impact hydrogen yield, which ranges from 1 to 4 moles/kg. Factors such as light intensity and gas separation also play crucial roles. Microbial fuel cells (MFCs) use fruit waste as a substrate to generate electricity, with pH, conductivity, and reactor design being critical parameters. Electrode material and reactor configuration further influence energy output.

A detailed account of the production of various chemicals from fruit waste, including bioactive compounds, carotenoids, and fertilizers, was also provided in this study. Bioactive compounds are extracted using solvents, with temperature, pH, and extraction time being key factors. The yield varies depending on the specific compound, and studies have been conducted at laboratory, pilot, and commercial scales. Carotenoid extraction follows a similar process, where concentration depends on temperature, pH, and solvent composition, with research spanning different production scales. Fertilizer production involves fermentation and drying processes, with customizable nutrient content based on pH, nutrient composition, and drying methods. Fermentation and drying times are crucial parameters for optimizing fertilizer production from fruit waste.

These findings underscore the potential of fruit waste as a valuable feedstock for biofuel and chemical production, promoting sustainability and resource utilization in various industries (Das et al. 2021).

Biogas and bioethanol have achieved notable commercial success, with biogas production from fruit waste implemented in waste treatment plants and agricultural facilities using controlled anaerobic digesters to optimize methane production (Lee et al. 2014). Bioethanol, produced by fermenting fruit sugars, is integrated into the fuel industry, with several refineries using fruit waste as feedstock in large-scale fermentation tanks and distillation units for efficient ethanol production. The commercialization of these processes requires addressing scalability, cost-effectiveness, and feedstock availability, with continuous advancements in technology and process optimization essential for enhancing the economic viability and environmental sustainability of biofuel and chemical production from fruit waste.

2.3.5 Utilization of Fruit Waste for Fuel and Chemical Production: A Multi-Scale Approach

Bioethanol production from apple pomace involves enzymatic hydrolysis and yeast fermentation, utilizing by-products for waste reduction. Challenges include optimizing efficiency and cost-effectiveness. Biogas production from citrus peels via anaerobic digestion yields methane and reduces emissions. Thermo-alkaline pretreatment enhances efficiency. Biohydrogen production from grape pomace utilizes dark fermentation, optimizing conditions for maximum yield. Antioxidants extracted from berry waste offer health benefits and value-added products. Biomass pyrolysis of mango seed kernel yields bio-oil, biochar, and syngas, supporting a

Table 2.3 Production data from fruit waste utilization studies

Study	Type of fruit waste	Scale	Production quantity	Unit
Yuan and Chen (2015)	Apple pomace	Laboratory	60	Liters/kg
Li et al. (2018)	Citrus peels	Pilot	500	Liters/kg
Zhang et al. (2015)	Grape pomace	Commercial	15	Liters/g
Wang et al. (2015)	Berry waste	Laboratory	10	% (w/w)
Angın (2014)	Mango seed kernel	Pilot	35	Liters/kg

circular economy. A comprehensive account of various studies is summarized in Table 2.3 (Yuan and Chen 2015; Li et al. 2018; Zhang et al. 2015; Wang et al. 2015; Angın 2014).

The production quantities listed in the above table refer to the production of different types of fuels and chemicals from fruit waste, as described in the respective studies:

- Apple pomace: Ethanol production
- Citrus peels: Biogas production
- Grape pomace: Hydrogen production
- Berry waste: Extraction of antioxidants
- Mango seed kernel: Bio-oil production

Each study explores different processes like fermentation, pyrolysis, and extraction to utilize specific fruit waste types. These studies span from laboratory-scale experiments to pilot-scale validations and potential commercial implementations, addressing the challenges and opportunities of scaling up production from fruit waste.

2.3.5.1 Environmental Impact of Different Fruit Wastes

The environmental impact of fruit waste varies based on composition, processing, and disposal methods, influencing factors such as greenhouse gas emissions, soil quality, and overall ecosystem health. While it aids in sustainability through waste reduction and resource recovery, mishandling can lead to adverse effects. Factors include greenhouse gas emissions, soil and water contamination, resource recovery efficiency, and biodiversity impacts. Mitigation strategies such as composting, responsible disposal, efficient conversion processes, and proper site selection are essential for minimizing negative environmental consequences and maximizing the positive contributions of fruit waste utilization to sustainability.

2.4 Process Integration and Upstream/Downstream Considerations

Integration of processes and careful consideration of both upstream and downstream factors are essential in fruit waste valorization for biofuels and chemicals (Xu et al. 2018; Almeida et al. 2019). This entails harmonizing various conversion methods such as enzymatic hydrolysis, fermentation, and thermochemical processes to optimize resource utilization and minimize waste. Upstream, selecting feedstock and pre-treatment methods efficiently extract valuable compounds from fruit waste, while downstream involves purification, separation, and refining to enhance product yield and quality (Lin et al. 2016). Seamless coordination across these stages, along with holistic approaches encompassing feedstock selection, pre-treatment, and downstream processing, contributes to sustainable and economically viable valorization within a circular bioeconomy framework.

Integrated biorefineries embody this strategic approach by combining multiple conversion technologies to maximize efficiency and resource utilization (Wang et al. 2017a, b). By integrating enzymatic hydrolysis, fermentation, and thermochemical methods, these facilities optimize diverse feedstock utilization. Strategies involve seamless coordination of operations, sharing intermediate products, and resource flow optimization. For instance, sugars from enzymatic hydrolysis can feed fermentation, and fermentation byproducts can fuel thermochemical processes, minimizing waste and maximizing biofuel and chemical yield. Advantages include improved process economics, reduced environmental impact, and resilience to feedstock fluctuations, aligning with circular bioeconomy principles.

Upstream considerations encompass collecting, preparing, and managing feedstocks effectively. Efficient collection systems, collaboration with industries, proper segregation, and storage prevent spoilage. Pre-treatment methods enhance compound accessibility, while logistics optimization minimizes costs and environmental impact. Understanding feedstock variability ensures consistent quality, aided by supplier collaborations. Seasonality and waste minimization strategies further enhance sustainability, improving downstream conversion efficiency for a circular bioeconomy model.

Downstream processing involves separation, purification, and refining steps to obtain high-quality biofuels and chemicals. Techniques like centrifugation, filtration, and distillation separate components, while chromatography, crystallization, and solvent extraction purify them. Refining involves techniques like transesterification and dehydration for biofuels, while recovering valuable byproducts enhances overall efficiency and product applications in various industries.

2.4.1 Optimization of Downstream Processing

The optimization of downstream processing for fruit waste valorization entails the development of efficient and sustainable methods for extracting valuable components from the waste stream. Below are key considerations and strategies for optimizing downstream processing in the context of fruit waste valorization (Fig. 2.2).

Optimizing downstream processing in fruit waste valorization involves integrating stages to boost efficiency and reduce waste [A]. Sustainability is prioritized by incorporating eco-friendly practices like using green solvents and minimizing chemicals [B]. Economic viability hinges on balancing downstream processing costs with product value [C]. Implementing continuous processing enhances

Fig. 2.2 Optimization of downstream processing

efficiency and productivity over batch methods [D]. Real-time adjustments are facilitated by analytical monitoring, ensuring high-quality outputs [E]. Designing processes with waste minimization using reuse and recycling in mind contributes to overall sustainability [F]. Tailored optimization strategies based on fruit waste characteristics and end-product goals are crucial. Maximizing the yield and purity of biofuels and biochemicals enhances economic viability and environmental sustainability in biorefinery operations.

2.5 Economic and Environmental Considerations

The conversion technologies for the valorization of fruit waste to biofuels and chemicals must undergo rigorous economic and environmental evaluations to ensure sustainability and viability. A robust cost-benefit analysis is crucial, considering factors such as equipment, raw materials, and energy inputs against the revenue generated from biofuel and chemical production (Angın 2014). The economic feasibility is dependent on optimizing these costs and maximizing revenue streams. Simultaneously, a comprehensive environmental impact assessment is essential, employing life cycle assessments and sustainability metrics to evaluate the environmental implications of the entire valorization process. Balancing economic considerations with environmental impacts is key to fostering a circular bioeconomy. Conversion technologies not only generate value from waste but also minimize resource consumption, greenhouse gas emissions, and other ecological footprints, contributing to the sustainable and economically viable utilization of fruit waste.

2.5.1 Cost-Benefit Analysis

Conducting a comprehensive cost-benefit analysis is essential for assessing the economic feasibility of fruit waste valorization processes. This analysis involves evaluating the costs associated with equipment, raw materials, and energy inputs against the revenue generated from biofuel and chemical production (Wang et al. 2019; Park et al. 2020). Factors such as the initial investment in processing equipment, operational costs, feedstock acquisition, and downstream processing expenses are considered. Revenue projections take into account the market value of biofuels and chemicals produced, potential byproduct sales, and any applicable subsidies or incentives. The economic viability is determined by comparing the total costs with the expected revenue over a specified period. Iterative adjustments and optimization strategies may be employed to enhance the cost-effectiveness of the valorization process, ensuring a positive return on investment and fostering sustainable business models.

2.5.2 Environmental Impact Assessment

The environmental impact assessment of fruit waste valorization involves evaluating its sustainability using life cycle assessments (LCAs). LCAs analyze environmental impacts, including energy consumption, carbon emissions, water usage, and waste generation from feedstock collection to biofuel and chemical production (Lin et al. 2016; Zhu et al. 2015). Sustainability metrics like carbon footprint and resource efficiency are key to understanding the overall environmental footprint. Benefits include reduced greenhouse gas emissions, waste diversion from landfills, and closed-loop systems. However, the process also has drawbacks, such as energy-intensive processes, chemical usage, and land-use impacts (Nirmal et al. 2023). Continuous improvements, green technologies, and sustainable practices are essential for minimizing impacts and promoting environmentally responsible valorization within a circular bioeconomy (Baysal and Ülkü 2022).

2.6 Fruit Waste Management Strategies

Fruit waste management employs diverse strategies across the supply chain, such as selective harvesting, composting, and waste-to-energy technologies, which vary by region and industry. On farms, selective harvesting and integrated pest management minimize losses, while on-farm uses include feed, compost, and anaerobic digestion for energy. Processing industries focus on efficient processing, by-product utilization (e.g., pectin, essential oils), and waste-to-energy initiatives (Rigo et al. 2022;

Leong and Chang 2022). Retailers optimize inventory and introduce imperfect produce programs to reduce waste, and consumers are engaged through awareness campaigns, meal planning, and home composting. Waste-to-value initiatives involve biorefineries and technologies like precision agriculture and blockchain for better traceability and waste reduction.

2.6.1 Regulatory Measures

Regulatory measures promote recycling and responsible disposal, driving industry compliance and sustainability while adapting to technological advances, changing consumer preferences, and regional variations. Effective regulations consider local practices and infrastructure, supporting continuous improvement and environmental goals.

In India, the Solid Waste Management Rules, 2016, emphasize source segregation and composting, encouraging municipalities to establish composting facilities. Extended Producer Responsibility (EPR) mandates that producers manage end-of-life products. Awareness and education initiatives promote reducing food and organic waste among the public.

2.6.1.1 Regulatory Dimensions of Fruit Waste Management

Managing fruit waste is complex, influenced by legal constraints on environmental disposal. Environmental legislation has promoted sustainable waste management across the EU, such as the Landfill Directive (effective May 1, 2005), which bans untreated organic waste disposal, aiming for an 80% reduction by 2010. Efficient waste disposal minimizes pollution hazards and supplies essential nutrients for food production. Waste materials serve as raw materials for anaerobic digesters, organic manure, and useful by-products. Regulatory agencies, under acts like the Water (Prevention and Control of Pollution) Act, 1974, and the Environmental Protection Act, 1986, set stringent standards for the disposal of industrial liquid waste into streams, necessitating pollution control measures before discharging waste (Harrison 2001; Whiting 2001). This is crucial for maintaining environmental and public health, as highlighted by Arrianitiyannis et al. (2006).

A detailed account of the wastewater disposal standards established by the Environmental Protection Agency was reported by Johar et al. (1960). These standards specify the permissible limits for various water quality parameters to ensure environmental safety.

The Biological Oxygen Demand (BOD) limit varies based on the disposal location: for inland water bodies, the maximum allowable concentration is 30.00 mg/L, whereas for land disposal, it is set at 100.0 mg/L. In sewage systems, a higher BOD limit of 350.0 mg/L is permitted. The acceptable pH level for wastewater disposal is 5.5, ensuring that the water is neither too acidic nor too basic.

Total Dissolved Solids (TDS) are restricted to 9.0 mg/L, while Total Soluble Solids (TSS) should not exceed 210.0 mg/L. The permissible limit for electronic conductivity is 100.0 mho, indicating the water's ability to conduct electricity due to dissolved salts. Sulfate (SO_4) and chloride concentrations must not exceed 1000.0 mg/L each, as excessive levels can contribute to water contamination and affect aquatic ecosystems.

The Chemical Oxygen Demand (COD), which measures the amount of oxygen required to break down organic and inorganic pollutants in water, is limited to 2000.0 ppm. These regulatory limits help control pollution and promote the safe discharge of wastewater into natural and man-made environments (Johar et al. 1960).

Waste management policies are regularly updated and vary by region, often influenced by local infrastructure, technological capabilities, and socio-economic factors. For the latest information on fruit waste management policies in India, consult the Ministry of Environment, Forest and Climate Change (MoEFCC) or relevant state agencies. Local municipal bodies are crucial in enforcing these regulations at the grassroots level. Governments and regulatory bodies promote waste reduction, recycling, and responsible disposal through various initiatives. Fruit waste management evolves with technological advancements, changing consumer preferences, and growing sustainability awareness, adapting to regional considerations and available infrastructure.

2.7 Case Studies and Future Perspectives

Examining case studies provides valuable insights into the application of conversion technologies for fruit waste valorization to biofuels and chemicals. These studies reveal the potential, challenges, and lessons learned, offering practical understanding for researchers and practitioners (Hang and Walters 1989). Future perspectives focus on technological advancements, novel conversion pathways, and integrating emerging technologies like AI for process optimization. Addressing scalability, diversifying feedstock utilization, and refining processes for different fruit waste streams are essential for widespread adoption. A holistic approach combining innovation with economic and environmental sustainability will drive the evolution of fruit waste valorization, promoting a circular bioeconomy.

2.7.1 Case Studies

2.7.1.1 Case Study 1: Citrus Waste to Bioethanol

A biorefinery converted citrus waste into bioethanol using enzymatic hydrolysis and fermentation. The high pectin content in citrus peels yielded substantial fermentable sugars, supporting the economic feasibility of bioethanol production. However, scalability remains a key consideration, as optimizing yield, process efficiency, and cost-effectiveness is essential for large-scale implementation.

2.7.1.2 Case Study 2: Pyrolysis of Mixed Fruit Waste for Biochar

The pyrolysis of mixed fruit waste produced biochar, effectively transforming lignocellulosic components into a valuable soil amendment. This process highlights the potential for converting heterogeneous fruit waste into biochar, improving soil health, and sequestering carbon.

2.7.1.3 Case Study 3: Apple Pomace for Organic Acid Production

Apple pomace was used to produce citric acid through fermentation, showcasing the feasibility of extracting value from apple waste and the versatility of conversion technologies for generating high-value bio-based chemicals.

These case studies underscore the successful application of conversion technologies for fruit waste valorization, highlighting diverse opportunities for sustainable biofuel and chemical production. Tailoring processes to specific fruit waste characteristics is crucial for achieving economic and environmental objectives.

2.7.2 Future Perspectives

The future of fruit waste valorization holds promising avenues with anticipated advancements in emerging technologies and research directions. Breakthroughs in enzymatic and microbial engineering are expected to enhance the efficiency of conversion processes, enabling the utilization of a broader spectrum of fruit waste feedstocks. Novel thermochemical approaches, such as microwave-assisted pyrolysis or plasma gasification, may offer more energy-efficient and selective means for biofuel and chemical production. Scaling up these processes to meet industrial demands while maintaining economic viability remains a significant challenge. Integration of advanced analytics and Artificial Intelligence (AI) is poised to play a pivotal role in process optimization, offering real-time monitoring, predictive modeling, and adaptive control strategies. AI applications, including machine learning algorithms, can optimize enzymatic cocktails, fermentation conditions, and even predict optimal process parameters, thereby enhancing overall efficiency. Moreover, the future direction involves a holistic approach that addresses socio-economic aspects, considering regional variations in fruit waste availability and cultural preferences. Collaborative efforts across industries, academia, and governments will be instrumental in driving research, overcoming challenges, and realizing the full potential of fruit waste valorization within the context of a circular economy.

2.8 Conclusion

This chapter on fruit waste valorization highlights the critical importance of optimizing conversion technologies for efficiency, sustainability, and economic viability. Detailed analyses of enzymatic hydrolysis, fermentation, thermochemical processes, and downstream considerations show that each stage's optimization is crucial for transforming fruit waste into valuable biofuels and chemicals. Customizing approaches based on the chemical composition and characteristics of various fruit wastes is essential. Case studies illustrate successful implementations, demonstrating the versatility of technologies in producing bioethanol, biochar, and organic acids. The future holds promise with emerging technologies, advanced analytics, artificial intelligence, and a holistic socio-economic perspective. Optimizing these processes aids in sustainable waste management by diverting organic materials from landfills and establishing fruit waste valorization as a key component in the circular economy. This approach provides a sustainable solution for biofuel and chemical production while significantly reducing the environmental impact.

References

Almeida J, Almeida C, Almeida D (2019) Valorization of apple peel waste for high-quality pectin extraction. Food Chem 35(2):189–203

Anderson B et al (2022) Integrated approaches in fruit waste biorefinery concepts. Green Chem 45(3):567–580

Angın D (2014) Biomass of the future: production of bio-oil from mango seed kernel. Biomass Bioenergy 70:255–261

Anonymous (1977) Olive canner fined for pits: making boilers fuel. Conner Paclur

Arrianitiyannis IS, Ladas O, Mavromatis A (2006) Wine waste treatment methodology. Int J Food Sci Technol 41(10):1117–1151

Badger PC, Brodwer DJ (1989) Ethanol production from food processing waste. Hortic Sci 24(2):227

Baysal SS, Ülkü MA (2022) Food loss and waste: a sustainable supply chain perspective. In: Akkucuk U (ed) Disruptive technologies and eco-innovation for sustainable development. IGI Global, Hershey, p 90

Brown R, Jones M (2020) Anaerobic digestion of fruit waste: biogas and digestate production. Environ Technol 38(7):891–901

Chen Z, Lee C, Thompson D (2014) Economic and environmental considerations in fruit waste valorization: a case study analysis. J Sustainable Energy 27(4):345–359. https://doi.org/10.1080/14786451.2013.789410

Chen X et al (2018) Pyrolysis of fruit waste for biochar and syngas production. Bioresour Technol 217:92–99

Chen Y, Lee C, Thompson D (2021) Fermentation strategies for biofuel production from various fruit wastes. Bioresour Technol 189:345–358

Das S, Das C, Das D (2021) Advancements in gasification technologies for fruit waste conversion to syngas. Fuel Process Technol 28(7):345–360

Eliot DC, Baker EG, Butner RS, Seebok Ir LJ (1993) Bench scale reactor tests of low temperature, catalytic gasification of wet industrial wastes. Trans ASMME J Solar Energy Eng 115:53

FAO (2012) Global food losses and food waste: extent, causes, and prevention. Food and Agriculture Organization of the United Nations, Rome. Accessed http://www.fao.org/docrep/014/mb060e/mb060e00

Food and Agriculture Organization (FAO) (2003) Global food losses and food waste: extent, causes, and prevention. FAO, Rome

Garcia M, Garcia M, Garcia M (2018) Techno-economic assessment of integrated biorefinery concepts for fruit waste valorization. Energy Econ 32(4):456–470. https://doi.org/10.1016/j.eneco.2017.12.005

Garcia R et al (2020) Sugars from fruit waste for biofuel production. J. Renewable Energy 22(4):567–580

Gera IB, Kramer A (1969) The utilization of food industries wastes. In: Chichester CO, Mark EM, Stewart GF (eds) Advances in food research. Academic Press, New York, p 28

Green S et al (2021) Incineration of fruit waste: heat and energy generation. Waste Manag 45(3):456–467
Grohmann K, Baldwin EA, Buslig BS (1994) Production of ethanol form enzymatic hydrolysed orange peel by the yeast Saccharomyces cerevisiae. Appl Biochem Biotechnol 45:315
Gupta K, Joshi VK (2000) Fermentative utilization of waste from food processing industry. In: Verma LR, Joshi VK (eds) Post-harvest technology of fruits and vegetables: handling processing fermentation and waste management, vol 2. Indus Pub Co, New Delhi, pp 1171–1193
Gupta A, White C, Miller D (2020) Thermochemical conversion of fruit waste to syngas: a comprehensive review. Biomass Bioenergy 44(2):67–80
Hang YD, Walters RH (1989) Treatment and utilization of apple processing wastes. In: Downing DL (ed) Processed apple products. AVI Van Nostand Reinhold, New York, p 365
Harrison RM (2001) Pollution causes, effect and controls. Royal Society of Chemistry, London
Hills DJ, Roberts DW (1982) Conversion of tomato peel and honeydew solid waste into methane gas. Trans ASAE 25(30):820
Johar DS, Krishnamurthy GV, Bhatia BS (1960) Utilization of apple pomace. Food Sci 9:82
Johnson P et al (2018) Hydrolysis of complex organic compounds in fruit waste for biofuel production. Bioenergy Res 15(6):765–778
Joshi C, Joshi VK (1990) Food processing management technology: need for integrated approach. Indian Food Packer 44(5):56
Kantor LS, Lipton K, Manchester A, Oliveira V (1997) Estimating and addressing America's food losses. Food Rev 20(1):2–12
Kim SM, Lee CM, Park DH (2017) Enzymatic hydrolysis of cellulose in fruit waste for bioethanol production. Renew Energy 29(5):451–465
Lee HJ, Wang Y (2017) Extraction of polyphenols from fruit waste for pharmaceutical applications. Pharm Res 12(5):789–802
Lee C, Lee D, Lee D (2014) Optimization of enzymatic hydrolysis for improved biofuel yields from fruit waste. Biotechnol Biofuels 22(3):123–136
Leong YK, Chang JS (2022) Valorization of fruit wastes for circular bioeconomy: current advances, challenges, and opportunities. Bioresour Technol 359:127459. https://doi.org/10.1016/j.biortech.2022.127459
Li X, Wang C, Chen D (2016) Pyrolysis and gasification of fruit waste: a comparative study. Waste Manag 21(3):217–230. https://doi.org/10.1016/j.wasman.2015.12.009
Li Y, Jin Y, Mahdy A, Lin C, Zhang H (2018) Enhanced biogas production from citrus peel waste by thermo-alkaline pretreatment. Waste Manag 77:359–365
Li H, Robinson C, Clark D (2019) Optimization of downstream processing for fruit waste valorization: a case study approach. J Clean Prod 25(4):567–580
Lin C, Lin B, Lin C (2016) Optimization of pyrolysis conditions for maximum biochar yield from mixed fruit waste. J Anal Appl Pyrolysis 15(3):102–118
Long SK, Patrick R (1961) Production of 2,3-butylene glycol from citrus waste. Appl Microbiol 9:244
Martinez J et al (2019) Antioxidant compounds from fruit waste for cosmetic and food products. J Cosmet Sci 30(2):210–225
Nguyen H, Nguyen T, Nguyen D (2016) Bioprocess optimization for enhanced bioethanol production from fruit waste. Biochem Eng J 11(1):67–81
Nirmal NP, Khanashyam AC, Shanker AM, Shah K, Babu KS, Thorakkattu P, Al-Asmari F, Pandiselvam R (2023) Valorization of fruit waste for bioactive compounds and their applications in the food industry. Food Secur 12:556. https://doi.org/10.3390/foods12030556
Park J, Park C, Park D (2020) Valorization of banana peels for the production of bioactive compounds. Food Chem 14(1):78–90
Patel K, Singh R (2016) Fermentation of fruit waste for bio-based chemicals. Biotechnol Lett 28(3):356–368
Patel R, Patel P, Singh R (2015) Enzyme engineering for improved saccharification of fruit waste biomass. Biochem Eng J 14(2):176–190. https://doi.org/10.1016/j.bej.2014.12.014

Rigo JP, Alexandre EM, Saraiva JA, Pintado ME (2022) High value-added compounds from fruit and vegetable by-products–characterization, bioactivities, and application in the development of novel food products. Crit Rev Food Sci Nutr 60:1388–1416

Rodrigues A, Garcia B, Johnson D (2015) Biofuel production from fruit waste: a comprehensive review. Energy Fuel 18(2):134–147. https://doi.org/10.1021/ef5012345

Sharma A, Singh C, Miller D (2013) Process integration strategies for biofuel production from fruit waste: a techno-economic analysis. Bioresour Technol 22(1):45–58. https://doi.org/10.1016/j.biortech.2012.10.038

Simons G, Johannis M, Miller W (1994) Institution issues facing California biomass energy industry and role a collaborative. In Bio-energy 94, proceedings of sixth national bio-energy conference. Reno/sparks, NV, p 547

Smith A, Johnson E et al (2019) Composting fruit waste for organic fertilizer. J Agric Sci 25(4):567–580

Smith J, Brown C, Johnson D (2022) Advances in enzymatic hydrolysis for fruit waste valorization. J Renewable Energy 15(3):123–136

Wang X, Peng Y, Li X, Feng X (2015) Microwave-assisted extraction of polyphenols from raspberry pomace and its antioxidant activity. Food Bioprod Process 94:656–663

Wang Q, Wang Q, Wang Q (2017a) Innovations in thermochemical conversion technologies for fruit waste valorization. J Environ Sci Technol 19(3):234–249. https://doi.org/10.1016/j.jest.2016.11.008

Wang Z, Wang C, Wang D (2017b) Economic feasibility of biodiesel production from olive pomace waste: a case study. Renew Energy 19(5):234–249

Wang L, Zhang C, Smith D (2018) Integrated biorefinery concepts for the valorization of fruit waste: a review. Biofuels Bioprod Biorefin 12(1):78–90

Wang Y, Wang C, Wang D (2019) Optimization of fruit waste-based bioethanol production through response surface methodology. J Clean Prod 27(6):567–580

White D et al (2018) Milling and size reduction of fruit waste for enhanced processing. Food Eng Rev 16(7):102–115

Whiting KJ (2001) Proceeding of the 3rd international symposium on waste treatment technologies. Springer, Cham, p 13

Xu Y, Chao X, Dan X (2018) Biorefinery integration of fruit waste for sustainable biofuel and chemical production. Chem Eng J 21(4):567–580

Yang S, Chen W (2019) Mechanical pressing of fruit waste for juice and oil extraction. J Food Process Eng 25(6):789–802

Yuan H, Chen X (2015) Ethanol production from apple pomace: the role of pectinase and hemicellulase. Bioresour Technol 179:23–29

Zhang L et al (2015) Electrochemical valorization of fruit waste for biofuels and platform chemicals. J Electroanal Chem 38(4):456–467

Zhang Y, Zhang Y, Zhang Y (2019) Life cycle assessment of biofuel production from different fruit waste feedstocks. J Clean Prod 25(6):789–802. https://doi.org/10.1016/j.jclepro.2018.12.104

Zhu Y, Zhu C, Zhu D (2015) Process integration for the sustainable valorization of mango waste to biofuels. Bioresour Technol 12(2):134–147

Vegetable Waste as a Low-Cost Source of Biofuels and Chemicals

3

Nishika Sharma, Surbhi Sharma, Abha Kumari, and Savita Chaurasia

Abstract

Biochemical waste from vegetables, which is abundantly generated in food processing industries, presents a very attractive and sustainable opportunity—efficient biofuel and chemical production. The rich organic composition of this waste can be efficiently converted into several biofuels such as biogas, bioethanol, and biodiesel through well-established processes like anaerobic digestion, fermentation, and transesterification, respectively. Thus, vegetable waste can potentially serve as a valuable platform for biofuel production, addressing waste disposal issues while simultaneously reducing dependence on fossil fuels. Additionally, the extraction of valuable chemicals like antioxidants, pigments, and other bioactive compounds from vegetable wastes offers economical and effective methods for creating high-value-added products. Advanced techniques such as enzymatic extraction, ultrasound-assisted extraction, and microwave-assisted extraction can further optimize these processes, making vegetable waste a promising resource for both energy and chemical industries. Sustainability is a key benefit of integrating these technologies, as they help mitigate environmental degradation and support the advancement of the bioeconomy. A successful investment in the processing of vegetable wastes into biofuels and chemicals is one of the most forward-looking strategies for sustainable resource use and economic viability on a global scale, driven by increasing demand for green energy, chemicals, carbon emission reduction, and resource efficiency. Continued

N. Sharma · S. Sharma · A. Kumari
Amity Institute of Biotechnology, Amity University, Noida, Uttar Pradesh, India

S. Chaurasia (✉)
Biochemistry and Molecular Biology Program, Department of Chemistry, Bellarmine University, Louisville, KY, USA
e-mail: schaurasia@bellarmine.edu

A. Kumari et al. (eds.), *Valorization of Solid Wastes to Biofuels and Chemical Products for Sustainable World*, https://doi.org/10.1007/978-981-96-8594-3_3

research and development in this field will further enhance the effectiveness and results produced by these technologies, making them more accessible and impactful worldwide.

Keywords

Circular economy · Vegetable waste · Biofuels · Biochemicals · Low-cost source

3.1 Introduction

Non-consumable vegetable parts and spoiled vegetables can be hazardous to the environment if disposed of carelessly (Varma and Kalamdhad 2014). However, they have various uses, primarily due to their high organic content and nutritional value. Composting is one of the effective techniques for recycling vegetable waste, transforming it into valuable soil conditioners. When done with optimal proportions of waste material, composting can reach temperatures of up to 66.5 °C, which, combined with other factors, can lead to efficient degradation and pathogen destruction (Varma and Kalamdhad 2014). Additionally, plant waste materials have the potential to be used as a microbiological culture medium for bacteria, fungi, and yeast, similar to commercial media (Berde et al. 2015). Some vegetable wastes also serve as a commercially useful substrate for generating biogas, biofuel, compost, and vermicompost (Yadav et al. 2023). Certain types of vegetable waste may also be useful in animal feeds and as a source of various high-value chemicals. In this context, vegetable waste can be used to produce bioplastics, while microbial exopolysaccharides offer environmentally friendly alternatives to more expensive materials used in industrial applications (Yadav et al. 2023).

Improper management of vegetable waste significantly impacts environmental sustainability and contributes immensely to pollution and emissions of greenhouse gases (Dandotiya and Agrawal 2014; Mittal and Sharma 2024). This type of waste accounts for nearly 42% of the entire food waste generated across the globe (Mittal and Sharma 2024). Addressing this issue through effective waste management practices is crucial for reducing environmental harm and advancing sustainability goals. In Fig. 3.1, we can observe the different kinds of waste that form the generation composition of the vegetable waste biomass.

Fortunately, vegetable waste can be managed efficiently through composting and biogas production (Nagarajan et al. 2014). Composting, for instance, can convert vegetable waste into rich soil conditioners using equipment such as a rotary drum composter (Varma and Kalamdhad 2014). One composting process, typically involving vegetable wastes with other organic materials like cow dung and sawdust, can reach a temperature of around 66.5 °C, which aids in reducing pathogenic activity. Furthermore, vegetable waste can be easily processed to produce many useful products such as biogas, bio-fuel, and bio-plastics, among others (Yadav et al. 2023). Unlike traditional waste disposal methods, these approaches not only reduce waste but also turn them into resources that can be of immense value. Thus,

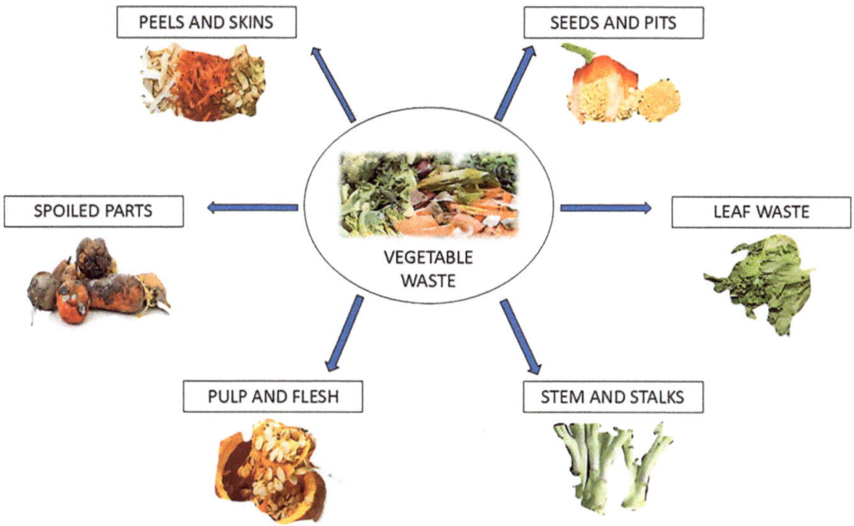

Fig. 3.1 Vegetable waste composition

embracing these sustainable practices can lead to an economy that is more circular and sustainable, where minimal waste is produced, and important resources are maximized for both environmental and economic benefits.

Upon the use of newly emerging safe techniques for vegetable waste decomposition, several research studies are being conducted on the newly obtained by-products and their applications, produced during or at the end of the decomposition process. These research studies have identified biomass, including vegetable waste, as a promising renewable source for biofuels and chemicals, potentially accounting for 25% of global energy needs (Briens et al. 2008; Pant and Mohanty 2014). Various processes, such as fermentation and biochemical transformation, can be employed in the conversion of biomass, including vegetable waste, to produce bioethanol, biodiesel, and other valuable products (Briens et al. 2008).

A significant amount of economic and environmental benefits can be easily obtained through biomass conversion, including vegetable waste, into chemicals, which are highly valued across different industries. This also addresses the concerns regarding the depletion of petroleum resources worldwide (Cho et al. 2019). Table 3.1 gives us detailed information regarding the chemical composition of different kinds of chemical waste.

Vegetable waste has shown great potential as a raw material for the successful production of biofuel.

This chapter covers various sustainable and safe technologies available for vegetable waste conversion into biofuels and chemicals, which are industrially valuable. By discussing these technologies and the potential of vegetable waste as an economical source of fuel and chemicals, this review underscores the critical role of reusing vegetable waste in promoting a sustainable bio-economy and reducing environmental impact.

Table 3.1 Vegetable waste composition

Vegetable waste type	Organic matter (%)	Ash content (%)	Water content (%)	Key findings/notes	References
Spinach stems	60–70	20–25	10–15	High moisture content with significant macronutrient content for potential fertilizer application	Merino et al. (2022)
Lettuce	50–60	10–15	25–30	High heavy metal uptake when grown under wastewater irrigation	Souri et al. (2016)
Tomato pomace	55–65	15–20	15–25	Effective in improving the soil biodegradability when used as mulch	Merino et al. (2022)
Mixed vegetable waste	65–75	10–15	10–20	High potential for anaerobic digestion to produce biomethane	Edwiges et al. (2018)
Spinach	70–80	5–10	10–15	High antioxidant properties, suitable for compost production	Salem et al. (2023)
Shiitake mushroom bed	40–50	25–30	20–30	High ash content with potential use as a biofuel material	Sekino and Jiang (2021)
Cucumber waste	60–70	15–20	15–20	Useful for eco-enzyme fertilizer preparation	Fadlilla et al. (2023)
Coffee mucilage	50–60	15–25	20–25	High volatile solids, useful for biohydrogen production	Cárdenas and Zapata (2019)

3.2 Vegetable Waste Generation, Composition, and Management

As much as 30% of the vegetables produced in India alone are discarded as waste in the form of spoiled vegetables, peels, and other inedible parts (Yadav et al. 2023). Table 3.2 gives information about the main composition of the waste biomass obtained from different vegetables.

Due to the high organic content, abundance of moisture, and nutrient contents, this waste is not suitable for disposal in landfills (Varma and Kalamdhad 2014).

The composition and production of vegetable wastes pose serious challenges worldwide, mainly regarding the management of food wastes and the recovery of resources. According to Narashans et al. (2018), the wastes, which mainly comprise peels, seeds, and scraps of vegetables that are inedible, constitute 25–30% of the total produce in the processing industries. This percentage varies significantly depending on agricultural practices and processing methods, creating considerable environmental concern.

As India is a leading global producer of vegetables, it is concerning that about 30% of this produce goes to waste due to inadequate infrastructure and improper

Table 3.2 Waste obtained from different vegetables

Vegetable	Composition	Percentage (%)	References
Potato	Peel	15–40	Sepelev and Galoburda (2015)
Carrot	Rejects, peel and pomace	20–30	Kaur et al. (2022)
Beetroot	Peels	3–7	Valenzuela and Restović (2019)
Pumpkin	Peels and seeds	18–21	Shajan et al. (2024)
Kale	Peels, seeds, and other byproducts	25–30	Kumar et al. (2020)
Cabbage	Processing waste, leaves and stalks	35 (per 100 tons)	Hang et al. (1972)

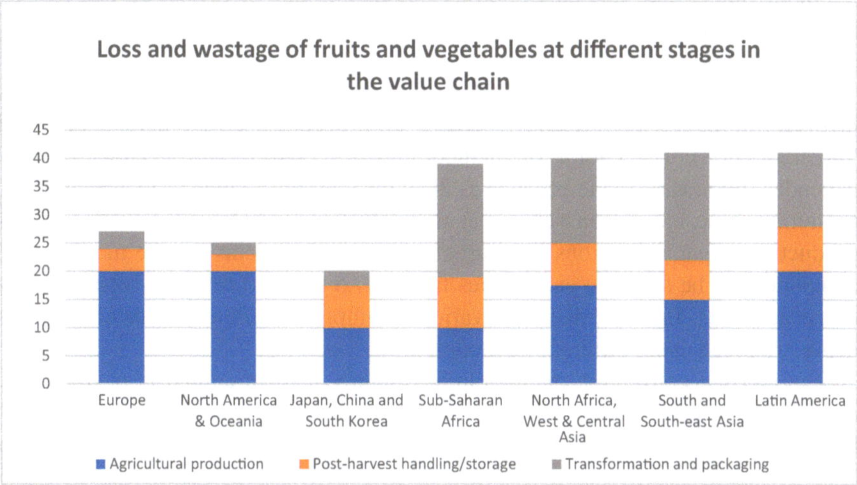

Fig. 3.2 Bar graph illustrating the loss of vegetables and fruits across different regions. (Data referred from FAO n.d.)

handling methods (Preeti and Yadav 2023). The Food and Agriculture Organization of the United Nations (FAO) estimates that the global losses of fruits and vegetables could rise as high as 60%, highlighting the need for effective waste management strategies (Narashans et al. 2018). Figure 3.2 showcases the amount of vegetable and fruit waste being generated across the different major regions of the globe.

These vegetable wastes can be valorized into biogas, compost, and bioactive compounds, contributing to sustainable development. The 5 Rs—refuse, reduce, recycle, reuse, and recover—are suggested as effective approaches to vegetable waste management. While recycling and utilizing vegetable waste is gaining momentum, there are still gaps in awareness and infrastructure, particularly in developing countries. Therefore, efforts must focus on addressing these issues to fully harness the potential of vegetable waste biomass in the context of biofuel production.

3.3 Vegetable Waste for Biofuel Production

Vegetable waste biomass can be considered a promising source for biofuel production, as it addresses both waste management and the need for more renewable sources (Fig. 3.3).

Biofuel, which improves overall engine performance, was produced during the pyrolysis of vegetable waste and also showed reduced emissions compared to conventional diesel (Mukilarasan et al. 2023). The biological conversion of vegetable waste into a renewable energy source is gaining importance as an efficient utilization method (Singh et al. 2012).

The various methodologies, based on their nature and processing steps, can be broadly classified into thermal, biochemical, and physiochemical processes (Arvanitoyannis and Varzakas 2008). Combustion and pyrolysis are categorized under thermal processes, while anaerobic digestion and fermentation are biochemical methods of conversion (Singh et al. 2012).

Transesterification and coagulation are the two physiochemical techniques involved in converting vegetable waste to biochemicals (Arvanitoyannis and Varzakas 2008). The choice of methodology for converting vegetable waste is dependent upon various factors such as the type of biomass being used, end product, economic value, etc. (Gnanasekaran et al. 2023). For example, in the case of biogas production, organic fractions like carbohydrates and proteins are the most crucial components. The methane yield varies from 207.5 to 346.3 mL/g of volatile solids (Hu et al. 2021). Table 3.3 discusses various examples of different kinds of bio-fuels being produced using different methodologies.

The thermal conversion methodologies mainly involve combustion, pyrolysis, and gasification (Kara et al. 2012). A particularly effective method of production of bio-oil has been observed through pyrolysis, occurring at 300–800 °C, from

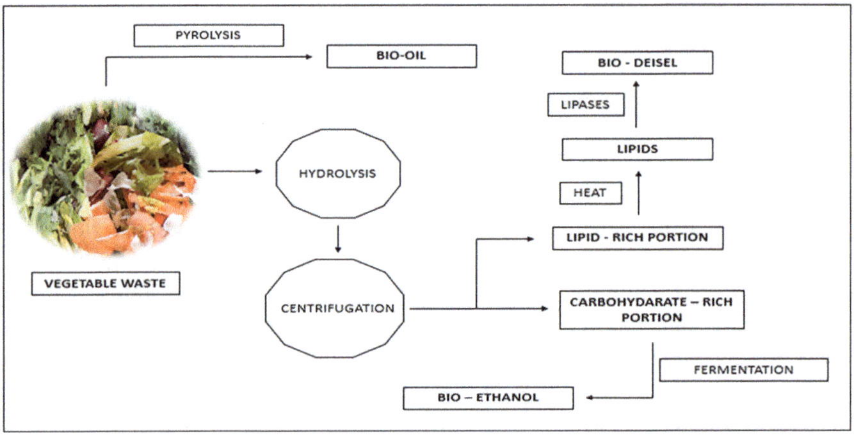

Fig. 3.3 Flow chart showcasing the production of the three main bio-fuels that can be generated from vegetable waste

Table 3.3 Different methods of biofuel production from vegetable wastes

Feedstock	Extraction method	Process description	Product description	References
Biogas				
Rumen digesta: vegetable waste: fruit waste (1:1:2, 1:2:1, 1:1.5:1.5, 1:0.5:0.5, 1:2:2 and 1.5:1.5:1)	Batch-type anaerobic digestion	Hydraulic retention period—30 days –3000 mL digester flask, –1500 mL water-cum-gas chamber, 1500 mL water collector flask Raw material—1200 g	Highest produced gas amount—35 mL Average gas production—17 mL/day	Anika et al. (2022)
Mixed vegetable waste and horse dung (34 kg) (30% VW and 30% horse dung, 20% VW and 40% horse dung)	Anaerobic digestion	Two sets of experiments with different ratios of vegetable waste and horse dung anaerobically digested for the period of 10 and 13 days	Primarily blue and yellow hints in the gas Combustion holding 3.22–7.22 First experiment—3–5.20 °C Second experiment—3–6.10 °C	German et al. (2023)
Biodiesel				
2000 L of waste vegetable oil (1:6 oil-alcohol ratio)	Neutralization and thermal drying, mixing of alcohol catalyst, transesterification, and decanting	Conversion rate: 81–85% by weight Optimal molar ratio (oil) was 1:6 Catalyst amount: 0.5% (NaOH) or 1% (KOH) of the oil weight	4000 l of biodiesel meeting ANP standards	Torres et al. (2013)
Bio-oil				
Oya husk—1 kg, rice husk—1 kg, cotton stalk—1 kg, bagasse—1 kg, and wood sawdust-sesame residues—1 kg	Pretreatment with sodium hydroxide solution and pyrolysis	– NaOH concentrations: 50 g/L, 100 g/L, 150 g/L – Tubular batch pyrolysis reactor – Temperature: 600 °C – Nitrogen gas injected—20 mL/min	Bio-oil production increased after pre-treatment Soya husk: 158% Rice husk: 80% Cotton stalk: 112% Bagasse: 269% Wood sawdust: 139%	Singh and Chouhan (2014)

(continued)

Table 3.3 (continued)

Feedstock	Extraction method	Process description	Product description	References
Bio-ethanol				
2000 mL of sugarcane molasses solution	Enzymatic hydrolysis and fermentation	– Amylase addition-temperature of 90 °C (2 h) – Saccharification stage enzyme-gluco-amylase—temperature of 65 °C (4 h) – Fermentation stage-variations of bread and yeast tape Time period—3 days	Maximum bioethanol content after fermentation—18.6% and reducing sugar of 10.98 g/L—Using 10% baker's yeast	(Jamil et al., 2016)
Bulk carrot waste (particle size reduced to 1–3 mm and stored at 4 °C)	Hydrothermal pretreatment and ABE fermentation	Pre-treated carrot pulp inoculated with specific strains of clostridium bacteria. The fermentation occurs under anaerobic conditions	With pretreatment Sugar recovery: 88% Butanol concentration: 69 g/kg carrot pulp ABE concentration: 101 g/kg carrot pulp	López-Linares et al. (2022)

vegetable biomass (Lewandowski et al. 2020). Studies on the pyrolysis of vegetable waste have indicated that it can yield biofuel suitable for compression ignition engines, with improved brake power and thermal efficiency (Mukilarasan et al. 2023). Typically, at temperatures of 700–1000 °C, the gasification process can convert vegetable waste into syngas (Lewandowski et al. 2020). Although combustion is considered as a common method of conversion, it is less environmentally friendly compared to pyrolysis and gasification, as it does not allow for the recovery or reuse of valuable chemical compounds (Lewandowski et al. 2020).

In the various biochemical pathways available for vegetable waste to biofuel production, anaerobic digestion (AD) is a promising pathway with significant potential for producing biofuels, particularly biomethane (Magama et al. 2022; Youcai and Ran 2021). Anaerobic digestion primarily involves the steps of *hydrolysis*, *acidogenesis*, *acetogenesis*, and *methanogenesis* (Zeng et al. 2022). A variety of products are obtained through anaerobic digestion, including methane, hydrogen, digestate, and volatile fatty acids, making it suitable for biorefinery operations (Magama et al. 2022). Fermentation, another biochemical method of conversion, can be used for the production of bioethanol and biohydrogen (Zeng et al. 2022). The conversion efficiency can be enhanced by 10–30% through pre-treatment before fermentation, and the efficiency of biochemical conversion of biomass can reach 90–95% (Saravanan et al. 2023).

The use of physicochemical methods, such as transesterification, to produce fuels like biodiesel is a promising solution to rising global energy demands and environmental concerns (Milano et al. 2018; Chafle et al. 2014). The process of transesterification involves the conversion of triglycerides to methyl esters using an alkali catalyst (typically potassium hydroxide) (Banihani 2016). Reaction temperature, catalyst concentration, methanol-to-oil ratio, and feedstock composition are key variables affecting the transesterification process of biomass (Banihani 2016).

Vegetable waste is a very economical and valuable source of organic material due to its composition and easy availability. It holds immense potential in the field of bio-fuel production. By leveraging this abundant resource, we can significantly reduce waste while contributing to sustainable energy solutions. Harnessing vegetable waste for biofuel production not only supports environmental sustainability but also offers a practical approach to meeting growing energy demands.

3.4 Vegetable Waste for Efficient and Sustainable Chemical Extraction

Vegetable waste is composed of various types of materials, including peels, seeds, and other inedible parts. These wastes are rich in valuable bioactive compounds that can be potentially used in various industries worldwide (Yadav et al. 2023; Sagar et al. 2018). Phytochemicals, vitamins, enzymes, and dietary fibers, known for their antioxidant, anti-inflammatory, and anticancer properties, have been discovered in vegetable waste (Rodríguez García and Raghavan 2021; Annegowda and Majumder 2021). These valuable compounds can be efficiently extracted through green extraction techniques, which offer higher yields and reduced solvent usage (Rodríguez García and Raghavan 2021).

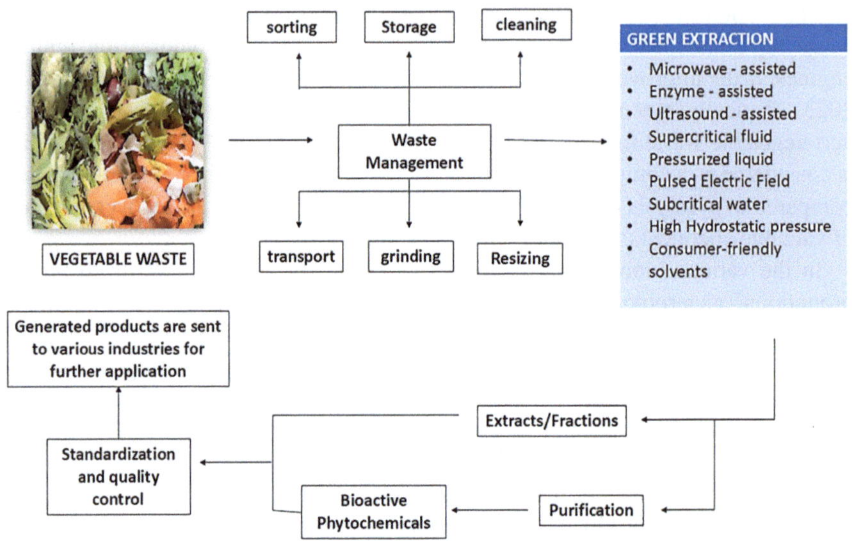

Fig. 3.4 Extraction of phytochemicals from fresh vegetable waste

The representation of the different kinds of green extraction methodologies used in chemical extraction from vegetable waste can be seen in Fig. 3.4.

The effective valorization of vegetable waste not only provides a cheaper source of these compounds but also helps reduce environmental contamination and protect endangered plant species (Annegowda and Majumder 2021). The application of these extracted biochemicals spans across many industries, such as food, pharmaceuticals, cosmetics, and textiles.

Various types of chemicals can be extracted from vegetable waste, with applications spread widely across different industries globally. These chemicals mainly include phenolic compounds, flavonoids, saponins, and alkaloids (Chandak et al. 2024; Salem et al. 2023). Phenolic compounds in vegetable waste have been identified in high concentrations and exhibit strong antioxidant activity and potential health benefits, with values ranging from 200 μg/ml to 250 μg/ml in studies (Chandak et al. 2024). Flavonoids are found abundantly in onion peels and contribute to the antioxidant and antimicrobial properties of vegetable waste, while saponins and alkaloids found in vegetable waste significantly contribute to enhancing immune response by exhibiting antimicrobial effects and other pharmacological activities (Salem et al. 2023).

There are many methods of extracting chemicals from vegetable waste, and one such method is solvent-based extraction, emphasizing sustainability and efficiency. Research has shown the potential of computational solvent screening to identify over 100 high-performing solvents for phenolic extractions (Gupta et al. 2023). A typical solvent-based extraction includes a pump, mixer, and settler to facilitate the extraction process; the pump introduces the solvent to the vegetable waste, and the mixer blends them together (Fredriksson et al. 2016). Common solvents used for solvent-based extraction include hexane, chloroform, and ethyl acetate (Babbar et al. 2014). Traditionally, ethanol and methanol have been used widely for solvent-based extraction, but now, further replacements have been discovered that are more effective (Gupta et al. 2023).

Another technique is the Ultrasound-Assisted Extraction (UAE), which utilizes high-frequency sound waves to create cavitation bubbles in the solvent that, upon collapse, generate localized high temperatures and pressures, facilitating the release of compounds from plant matrices (Goswami et al. 2024). Parameters such as ultrasound power, extraction time, and temperature, as demonstrated in the studies on Jerusalem artichoke tubers, can be optimized to enhance this extraction technique (Afoakwah et al. 2024). UAE has been successfully applied to various vegetable wastes with significant yields; for example, a substantial amount of lycopene can be extracted from tomato skins using volatile deep eutectic solvents (Goswami et al. 2024). It has been shown that the UAE can produce higher concentrations of phenolic compounds and antioxidants compared to traditional extraction techniques, as noted during the extraction process from Brosimum alicastrum (Aguilar-Piloto et al. 2023).

Extraction through a technique called supercritical fluid extraction is another sustainable technique for extracting valuable chemicals from vegetable waste. It operates above the critical temperature and pressure of carbon dioxide, allowing control over properties like density and viscosity to increase extraction efficiency (Perez-Vazquez et al. 2023). Both polar and non-polar compounds can be effectively extracted using this technique, including polyphenols, essential oils, and antioxidants from various plant residues (Herzyk et al. 2024; Yasuno 2022). The extraction of polar compounds can be enhanced in this technique by adding co-solvents like ethanol or water (Atwi-Ghaddar et al. 2023). SFE significantly contributes to waste valorization through the conversion of vegetable byproducts into high-value extracts for the food, cosmetics, and pharmaceutical industries (Herzyk et al. 2024; Yasuno 2022).

Microwave-assisted extraction (MAE) is a technique for chemical extraction from vegetable waste that directly employs microwave radiation to heat the biomass and solvent mixture, facilitating the rapid release of target compounds from the plant matrix (Trilokchandran 2024). There are various configurations of microwave systems that can be employed, which include features like closed vessels to ensure safety and efficiency (Ferrara et al. 2023). This technique is very energy-efficient and requires less time and solvent compared to conventional methods, making it more eco-friendly (Álvarez Viñas et al. 2023; López-Salazar et al. 2023). This innovative technique utilizes microwave energy to heat solvents in contact with samples, leading to improved extraction outcomes.

Another technique, which is innovative and environment-friendly, includes extraction through enzymes, also known as enzyme-assisted extraction (EAE), used for extracting valuable chemicals from vegetable waste biomass. Cellular structures are disrupted through enzymes such as cellulases, hemicellulases, and pectinases to facilitate the extraction of compounds from the plant matrix (Rizwan et al. 2024). A pre-treatment phase is often involved in this extraction technique, where enzymes are applied to the solid matrix, followed by conventional extraction methods, which improve yield and efficiency (Rizwan et al. 2024). The significantly lower energy requirements of this process give it recognition among the various extraction techniques, along with its higher extraction rate compared to traditional methods (Gao et al. 2024). EAE has been successfully employed to extract valuable compounds from various sources, such as olive leaves, achieving a significant increase in the total polyphenol and antioxidant activity (Vardakas et al. 2024).

Another technique, which is extraction through a pressurized liquid medium for extracting chemicals from vegetable biomass, operates below the critical point of solvents, maintaining them in a liquid state and altering their physicochemical properties for better penetration into the vegetable matrix (Barik et al. 2024; Barp et al. 2023). Automated and simultaneous extractions are possible with this technique, which significantly decreases energy consumption and extraction time (Jiménez-Soriano et al. 2024). The extraction can be efficiently increased and optimized for greater purity when the PLE extraction technique is combined with other extraction methods (Álvarez Alarcon et al. 2023). This technique is very effective in extracting valuable phytochemicals from vegetable waste, thus enhancing the recovery of antioxidants and other beneficial compounds (da Cunha Rodrigues et al. 2023).

Vegetable waste extraction has been increasingly recognized for its potential to enhance sustainability and reduce the negative environmental impact of this biomass. The development of various green extraction methodologies has enabled the efficient recovery of bioactive compounds from waste materials, particularly from fruits and vegetables. The use of sound waves and microwave energy in extraction methods is now leading to higher yields and concentrations of bioactive compounds such as carotenoids and polyphenols from vegetable waste (Valisakkagari et al. 2024). Green extraction methods are enabling researchers to extract valuable compounds such as essential oils from vegetable waste, offering a non-toxic alternative to traditional solvents (Dhenge et al. 2024).

As these innovative techniques continue to evolve, they hold significant promise for scaling up sustainable industrial applications, helping to minimize waste and maximize resource efficiency across multiple sectors. Table 3.4 discusses examples of various chemicals being obtained from vegetable waste through different green extraction methodologies.

3.5 Challenges and Future Prospects

There are a host of different technological, economic, and regulatory issues that come along with the use of vegetable waste biomass sources for chemical extraction and biofuel production that stand against the potential and sustainability of this input. Technologically, the conversion of vegetable waste into biofuels involves complex processes, such as enzymatic and chemical catalytic reactions, which may be inefficient and require optimization (Lee 2022). The high variability in the composition of vegetable waste complicates efforts at standardization, both for yield and quality of the biofuels produced (Adamu et al. 2023). Furthermore, high operational

Table 3.4 Table of different extraction methods and the products obtained using them from different vegetable wastes

Vegetable waste source	Extraction method	Chemical obtained	Yield description	Process description	References
Carrot peels, sweet potato, peppers	Supercritical fluid extraction	Carotene	Total carotenoid values were greater than 90% w/w for most samples (TCRs 88–100% w/w)	Temperature 59 °C, 350 bar, 15 g/min carbon dioxide, 15.5% (v/v) ethanol as co-solvent, and 30 min of extraction time	de Andrade Lima et al. (2019)
Tomato processing waste	Enzyme-assisted extraction	Lycopene	7–16 times more increased recovery of lycopene was observed, compared to the untreated TPW	Treatment optimized using pectinolytic and cellulolytic enzymes Temperature was 30 °C, the pressure 3.5 h	Zuorro et al. (2011)
Carrot residue	Ultrasound-assisted extraction	50 min of irradiation, 50 °C temperature, and 100 W power	64.66% maximum concentration yield of carotene	50 min of irradiation, 50 °C temperature, and 100 W power	Purohit and Gogate (2015)
Beetroot leaves and stems	Pressurized liquid extraction	Phenolic compounds	36.03 ± 0.04-ethanol: water (70/30) 23.24 ± 0.4 Ethanol: water (90/10)	40 °C temperature, 7.5 pressure, 10 and flow rate of 12.5 MPa at 3 mL/min	Lasta et al. (2019)
Red habanero chili waste	Supercritical fluid extraction with co-solvent	Capsaicinoids and carotenoids	14.9 mg of total capsaicinoids and 292.09 μg of total carotenoids per gram of sample	Various techniques used, such as Soxhlet, supercritical fluid with a co-solvent, and maceration assisted by ultrasound	Rojas et al. (2024)
Broccoli stems and leaves	Microwave-assisted extraction	Phenolic compounds	Leaves 1940.35 ± 0.794 μg GAE/g DW Florets 657.062 ± 0.771 μg GAE/g DW Stems 225.273 ± 0.897 μg GAE/g DW	Between the range of 73–75 °C temperature, 75–80% methanol concentration for 10–20 min was used for the different parts of the vegetables	Rodríguez García and Raghavan (2022)

costs, including those related to technology implementation, collection, and processing, hamper the economic viability of biofuels (Raina et al. 2024). Additionally, there is competition from fossil fuels and other energy sources, which results in fluctuating market prices, potentially discouraging investment in biofuel technologies (Negi 2024). From a policymaking perspective, inconsistent regulations and unsupportive policies hinder the development of biofuel industries, making it difficult for stakeholders to understand the legal stance on many issues regarding biofuels (Adamu et al. 2023). Despite these challenges, the potential of vegetable waste remains substantial for contributing to sustainable energy systems; hence, there is a need for follow-up studies and developments to overcome these barriers.

This guides us towards a promising vision of biofuel and chemical production using vegetable waste, which complements the current need for both sustainable energy solutions and effective waste management. The various technologies and methodologies used within the many approaches to vegetable waste conversion into valuable biofuels and chemicals can contribute towards the overall concept of a green and sustainable circular bioeconomy. Fermentation, anaerobic digestion, and pyrolysis are some of the conversion methods being optimized for better yield and efficiency (Francis and Makinde 2024). An integrated approach—combining thermochemical and biochemical processes in biorefinery systems—can result in improved productivity with reduced environmental impacts, making vegetable waste a promising feedstock for both biofuels and valuable chemicals (Afraz et al. 2023). Further investment in research and innovation will be essential to unlock the full potential of these technologies. With continued advancements, vegetable waste could emerge as a cornerstone in the transition towards a more sustainable future.

3.6 Conclusion

Organic waste from vegetables is one of the most promising feedstocks for biofuels and chemical production, addressing both waste management challenges and an increasing demand for renewable resources. Due to their organic content and nutrient-rich nature, vegetable wastes are highly valuable as raw material for biofuel production (Dhir 2024), and various conversion methods, including anaerobic digestion, fermentation, and transesterification, have been proven effective with these materials. Furthermore, the valorization of bioactive compounds from vegetable wastes through green technologies such as ultrasound-assisted extraction, supercritical fluid extraction, and enzyme-assisted extraction enhances their application in cosmetics, pharmaceutical, and food industries.

The application of green technologies contributes not only to reducing environmental impacts associated with waste disposal but also to fostering growth in the bioeconomy. It valorizes vegetable waste into value-added resources, thereby supporting a low-carbon environment and promoting the judicious use of natural resources—a critical consideration in the context of the emerging global consensus on sustainable development. Converting vegetable wastes into biofuels and chemicals represents an investment in the future (Afraz et al. 2023), addressing both environmental concerns and economic viability amidst increasing resource constraints. By prioritizing these sustainable practices, we can pave the way for a more resilient and eco-friendly future.

References

Adamu HM, Bello USB, Yuguda AU, Tafida UI, Jalam AM, Sabo A, Qamar M (2023) Production processes, techno-economic and policy challenges of bioenergy production from fruit and vegetable wastes. Renew Sustain Energy Rev. https://doi.org/10.1016/j.rser.2023.113686

Afoakwah NA, Tchabo W, Ansah PO (2024) Ultrasound-assisted extraction (UAE) of Jerusalem artichoke tuber bio-active ingredient using optimized conditions of box-Behnken response surface methodology. Heliyon. https://doi.org/10.1016/j.heliyon.2024.e25645

Afraz MT, Xu X, Adil M, Manzoor M, Zeng X-A, Han Z, Aadil RM (2023) Subcritical and supercritical fluids to valorize industrial fruit and vegetable waste. Food Secur 12:2417

Aguilar-Piloto G, Negrón-Díaz AC, Moo-Huchin VM, Ramírez-Sucre MO, Delgadillo-Díaz M, Cuevas-Glory L, Sauri-Duch E (2023) Ultrasound-assisted extraction (UAE) of phenolic compounds from Brosimum alicastrum fruit and their antioxidant capacity. Ecosistemas Recursos Agropecuarios 10:e3655

Álvarez Alarcon N, Ortiz-Ramírez JS, Ortega-Barbosa JP, Ballesteros-Vivas D, Rodríguez-Varela LI, Parada-Alfonso F (2023) Pressurized liquid extraction (PLE). In: Current and future developments in food science. Elsevier, Amsterdam, pp 357–395

Álvarez Viñas M, Sanz V, Rodríguez-Seoane P, López Hortas L, Flórez Fernández N, Torres MD, Domínguez H (2023) Microwave-assisted extraction (MAE). In: Current and future developments in food science. Elsevier, Amsterdam, pp 396–430

Anika T, Muhammad R, Mamun A, Anwar H, Towfiqur R, Janibul AS (2022) Comparison effect on biogas production from vegetable and fruit waste with rumen digesta through co-digestion process. Eur J Energy Res 2(1):1–7. https://doi.org/10.24018/ejenergy.2022.2.1.38

Annegowda HV, Majumder P (2021) Valuable bioactives from vegetable wastes. In: Valorization of agri-food wastes and by-products. Academic, New York

Arvanitoyannis IS, Varzakas T (2008) Vegetable waste treatment: comparison and critical presentation of methodologies. Crit Rev Food Sci Nutr 48:205–247

Atwi-Ghaddar S et al (2023) Supercritical fluid extraction (SFE) of polar compounds from Camellia sinensis leaves: use of ethanol/water as a green polarity modifier. Molecules 28(14):5485

Babbar N, Oberoi HS, Sandhu SK, Bhargav VK (2014) Influence of different solvents in extraction of phenolic compounds from vegetable residues and their evaluation as natural sources of antioxidants. J Food Sci Technol 51(10):2568–2575

Banihani FF (2016) Transesterification and production of biodiesel from waste cooking oil: effect of operation variables on fuel properties. J Chem Eng 4:154

Barik RC, Sugunan S, Shafri MAM (2024) Pressurized liquid extraction for the isolation of bioactive compounds. In: Methods and protocols in food science. Springer, Cham, pp 275–298

Barp L, Miklavčič Višnjevec A, Moret S (2023) Pressurized liquid extraction: a powerful tool to implement extraction and purification of food contaminants. Food Secur 12(10):2017

Berde CV, Berde VB, State M (2015) Vegetable waste as alternative microbiological media for laboratory and industry. Res J Pharm Technol 12(4):1521

Briens C, Piskorz J, Berruti F (2008) Biomass valorization for fuel and chemicals production—a review. Int J Chem React Eng 6:1

Cárdenas ELM, Zapata ADZ (2019) Biohydrogen production by co-digestion of fruits and vegetable waste and coffee mucilage. Rev Fac Nac Agron Medellín 72(1):1–10

Chafle SB, Parmar VK, Biya S (2014) Utilization of vegetable and fruit waste for bio-energy generation. J Autom Control Eng 2:143–145

Chandak AM, Mankar SA, Balpande RS (2024) Production of bioactive compounds from vegetable waste using microbial consortia. J Pure Appl Microbiol 18(3):1661–1673
Cho E, Trinh LT, Song Y, Lee Y, Bae H (2019) Bioconversion of biomass waste into high value chemicals. Bioresour Technol 298:122386
da Cunha Rodrigues L, Barbosa Bodini R, de Lima Caneppele F, Dacanal GC, Crevelin EJ, Beraldo de Moraes LA, Lopes de Oliveira A (2023) Pressurized liquid extraction (PLE) in an intermittent process as an alternative for obtaining passion fruit (Passiflora edulis) leaf hydroalcoholic extract (tincture). PRO 11(8):2308
Dandotiya P, Agrawal OP (2014) Stabilization of vegetable market waste through vermicomposting. In: Bioremediation and sustainable technologies for cleaner environment. Springer, Cham
de Andrade Lima M, Kestekoglou I, Charalampopoulos D, Chatzifragkou A (2019) Supercritical fluid extraction of carotenoids from vegetable waste matrices. Molecules 24:466
Dhenge R, Rinaldi M, Ganino T, Lacey K (2024) Recent and novel technology used for the extraction and recovery of bioactive compounds from fruit and vegetable waste. In: Emerging technologies in food science. CRC Press, Boca Raton, pp 162–187
Dhir B (2024) Biofuel production from agricultural waste: a global trend. In: Clean energy production technologies. Springer, Cham, pp 1–13
Edwiges T, Frare L, Mayer B, Lins L, Triolo JM, Flotats X, de Mendonça S, Costa MS (2018) Influence of chemical composition on biochemical methane potential of fruit and vegetable waste. Waste Manag 71:618–625
Fadlilla T, Budiastuti MS, Rosariastuti MR (2023) Potential of fruit and vegetable waste as eco-enzyme fertilizer for plants. Jurnal Penelitian Pendidikan IPA 9(1):1–10
FAO (n.d.) Global food losses and food waste. https://www.fao.org/4/mb060e/mb060e00.htm
Ferrara D, Beccaria M, Cordero C, Purcaro G (2023) Microwave-assisted extraction in closed vessel in food analysis. J Sep Sci 46:e2300390
Francis M, Makinde P (2024) Production of biofuels from agricultural waste. Asian J Agric Hortic Res 11(3):37–49
Fredriksson H, Saario R, Matinheikki J, Vesanen M, Vuortenväli H (2016) Extraction arrangement for solvent extraction
Gao Y, Ding Z, Liu Y, Xu Y-J (2024) Aqueous enzymatic extraction: a green, environmentally friendly and sustainable oil extraction technology. Trends Food Sci Technol 144:104315
German MAJ, Paiste RP, Rafanan CMD, Pilien VP (2023) Biogas generation from vegetable waste and horse dung through improvised anaerobic digester. Int J Environ Sci Technol Manag 32(1):307. https://doi.org/10.69566/ijestm.v32i1.307
Gnanasekaran L, Priya AK, Thanigaivel S, Hoang TKA, Soto-Moscoso M (2023) The conversion of biomass to fuels via cutting-edge technologies: explorations from natural utilization systems. Fuel 331:125668
Goswami MJ, Dutta U, Kakati D (2024) Ultrasound-assisted extraction for food, pharmacy, and biotech industries. In: Methods and protocols in food science. Springer, Cham, pp 103–128
Gupta Y, Bhattacharyya S, Vlachos DG (2023) Extraction of valuable chemicals from food waste via computational solvent screening and experiments. Sep Purif Technol 316:123719
Hang YD, Downing DL, Stamer JR, Splittstoesser DF (1972) Wastes generated in the manufacture of sauerkraut. J Milk Food Technol 35:432–435
Herzyk F, Piłakowska-Pietras D, Korzeniowska M (2024) Supercritical extraction techniques for obtaining biologically active substances from a variety of plant byproducts. Food Secur 13(11):1713–1713
Hu Y, Cai F, Wang L, Chen C, Liu G (2021) Compositional components and methane production potential of typical vegetable wastes. Environ Sci Pollut Res 28(38):54177–54186
Jamil TS, Fathy NA, El-Shafie AS, Bassiouny R (2016) Bioethanol production from sugarcane molasses using different chemically and biologically treated lignocellulosic biomasses. Renewable Energy, 92:355–363. https://doi.org/10.1016/j.renene.2016.02.050
Jiménez-Soriano Y, Andreu V, Picó Y (2024) Pressurized liquid extraction of organic contaminants in environmental and food samples. Trends Anal Chem 173:117624

Kara B, Emir Z, Kaygusuz K (2012) Thermal processing technologies for biomass conversion to energy. J Eng Res 1:55–62

Kaur P, Subramanian J, Singh A (2022) Green extraction of bioactive components from carrot industry waste and evaluation of spent residue as an energy source. Sci Rep 12:1–12

Kumar H, Bhardwaj K, Sharma R, Nepovimova E, Kuča K, Dhanjal DS, Verma R, Bhardwaj P, Sharma S, Kumar D (2020) Fruit and vegetable peels: utilization of high value horticultural waste in novel industrial applications. Molecules 25:2812

Lasta HFB, Lentz L, Rodrigues LGG, Mezzomo N, Vitali L, Ferreira SRS (2019) Pressurized liquid extraction applied for the recovery of phenolic compounds from beetroot waste. Biocatal Agric Biotechnol 21:101353

Lee J (2022) From fruit and vegetable waste to biofuel production: part II. In: Advances in biofuel production. Scrivener Publishing, Beverly, pp 81–124

Lewandowski WM, Ryms M, Kosakowski W (2020) Thermal Biomass Convers Rev Process 8(5):516

López-Linares JC, Coca M, Plaza PE, Lucas S, García-Cubero MT (2022) Waste-to-fuel technologies for the bioconversion of carrot discards into biobutanol. Renew Energy 202:362–369

López-Salazar H, Camacho-Díaz BH, Arenas Ocampo ML, Jiménez-Aparicio AR (2023) Microwave-assisted extraction of functional compounds from plants: a review. Bioresources 18(3):6614–6638

Magama P, Chiyanzu I, Mulopo JL (2022) A systematic review of sustainable fruit and vegetable waste recycling alternatives and possibilities for anaerobic biorefinery. Bioresour Technol Rep 18:101031

Merino D, Zych A, Athanassiou A (2022) Biodegradable and biobased mulch films: highly stretchable PLA composites with different industrial vegetable waste. ACS Appl Mater Interfaces 14(41):46920–46931

Milano J, Ong H, Masjuki HH, Silitonga AS, Kusumo F, Dharma S, Sebayang AH, Cheah MY, Wang C (2018) Physicochemical property enhancement of biodiesel synthesis from hybrid feedstocks of waste cooking vegetable oil and beauty leaf oil through optimized alkaline-catalysed transesterification. Waste Manag 80:435–449

Mittal V, Sharma A (2024) From scraps to solutions: harnessing the potential of vegetable and fruit waste in pharmaceutical formulations. Lett Funct Foods. https://doi.org/10.2174/0126669390271001231122051310

Mukilarasan N, Gopal K, De Poures MV, Sivanantham A, Poyyamozhi N, Venkatesh R, Naveen S, Vaishnavi M (2023) Sustainable production of bio-fuel derived from vegetable waste via pyrolysis technique. Environ Qual Manag 33:267–273

Nagarajan G, Rajakumar S, Ayyasamy P (2014) Vegetable wastes: an alternative resource for biogas and bio compost production through lab scale process. Int J Curr Microbiol App Sci 3:379–387

Narashans A, Sagar S, Pareek S, Sharma S, Yahia EM, Lobo MG (2018) Fruit and vegetable waste: bioactive compounds, their extraction, and possible utilization. Compr Rev Food Sci Food Saf 17(3):512–531

Negi S (2024) Current scenario and future prospects of enzymatic transformation of used vegetable cooking oil for biodiesel. In: Advances in biodiesel production. Elsevier, Amsterdam, pp 279–301

Pant KK, Mohanty P (2014) Biomass, conversion routes and products—an overview. Wiley, Hoboken

Perez-Vazquez A, Barciela P, Carpena M, Donn P, Seyyedi-Mansour S, Cao H, Fraga-Corral M, Otero P, Simal-Gandara J, Prieto MA, Cassani L (2023) Supercritical fluid extraction as a potential extraction technique for the food industry. Eng Proc 37(1):115

Preeti N, Yadav N (2023) Waste to worth: a review on utilization of vegetable waste. Int J Environ Climate Change 13(9):1385–1398

Purohit AJ, Gogate PR (2015) Ultrasound-assisted extraction of β-carotene from waste carrot residue: effect of operating parameters and type of ultrasonic irradiation. Sep Sci Technol 50:1507–1517

Raina N, Chuetor S, Charoenkool P, Jiradechakorn T, Sereenonchai C, Phojaroen J, Boonmee R, Pathak AK, Singh HM (2024) Opportunities and challenges in the production of biofuels from waste biomass. In: Advances in biofuels. Woodhead Publishing, New Delhi, pp 23–43
Rizwan D, Mir SA, Aziz SA, Masoodi FA (2024) Enzyme-assisted extraction of essential oils. In: Advances in essential oil research. Elsevier, Amsterdam, pp 217–233
Rodríguez García SL, Raghavan V (2021) Green extraction techniques from fruit and vegetable waste to obtain bioactive compounds—a review. Crit Rev Food Sci Nutr 62:6446–6466
Rodríguez García SL, Raghavan V (2022) Microwave-assisted extraction of phenolic compounds from broccoli (Brassica oleracea) stems, leaves, and florets: optimization, characterization, and comparison with maceration extraction. Recent Progr Nutr 2(2):1
Rojas JD, Vázquez-León LA, Palma M, Férnandez-Ponce MT, Casas L, Barbero GF, Rodríguez-Jimenes, G. C. (2024) Re-valorization of red habanero chili pepper (Capsicum chinense Jacq.) waste by recovery of bioactive compounds: effects of different extraction processes. Agronomy 14(4):660
Sagar NA, Pareek S, Sharma S, Yahia EM, Lobo MG (2018) Fruit and vegetable waste: bioactive compounds, their extraction, and possible utilization. Compr Rev Food Sci Food Saf 17(3):512–531
Salem MA, Mohamed OG, Mosalam EM, Elberri AI, Abdel-Bar HM, Hassan MT, Al-Karmalawy AA, Tripathi A, Ezzat SM, Abo Mansour HE (2023) Investigation of the phytochemical composition, antioxidant, antibacterial, anti-osteoarthritis, and wound healing activities of selected vegetable waste. Dental Sci Rep 13:13034
Saravanan A, Yaashikaa PR, Senthil Kumar P, Vickram AS, Karishma S, Kamalesh R, Rangasamy G (2023) Techno-economic and environmental sustainability prospects on biochemical conversion of agricultural and algal biomass to biofuels. J Clean Prod 414:137749
Sekino N, Jiang Z (2021) Fuel and material utilization of a waste shiitake (Lentinula edodes) mushroom bed derived from hardwood chips: characteristics of calorific value in terms of elemental composition and ash content. J Wood Sci 67:1–10
Sepelev I, Galoburda R (2015) Industrial potato peel waste application in food production: a review
Shajan AE, Dash KK, Hamid B, Bashir O, Shams R (2024) Comprehensive comparative insights on physico-chemical characteristics, bioactive components, and therapeutic potential of pumpkin fruit. Future Foods. https://doi.org/10.1016/j.fufo.2024.100312
Singh S, Chouhan A (2014) Experimental studies on enhancement of bio-oil production using agro waste materials pre-treated with alkaline solutions. Afr J Basic Appl Sci. https://doi.org/10.5829/idosi.ajbas.2014.6.1.2017
Singh A, Kuila A, Adak S, Bishai M, Banerjee R (2012) Utilization of vegetable wastes for bioenergy generation. Agric Res 1:213–222. https://doi.org/10.1007/s40003-012-0030-x
Souri M, Alipanahi N, Tohidloo G (2016) Heavy metal content of some leafy vegetable crops grown with wastewater in southern suburb of Tehran-Iran. Veg Sci 43(2):156–162
Torres EA, Cerqueira GD, Ferrer TM, Quintella CM, Raboni M, Torretta V, Urbini G (2013) Recovery of different waste vegetable oils for biodiesel production: a pilot experience in Bahia State. Brazil Waste Manag 33(12):2670–2674. https://doi.org/10.1016/j.wasman.2013.07.030
Trilokchandran B (2024) A review: microwave assisted extraction. Indian Sci J Res Eng Manag 8(2):1–10. https://doi.org/10.55041/ijsrem28772
Valenzuela ME, Restović F (2019) Chapter 13: valorization of tomato waste for energy production. In: Food chemistry, function and analysis. RSC, London. https://doi.org/10.1039/9781788016247-00245
Valisakkagari H, Chaturvedi C, Rupasinghe HPV (2024) Green extraction of phytochemicals from fresh vegetable waste and their potential application as cosmeceuticals for skin health. PRO 12(4):742. https://doi.org/10.3390/pr12040742
Vardakas A, Kechagias A, Penov N, Giannakas AE (2024) Optimization of enzymatic assisted extraction of bioactive compounds from Olea europaea leaves. Biomass 4(3):647–657. https://doi.org/10.3390/biomass4030035
Varma VS, Kalamdhad AS (2014) Bio-conversion of organic waste using composting technologies - a review. Int J Environ Technol Manag 17:483–507

Yadav P, Dhankhar S, Mehar R (2023) Waste to worth: a review on utilization of vegetable waste. Int J Environ Climate Change. https://doi.org/10.9734/ijecc/2023/v13i92368

Yasuno Y (2022) Recent advancements and challenges in supercritical fluid extraction methods and their applications in food industries. In: Proceedings from the international conference on food science and technology. Springer, Cham, pp 371–385. https://doi.org/10.1007/978-981-16-8125-7_18

Youcai Z, Ran W (2021) Anaerobic fermentation process for biomethane production from vegetable waste

Zeng J, Zeng H, Wang Z (2022) Review on technology of making biofuel from food waste. Int J Energy Res 46:10301–10319. https://doi.org/10.1002/er.7868

Zuorro A, Fidaleo M, Lavecchia R (2011) Enzyme-assisted extraction of lycopene from tomato processing waste. Enzym Microb Technol 49(6–7):567–573. https://doi.org/10.1016/j.enzmictec.2011.04.020

Conversion of Plastic Waste to Chemicals and Biofuels

4

Nalini Soni, Priya Gupta, Sangeeta Devendra Kumar Singh, Nikki Agrawal, Silambarasan Tamil Selvan, Sudheer D. V. N. Pamidimarri, Ravi Kant Singh, and Balasubramanian Velramar

Abstract

The environment is increasingly at risk due to the massive volumes of plastic waste generated, which have a negative influence on the ecosystem. One of the primary victims of this issue is the maritime environment. The amount of plastic waste produced worldwide is rising at a rate of about 3.5 MMT each year. Waste plastics composed of LDPE, HDPE, PVC, PS, and PET will amount to 25,000 MMT by 2040 if the current trend continues. Therefore, all researchers should concentrate on different ways to get rid of waste plastics and turn them into something useful. These devices are garbage-to-energy devices that physically turn waste into more useful forms, like biofuel and other industrially important chemicals like bioethanol, biobutanol, biogas, biohythane, CNG, syngas tar, ash, coke, char, and gas through various processes like combustion, pyrolysis, gasification, or biological treatments. Making use of these wastes would help preserve our ecosystem in addition to providing a sustainable supply of chemicals and

N. Soni · P. Gupta · S. D. K. Singh · B. Velramar (✉)
Amity Institute of Biotechnology, Amity University Chhattisgarh, Raipur, Chhattisgarh, India
e-mail: vbalasubramanian@rpr.amity.edu

N. Agrawal
Atal Bihari Vajpayee Vishwavidyalaya, Bilaspur, Chhattisgarh, India

S. T. Selvan
Helix Research Center, Department of General Medicine, School of Allied Health Sciences, Vinayaka Missions Research Foundation (Deemed University), Chennai, Tamil Nadu, India

Sudheer D. V. N. Pamidimarri
Discipline of Industrial Biotechnology, Gujarat Biotechnology University, Gandhinagar, Gujarat, India

R. K. Singh
Amity Institute of Biotechnology, Amity University, Noida, Uttar Pradesh, India

A. Kumari et al. (eds.), *Valorization of Solid Wastes to Biofuels and Chemical Products for Sustainable World*, https://doi.org/10.1007/978-981-96-8594-3_4

fuels. Numerous facets of using waste materials to generate clean, renewable energy for society's sustainable growth are crucial in the current plastic waste management, and they require quick attention.

Keywords

Plastic waste · Biofuel · Chemical · Environment · Sustainable growth

4.1 Introduction

Most of the plastic materials are made up of polyethylene (PE), polystyrene (PS), polypropylene (PP), polyvinyl chloride (PVC), polyurethane (PUR), and polyethylene terephthalate (PET) (Table 4.1). The most recent data from Plastics-Europe indicates that 348 million tons of plastic were produced worldwide in 2018 (Leal Filho et al. 2019). China and the European Union are the world's two largest consumers of plastic, using 29.4% and 18.5% of the material, respectively (Ru et al. 2020). Globally, the production of plastic garbage is rising at a rapid pace in tandem with rising plastic use. The production of plastic garbage is predicted to reach 26 billion tons by 2050, with almost half of that amount coming from disposed waste in landfills before entering lakes and oceans and causing significant environmental contamination. Consequently, plastic garbage has turned into a sinister representation of our wasteful civilization (Jambeck et al. 2015; Geyer et al. 2017). Researchers across the globe have embraced plastic trash as a valuable resource and are currently exploring novel technology that facilitates plastic recycling. This comprises developing chemical recycling techniques and doing research and development to produce new chemical raw materials so that waste can be turned into something useful. According to Peng et al. (2018), the three primary ways used today to dispose of plastic trash are burning, landfilling, and mechanical and chemical recycling. Due to its affordability and ease of use, landfilling is the primary method for disposing of plastic trash in many countries, particularly in developing nations. Nonetheless, a significant area of land is occupied by accumulated plastic waste. According to statistics, only 9–12% of the world's plastic garbage is recycled and burned, with the remaining 79% being disposed of in landfills or the environment. This suggests that finding new and creative ways to recycle plastic waste is imperative (Garcia and Robertson 2017; Peng et al. 2018).

4.2 Plastic Waste Problem: Current Situation with Future Problem

Plastics are widely employed due to their beneficial characteristics, which include their high impact resistance, shapeability, bacterial resistance, and low weight. However, due to its extensive use, a sizable amount of trash is generated. While most plastics are utilized for one-time (short-term) purposes, about one-fourth are

Table 4.1 Different types of chemicals with their symbols and applications in various industrial fields

S. No.	Types of plastics	Symbols	Uses	References
1	Polyethylene terephthalate (PET)	01 PET/PETE	Carbonated drinks bottles, textile fibers, luggage racks, mirror backs, food containers, films, polyester fibers for clothes, pillows, carpets, electrical parts, insulation equipment, and domestic applications	Liu et al. (2020), Olam (2022)
2	High-density polyethylene (HDPE)	02 HDPE	Oil bottles, cable insulations, plastic detergent bottles, plastic toys, shower soaps, milk jugs, motor oil containers, plant pots, rigid pipes, garden furniture, park benches	Rouch et al. (2022)
3	Polyvinyl chloride (PVC)	03 PVC	Blood containers, urine ostomy bags, flexible tubes, inhalation mask, oxygen mask, and PPE such as gloves. Floor and wall coatings. Packaging industry as food wraps, cosmetic packaging. Automotive industry as cable insulations	Lewandowski (2022)
4	Low-density polyethylene (LDPE)	04 LDPE	Packaging films, pipes, disposable plastic products, shopping bags, bread bags, irrigation pipes, beverage cups, garbage bags, coating for paper milk boxes	Alabi et al. (2022)
5	Polypropylene (PP)	05 PP	Fibers and fabrics like slit film, tape, mats, food storage containers, water bottles, houseware applications such as storage system, toys, can openers, blenders, disposable syringes, pesticide containers plastic tubs, water filters	Maddah (2016)
6	Polystyrene (PS)	06 PS	Vegetable trays, toys, CD cases, coat hangers, egg cartons, seed trays, plastic cutlery, disposable foam cups, floor laminating sheets, and fast-food trays	Alabi et al. (2022)
7	Polyethylene (PE) Acrylonitrile Butadiene styrene (ABS) Polyamide (PA) or Nylons Polybutylene terephthalate (PBT)	07 OTHER	Phone covers, toys, electrical products, windows, sheets, car rims, cables, aerospace applications, automobile industries, CDs, eyeglasses, dental sealants, baby bottles, and medical storage containers	Geueke et al. (2023)

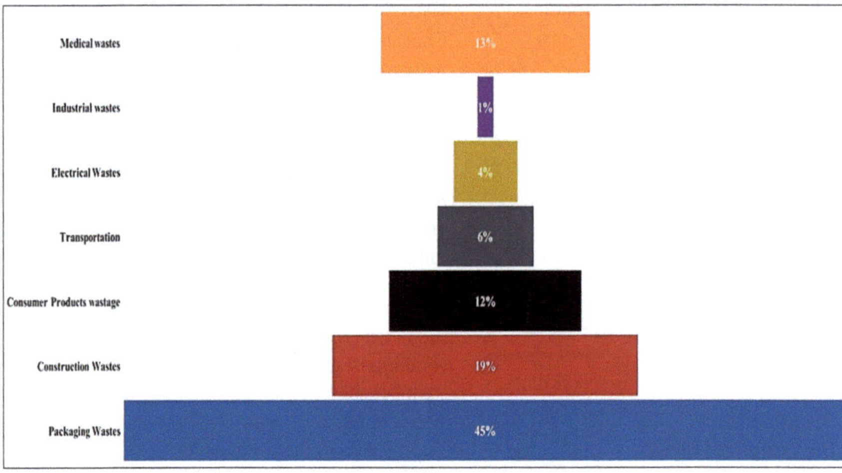

Fig. 4.1 Various sources of plastic waste dumping into the ecosystem

used for long-term applications like pipelines. Plastics' short lifespan has led to an increase in the amount of plastic waste generated each year and the amount that is improperly disposed of in the environment. The vast amount of plastic creates a sustainability challenge due to its high energy consumption and carbon emissions during manufacture, but disposing of it poses a significant environmental risk (Fig. 4.1). Because of inadequate management, plastic waste has been disposed of in areas such as the marine environment, where it is not only consumed by animals but also found in human diets. The non-biodegradable characteristics of the majority of plastic kinds make PW more difficult to dispose of than other solid wastes (Andrady 1994). Unfortunately, degradable materials require a prolonged duration, contingent upon the environmental factors they encounter (Kyrikou and Briassoulis 2007; Greenberg 2015). As a result, most PW end up in the environment, where they are transported to the marine environment by various precipitation events. Policymakers have been urged to suggest solutions to reduce plastic production and its usage; however, to address concerns regarding plastic manufacturing, waste, and recycling, it is crucial to understand how different stakeholders view the material (McNicholas and Cotton 2019).

4.3 Production of Chemicals from Plastic Wastes

The pyrolysis method is considered a stand-alone facility for the upgrading of plastic wastes in chemical recycling processes; it is particularly useful for converting polyolefins (POs), which account for two thirds of plastic waste, into gaseous or liquid fuels or raw chemicals, primarily light olefins and benzene, toluene, and xylene (BTX). Synthetic plastics are widely used in the packaging of products such as pharmaceuticals, food, cosmetics, chemicals, and detergents. Chemical recycling

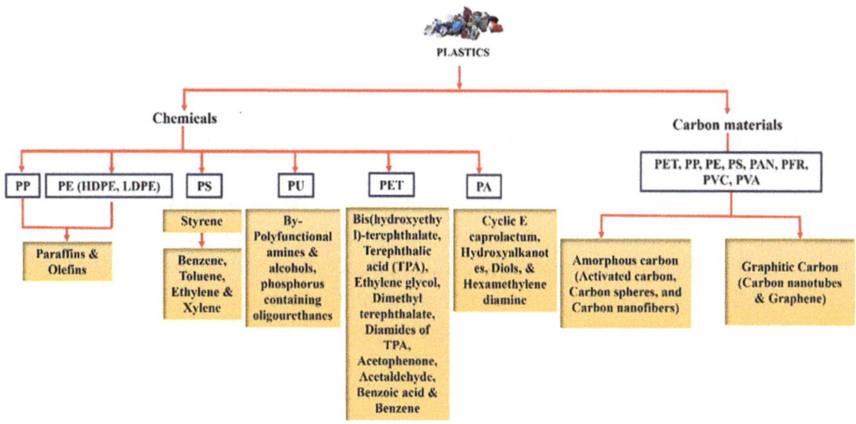

Fig. 4.2 Conversion of various plastic polymers into chemicals and carbon materials

can yield both original and new chemical raw materials while integrating sustainability principles (Fig. 4.2). As a result, the expenses associated with the transportation of plastic waste can be avoided (Hibbitts et al. 2016). The products of thermal pyrolysis of plastics at low and high temperatures. Paraffins, olefins, aromatics, and other hydrocarbons (HCs) can be produced by pyrolyzing plastic waste. Plastics' chemical composition, structure, and rate of breakdown all have an impact on these HCs' yields. Waste polyethylene (PE) and waste polypropylene (PP) can produce paraffins and olefins, while waste polystyrene (PS) can make styrene by refining the pyrolytic product. Additionally, methyl methacrylate, a monomer, is produced by thermal pyrolysis of polymethyl methacrylate; a 98% yield was recorded at 450 °C (Flaherty et al. 2014).

4.3.1 Polyethylene

When it comes to volume, polyethylene (PE) is the most utilized synthetic polymer, with polypropylene (PP) being among the remaining three most extensively utilized. These polymers are used in many applications where strong adhesion is necessary, and there are numerous scenarios in which issues can develop (Volanti et al. 2022). Commercial treatments were initially developed to enhance LDPE's adhesive properties, but they also work well on high-density polyethylene (HDPE) and polypropylene (PP). The development of industry has also made it possible to obtain more specialized and adaptable chemical structures to produce polymers with specific functions and qualities. Durable and electronic products are included in these technical applications besides packaging. In this review, three distinct polymer types were examined: bio-polyethylene (Bio-PE), bio-polypropylene (Bio-PP), and bio-poly (ethylene terephthalate) (Bio-PET). Their advantages and limitations are examined, as well as the latest advancements in product development, processing technologies, and applications (Siracusa and Blanco 2020). The literature provides excellent reports on all aspects of chemical, physical, and mechanical properties as well as their engineering applications. Copolymerization of ethylene with butene, hexane, or octane yields LLDPE (Pereira et al. 2016). The same equipment and plant can be used to process it. Bio-PE is clearly not biodegradable, a type of polymer extracted from biological sources, but it is nondegradable by nature. While Bio-PE could be naturally degraded compared to synthetic PE due to its chemical origins, it still needs specific conditions to degrade. The bagasse by-product is used as the primary fuel source in the sugar milling process because it burns to produce enough heat to meet the energy requirements; depending on the size of the plant, an excess of heat or electricity may also be generated; an alternative is to use the by-product of vinasse as fertilizer. The first companies to produce bio-PE on a commercial scale beginning in 2010 were the Brazilian company Braskem and the joint venture Dow and Crystalsev for use in food packaging, cosmetics, personal care, automotive, and toy applications (in this case in a joint venture with major Brazilian toy company Brinquedos Estrela), as well as agriculture and industry (Siracusa and Blanco 2020). Since the price of a barrel of sugar cane-derived ethanol (approximately \$115 US versus \$80 US; 1 L = 0.0085 barrel) is now competitive with the price of a barrel of crude oil, it was initially believed that the cost of producing bio-ethylene was not cost-competitive when compared to that of petrochemical-derived ethylene. One kilogram of bio-PE costs about 30% more than one kilogram of petrochemical PE (Nakajima et al. 2017).

4.3.2 Polypropylene

The olefin monomer propylene is the source of the downstream petrochemical product polypropylene (Tan et al. 2015). Addition polymerization is the process of connecting monomers to create polymers. This process combines monomers by adding heat, high-energy radiation, and an initiator or catalyst. As a result, propylene molecules polymerize into extremely lengthy chains or polymer molecules. Any polymer can be enhanced via one of four methods: bulk, gas-phase, solution, or suspension polymerization. Molecular weight and molecular weight distribution, copolymer components, process conditions, and polypropylene all affect its properties. Every carbon atom in polypropylene, a vinyl polymer, is joined to a methyl group, allowing it to be expressed (Jeon et al. 2021).

4.3.3 Polystyrene

PS is frequently found in cassette boxes, compact disks, food containers, food packaging, disposable plates, cups, and cutlery, and insulation for building supplies (Hwang et al. 2020). In 2013, almost 21 million tons of PS were produced worldwide. Plastics, including PS, have been used so extensively that they have accumulated in the environment and are now causing environmental pollution, human health problems, and ecosystem changes due to their toxicity and recalcitrant compounds (Kik et al. 2020). Although PS materials can be recycled, landfills end up with the majority of PS foam. Because PS foam is heavy and lightweight, a huge portion of its recycling is related to transportation costs. PS foam recycling rate in the United States increased from roughly 20% in 2008 to 28% in 2010. To lessen the pollution that the environment is subjected to, PS has implemented various solutions, including lowering the usage of PS products and utilizing modified PS materials that are environmentally biodegradable. PS foam's severe impact on waterways has led to its ban from sale and use in several locations, including New York, Paris, Washington, DC, Toronto, and San Francisco. Studies have been conducted to identify PS-digesting microbes in soil, landfills, and activated sludge environments (Ward 2010).

4.3.4 Polyurethane

The material polyurethane exemplifies the term "versatile." The link between the structure and properties of di-isocyanates and polyols allows the manufacturer a great deal of flexibility and customization. There are various varieties and uses for polyurethane. These include thermoplastic polyurethane, flexible and rigid polyurethane foam, polyurethane adhesives, and polyurethane coatings. Flexible polyurethane foam is used to make furniture, automobile seats, beds, and bedding. Because of its cushioning qualities, it provides support and comfort. Refrigerators, freezers, and structures are insulated using rigid polyurethane foam. It provides superior energy efficiency and thermal insulation. Thermoplastic Polyurethane (TPU): Found in automobile parts, medical equipment, and shoes. It is renowned for being pliable, strong, and abrasion-resistant. Wood, metal, and other surfaces are coated with polyurethane to prevent weathering, corrosion, and wear. Polyurethane Adhesives: Used to join different materials in the automotive, footwear, and construction sectors. The characteristics of polyurethane can vary greatly, ranging from rock-hard, unyielding construction material to soft-touch coatings. These mechanical, chemical, and biological qualities, along with their adaptability, have generated a great deal of attention from the scientific community and the relevant sectors (Das et al. 2020). A variety of additives and nanomaterials can be added, and the raw ingredients can be altered to improve the material. The main advantages of polyurethane have been its durability, toughness, and excellent chemical resistance. The development of inexpensive polyether polyols led to the production of flexible foams and other automotive and upholstery applications that are still relevant today. The result of rigorous research and continuous improvement in formulations, additives, and processing techniques has produced a wide range of polyurethane applications. Therefore, proper modifications in the raw material can produce a polyurethane suitable for every application (Yuan et al. 2020). The polymer can be as elastic as rubber and as durable as metal, which is why polyurethane is being used as a substitute for a variety of metals, plastics, rubber-based engineering, and commodity applications. Polyurethane is made when di/poly isocyanate reacts with a diol or polyol to form repeating urethane linkages in the presence of a chain extender and other additives. The building blocks of polyurethane can be broken down into two mother components: isocyanate and polyol. The final properties of the product are determined by both polyol and isocyanate. The structure-property relationship between polyol and isocyanate is essential for understanding and producing polyurethane products because changing either one can significantly alter the polyurethane's properties (Xu et al. 2021).

4.3.5 Polyamides

For about 80 years, polyamides have been available for purchase. The first polyamide products to be sold commercially were fibers composed of PA6 and PA66. In 1938, the first PA66-based polyamide fibers, known as nylon, hit the market. The primary uses of polyamides are in the manufacturing of polyamide fibers, which continues to be the world's largest consumer of polyamides, in the processing of polyamide engineering plastic, and in the processing of polyamides intended for film. The textile industry and carpet filaments are all made from polyamides. Manufacturers of undergarments, ornamental fabrics, clothing fabrics, technical textiles, upholstery, floor coverings, carpets, hoses, tire cords, safety belts, airbags, sewing threads, ropes, nets, sleeping bags, tarpaulins, tents, and other items are among the extensive list of consumers.

A polyamide is a polymeric material that is typically linear and has amide groups that repeat frequently as a necessary component of the polymeric chain. These amides are classified as secondary since there is only one hydrogen atom present in the nitrogen atom. This family of polymers, which includes polyamides and other polymers, is generated using the "step-growth" mechanism. In this mechanism, multifunctional (difunctional) molecules react with one another, creating new bonds as independent steps that neither cause nor depend on one another. Monomers, which are low molecular weight beginning materials, react with one another and with remaining monomers to form molecules of increasing length, resulting in a material with a desirable average molecular size (Xi et al. 2020). Polyamides are produced by this technique. In certain cases (like nylon 6 or PA 6), a "chain-growth" polymerization process occurs, in which the monomer (caprolactam) is constantly added to the expanding chain. But unlike the chain development of vinyl polymers, this mechanism is reversible and relatively slow (Shao et al. 2019). Nylons, also known as polyamides, are well-known for their use in fiber applications and are significant engineering plastics. Along with details on specifications, sales, and suppliers, this article discusses the qualities, production, processing, and applications of polyamide plastics. Among the materials studied are nylon-6,6, nylon-6,9, nylon-6,12, nylon-11, and nylon-12, as well as amorphous nylon and other semi-aromatic nylons. The following physical characteristics: crystallinity, electrical characteristics, moisture absorption, flammability, and thermal characteristics have been employed to characterize (Venoor et al. 2021). Chemical elements include things like solvents and chemical effects, hydrolysis/polycondensation, UV aging, oxidation, and thermal degradation. The manufacturing description covers solid-state polymerization, pultrusion, and compounding (Kausar 2019). An overview of compounding additives and modifications, including plasticizers, reinforcements, impact modifiers, flame retardants, stabilizers, lubricants, nanocomposites, and polymer blends. The following processing pathways have been explained in detail: extrusion, film, tubing and pipe, monofilament, rods and profiles, blow molding, roto molding, reaction injection molding, powder coating, wire and cable coating, and assembly techniques. Issues with recycling and economics are covered. Standards, quality control, and specifications are also examined, along with a recommendation for suitable testing techniques (Liu et al. 2020).

In addition, polyamides make up 17% of microplastics in contaminated areas and are the most prevalent plastic at five distinct drinking water locations, accounting for 33% of the total. Numerous reports of polyamide microplastics inside living things have also surfaced. Up to 46.11% of pig lungs contain polyamides, the most prevalent polymer, whereas up to 35.6% of 10 fish species in the English Channel include microplastics. In addition to harming marine life, these microplastics seriously damage ecosystems by altering the links in the food chain. Human health is also seriously threatened by microplastic waste, which can lead to issues with organs, delayed child development, obesity, and reproductive harm. Various studies have shown that it is urgent to solve the environmental pollution caused by polyamide microplastics.

4.4 Production of Biofuel from Plastic Wastes

Plastics come in a variety of forms, but not all of them are used to make plastic fuels. They need to be divided and categorized. Low-density polyethylene (LDPE), high-density polyethylene (HDPE), polypropylene (PP), polyvinyl chloride (PVC), polystyrene (PS), and polyethylene terephthalate (PET) make up the majority of plastic waste (Ouda et al. 2013). The polymer chain chemical structure is another factor used to categorize plastics. Acrylates, polyesters, silicones, polyurethanes, and halogenated plastics are a few significant categories. Because of its relative clarity, hardness, and flexibility, LDPE is used to produce bottles. It is used to manufacture films, dropper bottles, commercial and waste bags, and dropper bottles because of its strength and durability. Toys, tank lids, and several types of packaging are also made using it. Because PP has a high melting point, it is perfect for holding hot liquids (such as syrup, ketchup, and juice) that need to cool in a bottle. It is used to produce a stiff or flexible component. Containers for yoghurt, margarine, and takeaway made of polypropylene (PP). Bottle caps, medicinal bottles, and other items are made using it.

4.4.1 Torrefaction

Torrefaction is a safe and effective thermochemical method for thermal processing of biomass resources (Sarker et al. 2021). Torrefaction has been demonstrated to increase the effectiveness of thermochemical conversion processes such as pyrolysis, liquefaction, and gasification by removing moisture and partially decomposing the biomass (Basu et al. 2014). In addition to the great properties of torrefied and densified biomass, torrefaction has garnered a lot of attention due to the near-commercial readiness of the related technologies. The advantages of torrefied biomass over raw biomass include its higher heating value, low moisture content, hydrophobicity, grindability, and palletizability (Sarker et al. 2021). Torrefaction increases biomass's heating value by lowering its oxygen content and increasing its carbon content. The drying of feedstock, typically at a temperature of 200 to 300°C and in an inert atmosphere, is referred to as "torrefaction." According to Chen et al. (2015), the temperature range determines the three types of torrefactions: light (200–235 °C), mild (235–275 °C), and severe (275–300 °C). Another way to conceptualize it is as the gradual heating of biomass in an oxygen-free environment. Hot vapors produced during the torrefaction process are sent to the condenser for condensation. Water and carboxylic acids are present in the condensed liquid product of torrefaction, together with trace amounts of ethers, alcohols, ketones, aldehydes, and esters (Acharya et al. 2015). H_2, CO_2, CO, and CH_4 are present in trace amounts in the non-condensable gases. Due to their properties, the leftover torrefied solids are used as a green alternative to coal. The fact that torrefaction emits no greenhouse gases is one advantage. Instead, the carbon was retained for further thermochemical processing as densified and dehydrated biomass. Table 4.1 enumerates a number of significant studies on the process of turning waste biomass into solid fuels.

4.4.2 Pyrolysis

Pyrolysis is a technique that produces biofuels by subjecting the organic material in biomass to an irreversible thermochemical degradation reaction. There are three types of pyrolysis technologies: slow, fast, and flash pyrolysis methods. Fast heating rates and short vapor residence times induce the volatile hydrocarbon vapors to condense quickly into bio-oil, but slow pyrolysis produces more biochar because the biomass carbonizes more slowly due to slow heating rates and long vapor residence times (Zaimes et al. 2015). A typical flow diagram for the pyrolysis process is displayed in Fig. 4.3. By thermally decomposing biomass at temperatures between 300 and 700 °C without the presence of oxygen, a process known as pyrolysis, biochar, non-condensable gases (H_2, CH_4, CO, and CO_2), and bio-oil are produced. Both a watery and an organic phase are present in bio-oil. To provide clean transportation fuels, the organic phase of bio-oil refiners from tar and heavier, hydrocarbon-rich molecules. Using both catalytic and non-catalytic techniques, oxygen, nitrogen, and sulfur components must be eliminated from bio-oil in order to enhance it and stop NOx and SOx emissions during burning (Kumar and Strezov 2021). The heating value of the fuel is lowered as a result. The watery phase of the bio-oil contains water, esters, ethers, aldehydes, ketones, phenols, alcohols, acids, and other biochemicals (Singh et al. 2020). Through the catalytic Fischer-Tropsch synthesis,

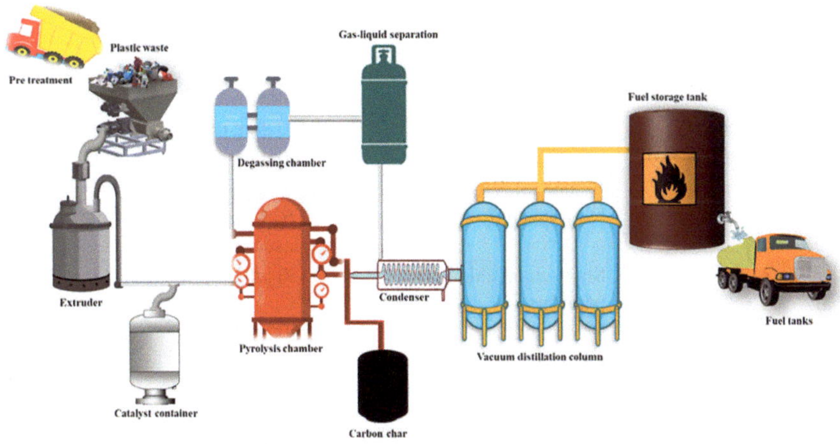

Fig. 4.3 Production of biofuel from plastic wastes

liquid hydrocarbons, specifically H_2 and CO, are produced from the gas phase. One technique for creating sustainable energy is pyrolysis, which employs a variety of lignocellulosic biomass. The properties of the pyrolysis products are influenced by temperature, heating rate, biomass composition, and residence duration. Food may be quickly and efficiently heated in a microwave, as is often known. The heating caused by microwaves is the result of material components directly absorbing microwave radiation. With wavelengths ranging from 300 MHz to 300 GHz, electromagnetic waves are referred to as "microwave radiation." It is recognized that microwave heating is a practical technique for the pyrolysis process (Jha et al. 2022). Many benefits of microwave-assisted pyrolysis over classical pyrolysis include a higher energy efficiency, a pyrolizer that reacts rapidly to start and stop, a faster heating rate, more consistent and rapid heating of large feedstocks, and low-aqueous bio-oil (Table 4.2).

4.4.3 Liquefaction

Another thermochemical method for converting biomass that primarily yields bio-crude oil is liquefaction. Bio-crude oil is produced through the hydrogenation and high-pressure thermal breakdown of biomass—hydrogenation is typically a process used after liquefaction (for upgrading), not directly in liquefaction. Liquefaction generally involves the thermal breakdown of biomass in the presence of water, without necessarily needing hydrogenation at the liquefaction stage. Hydrothermal liquefaction (HTL) requires catalysts and water to transform high-moisture solid waste into bio-crude oil. Figure 4.3 shows the flow diagram for the liquefaction process. Liquefaction takes place at temperatures between 240 and 380 °C and pressures between 5 and 30 MPa. The mechanism of the liquefaction reaction can be

Table 4.2 List of chemicals acquired from different types of waste plastics

S. No.	Plastics	Chemicals obtained	Concentration (g/kg of plastics)	References
1.	Polyvinyl chloride (PVC)	Diethylhexyl phthalate	300 g/kg plastic	Prajapati et al. (2021), Hahladakis et al. (2018)
		Dibutyl phthalate	300 g/kg plastic	
		Bisphenol A	1 g/kg plastic	
		Nonylphenol	2.6–4.2 g/kg plastic	
		Short-chain chlorinated paraffins	330 g/kg plastic	
		Medium-chain chlorinated paraffins	150 g/kg plastic	
		Tetralead trioxide sulphate	10–40 g/kg plastic	
		Lead distearate	10–40 g/kg plastic	
		Aluminum, zinc, cadmium	No data available	
2.	Polypropylene (PP)	2,4-DTBP	No data available	Geueke et al. (2023)
		2,6-di-tertbutylbenzoquinone (2,6-DTBQ)	No data available	
		Silver	No data available	
		BPA	No data available	
		Hexabromocyclododecane (HBCD)	No data available	
		Dechlorane plus	30–350 g/kg	
		Zinc, bromine	No data available	
3.	Polystyrene (PS)	Nonylphenol	0.9–1.8 g/kg	
		Hexabromocyclododecane (HBCD)	5–30 g/kg	
		Bromine metal	No data available	
4.	High-density polyethylene (HDPE)	Butane	4.3 g/kg	Sarker et al. (2012)
		Hexane	5.7 g/kg	
		Cycloheptene	0.2–4.1 g/kg	
		Pentane	0.1–4.3 g/kg	
5.	Low-density polyethylene (LDPE)	Ethyl benzene	3.24 g/kg	Htway et al. (2020)
		Pentachloroethane	0.79 g/kg	
		Tetrachloroethylene	0.30 g/kg	

summarized by the following basic pathways: (1) hydrolysis of biomass, smaller monomer, (2) smaller monomer, smaller compounds (due to cleavage and decarboxylation), and (3) recombination of the smaller fragments, new compounds (due to condensation, polymerization). Among the solvents commonly used in liquefaction are subcritical water and ethanol. In the dissolution of biomass, solvents are essential because they convert cellulose, hemicellulose, and lignin into volatile substances. Two phases are then produced, the tar phase and the aqueous phase, just like in pyrolysis. The watery phase that contains water and biochemicals can be recycled and utilized as a solvent in the process, as was previously mentioned (Hu et al. 2018). Compared to bio-oil produced by pyrolysis, bio-crude produced by liquefaction requires less upgrading through hydrotreating due to its low oxygen concentration, increased hydrocarbon yields, enhanced energy density, and flowability.

4.4.4 Gasification

Any carbonaceous material can be converted into syngas through a thermochemical process known as gasification (Makwana et al. 2019). Char is produced by using a range of feedstocks, and through gasification processes, the feedstock could be converted into produce syngas (such as H_2, CO, CO_2, and CH_4) in biofuel production. Syngas can be utilized to produce hydrogen, chemicals, power, heat, biofuel, and biomethane (Centi and Perathoner 2020). For gasification, a basic process flow diagram is shown in Fig. 4.4. Although hydrothermal gasification can operate at relatively low temperatures (400–700 °C), high temperatures (800–1200 °C) are necessary for thermal gasification of coal and complex biomass. Gasification yields the main gaseous components along with condensable liquids that are high in water and biochemicals. It is relatively easy to clean and recover the gas at higher temperatures because less tar is produced (Makwana et al. 2019). High gasification

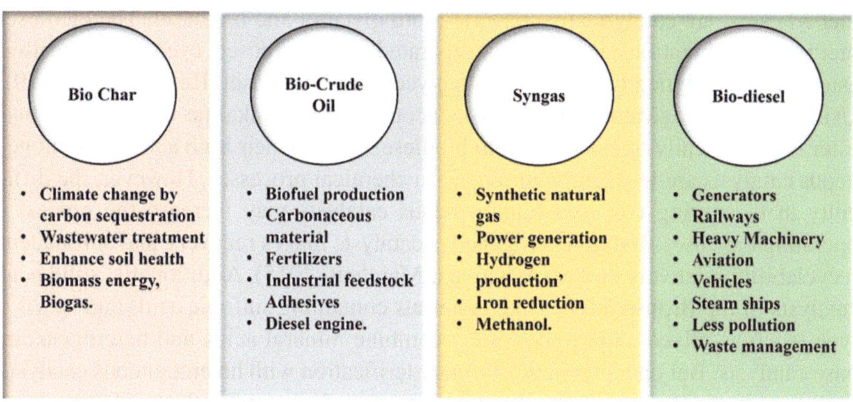

Fig. 4.4 Major biofuel products

temperatures also encourage endothermic processes, including steam reformation, methanation, and water-gas shift, which accelerate the rate of reaction and nearly completely convert biomass into gases. The syngas is converted into clean fuels and value-added products via Fischer-Tropsch catalysis. In hydrothermal gasification (HTG), syngas is produced by using subcritical or supercritical water as a reaction medium. Supercritical water occurs above the critical pressure of 22.1 MPa, whereas subcritical water occurs below the critical temperature of 374 °C. The primary benefit of using HTG is that it can efficiently convert biomass with a high moisture content, which often requires expensive and energy-intensive drying procedures before gasification (Pecchi and Baratieri 2019). However, converting wet biomass by pyrolysis, liquefaction, and gasification is inefficient because the heat of moisture evaporation surpasses the heat of biomass combustion. The HTG product has many uses, such as power generation, chemical synthesis, and fuel cells that produce hydrogen. It is challenging to increase the levels of CO and H_2 in syngas. Nonetheless, several strategies that can assist with this issue include the use of catalysts to promote water gas shift, steam reforming, tar reforming, and syngas redox processes (Nguyen et al. 2021). Additionally, by optimizing the gasification process parameters—such as temperature, pressure, reaction time, reaction medium, feedstock concentration, equivalency ratio, and catalyst loading—it is possible to increase gasification efficiency, carbon conversion efficiency, hydrogen selectivity, and syngas calorific value (Han et al. 2011).

4.4.5 Transesterification

Transesterification is the process of treating non-edible oils with alcohols (methanol and ethanol) to produce biodiesel. Because they are commonly available and reasonably priced, these alcohols are used. It is commonly known that transesterification facilitates the conversion of triglycerides into esters that are readily miscible in diesel and lowers the viscosity of inedible oils (Singh et al. 2020). Methanol and methyl esters are produced when alcohol combines with waste fat and oil. The methyl esters are catalytically converted into glycerol and biodiesel. The biodiesel methanol combination can be further separated and repurposed, even though methyl ester can be catalytically converted into glycerol and biodiesel (Rezania et al. 2019). Using both homogeneous and heterogeneous acid and alkaline catalysts, methyl ester is catalytically transformed into biodiesel. Due to their high activity, homogeneous catalysts are frequently employed in chemical processes. However, the difficulty in recovering the used homogeneous catalysts may increase the process's operating expenses. Using heterogeneous catalysts makes recovery and subsequent recyclability relatively easier (Avhad and Marchetti 2015). Additionally, solid acid catalysts of the Brønsted type (like materials containing sulfonic acid) and Lewis's type (such as mixed sulphated oxides) combine mineral acids and heterogeneous base catalysts. But one drawback of transesterification with heterogeneous catalysts is the production of soap, which complicates product purification and separation (Veljković et al. 2015). Alkali catalysts can increase yields and product selectivity while accelerating the transesterification process. Oil is catalytically converted to biodiesel using both base-catalyzed transesterification and acid-catalyzed esterification. For esterification or free fatty acids (FFA), acid catalysts are employed. Among the different kinds of acid catalysts are hydrochloric acid, sulfuric acid, phosphoric acid, and organic sulfonic acids (Borges and Díaz 2012). Acid catalysts are appropriate for glycerides with greater water and FFA levels, but transesterification is slower with them than with alkali catalysts. Materials used in transesterification include alkali catalysts such as NaOH, KOH, carbonates, and alkoxides such as sodium methoxide (CH_3NaO) and sodium ethoxide (C_2H_5ONa) (Jayakumar et al. 2021). Alkali-catalyzed transesterification is faster than that of acid catalysts. Whether catalytic or noncatalytic, the liquid products of transesterification must be separated from alcohol, oil, applied catalyst, and other impurities. Reduction of reactor corrosion, product selectivity, regeneration, and recyclability are some benefits of using solid acid catalysts (Mansir et al. 2017). However, high-quality oils can be produced using heterogeneous catalytic techniques due to their vulnerability to high FFAs. Heteropoly acids (HPAs) are solid acid catalysts that are effective in the production of biodiesel due to their strong acidity and exceptional resistance to high levels of FFA and water content (Negm et al. 2019). HPAs' exceptional dispersion, elastic structure, and appropriate interaction with support materials are further advantages. The catalyst's surface area and thermal stability are increased when HPAs are immobilized on the proper support material (Hanif et al. 2017). Because they can increase the catalyst's activity and effectiveness, the support material's surface area and porosity are carefully taken into account. When it comes to enzyme-catalyzed transesterification, the lipase enzyme serves as a bio-catalyst (Canet et al. 2016). Enzymes have a significant potential to reduce environmental problems and eliminate the need for high energy requirements in biodiesel refineries. In contrast to other acid- and base-catalyzed transesterification processes, enzymatic transesterification takes place at a comparatively lower temperature to avoid the loss of lipase activity. Since lipase-catalyzed reactions require a moderate temperature, enzymatic transesterification requires less energy. Enzymatic transesterification has several benefits, including high selectivity, mild reaction conditions, and a wide range of substrates. Nevertheless, the cost, stability, and low recyclability of enzymes are some of the disadvantages that call for more research (Jha et al. 2022).

4.5 Major Biofuel Products

4.5.1 Biochar

Thermochemical conversion of carbon-containing raw materials, such as woody biomass, crop residue, plastics, and plastic-biomass waste, using pyrolysis, gasification, hydrothermal carbonization, and torrefaction techniques, produces biochar, a solid material with a high carbon content (Adeniyi et al. 2023). Biochar is made up of carbon and a trace amount of minerals, depending on the kind of precursor that was utilized. Electrical conductivity, volatiles, cation exchange capacity, and pyrolysis temperature all affect how much biochar is produced. However, the output of biochar at high temperatures revealed a higher surface area and carbon content (Nanda et al. 2016). The impact of temperature on the yield of biochar using LDPE and sugarcane bagasse was examined in earlier research (Adeniyi et al. 2023). When plastics are used as a feedstock, they discovered that the generation of biochar increases by 16.67%, with the highest temperature occurring at 250 °C. Aside from these, biochar maintained characteristics such as a high percentage of carbon and cation exchange capacity, hydrophobicity, more surface area, and thermal stability (Bartoli et al. 2020). Because of its high carbon concentration, biochar generates a lot of heat, which is why it is used to make activated carbon. According to Tan et al. (2015), biochar is also utilized as a filter for pollutants, an absorbent to capture heavy metals and micropollutants, and a nutrient for soil remediation research. According to (Panwar et al. 2019), it also has a large surface area, pore size, cation exchange characteristics, and a functional group. Pb and 2,4-dinitrotoluene sorption is facilitated by this characteristic (Tanasupawat et al. 2016). These days, biochar is employed as a catalyst in processes such as hydrotreating, esterification, and hydrodeoxygenation to improve bio-oil (Kang et al. 2022). In metallurgical operations, biochar is a highly promising bioresource that can be used as a fuel and reducing agent in place of coal and cock (Koveria et al. 2020). Through carbon sequestration, wastewater treatment, soil health enhancement, biomass energy, and biogas production, biochar helps mitigate climate change (Fig. 4.4).

In addition to these, hydrochar is a carbon-rich product of hydrothermal modification procedures such as carbonization, gasification, and liquefaction. Hydrochar has more oxygenated functional groups and a higher H/C and O/C atomic ratio than biochar (Nanda et al. 2016). Hydrochar has a somewhat acidic composition, while biochar is alkaline. As a result, biochar is applied to acidic soil to balance its pH level (4.0–7.6) (Zhang et al. 2019).

4.5.2 Bio-crude Oil

Hydrothermal liquefaction, in which lignocellulosic feedstocks react with water at elevated pressure and temperature to produce bio-crude oil, is the energy-rich liquid product. In these processes, improved bio-crude oil production was noted when plastic waste, such as nylon-6, polyethylene, and polypropylene, was combined with biomass feedstock (Hongthong et al. 2020). The production of bio-crude oil also produces some undesirable qualities, such as high acidity, high moisture, and high corrosive properties; these are removed before using bio-crude oil as a practical fuel for transportation (Chand et al. 2019). Due to their lower rate of nitrogen recovery, microalgae have replaced algae in recent years (Chen et al. 2015). Catalytic and non-catalytic upgrading equipment eliminates the negative characteristics of bio-crude oil, such as its high viscosity, acidity, corrosivity, emission, high moisture content, and ash content (Saber et al. 2016). Carbonaceous material, fertilizers, industrial feedstock, adhesives, diesel engines, and biofuel can all be produced using bio-crude oil (Fig. 4.4).

4.5.3 Bio-oil

Bio-oil is produced by pyrolyzing lignocellulosic biomass. It is a breakdown of cellulose, lignin, and hemicellulose. The temperature, residence time, heating rate, and biomass composition all affect their yield and characteristics (Zhang et al. 2022). Since bio-oil is easier to transport and store than biogas and biochar, it can be utilized with existing pumping systems, equipment, and safety regulations (Lee et al. 2019). Hydrocarbon derivatives, aldehydes, sugars, acids, esters, and phenols make up bio-oil. Due to the high moisture content of biomass and the pyrolysis process's decarboxylation and dehydration, 30–40% of bio-oil is made of water (Salehi et al. 2011; Joshi and Lawal 2012; Gupta et al. 2019). Because of its high viscosity, corrosivity, presence of heteroatoms like nitrogen, oxygen, and sulfur, poor energy density, and low thermal stability, bio-oil has several main drawbacks that should be addressed before usage. Biomass and plastic waste are co-pyrolyzed to produce improved bio-oil (Dyer et al. 2021; Terry et al. 2021).

4.5.4 Syngas

One important byproduct of the steam gasification process for plastics is syngas. Among its gaseous constituents are H_2, CH_4, CO, and CO_2. Syngas has a high H_2/CO ratio and is rich in H_2 (Ruoppolo et al. 2012; Erkiaga et al. 2013). 40% H_2 concentration of syngas was created by gasifying polypropylene and polyethylene, which translates to a 4–3% weight percentage of H_2 production ratio (gH_2 100 g/plastics) (Shah et al. 2023). Syngas is utilized in the production of green fuels and chemicals (methanol and hydrocarbons) as well as in the production of electricity (Zhang 2010; Adnan et al. 2020). Syngas is used to create synthetic natural gas, which is then used to produce methanol, hydrogen, power, and iron reduction (Fig. 4.4). Hydrogen is used in various industrial applications like the production of medicines, aerospace, platform chemicals, metallurgy, maritime purposes, and electrical devices as an energy carrier. The production of hydrogen also gives several benefits in profitability, environment end users, and energy security (Abe et al. 2019). However, the large-scale production of hydrogen is difficult and most challenging, therefore government tries to accomplish several policies for the convenient, stable, and safe transportation of hydrogen energy (Tian et al. 2019; Jha et al. 2022).

4.5.5 Biodiesel

Most of the plastic trash is composed of highly volatile materials (>90 wt.% dry basis), apart from plastic film. This suggests that biofuels with high ignition, like syngas and bio-oil, can be produced. In addition, the comparatively low levels of S and N components in plastic wastes suggest that the high levels of C and H elements within the pyrolysis process will result in low NOx, HCl, and SO_2 emissions along with high heating value (HHV) gaseous products. Biodiesel exhibits biodegradability, non-toxicity, and the presence of minute amounts of aromatics and sulfur compounds (Alamer et al. 2023). Biodiesel is produced through the transesterification reaction with the help of homogeneous and heterogeneous catalysts. In these reactions, sodium hydroxide, hydrochloric acid, organic sulfonic acid, sodium methoxide, potassium methoxide, iron sulfate, and sulfuric acid are used as homogeneous alkali catalysts, and calcium oxide, calcium carbonate, and calcium hydroxide are used as heterogeneous catalysts. Heterogeneous catalysts are most effective, recyclable, and non-corrosive in nature as compared to homogeneous catalysts (Ling et al. 2019). Biodiesel has significant use in different areas, such as in generators, in railways, in heavy machinery, in aviation, in vehicles, in steam ships, for maintaining less pollution and for waste management (Fig. 4.4).

4.6 Research Gap and Future Prospects

Among the significant chemicals that are made from plastic trash are paraffins, olefins, naphthalene, benzene, toluene, xylene, carboxyl compounds, microcrystalline waxes, and lubrication and grease base stocks. The most cutting-edge operational condition-based processes show promise in the production of these chemicals. The disadvantages of large-scale applications include harsh reaction conditions, aggressive reagents, slow reaction rates, low chemical yields, and many purification procedures. Advanced technological options, including plasma pyrolysis, microwave treatment, and supercritical extractions using different solvents, need more investigation. The most challenging part is recovering the chemicals since it is hard to remove catalysts from products and purify them from other products. The next phase of polymerization requires a high-quality, pure monomer. Innovations like ionic liquids, deep-eutectic fluids, or nanoparticles acting as catalysts are considered. Both the monomer synthesis and the depolymerization reaction are accelerated by these catalysts. Waste plastic is used to make graphitic and amorphous carbon, among other carbon compounds. Carbon recovery from the commonly used polymers is lower. Large amounts of energy are wasted as a result, and the environment is severely harmed. To produce carbon materials from plastic waste, it is therefore imperative to increase carbon recovery to a high degree. It is common practice to discard plastics of various types and shapes, such as rubber, fabrics, electrical shells, fresh-keeping films, and plastic stacks. It is really challenging to identify and filter this garbage. For carbonization processes to be simple, affordable, and energy-efficient, plastic wastes must be treated without being sorted or cleaned. If the carbonization operations are successful on a larger scale, at least 1,000 metric tons of carbon materials will be produced. If we are successful in producing a significant amount of carbon materials, there will be an immediate need to explore new and practical applications. Adsorbents, electrode materials, and catalysts are examples of traditional applications. Potential applications for carbon materials include solar steam generation, soil remediation, and the production of biofilm precursors for the treatment of contaminated wastewater.

4.7 Conclusion

Waste biomass is an exceptional bioresource when transformed into useful energy carriers such as bio-oil, bio-crude oil, biodiesel, syngas, hydrogen, and charcoal. As covered in this book chapter on solid waste valorization, these waste materials were converted into biofuel and biochemical products by thermochemical processes as torrefaction, pyrolysis, gasification, liquefaction, and transesterification. Torrefaction results in biomass that is densified, has a lower moisture content, and a higher heating value. Bio-oil and bio-crude oil are produced by pyrolysis and liquefaction because the organic materials in biomass, such as cellulose, hemicellulose, and lignin, thermally break down in inert atmospheres and under the proper processing conditions to produce bio-oil and bio-crude oil, respectively. When gasification occurs in inert air or hydrothermal (steam and/or supercritical water) conditions, syngas—a mixture of H_2 and CO with CO_2 and CH_4—is produced. Biochar is a byproduct of pyrolysis and gasification that has a higher fixed carbon content. The aromaticity, surface area, porosity, calorific value, and thermal stability of biochar can all be improved at higher temperatures. Biodiesel is produced by trans esterifying the fatty acids and lipids in residual vegetable or animal fat, typically with the use of chemical or biological catalysts. Nevertheless, factors like biomass content and process parameters (temperature, pressure, reaction duration, reactor type, additional catalyst, etc.) have an impact on the process efficiency as well as the yield and quality of the biofuel products. Biofuels and biochemicals can be used directly or in combination with fossil fuels to reduce pollution, greenhouse gas emissions, global warming, and dependency on fossil fuels. transformation of solid waste into biofuels and chemicals for a sustainable future. This study describes the many plastic wastes that can be converted into chemicals and biofuels, as well as the research obstacles associated with this process. Additionally, this chapter offers a succinct summary of cutting-edge procedures to assist future system designers in the valorization of plastic waste. But new technology should be incorporated in the future studies to improve the waste management such as supercritical fluid technology (SCF) and Plasma pyrolysis technique. Over the past 65 years, the output of plastic has grown far more than that of any other industrial material. Similarly, plastic and electrical and electronic equipment have become essential components of modern life. The development of sustainable methods for recycling plastic and e-wastes is essential from the perspective of resource recovery and environmental preservation. By recovering its metal content, turning plastic garbage into useful items, and conserving the resources, it would be worthwhile to turn the waste into wealth. Supercritical fluid technologies may be able to address these problems in this way. It has been noted as an appealing and practical method for handling plastic and e-waste also because of its unique qualities. SCF and plasma-assisted pyrolysis have a number of benefits over traditional and microwave-assisted pyrolysis, including improved reaction kinetics, increased energy density, lower carbon emissions, and higher energy efficiency. However, further research is required to close this gap.

References

Abe JO et al (2019) Hydrogen energy, economy and storage: review and recommendation. Int J Hydrog Energy 44(29):15072–15086. https://doi.org/10.1016/j.ijhydene.2019.04.068

Acharya B, Dutta A, Minaret J (2015) Review on comparative study of dry and wet torrefaction. Sustain Energy Technol Assess 12:26–37. https://doi.org/10.1016/j.seta.2015.08.003

Adeniyi AG et al (2023) Thermochemical co-conversion of biomass-plastic waste to biochar: a review. Green Chem Eng. https://doi.org/10.1016/j.gce.2023.03.002

Adnan MA et al (2020) Gasification of wet microalgae to produce H2-rich syngas and electricity: a thermodynamic study considering exergy analysis. Renew Energy 147:2195–2205. https://doi.org/10.1016/j.renene.2019.10.027

Alabi AA et al (2022) Powder production from recycled low-density polyethylene waste and doum palm nuts for light weight engineering application: a circular economy approach. Cleaner Waste Syst 3:100044. https://doi.org/10.1016/j.clwas.2022.100044

Alamer NJ, Aldayel MF, Khalifa A (2023) Isolation and characterization of Brucella spp., low-density polyethylene (LDPE) plastic degrading bacteria in Al-Ahsa Region, Saudi Arabia. Appl Sci 13(7):4629. https://doi.org/10.3390/app13074629

Andrady AL (1994) Assessment of environmental biodegradation of synthetic polymers. J Macromolec Sci 34(1):25–76. https://doi.org/10.1080/15321799408009632

Avhad MR, Marchetti JM (2015) A review on recent advancement in catalytic materials for biodiesel production. Renew Sust Energ Rev 50:696–718. https://doi.org/10.1016/j.rser.2015.05.038

Bartoli M et al (2020) A review of non-soil biochar applications. Mater Ther 13(2):1–35. https://doi.org/10.3390/ma13020261

Basu P et al (2014) An experimental and theoretical investigation on torrefaction of a large wet wood particle. Bioresour Technol 159:215–222. https://doi.org/10.1016/j.biortech.2014.02.105

Borges ME, Díaz L (2012) Recent developments on heterogeneous catalysts for biodiesel production by oil esterification and transesterification reactions: a review. Renew Sust Energ Rev 16(5):2839–2849. https://doi.org/10.1016/j.rser.2012.01.071

Canet A et al (2016) Lipase-catalysed transesterification: viewpoint of the mechanism and influence of free fatty acids. Biomass Bioenergy 85:94–99. https://doi.org/10.1016/j.biombioe.2015.11.021

Centi G, Perathoner S (2020) Chemistry and energy beyond fossil fuels. A perspective view on the role of syngas from waste sources. Catal Today 342:4–12. https://doi.org/10.1016/j.cattod.2019.04.003

Chand R et al (2019) Evaluating the potential for bio-fuel upgrading: a comprehensive analysis of bio-crude and bio-residue from hydrothermal liquefaction of agricultural biomass. Appl Energy 254:113679. https://doi.org/10.1016/j.apenergy.2019.113679

Chen W-H, Peng J, Bi X (2015) A state-of-the-art review of biomass torrefaction, densification and applications. Renew Sust Energ Rev 44:847–866. https://doi.org/10.1016/j.rser.2014.12.039

Das M, Patra P, Ghosh A (2020) Metabolic engineering for enhancing microbial biosynthesis of advanced biofuels. Renew Sust Energ Rev 119:109562. https://doi.org/10.1016/j.rser.2019.109562

Dyer AC, Nahil MA, Williams PT (2021) Catalytic co-pyrolysis of biomass and waste plastics as a route to upgraded bio-oil. J Energy Inst 97:27–36. https://doi.org/10.1016/j.joei.2021.03.022

Erkiaga A et al (2013) Syngas from steam gasification of polyethylene in a conical spouted bed reactor. Fuel 109:461–469. https://doi.org/10.1016/j.fuel.2013.03.022

Flaherty DW, Hibbitts DD, Iglesia E (2014) Metal-catalyzed C–C bond cleavage in alkanes: effects of methyl substitution on transition-state structures and stability. J Am Chem Soc 136(27):9664–9676. https://doi.org/10.1021/ja5037429

Garcia JM, Robertson ML (2017) The future of plastics recycling. Science 358(6365):870–872. https://doi.org/10.1126/science.aaq0324

Geueke B et al (2023) Hazardous chemicals in recycled and reusable plastic food packaging. Cambridge Prisms Plast. https://doi.org/10.1017/plc.2023.7

Geyer R, Jambeck JR, Law KL (2017) Production, use, and fate of all plastics ever made. Sci Adv 3:e1700782

Greenberg R (2015) Premature termination in psychotherapy: strategies for engaging clients and improving outcomes. APA Books, Washington. https://doi.org/10.1037/14469-000

Gupta S, Gupta GK, Mondal MK (2019) Slow pyrolysis of chemically treated walnut shell for valuable products: effect of process parameters and in-depth product analysis. Energy 181:665–676. https://doi.org/10.1016/j.energy.2019.05.214

Hahladakis JN et al (2018) An overview of chemical additives present in plastics: migration, release, fate and environmental impact during their use, disposal and recycling. J Hazard Mater 344:179–199. https://doi.org/10.1016/j.jhazmat.2017.10.014

Han B, Gopalan V, Ezeji TC (2011) Appl Microbiol Biotechnol 91:565

Hanif MA, Nisar S, Rashid TDU (2017) Supported solid and heteropoly acid catalysts for production of biodiesel. Catal Rev 59:165–188. https://doi.org/10.1080/01614940.2017.1321452

Hibbitts DD, Flaherty DW, Iglesia E (2016) Effects of chain length on the mechanism and rates of metal-catalyzed hydrogenolysis of n-alkanes. J Phys Chem 120(15):8125–8138. https://doi.org/10.1021/acs.jpcc.6b00323

Hongthong S et al (2020) Co-processing of common plastics with pistachio hulls via hydrothermal liquefaction. Waste Manag 102:351–361. https://doi.org/10.1016/j.wasman.2019.11.003

Htway EE et al (2020) Determination of chemicals released from single use low density polyethylene plastic bags. IOP Conf Ser Earth Environ Sci 496(1):012012. https://doi.org/10.1088/1755-1315/496/1/012012

Hu T et al (2018) Improvement in bio-crude yield and quality through co-liquefaction of algal biomass and sawdust in ethanol-water mixed solvent and recycling of the aqueous by-product as a reaction medium. Energy Convers Manag 171:618–625. https://doi.org/10.1016/j.enconman.2018.06.023

Hwang J et al (2020) Potential toxicity of polystyrene microplastic particles. Sci Rep 10(1):1–12. https://doi.org/10.1038/s41598-020-64464-9

Jambeck JR et al (2015) Entradas de residuos plásticos desde la tierra al océano. Ciencia 347(6223):768–771. https://doi.org/10.1126/science.1260352

Jayakumar M et al (2021) Heterogeneous base catalysts: synthesis and application for biodiesel production - a review. Bioresour Technol 331:125054. https://doi.org/10.1016/j.biortech.2021.125054

Jeon J et al (2021) Evaluation of thermal properties and acetaldehyde adsorption performance of sustainable composites using waste wood and biochar. Environ Res 196:110910. https://doi.org/10.1016/j.envres.2021.110910

Jha S et al (2022) A review of thermochemical conversion of waste biomass to biofuels. Energies 15(17):1–23. https://doi.org/10.3390/en15176352

Joshi N, Lawal A (2012) Hydrodeoxygenation of pyrolysis oil in a microreactor. Chem Eng Sci 74:1–8. https://doi.org/10.1016/j.ces.2012.01.052

Kang K, Nanda S, Hu Y (2022) Current trends in biochar application for catalytic conversion of biomass to biofuels. Catal Today 404:3–18. https://doi.org/10.1016/j.cattod.2022.06.033

Kausar A (2019) Advances in carbon fiber reinforced polyamide-based composite materials. Adv Mater Sci 19(4):67–82. https://doi.org/10.2478/adms-2019-0023

Kik K, Bukowska B, Sicińska P (2020) Polystyrene nanoparticles: sources, occurrence in the environment, distribution in tissues, accumulation and toxicity to various organisms. Environ Pollut 262:114297. https://doi.org/10.1016/j.envpol.2020.114297

Koveria A et al (2020) Metallurgical coke production with biomass additives. Part 1. A review of existing practices. Can Metall Q 59(4):417–429. https://doi.org/10.1080/00084433.2021.1916293

Kumar R, Strezov V (2021) Thermochemical production of bio-oil: a review of downstream processing technologies for bio-oil upgrading, production of hydrogen and high value-added products. Renew Sust Energ Rev 135:110152. https://doi.org/10.1016/j.rser.2020.110152

Kyrikou I, Briassoulis D (2007) Biodegradation of agricultural plastic films: a critical review. J Polym Environ 15(2):125–150. https://doi.org/10.1007/s10924-007-0053-8

Leal Filho W et al (2019) An overview of the problems posed by plastic products and the role of extended producer responsibility in Europe. J Clean Prod 214:550–558. https://doi.org/10.1016/j.jclepro.2018.12.256
Lee SY et al (2019) Waste to bioenergy: a review on the recent conversion technologies. BMC Energy 1(1):1–22. https://doi.org/10.1186/s42500-019-0004-7
Lewandowski K (2022) A brief review of poly(vinyl chloride) (PVC) recycling. Polymers 14(15):3035
Ling JSJ et al (2019) A review of heterogeneous calcium oxide based catalyst from waste for biodiesel synthesis. SN Appl Sci 1(8):1–8. https://doi.org/10.1007/s42452-019-0843-3
Liu X et al (2020) The origin of memory effects in the crystallization of polyamides: role of hydrogen bonding. Polymer 188:122117. https://doi.org/10.1016/j.polymer.2019.122117
Maddah H (2016) Polypropylene as a promising plastic: a review. Am J Polym Sci 6:2163–1352. https://doi.org/10.5923/j.ajps.20160601.01
Makwana JP, Pandey J, Mishra G (2019) Improving the properties of producer gas using high temperature gasification of rice husk in a pilot scale fluidized bed gasifier (FBG). Renew Energy 130:943–951. https://doi.org/10.1016/j.renene.2018.07.011
Mansir N et al (2017) Investigation of heterogeneous solid acid catalyst performance on low grade feedstocks for biodiesel production: a review. Energy Convers Manag 141:171–182. https://doi.org/10.1016/j.enconman.2016.07.037
McNicholas G, Cotton M (2019) Stakeholder perceptions of marine plastic waste management in the United Kingdom. Ecol Econ 163:77–87. https://doi.org/10.1016/j.ecolecon.2019.04.022
Nakajima H, Dijkstra P, Loos K (2017) The recent developments in biobased polymers toward general and engineering applications: polymers that are upgraded from biodegradable polymers, analogous to petroleum-derived polymers, and newly developed. Polymers 9(10):1–26. https://doi.org/10.3390/polym9100523
Nanda S, Dalai AK et al (2016) Biochar as an exceptional bioresource for energy, agronomy, carbon sequestration, activated carbon and specialty materials. Waste Biomass Valoriz 7(2):201–235. https://doi.org/10.1007/s12649-015-9459-z
Negm N et al (2019) Clean transesterification process for biodiesel production using heterogeneous polymer-heteropoly acid nanocatalyst. J Clean Prod 238:117854. https://doi.org/10.1016/j.jclepro.2019.117854
Nguyen NM et al (2021) Biomass-based chemical looping gasification: overview and recent developments. Appl Sci 11(15):7069. https://doi.org/10.3390/app11157069
Olam M (2022) PET: production, properties and applications Complimentary Contributor Copy
Ouda OKM, Cekirge HM, Raza SAR (2013) An assessment of the potential contribution from waste-to-energy facilities to electricity demand in Saudi Arabia. Energy Convers Manag 75:402–406. https://doi.org/10.1016/j.enconman.2013.06.056
Panwar NL, Pawar A, Salvi BL (2019) Comprehensive review on production and utilization of biochar. SN Appl Sci 1(2):1–19. https://doi.org/10.1007/s42452-019-0172-6
Pecchi M, Baratieri M (2019) Coupling anaerobic digestion with gasification, pyrolysis or hydrothermal carbonization: a review. Renew Sust Energ Rev 105:462–475. https://doi.org/10.1016/j.rser.2019.02.003
Peng R et al (2018) Microbial degradation of polyurethane plastics. Chin J Biotechnol 34(9):1398–1409. https://doi.org/10.13345/j.cjb.170532
Pereira S et al (2016) Pereira et al. 2016 cmmi1686-sup-0001-SI
Prajapati R et al (2021) Potential chemicals from plastic wastes. Molecules 26(11):1–22. https://doi.org/10.3390/molecules26113175
Rezania S et al (2019) Review on transesterification of non-edible sources for biodiesel production with a focus on economic aspects, fuel properties and by-product applications. Energy Convers Manag 201:112155. https://doi.org/10.1016/j.enconman.2019.112155
Rouch DA et al (2022) Clarendon Policy & Strategy Group Plastic future
Ru J, Huo Y, Yang Y (2020) Microbial degradation and valorization of plastic wastes. Front Microbiol 11:1–20. https://doi.org/10.3389/fmicb.2020.00442

Ruoppolo G et al (2012) H2-rich syngas production by fluidized bed gasification of biomass and plastic fuel. Waste Manag 32(4):724–732. https://doi.org/10.1016/j.wasman.2011.12.004

Saber M, Nakhshiniev B, Yoshikawa K (2016) A review of production and upgrading of algal bio-oil. Renew Sust Energ Rev 58:918–930. https://doi.org/10.1016/j.rser.2015.12.342

Salehi E, Abedi J, Harding T (2011) Bio-oil from sawdust: effect of operating parameters on the yield and quality of pyrolysis products. Energy Fuel 25(9):4145–4154. https://doi.org/10.1021/ef200688y

Sarker M et al (2012) Low and high density polyethylene waste plastics conversion into liquid hydrocarbon fuel. Chem Today 30(6):42–43

Sarker TR et al (2021) A review of torrefaction technology for upgrading lignocellulosic biomass to solid biofuels. Bioenergy Res 14(2):645–669. https://doi.org/10.1007/s12155-020-10236-2

Shah HH et al (2023) A review on gasification and pyrolysis of waste plastics. Front Chem 10:1–38. https://doi.org/10.3389/fchem.2022.960894

Shao L et al (2019) One-pot synthesis of melamine-based porous polyamides for CO2 capture. Microporous Mesoporous Mater 285:105–111. https://doi.org/10.1016/j.micromeso.2019.05.005

Singh A et al (2020) Pyrolysis of Miscanthus and characterization of value-added bio-oil and bio-char products. Can J Chem Eng 99:23978. https://doi.org/10.1002/cjce.23978

Siracusa V, Blanco I (2020) Bio-polyethylene (bio-PE), bio-polypropylene (bio-PP) and bio-poly(ethylene terephthalate) (bio-PET): recent developments in bio-based polymers analogous to petroleum-derived ones for packaging and engineering applications. Polymers 12(8):1641. https://doi.org/10.3390/APP10155029

Tan X et al (2015) Application of biochar for the removal of pollutants from aqueous solutions. Chemosphere 125:70–85. https://doi.org/10.1016/j.chemosphere.2014.12.058

Tanasupawat S et al (2016) Ideonella sakaiensis sp. nov., isolated from a microbial consortium that degrades poly(ethylene terephthalate). Int J Syst Evol Microbiol 66(8):2813–2818. https://doi.org/10.1099/ijsem.0.001058

Terry LM et al (2021) Bio-oil production from pyrolysis of oil palm biomass and the upgrading technologies: a review. Carbon Resour Convers 4(June):239–250. https://doi.org/10.1016/j.crcon.2021.10.002

Tian M-W et al (2019) The multiple selections of fostering applications of hydrogen energy by integrating economic and industrial evaluation of different regions. Int J Hydrog Energy 44(56):29390–29398. https://doi.org/10.1016/j.ijhydene.2019.07.089

Veljković V, Bankovic-Ilic I, Stamenković O (2015) Purification of crude biodiesel obtained by heterogeneously-catalyzed transesterification. Renew Sust Energ Rev 49:97. https://doi.org/10.1016/j.rser.2015.04.097

Venoor V et al (2021) Understanding the effect of water in polyamides: a review. Polym Rev 61(3):598–645. https://doi.org/10.1080/15583724.2020.1855196

Volanti M et al (2022) Environmental impact of meals: how big is the carbon footprint in the school canteens? Foods 11(2):1–10. https://doi.org/10.3390/foods11020193

Ward OP (2010) Biodegradation

Xi X et al (2020) Soy protein isolate-based polyamides as wood adhesives. Wood Sci Technol 54(1):89–102. https://doi.org/10.1007/s00226-019-01141-9

Xu L, Du H, Zhang X (2021) Driving forces of carbon dioxide emissions in China's cities: an empirical analysis based on the geodetector method. J Clean Prod 287:125169. https://doi.org/10.1016/j.jclepro.2020.125169

Yuan J et al (2020) Science of the total environment microbial degradation and other environmental aspects of microplastics/plastics. Sci Total Environ 715:136968. https://doi.org/10.1016/j.scitotenv.2020.136968

Zaimes GG et al (2015) Biofuels via fast pyrolysis of perennial grasses: a life cycle evaluation of energy consumption and greenhouse gas emissions. Environ Sci Technol 49(16):10007–10018. https://doi.org/10.1021/acs.est.5b00129

Zhang W (2010) Automotive fuels from biomass via gasification. Fuel Process Technol 91(8):866–876. https://doi.org/10.1016/j.fuproc.2009.07.010

Zhang Z et al (2019) Insights into biochar and hydrochar production and applications: a review. Energy 171:581–598. https://doi.org/10.1016/j.energy.2019.01.035

Zhang K et al (2022) Biodegradation of polyester polyurethane by the marine fungus Cladosporium halotolerans 6UPA1. J Hazard Mater 437:129406. https://doi.org/10.1016/j.jhazmat.2022.129406

Valorization of Food Waste to Biofuels and Bioproducts

5

Ananya Verma, Sushant Gawali, Sarika Loni, Manju D. Tanwar, and Abhishek Guldhe

Abstract

Approximately one-third of the food that we consume are ultimately thrown away as food waste, a significant contributor to climate change, as the decomposition of organic waste releases potent greenhouse gases (GHGs). The food waste, along with inefficient food distribution and several economic factors, not only impacts the environment but also exacerbates global food insecurity. Implementing more sustainable techniques, such as composting and anaerobic digestion, can help in lowering the GHGs while producing valuable byproducts such as biofertilizer and biogas. Biofuel production from food waste can reduce fossil fuel reliance, while also providing a renewable source of energy for vehicles and machinery, promoting a more sustainable transportation sector. The bioproducts generated can be utilized in different industries, such as energy generation, pharmaceuticals, and agriculture. Additionally, the generation of bioproducts (biofuels, biopolymers, biofertilizers, biochemicals) from food waste

A. Verma · A. Guldhe (✉)
Amity Institute of Biotechnology, Amity University, Mumbai, Maharashtra, India
e-mail: asguldhe@mum.amity.edu

S. Gawali
Amity Institute of Biotechnology, Amity University, Mumbai, Maharashtra, India

Department of Chemical Engineering, State University of New York College of Environmental Science and Forestry, Syracuse, NY, USA

S. Loni
Amity Institute of Biotechnology, Amity University, Mumbai, Maharashtra, India

Organic Recycling Systems Ltd., Mumbai, India

M. D. Tanwar
Organic Recycling Systems Ltd., Mumbai, India

A. Kumari et al. (eds.), *Valorization of Solid Wastes to Biofuels and Chemical Products for Sustainable World*, https://doi.org/10.1007/978-981-96-8594-3_5

valorization can also create new employment opportunities and boost local economic growth, contributing to overall sustainable development. This chapter presents an overview of the numerous classification criteria for food waste and discusses biofuels and bioproducts derived from food waste valorization.

Keywords

Food waste valorization · Anaerobic digestion · Biogas production · Bioethanol · Organic acids

5.1 Introduction

Edible items that are either abandoned or lost at any stage in the food supply chain, which includes manufacturing, processing/refining, and consumption/utilization, are known as food waste. This includes both "food loss"—waste that accumulates throughout the stages of manufacturing, refinement, and distribution, and "food waste"—the waste that occurs when food is not consumed (Esteban-Lustres et al. 2022). It primarily consists of organic matter from a range of sources, including vegetables, fruits, dairy, cereals, and meat products.

Food waste is a major component of the solid waste generated by cities, accounting for around 30–60% of the whole urban residue (Esteban and Ladero 2018). The food and beverage sector, which includes beverages (26%), dairy (21%), fruit and vegetable cultivation and processing (14.8%), cereal processing (12.9%), and meat product processing and preservation (8%), vegetable and animal oil production (3.9%), accounts for 76.6% of food waste. Fish product processing and preservation contribute 0.4%, while the remaining 12.7% comes from other industries (Baiano 2014). According to the Food and Agriculture Organization (FAO), approximately one-third of global food production, equating to 1.3 billion tons, is wasted each year (Gustavsson n.d.). Europe produces a substantial quantity of food waste, which includes more than 29 million tons of dairy products each year (Sharma et al. 2020). Over 95% of the food waste is ultimately deposited in landfills, threatening the environment and our valuable resources that may be used for bioenergy production (Dhiman and Mukherjee 2021). These numbers indicate the importance of identifying sustainable approaches for the management, usage, and repurposing food waste for bioenergy. Inadequate harvesting strategies, infrastructure, processing, and packaging facilities together contribute to food waste in less developed nations. Conversely, in developed countries, a substantial portion of food waste stems from consumer actions, such as overpurchasing and subsequent disposal of unneeded food items.

Efficient food waste management involves using sustainable strategies to recover resources and create value products from items that would otherwise be discarded; the varied assortment of food waste, including dairy, vegetables, fruits, grains, meat, and seafood, adds to its intricacy. Food waste, while having high moisture content, has substantial value since it contains a large amount of nutrients and may be converted into bioenergy, which helps to decrease environmental impacts, alleviate GHGs, and foster a circular economy (Dahiya et al. 2018).

5.2 Classification of Food Waste

There are several ways to categorize food waste based on its function, origin, composition, and location along the food supply chain (Fig. 5.1). Food supply chain waste includes catering/kitchen waste, animal by-products/waste, expired packaged food waste, and mixed food waste, according to its usage. Domestic leftovers are another kind of waste. A wide variety of foods, such as dairy, vegetables, fruits, cereals, roots, tubers, oilseeds, and pulses, end up in the trash. Veggies, fatty foods, minerals, fibrous materials, starchy goods, and sludges are some of the common content categories used to describe food waste. Several useful resources, such as biofuels, enzymes, biofertilizers, biopolymers, and other biochemicals, may be extracted from food waste. The food supply chain encompasses several steps that result in food waste, such as production in agriculture, consumption, handling, and storage after harvest, distribution, and storage. This meticulous classification allows for a detailed understanding of the many components of food waste, which in turn allows for targeted strategies to minimize waste and maximize resource utilization.

Food waste valorization may result in the development of many kinds and subtypes of value-added bioproducts such as biofuels (bioethanol, biobutanol, bio-oil, biodiesel, biochar, biohydrogen, biogas or biomethane, and biohythane), biofertilizers, enzymes, biopolymers, and other biochemicals (Fig. 5.2), which will be touched upon in this chapter.

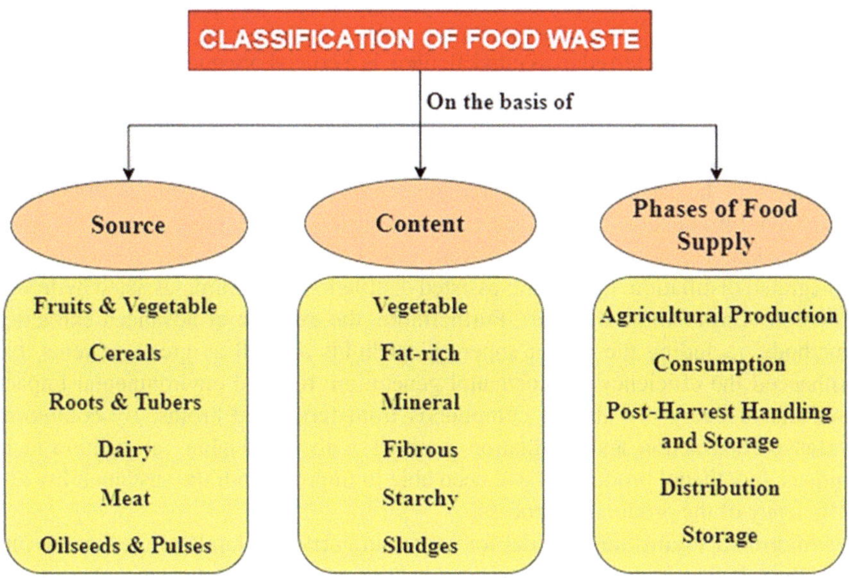

Fig. 5.1 Classification of food waste

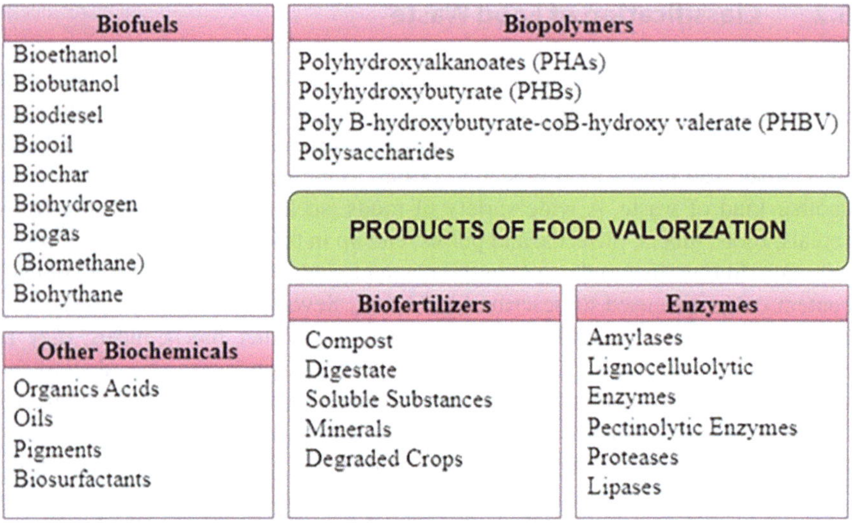

Fig. 5.2 Products of food waste valorization

5.3 Biofuels Obtained from the Valorization of Food Waste

5.3.1 Bioethanol

Bioethanol, often referred to as ethyl alcohol, is a crucial organic solvent utilized for color removal and the prevention of microbial contamination. Ethanol has the potential to be derived from several sources that possess sugar molecules, including corn, sugarcane, rice straw, sugar beet, sweet sorghum, and algae. Bioethanol production may be categorized into three generations: first generation, second generation, and third generation (Lee and Lavoie 2013).

Bioethanol production has evolved with energy-saving strategies like heat-integrated distillation, membrane-assisted distillation, and ohmic-assisted hydrodistillation (Gavahian et al. 2019). Furthermore, the exercise of advanced extraction methods, including the use of supercritical fluids as well as green solvents, has enhanced the efficiency of bioethanol generation, reduced environmental impact, and the recovery of valuable compounds from fermented broths. By combining extensive extraction and distillation methods with sustainable valorization techniques, bioethanol producers have been able to improve both the sustainability and efficiency of the production process.

Microbial strains are essential for converting different kinds of food waste into bioethanol. *Saccharomyces cerevisiae* (baker's yeast) is extensively used for the synthesis of bioethanol from different sources such as fruit pomaces (e.g., grape and apple pomace), sugarcane, maize, and starchy materials (Bibra et al. 2023). Lactic acid bacteria (LAB) undergo fermentation of waste from the dairy industry (such as whey) and other organic residues, resulting in the lactic acid production (Christensen et al. 2011). The lactic acid produced may then be converted to generate bioethanol (Kazemi Shariat Panahi et al. 2022). *Aspergillus* sp., including *Aspergillus niger*, aids in the enzymatic breakdown of lignocellulosic food waste, for the release of fermentable sugars that may be used for bioethanol production (Alabdalall et al. 2023). Furthermore, the bacterial and yeast components in microbial mixtures collaborate to break down constituents into sugars, thus providing a feasible method for producing bioethanol from a variety of food waste.

The type of conversion technology implemented, including enzymatic processes and fermentation, along with the source material used—such as sugarcane, rice, or any other elementary food waste—are the key parameters that influence the detritus generated and energy consumed during ethanol production. These factors also determine the bioproducts yielded, thereby affecting the overall environmental impact and efficiency of the process (de Farias Silva and Bertucco 2019). Consequently, it is critical to thoroughly analyze these parameters to optimize ethanol production approaches.

5.3.2 Biobutanol

Biobutanol, sometimes known as butyl alcohol, is widely regarded as an excellent alternative for petroleum-based fuel sources owing to its favorable characteristics, including reduced corrosiveness, non-hazardous nature, high biodegradability, and elevated energy content. Their isomers include *n*-butanol, *sec*-butanol, *tert*-butanol, and isobutanol.

Different types of food waste substrates, including lignocellulosic biomass (for example, rice straw), starchy materials (like maize and cassava), and feedstocks containing sugar (such as molasses and sugarcane juice), are used as raw materials for producing biobutanol using various microbial strains. Both *A. niger* and *Bacillus coagulans* play a role in the decomposition of food waste for the purpose of producing biobutanol; the choice of microbial strain relies on criteria such as the presence of substrate, cost, efficiency of fermentation, and desired characteristics of the final product. Furthermore, certain types of food waste, such as coffee grounds or fruit and vegetable waste, upon being acted upon by genetically engineered strains of *S. cerevisiae* and *Escherichia coli*, have led to the production of biobutanol (Santos et al. 2020).

Due to its favorable qualities like high energy content, low vapor pressure, and compatibility with current fuel infrastructure, biobutanol is often used as an alternative or an addition to conventional energy sources, especially in the field of transportation as a fuel for automobiles. It is not limited to just ground transportation—due to its low freezing point and high energy density, biobutanol also shows potential as an aviation fuel, providing the aviation sector with a renewable and sustainable alternative (López-Linares et al. 2023; Nimbalkar et al. 2018). Biobutanol functions as an environment-friendly solvent in diverse industrial procedures, and is used in chemical synthesis, paint manufacture, and cleaning operations.

5.3.3 Biodiesel

Biodiesel is a viable alternative to fossil diesel that is essentially obtained from a range of consumable oils and fats and is known for its compatibility with commercial diesel engines. Coming up with methods to generate biodiesel is the need of the hour since traditional raw materials may compete with water, land, and human feeding, necessitating fresh raw materials.

A variety of microbial strains plays a role in the generation of biodiesel from food waste, creating possibilities for environmentally friendly use of organic waste resources. *Rhodococcus opacus* effectively metabolizes food waste, such as potato peels and cheese whey, to accumulate triacylglycerols that may be converted into biodiesel (Chintagunta et al. 2021). *Aspergillus niger*, *Candida rugosa*, *Pseudomonas aeruginosa*, and *Yarrowia lipolytica* use lipases to break down triglycerides (Ali et al. 2017; Jang et al. 2012; Katre et al. 2018; Regner et al. 2019). The inclusion of microbial strains such as *Rhodosporidium toruloides*, *Cryptococcus curvatus*, *Rhodotorula glutinis*, and *Lipomyces starkeyi* showcases the diverse range of organisms involved in the synthesis of biodiesel from food waste (Close and Ojumu 2016; Wang et al. 2012). Microbial-driven techniques provide a great opportunity to use organic waste materials efficiently and sustainably in the manufacture of biodiesel. The composition of fatty acids in microbial oil biodiesel might vary according to the specific microbial strain and the kind of feedstock used. The fatty acid composition, catalyst used for transesterification, degree of purification, and transesterification efficiency in biodiesel are essential, as they can affect the fuel's properties—such as its viscosity, cetane number, and oxidative stability—which subsequently impact engine performance and emissions.

5.3.4 Bio-Oil

Bio-oil, a type of liquid fuel, consists of a diverse combination of organic molecules, such as oxygenated compounds and hydrocarbons, with characteristics that might differ within the different source materials and the methods used in production. It can be used as whole bio-oil, as fractionated bio-oil, or by extracting specific chemicals, and has promising potential as a sustainable source of energy for the production of electricity and heat, transportation fuels, and chemicals.

As one of the primary concerns is thermal stability, the enhancement of resilience to thermal deterioration is crucial to ensure that the bio-oil can withstand high temperatures during combustion or other uses without incurring substantial breakdown; thereby, enhancing its dependability and effectiveness as a feasible fuel source. Upgrading, which entails reducing the propensity of bio-oil to evaporate at typical temperatures, plays an important role in minimizing the instability of bio-oil, is necessary for storage as well as manipulation. Reduced volatility allows it to maintain a liquid state under normal operating circumstances, which improves its practicality and use.

Notable uses of bio-oil include the manufacture of improved biofuels using upgrading and refining methods, the synthesis of compounds like acetic acid, glycolaldehyde, and acetol, and the fabrication of surfactants and flotation agents. In addition, it is used as a wood preservative, agricultural pesticide, and can even be used as a replacement for petroleum products in asphalt emulsions.

5.3.5 Biohydrogen

Biohydrogen is hydrogen gas produced through biological processes by microorganisms like bacteria or algae and is recognized as a feasible source of energy that may be utilized for the generation of power using fuel cells and combustion processes. In addition to its function in energy production, it also acts as an eco-friendly transportation fuel, offering an alternative to non-renewable, prevailing fossil fuels and aiding in the reduction of GHG emissions. Biohydrogen is very adaptable in the chemical sector, where it plays a very important role in the production of essential chemicals like ammonia and methanol. When combined with renewable sources such as solar and wind energy, it results in uninterrupted and dependable electricity production. It is used in many industrial processes such as food processing, medicines, and wastewater treatment, serving as a versatile, renewable, and environment-friendly energy source.

During the early phase of dark fermentation (DF), microbes break down organic substances in the absence of oxygen and light, leading to the volatile fatty acids (VFAs) and alcohol production. These byproducts are useful substrates for the next stages of the conversion process. The second stage, photofermentation (PF), involves using the liquid portion acquired from DF to produce hydrogen (H_2) through the activity of photosynthetic bacteria such as *Rhodobacter sphaeroides*. In this stage, VFAs and alcohols undergo a conversion process into hydrogen gas (H_2) and carbon dioxide (CO_2), which increases the total energy output. In the last phase, anaerobic digestion (AD) is used to convert the solid residues, together with the effluent of PF, to produce methane by the activity of anaerobic microbes. This phase enhances the energy output even further by generating methane, a desirable kind of sustainable energy. By following this extensive three-step procedure, the efficiency of converting food waste is maximized, resulting in the extraction of valuable products such as VFAs, alcohols, hydrogen, and methane.

A primary obstacle in biohydrogen generation is the limited efficiency in generating hydrogen gas because this is often linked with other metabolic processes, such as the synthesis of organic acids or alcohols, which might vie for the accessible electrons and diminish the output of hydrogen gas. This phenomenon is often impeded by the availability of oxygen, resulting in the creation of water rather than hydrogen gas (Kumar and Mohan 2018).

5.3.6 Biogas

Biogas is generated by the biological decomposition of organic waste, which lacks oxygen. This process, known as anaerobic digestion or fermentation, involves biodegradable materials such as biomass, sewage, manure, green waste, municipal trash, energy crops, and plant components. The major components of biogas are methane (CH_4) and carbon dioxide (CO_2).

Biogas production is an intricate process that involves many microorganisms that aid in the decomposition of organic substances and the creation of methane via anaerobic digestion (Ghimire et al. 2015). Methanogenic archaea, including *Methanobacterium*, *Methanosarcina*, and *Methanococcus*, have a crucial function in the production of methane from a wide variety of organic sources, such as food waste, agricultural waste, and wastewater sludge (Dzulkarnain et al. 2022). LAB, such as *Lactobacillus* and *Enterococcus*, are often present and play a significant role in the process of fermentation during the early phases of anaerobic digestion (Detman et al. 2021). Moreover, *Clostridium* sp., including *Clostridium thermocellum* and *Clostridium cellulolyticum*, possess the capacity to break down cellulose and hemicellulose, which are often found in cellulosic food waste and agricultural leftovers (Li et al. 2014; Rambabu et al. 2020). Acidogenic Bacteria, such as *Acidithiobacillus* and *Acidogenic Clostridia*, have a significant impact during the acidogenesis phase by converting various organic molecules into VFAs and alcohols (Fuess et al. 2019; Sekoai et al. 2023). These bacteria work together to coordinate the complex steps of anaerobic digestion, emphasizing their importance in the environmentally friendly generation of biogas.

5.4 Biofertilizers Obtained from Food Waste Valorization

Besides biofuels, a range of technologies, encompassing both aerobic and anaerobic composting, chemical hydrolysis, direct burning, and in-situ degradation, have been developed and implemented on food waste to produce biofertilizers (Table 5.1). Biofertilizers are nutrient-rich preparations containing living microorganisms, which, when used for agricultural purposes (instead of chemical fertilizers), enhance soil fertility, crop productivity, and promote overall plant health.

Table 5.1 Biofertilizers obtained from various food waste

Food waste	Conversion technology	Biofertilizer
Food processing waste Wasted food material Agricultural waste	Aerobic composting	Compost
	Anaerobic composting	Digestate
	Chemical hydrolysis	Soluble substance
Food processing waste	Direct burning	Minerals
	In-situ degradation	Degraded crops

5.5 Enzymes Obtained from Food Waste Valorization

5.5.1 Amylases

Amylases are a class of enzymes that facilitate the starch breakdown by hydrolysis. Several enzymes catalyze the hydrolysis of starch or its derivative oligosaccharides. To optimize the production of amylase from food valorization, specific conditions and factors need to be considered. The choice of food waste as the primary carbon source in the fermentation medium was emphasized, and a blend of different food waste was recommended as an effective fermentation medium. Implementing a cell-substrate recycling system and optimizing amylolytic activity in aqueous extracts by adjusting the proportions of leftover food waste and additional agro-industrial waste were identified as essential steps. Additionally, supplementing the fermentation medium with a salt solution or corn steep liquor was suggested to enhance amylase production.

Amylase, an enzyme of great significance in the industrial sector, has a diversified range of uses in different sectors, highlighting its adaptability and importance (Benabda et al. 2019). In the food sector, it is used in starch liquefaction, baking, brewing, sweetener manufacture, and the transformation of starch into sugars to produce bread, beer, and other fermented goods. In the textile industry, it is used in desizing operations to facilitate the elimination of starch-based sizing agents from materials for effective dyeing and finishing procedures via enzymes. In the laundry business, it is a crucial component in detergents for stain removal by breaking down starch-based stains such as food residues and starchy soils. In the pharmaceutical industry, it has a role in generating glucose for intravenous feeding solutions and acts as an additive in tablet formulations. In addition, amylase is essential for bioethanol synthesis, since it plays a critical role in breaking down starch into sugars that may be fermented, which in turn facilitates the transformation of starchy substances into ethanol via enzymatic activities in fermentation procedures, thus providing sustainable energy options.

5.5.2 Lignocellulolytic Enzymes

Lignocellulosic biomass consists mostly of two major carbohydrate polymers, cellulose and hemicellulose, as well as lignin, a phenolic and non-carbohydrate polymer. Lignocellulolytic enzymes may break down hardwoods and serve a crucial ecological function as decomposers of lignocellulosic biomass. Moreover, these enzymes have diverse uses in many sectors such as textile, paper, food, detergent, cosmetic, environmental, pharmaceutical, biotechnological, and biofuel (Sánchez-Corzo et al. 2021).

Lignocellulolytic enzymes find utility in the textile industry, serving diverse purposes within this sector. Enzyme mixtures obtained from solid-state fermentation (SSF) can be employed in the pulp and paper industry for bio-pulping of lignocellulosic materials, contributing to the sustainable production of paper. The food industry can harness lignocellulolytic enzymes for various processes, including the production of products like fruit juices, wine, and beer. In the animal feed industry, these enzymes play a role in improving the digestibility of feed and reducing the presence of antinutritional factors, thereby improving overall nutritional quality. They play a vital role in the biofuel industry by catalyzing lignocellulosic biomass into fermentable sugars. This process is integral to bioethanol production, contributing to the advancement of sustainable and renewable fuel sources. The pharmaceutical sector benefits from lignocellulolytic enzymes in the production of various drugs and medicines, showcasing the versatility of these enzymes across different facets of pharmaceutical applications.

5.5.3 Pectinolytic Enzymes

Pectin, a complex polysaccharide, mostly consists of monomeric units of α-(1 $\rightarrow$ 4)-D-galacturonic acid, which are esterified with methoxy groups, forming primary constituent of plant cell walls, comprising about two-thirds of the dry weight and is extensively utilized in the food processing industry for its capacity to improve viscosity and retain water. Pectinases refer to a group of enzymes that break down pectic compounds, which are mostly found in microbes and higher plants and play a vital role in the cellular processes of cell wall metabolism, cell proliferation, senescence, fruit ripening, pathogenicity, and abscission.

The synthesis of pectinase from food waste has many obstacles, such as the fluctuating food waste composition, the existence of impurities, and the need for effective extraction techniques. Furthermore, the process of improving manufacturing conditions and extracting pectinase from intricate food waste matrices might pose difficulties. Nevertheless, the use of sophisticated methodologies like Response Surface Methodology (RSM) might facilitate the optimization of the manufacturing process (Mahesh et al. 2016). Moreover, the immobilization of pectinase onto an appropriate matrix is crucial for improving its stability and facilitating its retrieval and reusability, hence tackling some difficulties linked to the synthesis of free enzymes.

5.5.4 Proteases

Proteases or proteinases are proteolytic enzymes that can hydrolyze the peptide bonds in proteins. The manufacture of protease is a multifaceted process that presents several problems. A significant obstacle is the exorbitant manufacturing expenses, resulting from the costly raw materials, energy, and equipment necessary for the procedure. One further obstacle is the limited productivity of some sources, such as microbial cultures, which might restrict the quantity of protease that can be generated. The purification of proteases might pose challenges owing to the coexistence of other proteins and contaminants in the raw extract, which can impact the overall quality and activity of the end product. This may lead to a reduced output of functional protease and an increased manufacturing expense. Moreover, proteases may exhibit sensitivity to several environmental conditions, including pH, temperature, and substrate concentration, which might have an impact on their stability and activity (Tanasković et al. 2021). Attaining optimum protease activity and output necessitates meticulous regulation of the manufacturing conditions. These problems underscore the need for ongoing research and development in protease manufacturing to enhance output, decrease expenses, and optimize purifying and production settings.

Plant proteases may be used to extract high-value components from various byproducts, such as cheese whey, soy protein, keratinous materials, collagen, fish protein, gelatin, and bird feathers, via the process of hydrolysis. Furthermore, they might be used in the dehairing procedure within the leather business, providing a substitute for the utilization of detrimental chemicals. The above examples illustrate the wide variety of industrial by-products that may be efficiently processed and used with the assistance of plant proteases (David Troncoso et al. 2022).

5.5.5 Lipases

Lipases, also known as triacylglycerol acyl hydrolases, are water-soluble enzymes classified as ester hydrolases. They are characterized by their strong preference for nonpolar, water-insoluble ester substrates. These enzymes facilitate the breakdown of triglycerides by hydrolyzing carboxylic ester bonds. Found naturally in humans and other monogastric animals, lipases are predominantly produced in the pancreas and stomach, where they play a crucial role in fat and lipid digestion.

In the food sector, lipases are mostly used in the alteration of fats and oils, where they aid in the generation of reduced-calorie fats, structured lipids, and flavor compounds, whereas, in the pharmaceutical industry, they are necessary for chiral synthesis, which allows the creation of drugs with high enantiomeric purity; moreover, in the detergent industry they catalyze the breakdown of oils and fats, facilitating the removal of stains and dirt from garments. In the medical field, they are used in biological tests to measure the levels of triglycerides in the blood, whereas in the biodiesel sector, they facilitate the manufacturing process, which is in line with the increasing worldwide need for renewable energy sources (Salihu et al. 2012). In addition, lipases have a positive impact on the dairy industry by intensifying the taste of cheese and have a significant function in the bakery industry by strengthening the structural integrity and texture of dough. Lastly, in the forest products industry to control pitch, demonstrating their wide range of uses in many industries.

5.6 Biopolymers Obtained from Food Waste Valorization

5.6.1 Polyhydroxyalkanoate (PHA)

Polyesters of R-hydroxyalkanoic acids are the building blocks for a class of biodegradable and biocompatible plastics known as polyhydroxyalkanoates (PHAs). Their physical and mechanical characteristics vary significantly based on the carbon atom count in the monomer units that make up the polymer chain (Israni and Shivakumar 2020). PHAs are classified into two types: short-chain-length PHAs, which have monomer units of three to five carbon atoms, and medium-chain-length PHAs, which have monomer units of six to fourteen carbon atoms (Amstutz et al. 2019). Short-chain-length PHAs are typically rigid, crystalline, and brittle materials that possess high melting points. In contrast, medium-chain-length PHAs are generally flexible, elastomeric, and have lower melting points than short-chain-length PHAs.

PHAs are sustainable biopolymers that have attracted attention since they can be manufactured from food waste and agricultural leftovers utilizing different types of microbial strains. *Bacillus subtilis* has been recognized as a very promising organism for the manufacture of polyhydroxyalkanoates, indicating its significant promise in the field of biopolymer manufacturing (Vu et al. 2021). *Bacillus* sp. INT005, obtained from soil in a gas field, has the capacity to amass PHA, therefore contributing to the microbial variety involved in PHA synthesis (Basnett et al. 2017). The combination of these microorganisms, together with the use of various agricultural residues and food waste, highlights the adaptability and potential of PHA synthesis as a sustainable and environmentally beneficial option in the field of biopolymers.

5.6.2 Polyhydroxybutyrate (PHB)

It is a naturally biodegradable polymer, is produced by archaea and bacteria to store carbon and energy and is a typical component of PHA biopolymers. Its quick disintegration in natural situations makes it a possible alternative to biodegradable polymers.

Several studies have identified specific microbial strains and their corresponding food waste substrates as effective combinations for PHB synthesis. *B. megaterium* SRKP-3, for example, used dairy waste and saltwater as nutrition sources for PHB production (RamKumar Pandian et al. 2010). Studies have also delved into the mangrove rhizosphere, identifying *B. thuringiensis* and *Erythrobacter aquimaris* as microbial strains capable of PHB synthesis (Alarfaj et al. 2015; Mostafa et al. 2020). *Cupriavidus necator* DSM 7237 demonstrated its prowess by utilizing wine lees and crude glycerol from biodiesel production as feedstocks for PHB production (Dimou et al. 2015). Additionally, *B. cereus* MPTDC, *B. pumilus* MPTDFH, *E. faecium* MPTDF, *B. sonorensis* MPTD1, and *Enterobacter cloacae* SG1 utilized kitchen and food waste as carbon sources for production of PHB (Sindhu et al. 2020).

In a nutshell, PHB shows great promise in several industries, whether it be medicine, agriculture, nanotechnology, or packaging; in the field of medicine, they contribute to the creation of biodegradable medical implants, sutures, and drug delivery systems; and as biodegradable mulching films, controlled-release fertilizers, and biodegradable plant pots in agriculture. Its mechanical properties, like the capability to withstand high levels of tension and its rigidity, make it suitable for a variety of applications, and the strong oxygen and carbon dioxide barrier qualities of PHB make it appropriate for packaging applications that need utmost freshness and quality.

5.6.3 Poly(3-Hydroxybutyrate-Co-3-Hydroxyvalerate) (PHBV)

It is produced by microbial fermentation and is a form of polyester that is both biodegradable and biocompatible. It is known for its potential applications in different fields, including tissue engineering, biomedical applications, packaging, and other industrial sectors.

PHBV may be produced by a wide range of microbes, such as bacteria, archaea, and fungi. Notable bacterial strains used for PHBV production include *Cupriavidus necator*, *Ralstonia eutropha*, and *Pseudomonas putida*; Archaea strains, including *Haloferax mediterranei*, have also been used for PHBV synthesis (Han et al. 2015; Kachrimanidou et al. 2014). Additionally, *Aspergillus oryzae* has been utilized in SSF of sunflower meal (SFM) to produce nutrient-rich hydrolysates for PHB and P(3HB-co-3 HV) generation with glycerol and whole sunflower meal (SFM) hydrolysates serving as sole renewable sources using levulinic acid as a precursor. The SFM hydrolysates can be generated through a two-stage biological process, including SSF for enzyme production and submerged hydrolysis to produce nutrient-rich hydrolysates. Molasses and cheese whey have been used as a carbon source in the manufacture of PHBV (Melendez-Rodriguez et al. 2018).

In summary, several factors must be thoroughly evaluated when producing PHBV, such as the choice of substrate, downstream processing, environmental sustainability, microbial strain selection, process optimization, and product quality (Hassaini et al. 2017). By optimizing these factors, producers can achieve efficient and sustainable PHBV production for a variety of applications.

5.7 High-Value Products from Food Waste Valorization

5.7.1 Organic Acids

These represent a class of organic compounds that exhibit acidic characteristics and function as intermediates in the catabolic processes of amino acids, lipids, and carbohydrates. They display moderate acidity and are categorized according to the number of carboxylic groups they contain; however, carboxylic acids exhibit higher acidity in comparison to organic acids containing alcohol, enol, phenol, and thiol functional groups. Organic acids vary in structure depending on the number of hydroxyl or carboxyl functional groups and carbon-carbon double bonds. Acetic acid, lactic acid, butyric acid, citric acid, 3-hydroxypropionic acid, propionic acid, and succinic acid are the prominent organic acids that can be obtained from the valorization of food waste.

Lactic acid is often used in the culinary, cosmeceutical, and pharmaceutical sectors. Not only does it serve as a food preservative, flavor enhancer, pH adjuster, and hydrating ingredient in cosmetic goods, but also utilized for its antibacterial properties. Citric acid is a mild organic acid generated by the process of carbohydrate fermentation by the fungus *A. niger* (which is also utilized in the generation of various biofuels and bioproducts of significance). Besides serving as a flavor enhancer, preservative, and pH adjuster in food and beverage items, it functions as an antioxidant and chelating agent in medicinal preparations. 3-hydroxypropionic acid is an organic compound that can be generated from food waste via microbial fermentation and used as a fundamental component for creating biodegradable polymers and other chemical substances.

Short-chain fatty acids (SCFAs) are a collection of organic acids generated by the gut bacteria's fermentation of carbohydrates. These include acetic acid, propionic acid, and butyric acid. They provide a range of health advantages, such as enhancing gastrointestinal health, reducing inflammation, and regulating metabolic processes. Succinic acid is a SCFA precursor used in the synthesis of diverse compounds such as biodegradable polymers, solvents, and medicines. Through carbohydrate fermentation, *Actinobacillus succinogenes* is capable of producing succinic acid. Acetic acid, a type of organic acid, is often utilized in the food, beverage, and chemical sectors. It is generated by the process of carbohydrate fermentation by bacteria, such as *Acetobacter*. Acetic acid serves as a preservative, pH controller, and flavor enhancer in food and beverage items. Additionally, it finds use in the manufacturing of diverse compounds, such as solvents and polymers. Propionic acid, a significant compound employed as a mold inhibitor and pH regulator in the bakery industry, as a preservative in the food and animal feed sector, and in various medications in the pharmaceutical industry, is generally generated by bacterial (e.g., *Propionibacterium*) carbohydrate fermentation. Butyric acid has the potential to be used as a therapeutic intervention for a range of gastrointestinal illnesses, such as inflammatory bowel disease, and as a flavor enhancer in culinary items.

5.7.2 Pigments

Pigments are substances that alter light color by selectively absorbing specific wavelengths. They are used in various sectors like food, pharmaceuticals, cosmetics, and textiles due to their significant sensory attribute, which influences product aesthetic appeal and market reception (Kaur et al. 2020). Concern over the detrimental and deleterious impacts of synthetic pigments on animals, humans, and the environment has increased interest in natural pigments produced using microorganisms, otherwise known as bio-pigments. Various microbial strains have demonstrated their capacity for pigment synthesis, using a wide range of food waste sources as substrates (Charalampia et al. 2017). LAB strains have shown the capacity to produce red and pink pigments by using substrates such as whey and apple pomace (Mileriene et al. 2023).

A primary obstacle in pigment production is the insufficient abundance of pigments in food waste, resulting in challenging and ineffective extraction (Rodriguez-Lopez et al. 2020). Furthermore, conducting studies on innovative extraction methods and creating enhanced goods from food waste might help address the difficulties encountered in the manufacturing of pigments from natural sources.

5.7.3 Biochar

Biochar is a type of charcoal typically produced through pyrolysis, a process involving the decomposition of organic materials such as agricultural byproducts, wood, or manure, in a low-oxygen environment. With its versatile properties, Biochar proves beneficial across different applications. Mostly, its high porosity and surface area enable effective adsorption of substances like organic compounds and contaminants (Kumar et al. 2021). Additionally, it is a stable form of carbon, supporting carbon sequestration for climate change mitigation. In agriculture, it acts as a soil amendment, enhancing structure, water retention, and nutrient availability; and with its waste valorization potential, biochar contributes to the sustainable management of organic waste materials. Moreover, engineered activated biochar offers tailored characteristics for applications in energy storage, pollution control, and catalysis (Gabhane et al. 2020). Biochar emerges as a multifaceted solution for environmental sustainability, waste management, and agricultural improvement.

Producing biochar from food waste has several challenges and restrictions. An important obstacle is the elevated moisture level in food waste, which poses difficulties in processing and necessitates more energy for drying (Rago et al. 2018). Furthermore, the content of food waste might vary, which can impact the quality and uniformity of the resultant biochar. Another obstacle is the possibility of pollutants in the food waste, such as toxic metals or disease-causing microorganisms, which might impact the safety and quality of the biochar. Hence, it is crucial to meticulously choose and treat the food waste to guarantee its suitability for biochar production. Currently, the technology for creating biochar from food waste is in its infancy, and to improve the procedure, the efficiency, and the caliber of the biochar generated, more research is needed (Kane and Ryan 2022).

5.7.4 Biosurfactants

Biosurfactants are bioactive substances that are capable of decreasing the tension on surfaces and interfaces, possessing notable characteristics such as molecular self-assembly, reduced surface tension, and the ability for emulsion formation, which are either synthesized on the surface of microbial cells or secreted into their external environment. Microbial surfactants have several benefits in comparison to synthetic ones, these include low toxicity and exceptional biodegradability, and the maintenance of effectiveness under severe conditions of pH and salinity. Their notable characteristics include. They have distinct chemical compositions and molecular weights, serving as agents for cellular interaction, antagonism, and motility (Sharma et al. 2021).

Several promising techniques have been identified for biosurfactant production, each offering unique advantages in terms of cost-effectiveness and yield enhancement (Singh et al. 2019). These include the use of growth enhancers like lactones, incorporation of reliable nanoparticles composed of iron and manganese, coproduction with other industrial products, bioprospecting for strains, co-culture immobilization of organisms, and utilization of activated charcoal. Furthermore, the SSF process has emerged as a promising strategy, particularly in overcoming the foam production challenge encountered in submerged fermentations (Costa et al. 2017). Numerous microbial strains belonging to the *Bacillus* sp., *Pseudomonas* sp., *Candida* sp. and even Halobacteria have been employed for biosurfactant production using diverse substrates obtained from agro-industrial and food waste (Singh et al. 2019).

Biosurfactants play a key role in addressing the contamination of soil and water via environmental remediation and through the enhancement of the solubility and bioavailability of hydrophobic contaminants. These compounds also function as emulsifiers, stabilizers, and foaming agents in the food industry, as well as in shampoos, soaps, and other cosmetics in personal care products. They are used in agriculture as adjuvants to improve the efficacy of insecticides and herbicides in their formulations and play a crucial role in enhancing oil recovery (EOR) processes, hence enabling the oil industry to collect more oil from reservoirs (Lakatos et al. 2022). Furthermore, they may function as agents for delivering drugs and contribute to the process of wound healing in the medical field.

5.8 Challenges in Food Waste Valorization

Considering that there are several varieties of food waste in terms of their origins, compositions, and the physical and chemical transformations that must take place in order to create a desirable product, several challenges must be addressed in order to generate bioproducts from food waste. The variability in the composition of food waste from different sources affects the efficiency of the production process. Pretreatment of food waste introduces additional complexities and can produce inhibitors that affect fermentation. This needs to be kept in mind, especially when selecting fermentative organisms that may be sensitive to them (Abd-Alla et al. 2017). The separation of products of interest from the fermentation broth is challenging and expensive, and requires the availability of consistent and abundant food waste for year-round production. Converting agricultural waste into valuable products faces its own set of challenges (Awogbemi and Kallon 2022).

Limited knowledge among individuals leads to low-quality segregation of food waste, thereby producing compost of subpar quality. Market acceptance of products derived from agricultural waste may be hindered by consumer perception, competition with conventional products, and market demand. The absence of market demand for compost manufacturing is a significant hindrance since it might encounter rivalry from other fertilizer products (Karmee 2016). Environmental disputes, such as those over GHG emissions, unpleasant smells, and sanitation problems in composting zones, add complexity to the successful execution of procedures. Anaerobic digestion challenges revolving around material flow, location and pricing of facilities, and laws regarding the use of digestate may create uncertainty and impede its wider implementation. Bioenergy politics, policy-related obstacles, and regulatory compliance regarding waste management, product quality, and environmental considerations add another layer of complexity (Thi et al. 2016).

5.9 Conclusion

Understanding of the extensive classification and treatment of food waste generated on the basis of their forms, compositions, and sources, as well as the microbes that could be selectively employed to accelerate the process of food valorization, is necessary to discover sustainable methods of management, usage, and repurposing for energy generation. Collaborative efforts are essential for overcoming these obstacles, fostering innovation and sustainable practices in the valorization of agricultural waste. To tackle the difficulties associated with food waste valorization, a comprehensive strategy is needed that includes raising awareness, developing markets, establishing regulatory frameworks, and implementing sustainable policies. This will promote the efficient and eco-friendly use of food waste resources.

References

Abd-Alla MH, Zohri ANA, El-Enany AWE, Ali SM (2017) Conversion of food processing wastes to biofuel using clostridia. Anaerobe 48:135–143. https://doi.org/10.1016/j.anaerobe.2017.08.011

Alabdalall AH, Almutari AA, Aldakeel SA, Albarrag AM, Aldakheel LA, Alsoufi MH, Alfuraih LY, Elkomy HM (2023) Bioethanol production from lignocellulosic biomass using aspergillus Niger and aspergillus flavus hydrolysis enzymes through immobilized S. Cerevisiae. Energies 16(2):823. https://doi.org/10.3390/en16020823

Alarfaj AA, Arshad M, Sholkamy EN, Munusamy MA (2015) Extraction and characterization of polyhydroxybutyrates (PHB) from bacillus thuringiensis KSADL127 isolated from mangrove environments of Saudi Arabia. Braz Arch Biol Technol 58(5):781–788. https://doi.org/10.1590/S1516-891320150500003

Ali CH, Qureshi AS, Mbadinga SM, Liu JF, Yang SZ, Mu BZ (2017) Biodiesel production from waste cooking oil using onsite produced purified lipase from Pseudomonas aeruginosa FW_SH-1: central composite design approach. Renew Energy 109:93–100. https://doi.org/10.1016/j.renene.2017.03.018

Amstutz V, Hanik N, Pott J, Utsunomia C, Zinn M (2019) Tailored biosynthesis of polyhydroxyalkanoates in chemostat cultures. Methods Enzymol 627:99–123. https://doi.org/10.1016/bs.mie.2019.08.018

Awogbemi O, Kallon DV (2022) Valorization of agricultural wastes for biofuel applications. Heliyon 8(10):e11117. https://doi.org/10.1016/j.heliyon.2022.e11117

Baiano A (2014) Recovery of biomolecules from food wastes - a review. Molecules 19(9):14821–14842. https://doi.org/10.3390/molecules190914821

Basnett P, Ravi S, Roy I (2017) Natural bacterial biodegradable medical polymers: polyhydroxyalkanoates. In: Science and principles of biodegradable and bioresorbable medical polymers: materials and properties. Elsevier, Amsterdam, pp 257–277. https://doi.org/10.1016/B978-0-08-100372-5.00008-8

Benabda O, M'Hir S, Kasmi M, Mnif W, Hamdi M (2019) Optimization of protease and amylase production by Rhizopus oryzae cultivated on bread waste using solid-state fermentation. J Chem 3:1–9. https://doi.org/10.1155/2019/3738181

Bibra M, Samanta D, Sharma NK, Singh G, Johnson GR, Sani RK (2023) Food waste to bioethanol: opportunities and challenges. Fermentation 9(1):8. https://doi.org/10.3390/fermentation9010008

Charalampia D, Antonios KE, Constantina N, Haralabos KC (2017) Current trends and emerging technologies in biopigment production processes: industrial food and health applications. Int J Hortic 1(2):33. www.aipublications.com/ijhaf

Chintagunta AD, Zuccaro G, Kumar M, Kumar SPJ, Garlapati VK, Postemsky PD, Kumar NSS, Chandel AK, Simal-Gandara J (2021) Biodiesel production from lignocellulosic biomass using oleaginous microbes: prospects for integrated biofuel production. Front Microbiol 12:658284. https://doi.org/10.3389/fmicb.2021.658284

Christensen AD, Kádár Z, Oleskowicz-Popiel P, Thomsen MH (2011) Production of bioethanol from organic whey using Kluyveromyces marxianus. J Ind Microbiol Biotechnol 38(2):283–289. https://doi.org/10.1007/s10295-010-0771-0

Close D, Ojumu J (2016) Draft genome sequence of the oleaginous yeast Cryptococcus curvatus ATCC 20509. Genome Announc 4(6):16. https://doi.org/10.1128/genomeA.01235-16

Costa JAV, Treichel H, Santos LO, Martins VG (2017) Solid-state fermentation for the production of biosurfactants and their applications. In: Current developments in biotechnology and bioengineering: current advances in solid-state fermentation. Elsevier, Amsterdam, pp 357–372. https://doi.org/10.1016/B978-0-444-63990-5.00016-5

Dahiya S, Kumar AN, Shanthi Sravan J, Chatterjee S, Sarkar O, Mohan SV (2018) Food waste biorefinery: sustainable strategy for circular bioeconomy. Bioresour Technol 248:2–12. https://doi.org/10.1016/j.biortech.2017.07.176

David Troncoso F, Alberto Sánchez D, Luján Ferreira M (2022) Production of plant proteases and new biotechnological applications: an updated review. ChemistryOpen 11(3):17. https://doi.org/10.1002/open.202200017

de Farias Silva CE, Bertucco A (2019) Bioethanol from microalgal biomass: a promising approach in biorefinery. Braz Arch Biol Technol 62:1–14. https://doi.org/10.1590/1678-4324-2019160816

Detman A, Laubitz D, Chojnacka A, Kiela PR, Salamon A, Barberán A, Chen Y, Yang F, Błaszczyk MK, Sikora A (2021) Dynamics of dark fermentation microbial communities in the light of lactate and butyrate production. Microbiome 9(1):158. https://doi.org/10.1186/s40168-021-01105-x

Dhiman S, Mukherjee G (2021) Present scenario and future scope of food waste to biofuel production. J Food Process Eng 44(2):13594. https://doi.org/10.1111/jfpe.13594

Dimou C, Kopsahelis N, Papadaki A, Papanikolaou S, Kookos IK, Mandala I, Koutinas AA (2015) Wine lees valorization: biorefinery development including production of a generic fermentation feedstock employed for poly(3-hydroxybutyrate) synthesis. Food Res Int 73:81–87. https://doi.org/10.1016/j.foodres.2015.02.020

Dzulkarnain ELN, Audu JO, Wan Dagang WRZ, Abdul-Wahab MF (2022) Microbiomes of biohydrogen production from dark fermentation of industrial wastes: current trends, advanced tools and future outlook. Bioresour Bioprocess 9(1):16. https://doi.org/10.1186/s40643-022-00504-8

Esteban J, Ladero M (2018) Food waste as a source of value-added chemicals and materials: a biorefinery perspective. Int J Food Sci Technol 53(5):1095–1108. https://doi.org/10.1111/ijfs.13726

Esteban-Lustres R, Torres MD, Piñeiro B, Enjamio C, Domínguez H (2022) Intensification and biorefinery approaches for the valorization of kitchen wastes—a review. Bioresour Technol 360:127652. https://doi.org/10.1016/j.biortech.2022.127652

Fuess LT, Zaiat M, do Nascimento, C. A. O. (2019) Novel insights on the versatility of biohydrogen production from sugarcane vinasse via thermophilic dark fermentation: impacts of pH-driven operating strategies on acidogenesis metabolite profiles. Bioresour Technol 286:121379. https://doi.org/10.1016/j.biortech.2019.121379

Gabhane JW, Bhange VP, Patil PD, Bankar ST, Kumar S (2020) Recent trends in biochar production methods and its application as a soil health conditioner: a review. SN Appl Sci 2(7):1307. https://doi.org/10.1007/s42452-020-3121-5

Gavahian M, Munekata PES, Eş I, Lorenzo JM, Mousavi Khaneghah A, Barba FJ (2019) Emerging techniques in bioethanol production: from distillation to waste valorization. Green Chem 21(6):1171–1185. https://doi.org/10.1039/c8gc02698j

Ghimire A, Valentino S, Frunzo L, Trably E, Escudié R, Pirozzi F, Lens PNL, Esposito G (2015) Biohydrogen production from food waste by coupling semi-continuous dark-photofermentation and residue post-treatment to anaerobic digestion: a synergy for energy recovery. Int J Hydrog Energy 40(46):16045–16055. https://doi.org/10.1016/j.ijhydene.2015.09.117

Gustavsson (n.d.) Global food losses and food waste: extent, causes and prevention: study conducted for the International Congress 'Save Food!' at Interpack 2011 Düsseldorf, Germany

Han J, Wu LP, Hou J, Zhao D, Xiang H (2015) Biosynthesis, characterization, and hemostasis potential of tailor-made poly(3-hydroxybutyrate-co-3-hydroxyvalerate) produced by Haloferax mediterranei. Biomacromolecules 16(2):578–588. https://doi.org/10.1021/bm5016267

Hassaini L, Kaci M, Touati N, Pillin I, Kervoelen A, Bruzaud S (2017) Valorization of olive husk flour as a filler for biocomposites based on poly(3-hydroxybutyrate-co-3-hydroxyvalerate): effects of silane treatment. Polym Test 59:430–440. https://doi.org/10.1016/j.polymertesting.2017.03.004

Israni N, Shivakumar S (2020) Polyhydroxyalkanoate (PHA) biosynthesis from directly valorized ragi husk and sesame oil cake by bacillus megaterium strain Ti3: statistical optimization and characterization. Int J Biol Macromol 148:20–30. https://doi.org/10.1016/j.ijbiomac.2020.01.082

Jang MG, Kim DK, Park SC, Lee JS, Kim SW (2012) Biodiesel production from crude canola oil by two-step enzymatic processes. Renew Energy 42:99–104. https://doi.org/10.1016/j.renene.2011.09.009

Kachrimanidou V, Kopsahelis N, Papanikolaou S, Kookos IK, De Bruyn M, Clark JH, Koutinas AA (2014) Sunflower-based biorefinery: poly(3-hydroxybutyrate) and poly(3-hydroxybutyrate-co-3-hydroxyvalerate) production from crude glycerol, sunflower meal and levulinic acid. Bioresour Technol 172:121–130. https://doi.org/10.1016/j.biortech.2014.08.044

Kane S, Ryan C (2022) Biochar from food waste as a sustainable replacement for carbon black in upcycled or compostable composites. Compos Part C Open Access 8:100274. https://doi.org/10.1016/j.jcomc.2022.100274

Karmee SK (2016) Liquid biofuels from food waste: current trends, prospect and limitation. Renew Sust Energ Rev 53:945–953. https://doi.org/10.1016/j.rser.2015.09.041

Katre G, Raskar S, Zinjarde S, Ravi Kumar V, Kulkarni BD, RaviKumar A (2018) Optimization of the in situ transesterification step for biodiesel production using biomass of Yarrowia lipolytica NCIM 3589 grown on waste cooking oil. Energy 142:944–952. https://doi.org/10.1016/j.energy.2017.10.082

Kaur R, Panwar D, Panesar PS (2020) Biotechnological approach for valorization of whey for value-added products. In: Food industry wastes: assessment and recuperation of commodities. Elsevier, Amsterdam, pp 275–302. https://doi.org/10.1016/B978-0-12-817121-9.00013-9

Kazemi Shariat Panahi H, Dehhaghi M, Guillemin GJ, Gupta VK, Lam SS, Aghbashlo M, Tabatabaei M (2022) Bioethanol production from food wastes rich in carbohydrates. Curr Opin Food Sci 43:71–81. https://doi.org/10.1016/j.cofs.2021.11.001

Kumar AN, Mohan SV (2018) Acidogenic valorization of vegetable waste for short chain carboxylic acids and biohydrogen production: influence of pretreatment and pH. J Clean Prod 203:1055–1066. https://doi.org/10.1016/j.jclepro.2018.08.198

Kumar M, Dutta S, You S, Luo G, Zhang S, Show PL, Sawarkar AD, Singh L, Tsang DCW (2021) A critical review on biochar for enhancing biogas production from anaerobic digestion of food waste and sludge. J Clean Prod 305:127143. https://doi.org/10.1016/j.jclepro.2021.127143

Lakatos ES, Cioca LI, Szilagyi A, Vladu MG, Stoica RM, Moscovici M (2022) A systematic review on biosurfactants contribution to the transition to a circular economy. PRO 10(12):2647. https://doi.org/10.3390/pr10122647

Lee RA, Lavoie JM (2013) From first- to third-generation biofuels: challenges of producing a commodity from a biomass of increasing complexity. Anim Front 3(2):6–11. https://doi.org/10.2527/af.2013-0010

Li Y, Xu T, Tschaplinski TJ, Engle NL, Yang Y, Graham DE, He Z, Zhou J (2014) Improvement of cellulose catabolism in clostridium cellulolyticum by sporulation abolishment and carbon alleviation. Biotechnol Biofuels 7(1):25. https://doi.org/10.1186/1754-6834-7-25

López-Linares JC, Coca M, Plaza PE, Lucas S, García-Cubero MT (2023) Waste-to-fuel technologies for the bioconversion of carrot discards into biobutanol. Renew Energy 202:362–369. https://doi.org/10.1016/j.renene.2022.11.093

Mahesh M, Arivizhivendhan KV, Maharaja P, Boopathy R, Hamsavathani V, Sekaran G (2016) Production, purification and immobilization of pectinase from aspergillus ibericus onto functionalized nanoporous activated carbon (FNAC) and its application on treatment of pectin containing wastewater. J Mol Catal B Enzym 133:43–54. https://doi.org/10.1016/j.molcatb.2016.07.012

Melendez-Rodriguez B, Castro-Mayorga JL, Reis MAM, Sammon C, Cabedo L, Torres-Giner S, Lagaron JM (2018) Preparation and characterization of electrospun food biopackaging films of poly(3-hydroxybutyrate-co-3-hydroxyvalerate) derived from fruit pulp biowaste. Front Sustainable Food Syst 2:38. https://doi.org/10.3389/fsufs.2018.00038

Mileriene J, Serniene L, Kasparaviciene B, Lauciene L, Kasetiene N, Zakariene G, Kersiene M, Leskauskaite D, Viskelis J, Kourkoutas Y, Malakauskas M (2023) Exploring the potential of sustainable acid whey cheese supplemented with apple pomace and GABA-producing indigenous Lactococcus lactis strain. Microorganisms 11(2):436. https://doi.org/10.3390/microorganisms11020436

Mostafa YS, Alrumman SA, Otaif KA, Alamri SA, Mostafa MS, Sahlabji T (2020) Production and characterization of bioplastic by polyhydroxybutyrate accumulating Erythrobacter aquimaris isolated from mangrove rhizosphere. Molecules 25(1):179. https://doi.org/10.3390/molecules25010179

Nimbalkar PR, Khedkar MA, Chavan PV, Bankar SB (2018) Biobutanol production using pea pod waste as substrate: impact of drying on saccharification and fermentation. Renew Energy 117:520–529. https://doi.org/10.1016/j.renene.2017.10.079

Rago YP, Surroop D, Mohee R (2018) Assessing the potential of biofuel (biochar) production from food wastes through thermal treatment. Bioresour Technol 248:258–264. https://doi.org/10.1016/j.biortech.2017.06.108

Rambabu K, Show PL, Bharath G, Banat F, Naushad M, Chang JS (2020) Enhanced biohydrogen production from date seeds by clostridium thermocellum ATCC 27405. Int J Hydrog Energy 45(42):22271–22280. https://doi.org/10.1016/j.ijhydene.2019.06.133

RamKumar Pandian S, Deepak V, Kalishwaralal K, Rameshkumar N, Jeyaraj M, Gurunathan S (2010) Optimization and fed-batch production of PHB utilizing dairy waste and sea water as nutrient sources by bacillus megaterium SRKP-3. Bioresour Technol 101(2):705–711. https://doi.org/10.1016/j.biortech.2009.08.040

Regner EL, Salvatierra HN, Baigorí MD, Pera LM (2019) Biomass-bound biocatalysts for biodiesel production: tuning a lipolytic activity from aspergillus Niger MYA 135 by submerged fermentation using agro-industrial raw materials and waste products. Biomass Bioenergy 120:59–67. https://doi.org/10.1016/j.biombioe.2018.11.005

Rodriguez-Lopez AD, Melgar B, Conidi C, Barros L, Ferreira ICFR, Cassano A, Garcia-Castello EM (2020) Food industry by-products valorization and new ingredients: cases of study. Cases of study. In: Sustainability of the food system: sovereignty, waste, and nutrients bioavailability. Elsevier, Amsterdam, pp 71–99. https://doi.org/10.1016/B978-0-12-818293-2.00005-7

Salihu A, Alam MZ, AbdulKarim MI, Salleh HM (2012) Lipase production: an insight in the utilization of renewable agricultural residues. Resour Conserv Recycl 58:36–44. https://doi.org/10.1016/j.resconrec.2011.10.007

Sánchez-Corzo LD, Álvarez-Gutiérrez PE, Meza-Gordillo R, Villalobos-Maldonado JJ, Enciso-Pinto S, Enciso-Sáenz S (2021) Lignocellulolytic enzyme production from wood rot fungi collected in Chiapas, Mexico, and their growth on lignocellulosic material. J Fungi 7(6):450. https://doi.org/10.3390/jof7060450

Santos BAS, Azambuja SPH, Ávila PF, Pacheco MTB, Goldbeck R (2020) N-butanol production by Saccharomyces cerevisiae from protein-rich agro-industrial by-products. Braz J Microbiol 51(4):1655–1664. https://doi.org/10.1007/s42770-020-00370-6

Sekoai PT, Chunilall V, Ezeokoli O (2023) Creating value from acidogenic biohydrogen fermentation effluents: an innovative approach for a circular bioeconomy that is acquired via a microbial biorefinery-based framework. Fermentation 9(7):602. https://doi.org/10.3390/fermentation9070602

Sharma P, Gaur VK, Kim SH, Pandey A (2020) Microbial strategies for bio-transforming food waste into resources. Bioresour Technol 299:122580. https://doi.org/10.1016/j.biortech.2019.122580

Sharma J, Sundar D, Srivastava P (2021) Biosurfactants: potential agents for controlling cellular communication, motility, and antagonism. Front Mol Biosci 8:727070. https://doi.org/10.3389/fmolb.2021.727070

Sindhu R, Manju A, Mohan P, Rajesh RO, Madhavan A, Arun KB, Hazeena SH, Mohandas A, Rajamani SP, Puthiyamadam A, Binod P, Reshmy R (2020) Valorization of food and kitchen waste: an integrated strategy adopted for the production of poly-3-hydroxybutyrate, bioethanol, pectinase and 2,3-butanediol. Bioresour Technol 310:123515. https://doi.org/10.1016/j.biortech.2020.123515

Singh P, Patil Y, Rale V (2019) Biosurfactant production: emerging trends and promising strategies. J Appl Microbiol 126(1):2–13. https://doi.org/10.1111/jam.14057

Tanasković SJ, Šekuljica N, Jovanović J, Gazikalović I, Grbavčić S, Đorđević N, Sekulić MV, Hao J, Luković N, Knežević-Jugović Z (2021) Upgrading of valuable food component contents and anti-nutritional factors depletion by solid-state fermentation: a way to valorize wheat bran for nutrition. J Cereal Sci 99:103159. https://doi.org/10.1016/j.jcs.2020.103159

Thi NBD, Lin CY, Kumar G (2016) Waste-to-wealth for valorization of food waste to hydrogen and methane towards creating a sustainable ideal source of bioenergy. J Clean Prod 122:29–41. https://doi.org/10.1016/j.jclepro.2016.02.034

Vu DH, Wainaina S, Taherzadeh MJ, Åkesson D, Ferreira JA (2021) Production of polyhydroxyalkanoates (PHAs) by bacillus megaterium using food waste acidogenic fermentation-derived volatile fatty acids. Bioengineered 12(1):2480–2498. https://doi.org/10.1080/21655979.2021.1935524

Wang Q, Guo FJ, Rong YJ, Chi ZM (2012) Lipid production from hydrolysate of cassava starch by Rhodosporidium toruloides 21167 for biodiesel making. Renew Energy 46:164–168. https://doi.org/10.1016/j.renene.2012.03.002

Conversion of Paper and Packing Materials to Biofuels and Chemicals

6

Ashokkumar Veeramuthu and G. Flora

Abstract

The increasing production and usage of paper and packaging materials pose significant environmental challenges. Despite the digital advancements, paper consumption remains high, driven by various applications in daily life and industries. The traditional methods of paper disposal, predominantly landfilling and incineration, lead to environmental contamination and greenhouse gas emissions. Recycling, though beneficial, often results in downcycled products of lower quality. To address these issues, innovative conversion technologies are being explored to transform waste paper into valuable biofuels and chemicals. These technologies include thermochemical processes such as gasification and pyrolysis, as well as biochemical processes like anaerobic digestion and fermentation. The conversion of paper waste into bioethanol, biogas, cellulose-based products, and other biochemicals offers a sustainable alternative to conventional disposal methods. However, challenges such as contamination, the need for efficient pretreatment, and the economic feasibility of large-scale operations persist. Advancements in these technologies, combined with circular economy principles, could significantly reduce the environmental impact of paper waste, promoting a more sustainable and resource-efficient future.

A. Veeramuthu (✉)
Center for Waste Management and Renewable Energy, Saveetha Dental College and Hospitals, Saveetha Institute of Medical and Technical Sciences, Saveetha University, Chennai, Tamil Nadu, India

G. Flora
Department of Botany, St. Mary's College (Autonomous), Thoothukudi, Tamil Nadu, India

A. Kumari et al. (eds.), *Valorization of Solid Wastes to Biofuels and Chemical Products for Sustainable World*, https://doi.org/10.1007/978-981-96-8594-3_6

Keywords

Paper and packaging waste · Enzymatic hydrolysis · Fermentation · Anaerobic digestion · Circular economy

6.1 Introduction

Even if we are living in a more digitally advanced world, paper is still a common substance utilized in both our personal and professional daily lives. Approximately 414 million tons of paper are still produced and used annually in the UK (Business Waste 2024). Toilet paper, paperboard, envelopes, and writing and printing paper are all included in this. Paper is also used in newspapers, packaging, personal hygiene items, and cleaning supplies. It is made from the pulp of wood, grasses, or other fibrous materials. Every year, the world needs about 4 billion fewer trees to produce paper. Paper is 93% derived from trees. One ton of paper requires 24 trees. A4 paper requires 5 l of water to produce one sheet. Meanwhile, the amount of paper waste produced worldwide makes up nearly one-third of all solid garbage. In many organizations and companies, paper waste is a serious issue. Up to 70% of a company's waste can be attributed to paper due to printing errors, junk mail, bills, and packaging. An office worker uses roughly 10,000 sheets of paper annually on average. Companies use a variety of paper products, such as cardboard, envelopes, and wrappers, in addition to the paper used for printing (Baleforce 2024). Also, packaging materials are chosen by the food processing industry depending on factors such as cost-effectiveness, sustainability, strength, barrier qualities, and heat sealability. Paper, plastic, glass, aluminum, wood, and mixtures of these materials are common materials; each has advantages and disadvantages of its own. With 31% of the world's packaging market made up of paper and paperboard, these materials are widely used in primary and secondary packaging due to their favorable environmental effects. However, for better usefulness, lamination or additions are needed to improve plain paper's low heat sealability and weak barrier qualities. Milk cartons, microwave popcorn bags, ice cream cups, and fast-food containers are a few instances of this type of packaging. The manufacturing of these paper packing materials rises annually, contributing to environmental waste. Because of trims and off-cuts, the production of paper goods also generates some waste. The World Bank estimates global garbage production in cities to be at least two billion tons annually, with paper products accounting for a large portion of this waste. And if the current trends continue, this number is expected to increase by 70% by the year 2050 (Waste Control 2024; World Bank Group 2018). Paper waste is a serious problem that has an influence on the economy and ecology. It has an impact on waste management systems, forests, and air quality. Finding answers that satisfy both financial demands and environmental objectives is made easier when we are aware of its entire impact. Another important thing we must concentrate on is deforestation, which is a significant worldwide problem that is fueled by sectors like pulp and paper. It accelerates climate change by decreasing the Earth's capacity to absorb carbon, which results in the loss of biodiversity and the uprooting of indigenous communities. Paper goods are primarily landfilled or burned as trash, with just a small portion of this material being recycled. The amount of paper waste that ends up in landfills is a significant problem since it uses up valuable land resources and produces methane, a potent greenhouse gas, during its decomposition. The best way to get rid of waste paper is to recycle. Recycled paper can be used six or seven times, but because its fibers are shorter, it is frequently downcycled into materials of poorer quality. This lower-quality paper is often wasted as scrap, which is a source of cellulose that could be used more. Ethanol may be produced from this cellulose, which has advantages for the economy and ecology (Kolajo and Onovae 2023). For the management of paper waste, replanting trees contributes to a circular economy because trees are the source of almost all paper. These days, several biochemical compounds are produced from paper waste, including bioethanol, biogas, cellulose, lactic acid, acetic acid, butanol, biohydrogen, xylitol, furfural, levulinic acid, succinic acid, and enzymes. Significant research is being done to produce these chemicals at low cost as well. This chapter explains the recent techniques employed in the management of paper waste, focusing on the production of value-added chemicals, and discusses the associated challenges and prospective future directions.

6.2 Sources of Raw Materials

The process of making paper is intricate and resource-intensive, requiring the conversion of raw resources like wood, recycled paper, or other plant-based fibers into superior paper products. The majority of plant life naturally contains cellulose, which is the fiber component of wood. Paper is a material that is used widely in many industries, such as printing, packaging, personal care, and hygiene goods. Trees that are used to make paper are chosen for harvesting according to certain traits. The long fibers of softwood trees, including pine and spruce, are widely used to make paper that is stronger and more durable. Hardwood trees, on the other hand, like birch and eucalyptus, are used because their shorter strands enhance the smoothness and printing capabilities of some paper kinds (Faster Capital 2024).

Non-wood fibers are being used more and more in the making of paper due to their sustainability and adaptability. With its quick growth and high cellulose content, bamboo is a popular substitute for conventional wood fibers. A byproduct of processing sorghum or sugarcane, bagasse is utilized to make paper and other biodegradable goods. The raw material foundation for papermaking is further diversified by other materials like hemp, agricultural leftovers, corn stalks, cotton, kenaf, sisal, and jute. These commodities support industry innovation and sustainability while protecting forests and biodiversity (Abd El-Sayed et al. 2020).

To produce paper, waste paper has emerged as a vital secondary raw material. However, practical difficulties with collecting, sorting, and transportation prevent broad use. However, improvements in chemical and biological de-ink technologies are making it more viable. Because recycling discarded paper helps the environment and preserves trees and other cellulose resources for future use, it is becoming more and more popular (Defalque et al. 2021).

6.2.1 Types of Paper Waste

Paper garbage falls into two primary categories: pre- and post-consumer paper waste. Pre-consumer trash is material that was disposed of before it was suitable for consumer usage after it left the paper mill. Material disposed of after consumer use is known as post-consumer waste.

The paper waste is broadly classified as printing and writing paper, specialty paper, industrial and technical paper, packaging paper, tissue and hygiene paper, art and graft paper, recycled paper, scientific and technical paper, decorative and specialty papers, and eco-friendly and sustainable paper. Printing and writing paper include bond paper (Saloni Nagvekar 2024), coated paper, and uncoated paper (Scott Kravitz 2021). In the specialty paper category, the following type is included: acid-free paper (Preservation Equipment 2024), archival paper (Art Reproductions 2023), and security paper (Security Printing 2024). Under the industrial and technical paper type comes filter paper (Hawach Scientific 2024a), insulating papers (Paperpapers 2024), and release papers (LINTEC Corporation 2024). Paper waste in the packaging category includes kraft paper, paperboard, and corrugated paper (Ron Chen 2023). Paper waste in the hygiene category includes tissue paper, toilet paper, and paper towels (Green Seal 2024). Paper waste in the art and craft category includes watercolor paper, drawing paper, and scrapbooking paper (Altenew 2024). Paper waste in recycling is the secondary category that is further divided into two categories: post-consumer recycled paper and pre-consumer recycled paper (Strathmore 2025). Scientific and technical paper is a type of paper waste which includes laboratory filter paper and chromatography paper (Hawach Scientific 2024b). The decorative and specialty paper type consists of vellum paper (Papericious 2024), which is translucent paper used in crafts, and embossed paper, which is paper with a raised pattern. The final and environmentally friendly type of paper waste is called eco-friendly and sustainable paper, which comprises bamboo and hemp paper (Green Compostables 2024). The type of paper is depicted in Fig. 6.1.

6.2.2 Types of Packing Materials

Numerous paper packaging materials have been the subject of recent academic publications that have highlighted their uses as well as environmental concerns. Because of its strength and longevity, corrugated cardboard is frequently used to make shipping boxes. Paperboard, which is more robust than ordinary paper, is frequently used for product packaging, including cartons and cereal boxes (Deshwal et al. 2019). Kraft paper, which has a high tear resistance, is used to make void fill, paper bags, and wrapping paper (Primepac 2021). Paper-based cushioning materials, such as crumpled paper, paper shreds, and paper-based bubble wrap alternatives, serve as protective packaging (Charmaine Luk 2024). Laminated paper, which combines paper with plastic or aluminum for enhanced barrier properties, is frequently used in food packaging (Marsh and Bugusu 2007). Molded pulp, known for its

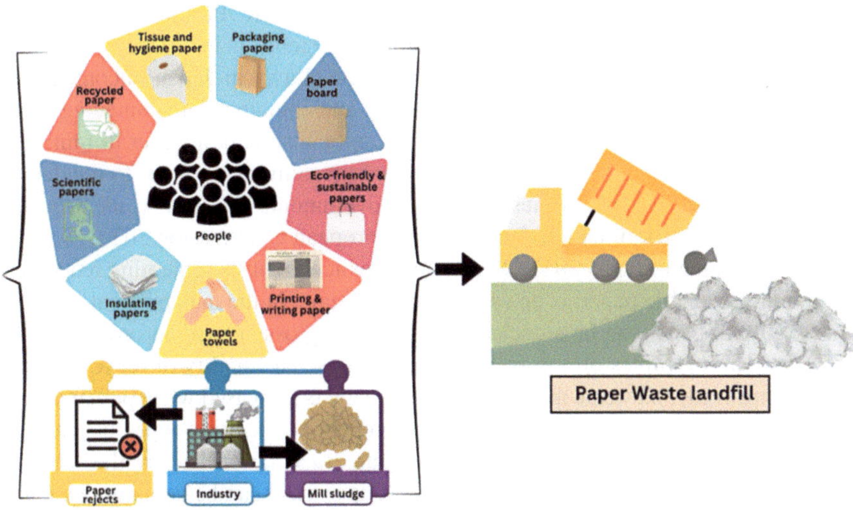

Fig. 6.1 Types of paper waste

cushioning properties, is used in protective packaging like egg cartons and electronics packaging (Zhang et al. 2022). Paper tubes and cores have been researched for their structural stability and recyclability because they are utilized as postal tubes and for winding materials like textiles and films. Recent developments in biodegradable coatings have improved waxed paper, which is used to wrap food items because of its ability to withstand moisture. Packaging uses honeycomb paperboard, which is lightweight and strong, for cushioning and structural support (Ficus Pax n.d.). Finally, paper bags are recognized for their recyclability and biodegradability; they are frequently used for grocery, retail, and takeout meals (Pactap n.d.).

6.3 Collection and Segregation Processes

Rag pickers and door-to-door collectors make up the informal sector, which is primarily in charge of managing the collection of post-consumer (PC) paper. Large traders dominate this industry, accounting for the collection of approximately 95% of the nation's waste paper. The value chain consists of zonal segregation centers run by semi-wholesalers, where garbage is baled and ready for delivery to end users, as well as direct collectors and small shops for preliminary sorting. The importance of this unorganized sector's involvement and structure in waste paper management has been highlighted by recent studies (Chintan Environmental Research and Action Group 2021). Waste paper collection is a well-organized industry regulated by a plethora of laws, including ordinances, directives, voluntary agreements, and procurement policy guidelines, in the majority of developed nations. Europe is a leader in packaging waste laws; to meet recycling requirements, Germany was the first to introduce extended producer responsibility (EPR) schemes. With over 130,000 businesses worldwide using the Green Dot logo, PROEUROPE, the umbrella organization for 33 National Producer Responsibility Systems, is in charge of the selective collection and recycling of packaging waste. Laws in Japan encourage the separated collection and recycling of packaging and containers, guaranteeing cooperation between businesses, local authorities, and consumers in order to minimize local solid waste. On the other hand, waste management laws are left to the discretion of individual municipalities and state governments in Canada, the US, and Mexico due to a lack of national legislation (European Commission 2020; Ministry of the Environment, Japan 2021). Segregation, which involves separating recyclable from non-recyclable paper, is an essential initial stage in the management of paper waste. Not all paper can be recycled; paper towels, glossy paper, and paper tainted with food are examples of non-recyclable materials. The quality of recycled products is improved, and waste management efficiency is increased when personnel are properly trained in the identification and segregation of certain types of garbage. Effective segregation needs to start at the source and be backed by the required infrastructure and staff awareness initiatives. Paper trash can be contaminated, and efficient source segregation can be established with the help of waste control specialists (Waste Control 2024).

6.4 Characterization of Paper and Packaging Materials

6.4.1 Physical Properties

Papers are used in many different applications, including newsprint, tissue paper, writing, printing, packaging, and specialty usage. Each of these uses requires a certain set of physical, mechanical, and morphological characteristics. Physical qualities deal with order, structure, and reaction to outside influences, whereas mechanical attributes include strength and flexibility under stress. These properties dictate suitability for various applications; for instance, packaging papers require high tear and bursting resistance, whereas writing and printing papers necessitate smoothness, excellent printability, optimal optical properties, dimensional stability, and adequate tensile strength (Pego et al. 2022).

Basis weight, thickness, density, two-sidedness, smoothness, permeability, handling, rigidity, roughness, and porosity are physical characteristics of paper sheets that are essential to their use. Properties including curl, strength, dimensional stability, softness, hardness, and compressibility are influenced by these qualities (Miao et al. 2023). Strength characteristics are frequently assessed since they are crucial for determining paper permanence and aging treatments. The MIT (Massachusetts Institute of Technology) folding endurance testing is used to quantify fold endurance, which evaluates the durability of book handling. Tear resistance is measured by the Elmendorf tear test, which looks at internal and external resistance to tearing, much like that of torn book pages. The Mullen test, which measures burst strength, is widely used in the paper industry but is less frequently applied in conservation. Standard span tensile testing is commonly used to determine tensile strength, which is important for interfiber bonding and network strength. It can also provide information about energy absorption, Young's modulus, and stress-strain relationships. Wet tensile tests, zero-span tests, and z-direction tests are additional measures that can be used to assess fiber strength and orientation (Caulfield and Gunderson 1988).

6.4.2 Chemical Composition

Paper is a very adaptable material with a wide range of applications that are sometimes overlooked in the world of digital screens. Different types of paper, including kraft, uncoated, copy, coated, paperboard, thermal, void filler, and continuous stationery, each have their own special qualities. Standardized test protocols are used by the paper industry and technical groups to evaluate the composition of paper, which is generally composed of 90–99% cellulose fibers. These fibers self-bond to generate a robust network that is essential for end-use properties. Filling and loading materials make up the majority of the inorganic percentage in paper, which affects its chemical and physical qualities (Worldpapermill 2020). Knowing the main chemicals utilized in the production of paper is the first step towards investigating the chemistry underlying its composition. The fundamental structural component of paper is cellulose, which is mostly sourced from plant materials such as wood or recycled paper fibers. Its complex carbohydrate structure, which is made up of repeating glucose units, gives paper its strength and rigidity. High-quality papers are frequently treated to lower the amount of lignin, a polymer that is also present in plant cell walls and adds to hardness. However, over time, lignin can cause paper disintegration and yellowing. With its branching structure, hemicellulose is also found in plant cell walls, which adds strength and flexibility. When it comes to colored paper, sizing agents like rosin and alum are employed both externally and internally to control ink interaction and improve water resistance (Rppaperimpex 2024).

6.4.3 Contaminants and their Effects

Paper products vary greatly in terms of their chemical makeup and physical characteristics. Many chemicals may be included at any point in their lifecycle, from production to consumer usage. Additives, including plasticizers, binders, strength-enhancing agents, coatings, pigments, inks, and solvents, are frequently added during manufacturing processes like printing and converting (e.g., binding, gluing, laminating, and labeling). The variety of chemicals involved may also be further expanded by adding other compounds at the usage phase of the paper, such as adhesives or inks, interactions with food, or cleaning procedures (Rizwan et al. 2022).

Wang et al. (2023) found out some sticky contaminants in recycled papers, the size, quantity, and weight of sticky contaminants in recycled fibers varied depending on the paper quality and source of collection. Curbside collected grades can have up to 15 wt.% stickies, while local paper recycling factories usually have 3–5 wt.% stickies. Depending on the quality, these impurities in recycled paper might range in size from 0.4 to 1.2 mm^2 on average. The main constituents of these stickies extracted using THF (tetrahydrofuran), according to chemical analysis using FT-IR and GC/MS, include polyvinyl acetate (PVA) polymers, styrene butadiene rubber (SBR), paraffin wax, resin, and polyamines.

Of the extractable chemicals, 101 were found in 132 paper-based food packaging materials studied in the Czech Republic. Of these compounds, 68 were quantified using tandem mass spectrometry in conjunction with gas and liquid chromatography. They included compounds from impregnation and coating operations (e.g., solvents, hydrocarbons), wastes from the manufacturing of paper pulp, and additives from printing inks or adhesives (e.g., photoinitiators, plasticizers, solvents). The most prevalent pollutants were aromatic hydrocarbons, photoinitiators, anthraquinone, and phthalate plasticizers. The substances with the highest concentrations were dibutyl phthalate, diisobutyl phthalate, and benzophenone, which were mostly derived from residues from regenerated pulp and materials used to modify functional qualities (Vápenka et al. 2016). Environmental deposition can be greatly impacted by the chemicals from paper-based materials that leak into soil and water during landfill disposal or are released into the atmosphere during combustion. This process opens the door for them to enter ecosystems' food webs, potentially endangering human health. Furthermore, aquatic organisms may be negatively impacted by these substances.

6.5 Pre-treatment Processes

Several challenges hinder the achievement of sustainable development, such as the exhaustion of natural resources and the impact of human activities on the ecosystem. The 2030 Agenda for Sustainable Development aims to enhance global quality of life and is underpinned by 17 Sustainable Development Goals (SDGs). SDG 12 focuses on promoting sustainable development by emphasizing responsible patterns of production and consumption, including improved waste management strategies like waste prevention, reduction, recycling, and reuse, alongside efficient resource utilization. For the past 27 years, greater attention has been placed on the manufacture of biochemicals and biofuels from cellulose-rich paper waste. Cellulose, a glucose biopolymer with renewable potential, is found in waste paper, a significant portion of solid waste. When converted to glucose, it has the potential to undergo fermentation for the generation of bioethanol and biopharmaceuticals. To use waste paper as a sustainable feedstock, the saccharification process must be optimized. Unfortunately, cellulose's structural makeup makes bioconversion ineffective and requires the creation of methods to increase the material's vulnerability to cellulase hydrolysis (Mokatse and van Wyk 2021). Hence, prior to the production of biochemicals, paper waste must undergo pretreatment. This section will comprehensively examine the various pretreatment methods utilized in this process.

6.5.1 Mechanical Pre-treatment

Physical procedures such as shredding, grinding, or milling are frequently utilized in mechanical pretreatment methods for waste paper to decrease the material's particle size and enhance its surface area. The goal of these techniques is to increase the availability of enzymes or other chemical treatments for use in fermentation or enzymatic hydrolysis, two more bioconversion processes. To improve the specific surface area available to microorganisms and decrease the size of biomass particles, a mechanical pre-treatment using a Hollander beater was assessed by Rodriguez et al. (2017). This increased the amount of biogas production and, in another study, showed the maximum bioconversion degradation for brown envelope paper was achieved by a combined chemical and physical pre-treatment using NaOH and sonication, followed by foolscap paper and office paper (Mokatse and van Wyk 2021). Capek-Ménard et al. 1992 conducted the experiment in a pressure vessel where waste office paper was handled with steam at a temperature of 170–220 °C. Enzymatic saccharification and fermentation to ethanol were then carried out. Acid-catalyzed steam treatment increased yields to 460 l/ton, a 29% increase above controls; however, steam alone did not significantly boost yields. Pretreating pulp and paper mill sludge with a variety of heating techniques, such as a hot air oven, autoclave, microwave, and water bath, increased hydrolysis. The solubilization of organic matter was enhanced by all approaches, as shown by chemical studies. Sludge solubilization was most significantly affected by the hot air oven treatment, which was applied for 90 min at 80 °C (Veluchamy and Kalamdhad 2017).

6.5.2 Chemical Pre-treatment

Phosphoric acid (H_3PO_4) and sulfuric acid (H_2SO_4) are frequently utilized for the chemical pretreatment because they are efficient and reasonably priced methods of hydrolyzing lignocellulose. Phosphoric acid is safe for the environment and has a softer effect. Compared to H_2SO_4, hydrochloric acid (HCl) is more volatile and easier to recover, which makes it a more effective means of breaking down biomass. High rates of conversion of cellulose to sugar are also observed in nitric acid (HNO_3) (Bensah and Mensah 2013). Chemical pretreatment techniques are used since cardboard and newspaper cannot be mechanically processed to produce methane. Newspaper lignin solubilization is greatly increased by adding nitric acid during the acetic acid pretreatment (Yuan et al. 2012). To produce fermentable sugar, waste paper underwent alkaline pretreatment before enzymatic hydrolysis (Ojewumi et al. 2018). The significant increase in cellulose and sugar yield, together with the efficient delignification, is performed by pretreatment with H_2O_2 (0.5% v/v) at 121 °C for 30 min was shown to be the most effective approach for waste paper (Annamalai et al. 2020).

6.5.3 Biological Pre-treatment

The viability of converting paper into biofuels and biochemicals hinges on advancements in pretreatment and hydrolysis techniques for the extraction of sugars from it. The benefits and difficulties of creating biological pretreatment methods to alter lignin for simpler enzymatic sugar extraction. Because of its inherent drawbacks, biological pretreatment has received less research attention than thermochemical approaches, but it has some advantages over thermochemical techniques, including being more energy-efficient, producing fewer inhibitory chemicals, producing more co-products, and being more ecologically friendly (Chen et al. 2010). To increase the production of methane, filter paper, office paper, newspaper, and cardboard were pretreated using a microbial consortium known as MC1. The hydrolysates' COD significantly increased before treatment, reaching a peak on day 7. The main volatile products produced during pretreatment are glycerol, butyric acid, ethanol, acetic acid, and propionic acid. MC1 showed good degrading capabilities, and all four substrates used had considerably higher methane generation rates and yields following pretreatment (Yuan et al. 2012). In another study to increase the production of methane, waste paper that mimicked cellulosic biomass was processed with rumen fluid. After the 6-h pretreatment, methane generation that was semicontinuous over a 20-day period produced the largest daily yield, 2.6 times more methane than untreated paper, or 73.4% of the theoretical yield (Baba et al. 2013). The enzymatic hydrolysis of office paper was enhanced by biological pretreatment using this combination strain of cellulose-hydrolyzing bacteria *Sphingomonas paucimobilis* MK1 and *Bacillus circulans* MK2, resulting in a reduction of processing time by more than a day and an increase in sugar recovery from 30% to 75%, and up to 94% under ideal conditions. Increased cellulose crystallinity following pretreatment was observed by X-ray diffraction, which was attributed to endoglucanase activity hydrolyzing amorphous regions (Kurakake et al. 2007). Using strong biocatalysts like xylanases, laccases, and cellulases to remove pitch, deink, and biobleach (the process of separating hemicelluloses from cellulose), paper waste is usually advantageous. Enzymes known as endoxylanases and β-xylosidases, together referred to as xylanases, are principally in charge of breaking down xylan, which is the main component of hemicellulose. β-xylosidases operate on these xylooligomers to produce xylose, while endoxylanases act on the homopolymeric backbone of 1,4-linked β-d-xylopyranose-producing xylooligomers. Xylans are found in pulp fibers and are closely related to cellulose and lignin. Lignin extraction is simple when enzymes break down the covalent link that binds lignin to xylan. During alkaline pulping, xylan that has reprecipitated on the fibers is hydrolyzed by xylanase enzymes (Chakraborty et al. 2019).

6.6 Conversion Technologies

Waste paper are very desirable material for the production of biochemicals because they are cheap, readily available lignocellulosic material with high holocellulose and low lignin concentrations. Numerous conversion technologies enable the transformation of waste paper into useful items. The goal of these technologies is to create new materials or products by recycling and upcycling waste paper. Thermochemical conversion (incineration, pyrolysis, and gasification), combustion, biochemical conversion, anaerobic digestion, fermentation, enzymatic hydrolysis, and hybrid processes (Fig. 6.2) are some of the key conversion technologies. Technologies for thermochemical conversion hold promise for managing trash and producing sustainable energy at the same time. Paper mill waste can be converted into useful energy vectors through the use of thermochemical methods such as incineration, gasification, and pyrolysis. Gasification is not a novel technique, but the pulp and paper industry is still in the early stages of its application's development. The process of gasification is the partial burning of organic materials to produce fuel gas, also known as product gas, which is mostly made up of CO, CO_2, H_2, CH_4, H_2O, N_2, and C_XH_Y and can be utilized for CHP (Combined Heat and Power) applications. Fazil et al. (2023) investigated the potential of paper-rich commercial refuse-derived fuel (RDF) with high ash content for gasification and co-gasification in a downdraft gasifier. Another study explores gasifying blends of pre-conditioned

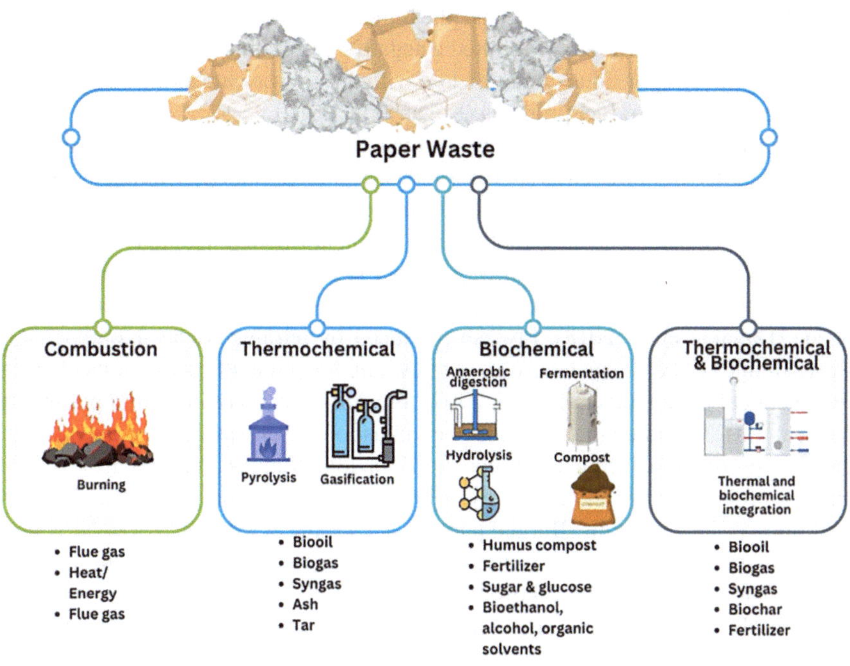

Fig. 6.2 Conversion process of paper waste into value-added products

paper mill rejects and de-inking sludge pellets with wood chips in a downdraft gasifier. Up to 80% of a brown paper mill's rejects can be successfully gasified with 20% wood chips, producing a gas suitable for a 10 kWe engine (Ouadi et al. 2013). By using a process called paper waste sludge (PWS) gasification, PWS can be transformed into greener gases, including CO, H_2, and CH_4. This type of energy reduces emissions and has the potential to be a sustainable option (Le Tan et al. 2023). Pyrolysis represents an additional methodology presently employed in the transformation of paper waste materials. Using the Microwave Assisted Pyrolysis approach, activated carbon (AC) was created from the waste paper. Utilizing microwave heating with a frequency of 2.45 GHz and several power outputs, the deinking paper waste was carbonized and activated. $ZnCl_2$ solution was used as an activating reagent during the activation process (Jauharatul et al. 2022). Another paper waste, specifically waste printing paper, was subjected to pyrolysis using a thermogravimetric analyzer combined with a Fourier transform infrared spectrometer to produce gaseous products (Ma et al. 2017). Lee et al. (2022) produced value-added products from the polyethylene-lined paper coffee cup waste by conventional slow pyrolysis. Pyrolysis is conducted within a Muffle furnace to facilitate the transformation of cardboard mixed biosolids into biochar (Zuhara et al. 2023). Melamine and urea formaldehyde-containing waste sheets were converted into pyrolytic oil, which has both commercial and environmental benefits (Yaman and Sinoplugil 2019). Angelo and Ridout (2016) investigated the pyrolytic conversion of paper waste sludge (PWSs) as a substitute for its valorization, with the rationale being the enhancement of exothermic reactions to achieve higher heating rates through the utilization of smaller pellet sizes. This phenomenon led to an evident rise in the non-condensable gas yield, which exhibited a negative correlation with the bio-oil yield. The paper recycling residue contaminated with plastic was processed using pyrolysis to recover valuable fuel products (Logan et al. 2019).

The process of combustion, which is a reactive oxidation, releases energy from chemical molecules utilized as fuels for industry, transportation, heating, and power generation. While the paper waste undergoes combustion, it produces heat energy that is used for many beneficial activities. Chemical Looping Combustion technology can be used to efficiently burn paper waste (Zainab et al. 2020). By using a cost-effective combustion process, paper waste and urea are converted into graphene (Dayakar et al. 2018). Fly ash produced by burning Malaysian paper mill waste sludge (PMWS) can be used as a substitute for cement because of its physical and chemical parameters (Yin et al. 2007). In addition to other industrial pollutants, Xu et al. (2023) studied the behavior of combustion paper waste. They also reduced the NOx production during the combustion by adding fungus and used tires. But the burning of waste paper and cardboard briquettes results in flue gas with a high concentration of heavy metals. Burned paper waste produces ash with a low NPK concentration (Gado et al. 2018). A small amount of mercury is found in the combust paper waste ashes (Kijo-Kleczkowska et al. 2023).

Anaerobic digestion and alcoholic fermentation are two processes used in the biochemical conversion of paper waste valorization, which transforms paper waste into different fuels. Anaerobic digestion produces methane and hydrogen, while fermentation produces ethanol. Anaerobic digestion is one of the most incredible techniques for handling high-strength organic effluents, such as recycled paper mill effluent (RPME), due to its low sludge output and energy-positive (production of methane) nature (Siti et al. 2022). One inexpensive method of producing bioplastic is to use waste paper as a resource for the anaerobic digestion process that produces polyhydroxyalkanoate (PHA). Huda et al. (2021) manufactured polyhydroxyalkanoate (PHA) from volatile fatty acids (VFAs) derived from waste paper. Biomethane is produced from paper industry waste by Thermophilic Anaerobic Digestion, some chemicals influence the production. Eda et al. (2023) studied the impact of two chemical molecules, 1-octanol and hexanal, originating from the alcohol and aldehyde groups, in relation to a thermophilic anaerobic process (55 ± 2 °C) that breaks down waste produced in the paper sector into methane. Anaerobic reactors that digest paper waste, the toxic effects of 1-octanol and hexanal would be better observed at higher concentrations, especially when their threshold levels are exceeded. As per their studies, the effect of the chemical compounds on methane generation has generally been concentration-dependent. Paper waste's resistant lignocellulosic structure creates challenges for anaerobic digestion, which can be overcome with the help of pretreatments such as air purging before feeding to reduce sulfide toxicity in the digester (Lin et al. 2014), mechanical pre-treatment with a Hollander beater, and anaerobic and microaerobic pretreatment (Chao et al. 2021).

Abduljabbar et al. (2022) used anaerobic fermentation to turn cardboard waste into bioethanol. By employing physical, chemical, and enzymatic treatments, they enhanced the two-stage saccharomyces and fermentation stage of glucose synthesis from cardboard waste and sped up the generation of bioethanol. Methane and carbon dioxide are the primary byproducts of biogas produced during the anaerobic fermentation process from the breakdown of lignin, cellulose, and hemicellulose, the main constituents of paper waste (Gonzalez-Estrella et al. 2017). Paper mill wastes can also be treated by anaerobic digestion, which also produces value-added bioproducts such as biogas, biomethane, biohydrogen, and biofertilizer. These are additional advantages of this process (Kumar et al. 2020). The thermophilic anaerobic membrane bioreactor (ThAnMBR), a revolutionary method for treating pulp and paper sludge with an emphasis on biogas production, was developed by Jiang (2018). The biological and membrane performance of a laboratory-scale ThAnMBR was assessed for 328 days during the study's operation. Higher hydraulic retention times (HRTs) enhanced performance, suggesting that ThAnMBR has the potential to be an efficient treatment for pulp and paper sludge.

Valorization of paper sludge and paper waste is a highly promising technique to produce renewable chemicals and fuels by using enzymatic hydrolysis. The microbial generation of bioplastic (Poly-3-hydroxybutyrate, or PHB) has a lot of potential when waste paper is used as a substrate. Office paper in a batch and fed-batch simultaneous saccharification and fermentation (SSF) method with *Cupriavidus necator* produced PHB. A shorter fermentation period was required in the fed-batch experiment, a good PHB yield was achieved with 10% paper loading utilizing H_2O_2 pretreatment in batch SSF. PHB exhibited traits that were comparable to those of conventional PHB, indicating that employing SSF to produce PHB from waste paper is a viable option (Sivakumar 2022).

6.7 Challenges and Future Directions Solutions

The paper waste generates a lot of solid and liquid waste, making it one of the most polluting materials. An industrialized nation's garbage output can reach up to ten million tons annually, and the processing of that waste requires significant financial and other resources. With a high percentage of lignin, cellulose, and hemicellulose, this discharge is a possible feedstock that can be utilized as a useful raw material for the manufacturing of biofuels, platform chemicals, and bioenergy (Chakraborty et al. 2019). Nevertheless, despite these benefits, there are a number of difficulties, such as handling contamination, making sure that collection and sorting are done quickly, and instructing customers on appropriate value-added product production procedures. There are many kinds of waste paper, such as cardboard, newspapers, office paper, and more. Because each variety has unique fibers, coatings, inks, and impurities, recycling and conversion procedures are made more difficult. To ensure a high-quality product and increase conversion efficiency, foreign elements and impurities must be removed during the cleaning step of waste paper recycling. In waste paper cleaning, physical, chemical, and biological cleaning techniques are frequently used; each has benefits and drawbacks of its own. High levels of contamination, issues with mechanical processing, the environmental effects of chemical cleaning, and the viability of biological approaches are some of the challenges associated with waste paper cleaning (Çiçekler and Tutuş 2024). Utilizing waste paper can provide high-value products such as chemicals, biofuels, and nanomaterials, but doing so needs sophisticated and sometimes expensive technology. Value-added product production may not be financially feasible due to the significant costs associated with gathering, classifying, shipping, and processing waste paper. Paper waste products with value additions must compete with items made from virgin materials and other waste streams, which are frequently less expensive.

Setting quality parameters for the final product, monitoring and adjusting these parameters at different stages of the process, and dynamically balancing process efficiency are the challenges in pre-treating and processing waste paper to value products (Markus et al. 2005). In addition, energy-efficient systems with high efficiency levels are needed for the treatment of waste fibers and products. Processing stages must use measurements for characteristic values, integrate chemical and physical advances, and alter chemicals, energy use, and waste disposal based on online evaluations to meet these issues (Rodriguez et al. 2017).

Though there are many obstacles to overcome, thermochemical conversion techniques like gasification, pyrolysis, and incineration have the potential to turn waste paper into goods of value. Environmental problems associated with incineration include ash disposal, toxic emissions, and greenhouse gases. While market growth obstacles and high costs restrict pyrolysis's economic feasibility, it is necessary to prepare feedstock uniformly and exercise precise process control to guarantee output quality. While scalability and system integration present additional challenges, gasification necessitates thorough feedstock pretreatment, optimal reaction conditions, and effective gas cleanup to prevent tar formation and contaminants. Critical factors for all approaches include market acceptance, regulatory compliance, high capital and operational expenses, and continual research and development for process improvements (Alao et al. 2022).

One of the challenges in biochemically converting paper waste is reducing cellulose crystallinity for enzymatic hydrolysis, which requires efficient pretreatment techniques (Pecar and Andreja 2015). Obstacles include the heterogeneous character of biomass sources, resistance to enzymatic attack, and the expensive, large-scale synthesis of enzymes that impede the enzymatic conversion of lignocellulosic resources into paper products (Gursharan et al. 2017). Optimizing enzymatic operations and overcoming inhibitors that may impede subsequent fermentation processes are issues associated with the enzymatic hydrolysis of paper waste to fermentable sugars for the manufacture of bio-ethanol (Modupe et al. 2022). These problems might be solved by hybrid processes that combine various conversion techniques, utilizing the advantages of each strategy to improve overall sustainability and efficiency.

The fastest-maturing vermicomposting method with adequate nutrient levels for plant application was achieved with newspaper bedding; however, the soil was contaminated due to the high Cu concentration (Chubo et al. 2021).

Paper trash will eventually be transformed into useful items by the use of environmentally friendly methods such as recycling and waste utilization. Circular economy principles are crucial since the pulp and paper sector produces large amounts of waste, including paper rejects (Hatem et al. 2023). High-value uses can be achieved through recycling waste paper to create cellulose-based goods such as biofuels, composites, and cellulose nanocrystals (Cynthia et al. 2018). Further demonstrating the possibility of turning garbage into useful materials is the conversion of waste paper into cellulose derivatives, nanocellulose, and aerogels (Mikhailidi and Kotelnikova 2021). This underscores the significance of sustainable waste management techniques in reducing environmental effects and fostering resource efficiency. The paper sector can help create a more sustainable future by cutting greenhouse gas emissions, water consumption, and manufacturing costs while producing useful goods from waste materials by concentrating on novel paths in recycling and waste utilization.

The transformation of paper waste into valuable products raises important policy and regulatory challenges that must be addressed for sustainable waste management. Significant waste streams, including paper rejects, are produced by the pulp and paper industry, which exacerbates environmental problems such as pollution and deforestation (Adnan et al. 2022). Reducing greenhouse gas emissions and resource consumption through recycling paper trash is crucial, yet there are certain constraints because of chemical usage and recycling cycles. Other uses for paper waste include making biofuels, nanocellulose, and building materials; these are sustainable options, but they need further investigation and development (Hatem et al. 2023). Furthermore, laws governing product take-back, such as extended producer responsibility (EPR), help to ensure that end-of-life products are properly recovered and recycled, even when changing recycling economics pose challenges to long-held beliefs. Initiatives to improve waste paper collecting and recycling are being put into place in nations like India to lessen their influence on the environment and their reliance on imported paper (Kamal et al. 2023).

6.8 Conclusion

Global pulp and paper consumption is projected to triple from 2010 to 2060, leading to a significant increase in waste paper, intensifying the existing pressure on the world's forests. Once considered a scarce and precious commodity, paper has now become ubiquitous, predominantly utilized for packaging purposes rather than for communication. A substantial amount of paper waste is disposed of in landfills, contributing to environmental contamination. The solid waste from the paper industry is a major source of water and air pollution. Every day, millions of individuals dispose of paper products, occupying significant space in landfills and occasionally encroaching on agricultural areas. Contemporary approaches to utilizing non-edible biomass entail converting it into glucose and then fermenting the resulting sugar to generate bioproducts. Due to its high cellulose content and low lignin levels, waste paper serves as an excellent raw material for bioconversion into useful goods like biofuel, biochar, organic acids, bioplastics etc., Diverse technologies exist for transforming waste paper into valuable substances, including thermochemical conversion (such as incineration, pyrolysis, and gasification), combustion, biochemical conversion, anaerobic digestion, fermentation, enzymatic hydrolysis, and hybrid processes. Each of these techniques presents unique obstacles. Consequently, further investigation is essential to devise economically feasible conversion strategies. Additionally, governmental and private sectors should collaborate to formulate policies that effectively tackle these challenges.

References

Abd El-Sayed ES, El-Sakhawy M, El-Sakhawy MA (2020) Non-wood fibers as raw material for pulp and paper industry. Nordic Pulp Pap Res J 35(2):215–230. https://doi.org/10.1515/npprj-2019-0064

Abduljabbar AT, Younes RW, Ibrahem AJ (2022) Converting of cardboard waste for bioethanol production using anaerobic fermentation. Iraqi J Ind Res 9(2):141–147. https://doi.org/10.53523/ijoirvol9i2id213

Adnan K, Sumeet M, Nisar A, Muhammad B, Farooq S, Vineet K, Luiz F, Romanholo F, Hafiz MN (2022) Treatment of pulp and paper industry waste effluents and contaminants. Elsevier, London, pp 349–370. https://doi.org/10.1016/b978-0-323-85835-9.00018-0

Alao MA, Popoola OM, Ayodele TR (2022) Waste-to-energy nexus: an overview of technologies and implementation for sustainable development. Cleaner Syst 3:100034. https://doi.org/10.1016/j.cles.2022.100034

Altenew (2024) Different types of paper for arts and crafts. https://altenew.com/blogs/paper-crafting-inspiration-and-tips/types-of-paper-for-crafting?srsltid=AfmBOooa2HtmVBzz3w5yR5vY4G5oxZaMxwbd4bXQSsCI2105RrVyhrHK

Angelo MCJ, Ridout J (2016) Valorisation of paper waste sludge using pyrolysis processing. https://core.ac.uk/download/pdf/188221453.pdf

Annamalai N, Al Battashi H, Anu SN et al (2020) Enhanced bioethanol production from waste paper through separate hydrolysis and fermentation. Waste Biomass Valoriz 11(1):121–131. https://doi.org/10.1007/s12649-018-0400-0

Art Reproductions (2023) What is archival paper? https://fineartprinting.com.sg/what-archival-painting/

Baba Y, Tada C, Fukuda Y, Nakai Y (2013) Improvement of methane production from waste paper by pretreatment with rumen fluid. Bioresour Technol 128:94–99. https://doi.org/10.1016/j.biortech.2012.09.077

Baleforce (2024) Paper waste: why does it matters? https://baleforce.com/paper-waste-why-does-it-matter-and-how-to-manage-it/

Bensah EC, Mensah M (2013) Chemical pretreatment methods for the production of cellulosic ethanol: technologies and innovations. Int J Chem Eng. https://doi.org/10.1155/2013/719607

Business Waste (2024) Paper waste facts and statistics. https://www.businesswaste.co.uk/your-waste/paper-recycling/paper-waste-facts-and-statistics/

Capek-Ménard E, Jollez P, Chornet E, Overend RP (1992) Pretreatment of waste paper for increased ethanol yields. Biotechnol Lett 14(10):985–988. https://doi.org/10.1007/BF01020642

Caulfield DF, Gunderson DE (1988) The book preservation guide. American Library Association, Chicago

Chakraborty D, Dahiya S, Amulya K, Srivastav V, Mohan SV (2019) Valorization of paper and pulp waste: opportunities and prospects of biorefinery. In: Prasad MNV, de Campos Favas PJ, Vithanage M, Mohan SV (eds) Industrial and municipal sludge. Butterworth-Heinemann, Oxford, pp 623–656. https://doi.org/10.1016/B978-0-12-815907-1.00027-1

Chao S, Li W, Li W, Cai F, Liu G, Chen C (2021) Anaerobic and microaerobic pretreatment for improving methane production from paper waste in anaerobic digestion. Front Microbiol 12:688290. https://doi.org/10.3389/fmicb.2021.688290

Charmaine Luk (2024) 4 sustainable paper-based protective packaging solutions. https://packmojo.com/blog/sustainable-protective-packaging-solutions/?srsltid=AfmBOorWW5GrRQNw2bf6FdVYDqaCIU-r4-vQlnhWIzv9FuiMKHgU2PQv

Chen S, Zhang X, Singh D, Yu H, Yang X (2010) Biological pretreatment of lignocellulosics: potential, progress and challenges. Biofuels 1(1):177–199. https://doi.org/10.4155/bfs.09.13

Chintan Environmental Research and Action Group (2021) Role and organization of the informal sector in waste paper management. https://www.chintan-india.org/research-and-publications

Chubo JK, Tengku A, Tengku Y-A, Marinovich M (2021) Paper wastes as beddings in vermicompost production. Borneo Sci 42(2):4469. https://doi.org/10.51200/bsj.v42i2.4469

Çiçekler M, Tutuş A (2024) The importance of cleaning stage in waste paper recycling. In *proceedings of the 2nd international conference on scientific and innovative studies*. Konya, Turkey. All Sciences Academy. https://as-proceeding.com/index.php/icsis

Cynthia A, Mark J, Vijay KT (2018) Exploring new horizons for paper recycling: a review of biomaterials and biorefinery feedstocks derived from wastepaper. Green Sustainable Chem 13:21–26. https://doi.org/10.1016/J.COGSC.2018.03.003

Dayakar CSB, Joseph V, Velmurugan S, Ravuri A, Nirmala G (2018) Combustion synthesis of graphene from waste paper for high performance supercapacitor electrodes. Int J Nanosci. https://doi.org/10.1142/S0219581X17600237

Defalque CM, Marins FAS, da Silva AF et al (2021) A review of waste paper recycling networks focusing on quantitative methods and sustainability. J Mater Cycl Waste Manage 23(1):55–76. https://doi.org/10.1007/s10163-020-01124-0

Deshwal GK, Panjagari NR, Alam T (2019) An overview of paper and paper-based food packaging materials: health, safety, and environmental concerns. J Food Sci Technol 56(10):4391–4403. https://doi.org/10.1007/s13197-019-03950-z

Eda Y, Yarsur I, Wåhlné H, Cigdem, Yangin-Gomec C (2023) Methane yield of paper industry waste in the presence of two compounds from alcohol and aldehyde groups during thermophilic anaerobic digestion. Environ Res Technol. https://doi.org/10.35208/ert.1212911

European Commission (2020) Packaging waste directive. https://ec.europa.eu/environment/topics/waste-and-recycling/packaging-waste_en

Faster Capital (2024) Exploring the raw materials of paper. https://fastercapital.com/topics/exploring-the-raw-materials-of-paper.html

Fazil A, Kumar S, Mahajani SM (2023) Gasification and co-gasification of paper-rich, high-ash refuse-derived fuel in downdraft gasifier. Energy 263(Part A):125659. https://doi.org/10.1016/j.energy.2022.125659

Ficus Pax (n.d.) Benefits of honeycomb paper packaging. https://ficuspax.com/blog/benefits-honeycomb-paper-packaging/#:~:text=Honeycomb%20paper%2C%20named%20after%20its,choice%20for%20various%20packaging%20applications. Accessed 12 January 2025

Gado IH, Ouiminga SK, Yonli AH, Sanogo O, Daho T, Koulidiati J (2018) Study of temperature fields and heavy metal content in the ash and flue gas produced by the combustion of briquettes coming from paper and cardboard waste. Recycling. https://doi.org/10.3390/recycling3030032

Gonzalez-Estrella J, Asato CM, Stone JJ, Gilcrease PC (2017) A review of anaerobic digestion of paper and paper board waste. Rev Environ Sci Biotechnol 16(3):569–590. https://doi.org/10.1007/s11157-017-9436-z

Green Compostables (2024) Eco-friendly paper—all you need to know. https://www.greencompostables.com/blog/eco-friendly-paper#:~:text=Eco%2Dfriendly%20papers%20are%20typically,processed%20into%20a%20reusable%20pulp

Green Seal (2024) GS1 sanitary paper products. https://greenseal.org/standards/gs-1-sanitary-paper-products/

Gursharan S, Shailendra KA, Vijaya G, Prince S (2017) Enzyme technology for lignocellulosic biomass conversion and recycling to valuable paper and other products: challenges ahead

Hatem A, Salma TG, Jiangyun M (2023) On the conversion of paper waste and rejects into high-value materials and energy. Sustain For 15(8):6915–6915. https://doi.org/10.3390/su15086915

Hawach Scientific (2024a) A brief introduction of filter paper. https://www.hawach.com/news/a--brief-introduction-of-filter-paper.html

Hawach Scientific (2024b) The difference between filter paper and chromatography. https://www.hawachfilterpaper.com/the-difference-between-filter-paper-and-chromatography/

Huda AB, Shatha A-K, Gupta VK, Nallusamy S (2021) Polyhydroxyalkanoate (PHA) production using volatile fatty acids derived from the anaerobic digestion of waste paper. J Polym Environ 29(1):250–259. https://doi.org/10.1007/S10924-020-01870-0

Jauharatul W, Achmad DK, Soeprijanto S, Pudjiastuti L, Daril RZ, Fajar PN, Nazilah S (2022) Synthesis of activated carbon from paper waste using microwave-assisted pyrolysis method. Key Eng Mater 937:157–164. https://doi.org/10.4028/p-bk445s

Jiang S (2018) Thermophilic anaerobic membrane bioreactor for pulp and paper sludge treatment

Kamal JS, Sameer A, Prashant G (2023) Waste paper trade and recycling policy in India: a review. Int J Ecol Environ Sci 49(7):2988. https://doi.org/10.55863/ijees.2023.2988

Kijo-Kleczkowska A, Gnatowski A, Tora B, Kogut K, Bytnar K, Krzywanski J, Makowska D (2023) Research on waste combustion in the aspect of mercury emissions. Materials. https://doi.org/10.3390/ma16083213

Kolajo TE, Onovae JE (2023) Biochemical conversion of waste paper slurries into bioethanol. Sci Afr 20:e01703. https://doi.org/10.1016/j.sciaf.2023.e01703

Kumar M, Morya R, Gupta A, Thakur IS (2020) Anaerobic biovalorization of pulp and paper mill waste. In: Biovalorisation of wastes to renewable chemicals and biofuels. Elsevier, Amsterdam, pp 41–61. https://doi.org/10.1016/B978-0-12-817951-2.00003-1

Kurakake M, Ide N, Komaki T (2007) Biological pretreatment with two bacterial strains for enzymatic hydrolysis of office paper. Curr Microbiol 54(5):424–428. https://doi.org/10.1007/s00284-006-0568-6

Le Tan NT, Le TL, Le NPT, Nguyen DQ (2023) Optimisation of syngas production from paper waste sludge by up-draft gasification: a response surface methodology approach. Chem Eng Trans 106:99. https://doi.org/10.3303/CET23106099

Lee H, Papari S, Bernardini G, Ciuffi B, Rosi L, Berruti F (2022) Value-added products from waste: slow pyrolysis of used polyethylene-lined paper coffee cup waste. Can J Chem Eng 101(3):1271–1285. https://doi.org/10.1002/cjce.24472

Lin C-J, Lin C-J, Zhang P, Pongprueksa P, Liu J, Evers SA, Hart PW (2014) Pilot-scale sequential anaerobic–aerobic biological treatment of waste streams from a paper mill. Environ Prog 33(2):359–368. https://doi.org/10.1002/ep.11785

LINTEC Corporation (2024) Release papers and films. https://www.lintec-global.com/products/process/#:~:text=Release%20papers%20are%20papers%20coated,Label%20and%20stickers%2Drelated

Logan A, Jeremy A, Brown A (2019) Valorisation of paper recycling residue contaminated with plastic into valuable fuel products by pyrolysis

Ma Y, Wang J, Zhang Y (2017) TG–FTIR study on pyrolysis of waste printing paper. J Therm Anal Calorim 129(2):1225–1232. https://doi.org/10.1007/s10973-017-6218-3

Markus, D, Volkmar, M, Thomas A, Runkler AS, Klaus V (2005) Method and installation for processing waste paper

Marsh K, Bugusu B (2007) Food packaging—roles, materials, and environmental issues. J Food Sci 72(3):R39–R55. https://doi.org/10.1111/j.1750-3841.2007.00301.x

Miao Y, Xiang S, Wei Y, Long X, Qiu J, Miao Y (2023) Physical properties of pulp and paper: a comparison of forming procedures. For Prod J 73(2):175–185. https://doi.org/10.13073/FPJ-D-23-00007

Mikhailidi AM, Kotelnikova NE (2021) Horizons for future sustainability: from trash to functional cellulose fibres. CRC Press, Boca Raton, pp 91–130. https://doi.org/10.1201/9780429347993-4

Ministry of the Environment, Japan (2021) Waste management and recycling. https://www.env.go.jp/en/recycle/recycling.html

Modupe E, Ojewumi M, Emetere OR, Damilola O, Elizabeth B, Ebuka G, Adimekwe (2022) Bio-conversion of waste paper into fermentable sugars—a review. Front Chem Eng 4:926400. https://doi.org/10.3389/fceng.2022.926400

Mokatse KMP, van Wyk JPH (2021) Physical and chemical pre-treatment of waste paper to increase saccharification by *Trichoderma viride* cellulase. Sefako Makgatho Health Sciences University, Ga-Rankuwa. https://doi.org/10.30955/gnj.002964

Ojewumi ME, Obielue BI, Emetere ME, Awolu OO, Ojewumi EO (2018) Alkaline pre-treatment and enzymatic hydrolysis of waste papers to fermentable sugar. J Ecol Eng 19(1):211–217. https://doi.org/10.12911/22998993/79404

Ouadi M, Brammer JG, Kay M, Hornung A (2013) Fixed bed downdraft gasification of paper industry wastes. Appl Energy 103:692–699. https://doi.org/10.1016/j.apenergy.2012.10.038

Pactap (n.d.) Why grocery stores are switching to paper bags. https://pactap.com/why-grocery-stores-are-switching-to-paper-bags/#:~:text=In%20the%20quest%20for%20sustainable,resonates%20with%20eco%2Dconscious%20consumers. Accessed 12 January 2025

Papericious (2024) Speciality paper. https://papericious.com/collections/speciality-paper?srsltid=AfmBOop7jv5ayZmiLWMlcpcm3O1FUJGY7TWfnbcH67cQbIlClmDVgUDd

Paperpapers (2024) What is insulating papers. https://www.paperpapers.com/news/what-is-insulating-paper/?srsltid=AfmBOopp2dBdkAxhP8sGxOOx616BynThIspr2VL7YtQn-dFrlxA3ZKR

Pecar D, Andreja G (2015) Enzymatic treatment of paper sludge. Chem Eng Trans 43:631–636

Pego MFF, Bianchi ML, Yasumura PK et al (2022) Physical and mechanical properties of papers produced from different cellulosic pulps blending (version 1). Res Sq. https://doi.org/10.21203/rs.3.rs-2257060/v1

Preservation Equipment (2024). https://www.preservationequipment.com/Catalogue/Conservation-Materials/Paper-Board/What-is-acid-free-paper

Primepac (2021) Types of kraft paper and their uses. https://info.primepac.co.nz/blog/types-of-kraft-paper-and-their-uses
Rizwan K, Rasheed T, Bilal M, Iqbal HMN (2022) Pulp and paper industry-based pollutants, and their adverse impacts. In: Bhat R, Kumar A, Nguyen TA, Sharma S (eds) Nanotechnology in paper and wood engineering. Elsevier, Amsterdam, pp 143–160. https://doi.org/10.1016/B978-0-323-85835-9.00005-2
Rodriguez C, Alaswad A, El-Hassan Z, Olabi AG (2017) Mechanical pretreatment of waste paper for biogas production. Waste Manag 68:157–164. https://doi.org/10.1016/j.wasman.2017.06.040
Ron Chen (2023) 4 popular types of paperboard grades for packaging. Pak factory blog. https://pakfactory.com/blog/4-popular-types-of-paperboard-grade-for-box-printing/?srsltid=AfmBOoqM46Yh4Iy7_p0AbACbXkpE-DGDCiIuu5A5bK2RmPI8MH4isZsD
Rppaperimpex (2024) Exploring the chemistry behind paper composition. https://www.rppaperimpex.com/exploring-the-chemistry-behind-paper-composition/
Saloni Nagvekar (2024) A comprehensive guide to paper types & printing essentials. https://www.printstop.co.in/blog/a-comprehensive-guide-to-paper-types-and-printing-essentials#:~:text=1.-,Bond% 20Paper,%2C% 20legal%20forms%2C%20and%20stationery
Scott Kravitz (2021) Coated vs. uncoated paper: when to use (or not use) each. https://ironmarkusa.com/coated-uncoated-paper/
Security Printing (2024). https://www.a1securityprint.com/security-paper/
Siti R, Hassan Y-T, Irvan D, Hamidi A, Aziz A (2022) Kinetic study of the anaerobic digestion of recycled paper mill effluent (RPME) by using a novel modified anaerobic hybrid baffled (MAHB) reactor. Water 14(3):390–390. https://doi.org/10.3390/w14030390
Sivakumar N (2022) High-solid loading enzymatic hydrolysis of waste office paper for poly-3-hydroxybutyrate production through simultaneous saccharification and fermentation. J Polym Environ 30(7):3045–3054. https://doi.org/10.1007/s10924-022-02392-7
Strathmore (2025) Recycled paper | pre vs. post consumer fibers. https://www.strathmoreartist.com/blog-reader/recycled-paper-pre-vs-post-consumer-fibers
Vápenka L, Vavrouš A, Votavová L, Kejlová K, Dobiáš J, Sosnovcová J (2016) Contaminants in the paper-based food packaging materials used in The Czech Republic. J Food Nutr Res 55(4):361–373
Veluchamy C, Kalamdhad AS (2017) Enhancement of hydrolysis of lignocellulose waste pulp and paper mill sludge through different heating processes on thermal pretreatment. J Clean Prod 168:219–226. https://doi.org/10.1016/j.jclepro.2017.09.040
Wang Y, Marcello C, Sawant N et al (2023) Identification and characterization of sticky contaminants in multiple recycled paper grades. Cellulose 30:1957–1970. https://doi.org/10.1007/s10570-022-05006-6
Wast Control (2024) Paper waste management: key considerations for businesses. https://wastecontrolinc.com/2024/02/28/paper-waste-management/
Waste Control (2024) Paper waste 101: building a waste management process with waste control. https://wastecontrolinc.com/2024/02/28/paper-waste/
World Bank Group (2018). https://www.worldbank.org/en/news/press-release/2018/09/20/global-waste-to-grow-by-70-percent-by-2050-unless-urgent-action-is-taken-world-bank-report
Worldpapermill (2020) 56 best chemicals used in pulp & paper mills (tried & tested). https://worldpapermill.com/chemicals-paper-mill/
Xu H, Li L, Tang W, Ang ZS, Chen Y, Sun G, Gu Q, Duan L (2023) Experimental study on the combustion behavior and NOx emission during the co-combustion of combustible industrial solid wastes. J Energy Inst. https://doi.org/10.1016/j.joei.2022.11.010
Yaman G, Sinoplugil H (2019) A pilot study for the pyrolysis of the urea formaldehyde-melamine formaldehyde resin paper waste. Therm Sci 23:253–266. https://doi.org/10.2298/TSCI181023039Y
Yin C-Y, Kadir SASA, Abdul-Malik AS, Syed-Ariffin SN, Mahazer SZ, Zamzuri Z (2007) Characterization of ash derived from combustion of paper mill waste sludge: comparison with municipal solid waste incinerator ash. ScienceAsia. https://doi.org/10.2306/scienceasia1513-

Yuan X, Cao Y, Li J, Wen B, Zhu W, Wang X, Cui Z (2012) Effect of pretreatment by a microbial consortium on methane production of waste paper and cardboard. Bioresour Technol 118:281–288. https://doi.org/10.1016/j.biortech.2012.05.058

Zainab T, Yaqub A, Bilainu A, Oboirien A (2020) Process modelling of chemical looping combustion of paper, plastics, paper/plastic blend waste, and coal. ACS Omega. https://doi.org/10.1021/ACSOMEGA.0C02880

Zhang Y, Duan C, Bokka SK, He Z, Ni Y (2022) Molded fiber and pulp products as green and sustainable alternatives to plastics: a mini review. J Bioresour Bioprod 7(1):14–25. https://doi.org/10.1016/j.jobab.2021.12.002

Zuhara S, Pradhan S, McKay GA (2023) Pyrolysis of biosolids with waste cardboard: effect of operating parameters, feedstock size and blending ratio. Int J Environ Sci Technol. https://doi.org/10.1007/s13762-023-04963-0

Conversion of Poultry Waste to Biofuels and Chemicals

7

Nishika Sharma, Shrestha Vashista, Surbhi Sharma, Abha Kumari, and Ayesha Sadaf

Abstract

Poultry waste is a substantial and mainly unexplored resource for the production of valuable products. This chapter investigates the viability of using several thermochemical and biochemical processes to transform poultry waste—which is mostly made up of feathers, manure, and other organic residues—into biofuels, biodiesel, biogas, and bio-oil. Poultry waste is a major source of proteins, carbohydrates, fats, that have the potential to be converted into biofuels like bio-oil, biogas (methane), bioethanol, biochar, and other chemicals like struvite and heme iron.

This chapter highlights the potential of poultry waste as a cost-effective and sustainable source for biofuel production. The efficiency and scalability of producing biofuel from poultry waste can be improved by further development of processing technologies and integration into existing waste management systems, which would contribute to a more sustainable future.

Keywords

Poultry waste valorization · Pyrolysis · Gasification · Transesterification · Hydrothermal liquefaction

N. Sharma · S. Vashista · S. Sharma · A. Kumari
Amity Institute of Biotechnology, Amity University, Noida, Uttar Pradesh, India

A. Sadaf (✉)
Interdisciplinary Biomedical Research Centre, School of Science and Technology, Nottingham Trent University, Nottingham, UK

A. Kumari et al. (eds.), *Valorization of Solid Wastes to Biofuels and Chemical Products for Sustainable World*, https://doi.org/10.1007/978-981-96-8594-3_7

7.1 Introduction

The manufacturing of biofuel from poultry waste has become an inventive method of producing renewable energy. Feathers, fat, and other components included in poultry manure are plentiful and can be difficult to dispose of, causing waste problems. These waste materials can be turned into valuable biofuels, such as biodiesel, biogas, or bioethanol, which can contribute to sustainable waste management and provide an environmentally responsible source of energy for the world (Seidavi et al. 2018).

Biofuels made from poultry by-products, like biogas, bioethanol, biodiesel, and bio-oil, offer a sustainable and renewable substitute to conventional fossil fuels, aiding in the decrease of greenhouse gas emissions and enhancing energy security. Techniques like anaerobic digestion, hydrothermal liquefaction, and fermentation have been fine-tuned to transform poultry waste efficiently into these various forms of biofuels (Kantarli et al. 2018). Incorporating such processes within the poultry waste management practices can foster a circular economic approach, where waste not only gets reduced but also serves as a resource for energy production, cutting down on environmental footprint, and generating economic benefits (Fig. 7.1).

The release of large amounts of waste such as excreta, bedding material, feathers, and offal is the major cause of significant amounts of land and water pollution (Thyagarajan et al. 2013). However, all these wastes can be processed into valuable products by different methodologies. Anaerobic digestion has been an efficient method for poultry waste management, contributing to biofuel production and nutrient recovery (Xu et al. 2015). Microbial bioconversion techniques utilizing enzymes such as lipases, proteases, and keratinases have also facilitated the transformation of

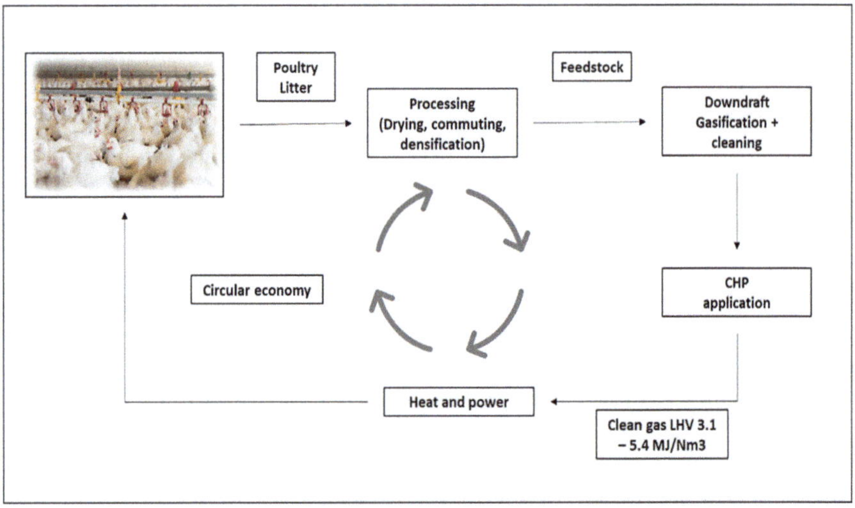

Fig. 7.1 Green circular economy implementation through the production of energy from poultry waste

poultry waste into valuable feed, fertilizers, and bioactive compounds (Brandelli et al. 2015). Fast pyrolysis of poultry waste into liquid biofuels has been considered in both non-catalytic and catalytic modes. Comparatively, poultry meal gives a higher yield of bio-oil compared to poultry litter, though the types of catalysts used, such as MgO and ZSM-5, can alter the composition of the product bio-oil (Kantarli et al. 2018). These techniques are thus viable solutions for the management of poultry waste in an environmentally friendly way while producing biofuels and valuable chemicals.

Some of the opportunities that a circular economy approach may create vis-à-vis poultry production are waste reduction, environmental impacts, and value creation. Various aspects of circularity have been studied concerning the poultry sector, relating to nitrogen flow analysis (Harchaoui et al. 2023), composting of wastes (Chiarelotto et al. 2021), and efficient technologies development in manure handling (Bartolacci et al. 2023). An efficient circular economy can be created based on three key strategies: improvement in the efficiency of nitrogen use, recycling manure and by-products, and use of biowaste streams in feed rations. However, composting poultry waste into agricultural input has its challenge of balancing the carbon-to-nitrogen ratio. The adoption of circular business models would go a long way in aiding small and medium-scale farms to reduce their emissions while converting the manure into energy and fertilizers. Poultry production can also be integrated with other types of agriculture, for example, bee-pollinated legumes, to derive a sustainable value chain that maintains biodiversity (Rauw et al. 2022).

There are various methods of disposal, but poultry waste management is still a major challenge to the industry. Windrow composting offers an environmentally safe method of disposing of poultry manure and mortalities (Vinodkumar et al. 2014). Nitrogen conservation and carbon loss reduction during decomposition in poultry wastes may be achieved by co-composting with paddy straw (Devi et al. 2012). Long-term application of poultry wastes to soil can lead to accumulation of heavy metals such as copper and zinc that may pose an environmental hazard (Han et al. 2000). In this respect, anaerobic digestion is also one of the feasible methods for poultry waste treatment; efficiency partly depends on how longer the accumulation time of wastes is. Longer accumulation periods are characterized by a decrease in the volatile solids content and further have less biogas production potential (Farias et al. 2012). Composting and anaerobic digestion processes significantly reduce pathogenic bacteria, including total and thermotolerant coliforms, in the poultry waste (Devi et al. 2012; Farias et al. 2012).

This chapter provides comprehensive learning on biofuel generation from poultry waste, wherein wastes like poultry fat, manure, and poultry food wastes are converted into bio-oil, crude oil, biogas like methane, and are also converted into chemicals like struvite, bioethanol, etc. The goal is to understand energy-rich chemicals from poultry waste using important conversion methods such as transesterification, pyrolysis, anaerobic digestion, hydrolysis, etc. These topics will be dealt with in detail in the upcoming sections.

7.2 Poultry Waste Generation and Composition

Poultry waste can be generated in various ways, which include poultry food waste, manure waste, farm waste, poultry fat, etc.

Poultry waste has been one of the major environmental problems in the world. In Nigeria, for example, a study estimated that 159,430 metric tons of poultry wastes, excluding slaughterhouse and hatchery wastes, are generated yearly in poultry farms in Minna (Adeoye et al. 2014). In the United States, an estimated 8.8 million tons of dry poultry fecal wastes were generated in 1989, aside from hatchery wastes and mortality birds (Pope and Cherry 1991).

India alone is estimated to produce 38.33 million tons/year of poultry manure that could fertilize 3.56 million hectares of farmland. Poultry waste is rich in high amounts of nutrients as well as pathogens (Prabakaran and Valavan 2021).

Wastes from Nigerian farms were found to contain nitrates as high as 206.75 mg/l and 48,214 CFU/100 mg of total coliform bacteria (Chukwu et al. 2011).

Raising chickens, processing eggs, and butchering birds are just a few of the processes that result in the production and processing of poultry, which also produces waste.

Waste from poultry encompasses all types of manure, litter, viscera, muscle, fat, bone, blood, and feathers (Fig. 7.2). Various valuable nutrients, in particular nitrogen and phosphorus, are contained within this type of waste; land application, therefore, is considered a realistic option for fertilizer as well as for soil amendment. There are, however, associated environmental concerns with such kinds of applications, like nutrient leaching and contamination (Table 7.1).

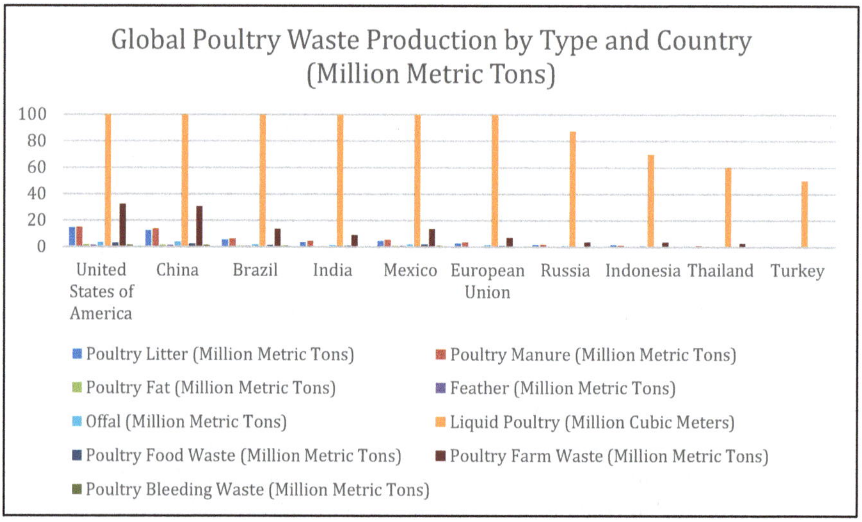

Fig. 7.2 Global poultry waste production by type and country (million metric tons) with data sourced from the FAO (Food and Agriculture Organization), USDA (United States Department of Agriculture), and World Bank reports

Table 7.1 Biofuel production from poultry waste

Type of poultry waste	Biofuel	Technique	Conditions	Yield	Reference
Poultry litter 1 ton	Biogas	Anaerobic digestion	Thermophilic conditions (50–55 °C)	20–60 m^3/ton	Bhatnagar et al. (2022)
Poultry fat (rendering waste)—1 ton	Biodiesel	Esterification and transesterification	Esterification around 70 °C, transesterification: 50–60 °C	800–950 kg/ton	Maf et al. (2022)
Feathers and offal (processing waste)—1 ton	Bio-oil (pyrolysis oil)	Pyrolysis	300 °C and 800 °C	300–400 kg/ton	Maf et al. (2022)
Poultry blood—100 kg	Hydrogen and syngas	Pyrolysis and gasification	700–900 °C	Hydrogen yield = 10–20 kg/h Syngas yield 70–90 kg/h	Tasnia Hassan Nazifa et al. (2022)
Poultry (wastewater) 1 m^3	Biogas (methane)	Anaerobic digestion	Thermophilic conditions: 52 °C	0.5–1.0 m^3/m^3	Gangani Rao et al. (2012)
Chicken feather—100 kg	Biodiesel	Transesterification	Chloroform methanol(2:1,V/V)	10–20 kg/100 kg	Purandas et al. (2018)
Poultry farm waste—100 kg	Bio-oil	Thermal depolymerization	200–350 °C (392–662 °F)	30–50 kg/100 kg	Laird et al. (2018)
Poultry manure and food waste—m^3	Bio-gas	Anaerobic digestion	Mesophilic range: 30–40° pH 6.8 and 7.5	0.5–1.0 m^3/m^3	Massé et al. (2018)
Poultry litter—1 kg	Ethanol production	Hydrolysis	pH: 4.8–5.5 Temp: 45–50 °C Time: 24–72 h	0.1–0.3 L/kg	Cantrell et al. (2008)
Poultry manure—100 kg	Bio-crude oil	Catalytic hydrothermal liquefaction	Temp: 250 and 350 °C Pressure: 10–20 MPa	30–50 kg/100 kg	Wang et al. (2013)

(continued)

Table 7.1 (continued)

Type of poultry waste	Biofuel	Technique	Conditions	Yield	Reference
Poultry litter—1 kg	Ethanol	Enzymatic hydrolysis	160–220 °C pressure (0.69–4 MPa)	0.1–0.3 L/kg	Sarkar et al. (2012)
Liquid poultry waste—100 kg	Bio-crude oil	Catalytic hydrothermal liquefaction	250–350 °C Pressure: 10–20 MPa	20–40 kg/100 kg	Chen et al. (2014)
Poultry litter—100 kg	Hydro-char	Hydrothermal carbonization	Temperature: 180–250 °C Pressure: 2–6 MPa	40–60 kg/100 kg	Libra et al. (2011)
Poultry litter—1 kg	Bioethanol	Enzymatic hydrolysis	Temp: 120–180 °C (15–30 min, 0.69–4 MPa)	0.05–0.15 L/kg	Kumar et al. (2009)

7.3 Products from Poultry Waste

7.3.1 Biogas

Biogas production from poultry waste is a research study that has been conducted, based on the concepts of sustainable waste management and energy production. It has been determined that poultry waste gives a higher yield of biogas compared to other substrates like banana peels and cow dung (Usman et al. 2023; Adeniran et al. 2014). Mixing the poultry waste with cattle feces in a 1:1 ratio improved the biogas production (El-Hadidi and Al-turki 2007). In an anaerobic digestion process, apart from biogas production, nutrient-rich effluent was produced, which is suitable for organic fertilizer (El-Hadidi and Al-turki 2007). Optimization of biogas production from poultry manure wastewater was done with the help of response surface methodology, and the optimum was found at 120 rpm agitation and a reaction time of 3.3 days (Chun et al. 2015). The parameters that were considered in this respect are pH, temperature, and substrate concentration (Usman et al. 2023). All these studies concluded that poultry waste is one of the promising feedstock for biogas production, not only from an environmental point of view but also from an economic one.

7.3.2 Biodiesel

Biodiesel from poultry waste was introduced as a substitute for fossil fuels in recent years. Production of biodiesel using chicken fat and skin wastes as raw materials through transesterification has been reported in various studies (Nagaraja et al. 2014; Moreira et al. 2010; Abid et al. 2018). The maximum production of biodiesel from chicken fat under optimum conditions using a 1:9 molar ratio of fat to methanol at a temperature of 35 °C, reaction time of 45 min, and 1% KOH as catalyst was 94.76% (Nagaraja et al. 2014). The biodiesel thus produced under these conditions usually met ASTM and EN 14214 standards with regard to the main properties such as density, viscosity, and flash points (Nagaraja et al. 2014; Moreira et al. 2010; Abid et al. 2018). Moreover, the production of biodiesel by poultry slaughter waste eliminates environmental pollution problems and can supply a considerable percentage of the diesel demand. For example, poultry slaughter wastes in Iran have the potential to annually produce around 112 million liters of biodiesel (Mahyari et al. 2021).

7.3.3 Bioethanol

Most recently, very promising results have been obtained by producing bioethanol from poultry waste. Hydrolysis of poultry manure using dilute sulfuric acid, under optimal conditions of 0.8M H_2SO_4 at 100 °C for 30 min, gave a maximum glucose concentration of 115.6 g/L (Woldesenbet et al. 2013). A glucose conversion of 7.1% was increased up to 27.6% by treating the substrate with different pretreatment methods, such as alkaline treatment using KOH (Champagne and Li 2008), followed by drying and grinding. Bio-oil was also produced through fast pyrolysis of poultry waste, with higher amounts produced from poultry meal as compared with poultry litter (Kantarli et al. 2018). The potential production of bioethanol from both agricultural and organic waste in Indonesia is very huge and estimated at about 11,880,641.29 and 72,511.2 kl from agricultural and organic waste, respectively (Susmiati 2018). To enhance the bioethanol production from these wastes, various technologies like separate hydrolysis and fermentation, simultaneous saccharification and fermentation, and consolidated bioprocessing are employed.

7.3.4 Hydrogen

During recent years, many scientists have been attempting to produce hydrogen from poultry waste through various techniques. It has been shown that the pretreatment and substrate concentration of sludge at an optimum condition can achieve high anaerobic fermentation production from poultry dung wastewater (Qian 2012). Co-digestion of sewage sludge, wine vinasse, and poultry manure resulted in a promising H_2 yield; the highest production was identified to be at 10 g/L poultry manure (Sillero et al. 2021). The anaerobic ponds of poultry slaughterhouses have been found to serve as efficient inoculum sources for the production of biohydrogen through the use of different substrates, sucrose, and glycerol (Cripa et al. 2019). In another study, chicken feather waste was utilized for bioenergy production through anaerobic digestion and bio-electrochemical systems (El Salamony et al. 2024). The feather hydrolysate showed promising methane and hydrogen production and electricity generation in microbial fuel cells and improved hydrogen production in microbial electrolysis cells. Such findings bring various methodologies being applied for poultry wastes in the interest of hydrogen production in the direction of sustainability.

7.4 Chemicals Generation from Poultry Waste

Poultry waste can be converted into value-added products by using a number of methods. Fast pyrolysis and hydrothermal conversion of chicken litter yielded a phenol-rich, high-oxygenated-compound-content, fatty acid-containing bio-oil (Pachón Gómez et al. 2022). Biochars from pyrolysis and hydrothermal carbonization of poultry litter have been found to contain proximate compositions like coal, and hydrothermal liquefaction of poultry meal yielded a promising bio-crude oil, likely suitable as a synthetic fuel. Composting of poultry waste with maize straw yielded a product meeting the minimum requirements of organic fertilizers. Hatchery waste meal and poultry by-product meal have been shown to be a good proteinous feed for animals (Prabakaran and Valavan 2021). The dried poultry manure has the potential to reduce the application of chemical fertilizers. Chicken fat can be utilized for the production of biodiesel by transesterification. The above processes are the ecology-friendly methods for poultry waste management and resource recovery.

7.4.1 Liquid Fertilizers

Liquid fertilizers from poultry wastes have the potential to represent a sustainable alternative for nutrient management in agriculture. Effluent produced during the thermophilic anaerobic digestion of poultry litter can be separated into solid and liquid fractions. The latter one would contain a very high level of nitrogen (Liedl et al. 2006). Subcritical water treatment of chicken feathers resulted in a liquid fertilizer with high levels of nitrogen and phosphorus, thus enhancing plant and algae growth (Zul et al. 2020). The liquid fertilizer obtained from bacterial fermentation of poultry litter and feathers was found to be rich in nitrogen in the form of amino acids and oligopeptides. The latter contributed to improved crop yield and drought tolerance (Sahoo et al. 2022). The liquid fertilizers obtained have been shown to give comparable or superior results compared with commercial fertilizers for various crops. Processing of poultry waste into liquid fertilizers offers an environmentally friendly solution for waste management, thereby reducing the cost of production and environmental risk resulting from poultry farming (Buzetti and Ivanov 2020).

7.4.2 Solid Fertilizers

Anaerobic digestion of poultry waste presents an efficient waste-to-resource approach in producing valuable fertilizers (Buzetti and Ivanov 2020; Silva et al. 2009). Both solid and liquid fractions are provided as end products, whereby phosphorus is present in the solid form and nitrogen in the liquid one. Application of such solid-type fertilizers derived from poultry wastes can result in reduced crop yield during the initial period, though it may increase soil phosphorus with time, while liquid effluents have produced results equal to or even superior to those of commercial fertilizers for many crops (Liedl et al. 2006). Treated poultry feather waste, because of its high content of organic matter and nitrogen, presents positive effects on plant growth, mainly in application rates starting at 12 t/ha (Joardar and Rahman 2018). These methods thus realize the efficacy in nutrient management and reduce the environmental risk attributed to untreated poultry wastes; hence, they are an environmentally friendly approach in managing wastes from the poultry industry (Buzetti and Ivanov 2020; Silva et al., 2009).

7.4.3 Biochar

Biochar derived from poultry waste is a potentially valuable material for a variety of agricultural applications. It can provide a bulking agent in composting, enhancing organic matter degradation and humification, and reducing losses of nitrogen (Dias et al. 2010). Biochar from crop residues modified with alum sludge gives good adsorption of nutrients from poultry wastewater and opens up possibilities for producing slow-release fertilizers (Thanh et al. 2023). Biochars derived from poultry manure are chemically composed of many elements and oxygen functionalities; hence, they can work both as fertilizer components and soil improvers (Dróżdż et al. 2020). However, they might contain high quantities of ammonium, volatile fatty acids, and nitrogen-containing organic compounds, which could potentially cause phytotoxicity. Such toxicities can be considerably reduced by solvent extraction or microbial treatment (Rombolà et al. 2019). Another important factor that would determine some of the properties of the biochar is the temperature of pyrolysis (Table 7.2). Generally, higher temperatures could lead to lower yields, but increasing the aromatization helps in maintaining the stability and nutrient retention capacity of the biochar (Dróżdż et al. 2020).

7.5 Challenges and Future Prospects

Harnessing poultry industry byproducts for the generation of sustainable fuel alternatives marks an innovative approach to renewable energy production. The abundant presence of feathers, fat, and other materials in poultry waste poses a disposal challenge. Converting these waste materials into biofuels, such as biodiesel, biogas, or bioethanol, offers a solution for waste management while providing a green energy source globally. It is estimated that poultry excreta can be used as an excellent source to produce nitrogen. Researchers have determined that approximately 3.29 million metric tons of nitrogen can be generated from the processing of meat chickens (based on an average of two estimates with differing assumptions of 2.65 and 3.94 million metric tons) (Seidavi et al. 2018) and 2.36 million metric tons from laying hens. The most common forms of poultry waste include poultry litter (manure and bedding), dissolved air flotation (DAF) sludge from poultry processing plants, and composts from bird carcasses and hatchery refuse. The primary poultry producers include broiler chicken, turkey, and egg operations, typically utilizing either floor/litter or caged pit housing systems.

Advancements in processing technologies and their integration with existing waste management practices could enhance the efficiency and scalability of biofuel production from chicken waste, contributing to a more sustainable energy future. With a number of developments and trends suggesting more practicality and uptake, the future of producing biofuel from chicken waste looks promising.

Research is continuously being conducted to optimize biofuel generation from poultry manure. Technological developments in anaerobic digestion, pyrolysis, gasification, and transesterification are anticipated to boost productivity and save expenses. Future systems may incorporate many processes, such as pyrolysis and anaerobic digestion, to enhance energy recovery from chicken waste and reduce residues. This idea of the biorefinery concept is gaining popularity. It involves using a single feedstock to create a variety of outputs, such as fertilizers, biogas, and biofuels. Carbon neutrality also plays a major role as biofuels made from poultry manure offer a carbon-neutral or carbon-negative energy alternative, particularly

Table 7.2 Chemical production from poultry waste

Type of waste	Chemical generated	Technique	Conditions	Yield	References
Poultry litter (manure and bedding)—1 ton	Biochar	Pyrolysis	No supply of oxygen, temperature range: 300–700 °C	300–400 kg/ton	Prabakaran and Valavan 2021
Poultry fat (rendering waste)—1 ton	Glycerol	Transesterification	Temperature 60–70 °C with a catalyst	100–120 kg of crude glycerol	Prabakaran and Valavan (2021)
Feathers and offal (processing waste)—1 ton	Keratin	(1) hydrolysis (2) enzymatic hydrolysis	Temperature 90–120 °C for 2–4 h At 50–60 °C	600–800 kg of keratin/1.000 kg of feathers	Kopeć et al. (2018)
Poultry blood—100 kg	Heme iron	Chemical treatment followed filtration process and drying	Separation and purification	30–50 kg of heme iron/1000 kg of fresh blood	Thyagarajan et al. (2013)
Poultry (wastewater) 1 m^3	Struvite (magnesium ammonium phosphate)	Precipitation	pH 8.0–9.5, ambient temperature is preferred	3–5 kg of struvite/1 m^3 of poultry wastewater	Kopeć et al. (2018)
Chicken feather –100 kg	Feather meal (high-protein animal feed)	Steam hydrolysis or rendering	150–180 °C pressure for 30 to 60 min	700–800 kg of feather meal per 1000 kg of chicken feathers	Prabakaran and Valavan (2021)
Poultry farm waste—100 kg	Ammonia	Ammonia stripping and scrubbing	pH 9.5 to 10.5, temperature 60–80 °C	5–10 kg of ammonia/1 metric ton of poultry manure	Prabakaran and Valavan (2021)
Poultry manure and food waste—m^3	Methane	Co-digestion (anaerobic digestion)	35–55 °C temperature for mesophilic conditions	30–50 m^3 of biogas per 1 metric ton of waste	Quiroga et al. (2010)

when contrasted with fossil fuels, due to the growing emphasis on reaching carbon neutrality.

7.6 Conclusion

The outlook for generating biofuels from poultry waste is promising, offering a considerable opportunity to support renewable energy objectives, enhance waste management practices, and advance sustainability efforts. The growth in this area is driven by ongoing technological progress, favorable regulations, and creative business strategies, all of which are crucial for transforming biofuels derived from poultry waste into a widely used energy option. The conversion of poultry waste into biofuels presents a compelling opportunity to address both waste management challenges and renewable energy needs. By utilizing advanced technologies such as anaerobic digestion, pyrolysis, and fermentation, poultry waste can be transformed into valuable biofuels like biogas, bioethanol, and biodiesel. This process not only reduces the environmental impact of waste disposal but also contributes to a sustainable energy future. Transforming poultry waste into biofuels within a circular economy context enhances the utility of resources, converting waste into a valuable commodity. This shift not only bolsters rural economies but also cuts down on greenhouse gases, contributing to worldwide carbon neutrality goals. Nonetheless, unlocking the full promise of poultry waste-to-biofuel processes requires ongoing research, innovation, and policy support. Tackling economic and technical barriers, refining these processes, and adhering to regulations will be essential for expanding these solutions. Therefore, transforming poultry waste into biofuels can play a crucial role in sustainable waste handling and renewable energy approaches, ushering in a cleaner, more sustainable future.

References

Abid M, Touzani A, Benhima R (2018) Synthesis of biodiesel from chicken's skin waste by homogeneous transesterification. Int J Sustain Eng 12:272–280

Adeniran KA, Ahaneku IE, Itodo IN, Rohjy HA (2014) Relative effectiveness of biogas production using poultry wastes and cow dung. Agric Eng Int CIGR J 16:126–132

Adeoye A et al (2014) New insights into valorization of agro-industrial wastes for production of citric acid: effects of mutation and optimization—a review. Eur J Sci Innov Technol 2(2):100–110. https://www.researchgate.net/publication/366205415

Bartolacci F, Marcantoni M, Usci R (2023) How to turn poultry manure into valuable resources: a circular business model for resilient and sustainable small and medium-sized farms. J Manage Sustain 13:2

Bhatnagar A et al (2022) Valorization of poultry waste through anaerobic digestion: a sustainable approach for waste management and energy production. Renew Sust Energ Rev 153:111763

Brandelli A, Sala L, Kalil SJ (2015) Microbial enzymes for bioconversion of poultry waste into added-value products. Food Res Int 73:3–12. https://doi.org/10.1016/j.foodres.2015.01.015

Buzetti K, Ivanov MV (2020) Production of liquid and granulated organic fertilizers from poultry farm wastes. Agr Sci. https://doi.org/10.32634/0869-8155-2020-335-2-61-63

Cantrell KB et al (2008) Livestock waste-to-bioenergy generation opportunities. Bioresour Technol 99(17):7941–7953
Champagne P, Li C (2008) Bio-product recovery from lignocellulosic materials derived from poultry manure. Bull Sci Technol Soc 28:219–226
Chen W-T, Zhang Y, Zhang J, Yu G, Schideman LC, Zhang P, Minarick M (2014) Hydrothermal liquefaction of mixed-culture algal biomass from wastewater treatment system into bio-crude oil. Bioresour Technol 152:130–139
Chiarelotto M et al (2021) Three months of composting are enough to eliminate the fowl typhoid agent salmonella Gallinarum from naturally infected broiler carcasses. Braz J Poul Sci 23(2):1441. https://www.scielo.br/j/rbca/a/R9ggWmn8YLQGMjtLYm8vKYH/
Chukwu O, Adeoye PA, Chidiebere IN (2011) Abattoir wastes generation, management and the environment: a case of Minna, north Central Nigeria. Int J Biosci 1:100–109
Chun C, Jamaludin NF, Zainol N (2015) Optimization of biogas production from poultry manure wastewater in 250 mL flasks. J Teknol 75(1):275–285
Cripa FB, Arantes MK, Sequinel R, Fiorini A, Rosado FR, Alves HJ (2019) Poultry slaughterhouse anaerobic ponds as a source of inoculum for biohydrogen production. J Biosci Bioeng 129:77–85
Devi S, Sharma CR, Singh K (2012) Microbiological biodiversity in poultry and paddy straw wastes in composting systems. Braz J Microbiol 43:288–296
Dias BD, Silva CA, Higashikawa FS, Roig A, Sánchez-Monedero MA (2010) Use of biochar as bulking agent for the composting of poultry manure: effect on organic matter degradation and humification. Bioresour Technol 101(4):1239–1246
Dróżdż D, Wystalska K, Malińska K, Grosser A, Grobelak A, Kacprzak M (2020) Management of poultry manure in Poland - current state and future perspectives. J Environ Manag 264:110327. https://doi.org/10.1016/j.jenvman.2020.110327
El Salamony DH, Hassouna MS, Zaghloul TI, He Z, Abdallah HM (2024) Bioenergy production from chicken feather waste by anaerobic digestion and bioelectrochemical systems. Microb Cell Factories 23:102
El-Hadidi YM, Al-turki A (2007) Organic fertilizer and biogas production from poultry wastes. J Food Agric Environ 5:228–233
Farias RM, Junior MA, Orrico AC, Garcia RG, Centurion SR, Fernandes AR (2012) Biodigestão anaeróbia de dejetos de poedeiras coletados após diferentes períodos de acúmulo. Cienc Rural 42:1089–1094
Han FX, Kingery W, Selim HM, Gerard PD (2000) Accumulation of heavy metals in a long-term poultry waste-amended soil. Soil Sci 165:260–268
Harchaoui S, Blazy V, Péchenart E, Wilfart A (2023) Challenges and opportunities for improving circularity in the poultry meat and egg sector: the case of France. Resour Conserv Recycl 193:106963
Joardar J, Rahman MM (2018) Poultry feather waste management and effects on plant growth. Int J Recycl Org Waste Agric 7:183–188
Kantarli IC, Stefanidis SD, Kalogiannis KG, Lappas AA (2018) Utilisation of poultry industry wastes for liquid biofuel production via thermal and catalytic fast pyrolysis. Waste Manag Res 37:157–167
Kopeć M, Gondek K, Mierzwa-Hersztek M, Antonkiewicz J (2018) Factors influencing chemical quality of composted poultry waste. Saudi J Biol Sci 25:1678–1685. https://doi.org/10.1016/j.sjbs.2016.09.012
Kumar R, Singh S, Singh OV (2009) Bioconversion of lignocellulosic biomass: biochemical and molecular perspectives. J Ind Microbiol Biotechnol 35(5):377–391. https://doi.org/10.1007/s10295-008-0327-8
Laird DA et al (2018) Impact of biochar amendments on the quality of a typical midwestern agricultural soil. Geoderma 158(3–4):443–449
Libra JA, Ro KS, Kammann C, Funke A, Berge ND, Neubauer Y, Titirici MM, Fühner C, Bens O, Kern J, Emmerich K (2011) Hydrothermal carbonization of biomass residuals: a compara-

tive review of the chemistry, processes, and applications of wet and dry pyrolysis. Biofuels 2(1):71–106. https://doi.org/10.4155/bfs.10.81
Liedl BE, Bombardiere J, Chatfield JM (2006) Fertilizer potential of liquid and solid effluent from thermophilic anaerobic digestion of poultry waste. Water Sci Technol 53(8):69–79
Maf NS et al (2022) Conversion of poultry waste into biochar: characterization and application for soil amendment. J Anal Appl Pyrolysis 161:105364
Mahyari ZF, Khorasanizadeh Z, Khanali M, Mahyari KF (2021) Biodiesel production from slaughter wastes of broiler chicken: A potential survey in Iran. SN Applied Sciences 3:57. https://doi.org/10.1007/s42452-021-04420-1
Massé V et al (2018) Treatment of swine and poultry slaughterhouse wastewater in an anaerobic sequencing batch reactor. Can Agric Eng 50(2):195–200
Moreira AL, Dias JM, Almeida MF, Alvim-Ferraz MC (2010) Biodiesel production through transesterification of poultry fat at 30 °C. Energy Fuel 24:5717–5721
Nagaraja Y, Venkatesh H, Vinay M (2014) Production of biodiesel from poultry waste. Int J Pharm Bio Sci 5:4
Nazifa TH et al (2022) Sustainable energy recovery from poultry waste: a review on the potential of anaerobic digestion and biochar production. Energies 15(3):987
Pachón Gómez EM, Domínguez RE, López DA, Téllez Bernal JF, Marino MD, Almada NS, Gange JM, Moyano EL (2022) Chicken litter: A waste or a source of chemicals? Fast pyrolysis and hydrothermal conversion as alternatives in the valorisation of poultry waste. J Anal Appl Pyrol 169:105796. https://doi.org/10.1016/j.jaap.2022.105796
Pope MJ, Cherry TE (1991) An evaluation of the presence of pathogens on broilers raised on poultry litter treatment. Poult Sci 70(4):750–755
Prabakaran G, Valavan SE (2021) Microbial degradation of poultry waste: a sustainable approach. J Environ Manag 287:112282
Purandas RB et al (2018) Biogas production from poultry litter waste: a case study of Baramati region. Int J Curr Microbiol App Sci 7(1):2108–2115
Qian Z (2012) Hydrogen production by anaerobic fermentation from wastewater in poultry farms. J Beijing Univ Technol 38(8):1258–1263
Quiroga G, Castrillón LM, Fernández-Nava Y, Marañón E (2010) Physico-chemical analysis and calorific values of poultry manure. Waste Manag 30(5):880–884. https://doi.org/10.1016/j.wasman.2009.12.016
Rao GG et al (2012) Utilization of poultry waste for production of protease by mutant strain of aspergillus oryzae (MTCC 5341) in solid state fermentation. J Environ Biol 33(4):617–620
Rauw WM et al (2022) Replacing soybean meal with narbon vetch (Vicia narbonensis L.) in pig diets: composition of subcutaneous fat and fresh loin and sensory attributes of dry-cured product. Animals 12(16):2061. https://www.researchgate.net/publication/353794574
Rombolà AG, Fabbri D, Baronti S, Vaccari FP, Genesio L, Miglietta F (2019) Changes in the pattern of polycyclic aromatic hydrocarbons in soil treated with biochar from a multiyear field experiment. Chemosphere 219:662–670
Sahoo S, Dash S, Rath B, Mondal KC, Mandal A (2022) Commercial initiation of feather hydrolysate as supreme fertilizer: a smart bio-cleaning strategy of poultry waste. Waste Biomass Valoriz 14:2151–2166
Sarkar N et al (2012) Bioethanol production from agricultural wastes: an overview. Renew Energy 37(1):19–27
Seidavi A, Zaker H, Scanes C (2018) Present and potential impacts of waste from poultry production on the environment. Worlds Poult Sci J 75:1–14
Sillero L, Solera R, Pérez M (2021) Anaerobic co-digestion of sewage sludge, wine vinasse and poultry manure for bio-hydrogen production. Int J Hydrog Energy 47:3667–3678
Silva MEF, Lemos LT, Cunha-Queda C, Nunes O (2009) Co-composting of poultry manure with low quantities of carbon-rich materials. Waste Manag Res 27(2):119–128. https://doi.org/10.1177/0734242X08096693
Susmiati Y (2018) The prospect of bioethanol production from agricultural waste and organic waste. Jurnal Teknologi dan Manajemen Agroindustri, Industria

Thanh NL, Hang VT, Thuy NT, Hien LPT, Huy NN (2023) Nutrient recovery from poultry wastewater by modified biochar: an optimization study. Curr Appl Sci Technol 23:6

Thyagarajan N, Shankar NU, Subrahmanyan R, Arcus W, Bernardi G, Bowman JD, Briggs F, Bunton JD, Cappallo RJ, Corey BE, deSouza L, Emrich D, Gaensler BM, Goeke RF, Greenhill LJ, Hazelton BJ, Herne D, Hewitt JN, Johnston-Hollitt M, Wyithe JSB (2013) A study of fundamental limitations to statistical detection of redshifted H i from the epoch of reionization. Astrophys J 776(1):6

Usman A, Ali T, Ibrahim KA, Magaji I (2023) Biogas production: A comparative study of chicken droppings (poultry waste) and banana peels as the gas source. Afr Sci Rep. https://doi.org/10.46481/asr.2023.2.1.71

Vinodkumar OR et al (2014) Bioconversion of poultry feather waste by keratinolytic bacteria and fungi. Int J Poult Sci 13(1):1–6

Wang T et al (2013) Thermochemical conversion of poultry litter and hardwood: TG-FTIR analysis and kinetic study. J Anal Appl Pyrol 100:1–7

Woldesenbet AG, Shiferaw G, Chandravanshi B (2013) Bio-ethanol production from poultry manure at Bonga poultry farm in Ethiopia. Afr J Environ Sci Technol 7:435–440

Zul S, Iwamoto K, Rahim MAM, Abdullah N, Mohamad SE, Shimizu K, Hara H (2020) Production of liquid fertilizer from chicken feather waste by using subcritical water treatment for plant and algal growth. IOP Conf Ser Earth Environ Sci 479:012033

Xu F, Wang B, Li Y, Bule M (2015) Anaerobic digestion of high-solid poultry litter with the aid of biochar and activated carbon. Bioresour Technol 196:144–150. https://doi.org/10.1016/j.biortech.2015.07.050

Conversion of Oils and Fat Waste to Chemicals and Biofuels

8

Biswajit Saha, Rishika Saha, Disha Patra, and Ranjan Kumar Mohapatra

Abstract

Excessive usage and high costs are making fossil fuels less popular. Over the past few decades, rising energy demand, severe environmental impacts, and resource depletion have driven research into alternative energy sources. Industrialized countries have begun to address the energy crisis with considerable measures and targets, but a growing population is driving energy demand at an unprecedented rate. Biofuels from renewable resources are needed to address growing issues. Biofuels from fats and oils are a sustainable, renewable energy source. It uses abundant natural resources to replace fossil fuels. This process converts lipid-rich feedstocks into biodiesel and renewable diesel, lowering greenhouse gas emissions and promoting a low-carbon economy. This technique uses natural oils to give hope for renewable fossil fuel alternatives. Biofuels may also be effortlessly integrated into existing infrastructure, making them a realistic and eco-friendly energy future option. Biofuels from oils could help create a greener energy future. Microalgae, oils derived from plants, along with fats from animals, are utilized to make these fuels. Biodiesel is made from transesterified vegetable oils, while bioethanol comes from corn and sugarcane. This study examines the complex process of turning lipid-rich substrates into biodiesel and sustainable diesel. An eco-friendly alternative to fossil fuels, these natural oils undergo transesterification and hydrotreatment. Transesterification, hydrotreatment, and pyrolysis are covered in this abstract for biofuel synthesis from fats

B. Saha · R. Saha · D. Patra
Amity Institute of Biotechnology, Amity University, Noida, Uttar Pradesh, India

R. K. Mohapatra (✉)
Department of Environmental and IT Convergence Engineering, Chungnam National University, Daejeon, South Korea
e-mail: mohapatrark@cnu.ac.kr; mranjankumar.biotech@gmail.com

A. Kumari et al. (eds.), *Valorization of Solid Wastes to Biofuels and Chemical Products for Sustainable World*, https://doi.org/10.1007/978-981-96-8594-3_8

and oils. Biofuel characteristics, environmental benefits, feedstock variety, and economic viability are examined. The article also discusses feedstock supply, technology, and policy issues. This report emphasizes the need for innovation and regulatory support to advance the oils and fats biofuel industry for a sustainable energy future.

Keywords

Biofuel · Renewable diesel · Transesterification · Hydrotreatment · Pyrolysis

8.1 Introduction

The transformation of used oils and fats into chemicals and biofuels has become a key scientific endeavor in the never-ending search for environmentally friendly energy alternatives. The need to reduce reliance on fossil fuels and lower the emission of greenhouse gas, along with the growing worldwide need for energy, has sparked extensive research into eco-friendly and alternative energy sources.

This transformative method, which is based on chemical and biochemical engineering concepts, turns valuable biofuels and chemical feedstocks from lipid-rich wastes from a range of sources, including municipal solid waste, industrial effluents, and agricultural residues. Researchers use state-of-the-art technologies to attempt to extract the energy that is naturally present in oils and fats by means of complex processes, including pyrolysis, hydrodeoxygenation, and transesterification.

Triglycerides, the main component of fats and oils, are converted at the molecular level by disassembling complicated hydrocarbon chains into simpler molecules. To reconfigure lipid molecules into biodiesel, a sustainable and cleaner-burning substitute for conventional fossil fuels, catalysts—both biological and chemical—are essential to these conversions. Moreover, the method produces a variety of useful chemical intermediates, increasing the range of possible uses outside the energy sector.

Investigating the intricacies of oil and fat waste conversion offers a compelling path for waste valorization in addition to bearing fruit for a sustainable energy future. This scientific endeavor not only corresponds with global sustainability goals but also highlights the necessity of innovation in solving the problems faced by our ever-evolving energy landscape by turning once-disregarded waste streams into high-value products.

8.2 Generation Status of Oils and Fat Waste Worldwide and their Potential as Feedstock

Waste is a growing global concern, particularly in nations such as India and China, along with Europe (Fig. 8.1). Waste can be categorized as solid urban, industrial, sanitary, or agricultural refuse, depending on its source. Solid urban leftovers can be

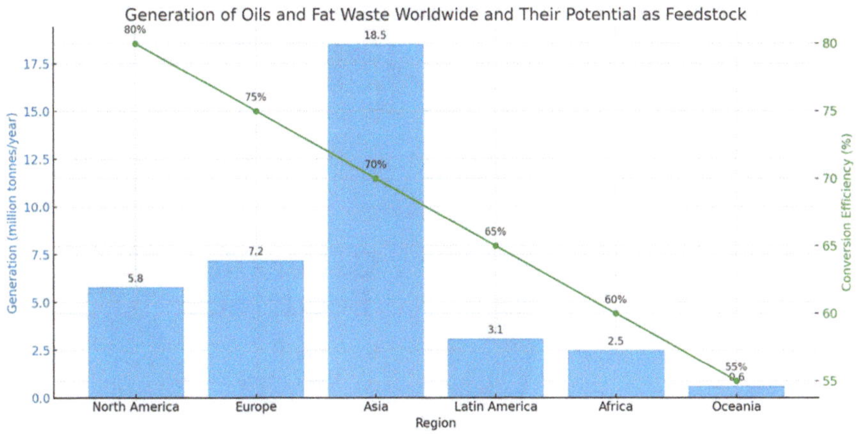

Fig. 8.1 Generation of oils and fat waste worldwide and their potential as feedstock

classified as glass, paper, metals, plastics, organic waste, and others. Food processing waste is an excellent illustration of pre-consumer waste that is created on a massive scale around the world. Oil and fat waste is a growing issue, accounting for almost 50% of total waste in some nations, with organic matter making up 65%.

Waste disposal is also growing more expensive. Between 2009 and 2011, the EU land tax directive resulted in an increase in land tax gate costs from £42–77 to £69–115. Management of oil and fat waste in traditional methods frequently causes sewer concerns. However, if properly used, this trash might be a beneficial means of energy and material recovery. Achieving a zero-waste sector will boost industry profitability, drive growth, and open up new prospects for innovation (Pfaltzgraff and Clark 2014). Some countries, such as the Czech Republic, have government-run initiatives to recycle fat waste. Beginning in 2020, municipalities must collect waste from cooking oils and fats separately (Van Fan et al. 2020).

Waste fats and oils are a valuable feedstock for energy generation and material recovery. This feedstock can be used in cogeneration facilities to generate thermal and electric energy, or directly in agricultural biogas plants without further treatment (Ortner et al. 2016). This fuel can be produced via transesterification, pyrolysis, hydrogenation, and hydrocracking techniques, which are widely used around the world, or by hydrotreating. Oil and fat wastes are an important raw material for biodegradable plastics, bio-based surfactants, soaps, and lubricants production, in addition to waste-to-energy solutions.(Maher & Bressler, 2007)

8.2.1 Oil and Fat Waste as Second-Generation Feedstock

Biofuels, including biodiesel, are often categorized into four generations based on feedstock and production technology. Feedstocks are typically referred to as raw materials or substances that are used as input for biofuel production. Biofuel feedstocks have evolved significantly since their initial generation. Biofuels eventually

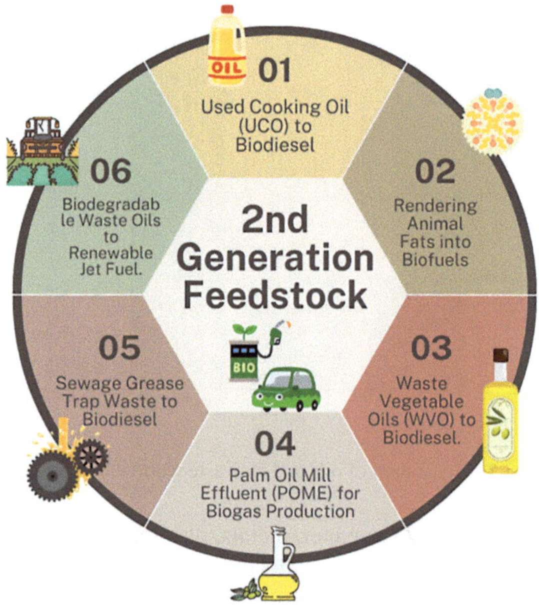

Fig. 8.2 Oil and fat waste as second-generation feedstock is the imperative need for stringent collection, processing, and conversion methods to ensure both environmental and economic viability

originate from sugar, starch, cellulose, animal or plant-based fats, and oils sourced from different livestock and agricultural feedstocks; nonetheless, a significant distinction among first, second, and third-generation biofuels has become frequently employed to indicate the initial feedstock source. Feedstocks are costly, competing with supply chains and the distribution of farmland. They also raised moral and sustainable growth.(Karmakar et al., 2012)

concerns, including rising staple-food prices in developing countries and unsustainable land use change from forested areas to farming. Furthermore, their use caused significant environmental issues, including contamination of water and soil, increased GHG emissions, biodiversity loss, and the propagation of crop diseases and pests.

First-generation feedstocks had comparable problems and were eventually replaced by fuels from the second and third generations. While certain initial-generation feedstocks may continue to provide biofuel for a while, they are becoming less relevant as better alternatives emerge. Second-generation biofuel differs from first-generation biofuels by using woodland or waste-based cellulose or lignocellulosic biomass as feedstocks, rather than food crops (Fig. 8.2).

8.3 Composition of Oil and Fat Waste and their Variation

Oil and fat waste have a wide makeup, with triglycerides being the most frequent component. This composition may also contain free fatty acids (FFAs) generated by hydrolysis or oxidation techniques, as well as glycerol, a consequence of

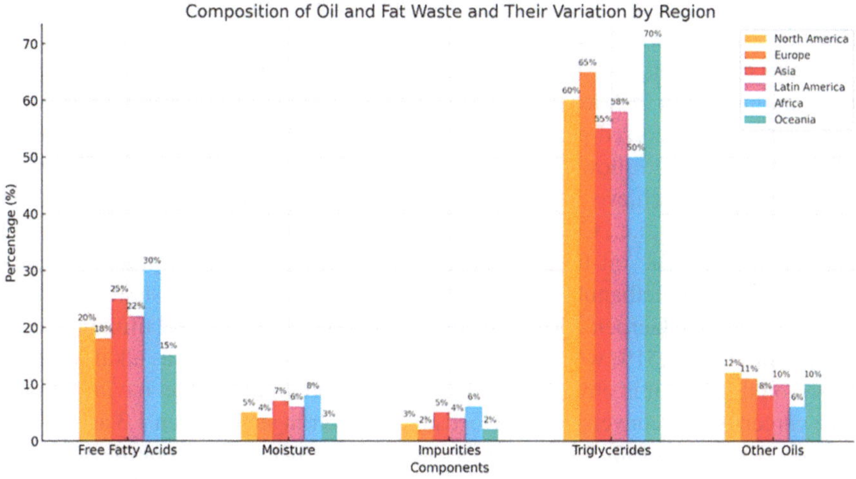

Fig. 8.3 Composition of oil and fat waste and their variation by region

triglyceride breakdown. Impurities like water, particle debris, and pollutants associated with processing may also exist. Processing procedures such as refining, hydrogenation, and extraction can all affect the fatty acid profile and other components. Furthermore, storage circumstances affect oxidative stability, which can lead to the development of oxidation products. Hydrolysis, mediated by variables such as high temperatures, water, or enzymes, can alter the composition by raising FFA concentrations. Some of the examples of oil and fat waste composition and their variation are given in Fig. 8.3.

8.4 Biofuel Production Technology from Oil and Fat Waste on Laboratory and Pilot Scale and Commercialization Detail

Triglycerides undergo three sequential reactions, resulting in a total of three molecules of fatty acid methyl ester, i.e., FAME, and one molecule of glycerol after passing via two intermediates (monoglycerides and diglycerides). Transesterification is a chemical process in which an ester molecule exchanges ester groups with an alcohol.

Direct synthesis of triglycerides into biodiesel utilizing inorganic alkalis (such as NaOH, KOH, or methylates) as catalysts for transesterification reactions is not feasible due to catalyst consumption during FFA neutralization, resulting in an insufficient output of FAME following the reaction (Van Gerpen 2005). Transesterification reactions using acid catalysts, such as hydrochloric, p-Toluenesulfonic acid, or phosphoric acid, are sluggish and need a significant amount of methanol, making the process economically disadvantageous. The choice of an appropriate catalyst is heavily influenced by the FFA content. Physical and chemical refining methods can be used, followed by transesterification employing homogenous alkali catalysts (Lam et al., 2010).

In the pilot-plant investigation, researchers utilized two types of fats obtained from animals (with FFA levels of 10% and 45%, each in turn) to produce biofuel. A two-stage acid-catalyzed pretreatment method utilizing sulfuric acid was used for FFA esterification, followed by the application of an alkali catalyst ($NaOCH_3$, 0.25% w/w) to achieve maximum ester production during transesterification. Another researcher, i.e., Chitra et al. (2005), scaled up biodiesel manufacturing by employing inorganic oil from Jatropha curcas (3.2% FFA concentration), NaOH (1.5% w/w), and methanol (21% w/w) as transesterification catalysts for reactions (Canakci and Van Gerpen 2001). The pilot plant produced biodiesel utilizing potassium hydroxide (1.51% w/w) and methanol and beef tallow (6:1) as alkali catalyst. Based on the data, it is obvious that they created a premium biofuel having a favorable percentage of conversion. The feedstock's acid ranged between 1.3 and 1.9 mg KOH/g (Espinosa et al. 2009). Torres et al. (2013) matched laboratory results to pilot-scale studies in their publication. The pilot-scale waste-to-biodiesel production from vegetable oils utilizing methanol, NaOH, and KOH as catalysts yielded similar results to laboratory-produced biodiesel. The overall yield for FAME was almost 90%. Abdullah et al. (2013) successfully manufactured biodiesel from used cooking oil on a trial scale utilizing a two-phase process (Devappa et al., 2010).

The first phase involved processing of fleshing oil and fats with sulfuric acid and methanol to lower the FFA level below 1%, that is, pre-treatment. Following that, alkali-catalyzed transesterification with KOH and methanol as catalysts resulted in a good FAME yield. In a preliminary scale investigation, Carlini et al. (2014) examined the conditions for producing biodiesel from household used cooking oil waste with an acidic value of 2.135 mg KOH/g. The study compared catalysts (NaOH and H2SO4) at various concentrations, as well as alcohol concentrations. The maximum reaction yield (94.5%) was achieved with 0.5% NaOH and 100% excess methanol.

Šánek et al. (2016) proposed a laboratory-scale refining melting technique followed by the transesterification of triglycerides using a potent organic catalyst, such as tetramethylammonium hydroxide (TMAH), to treat acidic waste fat. Using organic bases reduces ash content in glycerol compared to inorganic transesterification catalysts. The biodiesel synthesis process using rapeseed oil successfully converted triglycerides to FAME within 15–20 min under the specified conditions. Karavalakis et al. (2010) compared reaction settings to maximize methyl esters output from waste frying oil. They employed benzyl trimethyl and tetramethyl-ammonium hydroxides to be catalysts involved in the transesterification process. Panter et al. (2012) enhanced the reaction conditions for biodiesel catalyzed by TMAH from vegetable oil, achieving approximately 100% conversion in less than 2 h. In their work, Sánchez et al. (2013) identified the ideal parameters for the effective sunflower oil transesterification. These included a TMAH content of 1.77% w/w, 60 °C as the reaction temperature, a fat-to-methanol ratio of 1:6, and a reaction period of 2 h. This paper presents a pilot-scale biodiesel manufacturing process for waste frying oils (WFO) as well as animal fats (WAF) using a unique organic alkali catalyst.

8.5 Challenges and Prospects of the Production of Biofuels from Fats and Oils Wastes

Producing biofuels from oil and fat waste presents both challenges and prospects, which can significantly impact their feasibility and adoption.

8.5.1 Challenges of Production of Biofuels from Fat and Oil Wastes

Production of biofuels from fat and oil waste presents an array of formidable issues that need to be tackled to achieve their complete potential as a renewable energy source.

8.5.1.1 Feedstock Issue

The first issue, overbudget production, could be caused by poor feedstock selection. Feedstock prices should not exceed 55% of production costs. Refining and treating feedstock prior to processing may result in higher costs (50–80%). The establishment of a treatment facility for raw material would diminish the industry's overall profit (Anuar and Abdullah 2016).

8.5.1.2 Quality and Yield Issue

To effectively promote biodiesel as a fuel replacement, it is important to ensure that the quality meets global standards. The major problem throughout the manufacturing line is ensuring that the quality of the biodiesel produced meets international standards. Inadequate quality control for biodiesel can harm consumers, particularly transportation engines and industrial machines. Current manufacturing struggles to meet B100 criteria due to processing line problems. Producing 100% pure biodiesel from waste-based feedstock requires a robust processing line to produce a high-quality methyl ester product. To ensure commercial use, biodiesel manufactured must adhere to standards such as ASTM 6751, EN 1424, and ASTM 975. Failure to do so will result in useless fuel (Costenoble-Nen et al. n.d.).

8.5.2 Prospects of Biofuel Production from Oil and Fat Wastes

Despite these challenges, there are numerous promising opportunities for biofuel production from oil and fat waste.

8.5.2.1 Waste Management and Sustainability

Utilizing waste for the manufacturing of biofuel can increase resource efficiency by lowering the cost of waste disposal and environmental pollution caused by poor waste management practices. Global organizations are redesigning their sustainability programs to address challenges such as feedstock, geographic regions, and environmental effects of the biofuel sector. In 2011, the European Commission restructured the sustainability initiative into seven voluntary schemes that were directly applied to 27 Member States (Energy 2024).

8.5.2.2 Circular Economy

Integrating biofuel production from oil and fat wastes into a circular economy framework fosters resource efficiency and generates value from waste streams, helping to achieve sustainable development goals. According to the business and trade sector, circular economic transitions could result in 0.6 trillion euros in yearly economic activity benefits for the European Union's production industry. For several years, China has been leading the promotion of the concept of CE, having become the first nation globally to pass a circular economy law in 2008. Circular economy is linked with various concepts, with waste management being the most significant sub-sector. The aim is to repurpose and recycle materials into high-value goods and maintain their active use for as long as possible. Maintaining the intrinsic value of a product, material, or service can lead to a circular model that prioritizes resource efficiency and zero waste.

8.6 Feasibility and Economics of Biofuel Production from Waste of Fat and Oil

The economic viability of producing biofuel from oil and fat waste is influenced by several things, including feedstock acquisition as well as processing costs, government rebates and subsidies, and biofuel market prices. As demand for renewable fuels rises, opportunities for economically viable biofuel production from waste oils and fats are likely to expand (Table 8.1) (Alptekin et al., 2012).

Table 8.1 Key considerations in assessing the feasibility and economics of biofuel production from oil and fat waste

Aspect	Feasibility	Economics
Feedstock availability	Abundant, continuous supply	Varies based on local collection infrastructure and regulations
Resource sustainability	Renewable, reduces waste	May require initial investment in collection and processing infrastructure
Processing technology	Established methods (e.g., transesterification, hydroprocessing)	Initial setup costs, operational expenses
Product yield	High yield potential	Economies of scale may improve profitability
Carbon footprint	Carbon-neutral or negative	Cost of carbon credits, potential carbon tax implications
Market demand	Growing demand for renewable fuels	Market volatility, fluctuating oil prices

8.6.1 Commercialization Potential of Biodiesel Derived from Oil and Fat Wastes

The feedstock and resources necessary to transform fatty acids into fatty acid esters are the main factors influencing the cost of producing biodiesel. The cost of feedstock has a significant impact on the biodiesel economy's profitability, which accounts for 70.1–82.1% of the entire cost of manufacturing. Sustainability and feedstock availability are major issues in addition to cost.

Biodiesel from various feedstocks is currently more expensive (more than US$ 0.5/l) compared to US$ 0.36, which is per liter for normal diesel and is projected to increase threefold in underdeveloped nations (Baskar and Aiswarya 2016). To make biodiesel economically viable, governments must conduct thorough studies on new economical feedstocks and implement policies.

According to the European Biomass Industry Association, in 2016, 6.89 × 104 million gallons of biodiesel made from oil and fat wastes may replace 1.5% of diesel fuel usage in the EU-28 (Statista 2016). Additionally, oil and fat-based biodiesel has the possibility to replace 26.7% of the EU's total production of biodiesel. In 2018, 3814.12 million gallons of biofuel were produced using various feedstocks, including 19.89% from yellow grease. In the US, restaurants generate an average of 11,001 kg of grease trap waste (GTW) each year, with up to 45.1% lipid content (Wallace et al. 2017).

Because of its high FFA content and low cost, yellow and brown grease are frequently used as a feedstock for biodiesel. According to estimates, more than 400 million gallons of biofuel may be made annually in the US from GTW, which would have accounted for around 31.49% of all biodiesel generated in 2014 (Chisti 2008).

8.7 Chemical Production Technology from Fat and Oil Waste on a Laboratory and Pilot Scale and Commercialization

An ester molecule and an alcohol swap ester group in a chemical process known as transesterification. Figure 8.4 provides the general transesterification equation. This process is referred to as methanolysis if methanol is used.

A sequence of consecutive, reversible steps makes up transesterification. Diglycerides, monoglycerides, and glycerol are the products of the sequential metabolism of triglycerides, each of which releases a mole of ester. The equilibrium encourages the creation of glycerol and fatty acid ester, and the processes are reversible. The reactions are shown in Fig. 8.5.

Fig. 8.4 Transesterification of ester

$$\underset{\text{Ester}}{RCOOR^1} + \underset{\text{Alcohol}}{R^2OH} \rightleftharpoons \underset{\text{Ester}}{RCOOR^2} + \underset{\text{Alcohol}}{R^1OH}$$

Fig. 8.5 General equation for transesterification of triglycerides

$$\text{Triglyceride} + R^1OH \rightleftharpoons \text{Diglycerides} + RCOOR^1$$

$$\text{Diglycerides} + R^1OH \rightleftharpoons \text{Monoglyceride} + RCOOR^1$$

$$\text{Monoglyceride} + R^1OH \rightleftharpoons \text{Glycerol} + RCOOR^1$$

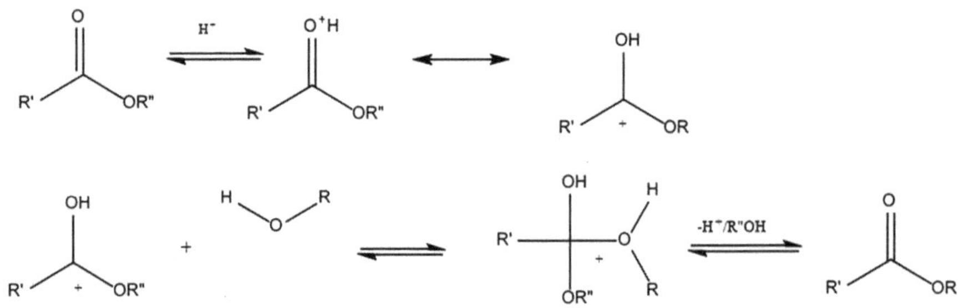

R" = Glycerides

R' = Carbon chain of fatty acid

R = Alkyl group of alcohol

Fig. 8.6 Mechanism of acid-catalyzed transesterification

8.7.1 Acid-Catalyzed Transesterification

Alcohols can esterify carboxylic acids in the absence of water using an acidic catalyst. The most widely used reagent for making methyl esters is 5% anhydrous hydrogen chloride in methanol.

Esterification typically involves dissolving the lipid sample in a 99.9-fold excess of methanolic hydrogen chloride and refluxing the solution for 2 h or overnight at 50.1 °C (30.1 °C for free acids alone). The mechanism for acid-catalyzed transesterification is shown in Fig. 8.6.

8.7.2 Base-Catalyzed Transesterification

In anhydrous methanol, sodium or potassium methoxide is the most effective basic trans esterifying agent. In the first step, the alkoxide ion attacks the carbonyl group of the triglyceride, forming a tetrahedral intermediate; in the second stage, the intermediate reacts with alcohol to form alkoxide ions. When an ester reacts with a base, such as an alcoholate anion, it produces an anionic intermediate that can either dissolve back into the original ester or produce a new ester.

The final stage involves the restructuring of the tetrahedral intermediate, resulting in an ester and diglyceride. Base-catalyzed transesterification is faster than acid transesterification and can be done at low temperatures (302–307 K) and pressures (0.11 MPa). It is shown in Fig. 8.7.

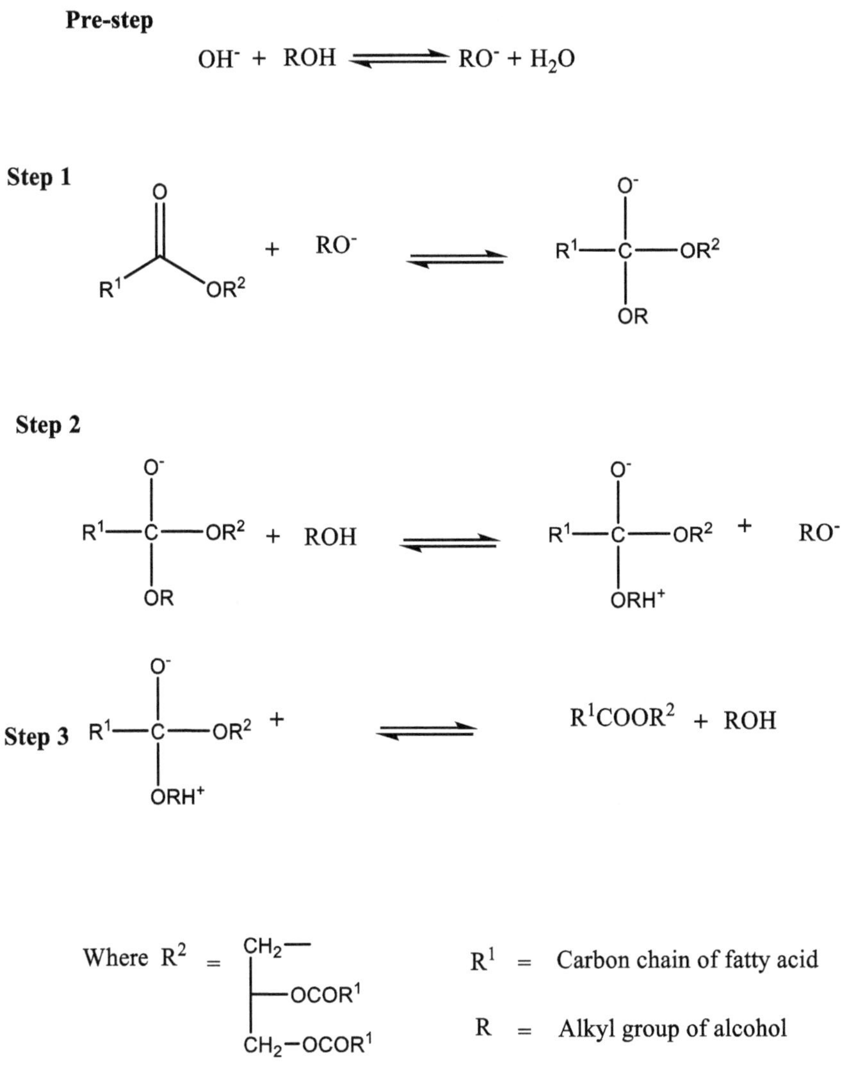

Fig. 8.7 Base-catalyzed transesterification

8.7.3 Acid-Base Transesterification

The esterification of natural fatty acids is made possible by acid catalysis, which decreases soap (glycerine, byproduct) generation and results in a higher yield of alkyl ester (biodiesel). Biodiesel is manufactured through transesterification, which involves reacting vegetable oil and alcohol with a basic or acidic catalyst. Because of its low cost and advantageous physical and chemical properties, methanol is the most widely used alcohol. The maximum biodiesel production was achieved at 3.1% catalyst weight, 60 °C, and 120 min of reaction time. Acid-base catalyzed synthesis produced higher yields of biodiesel than base-catalyzed production. Acid-base catalyzed biodiesel's kinematic viscosity met standard test requirements (Venkateswarulu et al. 2014).

8.7.4 Enzyme-Catalyzed Transesterification

Although transesterification is a cost-effective and efficient method for producing biodiesel, downstream processing issues such as catalyst separation, unreacted methanol, glycerol recovery, and inorganic salt removal arise. The potential of liberated acids or water contamination, combined with soap production, can make separation difficult.

Lipases, also known as triacyl glycerol acyl hydrolase (EC 3.1.1.3), are enzymes that break down and synthesize acylglycerols at the lipid-water interface (Sawangkeaw & Ngamprasertsith, 2013).

Using 7.5% (w/v) lipases and a 1:4 oil-to-methanol molar ratio, 86.1% of fatty acids were converted to fatty acid methyl ester (FAME) in 35 h at 45 °C. A new acyl acceptor, methyl acetate, has been created for biodiesel synthesis (Fig. 8.8).

8.7.5 Hydrocracking

Hydrocracking is a method that converts the elements of gas oil fractions (partially or totally) into lighter molecules using hydrogen H_2 and a catalyst. The hydrocracking mechanism consists of numerous stages, including the conversion of unsaturated bonds in triglyceride fatty acid chains are hydrogenated, resulting in saturated

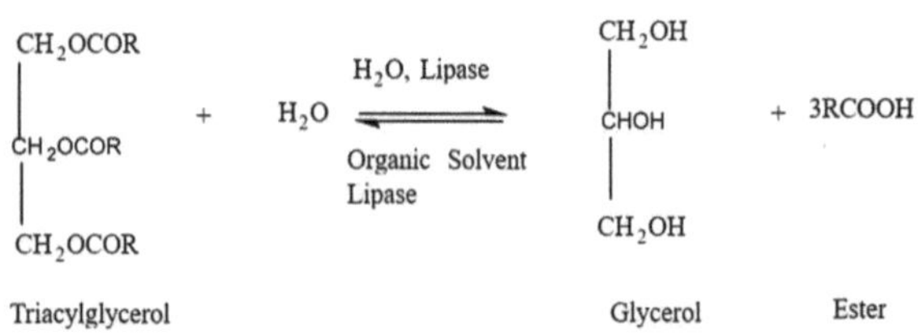

Fig. 8.8 Illustration of biodiesel generation through enzymatic transesterification using methyl acetate and the lipase enzyme as a catalyst (Venkateswarulu et al. 2014)

carbon-carbon double bonds. Hydro processing involves removing heteroatoms like sulfur and nitrogen, as well as breaking down carbon-carbon bonds to create premium diesel and gasoline molecules. Carbon-carbon bonds are broken, resulting in the fragmentation of fatty acid chains into tiny hydrocarbons. Catalyst regeneration during hydrocracking includes replenishing the catalyst, which may become deactivated over time, to keep it active and effective in the hydrocracking process. Key parameters for hydrocracking include temperature, LHSV, catalyst life, and other factors. Increasing temperature and decreasing LHSV lead to higher conversion and yield of biofuels through increased cracking activity (Bezergianni and Kalogianni 2009).

As hydrocracking temperatures rise, so does the removal of heteroatoms (sulfur, nitrogen, and oxygen), with de-oxygenation being most effective. Temperature does not favor saturation, indicating the need for a pre-treatment step before hydrocracking to ensure double bond saturation and remove heteroatoms.

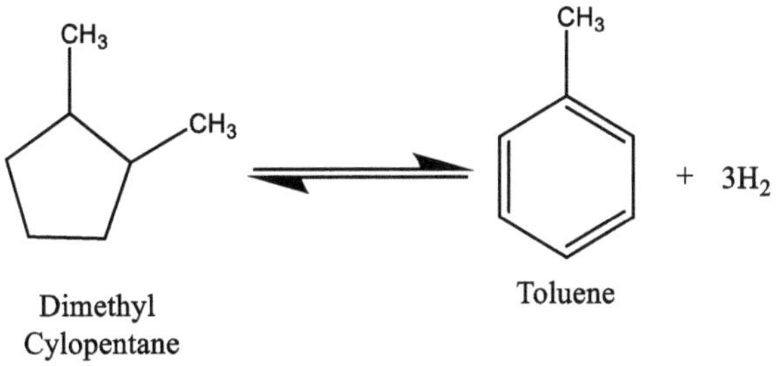

8.7.6 Laboratory or Pilot-Scale Biodiesel Production

Reactors can be made from materials such as glass, stainless steel, or even high-density polyethylene (HDPE), depending on volume and desired features, with typical capacities ranging from a few liters to tens of liters. To maintain the temperature, which is typically 50–60 °C, the reactor requires a heating mantle or band heater, and efficient stirring is necessary for proper mixing of the reactants (oil, alcohol, catalyst). Magnetic stirrers with a stir bar are commonly used in lab setups, while pilot plants might utilize mechanical or overhead stirrers. A temperature controller is essential for maintaining a constant reaction temperature for optimal conversion efficiency. For separating the two layers (biodiesel and glycerin) in a small-scale setup, a simple separation funnel can be used, which relies on gravity. In pilot-scale setups, centrifuges might be employed to accelerate the separation process, which uses centrifugal force to separate liquids with different densities, significantly reducing separation time compared to gravity settling. Crude biodiesel may still contain residual methanol, soap, and other impurities, which can be removed by washing with water in a separatory funnel and allowing the layers to separate again.

8.7.7 Commercialization Contemplations

1. Feedstock accessibility and fetched:
   (a) Securing dependable sources: Building long-term organizations with eateries, nourishment preparation offices, and oil trap collection companies for utilized cooking oil and animal fats is fundamental.
   (b) Collection and transportation coordination: Creating a productive and cost-effective framework for collecting and transporting the squander feedstock to the generation office is pivotal.
   (c) Maintainability concerns: Utilizing squander oils occupies them from landfills but does not address the root cause—the generation of these squander materials. Investigating non-food edible oils like jatropha or green growth offers a more feasible arrangement, but comes with its challenges, like utilization and water requirements.

2. Handle financial matters:
   (a) Capital costs: Setting up a large-scale biodiesel generation plant requires a significant investment in reactors, separators, refinement gear, and capacity offices.
   (b) Working costs: Adjusting the fetched of feedstock, handling chemicals (methanol, catalyst), utilities (energy for warming and blending), and labor is pivotal for productivity.
   (c) Glycerin advertisement: Finding steady markets for glycerin, a co-product of biodiesel generation, is imperative. Changes in glycerin costs can altogether affect the overall benefit.

3. Natural affect:
   (a) Life cycle appraisal (LCA): Conducting a careful LCA is vital to evaluate the natural impact of the whole generation preparation, including feedstock collection, preparation, and squander transfer.
   (b) Nursery gas emissions (GHG): Biodiesel, for the most part, offers lower GHG emissions compared to petroleum diesel, but this could change depending on feedstock source and handling strategies.

4. Controls and Arrangements:
   (a) Biofuel orders: Government arrangements that order at a minimum rate of biofuels (including biodiesel) mixed with ordinary diesel can make a steady advertise for biodiesel makers (P.N. Giannelos et al., 2005).

8.8 Challenges and Prospects of Chemical Production from Fat and Oil Waste

Cooking oil and other edible oils and fats that come from homes are now usually disposed of with municipal waste, including grease residues mixed in the waste, or just poured down the drain (Kulkarni & Dalai, 2023). While most of Europe has long required food industry companies to recycle this waste, household recycling is still not common, even though households contribute significantly to the total production (for example, households in South Korea account for about 18.9% of the total production).In some countries, the government has initiated programs aimed at recycling fat waste, such as the Czech Republic, where the municipalities are mandated to collect waste cooking oils and fats separately starting in 2020. Due to its physical characteristics and structure, fat waste is a unique area of waste management that requires special attention. Animal fats and vegetable oils both solidify at low temperatures and melt at high temperatures, but animal fats may appear in a semi-solid or paste state at room temperature, and vegetable oils tend to have a higher viscosity. Compared to other materials like glass, paper, or plastic, handling fat waste is more complicated (Callegari et al., 2019).

Production, disposal and collection, processing, and use are the four parts of the reverse supply chain for trash collecting. Each segment of this chain presents its own set of problems and logistical challenges, such as selecting the appropriate container vehicles for waste collectors. As a result, it is necessary to address each part of the chain individually, in addition to considering the entire chain as a whole.

8.9 Feasibility and Economics of Chemical Production from Oil and Fat Waste

The global community is increasingly focused on finding sustainable solutions, and this applies to the chemical industry as well. In 2014, 145 countries had a policy that targets renewable types of energy, with a total global annual investment of US $214.5 billion in renewable types of power as well as fuels. The top investment countries were China, the United States, Japan, the United Kingdom, and Germany. By 2012, renewables accounted for 18.9% of the world's total energy consumption, and the UAE is one of the countries with a renewable energy target, aiming to generate 6.99% of its power from renewable sources by 2020 to reduce its reliance on oil (Hussain et al. 2016). In Harare, visits to food outlets showed that 28.9% of the oils and fats taken were disposed of. An alkaline type of trans-esterification was utilized to convert WCO into biodiesel, with optimal operating conditions resulting in an average yield of 91.75%. An economic analysis indicated a payback period of 4.43 years with biodiesel selling at 0.95c/liter. To improve economic viability, it is recommended to recover process water, alcohol, and glycerin (Manyuchi et al. 2012).

8.10 Impact of Fat and Oil Waste on the Environment

Fat and oil waste, a seemingly innocuous byproduct of our daily lives, poses a significant threat to the environment. Although it may not be as visible as other forms of pollution, its impact can be just as devastating. This reveals the hidden dangers of improperly disposed of fat and oil waste, highlighting its detrimental effects on water resources, wildlife, and even our own infrastructure. The consequences for aquatic life are tragic, as oil coats feathers and fur, hindering their ability to survive. The impact on our infrastructure is also significant, with "fatbergs"—massive conglomerations of fat, oil, and debris—causing blockages and costly repairs.

8.10.1 Impact on Water Systems

- Clogged Sewer Lines: When fat and oil wastes are poured down drains, they cool and solidify, adhering to the insides of pipes. Over time, this buildup narrows the passage and can lead to complete blockages. These blockages cause sewage overflows, releasing untreated wastewater into streets, waterways, and basements, posing a serious health risk.
- Treatment Plant Strain: Even if fat and oil wastes do not cause blockages, they can still overwhelm wastewater treatment plants. Fat and oil waste disrupts the biological processes that break down waste, reducing treatment efficiency and potentially allowing harmful pollutants to enter waterways.

8.10.2 Impact on Ecosystems

- Habitat Disruption: When FOG enters waterways, it forms a film that reduces oxygen levels and hinders sunlight penetration, negatively impacting aquatic ecosystems and the plant and animal life that depend on these factors for survival.
- Wildlife Distress: FOG can adhere to birds, mammals, and other wildlife, interfering with their ability to regulate body temperature and seek sustenance. This can lead to hypothermia, starvation, and even fatalities.

8.10.3 Impact on Civil Engineering

- Fatberg Formation: FOG combines with other debris, such as wipes and sanitary products, to create massive blockages called "fatbergs." These behemoth accumulations can extend for hundreds of feet, causing sewage backups and necessitating expensive efforts to clear sewer lines.
- Infrastructure Erosion: FOG can coat and degrade sewer infrastructure, leading to accelerated corrosion and requiring costly repairs or replacements (Fig. 8.9).

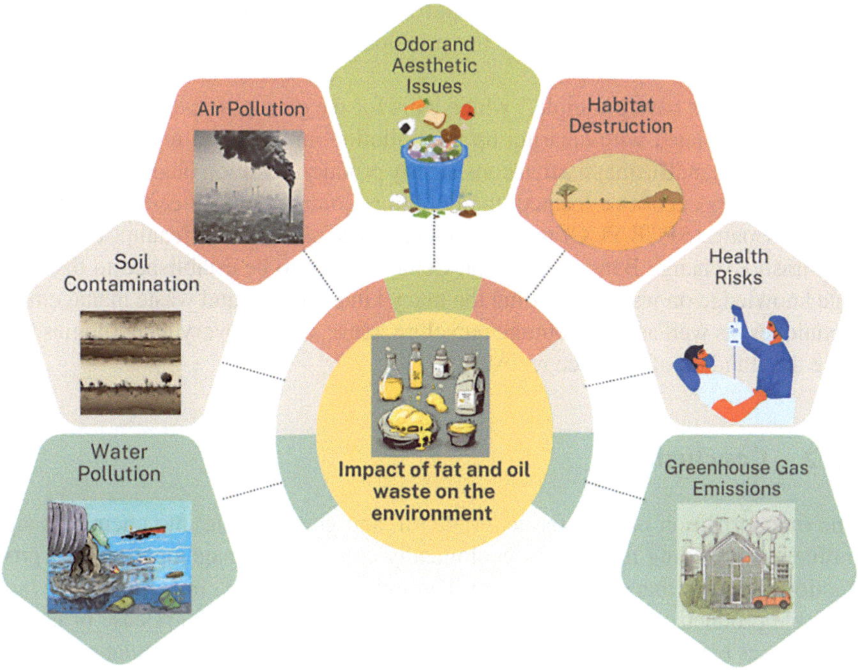

Fig. 8.9 Improper fat and oil disposal pollutes water, soil, and air, harming ecosystems and biodiversity

8.11 Policy for the Fat and Oil Waste Management

Waste made of fatty tissue, crude oil, and greasy (FOG) is a seemingly harmless consequence of everyday life, but it actually poses a serious risk to the environment and infrastructure.

Ineffective solid waste management is a major problem in India, especially in urban areas (Kumar et al. 2017). In 2013, there were 1249 million people on the planet, up from 1029 million in 2001. In urban regions, about one-third of the population resides. It is projected that by 2050, over half of India's population would reside in cities, and the country will produce 5.1% more waste annually.

Waste arising is projected to reach approximately 101 million metric tons (MMT), 163.9 MMT, and 437 MMT annually by 2021, 2031, 2050, respectively. The main causes of the increase in solid waste generation are changes in lifestyle, economic development, and population growth.

At the state level, solid waste management is the responsibility of urban local governments. Companies that generate hazardous waste must get permission from the appropriate SPCB.

The Environment (Protection) Act of 1986 established broad regulations for the management and handling of chemicals and hazardous materials, as well as for the protection and enhancement of the environment. This Act also incorporates a variety of guidelines, such as the Dangerous Materials (Managing and Handling) Rules of the year 2008, which are the principal standards for handling hazardous materials in India.

The “polluter pays principle,” which calls for the corrupt to pay both a fine and the costs associated with clearing up the pollution, must be the main focus of the remediation plan. Businesses that consistently produce pollution ought to be banned. If the polluters cannot be located, the Pollution Control Committee/State Pollution Control Board (SPCB/PCC) must establish a special fund for cleanup. In light of this, trash Exchange Banks/Collection Centers ought to be established to disseminate knowledge on waste, including the many kinds of trash and waste management techniques, as well as to encourage recycling, reuse, and recovery technologies that raise the standard of resource recovery.

8.12 Conclusion

The transformation of fatty tissue, crude oil, and greasy waste into chemicals and biofuels is a leading method for ecologically conscious management and the production of energy that can be renewed. Through the process of converting waste into valuable products, this innovative method not only decreases the release of greenhouse gases but also showcases a commitment to the ideals of the circular economy. The utilization of diverse catalysts, including metallic, zeolitic, and enzymatic catalysts, exemplifies ongoing technical advancements and provides enhanced selectivity and efficiency. The possibility of generating employment opportunities and positive impacts on local economies contributes to the social aspect of the overall sustainability profile. However, it is important to note that economic feasibility remains a crucial factor for the success of these conversion operations. Continuing to support these technologies will necessitate continual investigation and ingenuity to tackle challenges such as waste generation and catalyst deactivation. The achievement of a cleaner, more environmentally friendly, and economically efficient future relies heavily on the transformation of fatty tissue, crude oil, and greasy waste into chemicals and biofuels. This transformation process places great importance on the active involvement of all parties involved, strict compliance with sustainability standards, and ongoing enhancements.

References

Abdullah N, Sahibul-Anwar H, Ideris S, Hasuda T, Hitotsuyanagi Y, Takeya K, Diederich M, Choo CY (2013) Goniolandrene a and B from Goniothalamus macrophyllus. Fitoterapia 88:1–6. https://doi.org/10.1016/j.fitote.2013.03.028

Alptekin E, Canakci M, Sanli H (2012) Evaluation of leather industry wastes as a feedstock for biodiesel production. Fuel 95:214–220. https://doi.org/10.1016/j.fuel.2011.08.055

Anuar MR, Abdullah AZ (2016) Challenges in biodiesel industry with regards to feedstock, environmental, social and sustainability issues: a critical review. Renew Sustain Energy Rev 58:208–223. https://doi.org/10.1016/j.rser.2015.12.296

Baskar G, Aiswarya R (2016) Trends in catalytic production of biodiesel from various feedstocks. Renew Sustain Energy Rev 57:496–504. https://doi.org/10.1016/j.rser.2015.12.101

Bezergianni S, Kalogianni A (2009) Hydrocracking of used cooking oil for biofuels production. Bioresour Technol 100(17):3927–3932

Callegari A, Bolognesi S, Cecconet D, Capodaglio AG (2019) Production technologies, current role, and future prospects of biofuels feedstocks: a state-of-the-art review. Crit Rev Environ Sci Technol 50(4):384–436. https://doi.org/10.1080/10643389.2019.1629801

Canakci M, Van Gerpen J (2001) A pilot plant to produce biodiesel from high free fatty acid feedstocks. American Society of Agricultural and Biological Engineers, St. Joseph. https://doi.org/10.13031/2013.4209

Carlini M, Castellucci S, Cocchi S (2014) A pilot-scale study of waste vegetable oil transesterification with alkaline and acidic catalysts. Energy Procedia 45:198–206. https://doi.org/10.1016/j.egypro.2014.01.022

Chisti Y (2008) Biodiesel from microalgae beats bioethanol. Trends Biotechnol 26(3):126–131. https://doi.org/10.1016/j.tibtech.2007.12.002

Chitra P, Venkatachalam P, Sampathrajan A (2005) Optimisation of experimental conditions for biodiesel production from alkali-catalysed transesterification of Jatropha curcus oil. Energy Sustainable Dev 9(3):13–18. https://doi.org/10.1016/s0973-0826(08)60518-9

Costenoble-Nen O, Fischer J, Haupt-Agqm J (n.d.) Improvements needed for the biodiesel standard EN 14214. https://www.agqm-biodiesel.com/application/files/1915/2206/0269/BIOSCOPESlot1.pdf. Accessed 2 April 2024

Devappa RK, Makkar HPS, Becker K (2010) Jatropha toxicity—a review. J Toxicol Environ Health Part B 13(6):476–507. https://doi.org/10.1080/10937404.2010.499736

Energy (2024) Energy. https://energy.ec.europa.eu/index_en

Espinosa M, Krause LC, Silvana M, Faccini CS, Jacques RA, Almeida SR, Regina M, Caramão EB (2009) Beef tallow biodiesel produced in a pilot scale. Fuel Process Technol 90(4):570–575. https://doi.org/10.1016/j.fuproc.2009.01.001

Giannelos PN, Sxizas S, Lois E, Zannikos F, Anastopoulos G (2005) Physical, chemical and fuel related properties of tomato seed oil for evaluating its direct use in diesel engines. Ind Crop Prod 22(3):193–199. https://doi.org/10.1016/j.indcrop.2004.11.001

Hussain MN, Samad TA, Janajreh I (2016) Economic feasibility of biodiesel production from waste cooking oil in the UAE. Sustain Cities Soc 26:217–226. https://doi.org/10.1016/j.scs.2016.06.010

Karavalakis G, Anastopoulos G, Karonis D, Stournas S (2010) Biodiesel production using tetramethyl- and benzyltrimethyl ammonium hydroxides as strong base catalysts. Fuel Process Technol 91(11):1585–1590. https://doi.org/10.1016/j.fuproc.2010.06.006

Karmakar A, Karmakar S, Mukherjee S (2012) Biodiesel production from neem towards feedstock diversification: Indian perspective. Renew Sustain Energy Rev 16(1):1050–1060. https://doi.org/10.1016/j.rser.2011.10.001

Kulkarni MG, Dalai AK (2023) Waste cooking oil: an economical source for biodiesel: a review. Ind Eng Chem Res 45:2901–2913. https://doi.org/10.1021/ie0510526

Kumar S, Smith SR, Fowler G, Velis CA, Kumar S, Arya S, Rena K, R., & Cheeseman, C. (2017) Challenges and opportunities associated with waste management in India. R Soc Open Sci 4(3):160764. https://doi.org/10.1098/rsos.160764

Lam MK, Lee KT, Mohamed AR (2010) Homogeneous, heterogeneous and enzymatic catalysis for transesterification of high free fatty acid oil (waste cooking oil) to biodiesel: a review. Biotechnol Adv 28(4):500–518. https://doi.org/10.1016/j.biotechadv.2010.03.002

Maher KD, Bressler DC (2007) Pyrolysis of triglyceride materials for the production of renewable fuels and chemicals. Bioresour Technol 98(12):2351–2368. https://doi.org/10.1016/j.biortech.2006.10.025

Manyuchi M, Govha J, Nyanhongo M (2012) Biodiesel production from waste cooking oil: technical and economic feasibility

Ortner ME, Müller W, Schneider I, Bockreis A (2016) Environmental assessment of three different utilization paths of waste cooking oil from households. Resour Conserv Recycl 106:59–67. https://doi.org/10.1016/j.resconrec.2015.11.007

Panter R, Archer T, Schroetter A, Ricci J, Smith RB, Dooling TA, Tirla C (2012) TMAH-catalyzed biodiesel production for greater energy independence. In: Abstracts of papers of the American Chemical Society, vol 243. American Chemical Society, Washington

Pfaltzgraff LA, Clark JH (2014) Green chemistry, biorefineries and second generation strategies for re-use of waste: an overview. In: Elsevier EBooks. Elsevier, Amsterdam, pp 3–33. https://doi.org/10.1533/9780857097385.1.3

Sánchez BS, Mendow G, Levrand PG, Querini CA (2013) Optimization of biodiesel production process using sunflower oil and tetramethyl ammonium hydroxide as catalyst. Fuel 113:323–330. https://doi.org/10.1016/j.fuel.2013.05.085

Šánek L, Pecha J, Kolomazník K, Bařinová M (2016) Pilot-scale production of biodiesel from waste fats and oils using tetramethylammonium hydroxide. Waste Manag 48:630–637. https://doi.org/10.1016/j.wasman.2015.10.005

Sawangkeaw R, Ngamprasertsith S (2013) A review of lipid-based biomasses as feedstocks for biofuels production. Renew Sustain Energy Rev 25:97–108. https://doi.org/10.1016/j.rser.2013.04.007

Statista (2016) EU-28: diesel fuel sales 2016. https://www.statista.com/statistics/818096/diesel-fuel-sales-per-country-european-union-eu-28/

Torres EA, Cerqueira GS, Ferrer TM, Maria C, Raboni M, Torretta V, Urbini G (2013) Recovery of different waste vegetable oils for biodiesel production: a pilot experience in Bahia state. Brazil Waste Manage 33(12):2670–2674. https://doi.org/10.1016/j.wasman.2013.07.030

Van Fan Y, Klemeš JJ, Bertók B (2020) Implementing circular economy in municipal solid waste treatment system using P-graph. Sci Total Environ 701:134652–134652. https://doi.org/10.1016/j.scitotenv.2019.134652

Van Gerpen J (2005) Biodiesel processing and production. Fuel Process Technol 86(10):1097–1107. https://doi.org/10.1016/j.fuproc.2004.11.005

Venkateswarulu TC, Raviteja CV, Prabhaker KV, Babu DJ, Ranganadha Reddy A, Indira M, Venkatanarayana A (2014) Review on methods of transesterification of oils and fats in biodiesel formation. Int J ChemTech Res 6(4):2568–2576

Wallace T, Gibbons D, O'Dwyer M, Curran TP (2017) International evolution of fat, oil and grease (FOG) waste management—a review. J Environ Manag 187:424–435. https://doi.org/10.1016/j.jenvman.2016.11.003

Challenges and Prospects of Conversion of Municipal Solid Waste to Biofuel and Chemicals

9

Surbhi Sharma, Anshika Dedha, Abha Kumari, Prisha Gupta, and Manju M. Gupta

Abstract

This chapter explores the challenges and prospects of converting Municipal Solid Waste (MSW) into biofuels and chemicals, highlighting the technologies, their intricacies, and the potential of this renewable energy source. Despite the economic and environmental benefits, large-scale commercialization is still challenging because of several factors. These include high moisture content, compositional unpredictability, and the presence of toxic metals, which can hinder biofuel production. These issues, along with low quality and yield, further complicate the purification of biofuels. Furthermore, if waste is not correctly sorted, the economic and technological difficulties, such as high operating costs and the requirement for complex separation technologies, often outweigh the advantages. New technologies that have the potential to increase conversion efficiency and product quality include plasma gasification, anaerobic digestion, and so on. However, strong legislative backing, monetary rewards, and market mechanisms that encourage innovation and adoption are necessary for the effective commercialization of these technologies. Sustainable waste management also requires the integration of circular economy principles, which prioritize resource recovery, waste minimization, and byproduct valorization. The development of MSW conversion technologies and procedures depends on international collaboration, knowledge sharing, and community awareness and involvement. These initiatives can support the development of a waste management and renewable energy strategy that is more robust and sustainable. This chapter focuses on the recent

S. Sharma · A. Dedha · P. Gupta · M. M. Gupta (✉)
Amity Institute of Biotechnology, Amity University, Noida, Uttar Pradesh, India

A. Kumari
Department of Botany, University of Delhi, Sri Aurobindo College, New Delhi, India

A. Kumari et al. (eds.), *Valorization of Solid Wastes to Biofuels and Chemical Products for Sustainable World*, https://doi.org/10.1007/978-981-96-8594-3_9

technological developments, policy support, and community involvement required for overcoming the present obstacles and realizing the full potential of MSW as a valuable resource for the production of biofuel and chemicals.

Keywords

Pyrolysis · Gasification · Feedstock heterogeneity · Circular economy · Policy support · Resource recovery

9.1 Introduction

Waste output is rising at a rapid rate in emerging nations due to a combination of factors, including population growth, economic expansion, rapidly increasing urbanization, and improved living standards. These factors not only accelerate the generation of waste but also lead to various socioeconomic and environmental issues. Municipal solid waste (MSW) is one such waste generation that is significantly driven by population expansion. The population of the world, which is projected to be 8.2 billion in 2024, is predicted to rise for a further 50–60 years, peaking at 10.3 billion in the middle of the 2080s (UN DESA 2024). MSW production was 2.01 billion tons yearly in 2016, which by 2050 is anticipated to increase to 3.40 billion tons (World Bank 2018). This underscores increasing environmental issues brought on by the expanding world population. In India, the Central Pollution Control Board reported that the generation of waste per person has increased significantly. At the current rate of 1.2 kg per capita per day, 1300 million tons of MSW are generated annually (CPCB India 2018). According to the World Bank (2012), this amount is expected to increase to 2200 million tons per year by 2025, posing a significant threat to environmental sustainability. Generally, wealthier cities, especially those in Organisation for Economic Cooperation and Development countries, generate more than half of the world's urban waste. This includes 34 countries across Asia, the Pacific, Europe, North America (including Canada, Mexico, and the U.S.), and South America, with an average waste generation of about 2.2 kg per capita/day. In contrast, East Asia, Africa, and South Asia have lower per capita generation rates, averaging 0.95, 0.65, and 0.45 kg/day, respectively (Hoornweg et al. 2015). This explains the gravity of the problem and the need for effective management.

Waste can be mainly divided into three categories based on its physical state—solid, liquid, and gaseous waste. Solid wastes can further be classified into municipal wastes, hazardous wastes, medical wastes, and radioactive wastes (Dehghani et al. 2021). MSW refers to waste produced by residential, commercial, or institutional operations and is typically not dangerous. The dynamic nature of waste composition and the wide range of waste generation rates in India make the development of a sustainable waste management system a challenging task. In India, organic matter makes up 51% of urban MSW, recyclables make up 17.5%, and the rest 31% is categorized as inert waste. Urban MSW has an average moisture content of 47% and a calorific value of 1751 kcal/kg (Sharma et al. 2019). Due to its diverse makeup, MSW can store energy in the form of chemical bonds between molecules of carbon, hydrogen, and oxygen. When these chemical bonds break, a large amount of energy is released, creating gases, liquids, or solids known as biofuel. With a heating value of about 20.57 MJ/kg, MSW is useful for several petrochemical, transportation, and power production industries (Ram et al. 2021).

Solid waste has different properties and compositions in the commercial, industrial, and residential domains. Solid waste management is a major challenge in India, where approximately 68.8 million tons of MSW are generated annually (Ahmad et al. 2024). Of this, 47.2 million tons are produced by 366 cities, which account for 70% of India's urban population. If current trends continue, these cities are expected to create a staggering 161 million tons of MSW by 2041, marking a fivefold increase in just four decades (Ahmad et al. 2024; Rawat and Mohanty 2021). To implement the right technology for solid waste management, it is critical to understand the composition of the waste. The Indian MSW has unique characteristics and composition compared to those of the other industrialized nations. It contains a large fraction (51%) of food and biodegradable waste, while paper constitutes only 7% of the waste (Ahluwalia and Patel 2018). This concerning prediction highlights the critical need for proactive approaches to waste management plans, with a focus on sustainable practices and creative solutions to mitigate future problems for society and the environment. The practice of segregating household, plastic, commercial, industrial, and e-waste is almost non-existent in India. Open dumping of MSW leads to chemical and biological degradation of soil and water resources, as well as causing air pollution. Currently, only 20% (27,000 tons/day) of the total MSW collected is processed; the remaining 80% (110,000 tons/day) is left untreated (MoHUA 2016), which results in its disposal in landfills and aquatic bodies, thereby polluting soil and water (Siddiqua et al. 2022). Since waste composition changes with place, time, and societal factors, we need tailored approaches for effective waste management in different contexts.

Landfilling and incineration continue to be the principal methods of MSW disposal, even in economically developed countries. Incineration is frequently combined with heat recovery to provide energy, but it produces substantial emissions, such as nitrous oxide and carbon dioxide. Furthermore, compared to landfilling, it necessitates more advanced infrastructure, limiting its use in low-income countries (Eriksson and Finnveden 2017). Composting is an alternative method of waste management in addition to incineration, which has drawbacks such as high initial and ongoing costs and the requirement for extra caution in the secure disposal of the fly ash produced. By stabilizing organic materials with the help of local microbes, composting produces green fertilizer. However, proper composting techniques are crucial to prevent problems like overpowering odor and the possible release of greenhouse gases (GHGs) (Ayilara et al. 2020). There is growing interest in valorizing waste by creating useful chemicals and fuels, both in the scientific community and beyond. This approach addresses waste management challenges while producing valuable resources. Innovative and comprehensive waste management systems are desperately needed globally to handle the growing waste volumes and reduce the negative environmental effects of inadequate waste management. These technologies can support environmental sustainability by facilitating resource recycling and managing growing waste volumes effectively.

Waste-to-energy (WtE) technologies are emerging as one of the most promising choices for producing renewable energy, given the current environmental challenges. By effectively converting waste into heat and electricity, these cutting-edge technologies can address numerous environmental problems related to the management of solid waste (Kumari et al. 2022a). Additionally, they aid in lowering the consumption of fossil fuels, which contribute to climate change and global warming through the release of greenhouse gases (GHGs) (Alqattan et al. 2018; Gupta et al. 2024). WtE technologies also offer significant economic and ecological benefits to society. MSW can be converted into a variety of chemicals and biofuels using different conversion routes. These consist of chemical reactions like hydrolysis, thermochemical reactions like pyrolysis and combustion, and biological reactions like fermentation and anaerobic digestion (Dabe et al. 2019; Rizwan et al. 2018). Each of these conversion pathways has unique advantages, enabling the conversion of MSW into a wide array of chemicals and biofuels. Thermochemical disposal techniques like pyrolysis and gasification are particularly effective and versatile for recovering fuels and value-added chemicals from MSW, including waste plastics. Pyrolysis, for example, involves the thermal decomposition of organic substances in the absence of oxygen to produce liquids (condensable), solids, and non-condensable gases like hydrogen (H_2), carbon monoxide (CO), carbon dioxide (CO_2), etc. (Maluf Filho et al. 2022 ; Gvero et al. 2017; Vershinina et al. 2022). This process provides an efficient way to convert waste plastics in MSW into valuable energy resources.

The conversion of MSW into biofuels and chemicals is influenced by several factors, including waste management practices, policies, energy availability, energy use, and accessible technology. Effective planning of MSW management (MSWM) operations requires an understanding of the current state of affairs, the ability to recognize concerns and challenges, and the capacity to highlight ongoing initiatives. This review aims to explore the main challenges and future possibilities of WtE technologies for the production of energy. Through examining these facets, the aim is to offer perspectives on how these technologies might be maximized to improve waste management and make a positive impact on the production of renewable energy and resources, which will eventually promote environmental sustainability.

9.2 Municipal Solid Waste Management Operations

MSWM encompasses a group of activities including collection, transfer, resource recovery, recycling, and treatment. These activities can vary according to the composition of municipalities, cities, countries, etc. The main operations needed for the proper and safe management of MSW generally involve three steps: segregation of waste, collection and recycling, and safe disposal. The segregation of wastes is a crucial stage in MSWM to properly separate the many different kinds of waste materials depending on many different factors such as size, biodegradability, flammability, etc. which is very helpful for sorting out non-biodegradable materials such as plastics, glass, etc. as they can act as obstacles in the treatment process needed for the other waste materials like paper, styrofoam, etc. and are essentially more suited for other kinds of treatment processes. However, this stage of segregation, while necessary, can also drive up the downstream processing costs for municipal solid waste management (Nanda and Berruti 2021). In India, this process is very unorganized, with low efficacy, occurring in extremely dangerous and hazardous conditions, with only a few of the more economically viable constituents being separated from the waste stream (Kaushal et al. 2012; Joshi and Ahmed 2016). The general advice to individual households, organizations, and industries to be active participants in the segregation of MSW at the source and the governmental agencies responsible for waste management to help the citizens by providing disposal containers/bags with distinguishable colors and text to make this process easier (Nanda and Berruti 2021) is not of much use.

The next stage in MSWM following segregation is the collection of wastes, which involves the transportation of the generated and sorted wastes from commercial, residential, and special waste sources (Zhang et al. 2022). Household waste is often disposed of in communal containers built of concrete, metal, or a combination of the three materials. Along with household waste, several important commercial organizations in the area also use these communal waste bins, except for certain commercial complexes and industrial units, which pay the municipal authorities to have their waste transferred to the disposal site (Joshi and Ahmed 2016). The industrial organization handles different types and volumes to meet the specific needs of businesses. Bulky goods, toxic materials, and e-waste are among the items that are the focus of special waste collection since they have unique disposal needs and must be handled separately (Ferronato and Torretta 2019). In India, waste collection is incredibly disorganized. The majority of urban areas notably lack MSW storage at the source. The collection containers that are used in different cities are not appropriately maintained, placed, or designed in the country (Gupta et al. 2015).

This is followed by efficient recycling and safe disposal of MSW, which are very critical stages in MSWM, aiming to reduce the use of landfills and their overall environmental impact (Liikanen et al. 2018). Recycling is the act of collecting, transforming, and creating new products out of resources, including glass, metal, plastic, and paper. Curbside recycling programs, drop-off locations, and buy-back facilities help to facilitate this process. Efficient recycling minimizes pollution, conserves energy, and lowers the need for raw materials. However, waste that cannot be recycled or reused is disposed of, mostly through incineration and landfilling (Buffington 2015). While incineration lowers waste volume and has the potential to produce electricity, modern landfills are constructed with liners and leachate systems to save groundwater. When combined, these techniques provide sustainable waste management that strikes a balance between resource recovery and environmental protection (Siddiqua et al. 2022). Around the world, a mere 25% of the collected municipal solid waste is directed towards waste management techniques like composting, anaerobic digestion, and recycling. Therefore, as waste recycling patterns and practices vary throughout countries, it is crucial to determine the most effective method for the disposal or treatment of MSW worldwide (Nanda and Berruti 2021).

Protecting public health, enhancing environmental quality, ensuring sustainability, and boosting economic output are the primary objectives of MSWM (Peri et al. 2018). To accomplish these goals, local governments must fully adopt sustainable solid waste management systems and collaborate with the public and commercial sectors. In India, the Solid Waste Management Rules (2016) regulate MSWM. Under these regulations, all municipal authorities known as urban local bodies are responsible for the creation, on-site storage, collection, transportation, processing, and treatment of MSW in an environmentally sound manner. These activities must meet the criteria of economy, energy efficiency, aesthetics, and environmental sustainability. The cost of solid waste management in India is borne by urban local bodies at a rate of INR 500 to INR 1500 per metric tonne. Of this, 60%–70% is allocated for waste collection, 20%–30% for transportation, and less than 5% for waste treatment (Radhakrishnan and Pant 2021). This breakdown of spending demonstrates how minimal emphasis is currently placed on waste disposal within MSWM procedures.

There are various methods to reduce the amount of MSW, such as anaerobic digestion, incineration, composting, waste minimization at the source, and the 3 Rs plan (Reduce, Reuse, Recycle). Figure 9.1 depicts the hierarchy of solid waste management. However, unscientific or unclean landfilling is still the most popular method for disposing of MSW in India (Kulkarni and Anantharama 2024). Choosing the right procedures for waste processing and disposal requires prior knowledge of the composition of MSW. The amount and composition of MSW differ greatly depending on variables, including dietary preferences, lifestyle choices, cultural customs, socioeconomic status, and climate. Therefore, understanding the characteristics of waste in each place is necessary for selecting appropriate waste management techniques.

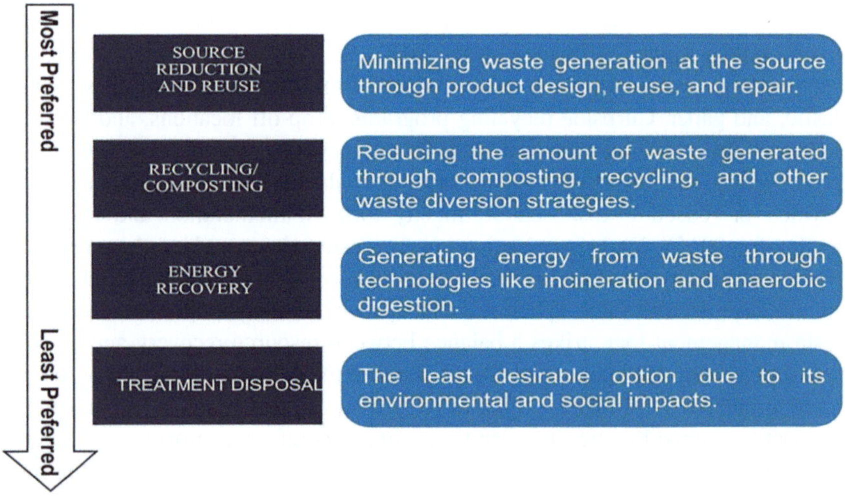

Fig. 9.1 Integrated solid waste management hierarchy

9.3 Challenges in the Conversion of MSW to Biofuels

The demand for biofuels is rising globally due to their appealing and ecologically sustainable benefits, such as being renewable, improving carbon fixation and lowering carbon dioxide emissions indirectly, promoting local economic growth, lowering air pollution from burning and decomposing biomass in fields, bringing energy security and lowering reliance on imported oil, and supporting employment opportunities for scientists, engineers, specialists, and process bioengineers (Kumari et al. 2022b). However, the conversion of MSW into fuels is fraught with difficulties for the renewable energy sector, such as feedstock heterogeneity, low energy density, variable moisture content, contaminants, infrastructure and supply chain logistics, handling regulations, and a dearth of innovative MSW characterization techniques. These challenges highlight the need for innovative solutions for the effective conversion of MSW into biofuels and chemicals. In Fig. 9.2, an overview of challenges and prospects for the conversion of MSW to biofuels and chemicals is presented.

9.3.1 Variability in Feedstock Content

MSW as a feedstock presents several challenges, including high moisture content, compositional variability, particle size and shape, density, and ash content. Normalization of the feedstock through physical or chemical pre-treatment is still required in most situations, even if MSW processing facilities include automated systems to separate and disinfect commingled waste (Mantovani et al. 2021). Converting MSW into biofuels is difficult due to its heterogeneous nature, which includes a wide range of materials like newspapers, textiles, rubber, batteries, and

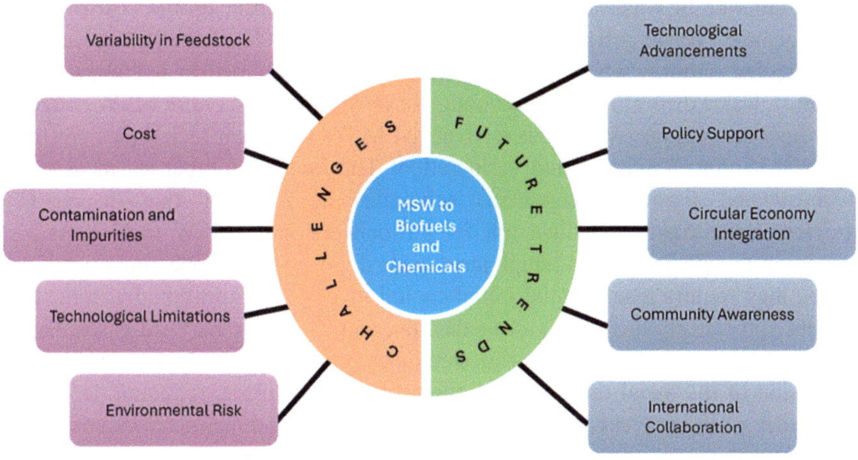

Fig. 9.2 Overview of challenges and prospects for the conversion of biofuels and chemicals from MSW

circuit boards. Certain components may not be appropriate for the manufacture of biofuel and require various treatment techniques.

During the conversion process, high moisture content enhances the generation of gas and decreases the output of liquid biofuel, or bio-oil (Dong et al. 2016). Additionally, the moisture complicates the recovery of recyclable items, increases leachate in landfills, and lowers the net caloric value of MSW for incineration (Khan et al. 2022). Furthermore, the water content of MSW might contaminate the yield of bio-oil and make the waste vulnerable to microbial contamination, posing health risks due to pathogens. The high moisture content also makes MSW heavier and more voluminous, which necessitates careful drying at high temperatures. However, the drying process is expensive and energy-intensive, making it a strain on the economy (Nanda and Berruti 2021; Logan and Visvanathan 2019). To address this issue, researchers suggest a decentralized approach where low-volume facilities use MSW locally, reducing the need for long-distance transportation and drying (Kang et al. 2021; Maluf Filho et al. 2022; Molina-Peñate et al. 2022). Moreover, a sequence of filtration and precipitation procedures is required to extract the water content from the generated biofuel.

The chemical characteristics of MSW, including ash, moisture, total solids, volatile solids, CHNSO, heavy metals, and heating values, significantly impact energy conversion efficiency. For example, the generation of biogas in MSW landfills requires a moisture level of more than 20% (wt/wt) to promote waste biodegradation by microorganisms, which in turn produces methane (Chai et al. 2016; Pearse 2019). Comprehending these waste attributes is vital for efficient waste handling and enhancing energy conversion procedures.

9.3.2 Cost

The technique of waste treatment used has a major impact on the cost of MSW management. Waste collection alone accounts for 80–90% of the MSW budget in low-income nations, where the majority of waste is dumped in open spaces. By contrast, waste disposal usually accounts for only 10% of the entire MSW budget in high-income nations, where more advanced MSW treatment techniques are used, such as sanitary dumping and incineration. The increased expenses incurred by sophisticated waste management techniques account for this discrepancy. Implementing Anaerobic digestion for the conversion of MSW to biofuels also poses certain challenges. Although large-scale digesters offer appealing operational costs, they require a substantial initial capital investment. Additionally, some hazardous substances, such as alkali metals, tar, and extremely harmful gases. Heavy metals might not be completely digested during the process, and as a result, rapid corrosion of the metal tubes used in the process takes place, reducing the overall efficiency of the process and increasing the operation and maintenance costs. This influences the operating cost as well as the maintenance cost. The financial feasibility of anaerobic digestion for MSW conversion to biofuels is thereby impacted by these factors. Though the process offers significant environmental benefits, its economic feasibility is still a matter of concern. The difficulties associated with anaerobic digestion highlight the intricacy and expense of turning MSW into biofuels (Moura and Pereira 2018).

9.3.3 Contamination and Impurities

MSW is contaminated with a variety of substances, including halogens, siloxanes, and sulfur compounds (Ortiz-Ardila et al. 2024). These contaminants taint the biogas generated during the conversion process and persist in trace amounts even after purification. Such pollutants corrode engines and other metallic components, necessitating part replacements or higher maintenance expenses (Aytav et al. 2020). Impurities also lead to subsequent thermal cracking of waste, which has a significant impact on the yield of the final product (Macor and Benato 2020). The presence of fine particles in MSW affects the quality of the biofuel produced. Only 15%–20% of all plastic scrap types can be recycled using conventional mechanical recovery procedures like sorting, grinding, washing, and extrusion (Jeswani et al. 2021; Vogt et al. 2021). Above this point, things like dirt, dust, aluminum foil, food scraps, and paper labels contaminate plastics. Thus, employing ceramic membranes or conventional upgrading methods could ensure high-quality production of biofuels like bio-oil (Mishra et al. 2023).

9.3.4 Technological Challenges

From a technological standpoint, the expenses associated with material separation can easily outweigh the value of the finished bioproducts if waste is not properly sorted. Separation technologies are needed to remove cellulose, antioxidants, amino acids, or any other undesirable materials from the refinery process chain (Kolakoti et al. 2022). In petroleum refineries, standard distillation can be employed to separate products. However, chemical components derived from biomass have lower volatility than conventional fuels, resulting in drastically different evaporation and combustion characteristics (Qamar et al. 2023; Pandey et al. 2012). Thus, in the bioeconomy, emphasis will need to be placed on more comprehensive waste sorting programs rather than only isolating organic fractions from MSW (Edo et al. 2022) and developing suitable techniques such as solvent extraction, membrane, reactive adsorption, and chromatography. Meanwhile, certain biogenic wastes are challenging to treat using the traditional biological method due to their high chemical oxygen demand, dissolved salt concentration, and micro-pollutant content (Shah et al. 2020). It might therefore be necessary to investigate and assess new and more sophisticated anaerobic treatments for the production of biogas and material recycling.

The arduous process of anaerobic digestion necessitates a hydraulic retention time of more than 30 days. This results in higher expenses because a bigger digester volume is required (Parajuli et al. 2022). Slow recovery and low loading rates are further constraints on biogas generation. The ideal temperature for biogas production is approximately 37 °C, so effective heat management systems are needed to keep this temperature constant and guarantee optimal production. Reduced biogas productivity can be caused by year-round temperature variations, particularly in the absence of adequate heat control. In colder climates, it might be difficult to maintain ideal temperatures for anaerobic digestion since wintertime temperature dips can cause a decrease in biogas generation, which lowers the process's overall efficiency (Mahmudul et al. 2021). It is challenging for areas with cold climates to effectively embrace this technology because of this constraint.

9.3.5 Environmental Risk

Heavy metals like chromium (Cr), lead (Pb), zinc (Zn), copper (Cu), and mercury (Hg) are abundant in MSW components (Gujre et al. 2021). These metals are converted and then fed into gasification plants, where they are distributed into slag, fly ash, and exhaust gas. These substances eventually build up in sediments, groundwater systems, surface water, or soil, causing serious environmental problems (Wang et al. 2022). Water contamination may result from improper management of MSW and its byproducts. Inadequate leachate management, specifically the failure of containment liners, collection drains, and treatment systems, may lead to contamination of aquatic habitats and drinking water sources by leaking into the soil and groundwater (Khan et al. 2022; Mukherjee et al. 2015). Furthermore, contaminants from MSW conversion facilities' discharge may enter adjacent bodies of water, worsening the condition of the water and endangering aquatic life (Mathana et al. 2021). Furthermore, burning biofuels made from MSW has the potential to release greenhouse gases like carbon dioxide (CO_2) into the environment. Although the production and use of biofuels are thought to be more environmentally friendly than fossil fuels, the net greenhouse gas emissions that result from these processes vary depending on several factors, such as the efficiency of the conversion process, the source of the feedstock, and the management of emissions and byproducts. Degradation of the soil, erosion, and loss of ecosystem services are further consequences of converting natural habitats to agricultural land for the production of biofuel (Sebos 2022; Rawat and Mohanty 2021).

The conversion of MSW to biofuels and chemicals faces major issues like the heterogeneity and high moisture content of feedstock, which makes the conversion process complicated and reduces the yield of biofuels. High initial capital investment expenses and the high cost of advanced waste management and treatment technology hamper the economic viability. The quality and longevity of biogas and biofuel production are impacted by contamination from different contaminants present in MSW. Technological hurdles, like the requirement for efficient material separation and temperature control, make the process even more difficult. Furthermore, environmental hazards such as greenhouse gas emissions and heavy metal contamination highlight the necessity for creative and sustainable approaches to MSW to biofuel conversion.

9.4 Future Prospects and Emerging Trends

Biofuels are expected to complement non-renewable resources like fossil fuels on a commercial scale due to their economic benefits, ease of transportation, flexibility, and promotion of circular economy principles aimed toward the use of waste as a valuable resource. However, there is still a long way to go as bioenergy expenses cannot be generalized due to their complex nature, including high costs, large-scale upkeep, and other challenges that were discussed above (Rajmohan et al. 2021). The future of the production of biofuels and chemicals from MSW holds a lot of promise, with growing interest in the field with technological advancements, regulatory backing, and rising environmental awareness aimed towards addressing challenges that ultimately act as an obstacle in the commercialization of biofuels and chemicals. Continuous efforts have been made, especially in the past couple of years, to drive innovation in MSW conversion technology, resulting in improved effectiveness, cost-effectiveness, and scalable processes. Some of these research efforts involve the exploration of newer conversion methods, catalyst research, reactor design, and process optimization to improve efficiency, yield, and product quality (Bashir et al. 2022). Recent technologies that are being studied for the conversion of MSW to biofuels and chemicals include anaerobic digestion, microwave irradiation-assisted pyrolysis, supercritical water gasification, plasma gasification, and others (Negi et al. 2019; Al Rayaan 2021). There is also a need to develop pretreatment technologies with the objective of low toxicity and minimal to no inhibitor generation through the integration of pretreatment and detoxification methods, and their further evaluation to determine their technical, environmental, and economic benefits (Ebrahimian et al. 2023).

For the scaling up and commercialization of the innovative methods and technologies being researched for the conversion of biofuels and chemicals, there is also the requirement for the establishment of supportive legal frameworks, financial incentives, and market mechanisms to encourage MSW conversion as part of sustainable waste management and renewable energy mandates through the collaboration of these facilities with consumers, including homes, enterprises, and educational institutions, to understand the impact of individual decisions on the industry (Negi et al. 2019; Batista et al. 2021). The following discusses the prospects, recommendations, and recent trends that are being seen for the large-scale production and use of biofuels and chemicals from MSW through technological advancements, policy support, incentives, resource recovery, community-based cooperative models, integration of circular economy principles, etc. Figure 9.3 represents an overview of the prospects and recent trends happening around the world, maximizing the effective conversion of MSW to biofuels and chemicals.

9.4.1 Recent Technological Advancements

In the past few years, there have been significant technological breakthroughs in the conversion of MSW to biofuels and chemicals, acting as critical drivers for overcoming current hurdles and realizing the full potential of waste-to-energy and

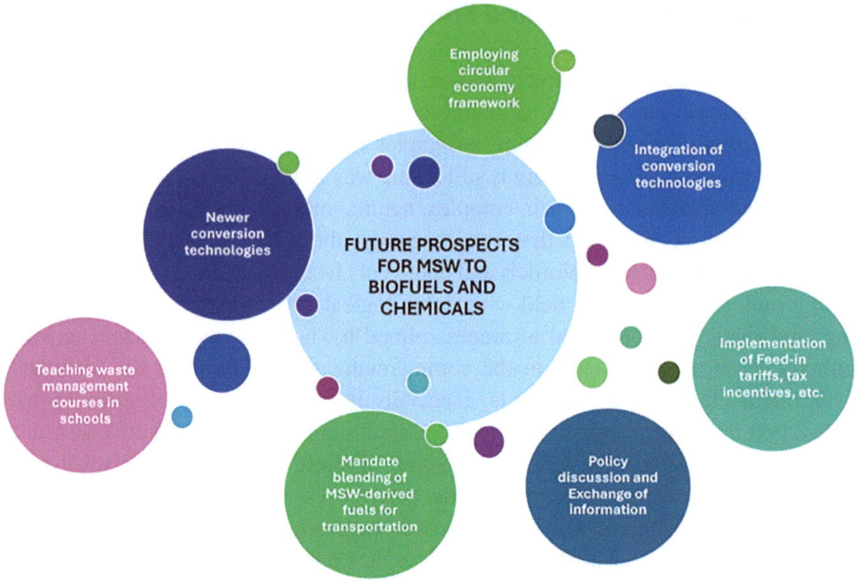

Fig. 9.3 Overview of future prospects for the conversion of biofuels and chemicals from MSW

waste-to-chemical operations. These breakthroughs include several areas of technology development, process optimization, and innovation. These breakthroughs include the study of new conversion technologies like anaerobic digestion that involve the set of metabolic events in which microorganisms degrade the organic matter of any substrate to produce a gaseous mixture (CH_4, CO_2, H_2, H_2S, etc.) in the absence of free oxygen (Uddin and Wright 2023), Microwave irradiation-assisted pyrolysis is a thermal degradation technique that utilizes microwave radiation to pyrolyze organic materials such as plastics, paper, and biomass under high temperatures (573–1073 K) facilitating the breakdown of complex compounds into valuable products like bio-oil, syngas, and char (Lam et al. 2019; Al-Rumaihi et al. 2022), Plasma gasification that is an allothermal gasification process in which the heat necessary by the endothermic reactions is provided by thermal plasma, commonly generated by direct current (DC) non-transferred arc plasma torches (Mazzoni and Janajreh 2017), etc. providing excellent environmental performance, resource recovery, and energy efficiency. There are also studies being conducted to incorporate conversion technologies, such as catalytic microwave torrification and pyrolysis, to treat waste and remove components such as hemicellulose while partially decreasing cellulose and lignin. This strategy seeks to increase the quality and yield of biofuels derived from biogenic wastes (Kang et al. 2022; Ingole et al. 2016). In addition to that, there is also a rise in studies surrounding the use of carbon nanotubes and metal oxide nanoparticles of iron, zinc, calcium, silica, etc. to act as nanocatalysts in combination with other processes to enhance biofuel production due to their size, catalytic activity, high surface area, stability, and adsorption capacity, etc. (Arya et al. 2021; Dikshit et al. 2021). Metal oxides like Fe_2O_3 and ZnO are very effective in accelerating reactions during the conversion of biomass to biofuels, such as transesterification and pyrolysis (Banerjee et al. 2019). Iron oxide, for example, can accelerate the breakdown of complex organic compounds and contribute to the production of bio-oil. Zinc oxide aids in the reduction of undesirable byproducts produced during the catalytic cracking of bio-oil, hence increasing production and quality. Calcium oxide (CaO) and silica (SiO_2) are stable and have high adsorption capability, enhancing the overall efficiency of biofuel production processes (Pandit et al. 2023).

There are also studies associated with the incorporation of several conversion routes, such as thermochemical (e.g., pyrolysis, gasification) and biochemical (e.g., fermentation, anaerobic digestion) processes, and based on different types of feedstocks, wastes, products, byproducts, platforms, processes, and industrial infrastructure into integrated biorefinery ideas that allow for synergistic interactions, waste valorization, and resource optimization (Ochieng et al. 2022). However, further research is needed to develop technologies and optimization methods that address feedstock variability and enhance process recovery.

9.4.2 Policy Support and Market Incentives

Due to the increasing amount of waste and shortage of areas for disposal leading to increasing environmental concerns, there is always a strict need for governments to take political actions toward the reduction of MSW, their management and recycling, and incentives promoting the production of bioenergy from these wastes and other important chemicals on a wider scale (Matsakas et al. 2017). In the sphere of MSW management, the EU has enacted numerous laws including the Waste Framework Directive (2008/98/EC) establishing its member nations' targets to fulfil regarding repurposing and deflecting MSW from ending up in disposal sites, intending to recycle 50% of household trash by 2020 with the top priority is to minimize the amount of new waste generation, followed by reuse and recycling (Valenćiková and Fandel 2023; Hemidat et al. 2022). The above-mentioned scheme is represented by the waste management hierarchy scheme in Fig. 9.1. The governmental institutions of any country, developed or developing, should ultimately aim toward the promotion of different technologies used for biofuels and chemical production from waste by supplying information and instruction on treatment systems, fast-tracking applications for new WtE plants, and providing financial incentives like subsidies/tax deductions for the achievement of waste reduction and other WtE goals (Malav et al. 2020). Some of these policies include the "National Policy on Biofuels," which was published by India in 2018. It was amended in 2022 to reduce petroleum fuel imports by increasing domestic biofuel production (IEA 2023), the United States Energy Policy Act of 2005 that promotes biofuel research and development through a variety of economic incentives such as grants, income tax credits, subsidies, and loans (US EPA 2024), etc. These policies can ultimately complete their set goals through the implementation of feed-in-tariffs, renewable portfolio standards, tax incentives, tax on greenhouse emissions or usage of fossil fuels, ban or restriction of landfill disposal of waste to only under dire circumstances, R&D grants and loans, renewable fuel standards to mandate the blending of renewable fuels like MSW-derived fuel to transportation fuel, etc. as a way to essentially encourage investors and developers in the recovery of MSW and the large-scale production and infrastructure of biofuels and chemicals from these wastes.

9.4.3 Integration of Circular Economy Principles

The purpose of current and future policies being implemented by governmental institutions should also be to integrate into the circular economy framework, with an emphasis on resource recovery, waste minimization, and byproduct valorization (Iacovidou et al. 2017). Circular economy is a production and consumption paradigm that emphasizes sharing, leasing, reusing, repairing, refurbishing, and recycling existing materials and products as feasible. This extends the product's life cycle, aiming towards reducing waste through repurposing it into valuable resources and reusability of byproducts from these resources again and again, creating additional value (Hemidat et al. 2022). The policymakers should take these principles into understanding during the introduction of policies, applying them to MSW conversion processes maximizing recovery of wastes and byproducts, their valorization, segregation, and recycling of said wastes, employing circular supply chains like integrated biorefineries, and continuously promoting improvement and innovation through the utilization of these principles in the field of MSW to biofuels and chemical systems around the world.

9.4.4 Community Awareness

Community awareness is critical in increasing understanding, involvement, and support for efforts to turn municipal solid waste, or MSW, into biofuels and chemicals (Tsui and Wong 2019). The public's awareness of the environmental concerns faced due to improper disposal can essentially help in the implementation of policies for proper waste management and their conversion in a swift manner. People should also learn to properly characterize and segregate wastes for proper waste management and more effective bioenergy production, with facilitations provided for this purpose in locations of disposal, and training courses being provided for a shorter period aimed towards understanding the importance of segregation of wastes into inert, biodegradable, and recyclable wastes. The governing bodies should also take the appropriate steps to encourage scientific and professional schools to integrate hands-on waste management courses in their curricula. The academic community should assist with trash classification, especially if it occurs near them (Malav et al. 2020). Another method to spread the word about the value of MSW conversion and sustainable waste management can be through the launch of multimedia campaigns through the medium of television, newspapers, radio, social media, etc. One great example of this level of campaign is the "Swachh Bharat

Mission," which is a country-wide campaign launched in 2014 by the Indian government to eliminate open defecation by 2019 and to promote the improvement of solid waste management and production of biogas from these wastes (SBMG n.d.). This is assuredly helpful as stakeholders may develop cooperation, trust, and support for MSW conversion initiatives by placing a high priority on community knowledge and involvement. This will also instill a sense of shared responsibility and ownership for building a more resilient and sustainable community.

9.4.5 Knowledge Sharing and International Collaborative Efforts

The field of turning MSW into biofuels and chemicals also ultimately requires advancement through worldwide collaboration and knowledge sharing. The sharing of best practices, lessons discovered, and success stories from MSW conversion programs worldwide is ultimately made possible by international collaboration. Researchers, legislators, business executives, and community leaders can exchange ideas, present their findings, and work together on projects through venues like conferences, workshops, and online discussion boards. In the fields of waste-to-energy and waste-to-chemicals, this kind of collaboration between research centers, academic institutions, and industrial players stimulates interdisciplinary cooperation, innovation ecosystems, and intellectual interchange (Malav et al. 2020). This ultimately encourages policy discussion, standardization initiatives, and information exchange on standards, best practices, and regulatory frameworks about waste management, renewable energy, and environmental protection. This can be done through engaging in global accords, campaigns, and associations, like the Paris Agreement, Sustainable Development Goals (SDGs), and Circular Economy Action Plan, which offers nations the chance to exchange insights, establish mutual objectives, and synchronize efforts to tackle worldwide predicaments concerning waste handling and climate change.

9.5 Conclusion

The exponential increase in waste production poses serious environmental and economic problems, especially in developing countries like India. The urgent need for creative waste management solutions is highlighted by the anticipated rise in the production of MSW. Waste-to-energy technologies, including pyrolysis, gasification, torrefaction, etc., present viable ways to transform waste into useful energy sources, lowering the dependency on fossil fuels and subsequently reducing environmental consequences. To efficiently commercialize biofuels and chemicals derived from MSW, however, issues including high costs and technological constraints need to be resolved. Market processes, financial incentives, and supportive regulatory frameworks are critical to overcoming these obstacles. Promoting sustainable waste management techniques requires cooperation between consumers and trash management facilities. We can realize the promise of WtE technologies and transition to a more ecologically friendly and sustainable method of waste management and energy production by promoting innovation, putting in place sensible policies, and encouraging teamwork.

References

Ahluwalia IJ, Patel U (2018) Solid waste management in India: an assessment of resource recovery and environmental impact. Indian Council for Research on International Economic Relations Working Paper No. 356

Ahmad W, Hassan M, Masud SFB, Amjad MS, Samara F, Anwar M, Nawaz T (2024) Socio-economic benefits and policy implications of generating sustainable energy from municipal solid waste in Pakistan. Energy Clim Change 5:100124. https://doi.org/10.1016/j.egycc.2023.100124

Al Rayaan MB (2021) Recent advancements of thermochemical conversion of plastic waste to biofuel-A review. Cleaner Eng Technol 2:100062. https://doi.org/10.1016/j.clet.2021.100062

AlQattan N, Acheampong M, Jaward FM, Ertem FC, Vijayakumar N, Bello T (2018) Reviewing the potential of waste-to-energy (WTE) technologies for sustainable development goal (SDG) numbers seven and eleven. Renew Energy Focus 27:97–110. https://doi.org/10.1016/j.ref.2018.09.005

Al-Rumaihi A, Shahbaz M, Mckay G, Mackey H, Al-Ansari T (2022) A review of pyrolysis technologies and feedstock: a blending approach for plastic and biomass towards optimum biochar yield. Renew Sust Energ Rev 167:112715. https://doi.org/10.1016/j.rser.2022.112715

Arya I, Poona A, Dikshit PK, Pandit S, Kumar J, Singh HN, Kumar S (2021) Current trends and future prospects of nanotechnology in biofuel production. Catalysts 11(11):1308. https://doi.org/10.3390/catal11111308

Ayilara MS, Olanrewaju OS, Babalola OO, Odeyemi O (2020) Waste management through composting: challenges and potentials. Sustain For 12(11):4456. https://doi.org/10.3390/su12114456

Aytav E, Koçar G, Teksan AE, Şahin M (2020) Experimental investigation of biogas source H2SO4 on engine oil and bearing materials in gas engines. Int J Energy Stud 5(2):71–87

Banerjee S, Rout S, Banerjee S, Atta A, Das D (2019) Fe2O3 nanocatalyst aided transesterification for biodiesel production from lipid-intact wet microalgal biomass: a biorefinery approach. Energy Convers Manag 195:844–853. https://doi.org/10.1016/j.enconman.2019.05.060

Bashir MA, Wu S, Zhu J, Krosuri A, Khan MU, Aka RJN (2022) The recent development of advanced processing technologies for biodiesel production: a critical review. Fuel Process Technol 227:107120. https://doi.org/10.1016/j.fuproc.2021.107120

Buffington J (2015) The recycling myth: disruptive innovation to improve the environment. Bloomsbury Publishing, London

Batista M, Caiado RGG, Quelhas OLG, Lima GBA, Leal Filho W, Yparraguirre ITR (2021) A framework for sustainable and integrated municipal solid waste management: Barriers and critical factors to developing countries. J Clean Prod 312:127516. https://doi.org/10.1016/j.jclepro.2021.127516

Chai X, Tonjes DJ, Mahajan D (2016) Methane emissions as energy reservoir: context, scope, causes and mitigation strategies. Prog Energy Combust Sci 56:33–70. https://doi.org/10.1016/j.pecs.2016.05.001

CPCB India (2018) MSW-generation & composition. Accessed 9 March 2024

Dabe SJ, Prasad PJ, Vaidya AN, Purohit HJ (2019) Technological pathways for bioenergy generation from municipal solid waste: renewable energy option. Environ Prog Sustain Energy 38(2):654–671. https://doi.org/10.1002/ep.12981

Dehghani MH, Omrani GA, Karri RR (2021) Solid waste—sources, toxicity, and their consequences to human health. In: Soft computing techniques in solid waste and wastewater management. Elsevier, Amsterdam, pp 205–213. https://doi.org/10.1016/B978-0-12-824463-0.00013-6

Dikshit PK, Kumar J, Das AK, Sadhu S, Sharma S, Singh S, Kim BS (2021) Green synthesis of metallic nanoparticles: applications and limitations. Catalysts 11(8):902. https://doi.org/10.3390/catal11080902

Dong J, Chi Y, Tang Y, Ni M, Nzihou A, Weiss-Hortala E, Huang Q (2016) Effect of operating parameters and moisture content on municipal solid waste pyrolysis and gasification. Energy Fuel 30(5):3994–4001. https://doi.org/10.1021/acs.energyfuels.6b00042

Ebrahimian F, Denayer JF, Mohammadi A, Khoshnevisan B, Karimi K (2023) A critical review on pretreatment and detoxification techniques required for biofuel production from the organic fraction of municipal solid waste. Bioresour Technol 368:128316. https://doi.org/10.1016/j.biortech.2022.128316

Edo C, Fernández-Piñas F, Rosal R (2022) Microplastics identification and quantification in the composted organic fraction of municipal solid waste. Sci Total Environ 813:151902. https://doi.org/10.1016/j.scitotenv.2021.151902

El-Mathana ME, Mostafa NG, Galal MM, Elawwad A (2021) Assessment and simulation of a solid waste dumpsite impact on the surrounding water resources: A case study in Abu Zaabal, Egypt. Heliyon 7(11)

Eriksson O, Finnveden G (2017) Energy recovery from waste incineration—the importance of technology data and system boundaries on CO_2 emissions. Energies 10(4):539. https://doi.org/10.3390/en10040539

Ferronato N, Torretta V (2019) Waste mismanagement in developing countries: a review of global issues. Int J Environ Res Public Health 16(6):1060. https://doi.org/10.3390/ijerph16061060

Gujre N, Mitra S, Soni A, Agnihotri R, Rangan L, Rene ER, Sharma MP (2021) Speciation, contamination, ecological and human health risks assessment of heavy metals in soils dumped with municipal solid wastes. Chemosphere 262:128013. https://doi.org/10.1016/j.chemosphere.2020.128013

Gupta N, Yadav KK, Kumar V (2015) A review on current status of municipal solid waste management in India. J Environ Sci 37:206–217. https://doi.org/10.1016/j.jes.2015.01.034

Gupta MM, Dutta R, Kumari A, Bala K (2024) State-of-the-art of microbial whole-cell catalysts for biofuel production. In: Whole-cell biocatalysis. Apple Academic Press, Palm Bay, pp 355–375. https://doi.org/10.1201/9781003413134-14

Gvero P, Mujanić I, Papuga S, Vasković S, Anatunović R (2017) Review of synthetic fuels and new materials production based on pyrolysis technologies. Adv Appl Ind Biomater 65-85. https://doi.org/10.1007/978-3-319-62767-0_4

Hemidat S, Achouri O, El Fels L, Elagroudy S, Hafidi M, Chaouki B, Guo J (2022) Solid waste management in the context of a circular economy in the MENA region. Sustain For 14(1):480. https://doi.org/10.3390/su14010480

Hoornweg D, Bhada-Tata P, Kennedy C (2015) Peak waste: when is it likely to occur? J Ind Ecol 19(1):117–128. https://doi.org/10.1111/jiec.12165

Iacovidou E, Millward-Hopkins J, Busch J, Purnell P, Velis CA, Hahladakis JN, Brown A (2017) A pathway to circular economy: developing a conceptual framework for complex value assessment of resources recovered from waste. J Clean Prod 168:1279–1288. https://doi.org/10.1016/j.jclepro.2017.09.002

IEA (2023). National policy on biofuels. https://www.iea.org/policies/17006-national-policy-on-biofuels-2022-amendment. Accessed 10 March 2024

Jeswani H, Krüger C, Russ M, Horlacher M, Antony F, Hann S, Azapagic A (2021) Life cycle environmental impacts of chemical recycling via pyrolysis of mixed plastic waste in comparison with mechanical recycling and energy recovery. Sci Total Environ 769:144483Ingole PM, Ranveer AC, Deshmukh SM, Deshmukh SK (2016) Microwave assisted pyrolysis of biomass: a review. Int J Adv Technol Eng Sci 4(6):78–84

Joshi R, Ahmed S (2016) Status and challenges of municipal solid waste management in India: a review. Cogent Environ Sci 2(1):1139434. https://doi.org/10.1080/23311843.2016.1139434

Kang K, Klinghoffer NB, ElGhamrawy I, Berruti F (2021) Thermochemical conversion of agroforestry biomass and solid waste using decentralized and mobile systems for renewable energy and products. Renew Sust Energ Rev 149:111372. https://doi.org/10.1016/j.rser.2021.111372

Kang K, Nanda S, Hu Y (2022) Current trends in biochar application for catalytic conversion of biomass to biofuels. Catal Today 404:3–18. https://doi.org/10.1016/j.cattod.2022.06.033

Kaushal RK, Varghese GK, Chabukdhara M (2012) Municipal solid waste management in India-current state and future challenges: a review. Int J Eng Sci Technol 4(4):1473–1489

Khan S, Anjum R, Raza ST, Bazai NA, Ihtisham M (2022) Technologies for municipal solid waste management: current status, challenges, and future perspectives. Chemosphere 288:132403. https://doi.org/10.1016/j.chemosphere.2021.132403

Kolakoti A, Setiyo M, Rochman ML (2022) A green heterogeneous catalyst production and characterization for biodiesel production using RSM and ANN approach. Int J Renew Energy Dev 11(3):43627. https://doi.org/10.14710/ijred.2022.43627

Kulkarni BN, Anantharama V (2024) Comparative analysis of soil contamination caused by existing municipal solid waste management facilities. Environ Nanotechnol Monitor Manage 22:100979. https://doi.org/10.1016/j.enmm.2024.100979

Kumari A, Dutta R, Gupta MM, Rai MP, Srivastava S, Chandra S, Nimkar V (2022a) Wastewater and solid waste as feedstock for energy production. In: Novel feedstocks for biofuels production. Springer, Singapore, pp 219–270. https://doi.org/10.1007/978-981-19-3582-4_8

Kumari A, Kundu PK, Gupta MM, Bala K, Chandra S, Dutta R, Das A (2022b) Role of microorganisms in production of biofuels. In: Novel feedstocks for biofuels production. Springer, Singapore, pp 65–116. https://doi.org/10.1007/978-981-19-3582-4_4

Lam SS, Mahari WAW, Ok YS, Peng W, Chong CT, Ma NL, Tsang DC (2019) Microwave vacuum pyrolysis of waste plastic and used cooking oil for simultaneous waste reduction and sustainable energy conversion: recovery of cleaner liquid fuel and techno-economic analysis. Renew Sust Energ Rev 115:109359. https://doi.org/10.1016/j.rser.2019.109359

Liikanen M, Havukainen J, Viana E, Horttanainen M (2018) Steps towards more environmentally sustainable municipal solid waste management–a life cycle assessment study of São Paulo, Brazil. J Clean Prod 196:150–162. https://doi.org/10.1016/j.jclepro.2018.06.005

Logan M, Visvanathan C (2019) Management strategies for anaerobic digestate of organic fraction of municipal solid waste: current status and future prospects. Waste Manag Res 37(1_suppl):27–39. https://doi.org/10.1177/0734242X18816793

Macor A, Benato A (2020) Regulated emissions of biogas engines—on site experimental measurements and damage assessment on human health. Energies 13(5):1044. https://doi.org/10.3390/en13051044

Mahmudul HM, Rasul MG, Akbar D, Narayanan R, Mofijur M (2021) A comprehensive review of the recent development and challenges of a solar-assisted biodigester system. Sci Total Environ 753:141920. https://doi.org/10.1016/j.scitotenv.2020.141920

Malav LC, Yadav KK, Gupta N, Kumar S, Sharma GK, Krishnan S, Bach QV (2020) A review on municipal solid waste as a renewable source for waste-to-energy project in India: current practices, challenges, and future opportunities. J Clean Prod 277:123227. https://doi.org/10.1016/j.jclepro.2020.123227

Maluf Filho D, Coelho ST, Perecin D (2022) Opportunities and challenges of gasification of municipal solid waste (MSW) in Brazil. Energies 15(8):2735. https://doi.org/10.3390/en15082735

Mantovani L, Tribaudino M, Matteis CD, Funari V (2021) Particle size and potential toxic element speciation in municipal solid waste incineration (MSWI) bottom ash. Sustain For 13(4):1911. https://doi.org/10.3390/su13041911

Matsakas L, Gao Q, Jansson S, Rova U, Christakopoulos P (2017) Green conversion of municipal solid wastes into fuels and chemicals. Electron J Biotechnol 26:69–83. https://doi.org/10.1016/j.ejbt.2017.01.004

Mazzoni L, Janajreh I (2017) Plasma gasification of municipal solid waste with variable content of plastic solid waste for enhanced energy recovery. Int J Hydrog Energy 42(30):19446–19457. https://doi.org/10.1016/j.ijhydene.2017.06.069

Mishra RK, Saini R, Kumar DJP, Sankannavar R, Binnal P, Dwivedi N, Kumar P (2023) Thermo-catalytic pyrolysis of Azadirachta indica seeds over CaO and CuO: pyrolysis kinetics, impact of catalysts on yield, fuel properties and its chemical compositions. J Energy Inst 111:101366. https://doi.org/10.1016/j.joei.2023.101366

MoHUA (2016) Annual report. Ministry of housing and urban Affairs, government of India. https://mohua.gov.in/upload/uploadfiles/files/annual%20report%20English%20-%20print.pdf

Molina-Peñate E, Artola A, Sánchez A (2022) Organic municipal waste as feedstock for biorefineries: bioconversion technologies integration and challenges. Rev Environ Sci Biotechnol 21(1):247–267. https://doi.org/10.1007/s11157-021-09605-w

Moura I, Pereira D (2018) Anaerobic digestion of MSW: challenges of a high cost technology. In: Modeling innovation sustainability and technologies: economic and policy perspectives. Springer, Cham, pp 295–300. https://doi.org/10.1007/978-3-319-67101-7_22

Mukherjee S, Mukhopadhyay S, Hashim MA, Sen Gupta B (2015) Contemporary environmental issues of landfill leachate: assessment and remedies. Crit Rev Environ Sci Technol 45(5):472–590. https://doi.org/10.1080/10643389.2013.876524

Nanda S, Berruti F (2021) Municipal solid waste management and landfilling technologies: a review. Environ Chem Lett 19(2):1433–1456. https://doi.org/10.1007/S10311-020-01100-Y

Negi H, Agrawal R, Verma A, Goel R (2019) Municipal solid waste to bioenergy: current status, opportunities, and challenges in Indian context. In: New and future developments in microbial biotechnology and bioengineering. Elsevier, Amsterdam, pp 191–203. https://doi.org/10.1016/B978-0-444-64191-5.00014-6

Ochieng R, Gebremedhin A, Sarker S (2022) Integration of waste to bioenergy conversion systems: a critical review. Energies 15(7):2697. https://doi.org/10.3390/en15072697

Ortiz-Ardila AE, Restrepo JD, Angenent LT, Usack JG, Labatut RA (2024) Protecting human health and the environment against siloxanes: the role and effectiveness of wastewater treatment technologies. Crit Rev Environ Sci Technol 54(1):68–94. https://doi.org/10.1080/10643389.2023.2221156

Pandey RK, Rehman A, Sarviya RM (2012) Impact of alternative fuel properties on fuel spray behavior and atomization. Renew Sust Energ Rev 16(3):1762–1778. https://doi.org/10.1016/j.rser.2011.11.010

Pandit C, Banerjee S, Pandit S, Lahiri D, Kumar V, Chaubey KK, Joshi SJ (2023) Recent advances and challenges in the utilization of nanomaterials in transesterification for biodiesel production. Heliyon 9(4):e15475. https://doi.org/10.1016/j.heliyon.2023.e15475

Parajuli A, Khadka A, Sapkota L, Ghimire A (2022) Effect of hydraulic retention time and organic-loading rate on two-staged, semi-continuous mesophilic anaerobic digestion of food waste during start-up. Fermentation 8(11):620. https://doi.org/10.3390/fermentation8110620

Pearse LF (2019) Biochemical methane potential of landfilled municipal solid waste using a non-slurry approach. Department of Civil Engineering

Peri G, Ferrante P, La Gennusa M, Pianello C, Rizzo G (2018) Greening MSW management systems by saving footprint: the contribution of the waste transportation. J Environ Manag 219:74–83. https://doi.org/10.1016/j.jenvman.2018.04.098

Qamar OA, Jamil F, Hussain M, Alaa H, Inayat A, Waris A, Park YK (2023) Feasibility-to-applications of value-added products from biomass: current trends, challenges, and prospects. Chem Eng J 454:140240. https://doi.org/10.1016/j.cej.2022.140240

Radhakrishnan G, Pant A (2021) PPPs in solid waste management in India. JGBS working papers, pp 1–15

Rajmohan KS, Ramya C, Varjani S (2021) Trends and advances in bioenergy production and sustainable solid waste management. Energy Environ 32(6):1059–1085. https://doi.org/10.1177/0958305X19882415

Ram C, Kumar A, Rani P (2021) Municipal solid waste management: a review of waste to energy (WtE) approaches. Bioresources 16(2):4275

Rawat P, Mohanty S (2021) Utilization of municipal solid waste as backfill material. In: Proceedings of the Indian geotechnical conference 2019: IGC-2019 volume II. Springer, Singapore, pp 49–62. https://doi.org/10.1007/978-981-33-6370-0_5

Rizwan M, Saif Y, Almansoori A, Elkamel A (2018) Optimal processing route for the utilization and conversion of municipal solid waste into energy and valuable products. J Clean Prod 174:857–867. https://doi.org/10.1016/j.jclepro.2017.10.335

SBMG (n.d.). https://swachhbharatmission.ddws.gov.in/. Accessed 10 March 2024

Sebos I (2022) Fossil fraction of CO_2 emissions of biofuels. Carbon Manage 13(1):154–163. https://doi.org/10.1080/17583004.2022.2046173

Shah AI, Dar MUD, Bhat RA, Singh JP, Singh K, Bhat SA (2020) Prospectives and challenges of wastewater treatment technologies to combat contaminants of emerging concerns. Ecol Eng 152:105882. https://doi.org/10.1016/j.ecoleng.2020.105882

Sharma A, Ganguly R, Gupta AK (2019) Municipal solid waste characterization and analysis in Himachal Pradesh (Doctoral dissertation, Jaypee University of Information Technology, Solan, HP)

Siddiqua A, Hahladakis JN, Al-Attiya WAK (2022) An overview of the environmental pollution and health effects associated with waste landfilling and open dumping. Environ Sci Pollut Res 29(39):58514–58536. https://doi.org/10.1007/s11356-022-21578-z

Tsui TH, Wong JW (2019) A critical review: emerging bioeconomy and waste-to-energy technologies for sustainable municipal solid waste management. Waste Disp Sustain Energy 1:151–167. https://doi.org/10.1007/s42768-019-00013-z

Uddin MM, Wright MM (2023) Anaerobic digestion fundamentals, challenges, and technological advances. Phys Sci Rev 8(9):2819–2837

UN DESA (2024). https://desapublications.un.org/publications/world-population-prospects-2024-summary-results. Accessed 10 March 2024

US EPA (2024). https://www.epa.gov/environmental-economics/economics-biofuels#:~:text=U.S.%20policy%20approaches%20to%20support%20biofuel%20production,-The%20Energy%20Policy&text=EISA%20expanded%20the%20Renewable%20Fuel,from%20feedstocks%20other%20than%20cornstarch. Accessed 10 March 2024

Valenćiková M, Fandel P (2023) Assessing waste management efficiency in the European Union: a focus on the Slovak Republic. Ecocycles 9(2):7–25

Vershinina K, Nyashina G, Strizhak P (2022) Combustion, pyrolysis, and gasification of waste-derived fuel slurries, low-grade liquids, and high-moisture waste. Appl Sci 12(3):1039. https://doi.org/10.3390/app12031039

Vogt BD, Stokes KK, Kumar SK (2021) Why is recycling of postconsumer plastics so challenging? ACS Appl Polym Mater 3(9):4325–4346. https://doi.org/10.1021/acsapm.1c00648

Wang S, Han Z, Wang J, He X, Zhou Z, Hu X (2022) Environmental risk assessment and factors influencing heavy metal concentrations in the soil of municipal solid waste landfills. Waste Manag 139:330–340. https://doi.org/10.1016/j.wasman.2021.11.036

World Bank (2012) What a waste: a global review of solid waste management. Urban development series knowledge papers. https://documents.worldbank.org/en/publication/documents-reports/documentdetail/302341468126264791/what-a-waste-a-global-review-of-solid-waste-management. Accessed 10 March 2024

World Bank (2018) What a waste 2.0: a global snapshot of solid waste management to 2050. World Bank data. https://data.worldbank.org/indicator/SP.POP.TOTL?locations¼IN. Accessed 9 March 2024

Zhang X, Liu C, Chen Y, Zheng G, Chen Y (2022) Source separation, transportation, pretreatment, and valorization of municipal solid waste: a critical review. Environ Dev Sustain. https://doi.org/10.1007/S10668-021-01932-W

Part II

Agriculture Solid Waste to Biofuels and Chemical Products

Agriculture Solid Waste as Feedstocks for the Production of Biofuels and Chemicals

10

Aryan Prajapati, Kavya Adlakha, Bhawna Jain, Meenal Rastogi, and Smriti Shrivastava

Abstract

Ongoing deterioration of the environment and the instability of the fossil fuel market have increased the pace of the hunt for alternative fuels. Owing to the phenomenal increase in the world's population, better farming, business, and business endeavors, improved farming and other food production, the production of agricultural waste, the use of renewable fuels, and the emission of harmful gases. There has never been a more pressing demand for agricultural waste to biofuels that is affordable, easily accessible, and safe for the environment. Biofuels are inexpensive, environmentally benign, renewable, and biodegradable fuels made from leftover lignocellulosic biomass by microorganisms. This paper examines the process of turning agricultural wastes into biofuels, focusing on the many applications of bioethanol, biohydrogen, butanol, biomethane, and biodiesel. It is reasonable to draw the conclusion that wastes produced by agricultural processes and activities have value and can be used to help reach the worldwide renewable energy target in an accessible and cost-effective manner. The investigation's findings will advance our body of knowledge and offer fresh approaches and methods for using agricultural waste. More multidisciplinary research is required to develop cutting-edge, environmentally responsible processes. The present chapter gives a detailed description of various agricultural residues that can be used to produce diverse value-added products.

A. Prajapati · K. Adlakha · B. Jain · S. Shrivastava (✉)
Amity Institute of Biotechnology, Amity University, Noida, Uttar Pradesh, India

M. Rastogi
CSIR-Institute of Microbial Technology, Chandigarh, Punjab, India

A. Kumari et al. (eds.), *Valorization of Solid Wastes to Biofuels and Chemical Products for Sustainable World*, https://doi.org/10.1007/978-981-96-8594-3_10

Keywords

Agricultural wastes · Lignocellulosic biomass · Biofuel · Bio methane · Waste conversion · Biodiesel

10.1 Introduction

Worldwide, fossil fuels are utilized to provide energy, yet they have serious drawbacks, including global warming. As biomass is one of the more promising environmentally friendly energy resources, biomass-based fuels offer a tremendous potential alternative to minimize the usage of fossil fuels (Wang et al. 2021a). To feed a growing population, agricultural and industrial activities have intensified, resulting in a notable increase in biomass waste; nevertheless, disposal, utilization, and management strategies are not effectively implemented (Tripathi et al. 2019).

Renewable energy sources, for instance, solar, wind, hydro-power, and biomass, are increasingly pursued as alternatives to mitigate pollution attributed to fossil fuel usage. Biomass derived from diverse sources (agricultural, forestry, and aquatic) can be utilized for the production of varied biofuels (solid, liquid, or gaseous). Currently, Brazil and the United States are leading producers of biofuels (Food and Agriculture Organization of the United Nations (FAO) 2016, 2017a, b; European Commission 2015), with Europe and Eurasia holding the largest share in global power generation (United Nations 2019).

Second-generation biofuels have immense application potential. Various surveys have predicted that only 10% of the global residues are sufficient to address 50% of the entire biofuel demand tree by 2030. Extensive research is ongoing in algal engineering for enhanced sustainable biofuel production and exploring algal biodiversity to identify highly efficient fuel-producing species (European Commission 2015). To date, ethanol accounts for up to 75% of total biofuel consumption (Bracco et al. 2018). Global bioethanol production has risen significantly since 2007, from 13.1 to 25.7 billion gallons in 2015, with the United States and Brazil predominantly using simple feedstocks such as corn and sugarcane, respectively (McCormick and Kautto 2013; Scarlat et al. 2015). India, being the world's second-largest sugarcane producer, contributes to only 2% of global bioethanol production.

Biofuels have gained recognition as viable substitutes for liquid fuels, such as biobutanol, biogas, and bioethanol (Amiri et al. 2015). One practical means of reducing the undesirable environmental effects of the extraction, processing, and use of fossil fuels is the utilization of renewable energy for a variety of purposes. This has inspired governments, environmentalists, and international organizations to keep pushing for more funding, investments, policies, and human resources to support fuel refiners and researchers in their efforts to create affordable, environmentally friendly energy sources, particularly biofuel.

The materials produced by agricultural processes and activities through the value chain are known as agricultural wastes. They can be produced as by-products, finished products, or raw materials from a variety of processes and activities. The materials are discarded as waste since they are no longer suitable for use (Ylä-Mella et al. 2022). Off-cuts, crop residues, waste from factories, animal wastes, and wastes connected to food are the different types of agricultural wastes (Pattanaik et al. 2019).

Alternate and sustainable fuel sources are major concerns, and bioethanol from various agro-residues is gaining a lot of importance. As an agricultural economy, India has abundant substrates to be converted into products. In addition to basic techniques, high-throughput methods involving next-generation sequencing, metagenomics, and others play a crucial role in improving product quality. Our research focuses on optimizing processes towards efficient saccharification of agricultural biomass for the extraction of monosaccharides, which are subsequently fermented and hence boost ethanol production from varied agricultural substrates (Shrivastava et al. 2014).

10.2 Agriculture Solid Waste

The agricultural sector is one of the major contributors to the generation of agricultural solid waste, which can accumulate indiscriminately and pose significant risks to global health and food security if left unmanaged. Alternatively, these wastes can serve as valuable raw materials for the bio-economy (Rastogi and Shrivastava 2017; Aro 2016; Awasthi et al. 2015). Recycling agricultural waste offers several benefits, including reduced greenhouse gas emissions, decreased dependency on fossil fuels, support for green market development, employment opportunities, bioenergy production, and bio-conversion of agricultural solid wastes to animal feed (Renewable Fuels Association 2016; Cheng and Timilsina 2011). Global food production has been estimated to decline by about 42% if pesticide use is completely halted (Zhao et al. 2012). Agricultural solid waste generation is reported to grow continually at an annual rate of approximately 7.5% (Scully and Orlygsson 2015; Plácido and Capareda 2015). In many developing countries, these wastes are indiscriminately dumped or burned in public areas, leading to air pollution, soil contamination, generation of harmful gases, smoke, and dust, and water pollution through runoff, thereby harming both terrestrial and aquatic ecosystems (Martin et al. 2002; Asgher et al. 2013; Gutiérrez et al. 2012).

The development and use of biofuels have drawn more attention due to factors such as the depletion of the world's oil reserves, the volatility of crude oil prices, and the environmental effects of using fossil fuels. Biofuels are seen to be a viable substitute for non-renewable fuels and alleviate greenhouse gas emissions, increase energy security, and lessen dangerous emissions from transportation engines. The production of biofuels has increased globally. According to reliable sources, the production of biofuel was 59 million oil equivalent tons (metric tons) in 2010, increased to 96 metric tons in 2010, and is expected to reach 98 metric tons by 2022 (United Nations 2019). Biofuel is a naturally occurring bio-based fuel that is produced by living things or other processes from a variety of feedstocks (Awogbemi et al. 2021). Biodiesel, bioethanol, biohydrogen, biogas, biomethane, and green diesel are typical forms of biofuel. Biofuels are produced from plants, animal waste,

Table 10.1 Classification of agricultural solid waste

Agricultural solid waste			
Crop residues	Livestock wastes	Fruit wastes	Industrial processing wastes
• Rice straw • Wheat straw • Barley straw • Leaves • Corn stover	• Cattle manure • Animal fat • Bird feathers • Pig manure	• Cabbage • Mango • Lettuce • Tomato	• Rice husk • Rice barn • Banana peel • Sugarcane bagasse • Egg shells

manure, sludge, and other waste materials. They are biodegradable, sustainable, affordable, and environmentally benign. The issues with using edible feedstocks for biofuel production have been resolved by using lignocellulosic waste, wastewater, and other inedible feedstocks (Table 10.1).

Engine wear and NOx emissions are increased when internal combustion engines run on biodiesel, a type of liquid biofuel. Moreover, methane, a vital component of biogas, is a substantial contributor to global climate change, and the production and use of biogas, a gaseous biofuel, exacerbates ozone layer depletion (Awogbemi et al. 2021). Even with its unquestionable downside, biofuel is still a cost-effective, clean, and ecological energy source. Utilizing biofuel has the potential to mitigate environmental deterioration and protect the environment for future generations. There is potential for carbon mitigation and CO_2 reduction through the use of biofuels in a variety of applications. According to Mukhtar et al. (2021) and Nakamya (2022), biofuels can promote and aid in social inclusion, employment creation, poverty reduction, and economic development.

10.2.1 Composition of Agricultural Solid Waste

Agricultural wastes are categorized as lignocellulosic biomass, which is made up of complex molecules including cellulose, hemicellulose, lignin, ash, and a few extractives, primarily protein. Before being converted into new configurations, the wastes' structures first break down into simple monomers during degradation (Stefanidis et al. 2014). The majority of agricultural wastes include a higher percentage of cellulose than hemicellulose or lignin combined. Cellulosic properties of biomass determine its mechanical stability, strength, and chemical fingerprint (Dhyani and Bhaskar 2018).

The primary link between cellulose and lignin is hemicellulose, which is the random, heterogeneous structure of branching polysaccharides found in cellulose. It is made of arabans, xylans, galactans, and various sugars derived from a heterogeneous mixture of sugars such as d-xylose, d-mannose, and d-galactose (Peng et al. 2012). The amount of hemicellulose in biomass determines how well it can be converted to biofuel (Kumari and Singh 2018).

Plants require lignin, which is second only to cellulose in importance. The lignin concentration of the woody biomass water controls the cell wall formation, stiffness, water resistance, as well as defense against physical, chemical, and microbial

Table 10.2 Certain agricultural solid waste's composition

S. No.	Agricultural waste	Compositional content dry basis (%)			References
		Cellulose	Hemicellulose	Lignin	
1.	Wheat straw	35.69	29.68	18.80	
2.	Corn cobs	45	35	15	
3.	Rice straw	38.82	27.59	19.55	Wang et al. (2021b)
4.	Sugar cane	37.72	22.95	22.34	
5.	Olive tree	36.5	21.3	24.1	
6.	Banana stems	33.2	18.2	5.5	
7.	Cocoa pods	26.1	4.82	21.29	

attack. It gives the plant structural support as it grows and aids in the plant's ability to transport water from the soil (Dhyani and Bhaskar 2018). On a dry basis, the majority of crop residues and agricultural wastes are made up of 15–20% lignin, 35–50% hemicellulose, 15–20% cellulose, and 15–20% extractives (protein, ash, etc.) (Table 10.2).

10.3 Biofuels from Agricultural Wastes

10.3.1 Bioethanol

Bioethanol, a renewable liquid biofuel, is typically produced through the fermentation of sugar and starch components found in plant derivatives and agricultural wastes. The most widely used liquid biofuel in the world is called bioethanol, or just ethanol. The need for bioethanol has persisted in driving up output. Global demand for bioethanol is projected to increase from an estimated 100.2 billion liters (2016) to 134.5 billion liters by 2024 (Bušić et al. 2018). In addition to its use as a fuel, bioethanol is also used to make drinks, personal-care items, disinfectants, and serves as feedstock for various pharmaceutical and chemical sectors. The global ethanol market, valued at USD 89.1 billion in 2019, is projected to surpass USD 155.6 billion by 2030. The principal application for bioethanol is fuel blend with petrol for internal combustion engines. Customers can profit financially and perform better with this strategy (Table 10.3). According to, the usage of bioethanol/gasoline blend in transportation engines results in emissions of 90% CO_2, 60–80% SO_2, and roughly 40% particulate matter.

10.3.2 Biohydrogen

Biohydrogen is an advanced biofuel, produced from biomass using thermochemical and biological processes. It is clean, non-toxic, and free from carbon emissions. It is a colorless, odorless, tasteless, highly combustible, renewable fuel that can sustainably replace fossil fuels. Because of its high calorific value (143 GJ/ton), it is the most lucrative and favored type of biofuel. According to Kumari and Singh (2018),

Table 10.3 Bioethanol production from some agricultural wastes

S. no.	Feedstock	Methodology	Microorganism	Yield	References
1.	Rice straw	Fermentation	*Saccharomyces cerevisiae*	61.55%	
2.	Corn cobs	Simultaneous saccharification and fermentation	*Saccharomyces cerevisiae* BY4743	42.24 g/L	
3.	Banana peel	Fermentation	*Saccharomyces cerevisiae*	4.24 g/L	
4.	Sugarcane bagasse	Hydrolysis and fermentation	*Saccharomyces cerevisiae* ATCC 9763	63.874 ppm	
5.	Apple wood	Fermentation	*Saccharomyces cerevisiae* M3013	70%	
6.	Potato peel	Fermentation	*Wickerhamia* sp.	21.7 g/L	

Table 10.4 Production of biohydrogen from some agricultural wastes

S. No.	Feedstock	Methods	Microorganism	Yield	References
1.	Sugarcane bagasse	Photo fermentation	*Rhodobacter capsulatus*-PK	513 mL/g	
2.	Corn stalk	Dark fermentation	Cow dung	168.9 mL/g	
3.	Corn cobs	Photo fermentation	*Clostridium acetobutylicum*	132 L/kg	Medina-Morales et al. (2021)
4.	Sorghum stover	Anaerobic fermentation	*Trichoderma asperellum*	402.01 mL/g	
5.	Potato peel	Dark fermentation	Sewage sludge inoculum	2.63 mL/g	
6.	Rice husk	SSF (solid state fermentation)	*Clostridium termitidis* and *clostridium intestinale*	5.9 mL/g	
7.	Raw palm oil mill effluent	Dark fermentation	POME sludge	28.47 mL/g	

POME palm oil mill effluent

biohydrogen has diverse applications in transportation, energy, medical, food and beverage, and industrial sectors. It also has positive social and economic effects and plays a significant role in the circular economy.

According to current government policies and targets on the shift to clean energy, the total global output of biohydrogen in 2015 was 1.8 million metric tons, which is expected to rise to 19 million metric tons (2030) and approximately 65 million metric tons by 2050 (Table 10.4).

In a similar vein, the market share of worldwide biohydrogen production, which was valued at roughly USD 103 billion (2017), is predicted to grow to USD 183 billion (2023) and reach USD 207 billion by 2026.

10.3.3 Biomethane

According to, the region of Asia and Oceania has the greatest potential for producing biomethane in the future (165 MTPA), followed by North America (105 MTPA) and the United States and Canada (125 MTPA). Comparably, it is anticipated that the worldwide biomethane market, which was valued at USD 1.9 billion in 2020, will reach USD 4 billion by 2031. This trend arises from increasing environmental concerns, a surge in biomethane usage within the transportation and maritime industries, and the implementation of supportive regulations promoting biomethane production and use. Additionally, biomethane will be used more in renewable energy since it is a net-zero fuel (Medina-Morales et al. 2021).

A substantial investment is planned for the anticipated deployment of biomethane as a substitute for natural gas in chemical, food and beverage, steel, and other sectors to meet the predicted demand for biomethane. Significant resources and efforts have been dedicated to ensuring cost-effective and environmentally sustainable processes for biomethane production, owing to its high energy content (36 MJ/m3) and net-zero emission properties. The utilization of agricultural wastes for production is one of those approaches. The most crucial element influencing a feedstock's anaerobic digestion is its C/N ratio. Recently, the viability of using sugar beetroot leaves as substrate for biomethane production has been reported by Undiandeye et al. (2022). An enzymatic pre-treatment based on *Aspergillus* and *Neurospora* enhanced the substrate's anaerobic digestibility and guaranteed a 516 mL/g yield.

10.3.4 Biobutanol

Biobutanol is a four-carbon biofuel produced through microbial acetone-butanol-ethanol (ABE) fermentation using renewable feedstocks, typically starch, sugars, or cellulosic biomass. It is a more sophisticated biofuel that vaporizes at a higher temperature than petrol. In addition, biobutanol has been used in a variety of industrial and chemical fields, such as herbicides, detergent formulations, paints and coatings, cosmetics, and chemical intermediates. In the food and pharmaceutical industries, as well as resins and plasticizers, it finds application. Shows that the volume of the global biobutanol market was 4.39 million metric tons in 2015, increased to 5.23 million metric tons in 2021, and is expected to reach 6.72 million metric tons by 2029. The biobutanol industry, which was estimated to be worth USD 90 million (2020), is expected to grow to USD 114.7 million by 2026.

To promote the economic sustainability of biobutanol production, various bio-based biomass are being utilized as feedstocks, which include agricultural waste

Table 10.5 Production of biodiesel from some agricultural waste

S. No.	Feedstocks	Method	Yield	References
1.	Microalgae grown on banana peel	Transesterification	420 mg/L	
2.	Corn stover	Fermentation	1.3 g/L	
3.	Microalgae grown on sugarcane bagasse	Transesterification	112 ± 5.2 mg/L	Aro (2016)
4.	Pineapple waste	Transesterification	13 ml/L	
5.	Rice bran oil	Ultrasound assisted transesterification	94.12%	
6.	Cassava starch	Fermentation	0.187 g/g	

and crop residues like cassava peel, corn, sugarcane bagasse, wheat, and molasses. Agricultural solid waste for biobutanol production includes sugarcane molasses, sugarcane bagasse, rice straw, corncobs, wheat straw, etc., which are processed through ABE fermentation methods using numerous microorganisms (*Clostridium* sp., *Pseudomonas, Ruminococcus,* etc.) with very high butanol yield Khalil and Amiri 2020).

10.3.5 Biodiesel

A renewable, biodegradable, and environmentally beneficial type of liquid biofuel is biodiesel. It is typically made from animal fats, vegetable oils, recovered restaurant grease, and spent cooking oil and is composed of long-chain fatty acid esters (Onuh et al. 2021). Biodiesel is an environmentally friendly substitute for fossil fuels that has been shown to extend engine life and produce fewer greenhouse gases and other harmful pollutants. In 2019, worldwide biodiesel production reached 42 billion liters, continually increasing to 43 billion liters (2021), and is expected to average 46 billion liters between 2023 and 2025. In a similar vein, it is anticipated that the worldwide biodiesel market value, which stood at USD 46.79 billion as reported for 2021, will increase to USD 51.48 billion by 2026. The product satisfies international requirements, with the authors reporting biodiesel yields for banana peel and sugarcane bagasse of 420 mg/L and 400 mg/L, respectively. In a similar vein, Aro (2016) produced biodiesel by growing oleaginous microalgae on sugarcane bagasse extract, with a yield of 5.2 mg/L (Table 10.5).

10.4 Conclusion

The world's population growth, agricultural practices, and food processing have led to an increase in the generation of agricultural wastes, sanitation issues, and the release of harmful gases from decomposing materials. This research has offered environmentally sound and sustainable methods for managing agricultural waste, as well as bioengineering techniques for waste reduction, conversion, and utilization. Moreover, it has been demonstrated that agricultural wastes can be processed to produce biodiesel, bioethanol, biomethane, biohydrogen, and butanol for use in the transportation and energy production industries. While using the produced biofuels as a transport engine lowers greenhouse gas emissions, ensures better engine performance, promotes smoother engine running, and extends engine lifespan, valorization of agricultural waste towards biofuel production provides ecological, economic, sanitary, industrial, and technical benefits.

Given the significance of employing agricultural waste as a feedstock for the production of biofuels, it is reasonable to speculate on the direction of future research. To ensure greater breakdown and conversion of crop residues and other agricultural waste into valuable products, new and improved pretreatment procedures are required. Further interdisciplinary research on techno-economic assessment, life cycle assessments, energy and exergy analyses, cost analyses, and sensitivity analyses is required to improve our comprehension of the intricate molecular kinetics and complex processes involved in waste degradation and conversion.

Conversion of agricultural residues to products generates huge amounts of revenue, contributing significantly to the bioeconomy. Recycling of agricultural residues finds opportunity as recycling of agricultural solid wastes into useful products is a great initiative towards sustainable development.

References

Amiri H et al (2015) Biobutanol from lignocellulosic wastes. In: Lignocellulose based bioproducts. Springer, Cham, pp 289–324

Aro EM (2016) From first generation biofuels to advanced solar biofuels. Ambio 45(1):24–31

Asgher M, Ahmad Z, Iqbal HMN (2013) Alkali and enzymatic delignification of sugarcane bagasse to expose cellulose polymers for saccharification and bio ethanol production. Ind Crop Prod 44:488–495

Awasthi P, Shrivastava S, Kharkwal AC, Varma A (2015) Biofuel from agricultural waste: a review. Int J Curr Microbiol Appl Sci 4(1):470–477

Awogbemi O et al (2021) An overview of the classification, production and utilization of biofuels for internal combustion engine applications. Energies 14(18):5687

Bracco S, Calicioglu O, Juan MGS, Flammini A (2018) Assessing the contribution of bioeconomy to the total economy: a review of national frameworks. Sustain For 10:1698. https://doi.org/10.3390/su10061698

Bušić A et al (2018) Bioethanol production from renewable raw materials and its separation and purification: a review. Food Technol Biotechnol 56(3):289–311

Cheng JJ, Timilsina GR (2011) Status and barriers of advanced biofuel technologies: a review. Renew Energy 36(12):3541–3549

Dhyani V, Bhaskar T (2018) A comprehensive review on the pyrolysis of lignocellulosic biomass. Renew Energy 129:695–716

European Commission (2015) EIP-AGRI workshop opportunities for agriculture and forestry in the circular economy. Workshop report 28-29 October 2015. Brussels, Belgium. https://ec.europa.eu/eip/agriculture/sites/agri-eip/files/eip-agri_ws_circular_economy_final_report_2015_en.pdf

Food and Agriculture Organization of the United Nations (FAO) (2016) The state of food and agriculture. Climate change, agriculture and food security http://www.fao.org/3/a-i6030e.pdf. Accessed 13 August 2020

Food and Agriculture Organization of the United Nations (FAO) (2017a) Strategic work of FAO for sustainable food and agriculture. http://www.fao.org/3/a-i6488e.pdf

Food and Agriculture Organization of the United Nations (FAO) (2017b). The state of food and agriculture: leveraging food systems for inclusive rural transformation. http://www.fao.org/3/a--i7658e.pdf

Gutiérrez A, Rencoret J, Cadena EM, Rico A, Barth D, del Río JC et al (2012) Demonstration of laccase-based removal of lignin from wood and non-wood plant feedstocks. Bioresour Technol 119:114–122

Kumari D, Singh R (2018) Pretreatment of lignocellulosic wastes for biofuel production: a critical review. Renew Sust Energ Rev 90:877–891

Martin C, Galbe M, Wahlbom CF, Hahn-Hägerdal B, Jönsson LJ (2002) Ethanol production from enzymatic hydrolysates of sugarcane bagasse using recombinant xylose-utilising Saccharomyces cerevisiae. Enzym Microb Technol 31(3):274–282

McCormick K, Kautto N (2013) The bioeconomy in Europe: an overview. Sustain For 5:2589–2608. https://doi.org/10.3390/su5062589

Medina-Morales MA et al (2021) Biohydrogen production from thermochemically pretreated corncob using a mixed culture bioaugmented with Clostridium acetobutylicum. Int J Hydrog Energy 46(51):25974–25984

Mukhtar A et al (2021) Mitigation of CO2 emissions by transforming to biofuels: optimization of biofuels production processes. Renew Sust Energ Rev 150:111487

Nakamya M (2022) How sustainable are biofuels in a natural resource-dependent economy? Energy Sustain Dev 66:296–307

Onuh EI, Inambao F, Awogbemi O (2021) Performance and emission evaluation of biodiesel derived from waste restaurant oil and Moringa oleifera: a comparative study. Int J Ambient Energy 42(8):912–919

Pattanaik L et al (2019) Biofuels from agricultural wastes. In: Second and third generation of feedstocks. Elsevier, Amsterdam, pp 103–142

Peng F et al (2012) Fractional purification and bioconversion of hemicelluloses. Biotechnol Adv 30(4):879–903

Plácido J, Capareda S (2015) Ligninolytic enzymes: a biotechnological alternative for bioethanol production. Bioresour Bioprocess 2(23):1–12

Rastogi M, Shrivastava S (2017) Recent advances in second generation bioethanol production: an insight to pretreatment, saccharification and fermentation processes. Renew Sust Energ Rev 80:330–340. https://doi.org/10.1016/j.rser.2017.05.225

Renewable Fuels Association (2016) Industry statistics for World Fuel Ethanol Production. http://ethanolrfa.org/resources/industry/statistics/#1454099788442-e48b2782-ea53. Accessed 20 October 2016

Scarlat N, Dallemand JF, Monforti-Ferrario F, Nita V (2015) The role of biomass and bioenergy in a future bioeconomy: policies and facts. Environ Dev 15:3–34. https://doi.org/10.1016/j.envdev.2015.03.006

Scully SM, Orlygsson J (2015) Recent advances in second generation ethanol production by thermophilic bacteria. Energies 8(1):1–30

Shrivastava S, Tekriwal KG, Kharkwal AC, Varma A (2014) Bio-ethanol production by simultaneous saccharification and fermentation using microbial consortium. Int J Curr Microbiol Appl Sci 3(7):505–511

Stefanidis SD et al (2014) A study of lignocellulosic biomass pyrolysis via the pyrolysis of cellulose, hemicellulose and lignin. J Anal Appl Pyrolysis 105:143–150

Tripathi N et al (2019) Biomass waste utilisation in low-carbon products: harnessing a major potential resource. NPJ Clim Atmos Science 2(1):35

Undiandeye J et al (2022) Effect of novel Aspergillus and Neurospora species-based additive on ensiling parameters and biomethane potential of sugar beet leaves. Appl Sci 12(5):2684

United Nations (2019) Programme performance report 2018. UN Environment Programme. https://wedocs.unep.org/bitstream/handle/20.500.11822/27734/PPR_2018_FINAL.pdf?sequence=1&isAllowed=y. Accessed 13 August 2020

Wang J et al (2021a) A review on polyhydroxyalkanoate production from agricultural waste biomass: development, advances, circular approach, and challenges. Bioresour Technol 342:126008

Wang T et al (2021b) Adsorption characteristics and mechanisms of Pb2+ and Cd2+ by a new agricultural waste–Caragana korshinskii biomass derived biochar. Environ Sci Pollut Res 28:13800–13818

Ylä-Mella J, Keiski RL, Pongrácz E (2022) End-of-use vs. end-of-life: when do consumer electronics become waste? Resources 11(2):18

Zhao XQ, Zi LH, Bai FW, Lin HL, Hao XM, Yue GJ et al (2012) Bioethanol from lignocellulosic biomass. Adv Biochem Eng Biotechnol 128:25–51

Valorization of Wheat Residue to Biofuels and Chemicals

11

Shruti S. Raut, Arpit Sharma, and Abha Mishra

Abstract

The escalating challenges of environmental pollution and energy depletion have underscored the urgent need for renewable and sustainable biofuels as viable alternatives to traditional fossil fuels. The increasing global demand for nonrenewable fossil fuels is driven by the challenges of rapid population growth and accelerating economic development. As nations strive to sustain expanding industries and urbanization, the reliance on finite energy resources continues to pose significant environmental and economic challenges, necessitating urgent exploration of sustainable alternatives. The conversion of abundantly available biomass into biofuels is attracting significant attention, and it is seen as an innovative technique for addressing environmental problems and attaining net-zero carbon emissions involves balancing carbon output with removal strategies, ensuring that anthropogenic CO_2 emissions are fully offset through carbon sequestration, renewable energy adoption, and advanced mitigation technologies. The utilization of a wide range of raw materials as feedstock for biofuel production is dependent on factors such as biomass availability, cost-effectiveness, and geographic location. Lignocellulosic (LC) biomass has received significant attention from researchers worldwide due to its diverse range of basic compo-

S. S. Raut · A. Mishra (✉)
Biomolecular Engineering Laboratory, School of Biochemical Engineering, Indian Institute of Technology (BHU), Varanasi, Uttar Pradesh, India
e-mail: abham.bce@itbhu.ac.in

A. Sharma
Biomolecular Engineering Laboratory, School of Biochemical Engineering, Indian Institute of Technology (BHU), Varanasi, Uttar Pradesh, India

Department of Pharmacology, Institute of Medical Science-Banaras Hindu University (IMS-BHU), Varanasi, Uttar Pradesh, India

A. Kumari et al. (eds.), *Valorization of Solid Wastes to Biofuels and Chemical Products for Sustainable World*, https://doi.org/10.1007/978-981-96-8594-3_11

nents, and it is suitable for second-generation biofuel production. This chapter presents an in-depth exploration of the fundamental principles and advanced applications of thermochemical and biochemical conversion processes. It explores thermochemical processes such as pyrolysis, torrefaction, hydrothermal liquefaction, and gasification, along with biochemical approaches, including pretreatment methods, enzymatic breakdown, and microbial conversion techniques. The discussion integrates process efficiencies, reaction mechanisms, and technological advancements, offering a comprehensive perspective on biomass valorization. These processes play a pivotal role in the biomass biorefinery of lignocellulosic materials and offer substantial potential for biofuel production. The technical aspects of lignocellulosic biofuel production are examined, with a focus on distinct processing strategies, including separate hydrolysis and fermentation (SHF), simultaneous saccharification and fermentation (SSF), simultaneous saccharification and co-fermentation (SSCF), and integrated consolidated bioprocessing (CBP).

Keywords

Biorefinery · Lignocellulosic (LC) biomass · Biofuel · Pretreatment · Biochemical processing

11.1 Introduction

The global community has been compelled by recent fuel price increases to produce biofuels from renewable sources, including lignocellulosic crops. Nevertheless, it is imperative to keep in mind that the accessibility of renewable agricultural biomass differs significantly by geographical region, with examples including maize in the United States and sugarcane in Brazil (Qureshi et al. 2010). Bioenergy, derived from biomass, has been utilized for thousands of years, dating back to the practice of burning wood for cooking and heating purposes. Presently, wood remains the most abundant biomass resource employed for bioenergy production. Biomass is a comprehensive term that includes a wide range of organic materials, such as agricultural and forestry by-products, municipal solid waste (MSW), industrial effluents, animal waste, aquatic vegetation, phytoplankton, and municipal biosolids. These renewable biological resources play a pivotal role in sustainable energy production, bioresource valorization, and circular bioeconomy frameworks (Demirbas 2009). Biomass serves as a sustainable and environmentally friendly energy resource that has the potential to significantly enhance our ecological conditions, economic stability, and energy independence. Energy can be derived by the process of direct combustion of biomass, wherein dry organic materials such as agricultural residue are burned (Demirbas 2009).

The utilization of biofuels has emerged as a possible alternative to conventional fossil fuels (Luque et al. 2008; Jeswani et al. 2020). Numerous nations, such as the European Union, India, Brazil, China, Thailand, and Argentina, have already pledged to develop biofuel production as a means of decreasing the dependency on fossil fuels (Robak and Balcerek 2020). Biofuels are derived from organic matter, encompassing both direct extraction from plants and indirect source from agricultural by-products, among other sources. Biomass feedstocks can be systematically classified based on their composition into herbaceous and lignocellulosic sources. Nonwoody biomass includes oilseed crops (e.g., rapeseed), sugar- and starch-rich plants (e.g., sugarcane, maize), agricultural residues, animal-derived waste, and by-products from food processing industries. In contrast, lignocellulosic biomass comprises forest-derived materials, fast-growing energy crops such as willow and short-rotation coppice, and unprocessed wood-based residues, all of which contribute significantly to bioenergy and bioproducts development (Hammond and McCann 2023). The emission of carbon dioxide (CO_2) resulting from the generation of energy through biomass is roughly compensated by the amount of CO_2 absorbed during the production of the fuel. Therefore, it is occasionally considered a "carbon-neutral" procedure, and its utilization presents the potential to contribute toward achieving CO_2 reduction goals (Hammond and Mansell 2018).

All biofuels are categorized into four generations based on their respective core resources, including technical advancements. First-generation biofuels are synthesized from edible feedstocks traditionally cultivated for human consumption. These biofuels are primarily derived from starch- and sugar-rich crops, including cassava, wheat, maize, and sugarcane, which undergo fermentative and catalytic conversion processes to produce bioethanol and biodiesel. However, their reliance on food-based resources raises concerns regarding food security, land use competition, and sustainability within the bioenergy sector. Consequently, the development of these biofuels gives rise to conflicts over the availability of food and feed resources, as well as competition for arable land (Qureshi et al. 2010). It is noted that in the year 2020 first-generation biofuels constituted a substantial majority of 96% of the global output of biofuels. Lignocellulosic-based biofuels, also known as second-generation biofuels, have the potential to serve as an economical substitute for petroleum in the coming years. Algal biomass serves as a versatile feedstock for the production of third- and fourth-generation biofuels, with the latter specifically derived from genetically engineered microalgae optimized for enhanced lipid accumulation. These advanced biofuels leverage biotechnological innovations to improve biomass productivity, carbon sequestration efficiency, and biofuel yield, positioning algae as a sustainable and high-potential resource for next-generation bioenergy solutions. The United States, the European Union, Brazil, China, and Canada are the primary contributors to global bioethanol production, accounting for 56%, 5%, 28%, 4%, and 2%, respectively.

Lignocellulosic biomass consists mainly of covalently bonded carbohydrate polymers, such as cellulose and hemicellulose, along with lignin, a complex noncarbohydrate structural component. Typically, its composition varies depending on the feedstock type, with cellulose accounting for 30–50%, hemicellulose making up 20–40%, and lignin comprising 10–20%. However, it is important to note that the proportion of carbohydrates may vary depending on the specific subtype of biomass (Karagoz et al. 2019). The increasing focus on the biochemical conversion of lignocellulosic waste into biofuel, particularly through fermentation, is driven by the favorable operating conditions it offers in comparison to capital-intensive thermochemical procedures such as pyrolysis. Numerous issues concerning the primary stages of processing, including pretreatment, enzymatic hydrolysis, and fermentation, are examined in the context of biotechnological technologies employed in biofuel generation (Robak and Balcerek 2020).

11.2 Composition of Different Straw Materials

Straw is a type of agricultural waste consisting of the desiccated stems of cereal plants after the removal of the grain and chaff (Hammond and Mansell 2018). The chemical composition and structural characteristics of lignocellulose have undergone evolutionary changes to establish a protective barrier against microbial infections in plants. Straw, like other lignocellulosic substrates, is a diverse and complex substance primarily composed of three prominent plant polymers: hemicellulose, cellulose, and lignin. Cellulose is a polysaccharide made up of β-D-glucose monomers connected through β-1,4-glycosidic linkages, with cellobiose acting as its basic disaccharide unit. These β-1,4 linkages enable tight molecular packing, contributing to the formation of highly ordered crystalline fibrils within the cellulose structure. Straw is primarily composed of hemicellulose in monocotyledonous plants, consisting of branched polysaccharides with a xylan backbone and arabinose and glucuronic acid side chains, the latter of which is often methylated. The cellulose chains present in the epidermis of wheat straw exhibit an orientation that aligns with the growing direction of the straw. The cellulose molecules undergo compaction, resulting in the formation of elongated polymer chains known as "microfibrils." These microfibrils are interconnected by short-chain hemicellulose polymers (Talebnia et al. 2010). Hemicelluloses exhibit an amorphous, heterogeneous, and randomly branched structure, making them more soluble and less crystalline than cellulose. Crosslinking cellulose fibrils and lignin construct a network within the plant cell wall. Unlike cellulose and hemicellulose, lignin is not classified as a polysaccharide but rather as a complex aromatic polymer composed of cross-linked phenylpropanoid units. Instead, it is composed of phenylpropanoid units that combine to create a complex three-dimensional structure. Lignin exhibits hydrophobic properties, contributing to the rigidity of the cell wall and conferring protection against both insect and plant infections. Several studies have reported that the cellulose content of wheat straw can reach around 30–50%. In addition to organic polymers, straw also has inorganic chemicals that persist as ash post-combustion. The structural composition of wheat straw is given in Table 11.1 (Balat 2011).

11.3 Pretreatment of Wheat Straw

The degree to which the pretreatment method enhances the biodegradability of the polysaccharides in the straw, particularly cellulose, and hemicellulose, is decisive in determining whether bioethanol production from wheat straw is successful or

Table 11.1 Wheat straw composition

Component	Dry solids (% w/w)
Neutral detergent fiber	86.56 ± 0.09
Acid detergent fiber	58.86 ± 0.04
Crude fiber	45.86 ± 0.20
Cellulose	48.57 ± 0.30
Hemicellulose	27.70 ± 0.12
Ash	6.68 ± 0.01
Lignin	8.17 ± 0.90
Crude fat	0.47 ± 0.01
Crude protein	3.48 ± 0.09

unsuccessful. The objective of pretreatment is to induce a disruption in the crystalline structure of microfibrils and degrade the lignin content, thereby facilitating the release of the cellulose and hemicellulose polymer chains (Kumar et al. 2012).

The pretreatment process must satisfy the subsequent criteria:

1. Improve the efficiency of sugar production or carbohydrate breakdown through hydrolysis.
2. Prevent the loss or degradation of carbohydrates.
3. Prevent the subsequent fermentation and hydrolysis from producing inhibitory metabolites.
4. Must be economically efficient.

Pretreatment techniques can be classified into various categories:

1. Physical methods: includes milling and grinding to reduce particle size and enhance surface area.
2. Physicochemical methods: encompasses techniques such as steam pretreatment/autohydrolysis, hydrothermolysis, and wet oxidation, which combine thermal and chemical effects.
3. Chemical methods: utilizes alkali, dilute acid, oxidizing agents, and organic solvents to break down lignocellulosic structures and improve accessibility.
4. Biological methods: utilizes microorganisms or enzymes to target and break down lignin and hemicellulose, enhancing the hydrolysis of cellulose.

11.3.1 Physical Pretreatment

11.3.1.1 Mechanical Disintegration

A combination of chipping, grinding, and milling can be utilized to disrupt the crystalline structure of cellulose, thereby enhancing the breakdown of lignocellulosic biomass and improving its susceptibility to enzymatic or chemical hydrolysis. Following decomposition, the dimensions of the materials are typically 10–30 mm;

however, grinding or milling reduces them to 0.2–2 mm. The power needed for the mechanical breakdown of agricultural materials is determined by the ultimate particle size and biomass characteristics (Talebnia et al. 2010). Research suggests that optimizing the final particle size within the 3–6 mm range can significantly lower energy requirements, reducing the grinding energy consumption to below 30 kWh/ton of biomass. Typically, excessive energy consumption occurs when the quantity of energy utilized surpasses the biomass's theoretical energy content.

11.3.1.2 Pyrolysis

Pyrolysis has been employed as a method for the processing of lignocellulosic materials as well. When biomass is subjected to temperatures exceeding 300 °C, cellulose undergoes fast decomposition, resulting in the formation of gaseous by-products and residual char. Reduced temperatures result in a significantly delayed decomposition rate and less volatile by-products. According to a study by Fan et al. moderate acid hydrolysis (1N H_2SO_4) at 97 °C for 2.5 h on pyrolysis, pretreatment by-products resulted in the formation of reducing sugars (including glucose, which comprises more than half of the cellulose) (Antonopoulou and Lyberatos 2013). The presence of oxygen alters the pyrolysis process, shifting it toward oxidative decomposition rather than pure thermal degradation. Zwart et al. documented the utilization of biomass in a biomass-to-liquids (BtL) process to produce transportation fuels. This method involves the conversion of biomass to syngas, which is then utilized to synthesize Fischer–Tropsch (FT) fuels of superior quality (Zwart et al. 2006).

11.3.2 Physicochemical Pretreatment

11.3.2.1 Steam Explosion

The predominant technique employed for the pretreatment of lignocellulosic materials is steam explosion (Sun and Cheng 2002). This method exposes biomass to high-pressure saturated steam, followed by a sudden pressure release, causing rapid decompression that facilitates structural disruption and enhances biomass breakdown. Typically, steam explosions are initiated at a temperature range of 160–260 °C (0.69–4.83 MPa) before the material is exposed to air pressure for a few seconds to minutes. The biomass and steam mixture is subjected to a prolonged duration to facilitate the hydrolysis of hemicellulose, and the procedure is concluded through a sudden and forceful decompression. The elevated temperature induces the degradation of hemicellulose and the modification of lignin, thereby enhancing the cellulose hydrolysis potential. It is hypothesized that hemicellulose undergoes hydrolysis as a result of the action of acetic and other acids generated during steam explosion pretreatment. An enzyme-accessible cellulose surface is hypothesized to result from the elimination of hemicelluloses from the cellulose microfibrils. The removal of lignin during the pretreatment process is only partially achieved as it is redistributed on the surfaces of the fibers due to the occurrence of melting and depolymerization/repolymerization events. The volume of the pretreatment sample is increased through the process of removing and redistributing hemicellulose and lignin. The fragmentation of material is facilitated by rapid fluctuations in atmospheric pressure and the presence of turbulent flow, increasing the available surface area. At elevated temperatures, water exhibits acidic properties. The inclusion of H_2SO_4 (or SO_2) or CO_2, typically at concentrations ranging from 0.3% to 3% (w/w), during steam explosion processes has been shown to reduce both the time and temperature required for the process. This addition has been shown to enhance hydrolysis efficiency, decrease the formation of inhibitory compounds, and facilitate the complete removal of hemicellulose. The utilization of steam as a means to efficiently elevate the temperature of cellulosic materials to the desired level, while minimizing the undesirable effect of excessive dilution of the resultant sugars, is facilitated by steam. A sudden pressure drop efficiently lowers the temperature, halting the reaction once the pretreatment process is complete. This rapid thermal expansion leads to the structural expansion of biomass particles, enhancing their accessibility for further processing. However, a weak correlation can be observed between this physical phenomenon and the pretreated solid's improved cellulose digestibility. Steam explosion pretreatment is influenced by a variety of variables, such as residence duration, temperature, fragment size, and moisture content. The achievement of optimal solubilization and hydrolysis of hemicellulose can be attained by two distinct approaches: one involving high temperature and short residence time (specifically, 270 °C for 1 min), and the other involving lower temperature and longer residence time (specifically, 190 °C for 10 min). Kobayashi et al. investigated the expansion of steam explosion technology, focusing on its potential application in enhancing the fermentation process for bamboo-to-methane conversion. The microbial seed for the methane fermentation process was procured from decomposing sludge acquired from a sewage treatment facility. It was discovered by researchers that unprocessed bamboo could only generate a negligible quantity of methane. However, it was observed that the introduction of steam detonation significantly increased methane production. Through the application of steam pressure of 3.53 MPa for 5 min on 1 g of pulverized bamboo, an approximate methane yield of 215 mL was produced. A negative correlation was observed between the quantity of steam-exploded bamboo utilized in methane fermentation and the quantity of high-molecular-weight lignin, or Klason lignin. One of the notable benefits associated with steam explosion pretreatment is its relatively low energy demand in comparison to mechanical comminution. Additionally, this method does not entail any recycling or environmental expenditures. Traditional mechanical size reduction techniques require approximately 70% more energy than steam explosion to achieve a comparable particle size reduction. However, steam explosion presents certain limitations that must be considered. One such issue is the insufficient disruption of the lignin–carbohydrate matrix, which results in the depletion of a portion of the xylan fraction. The formation of compounds that may inhibit the activity of microorganisms utilized in subsequent processes is another. Inhibitory compounds, water-soluble hemicellulose, and breakdown products that impede microbial growth, fermentation, and enzymatic hydrolysis are eliminated by rinsing pretreated biomass with water. The water-cleanse process eliminates soluble sugars, including those produced during the hydrolysis of hemicellulose, which results in a decrease in the overall saccharification yield.

11.3.2.2 Ammonia Fiber Explosion (AFEX)

In the AFEX physicochemical pretreatment process, lignocellulosic biomass is exposed to liquid ammonia under elevated temperature and pressure for a specific duration, followed by a rapid pressure release, which facilitates structural disruption and enhances enzymatic accessibility. In a conventional ammonia fiber expansion (AFEX) procedure, the liquid ammonia is administered at a dosage of 1–2 kg/kg of dry biomass. The process is conducted at a temperature of 90 °C, with a residence period of 30 min. In the pretreatment phase, only a little quantity of the solid material undergoes solubilization, resulting in the retention of nearly all hemicellulose and lignin components. The depolymerization of hemicellulose into oligomeric sugars, followed by deacetylation, is a potential factor contributing to its limited solubility (Gollapalli et al. 2002).

11.3.3 Biological Pretreatment

Biopretreatment can be employed as an alternative approach to facilitate the degradation of biomass through the utilization of microorganisms. A wide range of bacterial and fungal species possess the ability to enzymatically break down cellulose and hemicellulose into its constituent monomers. Several species of *Clostridia*, such as *Clostridium thermocellum*, have undergone thorough examination about their ability to degrade cellulose in anaerobic conditions. Under aerobic circumstances, cellulose and hemicellulose can be degraded by actinomycetes and other fungi. An additional approach to enhance the susceptibility of polysaccharides to enzymatic degradation is through microbial lignin degradation. Efforts have been made to achieve lignification through the utilization of white-rot fungi (e.g., *Phanerochaete chrysosporium*) or isolated lignin-degrading enzymes (e.g., laccases). When living fungi are utilized, this type of pretreatment is frequently time-consuming and necessitates additional procedures, such as detoxification. To achieve optimal liberation of sugar monomers, it is imperative to employ enzymes that possess the capability to degrade the three primary polymers, namely cellulose, hemicellulose, and lignin. Cellulose-degrading enzymes are produced by fungi, anaerobic, and aerobic bacteria, as well as thermophilic and mesophilic bacteria. Cellulolytic enzymes utilized in commercial applications are often sourced from diverse fungal species. The cellulolytic enzyme systems that have been subject to the most comprehensive research are derived from the ascomycete *Trichoderma reesei*, also known by its teleomorph name *Hypocrea jecorina*. Originally, the degradation of cellulose has relied on the utilization of three enzyme activities that work in conjunction: endoglucanases, exoglucanases, and β-glucosidases. Endoglucanases (endo-1,4-β-D-glucanases, EC 3.2.1.4) function as catalytic enzymes that break β-1,4 glycosidic linkages within the amorphous regions of cellulose, generating both reducing and nonreducing chain ends to facilitate further biodegradation. Exoglucanases, alternatively referred to as cellobiohydrolases (cellulose 1,4-β-cellobiosidases), are processive enzymes that catalyze the release of cellobiose from the cellulose molecule's reducing (EC 3.2.1.176) or nonreducing (EC 3.2.1.91) end. β-Glucosidases convert cellobiose or cellooligosaccharides into glucose via hydrolysis (EC 3.2.1.21) (Obeng et al. 2017; Payne et al. 2015). Thermal pretreatment typically enhances the solubility of hemicelluloses, whereas acid-based pretreatment hydrolyzes sufficient sugar monomers, enabling their utilization as a substrate for microbial fermentation in subsequent processes. Nevertheless, it has been noted that thermochemical pretreatment frequently results in the formation of oligosaccharides that are unassimilable to the majority of essential fermentation organisms. Consequently, hemicellulase treatment can significantly enhance the performance of lignocellulose-based processes. Equation (11.1) represents glucose recovery from wheat straw.

$$\text{Glucose Recovery}\left(\%\right)=\frac{\text{Total glucose}_{\text{Solid fraction}}+\text{Total glucose}_{\text{liquid fraction}}}{\text{Total glucose}_{\text{Straw}}}\times 100 \quad (11.1)$$

11.4 Utilizing Wheat Straw as a Feedstock to Produce Biofuel and Other Bioproducts

Wheat straw is an excellent substrate for producing mushrooms, animal fodder, and firewood, in addition to its many other applications (Townsend et al. 2017). A gradual reduction in soil carbon content occurs as straw is entirely extracted from fields (Karlsson et al. 2017; Townsend et al. 2017). Even with all of these potential applications, there is still more straw than can be utilized in a biorefinery. Please note that the annual availability of straw may fluctuate due to various factors, including the weather conditions, the specific cultivars chosen, and other comparable variables (Talebnia et al. 2010; Scarlat et al. 2010). Biofuels such as ethanol, butanol, biodiesel, and methane (biogas) can be produced from the monosaccharides obtained during the pretreatment of straw (Chandel et al. 2018).

11.4.1 Bioethanol Production from Wheat Straw

The synthesis of bioethanol from sugars produced from lignocellulose (wheat straw) has been extensively studied and researched. For efficient large-scale bioethanol production, the use of microorganisms with key attributes is crucial. These include the ability to utilize diverse substrates, high ethanol yield and productivity, resistance to hydrolysate-derived inhibitors, tolerance to elevated ethanol concentrations, cellulolytic capabilities, and the ability to ferment sugars under extreme temperatures. Among them, *Saccharomyces cerevisiae* is a widely recognized ethanol-producing microorganism, efficiently converting glucose and other hexose sugars into ethanol. *Saccharomyces cerevisiae* and *Zymomonas mobilis* are microbial strains recognized for their high ethanol production efficiency, achieving yields between 90% and 97% of the theoretical maximum. Additionally, both strains exhibit a remarkable ethanol tolerance of approximately 10% (w/v) in the fermentation medium. The primary pentose sugar found in hemicellulose, xylose, cannot be metabolized by the native *S. cerevisiae* and *Z. mobilis* strains. Several microorganisms, such as *enteric bacteria*, *Pichia stipitis*, *Candida shehatae*, and *Pachysolen tannophilus*, have the ability to convert xylose into ethanol through fermentation processes. Nevertheless, these microorganisms are notable for their inefficient ethanol production rates and their tendency to reassimilate the ethanol they generate (Chandel et al. 2018). To address the issue, researchers have successfully engineered genetically modified variants of *S. cerevisiae* that possess the ability to ferment both hexoses and pentoses. The xylose transporters that are implicated in the pentose-phosphate and xylose assimilation pathways are overexpressed in these strains. Either the xylose-fermenting yeast *Scheffersomyces stipitis*, which contains xylose reductase/xylitol dehydrogenase, or *Clostridium phytofermentans*, which contains a codon-optimized xylose isomerase, may serve as these transporters. In addition, a notable limitation exists regarding the extensive adoption of recombinant ethanologens as the host organism frequently exhibits rejection of plasmids containing xylose conversion genes. Zayed and Meyer (1996) established a highly efficient single-batch fermentation method that utilized two phylogenetically distinct microorganisms, resulting in a yield of 86% ethanol. In this study, delignified wheat straw was enzymatically hydrolyzed into simple sugars using the fungus *Trichoderma viride*. Subsequently, the resulting xylose and glucose were fermented into ethanol using the aerotolerant yeast *Pachysolen tannophylus*. With thermophiles, the ethanol productivity achieved using *Thermoanaerobacter* sp. cultivated on pretreated wheat straw exhibited significantly lower levels compared to the ethanol productivity achieved using yeasts and fungi. A pilot facility, established by the Danish company Inbicon, possesses the capability to produce 576 kg (730 l) of ethanol hourly from wheat straw. The second-generation ethanol facility situated in Crescentino, Italy, stands as the inaugural commercial-scale example of its kind worldwide. Annually, around 40,000 metric tons of ethanol are generated from 270,000 tons of biomass, which includes wheat straw, rice straw, and the energy crop giant cane. The commencement of construction for an additional ethanol plant, known as Clariant Sunliquid, in Romania, was initiated in the year 2018. It is anticipated that the previous facility, which is financed through the seventh Framework Programme of the European Union, will possess the capability to generate 50,000 metric tons of cellulosic agricultural waste per year. Multiple commercial-scale second-generation ethanol facilities have been established in the United States and Brazil, with a combined annual production capacity of approximately 530 million liters of ethanol. Straw makes up a significant portion of the feedstock mixture used by these plants.

11.4.1.1 Different Approaches for Fermentation of Ethanol

Several methodologies have been considered for the fermentation process of bioethanol. Simultaneous saccharification and fermentation (SSF), separate hydrolysis and fermentation (SHF), and direct microbial conversion (DMC) are key bioconversion strategies employed in bioethanol production. SHF involves the execution of hydrolysis and fermentation in distinct outlets each operating under its optimal conditions. Nevertheless, this approach is accompanied by challenges related to contamination and end-product inhibition of enzyme activity. A new strategy, termed simultaneous saccharification and fermentation (SSF), has been developed to overcome the challenges linked to the separate hydrolysis and fermentation (SHF) process. The sugars generated by the process of hydrolysis undergo immediate fermentation into ethanol, thereby overcoming issues related to sugar buildup, enzyme inhibition, and contamination. An added advantage of simultaneous saccharification and fermentation (SSF) over separate hydrolysis and fermentation (SHF) is the potential cost reduction resulting from the use of a single-reactor system. The primary limitation of simultaneous saccharification and fermentation (SSF) is the temperature mismatch between enzymatic hydrolysis and microbial fermentation. While hydrolyzing enzymes function optimally at 50 °C, most fermenting yeasts achieve peak performance within a 30–35 °C range, creating a challenge in process optimization. Additionally, discovered that the utilization of SHF instead of SSF resulted in a greater ethanol yield from washed steam-pretreated wheat straw. Incorporating a washing step into the ethanol production process would lead to increased production costs and a significant loss of fermentable sugars, which cannot be effectively recovered in large-scale operations. An alternative method for reducing the concentration of inhibitors before hydrolysis involves the implementation of a detoxification stage. Nevertheless, it should be noted that this factor may account for as much as 22% of the whole cost of ethanol production, making it advisable to steer clear of it. In the given study, assessed the efficacy of SHF and SSF by utilizing recombinant *E. coli* (FBR5) and pretreated wheat straw with diluted acid. The SHF approach demonstrated superior performance compared to the SSF strategy, resulting in increased ethanol production within a shorter fermentation period. The implementation of the overliming approach for detoxification resulted in a significant reduction in fermentation time during the simultaneous saccharification and fermentation (SSF) process, while also leading to increased ethanol production in the context of solid-state fermentation (SSF). Also observed that the SHF strategy exhibited superior efficacy compared to the SSF approach while utilizing the same strain (*E. coli*) and alkaline peroxide-prepared wheat straw. Figure 11.1 shows the complete production process of bioethanol (Balat 2011).

11.4.2 Biogas Production from Wheat Straw

An alternative approach to biofuel production involves generating methane (biogas) from wheat straw. Biogas, a sustainable energy source, is mainly composed of methane (CH_4) and carbon dioxide (CO_2) and can be harnessed for both thermal energy and electricity production. Additionally, beyond its conventional applications, biogas has the potential to serve as a transportation fuel. Once upgraded to biomethane, it can be seamlessly integrated into the existing natural gas

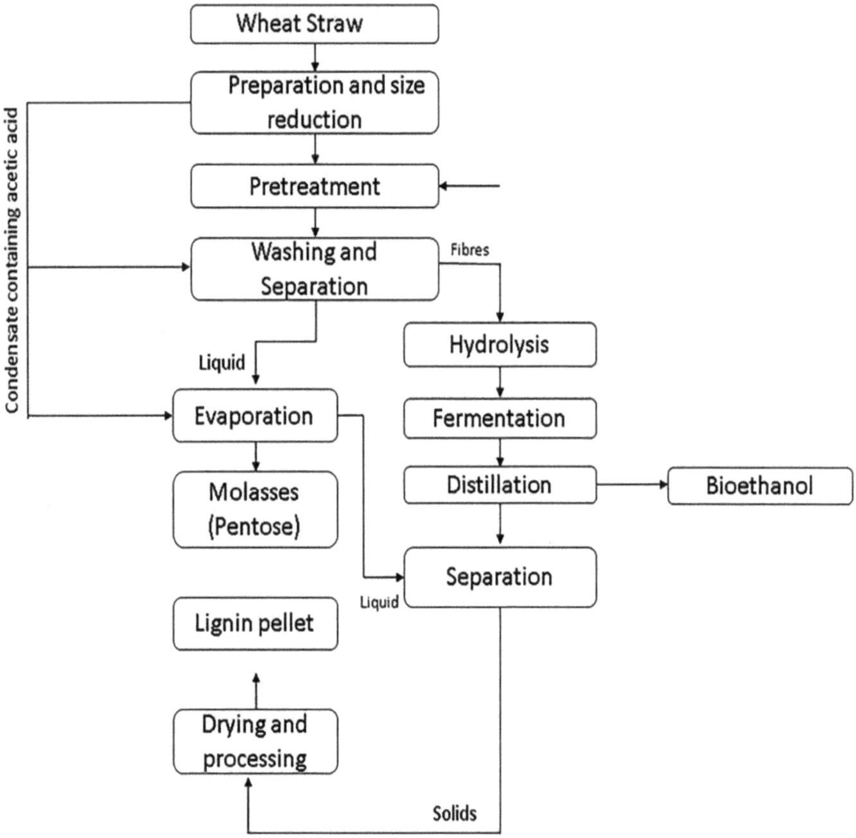

Fig. 11.1 Complete overview of bioethanol production

infrastructure, enhancing its versatility and sustainability. The most common biomass source in Europe is unquestionably wheat straw, but rice straw is the second most common worldwide (Kim and Dale 2004; Talebnia et al. 2010; Horn et al. 2011). As a result of its chemical constituents and rigid structure, wheat straw frequently has a limited capacity for anaerobic biodegradation. To begin with, it is important to note that wheat straw contains significant quantities of lignin, a substance that is resistant to degradation in anaerobic conditions (Fernandes et al. 2009). Accessing lignin is challenging for microbes due to the intricate structure it generates through intermolecular bonds with hemicellulose and cellulose. Anaerobic digestion generates biogas with superior biomass conversion, lower greenhouse gas emissions, and higher energy efficiency compared to ethanol production (Börjesson and Mattiasson 2008). Anaerobic digestion consists of four sequential stages: biopolymer hydrolysis, acidogenesis, acetogenesis, and methanogenesis. The final products of the hydrolysis process, which degrades a wide range of polymers, are monosaccharides, amino acids, and short-chain fatty acids. These consist of lipids, proteins, and carbohydrates (i.e., celluloses and hemicelluloses found in lignocellulosic materials). By further conversion of the aforementioned components, organic acids, or volatile fatty acids (VFAs), are created during acidogenesis. In conjunction with the primary products, the chemical process generates several compounds as by-products, such as carbon dioxide and ammonia. The process of acetogenesis involves the conversion of volatile fatty acids (VFAs) into acetic acid, hydrogen, and carbon dioxide, which serve as key intermediates in anaerobic digestion. Methanogenesis is the conversion of acetate to methane and carbon dioxide (Börjesson and Mattiasson 2008). The carbon-to-nitrogen (C/N) ratio, which periodically surpasses 90 (w/w), is a critical factor hindering the anaerobic biodegradation of wheat straw (Ferreira et al. 2013; Pohl et al. 2013). According to prior research, anaerobic digestion involves microbes that consume more carbon than nitrogen. Laboratory studies indicate that the optimal carbon-to-nitrogen (C/N) ratio for anaerobic digestion generally ranges between 20 and 30 (w/w). C/N ratios that fall below the optimal range may lead to an overabundance of volatile fatty acids (VFAs) accumulation and/or increased concentrations of total ammonium nitrogen. The overall anaerobic digestion process may incur failure during the methanogenesis phase due to the aforementioned factors. On the contrary, an excessively high carbon-to-nitrogen (C/N) ratio could impede microbial growth and diminish the biodegradability of the organic material. Variations in the carbon-to-nitrogen (C/N) ratio beyond the optimal range for anaerobic digestion (AD) have an adverse impact on both substrate biodegradability and biogas production (Sawatdeenarunat et al. 2015). Biogas production is generally considered more sustainable, generating lower greenhouse gas emissions and fewer harmful pollutants compared to open-field straw burning or fossil fuel combustion. However, if not properly managed, biogas production can still contribute to greenhouse gas emissions and global warming, particularly through methane leaks and inefficient gas utilization. Key factors in mitigating adverse effects associated with biogas production in China include the optimization of purification processes, proper disposal of biogas residue, and overall power consumption (Wang et al. 2016).

11.4.3 Biobutanol Production from Wheat Straw

Biobutanol is an ideal alternative fuel due to its high energy content, low volatility, and the absence of hygroscopic and corrosive properties (Ni and Sun 2009; Rakopoulos et al. 2011). In the absence of any modifications to existing engine or vehicle technologies, it is also capable of substituting gasoline (Campos-Fernández et al. 2012; Lee et al. 2008). Butanol (n-butanol and iso-butanol) exhibits several advantages over ethanol as a fuel, including superior energy density, reduced corrosiveness, and enhanced engine compatibility. The acetone–butanol–ethanol fermentation process, which utilizes *Clostridium acetobutylicum*, is the prevailing approach employed in the production of biobutanol (Zverlov et al. 2006). In terms of fermentation volume, biobutanol production ranks second after bioethanol as butanol serves as both a crucial industrial solvent and a promising alternative energy source. The annual production of butanol is estimated to be between 5 and 6 million tons, according to recent research. The global butanol market is valued at $7.2 billion (Lee et al. 2008). Furthermore, even if butanol can be produced at a lower cost, market demand is anticipated to increase substantially (Dürre 1998). One of the strategies to lower the cost of biobutanol production is to use renewable and affordable feedstock, such as abundant and sustainable lignocellulosic materials (Kim and Dale 2004; Kumar et al. 2012). Butanol is extremely toxic to cells, so the highest butanol titers in batch culture were below 13 g/l (Mariano et al. 2011). As potential methods for converting lignocellulosic materials to butanol, metabolically modified *Clostridia*, cellulolytic enzymes added to fermentations, and mixed cultures of solventogenic *Clostridia* and cellulolytic bacteria have all been examined. To illustrate, the utilization of rice straw could result in the production of 5.5 g/l of butanol, while corncobs could yield 10.9 g/l. Although initial results were encouraging, the final butanol titers and production rates were ultimately insufficient to be economically viable (Jiang et al. 2017, 2018). The synthesis of butanol was examined by Qureshi et al. (2007), who utilized wheat straw hydrolysates (WSH) as feedstock and *C. beijerinckii*. The study achieved a yield of 0.41 and an acetone–butanol–ethanol (ABE) productivity of 0.28 g L^{-1} h^{-1}. At the conclusion of fermentation, the concentrations of acetone, butanol, and ethanol were 6.7 g L^{-1}, 12.7 g L^{-1}, and 0.6 g L^{-1}, respectively. Additionally, butyric acid (1.9 g L^{-1}) and acetic acid (4.9 g L^{-1}) were generated. A culture supplemented with 100 g L^{-1} glucose produced 23.5 g L^{-1} ABE and 4.58 g L^{-1} organic acids. When the glucose concentration was increased to 150 g L^{-1} and 200 g L^{-1}, the resulting ABE production was 22.8 g L^{-1} and 14.7 g L^{-1}, respectively. At an initial 200 g L^{-1} glucose concentration, the culture yielded 41.7 g L^{-1} ABE, with acetone (21.3 g L^{-1}), butanol (19.5 g L^{-1}), and ethanol (0.9 g L^{-1}). A maximum productivity of 0.60 g L^{-1} h^{-1} was recorded in this study. In the control batch fermentation, 48.9 g L^{-1} glucose resulted in 20.1 g L^{-1} ABE, achieving a yield of 0.41 and a productivity of 0.28 g L^{-1} h^{-1}. Under similar conditions using WSH, the culture exhibited a yield of 0.42, producing 25 g L^{-1} ABE with an improved productivity of 0.60 g L^{-1} h^{-1}. The productivity increased by 214% compared to the control trial, and the outcomes are better. This provided evidence that this strain can efficiently employ WSH to generate ABE. Processing glucose seems to result in a much higher cost than WS sugars (cost of WS plus hydrolysis). Furthermore, it was mentioned that the *C. beijerinckii* P260 strain is an excellent choice since it can make ABE from WSH in excess of 28 g L^{-1} (and in one instance 41.7 g L^{-1} from glucose) (Qureshi et al. 2007).

11.4.4 Biodiesel Production from Wheat Straw

Biodiesel is presently the second most prevalent biofuel in the world, following bioethanol. Palm, rapeseed, and soya oil are a few examples of vegetable oils used to create biodiesel. The cultivation and processing of vegetable oil can contribute substantially to greenhouse gas emissions. To illustrate, one metric ton of palm or soya oil generates 2000 kg of CO_2 equivalents that are discharged into the atmosphere. Microbial lipids derived from lignocellulose (e.g., straw) serve as a sustainable alternative to vegetable oils. They offer numerous advantages over many of the currently used feedstocks, including low cost, high availability, and minimal environmental impact (Balat 2011). It has been demonstrated that oleaginous yeasts store more than 70% of their biomass as lipids, with some strains accumulating over 20%. Yeasts are regarded as potentially useful due to their rapid growth rate, ability to thrive on a wide range of substrates (including pentose sugars), high cell density, decreased vulnerability to viral infection, and the potential to regulate bacterial contamination by using low pH conditions. In the presence of carbon excess, straw hydrolysates frequently have high C/N ratios, which may result in lipid formation. In the presence of an excess of carbon, oleaginous yeasts cease citrate production and expel it from the mitochondria. Citric acid is enzymatically converted into oxaloacetate and acetyl-CoA within the cytoplasm by a protein known as ATP citrate lyase. During fatty acid biosynthesis, acetyl-CoA serves as a precursor molecule, while NADPH is utilized as a reducing agent. This process is facilitated by acetyl-CoA carboxylase and the fatty acid synthase (FAS) complex. Meanwhile, a portion of acetyl-CoA is transported back to the mitochondria for further metabolic processes. The conversion of numerous hexoses, pentoses, and organic acids generated during the pretreatment of lignocellulose to lipids can be achieved through the utilization of multiple oleaginous yeasts. The hydrolysate contains inhibitors that negatively impact both oleaginous yeasts and the feedstock dry matter, thereby limiting yeast growth and, consequently, reducing product yield. However, advancements in feeding strategies have the potential to enhance the tolerance of lipid-accumulating yeast strains to these inhibitors, improving overall bioprocess efficiency. Producing microbial biodiesel from lignocellulose could potentially yield an energy balance similar to that of bioethanol. The most productive lignocellulose hydrolysate-fed-batch cultures were oleaginous *Rhodosporidium* spp., which have been accurately identified as *Rhodotorula*. This finding suggests that strains of *Rhodosporidium* may develop resistance to inhibitors of fermentation (Wang et al. 2016). Producing valuable biodiesel and by-products could assist in achieving competitive production costs. The potential by-products of catalytic hydrogenation following the fermentation of maize stover with *Rhodotorula toruloides* are hydrocarbons with chemical similarities to fossil fuel. Additionally, wheat straw by-products can be refined into furfural, an additional valuable chemical. The pentose-derived by-product generated through dehydration acts as a crucial intermediate in the synthesis of biofuels, petroleum additives, and a range of value-added chemical compounds. Being a consequence of thermochemical pretreatment, it impedes the fermentation process. At present, synthetic furfural cannot be produced; lignocellulose is the only material that can be utilized in its synthesis. Regrettably, the present furfural synthesis procedures result in cellulose degradation, which renders the glucose monomers unusable as biofuels. A new method for furfural extraction has been innovated at the State Institute of Wood Chemistry in Latvia. The furfural is extracted from lignocellulosic materials using this method, which ensures minimal damage to the cellulose fraction. A more recent study demonstrated that ethanol or lipids can be co-produced with furfural and straw (111 g ethanol, or 33 g lipids, and 110 g furfural, or

69% of the theoretical maximum) from 1-kg straw. Theoretically, the pentoses underwent a complete transformation into furfural, and the hydrolysate contained glucose, which was easily fermentable. Inhibitors that stop fermentation were present in trace quantities in the hydrolysate. Additional potential applications exist for the lipids produced by oleaginous yeasts; for instance, they could be utilized as a substitute for vegetable oils such as palm oil in fish feed. Given the environmental impact of sustainable palm oil production, this strategy, which does not require oil extraction or transesterification, may be a valuable step toward a sustainable economy.

11.4.5 Bioplastics Production from Wheat Straw

An enormous quantity of toxins has been generated as a result of the widespread use of plastic products; these toxins are frequently deposited in landfills, where they cause ongoing environmental degradation and damage to natural resources and human health (Thompson et al. 2009). Bioplastics have gained renewed attention as a sustainable alternative to petroleum-based plastics due to concerns over fossil fuel depletion and environmental pollution. They offer several advantages, including biodegradability, lower greenhouse gas emissions, and the use of renewable resources, making them a promising solution for reducing plastic waste and environmental impact. These compounds have the potential to conserve valuable natural resources, making them a remarkable advancement in the realm of environmental sustainability (Thompson et al. 2009; Harding et al. 2017; Ahmed et al. 2018). They are manufactured with biopolymers, proteins, vegetable oil, carbohydrates derived from legumes and corn, lignocellulose, and sustainable biomass feedstock. Comprising the preponderance of biomass are lignocellulosic feedstocks that are indigestible. Regularly, substantial quantities of lignocellulosic waste are generated by various industries, including agriculture, food processing, alcohol, wood, and sugar. They are typically discarded in an unregulated manner. An extensive array of aesthetic, environmental, health, and safety concerns arise as a result of their accumulation (Patel et al. 2003). Utilizing lignocellulosic wastes is the most optimal and efficient approach to address these issues. There are numerous potential applications for these agrowastes, including the improvement of soil fertility, paper production, and synthetic fuels. However, only a small portion of all waste is accounted for by these applications. Utilizing this lignocellulosic biomass for the creation of sustainable products is ideal in accordance with green protocols. This extremely prospective field has the potential to yield an abundance of valuable products from lignocellulosic waste (Al-Battashi et al. 2019; Adekunle et al. 2016). Recent interdisciplinary interest in biomaterials has been sparked by their numerous advantageous characteristics, such as biodegradability, biocompatibility, affordability, and environmental tolerance. Numerous scientific studies on the potential uses of lignocellulosic waste have contributed to the advancement of bioplastics (Yang et al. 2019; Brinchi et al. 2013). Novel plastic technologies have been used to enhance lignocellulosic biomass in a variety of industries, including fuels, bioplastics,

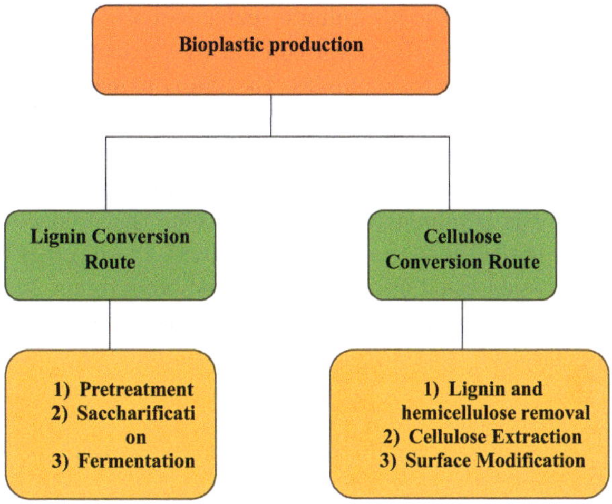

Fig. 11.2 Key pathways for bioplastic production from lignocellulosic biomass

lubricants, solvents, medications, cosmetics, surfactants, animal feed, and nutraceuticals (Yang et al. 2019). Polyhydroxybutyrate (PHB), a type of polyhydroxyalkanoate (PHA), is synthesized by certain microorganisms as an intracellular storage compound. PHAs have the potential to serve as bioplastics instead of plastics extracted from mineral oil, which are carbon-based materials. The industry desperately requires new technology as using lignocellulosic waste routes to create bioplastics is not without its challenges, including costly manufacturing routes. According to research by Kim et al. the utilization of lignocellulosic biomass and organic waste in place of cone stover could substantially reduce the cost of manufacturing bioplastics (Kim et al. 2020). The steps in the production of bioplastics from biomass materials include pretreatment, saccharification, liquid detoxification, fermentation, plastic, and biocomposite development (Brodin et al. 2017). Various routes lead to a variety of bioplastics with properties suitable for packaging, biomedical, water treatments, or other high-end products (Abdul-Latif et al. 2020). The production of bioplastics from lignocellulosic biomass entails two distinct conversion routes, one for cellulose and one for lignin (Fig. 11.2).

11.5 Conclusion and Future Perspectives

Straw is a lignocellulosic agricultural residue that is an excellent choice for use as a feedstock in biotechnological production processes. Co-manufactured alongside cereal grain and other food basic materials, this product's production has no discernible effect on food production or land utilization. In recent years, enormous research has been devoted to determining how to transform this feedstock into biofuels and other valuable compounds. An ever-growing variety of intriguing compounds can be synthesized from straw materials (Chandel et al. 2018). The present moment is characterized by an extreme reliance of the global economy on fossil resources, which renders alternative methods for producing fuels, chemicals, and food critical. The agricultural sector will also benefit from the production of high-value chemicals from agricultural by-products, thereby enhancing the economic appeal of green technologies. The ability to manufacture products that are optimally suited to local conditions and market demands will become possible with the advancement of technologies that enable the production of a variety of fuels and chemicals from straw. As is the case with biodiesel and biogas or ethanol and biogas, it is common practice to combine the production of two or more biofuels (Karlsson et al. 2016; Lantz et al. 2018). On the other hand, process energy is commonly generated through the combustion of lignin. Identifying enzymes capable of decomposing lignin and repurposing it is a current endeavor, given its underutilized status as a resource. Returning lignin to the soil can also contribute to the retention of soil organic carbon levels and ultimately mitigate greenhouse gas emissions (Karlsson et al. 2016; Lantz et al. 2018). To identify a sustainable substitute for fossil fuels, it is imperative to evaluate the prospective ecological and economic ramifications associated with every possible application. The lignocellulosic (LC) biomass biorefinery employs a diverse range of conversion technologies to generate large-scale liquid fuels and high-value compounds, addressing the growing global energy demand. Nevertheless, biorefinery technologies still face various challenges that must be overcome for further advancement. Challenges in the thermochemical conversion of lignocellulosic biomass encompass suboptimal overall energy efficiency, variability and inconsistency in bio-oil quality, limitations in reactor reliability and scalability, and the stringent product quality requirements set by both producers and consumers. Advances in reactor design and dependability, catalysts for bio-oil refining, substrate co-blending, and integration with other renewable energy sources are a few of the numerous potential research avenues. Chemical-looping conversion, integrated gasification combined cycle (IGCC), hydrothermal conversion, microwave-assisted conversion, pyrolysis, and the integration of multiple conversion technologies are among the emerging conversion methods currently in the research and development phase, with the potential for future industrial applications. In contrast, the weak reaction rates and low product concentration of biological conversion processes increase the expenses associated with downstream processing. Exploring catalyst advancements for bio-oil upgrading, feedstock co-processing, and synergistic integration with renewable energy sources represents just a few of the many promising research directions in this field. Emerging conversion technologies, such as chemical-looping gasification, integrated gasification combined cycle (IGCC), hydrothermal pyrolysis, microwave-assisted pyrolysis, supercritical water gasification, and hybrid conversion systems, are currently in the research and development phase but hold significant potential for industrial-scale implementation. In contrast, the high downstream processing costs are primarily attributed to the dilute product streams and intrinsically slow reaction kinetics characteristic of biological conversion processes (Antunes et al. 2019). Recent developments in fields including systems biology engineering, metabolic engineering, strain development, genetic engineering, genome editing and sequencing, high-throughput screening, in silico model prediction, and evolutionary engineering have also revolutionized the biological conversion technologies of LC biomass. Systemic biology is required to design and reconstruct a more effective strain for consolidated bioprocessing. This strategy integrates a range of genome-wide analytical techniques, including transcriptomics, metabolomics, and proteomics, in conjunction with computational modeling approaches such as kinetic models, population interaction models, and genome-scale metabolic network models. The objective is to bring a systems-level perspective and analyses of dynamic physiological data, gene expression, and cellular metabolism (Hasunuma and Kondo 2012).

References

Abdul-Latif NIS, Ong MY, Nomanbhay S, Salman B, Show PL (2020) Estimation of carbon dioxide (CO2) reduction by utilization of algal biomass bioplastic in Malaysia using carbon emission pinch analysis (CEPA). Bioengineered 11(1):154–164. https://doi.org/10.1080/21655979.2020.1718471

Adekunle A, Orsat V, Raghavan V (2016) Lignocellulosic bioethanol: a review and design conceptualization study of production from cassava peels. Renew Sust Energ Rev 64:518–530. https://doi.org/10.1016/j.rser.2016.06.064

Ahmed T, Shahid M, Azeem F, Rasul I, Shah AA, Noman M, Hameed A, Manzoor N, Manzoor I, Muhammad S (2018) Biodegradation of plastics: current scenario and future prospects for environmental safety. Environ Sci Pollut Res 25(8):7287–7298. https://doi.org/10.1007/s11356-018-1234-9

Al-Battashi HS, Annamalai N, Sivakumar N, Al-Bahry S, Tripathi BN, Nguyen QD, Gupta VK (2019) Lignocellulosic biomass (LCB): a potential alternative biorefinery feedstock for polyhydroxyalkanoates production. Rev Environ Sci Biotechnol 18(1):183–205. https://doi.org/10.1007/s11157-018-09488-4

Antonopoulou G, Lyberatos G (2013) Effect of pretreatment of sweet sorghum biomass on methane generation. Waste Biomass Valoriz 4(3):583–591. https://doi.org/10.1007/s12649-012-9183-x

Antunes FAF, Chandel AK, Terán-Hilares R, Ingle AP, Rai M, dos Santos Milessi TS, da Silva SS, dos Santos JC (2019) Overcoming challenges in lignocellulosic biomass pretreatment for second-generation (2G) sugar production: emerging role of nano, biotechnological and promising approaches. Biotech 9(6):13205. https://doi.org/10.1007/s13205-019-1761-1

Balat M (2011) Production of bioethanol from lignocellulosic materials via the biochemical pathway: a review. Energy Convers Manag 52(2):858–875. https://doi.org/10.1016/j.enconman.2010.08.013

Börjesson P, Mattiasson B (2008) Biogas as a resource-efficient vehicle fuel. Trends Biotechnol 26(1):7–13. https://doi.org/10.1016/j.tibtech.2007.09.007

Brinchi L, Cotana F, Fortunati E, Kenny JM (2013) Production of nanocrystalline cellulose from lignocellulosic biomass: technology and applications. Carbohydr Polym 94(1):154–169. https://doi.org/10.1016/j.carbpol.2013.01.033

Brodin M, Vallejos M, Opedal MT, Area MC, Chinga-Carrasco G (2017) Lignocellulosics as sustainable resources for production of bioplastics—a review. J Clean Prod 162:646–664. https://doi.org/10.1016/j.jclepro.2017.05.209

Campos-Fernández J, Arnal JM, Gómez J, Dorado MP (2012) A comparison of performance of higher alcohols/diesel fuel blends in a diesel engine. Appl Energy 95:267–275. https://doi.org/10.1016/j.apenergy.2012.02.051

Chandel AK, Garlapati VK, Singh AK, Antunes FAF, da Silva SS (2018) The path forward for lignocellulose biorefineries: bottlenecks, solutions, and perspective on commercialization. Bioresour Technol 264:370–381. https://doi.org/10.1016/j.biortech.2018.06.004

Demirbas A (2009) Biofuels from agricultural biomass. Energy Sour Part A Recov Utiliz Environ Effects 31(17):1573–1582. https://doi.org/10.1080/15567030802094011

Dürre P (1998) New insights and novel developments in clostridial acetone/butanol/isopropanol fermentation. Appl Microbiol Biotechnol 49(6):639–648. https://doi.org/10.1007/s002530051226

Fernandes TV, Klaasse Bos GJ, Zeeman G, Sanders JPM, van Lier JB (2009) Effects of thermo-chemical pre-treatment on anaerobic biodegradability and hydrolysis of lignocellulosic biomass. Bioresour Technol 100(9):2575–2579. https://doi.org/10.1016/j.biortech.2008.12.012

Ferreira LC, Donoso-Bravo A, Nilsen PJ, Fdz-Polanco F, Pérez-Elvira SI (2013) Influence of thermal pretreatment on the biochemical methane potential of wheat straw. Bioresour Technol 143:251–257. https://doi.org/10.1016/j.biortech.2013.05.065

Gollapalli LE, Dale BE, Rivers DM (2002) Predicting digestibility of ammonia fiber explosion (AFEX)-treated rice straw. Appl Biochem Biotechnol 98(1):23–35. https://doi.org/10.1385/ABAB:98-100:1-9:23

Hammond GP, Mansell RVM (2018) A comparative thermodynamic evaluation of bioethanol processing from wheat straw. Appl Energy 224(April):136–146. https://doi.org/10.1016/j.apenergy.2018.04.123

Hammond GP, McCann N (2023) Bioethanol processing from wheat straw: investment appraisal of a full-scale UK biofuel refinery. Biofuels 14(3):267–277. https://doi.org/10.1080/17597269.2022.2132722

Harding KG, Gounden T, Pretorius S (2017) "Biodegradable" plastics: a myth of marketing? Proc Manuf 7:106–110. https://doi.org/10.1016/j.promfg.2016.12.027

Hasunuma T, Kondo A (2012) Development of yeast cell factories for consolidated bioprocessing of lignocellulose to bioethanol through cell surface engineering. Biotechnol Adv 30(6):1207–1218. https://doi.org/10.1016/j.biotechadv.2011.10.011

Horn SJ, Nguyen QD, Westereng B, Nilsen PJ, Eijsink VGH (2011) Screening of steam explosion conditions for glucose production from non-impregnated wheat straw. Biomass Bioenergy 35(12):4879–4886. https://doi.org/10.1016/j.biombioe.2011.10.013

Jeswani HK, Chilvers A, Azapagic A (2020) Environmental sustainability of biofuels: a review: environmental sustainability of biofuels. Proc R Soc A Math Phys Eng Sci 476(2243):351. https://doi.org/10.1098/rspa.2020.0351

Jiang Y, Xin F, Lu J, Dong W, Zhang W, Zhang M, Wu H, Ma J, Jiang M (2017) State of the art review of biofuels production from lignocellulose by thermophilic bacteria. Bioresour Technol 245:1498–1506. https://doi.org/10.1016/j.biortech.2017.05.142

Jiang Y, Zhang T, Lu J, Dürre P, Zhang W, Dong W, Zhou J, Jiang M, Xin F (2018) Microbial co-culturing systems: butanol production from organic wastes through consolidated bioprocessing. Appl Microbiol Biotechnol 102(13):5419–5425. https://doi.org/10.1007/s00253-018-8970-0

Karagoz P, Bill RM, Ozkan M (2019) Lignocellulosic ethanol production: evaluation of new approaches, cell immobilization and reactor configurations. Renew Energy 143:741–752. https://doi.org/10.1016/j.renene.2019.05.045

Karlsson H, Ahlgren S, Sandgren M, Passoth V, Wallberg O, Hansson PA (2016) A systems analysis of biodiesel production from wheat straw using oleaginous yeast: process design, mass and energy balances. Biotechnol Biofuels 9(1):1–13. https://doi.org/10.1186/s13068-016-0640-9

Karlsson H, Ahlgren S, Sandgren M, Passoth V, Wallberg O, Hansson PA (2017) Greenhouse gas performance of biochemical biodiesel production from straw: soil organic carbon changes and time-dependent climate impact. Biotechnol Biofuels 10(1):1–15. https://doi.org/10.1186/s13068-017-0907-9

Kim S, Dale BE (2004) Global potential bioethanol production from wasted crops and crop residues. Biomass Bioenergy 26(4):361–375. https://doi.org/10.1016/j.biombioe.2003.08.002

Kim H, Lee S, Ahn Y, Lee J, Won W (2020) Sustainable production of bioplastics from lignocellulosic biomass: technoeconomic analysis and life-cycle assessment. ACS Sustainable Chem Eng 8(33):12419–12429. https://doi.org/10.1021/acssuschemeng.0c02872

Kumar M, Goyal Y, Sarkar A, Gayen K (2012) Comparative economic assessment of ABE fermentation based on cellulosic and non-cellulosic feedstocks. Appl Energy 93:193–204. https://doi.org/10.1016/j.apenergy.2011.12.079

Lantz M, Prade T, Ahlgren S, Björnsson L (2018) Biogas and ethanol from wheat grain or straw: is there a trade-off between climate impact, avoidance of iLUC and production cost? Energies 11(10):2633. https://doi.org/10.3390/en11102633

Lee SY, Park JH, Jang SH, Nielsen LK, Kim J, Jung KS (2008) Fermentative butanol production by clostridia. Biotechnol Bioeng 101(2):209–228. https://doi.org/10.1002/bit.22003

Luque R, Herrero-Davila L, Campelo JM, Clark JH, Hidalgo JM, Luna D, Marinas JM, Romero AA (2008) Biofuels: a technological perspective. Energy Environ Sci 1(5):542–564. https://doi.org/10.1039/b807094f

Mariano AP, Qureshi N, Filho RM, Ezeji TC (2011) Bioproduction of butanol in bioreactors: new insights from simultaneous in situ butanol recovery to eliminate product toxicity. Biotechnol Bioeng 108(8):1757–1765. https://doi.org/10.1002/bit.23123

Ni Y, Sun Z (2009) Recent progress on industrial fermentative production of acetone-butanol-ethanol by clostridium acetobutylicum in China. Appl Microbiol Biotechnol 83(3):415–423. https://doi.org/10.1007/s00253-009-2003-y

Patel DM, Bastioli DC, Marini DL, Würd-inger D-GE (2003) Life-cycle assessment of bio-based polymers and natural fibres. Biopolymers 10:14

Pohl M, Heeg K, Mumme J (2013) Anaerobic digestion of wheat straw - performance of continuous solid-state digestion. Bioresour Technol 146:408–415. https://doi.org/10.1016/j.biortech.2013.07.101

Qureshi N, Saha BC, Cotta MA (2007) Butanol production from wheat straw hydrolysate using Clostridium beijerinckii. Bioprocess Biosyst Eng 30(6):419–427. https://doi.org/10.1007/s00449-007-0137-9

Qureshi N, Saha BC, Hector RE, Dien B, Hughes S, Liu S, Iten L, Bowman MJ, Sarath G, Cotta MA (2010) Production of butanol (a biofuel) from agricultural residues: part II - use of corn Stover and switchgrass hydrolysates. Biomass Bioenergy 34(4):566–571. https://doi.org/10.1016/j.biombioe.2009.12.023

Rakopoulos CD, Dimaratos AM, Giakoumis EG, Rakopoulos DC (2011) Study of turbocharged diesel engine operation, pollutant emissions and combustion noise radiation during starting with bio-diesel or n-butanol diesel fuel blends. Appl Energy 88(11):3905–3916. https://doi.org/10.1016/j.apenergy.2011.03.051

Robak K, Balcerek M (2020) Current state-of-the-art in ethanol production from lignocellulosic feedstocks. Microbiol Res 240:126534. https://doi.org/10.1016/j.micres.2020.126534

Sawatdeenarunat C, Surendra KC, Takara D, Oechsner H, Khanal SK (2015) Anaerobic digestion of lignocellulosic biomass: challenges and opportunities. Bioresour Technol 178:178–186. https://doi.org/10.1016/j.biortech.2014.09.103

Scarlat N, Martinov M, Dallemand JF (2010) Assessment of the availability of agricultural crop residues in the European Union: potential and limitations for bioenergy use. Waste Manag 30(10):1889–1897. https://doi.org/10.1016/j.wasman.2010.04.016

Talebnia F, Karakashev D, Angelidaki I (2010) Production of bioethanol from wheat straw: an overview on pretreatment, hydrolysis and fermentation. Bioresour Technol 101(13):4744–4753. https://doi.org/10.1016/j.biortech.2009.11.080

Thompson RC, Moore CJ, Saal FSV, Swan SH (2009) Plastics, the environment and human health: current consensus and future trends. Philos Trans R Soc B Biol Sci 364(1526):2153–2166. https://doi.org/10.1098/rstb.2009.0053

Townsend TJ, Sparkes DL, Wilson P (2017) Food and bioenergy: reviewing the potential of dual-purpose wheat crops. GCB Bioenergy 9(3):525–540. https://doi.org/10.1111/gcbb.12302

Wang QL, Li W, Gao X, Li SJ (2016) Life cycle assessment on biogas production from straw and its sensitivity analysis. Bioresour Technol 201:208–214. https://doi.org/10.1016/j.biortech.2015.11.025

Yang J, Ching YC, Chuah CH (2019) Applications of lignocellulosic fibers and lignin in bioplastics: a review. Polymers 11(5):1–26. https://doi.org/10.3390/polym11050751

Zverlov VV, Berezina O, Velikodvorskaya GA, Schwarz WH (2006) Bacterial acetone and butanol production by industrial fermentation in the Soviet Union: use of hydrolyzed agricultural waste for biorefinery. Appl Microbiol Biotechnol 71(5):587–597. https://doi.org/10.1007/s00253-006-0445-z

Zwart RWR, Boerrigter H, van der Drift A (2006) The impact of biomass pretreatment on the feasibility of overseas biomass conversion to Fischer-Tropsch products. Energy Fuels 20(5):2192–2197. https://doi.org/10.1021/ef060089f

Rice Residue as a Feedstock for Biofuel and Chemicals

12

Azhagu Saravana Babu Packirisamy, Vithyasree Srikanth, Vajiha Aafrin Basheer, and Sai Nandhini Ravi

Abstract

This study focuses on the utilization of rice residue as a feedstock for biofuels and chemicals. Bioethanol is a renewable energy source that has been attracting attention in recent years as a substitute to fossil fuels. Numerous bacteria naturally produce polyhydroxyalkanoate (PHA) biopolymer that has garnered a lot of attention because of its biodegradability, lack of toxicity, and biocompatibility. Because rice residues, such as rice straw and rice husk, are abundant and a renewable resource; therefore, it is a good choice for the production of PHA and bioethanol. Among its many qualities is a high cellulose and hemicellulose content, which requires pretreatment and enzymatic hydrolysis and then transforms into fermentable sugars. The selection of a suitable pretreatment technique is a major challenge in the development of technology for the production of bioethanol and PHA from rice residue. The selection of pretreatment methods is essential in order to maximize the effectiveness of the enzyme saccharification process, thus making the entire process cost-effective. The present study examines the various technologies available for the production of bioethanol and PHA using rice residue as feedstock.

A. S. B. Packirisamy (✉) · V. Srikanth · S. N. Ravi
Department of Biotechnology, Vel Tech Rangarajan Dr. Sagunthala R&D Institute of Science and Technology, Chennai, Tamil Nadu, India
e-mail: drazhagusaravanababu@veltech.edu.in

V. A. Basheer
Department of Biotechnology, Vel Tech Rangarajan Dr. Sagunthala R&D Institute of Science and Technology, Chennai, Tamil Nadu, India

Centre for Food Technology, Anna University, Chennai, Tamil Nadu, India

A. Kumari et al. (eds.), *Valorization of Solid Wastes to Biofuels and Chemical Products for Sustainable World*, https://doi.org/10.1007/978-981-96-8594-3_12

Keywords
Bioethanol · Polyhydroxyalkanoate · Rice residue · Pretreatment · Feedstock

12.1 Introduction

Biofuel is a form of energy derived from biomass, commonly referred to as organic matter. The two most popular biofuels are bioethanol and biodiesel (Tse et al. 2021). The hunt for renewable energy sources that lower carbon dioxide emissions is increasingly emphasized as people become more aware of the effects of global warming and how reliant we are on fossil fuels. One strategy to lower the amount of environmental pollution and crude oil consumption is to produce bioethanol from biomass. The utilization of bioethanol-based blended fuel for cars has the potential to drastically decrease reliance on petroleum resources and the emissions of greenhouse gases. Cellulose feedstocks for the production of bioethanol can be derived from a variety of sources, which include municipal solid waste, sugar cane bagasse, maize stalks, pulpwood, rice straw, and switchgrass (Balat 2009).

Petrochemical is a broad term that includes a wide range of commercial polymers and chemicals. One such naturally produced microbial biopolymer is polyhydroxyalkanoate (PHA). PHA is bound to be one of the essential biopolymers that could be a great alternative to the traditional petrochemical-based plastics that pose a threat to the global environment (Sabapathy et al. 2020). PHA is primarily synthesized by the metabolic process of microorganisms and is stored in cells as an energy storage element. Due to their direct use, PHAs have a higher degree of biodegradability compared to other biopolymers (Novelli et al. 2021).

Rice residue is a significant crop residue that is cultivated extensively in Asia. Rice residues are renewable resources, which may be utilized to make a variety of environmentally beneficial and high-value products (Huang and Lo 2019). Rice straw is a primary by-product of rice cultivation and an excellent bioresource. Rice straw, a widely utilized waste material due to its high lignin concentration, is employed in various applications such as paper manufacturing, biofuel production, and animal feed (Roberto et al. 2003). Due to the siliceous nature of rice crop residues, rice straw has the potential to alter the electrochemical characteristics of acidic soils, resulting in a decrease in phosphorus fixation while increasing base retention and soil pH (Mandal et al. 2004).

The chemical makeup of the rice straw determines its nutritious content. According to research by Roberto et al. (2003), the primary components of rice straw are hemicellulose, lignin, cellulose, and ash. Hemicelluloses containing xylose are predominantly composed of pentoses, with xylose being the most prominent sugar, arabinose and hexose are also present. Rice straw carbohydrates include galactose (0.4%), mannose (1.8%), glucose (41–43.4%), arabinose (2.7–4.5%), and xylose (14–20.2%).

With over 750 million tons produced globally each year, rice residue produces a significant quantity of biomass waste, including bran, husk, and straw, which is an excellent feedstock for industries that use biomass conversion (Singh and Patel

2022; Zhang et al. 2023). Rice residues can be transformed into biochar and biofuels, as well as valuable chemicals (like glucose) and platform molecules.

12.2 Carbon Sources of Rice Straw

12.2.1 Cellulose

Cellulose is the most common polysaccharide found in plant biomass. Cellulosic polymer is characterized by its two-phase structure, which consists of a crystalline phase and an amorphous phase. The crystalline structure of cellulose limits its accessibility for enzymatic breakdown by making it resistant to enzymatic hydrolysis (Park et al. 2010). Cellulose is a highly abundant natural resource and has been utilized in a variety of industrial processes, including the production of paper and fibrous material and fermentation. Cellulose has been recognized as a key contributor to the advancement of bioenergy due to its polysaccharide structure (Park et al. 2018; Yang et al. 2007; Kumar et al. 2009; Taher et al. 2014; Fatma et al. 2018). Currently, microbial fermentation is the most popular form of renewable energy research (Lu et al. 2017; Lynd et al. 2002; Lee and Jin 2017).

12.2.2 Hemicellulose

A fundamental component of plant cell walls is hemicellulose, a branched polysaccharide that is different from cellulose (Hu LiSong et al. 2018; Saha 2003). Hemicellulose is a noncrystalline, hetero-polysaccharide. It is composed of furanoses and pyranoses sugar, including xylan (the building blocks of plant cell wall) and xyloglucan, as well as manans and other compounds related to the β-glucan linkage. Hemicellulose has two main groups of acids, which makes it really hydrophilic and alkaline-soluble (Patel and Parsania 2017). Hemicellulose hydrolysis has the potential to yield xylans, which are widely utilized in a variety of commercial products, including biofuels, pharmaceuticals, and foodstuffs (Sabiha-Hanim and Siti-Norsafurah 2012). Extraction of hemicellulose from plant cells can be accomplished via an organic solvent, ionic liquid-based extraction, liquid hot water treatment, and mercerization. The mercerization is the most widely employed in industries and laboratories due to its effectiveness. The hardwood is hydrolyzed using potassium hydroxide, while the soft plant cell wall is typically hydrolyzed using sodium hydroxide (Sun et al. 2013; Froschauer et al. 2013; Peng et al. 2011). Hemicelluloses are produced when ester bonds are broken during hydrolysis. Following the alkali treatment, hemicelluloses can be precipitated using ethanol (Stoklosa and Hodge 2012). Ultrasound can assist in decreasing the extraction time to approximately 1.5–2.5 h by increasing the efficiency of hemicellulose. Rice straw cells are easily damaged by the ultrasound waves as they disrupt the biomass cells and aid in the extraction of hemicelluloses (Wang et al. 2018; Dinh Vu et al. 2017; Minjares-Fuentes et al. 2016).

12.2.3 Lignin

Lignin is composed of three primary components and is an amorphous three-dimensional polymer. Globally, it is the second largest source of biomass after cellulose. Due to its benefits, which have been thoroughly investigated over the last 10 years, it is thought to be a possible value-added chemical. These benefits include antibacterial activity, environmentally friendly, heat-resistant, and adhesive (Binod et al. 2010; Figueiredo et al. 2018). Lignin has an aromatic backbone and is one of the naturally occurring polymer having a high molecular weight. This makes it an excellent raw material for biofuel, chemical, and fossil-derived materials. High concentrations of lignin can be obtained from lignocellulosic residues that are left over after processing biomass from forests or farms (Rosado et al. 2021). All vascular plants have lignin, a complex aromatic polymer, accumulated in their secondary cell wall (Sarkar et al. 2009; Bonawitz and Chapple 2010). It is closely interconnected with other components of the cell wall and can therefore be thought of as a "cellular adhesive" (Rubin 2008). It also plays a protective role in plants, providing protection against environmental factors such as pathogens and insects (Moura et al. 2010), as well as providing stability for xylem vessels to facilitate efficient water circulation (Voelker et al. 2011). Agricultural by-products are estimated to produce over 2×10^{11} tons of lignocellulosic material annually, which includes bagasse, straw, shells, roots, and husks (Tuck et al. 2012). Recently, lignocellulose has been recognized as a major source of raw materials for the production of bioenergy (Tuck et al. 2012; Somerville et al. 2010).

12.3 Method for Producing Bioethanol Using Lignocellulosic Biomass from Rice Residue

Bioethanol is produced by converting biomass into bioethanol. There are three main steps in the bioethanol process: pretreatment (using chemicals to treat the biomass), hydrolysis (using enzymes to break down the biomass), and fermentation (using enzymes to ferment the biomass). Each step can be done on its own or in combination with several other steps.

12.3.1 Pretreatment Method

The pretreatment process is regarded as an important step in the biochemical conversion of LC materials with additional value (Koupaie et al. 2019). Pretreatment disrupts the lignocellulosic structure and primarily breaks down lignin and exposes cellulose and hemicellulose for enzymatic hydrolysis. These exposed polysaccharides are then hydrolyzed by enzymes to produce sugars (Kumar and Sharma 2017). Therefore, the pretreatment is carried out prior to acid/enzyme hydrolysis (Sasmal et al. 2012). A variety of pretreatment techniques can be applied for the production of fermentable sugars (Bhatia et al. 2019; Rocha-Meneses et al. 2020).

12.3.2 Enzymatic Hydrolysis

In general, the categories for biomass hydrolysis strategies are enzymatic and chemical. Fermentable sugars are released as a result of chemical hydrolysis, but this process is typically hazardous, corrosive, and poisonous, and it requires reactors that are resistant to corrosion, which makes it a labor-intensive procedure overall (Dubey et al. 2012). Thus, one of the fundamental requirements that should be met by the bioethanol sector in the future is advancements in hydrolysis.

12.3.3 Fermentation

The fermentation process in the bioethanol sector has been greatly aided by fermentation-related microorganisms. In order to enhance yeast/bacteria fermentation efficiency, a number of studies are currently concentrating on novel fermentation pathways. These studies employ a variety of techniques, such as nutrition optimization, metabolic engineering, adaptive laboratory development, and strain selection and isolation from natural or industrial environments. Conventional yeast used in the production of ethanol cannot convert pentoses to ethanol. Only a small number of natural yeasts, including *Candida shehatae* (a type of brewer's yeast), *Pichia stipitis* (a type of spore-forming yeast), and *Pachysolen tannophilus* (a type of yeast that aids in the metabolism of xylose to ethanol), can metabolize pentoses (Berłowska et al. 2016; Zabed et al. 2017). Additionally, to optimize the utilization of xylose contained in hydrolyzate, it may be obtained by introducing genes that encode xylose into the metabolism of a specific microbe (Robak and Balcerek 2018).

12.4 Microbe and Method for Producing Lignocellulosic Biomass

12.4.1 Microbe for Producing PHA Using Cellulose as Carbon Source

Cellulose is a polysaccharide that is widely distributed in the marine environment. *Saccharophagus degradans*, called (Sde2-40), has a remarkable capacity to break down cellulose (Ekborg et al. 2005; Taylor et al. 2006). The microbe was collected from a decomposing salt marsh grass in Chesapeake Bay, Virginia, United States (Andrykovitch and Marx 1988). Sde2–40 has been identified as a bacterium that is highly versatile in its ability to degrade polysaccharides, as evidenced by a comprehensive genomic sequence analysis (Weiner et al. 2008) and metabolomics and transcriptomics studies (Shin et al. 2010a; Shin et al. 2010b; Zhang and Hutcheson 2011). The metabolism and degradation of these polysaccharides via Sde2–40 have been elucidated by the characterization of enzymes that hydrolyze carbohydrates, such as xylanase and cellulase (Taylor et al. 2006; Zhang and Hutcheson 2011; Watson et al. 2009). Sde2–40 has been sequenced and identified with the genes encoding enzymes, such as PhaA, PhaB, and PhaC, that are responsible for PHA synthesis (Kalia et al. 2003).

12.4.2 Method for Producing PHA Using Cellulose as a Carbon Source

Sawant et al. (2017) used 2 L of sterile minimal media and glucose, then the 5 L bioreactor (KoBio Tech, Korea) was used to produce PHA by Sde2-40. The temperature and pH were adjusted by adding HCl or NH_4OH, while aeration and agitation were set at 250 rpm and 1 vvm, respectively. When the pH of the medium reached 7.5, a 100 mL inoculum was added to the bioreactor. A 20 g sterile glucose solution was supplied every 16 h in fed-batch culture in order to get the glucose concentration back to 10 g/L. Every 48 h, a 20 g sterile avicel solution was supplied to the fed-batch culture, which used the avicel as a carbon source. In the fed-batch culture, the total amount of avicel and glucose used were 40 g/L and 50 g/L, respectively. Samples were collected and examined for PHA concentration and protein content.

12.4.3 Microbe for Producing PHA Using Hemicellulose as a Carbon Source

One of the most researched native PHA-producing bacteria is *Burkholderia cepacia*, which can consume a wide range of carbon sources for the production of SCL-PHA, including homopolymer-like substances polyhydroxyvalerate (PHV) (Wendy et al. 2022). Dietrich et al. (2013) showed that *B. cepacia* also has the ability to use the carbohydrates mannose, arabinose, and other derived carbohydrates from lignocelluloses as substrates for PHA synthesis.

12.4.4 Method for Producing PHA Using Hemicellulose as a Carbon Source

The microorganism *P. cepacia* strain ATCC 17759 was cultivated using $(NH_4)_2SO_4$ as the nitrogen source and either glucose or xylose as the carbon source at neutral pH for shake-flask tests in a mineral salts medium under nitrogen-limited conditions. The cultivation process was conducted in Erlenmeyer flasks with agitation for 60 h at 200 rpm and 30 °C. A prior paper (Ramsay et al. 1990) described the mineral salts medium's composition. The mineral salts medium was used, together with $(NH_4)_2SO_4$ and xylose at 30–35 °C, for batch fermentations carried out in a Multigen F-2000, 20% (w/v) NaOH and 20% (v/v) H_2SO_4 were used to maintain a neutral pH. At all times, the agitation and aeration were regulated.

Ramsay et al. determined the biomass by measuring the dry mass (Ramsay et al. 1990). Protein was examined using the biuret method, with a standard of bovine serum albumin. Braunegg et al. (1978) reported the methylation process for PHB samples, while Ramsay et al. (1990) outlined the requirements for gas chromatography quantification. NH_4^+ was determined using a gas ammonia probe and standards consisting of $(NH_4)_2SO_4$ solutions (Ramsay et al. 1995).

12.4.5 Microbe for Producing PHA Using Lignin as a Carbon Source

The aerobic bacterium *Pseudomonas putida*, which is Gram-negative, has a wide range of benefits that make it an attractive microbial host for a variety of biotechnology applications. It is characterized by its meticulous growth, reduced need for nutrients, its capacity to tolerate extreme environmental conditions, and its high ability to provide redox energy (Oliva-Arancibia et al. 2017). *P. putida* is typically found in soil and aquatic environments. However, due to their highly adaptable metabolism, members of the *P. putida* family can survive in a variety of environments with varying nutritional conditions (Molina et al. 2014). These are so-called "saprophagous soil colonizers" that have been collected from all over the globe (Ramos et al. 2015). *P. putida* NX-1 has been studied because it has different metabolic properties than other *P. putida* strains (Xu et al. 2018). The *P. putida* has been identified as an ideal microorganism for the transformation of aromatics derived from lignin into bio-based products via metabolic engineering (Weng et al. 2021). *P. putida* is a natural producer of PHA and is widely regarded as a specimen bacterium for the purpose of biodegradation research and as a production platform. This is largely due to its ability to tolerate the presence of solvents as well as oxidative stress.

12.4.6 Method for Producing PHA Using Lignin as a Carbon Source

In order to generate PHA from lignin, *P. putida* strain NX-1 cells that had been kept in glycerol were injected to LB medium and cultured. Centrifugation was used to gather the grown cells, and after 10 min at 4000 rpm, the resulting cell pellets were twice cleaned in a phosphate buffer. Following cleaning, the cells were inoculated in two-liter duplicate flasks that held 600 ml of modified medium with lignin, and incubated at 30 °C and 200 rpm for 7 days. Samples were taken every 24 h at a regular interval of 60 mL for the first 3 days and 30 mL for the final 4 days. The sample was centrifuged for 10 min at 4000 rpm. The resultant cell pellets were collected, dried, and twice washed with equi-volumetric deionized water in order to calculate the cell dry weight. Chloroform was mixed with *P. putida* strain NX-1 cells and incubated in a water bath for 4 h in order to extract the intracellular PHA. The chloroform solution was then concentrated by evaporation after being filtered to exclude cell debris. PHA solids were collected when the ethanol and chloroform mixture was removed during centrifugation. After being gathered, the PHA was dried at

30 °C in a vacuum dryer until all solvent residue was gone. PHA content = extracted PHA/dry cell weight was the formula used to determine the intracellular PHA content (Xu et al. 2021).

12.5 Production of Bioethanol from Lignocellulosic Biomass

Cell wall modifications and also genome editing are the most recent developments in feedstock modification. Until recently, there was no industrially viable pretreatment method that could remove lignin at a low cost and achieve financial viability. Lamichhane et al. (2021) covered the current energy demand scenario, technological advancements in bioethanol, and associated issues. The bioethanol industry should prioritize future research aimed at primarily changing the composition of LC biomass, developing a strong enzyme that maximizes fermentable sugar yield and facilitates hydrolysis, and developing genetically modified microorganisms, such as bacteria and yeast that can efficiently ferment sugars at high rates through simultaneous fermentation.

12.6 PHA Production from Lignocellulosic Biomass

12.6.1 PHA Production from Cellulose

Sde2–40 is a bacterium that can produce PHA from lignocellulosic biomass. Sawant et al. (2017) examined the development and synthesis of PHA in Sde2-40 cells that was supplemented with β-1, 4-glucan (1% different). By using Nile blue, the biopolymer that had accumulated in Sde2-40 cells was determined. FT-IR and GC analysis revealed that the type and nature of the PHA. PHA content was 11.80%, 12.8%, 13.7%, and 14.6% of dry cell weight in flask cultures using α-cellulose, carboxymethyl cellulose (CMC), sigmacell 101, and avicel PH101 as carbon sources, respectively.

12.6.2 PHA Production from Hemicellulose

Ramsay et al. (1995) determined whether *P. cepacia* could use xylose, a key hemicellulosic sugar found in hardwoods, to produce the thermoplastic, biodegradable poly(P-hydroxybutyrate) (PHB). Grown on mineral salts medium with ammonium limitation containing xylose, the culture yielded biomass containing PHB in shake flask. According to batch fermentation data, growth of the microorganism and the kinetics of PHB production on xylose matched those of those that had been previously published for it on fructose. The hydrolyzed hemicellulose substrate cost for PHB synthesis was calculated to be half that of bulk glucose and comparable to cane molasses using YPHBIS.

12.6.3 PHA Production from Lignin

According to the genome information, *P. putida* NX-1 possesses a biosynthesis pathway for the production of PHA, which includes ligninolytic enzymes and the second-class poly(R)-hydroxyalkanoic acid synthase, poly(hydroxyalkanoate) granule-associated protein, poly(3-hydroxyalkanoate) granule-associated protein PhaI, and polyhydroxyalkanoic acid system protein. Based on Kumar et al. (2018), Lin et al. (2016), Salvachúa et al. (2020), it appears that *P. putida* strain NX-1 possesses the ability to convert lignin into PHA. In order to validate this hypothesis, *P. putida* strain NX-1 was cultured aerobically with lignin as a carbon source and limited nitrogen supply was provided during the incubation. This occurrence was most likely caused by *P. putida* NX-1's ability to quickly use aromatic monomers generated from lignin and low-molecular-weight lignin fragments. According to earlier research, fungi mostly depolymerized the high-molecular-weight lignin, and bacteria in the natural world were able to effectively use the resulting low-molecular-weight portion (Abdelaziz et al. 2016; Kamimura et al. 2019). In the end, the intracellular PHA concentration was found. Using Nile red dye and fluorescence microscopy to visualize intracellular granules, the PHA buildup was further verified. Strong intracellular fluorescent spots indicated that PHA granules were present (Xu et al. 2021).

12.7 Summary

Lignocellulose biomass from rice straw appears to be an excellent substrate for a variety of biotechnology and industrial applications since exploiting these resources could reduce our reliance on petroleum and address concerns related to its fluctuating availability and rising prices. The concept of industrial biorefineries that convert biomass into a variety of value-added products is appealing despite the challenges associated with using lignocellulosic biomass for bio-based processes, which is gaining traction and can be expected to grow in importance over the coming decades. PHAs could be a significant product of bio-based refineries, which could help these valuable eco-friendly plastics compete with, and in some cases, partially substitute petrochemical plastics in their respective applications.

References

Abdelaziz OY, Brink DP, Prothmann J, Ravi K, Sun M, García-Hidalgo J, Gorwa-Grauslund MF (2016) Biological valorization of low molecular weight lignin. Biotechnol Adv 34(8):1318–1346

Andrykovitch G, Marx I (1988) Isolation of a new polysaccharide-digesting bacterium from a salt marsh. Appl Environ Microbiol 54(4):1061–1062

Balat M (2009) Bioethanol as a vehicular fuel: a critical review. Energy Sour Part A 31(14):1242–1255

Berłowska J, Pielech-Przybylska K, Balcerek M, Dziekońska-Kubczak U, Patelski P, Dziugan P, Kręgiel D (2016) Simultaneous saccharification and fermentation of sugar beet pulp for efficient bioethanol production. Biomed Res Int 2016(1):3154929

Bhatia L, Garlapati VK, Chandel AK (2019) Scalable technologies for lignocellulosic biomass processing into cellulosic ethanol. In: Horizons in bioprocess engineering. Springer, Cham, pp 73–90

Binod P, Sindhu R, Singhania RR, Vikram S, Devi L, Nagalakshmi S, Pandey A (2010) Bioethanol production from rice straw: an overview. Bioresour Technol 101(13):4767–4774

Bonawitz ND, Chapple C (2010) The genetics of lignin biosynthesis: connecting genotype to phenotype. Annu Rev Genet 44(1):337–363

Braunegg G, Sonnleitner BY, Lafferty RM (1978) A rapid gas chromatographic method for the determination of poly-β-hydroxybutyric acid in microbial biomass. Eur J Appl Microbiol Biotechnol 6:29–37

Dietrich D, Illman B, Crooks C (2013) Differential sensitivity of polyhydroxyalkanoate producing bacteria to fermentation inhibitors and comparison of polyhydroxybutyrate production from Burkholderia cepacia and Pseudomonas pseudoflava. BMC Res Notes 6:1–4

Dinh Vu N, Thi Tran H, Bui ND, Duc Vu C, Viet Nguyen H (2017) Lignin and cellulose extraction from Vietnam's rice straw using ultrasound-assisted alkaline treatment method. Int J Polym Sci 2017(1):1063695

Dubey AK, Gupta PK, Garg N, Naithani S (2012) Bioethanol production from waste paper acid pretreated hydrolyzate with xylose fermenting Pichia stipitis. Carbohydr Polym 88(3):825–829

Ekborg NA, Gonzalez JM, Howard MB, Taylor LE, Hutcheson SW, Weiner RM (2005) Saccharophagus degradans gen. Nov., sp. nov., a versatile marine degrader of complex polysaccharides. Int J Syst Evol Microbiol 55(4):1545–1549

Fatma S, Hameed A, Noman M, Ahmed T, Shahid M, Tariq M, Tabassum R (2018) Lignocellulosic biomass: a sustainable bioenergy source for the future. Protein Pept Lett 25(2):148–163

Figueiredo P, Lintinen K, Hirvonen JT, Kostiainen MA, Santos HA (2018) Properties and chemical modifications of lignin: towards lignin-based nanomaterials for biomedical applications. Prog Mater Sci 93:233–269

Froschauer C, Hummel M, Iakovlev M, Roselli A, Schottenberger H, Sixta H (2013) Separation of hemicellulose and cellulose from wood pulp by means of ionic liquid/cosolvent systems. Biomacromolecules 14(6):1741–1750

Hu LiSong HL, Du MengHao DM, Zhang JinPing ZJ (2018) Hemicellulose-based hydrogels present status and application prospects: a brief review. Open J For 8:15–28

Huang YF, Lo SL (2019) Utilization of rice hull and straw. In: Rice. AACC International Press, St. Paul, pp 627–661

Kalia VC, Chauhan A, Bhattacharyya G, Rashmi. (2003) Genomic databases yield novel bioplastic producers. Nat Biotechnol 21(8):845–846

Kamimura N, Sakamoto S, Mitsuda N, Masai E, Kajita S (2019) Advances in microbial lignin degradation and its applications. Curr Opin Biotechnol 56:179–186

Koupaie EH, Dahadha S, Lakeh AB, Azizi A, Elbeshbishy E (2019) Enzymatic pretreatment of lignocellulosic biomass for enhanced biomethane production-a review. J Environ Manag 233:774–784

Kumar AK, Sharma S (2017) Recent updates on different methods of pretreatment of lignocellulosic feedstocks: a review. Bioresour Bioprocess 4(1):7

Kumar P, Barrett DM, Delwiche MJ, Stroeve P (2009) Methods for pretreatment of lignocellulosic biomass for efficient hydrolysis and biofuel production. Ind Eng Chem Res 48(8):3713–3729

Kumar M, Verma S, Gazara RK, Kumar M, Pandey A, Verma PK, Thakur IS (2018) Genomic and proteomic analysis of lignin degrading and polyhydroxyalkanoate accumulating β-proteobacterium Pandoraea sp. ISTKB Biotechnol Biofuels 11:1–23

Lamichhane G, Acharya A, Poudel DK, Aryal B, Gyawali N, Niraula P, Parajuli N (2021) Recent advances in bioethanol production from lignocellulosic biomass. Int J Green Energy 18(7):731–744

Lee WH, Jin YS (2017) Evaluation of ethanol production activity by engineered Saccharomyces cerevisiae fermenting cellobiose through the phosphorolytic pathway in simultaneous saccharification and fermentation of cellulose. J Microbiol Biotechnol 27(9):1649–1656

Lin L, Cheng Y, Pu Y, Sun S, Li X, Jin M, Yuan JS (2016) Systems biology-guided biodesign of consolidated lignin conversion. Green Chem 18(20):5536–5547

Lu Y, Li GS, Lu YC, Fan X, Wei XY (2017) Analytical strategies involved in the detailed componential characterization of biooil produced from lignocellulosic biomass. Int J Anal Chem 2017(1):9298523

Lynd LR, Weimer PJ, Van Zyl WH, Pretorius IS (2002) Microbial cellulose utilization: fundamentals and biotechnology. Microbiol Mol Biol Rev 66(3):506–577

Mandal KG, Misra AK, Hati KM, Bandyopadhyay KK, Ghosh PK, Mohanty M (2004) Rice residue-management options and effects on soil properties and crop productivity. J Food Agric Environ 2:224–231

Minjares-Fuentes R, Femenia A, Garau MC, Candelas-Cadillo MG, Simal S, Rosselló C (2016) Ultrasound-assisted extraction of hemicelluloses from grape pomace using response surface methodology. Carbohydr Polym 138:180–191

Molina L, Udaondo Z, Duque E, Fernández M, Molina-Santiago C, Roca A, Ramos JL (2014) Antibiotic resistance determinants in a pseudomonas putida strain isolated from a hospital. PLoS One 9(1):e81604

Moura JCMS, Bonine CAV, de Oliveira Fernandes Viana J, Dornelas MC, Mazzafera P (2010) Abiotic and biotic stresses and changes in the lignin content and composition in plants. J Integr Plant Biol 52(4):360–376

Novelli LDD, Sayavedra SM, Rene ER (2021) Polyhydroxyalkanoate (PHA) production via resource recovery from industrial waste streams: a review of techniques and perspectives. Bioresour Technol 331:124985

Oliva-Arancibia B, Órdenes-Aenishanslins N, Bruna N, Ibarra PS, Zacconi FC, Pérez-Donoso JM, Poblete-Castro I (2017) Co-synthesis of medium-chain-length polyhydroxyalkanoates and CdS quantum dots nanoparticles in pseudomonas putida KT2440. J Biotechnol 264:29–37

Park S, Baker JO, Himmel ME, Parilla PA, Johnson DK (2010) Cellulose crystallinity index: measurement techniques and their impact on interpreting cellulase performance. Biotechnol Biofuels 3:1–10

Park J, Kim B, Son J, Lee JW (2018) Solvo-thermal in situ transesterification of wet spent coffee grounds for the production of biodiesel. Bioresour Technol 249:494–500

Patel JP, Parsania PH (2017) Characterization, testing, and reinforcing materials of biodegradable composites. In: Biodegradable and biocompatible polymer composites: processing, properties and applications. Elsevier, Amsterdam, pp 55–79

Peng P, Peng F, Bian J, Xu F, Sun RC, Kennedy JF (2011) Isolation and structural characterization of hemicelluloses from the bamboo species Phyllostachys incarnata Wen. Carbohydr Polym 86(2):883–890

Ramos JL, Sol Cuenca M, Molina-Santiago C, Segura A, Duque E, Gómez-García MR, Roca A (2015) Mechanisms of solvent resistance mediated by interplay of cellular factors in pseudomonas putida. FEMS Microbiol Rev 39(4):555–566

Ramsay JA, Berger E, Ramsay BA, Chavarie C (1990) Recovery of poly-3-hydroxyalkanoic acid granules by a surfactant-hypochlorite treatment. Biotechnol Tech 4:221–226

Ramsay JA, Hassan MCA, Ramsay BA (1995) Hemicellulose as a potential substrate for production of poly (β-hydroxyalkanoates). Can J Microbiol 41(13):262–266

Robak K, Balcerek M (2018) Review of second generation bioethanol production from residual biomass. Food Technol Biotechnol 56(2):174

Roberto IC, Mussatto SI, Rodrigues RC (2003) Dilute-acid hydrolysis for optimization of xylose recovery from rice straw in a semi-pilot reactor. Ind Crop Prod 17(3):171–176

Rocha-Meneses L, Ferreira JA, Mushtaq M, Karimi S, Orupõld K, Kikas T (2020) Genetic modification of cereal plants: a strategy to enhance bioethanol yields from agricultural waste. Ind Crop Prod 150:112408

Rosado MJ, Rencoret J, Marques G, Gutiérrez A, Del Río JC (2021) Structural characteristics of the guaiacyl-rich lignins from rice (Oryza sativa L.) husks and straw. Front Plant Sci 12:640475
Rubin EM (2008) Genomics of cellulosic biofuels. Nature 454(7206):841–845
Sabapathy PC, Devaraj S, Meixner K, Anburajan P, Kathirvel P, Ravikumar Y, Qi X (2020) Recent developments in polyhydroxyalkanoates (Phas) production–a review. Bioresour Technol 306:123132
Sabiha-Hanim S, Siti-Norsafurah AM (2012) Physical properties of hemicellulose films from sugarcane bagasse. Proc Eng 42:1390–1395
Saha BC (2003) Hemicellulose bioconversion. J Ind Microbiol Biotechnol 30(5):279–291
Salvachúa D, Rydzak T, Auwae R, De Capite A, Black BA, Bouvier JT, Guss AM (2020) Metabolic engineering of pseudomonas putida for increased polyhydroxyalkanoate production from lignin. Microb Biotechnol 13(1):290–298
Sarkar P, Bosneaga E, Auer M (2009) Plant cell walls throughout evolution: towards a molecular understanding of their design principles. J Exp Bot 60(13):3615–3635
Sasmal S, Goud VV, Mohanty K (2012) Ultrasound assisted lime pretreatment of lignocellulosic biomass toward bioethanol production. Energy Fuel 26(6):3777–3784
Sawant SS, Tran TK, Salunke BK, Kim BS (2017) Potential of Saccharophagus degradans for production of polyhydroxyalkanoates using cellulose. Process Biochem 57:50–56
Shin MH, Lee DY, Liu KH, Fiehn O, Kim KH (2010a) Evaluation of sampling and extraction methodologies for the global metabolic profiling of Saccharophagus degradans. Anal Chem 82(15):6660–6666
Shin MH, Lee DY, Skogerson K, Wohlgemuth G, Choi IG, Fiehn O, Kim KH (2010b) Global metabolic profiling of plant cell wall polysaccharide degradation by Saccharophagus degradans. Biotechnol Bioeng 105(3):477–488
Singh R, Patel M (2022) Effective utilization of rice straw in value-added by-products: a systematic review of state of art and future perspectives. Biomass Bioenergy 159:106411
Somerville C, Youngs H, Taylor C, Davis SC, Long SP (2010) Feedstocks for lignocellulosic biofuels. Science 329(5993):790–792
Stoklosa RJ, Hodge DB (2012) Extraction, recovery, and characterization of hardwood and grass hemicelluloses for integration into biorefining processes. Ind Eng Chem Res 51(34):11045–11053
Sun SL, Wen JL, Ma MG, Sun RC (2013) Successive alkali extraction and structural characterization of hemicelluloses from sweet sorghum stem. Carbohydr Polym 92(2):2224–2231
Taher H, Al-Zuhair S, Al-Marzouqi AH, Haik Y, Farid M (2014) Effective extraction of microalgae lipids from wet biomass for biodiesel production. Biomass Bioenergy 66:159–167
Taylor LE, Henrissat B, Coutinho PM, Ekborg NA, Hutcheson SW, Weiner RM (2006) Complete cellulase system in the marine bacterium Saccharophagus degradans strain 2-40T. J Bacteriol 188(11):3849–3861
Tse TJ, Wiens DJ, Reaney MJ (2021) Production of bioethanol—a review of factors affecting ethanol yield. Fermentation 7(4):268
Tuck CO, Pérez E, Horváth IT, Sheldon RA, Poliakoff M (2012) Valorization of biomass: deriving more value from waste. Science 337(6095):695–699
Voelker SL, Lachenbruch B, Meinzer FC, Kitin P, Strauss SH (2011) Transgenic poplars with reduced lignin show impaired xylem conductivity, growth efficiency and survival. Plant Cell Environ 34(4):655–668
Wang C, Tallian C, Su J, Vielnascher R, Silva C, Cavaco-Paulo A, Fu J (2018) Ultrasound-assisted extraction of hemicellulose and phenolic compounds from bamboo bast fiber powder. PLoS One 13(6):e0197537
Watson BJ, Zhang H, Longmire AG, Moon YH, Hutcheson SW (2009) Processive endoglucanases mediate degradation of cellulose by Saccharophagus degradans. J Bacteriol 191(18):5697–5705
Weiner RM, Taylor LE, Henrissat B, Hauser L, Land M, Coutinho PM, Hutcheson S (2008) Complete genome sequence of the complex carbohydrate-degrading marine bacterium, Saccharophagus degradans strain 2-40T. PLoS Genet 4(5):e1000087

Wendy YD, Fauziah MN, Baidurah YS, Tong WY, Lee CK (2022) Production and characterization of polyhydroxybutyrate (PHB) BY Burkholderia cepacia BPT1213 using waste glycerol as carbon source. Biocatal Agric Biotechnol 41:102310

Weng C, Peng X, Han Y (2021) Depolymerization and conversion of lignin to value-added bioproducts by microbial and enzymatic catalysis. Biotechnol Biofuels 14:1–22

Xu Z, Qin L, Cai M, Hua W, Jin M (2018) Biodegradation of Kraft lignin by newly isolated Klebsiella pneumoniae, pseudomonas putida, and Ochrobactrum tritici strains. Environ Sci Pollut Res 25:14171–14181

Xu Z, Xu M, Cai C, Chen S, Jin M (2021) Microbial polyhydroxyalkanoate production from lignin by pseudomonas putida NX-1. Bioresour Technol 319:124210

Yang H, Yan R, Chen H, Lee DH, Zheng C (2007) Characteristics of hemicellulose, cellulose and lignin pyrolysis. Fuel 86(12–13):1781–1788

Zabed H, Sahu JN, Suely A, Boyce AN, Faruq G (2017) Bioethanol production from renewable sources: current perspectives and technological progress. Renew Sust Energ Rev 71:475–501

Zhang H, Hutcheson SW (2011) Complex expression of the cellulolytic transcriptome of Saccharophagus degradans. Appl Environ Microbiol 77(16):5591–5596

Zhang W, Xu X, Yuan Y, Wang Z (2023) Sustainable application of rice-waste for fuels and valuable chemicals-a mini review. Front Chem 11:1225073

Conversion of Corn Residue to Chemicals and Biofuels

13

Alla Singh, Mamta Gupta, Sunil Kumar, Gurpinder Singh,
D. P. Chaudhary, and H. S. Jat

Abstract

This chapter provides an overview of corn residue conversion to chemicals and biofuels, highlighting maize's growing importance in industrial applications and bioenergy production. The structural composition of corn residue, detailing cellulose, hemicellulose, and lignin characteristics are discussed. Various pretreatment methods for corn residue processing, including milling, extrusion, microwave irradiation, and chemical treatments, have been described. The chapter explores the conversion of pretreated corn residue into biofuels, particularly bioethanol and biomethane, and discusses the production of high-value chemicals from corn residue, categorizing them into furans, polyols, and organic acids. The present work reviews the growth trends of corn cultivation, basic biology, and existing and upcoming innovations in corn residue processing. In conclusion, corn residue's potential in the bio-based economy is emphasized.

Keywords

Maize stover · Lignin valorization · Lignocellulosic biomass · Pretreatment · Waste valorisation

13.1 Introduction

Maize or corn (*Zea mays* L.), also referred to as the "queen of cereals," is an important crop for various industrial applications. Many of its products, including silage, bioethanol, corn starch, and bioplastics, are of industrial importance. Recently, there has been a

A. Singh (✉) · M. Gupta · S. Kumar · G. Singh · D. P. Chaudhary · H. S. Jat
ICAR-Indian Institute of Maize Research, Ludhiana, Punjab, India
e-mail: alla.singh@icar.org.in

A. Kumari et al. (eds.), *Valorization of Solid Wastes to Biofuels and Chemical Products for Sustainable World*, https://doi.org/10.1007/978-981-96-8594-3_13

renewed interest in its utilization for bioenergy production. The launching of Global Biofuel Alliance in 2023 implies the aspirations of the Global South for energy resilience. A lot of research has gone into conversion of corn residues to biofuels and biochemicals. Kramer (2022) reviewed the commercial ventures that attempted or did produce cellulosic ethanol globally. The author mentioned that most of the plants failed to produce the expected results in the United States. The 2007 Renewable Fuel Standard (RFS) target of 16 billion gallons of cellulosic ethanol turned out to be a gross overestimation. The commercial-scale second-generation bioethanol plants started in the mid-2010s in the United States by DuPont, Poet-DSM, and Abengoa are not operational now. Besides, the Environmental Protection Agency, which oversees the RFS program, regularly waives off the cellulosic component of RFS to meet realistic data. In contrast, Raizen, a Brazilian firm, has come up as a success story. The global and regional availability of maize differentially influences different industries that depend on it. Apart from technical considerations in its conversion, the transportation of corn residue influences the economic potential of a particular value-added chemical derived from it.

13.2 Global Corn Production Trends

Erenstein et al. (2022) have reviewed the production, consumption, and trade trends of maize globally. The authors mention that in comparison to rice and wheat, maize is more versatile due to its multiple uses. The percent relative change in production of maize from 1995 to 2019 is 118%, while the same is 41% and 39% for paddy and wheat, respectively (Fig. 13.1a). In addition, a study covering more than a hundred years of crops data in France concluded that maize is one of the few crops that has not shown yield plateau, while other major crops are already at the saturation level (Schauberger et al. 2022). In terms of average annual growth rate, Asia region shows higher growth in area and production compared to other parts of the world. Maximum yield increase is observed in the America. Deutsch et al. (2018) demonstrated the likely impact of a temperature rise scenario of different crops in various parts of the world. In Southeast Asian region, maize is likely to be climate resilient, with a mere 7% loss in production, while the same is a whopping 59% for rice. This shows that maize has the potential for climate resilience in the vulnerable regions of the Global South and is expected to play an important part if agricultural policy in the region. Additionally, maize has lesser water footprint than other cereals, requiring about 1200 l of water for every kg of grain produced. Due to this, it is at the forefront of agricultural policies incentivizing renewable efforts.

13.3 Corn Residue Fractions and Its Valorization

Figure 13.1a provides a schematic representation of the lignocellulose structure, with cellulose forming the microfibrils, hemicellulose wrapping around the microfibrils, and lignin network providing overall strength to the framework and shielding the inner polysaccharide component from intrusion by external agents. Cellulose is the most abundant natural substrate having D-glucose units connected to each other by $\beta 1 \rightarrow 4$ glycosidic linkage. Van der Waals interactions and H-binding between

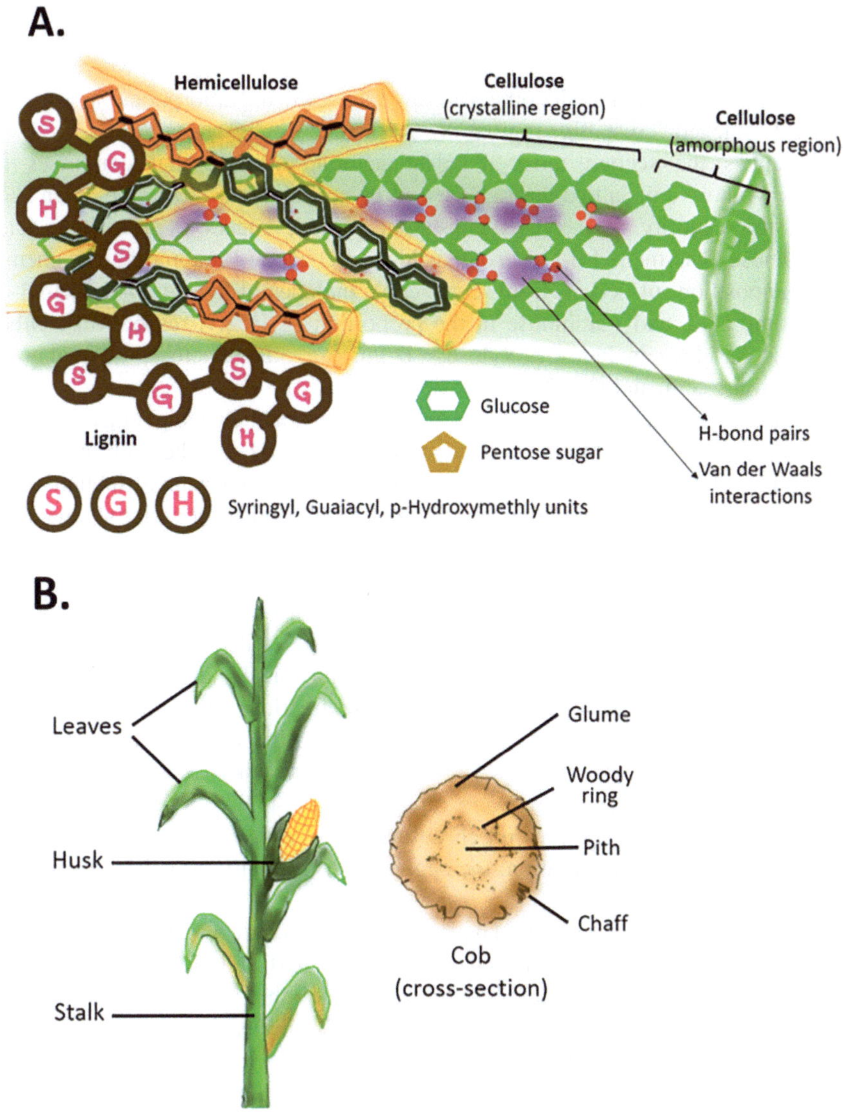

Fig. 13.1 Lignocellulose microstructure and corn residue fractions. (**a**) Cellulose, hemicellulose, and lignin form the lignocellulose framework that provides strength to the plant cell walls. (**b**) Corn residue consists of leaves, husks, stalks, and cobs. Cob consists of an outer region and an inner region, pith. Different fractions possess varying amounts of structural components, which influence their selection for tailored biorefinery

exposed hydroxyl groups keep the linear chains of cellulose stacked with each other. This results in "crystallinity" in plant fibers, where long chains of cellulose align with each other, providing good strength in return. There are regions where the cellulose chains do not align with others and are referred to as "amorphous" as opposed to highly stacked "crystalline regions" (Fig. 13.1a).

Xylan is the second most abundant natural substrate that comprises of D-xylopyranose units allied through $\beta 1 \rightarrow 4$ glycosidic bonds. The side-chain sugars in the majority of d-xylans consist of pentose sugars D-glucuronic acid, L-arabinose, 4-O-methyl-d-glucuronic acid, and O-acetyl-l-arabinose. The chemical composition of xylan from corn cobs includes d-xylose, 4-O-methyl-d-glucuronic acid, and l-arabinose in the proportion of 19:2:7, respectively. Other monosaccharides include D-xylose, D-glucose, D-galactose, D-mannose, D-glucuronic acid, and D-galacturonic acid.

Lignin is the most prevalent natural aromatic substrate. On weight basis, it comprises around 10–35% of biomass content. It is characterized by heavy branching in a network, whose building blocks are methoxylated phenylpropanoid subunits. It is primarily composed of three units referred to as S, G, and H for syringyl, guaiacyl, and p-hydroxymethyl units. It covers the cellulose and hemicellulose polymers and imparts strength to plant cell wall.

Figure 13.1b shows the different fractions of corn residue, with leaves, husks, and stalk collectively referred to as corn stover and the shelled shank, also known as corn cob. Woźniak et al. (2021) characterized the chemical and structural composition of different fractions of the corn residue. The fractions taken for analysis were stalks, cobs, husks, and leaves. The degree of crystallinity, carbon:nitrogen ratio, cellulose, and lignin contents, were the highest in corn stalks. Cellulose contents in stalks, leaves, husks, and cobs were reported to be 44.9%, 39.3%, 38.1%, and 37.9%, respectively. Variation in cellulose content with change of geographical region was observed. In addition, the degree of crystallinity of cellulose varies in different fractions.

Similarly, Gandam et al. (2022) have characterized the two regions of corn cobs, viz., corn cob outer region composed of glume and chaff, and the inner region, pith. The authors demonstrated that cob pith region had higher content of carbohydrate than in cob outer region (83.3% versus 79.9%). Similarly, the lignin content was lesser in pith (13.6%) compared to cob outer region (17.1%). The syringyl/guaiacyl ration, an indicator of lignin digestibility, was also higher in corn pith (1.34) and the outer region (1.28). The authors proposed that corn pith can be used for enzymatic production of xylooligosacchharides without pretreatment, while the outer region can be pretreated prior to valorization. The lowered requirement of pretreatment is aimed at enhancing the overall productivity of utilizing the corn cob biomass. Corn cobs have the benefit of higher bulk density than corn stalk, which is an important consideration for biomass transportation to industries and refineries.

13.4 Corn Residue Processing

Whether fractionated or not, the corn residue must be pretreated by physical–chemical and/or biological approaches to break the lignocellulosic structure. Both lignin and cellulose, especially the crystalline portion, contribute to "recalcitrance" of biomass, restricting the access of catalysts, both chemical and biological. Lignin is primarily responsible for reduced accessibility of enzymes to the cellulose and hemicellulose content. These strength-giving polymers must be broken efficiently and depending on the desired product, sometimes selectively in order to utilize lignocellulosics economically. Most of the pretreatment methods like steam explosion, microwave irradiation, milling, and extrusion break the lignocellulose structure nonselectively, while some methods like acid and alkali treatment, and biological pretreatment do achieve a certain degree of selectivity. Zentou et al. (2021) developed a mathematical model to assess the inhibitory effect of different compounds, viz., acetic acid, glycerol, succinic acid, and lactic acid in fermentation. The authors demonstrated that the inhibitors significantly inhibit specific growth rate, rather than half-velocity constant. Earlier, Pan et al. (2006) had demonstrated the reasons behind the poor digestibility of acetylated cellulosic substrates. The authors implicate poor binding of the cellulose binding domain of the cellulase enzymes as the main reason for poor digestibility of cellulose. Either the acetylated cellulose interferes with productive enzyme active site inside the cellulases or acetylation inhibits the general binding of cellulose with the enzyme. These considerations influence the selection of pretreatment approach for a particular lignocellulosic material. The pretreatment methods have been extensively reviewed by Mankar et al. (2021) and Den et al. (2018).

1. *Milling*: This is the most frequently used technique to deconstruct the lignocellulose structure. Reduction in particle size results in better mass transfer and improved accessibility of other reagents. Additionally, cellulose crystallinity also decreases. Depending on the specifications of the milling process, a range of particle size distribution can be obtained—from 0.2–2 mm to 10–30 mm. However, milling is an energy-expensive process. Hence, efforts have been made to reduce the energy requirements of milling. One process described is Wet Disk Milling (Hideno et al. 2009). For rice, wet disc milling achieved similar particle size distribution with ten times lesser energy than ball milling. Milling strategies have been combined with other pretreatment options. Peterson et al. (2024) have demonstrated the utility of deacetylated disc-refining as a pretreatment strategy to solve issues associated with steam- and dilute acid-based pretreatments. It has been demonstrated while acetic acid is a good choice for breaking lignocellulosic structure, it is not the best option when trying to ferment the acetylated cellulose, thus formed. In order to circumvent this, deacetylation has been considered as an option (Peterson et al. 2024). Combining it with disc-refining, the process was shown to be efficient at preventing inhibitor formation. Disc-refining opens up the macrocellular structure through a combination of cutting, shearing, and compressive forces. Deacetylation disc-refining effectively removes acetyl

groups and lignin from biomass without generating inhibitors like hydroxymethylfurfural (HMF) and furfural, which are common in steam and dilute acid pretreatments This is an example whereby milling was combined with another pretreatment to release sugars for fuel production.

2. *Extrusion*: A thermomechanical pretreatment strategy, extrusion, deconstructs the lignocellulosic material by employing high shear forces between a rotating screw and biomass under a set temperature. While increasing the screw speed was shown to result in better breakdown of biomass (Karunanithy and Muthukumarappan 2011), increasing the temperature beyond a certain limit is known to affect the glucose yields negatively. Extrusion has been combined with steam explosion (Zheng et al. 2014) and alkaline soaking (Karunanithy and Muthukumarappan 2011) for deconstructing lignocellulosic materials.
3. *Microwave irradiation*: A thermal process for pretreating biomass, microwave irradiation benefits from the presence of different biologically conductive regions in lignocellulose, including the crystalline and amorphous regions. Especially the polar regions form hotspots within the material, which get heated up due to irradiation. This also generated pressure inside the material components, resulting in its deconstruction. Microwave treatment causes fiber swelling and fragmentation, making the regions accessible for enzymatic hydrolysis. Tatijarern et al. (2013) used a two-step microwave-assisted sodium hydroxide pretreatment strategy to break the lignocellulose. Under milder conditions, microwave assists nonthermal solvolysis, while in the absence of oxygen, they assist thermal pyrolysis. However, the technology involves high capital costs at present.
4. *Acid pretreatment*: Lignocellulose can be deconstructed using either concentrated acid at low temperature or dilute acid at high temperature. Sulfuric acid, phosphoric acid, hydrochloric acid, nitric acid, and acetic acid are the most common acids used for this purpose. Mancini et al. (2016) demonstrated that concentrated phosphoric acid converts cellulose to amorphous form and retains high reactivity for subsequent action by cellulases. The process involves the use of a corrosive reagent, which necessitates special reactor construction material and mandatory downstream neutralization and makes it a negative aspect.
5. *Alkali pretreatment*: Lignin and various polysaccharides can be removed using a number of bases like sodium hydroxide, potassium hydroxide, calcium hydroxide, and sodium carbonate. Sodium hydroxide is the most common base, with use at either low concentration with high temperature and pressure, or use at high concentration with low temperature and pressure. While the former strategy removes lignin and hemicellulose, the latter strategy dissolves cellulose. The main disadvantages of alkali pretreatment are the formation of inhibitory compounds and downstream pH neutralization step requirement.
6. *Other strategies*: Many other strategies have been tested to deconstruct the lignocellulose in biomass. These include the use of Ionic liquids, which are salts that become liquids at room temperature and release their ions. They penetrate lignin and deconstruct the biomass. Their main advantage is the ability to dissolve large amounts of cellulose with low energy input. However, the main

issues in this approach are the expense and restricted sugar release from treated material, apart from inhibitory effect on hydrolytic enzymes.

Greener oxidative processes have also been employed for lignocellulose deconstruction. One prominent and environment-friendly method is ozonolysis. Hydroxy radicals are highly reactive and powerful oxidizers that are generated as a result of disintegration of ozone in basic medium. Ozone mineralizes the lignin component. Al jibouri et al. (2015) demonstrated a two-step process with wheat straw, whereby an intermediate washing step was incorporated between two ozone treatments. The washing step re-exposed functional groups of lignin by removing the organic acid molecules that got generated after the first ozone treatment. Similarly, Perrone et al. (2016) demonstrated the effectiveness of a multistep protocol that employed ozone treatment, alkaline washing, and ultrasound radiation. While ozone can be generated on-site, it is expensive and suffers from poor production energy balance.

Apart from this, photocatalysis, oxidative catalysis, and electrochemical processes have been employed to varying degrees. The use of white-rot and brown-rot fungi makes the biological pretreatment an environmentally friendly option. While white-rot fungi majorly degrade the lignin component, the brown-rot fungi degrade all the components of the lignocellulose structure.

13.5 Conversion to Biofuels and Chemicals

Biomass deconstruction is the first step for conversion of corn residue to biofuels and chemicals. The major focus is on delignification, which leads to an easier access of the hydrolytic agents or enzymes to the inner cellulose and hemicellulose fractions. Either of these or both the fractions can be converted to monomeric sugars. Figure 13.2 shows the schematic representation of conversion of corn residue to bioethanol. The process started with delignification, which can either be a physical–chemical method of biological delignification or a combination of both. Delignification allows access of cellulase enzyme to cellulose fibrils. Cellulase, which is a combination of different enzymes, endo- and exo-glucanases, and β-glucosidase, breaks the cellulose to monomeric glucose units. The glucose is then fermented to ethanol using the yeast *Sachharomyces cerevisiae*. Other than *Sachharomyces*, various other yeasts, fungi, or bacteria can be utilized to convert C6 and C5 sugars to bioethanol.

In contrast, biomethane can also be produced from corn residue, which is one of the most preferred residues for biogas manufacturing. For biogas production, the corn residue is anaerobically digested, and it goes through different phases of acidogenesis, acetogenesis and methanogenesis, which lead to breakdown of biomaterial to volatile fatty acids, then to acetic acid and finally to methane and carbon dioxide. Yet another route of value addition in corn residue is to anaerobically ferment it to lactic acid in enclosed bales, resulting in silage. Silage is a good source of nutrition for ruminants, and silage manufacturing can be undertaken at different scales, depending on local or global demand.

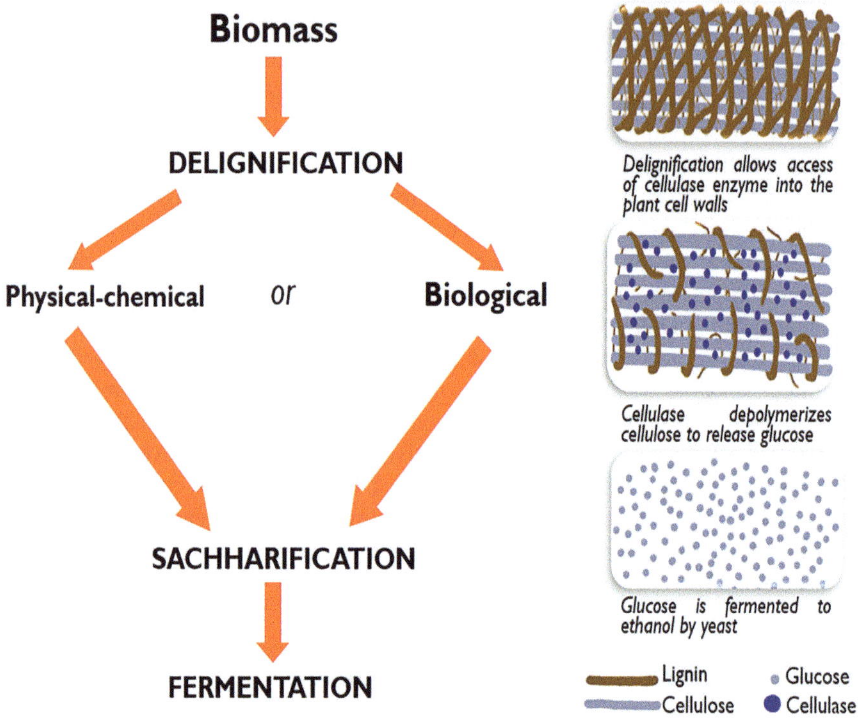

Fig. 13.2 Schematic representation of conversion of corn residue to bioethanol

Apart from biofuels, various pathways for conversion of corn residue to different chemicals are known. The three major categories of chemicals that can be produced are furans, polyols, and organic acids.

Furans: Xylose can be dehydrated to furfural, while glucose is typically converted to 5-hydroxymethylfurfural (HMF).

The market for both FF and HMF is already established, with both being important players in sustainable chemistry. Through different reactions, these can be further transformed into a number of other chemicals (Fig. 13.2). These compounds are processed into high-value chemicals through either oxidation or reduction. Both maleic anhydride (MAH) and maleic acid (MA) can be produced by oxidation of FF and HMF. Furfural can be oxidized to furoic acid or further oxidized to maleic acid/maleic anhydride, which can then be isomerized to fumaric acid. 2,5-furandicarboxylic acid (FDCA) can be produced by both FF and HMF. The conversion of HMF to 2,5-diformylfuran involves selective oxidation of the hydroxymethyl (–CH_2OH) group of HMF into an aldehyde (–CHO) while preventing overoxidation to 2,5-furandicarboxylic acid (FDCA). Furan-2,5-dimethylcarboxylate (FDMC) can be produced through HMF in the presence of low-molecular-weight alcohols and can be further hydrolyzed to FDCA.

Similarly, both FF and HMF can be reduced to yield a number of compounds. FF can be reduced to furfural alcohol (FFA) and later to different compounds, viz., tetrahydrofurfuryl alcohol (THFA), cyclopentanone (CPO), methylfuan (MF), and 1,5-pentanediol (PeD). HMF can be reduced to 2-methylfurna (2-MF), 2,5-dihydroxymethylfuran (DHMF), and 2,5-dimethylfuran (DMF).

Polyols: Different polyols (C2–C6) can be synthesized by hexoses and pentoses that result from hydrolytic cleavage of cellulose and hemicellulose, respectively. Hexoses (e.g., glucose, mannose, galactose) can be converted into various polyols and glycols through different catalytic and microbial processes. Sorbitol and mannitol are produced via catalytic hydrogenation using Ni, Ru, or Pt catalysts at 80–150 °C under 50–101 MPa H_2 (Werpy and Petersen 2004). Ethylene glycol, propylene glycol, and glycerol are obtained through hydrogenolysis of hexoses or sorbitol using Cu/Cr_2O_3, Ni/Al_2O_3, or Ru-based catalysts at 180–220 °C under 20–50 bar H_2 (Werpy and Petersen 2004). Erythritol, threitol, and ribitol can be obtained from microbial fermentation of glucose by *Moniliella* sp. or *Candida* sp. under aerobic conditions (27–32 °C, pH 3–6, preferably 4–5) (Moon et al. 2010). Butanediol (mainly 2,3-butanediol) is produced via bacterial fermentation (e.g., *Klebsiella pneumoniae*, *Enterobacter aerogenes*) at 30–37 °C (Ji et al. 2011).

Pentoses (e.g., xylose, arabinose) can be converted to polyols such as erythritol, threitol, arabitol, xylitol, and ribitol. Xylitol is produced by catalytic hydrogenation of xylose using Ni or Ru catalysts at 80–140 °C under 30–50 bar H_2 (Ur-Rehman et al. 2015). Arabitol, erythritol, and xylitol are obtained via microbial fermentation by yeasts like *Candida* sp., *Moniliella* sp., and *Yarrowia lipolytica* under aerobic conditions (27–32 °C, pH 3–6, preferably 4–5) (Kordowska-Wiater 2015; Moon et al. 2010; Tomaszewska et al. 2012)

Organic acids: A number of organic acids, viz., gluconic acid, lactic acid, xylonic acid, adipic acid, glucaric acid, levulinic acid, fumaric acid, succinic acid, etc., can be prepared from hexoses and pentoses, obtained after biomass deconstruction.

Similarly, lignin valorization approaches have been reviewed in Sethupathy et al. (2022). Currently, the two main industrial applications of lignin are in the recycling of inorganic compounds used in the pulping process and in combustion for the production of electricity along with some proportion in biofuels, chemicals, and bioplastics. The review covers recent developments in lignin applications, including lignin-based hydrogels, surfactants, three-dimensional printing materials, and electrodes. It also addresses chemical depolymerization for producing platform chemicals. Microbial and enzymatic degradation of lignin are presented as environmentally friendly approaches, while thermochemical methods like gasification are mentioned for transforming low-quality lignin into syngas.

13.6 Conclusion

Corn residue will remain available in large quantities due to the high demand of both corn grain as well as the residue. While corn grain is already in demand for a number of industrial processes, including first-generation bioethanol, corn residue finds use in silage for ruminant nutrition. Value chains around the usage of corn grain in feed, food, and bioenergy are already in operation, while corn silage value chain is also well-established. A number of sectors, especially polymer industry, are looking for sustainability in their operations. As such, biomass-derived chemicals provide an opportunity. The experience with cellulosic ethanol has been mixed, but a lot of learning remains to be implemented. A comparison of US and Brazilian cellulosic ethanol demonstrated the importance of residue supply and transportation chains to the overall economy of cellulosic ethanol. The success stories encourage a renewed look at the bio-based economy, with corn residue at the center stage. Development of benign pretreatment technologies, more efficient catalysts, higher selectivity and control in process operations and an overall favorable regulatory policy framework is needed to develop large commercial scale ventures around different bio-based fuels and chemicals centered on corn residue. The favorable orientation of different government policies, intergovernmental collaboration, and heightened awakening toward climate change, together with new chemistries and biological resources screening, are expected to lead to tailored valuation approaches, which would aim for cost-effectiveness, selectivity, and overall efficiency in integrated biorefinery approaches.

References

Al Jibouri AK, Turcotte G, Wu J, Cheng CH (2015) Ozone pretreatment of humid wheat straw for biofuel production. Energy Sci Eng 3(6):541–548

Den W, Sharma VK, Lee M, Nadadur G, Varma RS (2018) Lignocellulosic biomass transformations via greener oxidative pretreatment processes: access to energy and value-added chemicals. Front Chem 6:141

Deutsch CA, Tewksbury JJ, Tigchelaar M, Battisti DS, Merrill SC, Huey RB, Naylor RL (2018) Increase in crop losses to insect pests in a warming climate. Science 361(6405):916–919

Erenstein O, Jaleta M, Sonder K, Mottaleb K, Prasanna BM (2022) Global maize production, consumption and trade: trends and R&D implications. Food Secur 14(5):1295–1319

Gandam PK, Chinta ML, Gandham AP, Pabbathi NP, Konakanchi S, Bhavanam A, Atchuta SR, Baadhe RR, Bhatia RK (2022) A new insight into the composition and physical characteristics of corncob—substantiating its potential for tailored biorefinery objectives. Fermentation 8(12):704

Hideno A, Inoue H, Tsukahara K, Fujimoto S, Minowa T, Inoue S, Endo T, Sawayama S (2009) Wet disk milling pretreatment without sulfuric acid for enzymatic hydrolysis of rice straw. Bioresour Technol 100(10):2706–2711

Ji XJ, Huang H, Ouyang PK (2011) Microbial 2,3-butanediol production: a state-of-the-art review. Biotechnol Adv 29(3):351–364

Karunanithy C, Muthukumarappan K (2011) Optimization of alkali soaking and extrusion pretreatment of prairie cord grass for maximum sugar recovery by enzymatic hydrolysis. Biochem Eng J 54(2):71–82

Kordowska-Wiater M (2015) Production of arabitol by yeasts: current status and future prospects. J Appl Microbiol 119(2):303–314

Kramer D (2022) Whatever happened to cellulosic ethanol? Phys Today 75(7):22–24

Mancini G, Papirio S, Lens PN, Esposito G (2016) Solvent pretreatments of lignocellulosic materials to enhance biogas production: a review. Energy Fuel 30(3):1892–1903

Mankar AR, Pandey A, Modak A, Pant KK (2021) Pretreatment of lignocellulosic biomass: a review on recent advances. Bioresour Technol 334:125235

Moon HJ, Jeya M, Kim IW, Lee JK (2010) Biotechnological production of erythritol and its applications. Appl Microbiol Biotechnol 86(4):1017–1025

Pan X, Gilkes N, Saddler JN (2006) Effect of acetyl groups on enzymatic hydrolysis of cellulosic substrates. Holzforschung. https://doi.org/10.1515/HF.2006.062

Perrone OM, Colombari FM, Rossi JS, Moretti MM, Bordignon SE, Nunes CD, Gomes E, Boscolo M, Da-Silva R (2016) Ozonolysis combined with ultrasound as a pretreatment of sugarcane bagasse: Effect on the enzymatic saccharification and the physical and chemical characteristics of the substrate. Bioresour Technol 218:69–76

Peterson DJ, Paek C, Tao L, Davis R, Chen X, Brunecky R, Fowler M, Elander R (2024) Feedstock/pretreatment screening for bioconversion of sugar and lignin streams via deacetylated disc-refining. Biotechnol Biofuels Bioprod 17(1):52

Schauberger B, Kato H, Kato T, Watanabe D, Ciais P (2022) French crop yield, area and production data for ten staple crops from 1900 to 2018 at county resolution. Sci Data 9(1):38

Sethupathy S, Morales GM, Gao L, Wang H, Yang B, Jiang J, Sun J, Zhu D (2022) Lignin valorization: status, challenges and opportunities. Bioresour Technol 347:126696

Tatijarern P, Prasertwasu S, Komalwanich T, Chaisuwan T, Luengnaruemitchai A, Wongkasemjit S (2013) Capability of Thai mission grass (Pennisetum polystachyon) as a new weedy lignocellulosic feedstock for production of monomeric sugar. Bioresour Technol 143:423–430

Tomaszewska L, Rywińska A, Gładkowski W (2012) Production of erythritol and mannitol by Yarrowia lipolytica yeast in media containing glycerol. J Ind Microbiol Biotechnol 39(9):1333–1343

Ur-Rehman S, Mushtaq Z, Zahoor T, Jamil A, Murtaza MA (2015) Xylitol: a review on bioproduction, application, health benefits, and related safety issues. Crit Rev Food Sci Nutr 55(11):1514–1528

Werpy T, Petersen G (2004) Top value added chemicals from biomass: volume I – results of screening for potential candidates from sugars and synthesis gas. National Renewable Energy Laboratory, Golden, p 76

Woźniak M, Ratajczak I, Wojcieszak D, Waśkiewicz A, Szentner K, Przybył J, Borysiak S, Goliński P (2021) Chemical and structural characterization of maize stover fractions in aspect of its possible applications. Materials 14(6):1527

Zentou H, Zainal Abidin Z, Yunus R, Awang Biak DR, Abdullah Issa M, Yahaya PM (2021) A new model of alcoholic fermentation under a byproduct inhibitory effect. ACS Omega 6(6):4137–4146

Zheng J, Choo K, Bradt C, Lehoux R, Rehmann L (2014) Enzymatic hydrolysis of steam exploded corncob residues after pretreatment in a twin-screw extruder. Biotechnol Rep 3:99–107

Transfer of Sawdust to Chemicals and Biofuels

14

Simran Neeraj, Yukta Menon, and Anirudh Sharma

Abstract

This chapter explores the valorization of sawdust, a lignocellulosic waste from the wood processing and furniture industries, into value-added chemicals and biofuels via thermochemical and biochemical conversion pathways. Pyrolysis converts sawdust into bio-oils and biochar, while anaerobic digestion yields biogas for energy applications. Acid hydrolysis liberates fermentable sugars for bioethanol production. Chemical activation using phosphoric acid transforms sawdust into activated carbon valuable for adsorption processes. Submerged fermentation mediated by microbes like *Streptomyces lazureus* produces glucose powder. Deacetylation enables acetic acid extraction, with applications spanning textiles to pharmaceuticals. The chapter highlights the challenges of overcoming the recalcitrance of lignocellulosic components, developing efficient pretreatment methods, selectively degrading lignin, and establishing economic viability through cost-effective production and market competitiveness. However, utilizing sawdust as a sustainable feedstock avoids competition with food crops, reduces environmental impacts through waste valorization, and promotes a circular bioeconomy. Sawdust-derived products offer solutions for carbon sequestration, soil amendment, and remediation. Harnessing sawdust's potential through research, strategic collaborations, supportive policies, and investments can catalyze the transition toward a circular bioeconomy, fostering resource efficiency, energy security, and environmental sustainability globally. The valorization of this abundant waste into high-value chemicals and biofuels harmonizes economic growth with environmental stewardship.

S. Neeraj · Y. Menon · A. Sharma (✉)
Department of Biotechnology, Jaypee Institute of Information Technology, Noida, Uttar Pradesh, India
e-mail: anirudh.sharma@mail.jiit.ac.in

A. Kumari et al. (eds.), *Valorization of Solid Wastes to Biofuels and Chemical Products for Sustainable World*, https://doi.org/10.1007/978-981-96-8594-3_14

Keywords
Sawdust · Feedstock · Lignocellulose biomass · Chemicals · Biofuels

14.1 Introduction

The depletion of fossil fuel reserves and rising global energy demands have driven research into sustainable alternatives. Overreliance on natural resources leads to "natural resource dependence," causing economic and social issues. The "resource curse" shows that resource-rich nations often grow slower economically than those without such resources (Nath 2015). Communities dependent on fossil fuels face risks like food insecurity, disease transmission, heat-related illnesses, and poor mental health, alongside future fatal temperature increases (Hussain and Reza 2023). On-site power generation offers local government renewable energy access, mitigating financial risks and improving power quality. Utilizing lignocellulosic waste like sawdust, abundant in wood processing and furniture industries, has emerged as a promising renewable feedstock. Sawdust, often discarded or underused, contains cellulose, hemicellulose, and lignin, allowing conversion into biofuels like bioethanol and biochemicals like furfural through pretreatment and conversion technologies (Al-Dhabaan 2022). Exploring sawdust valorization can establish integrated bio-refineries, producing diverse products and reducing environmental impacts (Vu et al. 2020). Research on product profiles, optimizing conversion efficiencies, techno-economic assessments, and lifecycle analyses is crucial for developing commercially viable sawdust-to-biofuels processes. These efforts support the global shift toward a sustainable, circular bio-economy, maximizing resource utilization and minimizing waste and environmental footprints. This chapter explores converting sawdust from the furniture industry into value-added chemicals and biofuel products through innovative pathways.

14.2 Generation Status of Sawdust Worldwide and their Potential as Feedstock

Sawdust, a by-product of the wood processing industry, is generated in substantial quantities worldwide. In India alone, an estimated 18 million tons of sawdust is generated annually, with an average sawmill capable of producing up to 822.4 tons. Indonesia's log production in 2019 reached an impressive 48.0 million m^3, generating an estimated 2 million m^3 of sawdust. These staggering figures highlight the significant volume of sawdust generated globally (Pitti et al. 2019). The technological advancements in sawdust utilization have unveiled its remarkable potential as a versatile feedstock across various industries. Natural cellulose fibers derived from sawdust offer numerous advantages due to their affordability, abundant availability, and eco-friendly nature. Sawdust transforms from workshop waste into a versatile fuel source across various sectors (Emenike et al. 2024). In households, specially designed stoves and fireplaces burn sawdust directly, providing both warmth and cooking capabilities. The agricultural industry utilizes sawdust for drying and curing crops, space heating in greenhouses, and even warming chicken brooders (Palange 2009). Sawdust's industrial applications are just as diverse. Blacksmiths, brickmakers, and poultry producers all take advantage of its efficient burning properties to fuel their respective processes. Sawdust can be utilized as a feedstock for bio-oil generation via pyrolysis processes. The resulting bio-oil can be further utilized for heating applications such as furnaces, diesel engines, boilers, gas turbines, and more (Vamvuka 2011). Pyrolysis processes offer the potential for high bio-oil yield from sawdust. It has proven its potential as a raw material for improvement in a wide range of materials and chemicals, including methanol production (Rominiyi et al. 2017). Additionally, it serves as an additional raw material for the production of chipboards and fibreboards. In several pulp plants in the United States, processed sawdust waste is already being utilized to produce other chemicals and beverages, including quality alcohol. The economic viability of sawdust as a feedstock is particularly promising in the energy, manufacturing, and agricultural industries (Alizadeh et al. 2023). With the substantial global generation of sawdust, a sustainable waste management strategy is necessary to valorize this resource effectively. The substantial global generation of sawdust presents a valuable opportunity to harness its potential as a versatile feedstock across various industries, contributing to sustainable waste management, energy generation, material production, and resource utilization.

14.3 Composition of the Sawdust

The composition of sawdust is contingent upon the specific type of wood and the processing techniques employed. From a chemical standpoint, sawdust is predominantly constituted of organic compounds such as cellulose, hemicellulose, and lignin. These constituents contribute to the overall structural integrity and attributes of the material. Moreover, it contains extractives, which are low-molecular-weight compounds that can be isolated from wood through the application of solvents (Atiwesh 2022) (Fig. 14.1).

The physical characteristics of sawdust are equally pivotal as they influence its behavior and potential applications. Factors such as moisture content, apparent specific gravity, porosity, water retention capacity, and water drainage rate play a crucial role in determining its suitability for various purposes. Sawdust exhibits a high degree of porosity, which facilitates efficient water retention and drainage (Abd El Halim and El Baroudy 2014). This property renders it an ideal material for applications such as composting, animal bedding, or a soil amendment. Its moisture content and apparent specific gravity also impact its handling, transportation, and storage characteristics (Tran 2005).

Furthermore, the chemical composition of sawdust, particularly its carbon, hydrogen, and oxygen content, contributes to its potential utilization as a fuel source or for the production of bio-based products (Mansur et al. 2020). Ultimately, the unique amalgamation of physical and chemical properties renders sawdust a

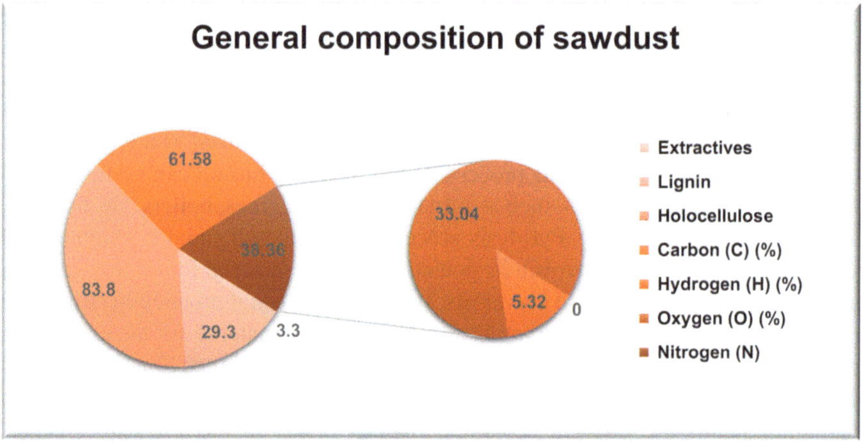

Fig. 14.1 Chemical composition of sawdust

valuable resource with diverse applications across various industries, including agriculture, construction, and energy production (Bamidele Charles et al. 2023). Comprehending these properties is essential for optimizing its utilization and minimizing any potential environmental impacts associated with its disposal or processing (Rusanen et al. 2019).

14.4 Biofuel Production Technology from Sawdust

14.4.1 Bio-Oil

Firstly, the process begins with size reduction, where sawdust undergoes mechanical treatment to decrease its size, thereby augmenting its surface area and enhancing heat transfer during pyrolysis (Ahmed et al. 2020). This step is crucial as it sets the stage for subsequent processes. Secondly, pyrolysis takes place, involving the rapid heating of the reduced sawdust to temperatures exceeding 400 °C, achieved either in the absence or near absence of oxygen (Sichone 2013). This high-temperature environment induces thermal decomposition, leading to the creation of bio-oil vapors, syngas, and biochar (Abou Rjeily et al. 2023). The third stage involves condensation, where the bio-oil vapors produced during pyrolysis are cooled and condensed. This is typically accomplished through the utilization of a series of condensers or a quenching system, resulting in the formation of liquid bio-oil (Vamvuka 2011). Finally, in the collection and refining phase, the condensed bio-oil is gathered and subjected to further refining processes. These processes may include solvent extraction or catalytic upgrading, aimed at enhancing the quality and properties of the bio-oil for various applications (Chan et al. 2020). Each of these steps plays a vital role in the overall conversion of sawdust into valuable bio-oil products (Fig. 14.2).

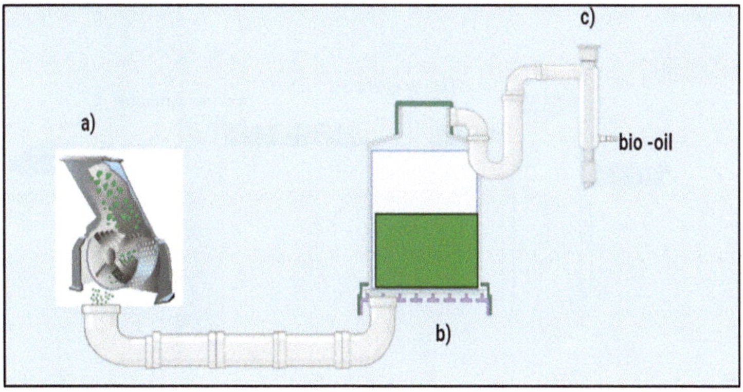

Fig. 14.2 Bio-oil production illustration, where (**a**) symbolizes size reduction process, (**b**) pyrolysis, and (**c**) condensation

14.4.2 Biogas

Sawdust undergoes various pretreatment processes to enhance its biodegradability and improve the efficiency of anaerobic digestion (Hashemi et al. 2021). These processes include alkaline or acidic treatment, steam explosion, and biological pretreatment (Kumar and Sharma 2017). After pretreatment, the sawdust is subjected to hydrolysis, where enzymes or acids break down the complex carbohydrates into simpler sugars (Karuppiah and Azariah 2019). This breakdown facilitates the subsequent fermentation process. The hydrolyzed sawdust is then fed into an anaerobic digester, where methanogenic bacteria convert the organic matter into biogas, primarily consisting of methane and carbon dioxide, under anaerobic conditions (Issah et al. 2020). The produced biogas can undergo upgrading processes, such as carbon dioxide and moisture removal, to improve its quality and calorific value for specific applications (Kapoor et al. 2019).

14.4.3 Solid Biochar

During the pyrolysis process, sawdust is heated to temperatures below 400 °C in an environment with minimal or no oxygen. This controlled heating results in the decomposition of the sawdust into three main products: biochar, bio-oil, and. The absence of oxygen is crucial because it prevents combustion, allowing the organic material to break down through thermal decomposition instead. The biochar produced in this step is a carbon-rich solid, while bio-oil is a liquid mixture of various organic compounds, and syngas is a gaseous mixture primarily composed of hydrogen, carbon monoxide, and carbon dioxide syngas (Elhenawy et al. 2024) (Fig. 14.3).

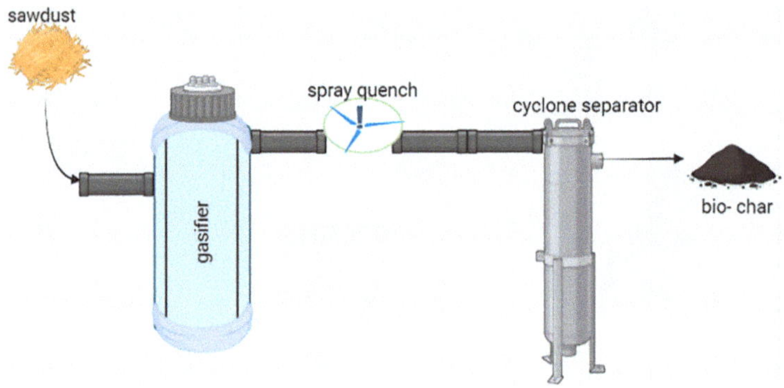

Fig. 14.3 Biochar production illustration

14.5 Chemical Production Technology for Sawdust

14.5.1 Ethanol Production from Acid Hydrolysis of Sawdust

To prepare the sawdust for further processing, it undergoes a pretreatment phase where it is treated with dilute sulfuric acid. Specifically, the sawdust is immersed in a solution with a sulfuric acid concentration of 1.5%. This mixture is then heated to a temperature of 120 °C and maintained at this temperature for 30 min. The heating process helps break down the lignocellulosic structure of the sawdust, making it more amenable to subsequent processing (Tulashie et al. 2023). After heating, the mixture is filtered to separate the solid insoluble components from the liquid portion, which contains soluble substances released during the pretreatment. Following the pretreatment, the insoluble component of the sawdust undergoes hydrolysis. In this phase, the solid residue from the pretreatment is treated with diluted sulfuric acid under varying conditions of acid concentration, temperature, and duration (Megawati et al. 2015). The purpose of this hydrolysis step is to break down the complex carbohydrates in the sawdust into simpler fermentable sugars. By adjusting the concentration of sulfuric acid, the temperature, and the time of hydrolysis, the process can be optimized to maximize the yield of these sugars. The fermentable sugars produced from hydrolysis are then subjected to fermentation. In this stage, *Saccharomyces cerevisiae*, commonly known as baker's yeast, is used as the fermenting agent (Chukwuemeka et al. 2020). The fermentation takes place in a medium that contains dextrose, yeast extract, urea, and other necessary nutrients to support yeast growth and activity. The pH of the fermentation medium is adjusted to 5 to create an optimal environment for the yeast. The mixture is then incubated at 30 °C for a period of 72 h, during which the yeast converts the fermentable sugars into ethanol and carbon dioxide. After fermentation, the mixture contains ethanol along with other by-products and residual nutrients. To isolate the ethanol, the fermented product undergoes distillation. The distillation process is conducted at a temperature of 85 °C, which allows the ethanol to vaporize and be collected separately from the other components in the mixture (Otulugbu 2012). The collected

Fig. 14.4 Illustration of pretreatment of sawdust for ethanol production

ethanol is then condensed back into liquid form. Through a series of optimization experiments, the conditions for achieving the maximum ethanol yield were determined. It was found that using a sulfuric acid concentration of 1.5%, a hydrolysis temperature of 123.4 °C, and a hydrolysis time of 30 min resulted in the highest yield of ethanol. Under these optimized conditions, the process achieved an ethanol yield of 47.2%, making it an efficient method for converting sawdust into bioethanol (Tulashie et al. 2023) (Fig. 14.4).

There are numerous uses for the ethanol made from sawdust via acid hydrolysis. It functions as a renewable fuel that spark ignition engines can use in place of gasoline, lowering energy prices and allowing renewable energy sources to take the place of non-renewable ones. Ethanol serves as a flexible solvent in a range of chemical processes in an industrial context. It serves as a preservative in the food sector and is an effective cleaning agent due to its disinfecting qualities (Gorman and Scott 2004). Ethanol is also a biofuel; it may be converted into sustainable aviation fuel and powers machinery and vehicles intended for biofuels. Sawdust-to-ethanol manufacturing promotes economic growth and employment creation, both of which are important aspects of community development. Additionally, using ethanol as fuel helps lower greenhouse gas emissions, boosting environmental sustainability (Demirbas 2009).

14.5.2 Activation of Carbon

Activated carbons are highly porous materials with extensive applications in adsorption processes due to their large specific surface area and pore volume. The production of activated carbons from renewable resources like sawdust is gaining attention as an eco-friendly and cost-effective method. Here, we explore the process of creating activated carbons from sawdust utilizing chemical activation with H_3PO_4, investigating the effects of different operational factors on the final activated carbons' adsorption properties, including impregnation ratio, activation temperature, and time (Shrestha et al. 2018). The potential of sawdust as a precursor for premium-activated carbons appropriate for gas-phase and liquid-phase adsorption applications is demonstrated, along with the ideal conditions for attaining better nitrogen adsorption capabilities. Our goal is to present a thorough explanation of the chemical activation process and how it affects the development of porosity in activated carbons made from sawdust (Hamad and Idrus 2022). One of the many uses for this adaptable adsorbent is activated carbon. The renewable resource sawdust can be used to produce activated carbon, providing a sustainable substitute for conventional sources like coal or petroleum-based goods.

Water and air purification is one of the main applications for activated carbon made from sawdust. It is a superior adsorbent for eliminating pollutants and impurities due to its high porosity and surface area. It improves the taste, odor, and color of water by efficiently removing organic compounds, chlorine, and other contaminants (Rodriguez-Narvaez 2023). It improves indoor air quality through air purification by adsorbing other contaminants and volatile organic molecules. Activated carbon made from sawdust is essential for gas separation. In order to absorb harmful gases and produce clean air, it is utilized in gas masks and respirators. It is also used in industrial procedures for purifying and separating gases. Activated carbon's high porosity and surface area also make it a good choice for energy storage applications (Durán et al. 2018). It is a component of supercapacitors, which are long-lasting energy storage devices with a high power density. Sawdust-derived activated carbon efficiently eliminates dyes from wastewater, assisting sectors like textile, leather, food, and cosmetics. Wastewater is made less hazardous by adsorbing dye particles (Vieira et al. 2022). Activated carbon is used in the food and beverage sector to decolorize and deodorize items, eliminating undesirable colors and tastes and enhancing the overall quality of the product. Moreover, medical and laboratory equipment is made using sawdust-derived activated carbon. It is necessary for many medical devices filters and purification systems (Udokpoh and Nnaji 2023) (Figs. 14.5 and 14.6).

14.5.3 Glucose Powder Formation

Streptomyces lazureus biodegrades sawdust through submerged fermentation, aiming to produce cellulase and ligninase enzymes. These enzymes play a crucial role in breaking down the complex polysaccharides in sawdust, ultimately yielding glucose. *Streptomyces lazureus* is a species of actinobacteria within the genus

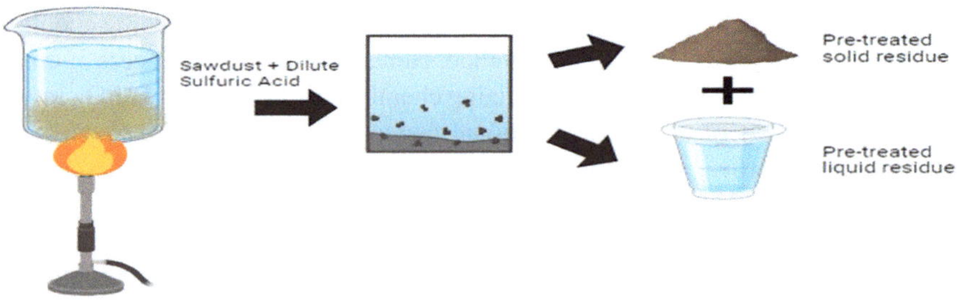

Fig. 14.5 Illustration of activation of carbon from sawdust

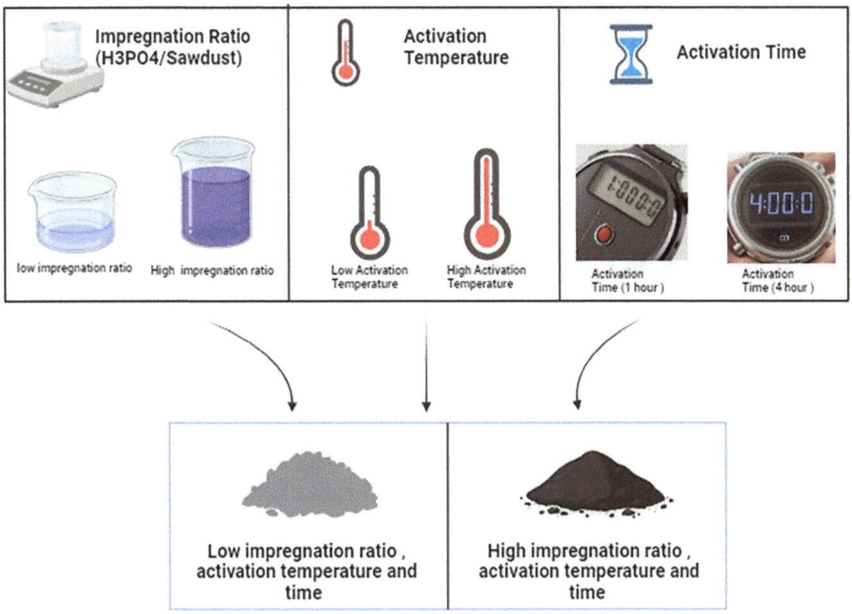

Fig. 14.6 Parameter optimization for the activation of carbon

Streptomyces. These bacteria are known for their complex life cycle, which includes both filamentous growth and spore formation (Salwan and Sharma 2020). Streptomyces species are notable for their ability to produce a wide range of secondary metabolites, including many antibiotics, enzymes, and other bioactive compounds.

Streptomyces lazureus, like other members of its genus, is found predominantly in soil and is involved in the decomposition of organic materials. This particular species has garnered interest for its ability to produce cellulase and ligninase enzymes (Dhillon et al. 2017). These enzymes are significant because they can break down cellulose and lignin, which are major components of plant biomass, into simpler molecules such as glucose. This capability makes *Streptomyces lazureus* valuable in biotechnological applications, including biodegradation, bioconversion of agricultural waste, and potentially in the production of biofuels. The most effective isolate was found to be *Streptomyces lazureus* (St-9), after 40 *Streptomycetes* were isolated from soil samples and 3 of them showed cellulolytic-ligninolytic activity (Al-Dhabaan 2022). In order to maximize enzyme activity, researchers adjusted temperature, pH, inoculum size, incubation time, and optimum sources of carbon and nitrogen. The enzymes cellulase and ligninase were isolated and purified by means of gel filtration column chromatography. The final product of the fermentation process was glucose powder, which underwent partial characterization to ascertain its solubility and melting point gathered soil samples and, using a particular agar medium, separated the *Streptomyces* bacteria. In order to prepare sawdust for fermentation, they prepared it and put it through acid hydrolysis (Rafiqul and Sakinah 2012). They determined that *S. lazureus* was the most potent *Streptomyces* isolate, St-9, using morphological and genomic techniques. Following their discovery of the ideal circumstances for the synthesis of lignin and cellulase, the enzymes were isolated and purified. They then analyzed the glucose that was created by the sawdust's submerged fermentation (Al-Dhabaan 2022) (Fig. 14.7).

14.5.4 Acetic Acid Formation Using Sawdust

Acetic acid can be produced from sawdust through deacetylation, which involves treating sawdust with an alkali solution like aqueous ammonia (NH_4OH). The acetic acid formation depends on the NH_4OH loading and the solid-to-liquid ratio. In one study, treating sawdust with 2–10% NH_4OH (w/v) at a solid-to-liquid ratio of 1:4–1:10 and heating at 121 °C for 60 min yielded a maximum acetic acid concentration of 8.16 g/L with 2% NH_4OH and a 1:4 ratio. Following deacetylation, the

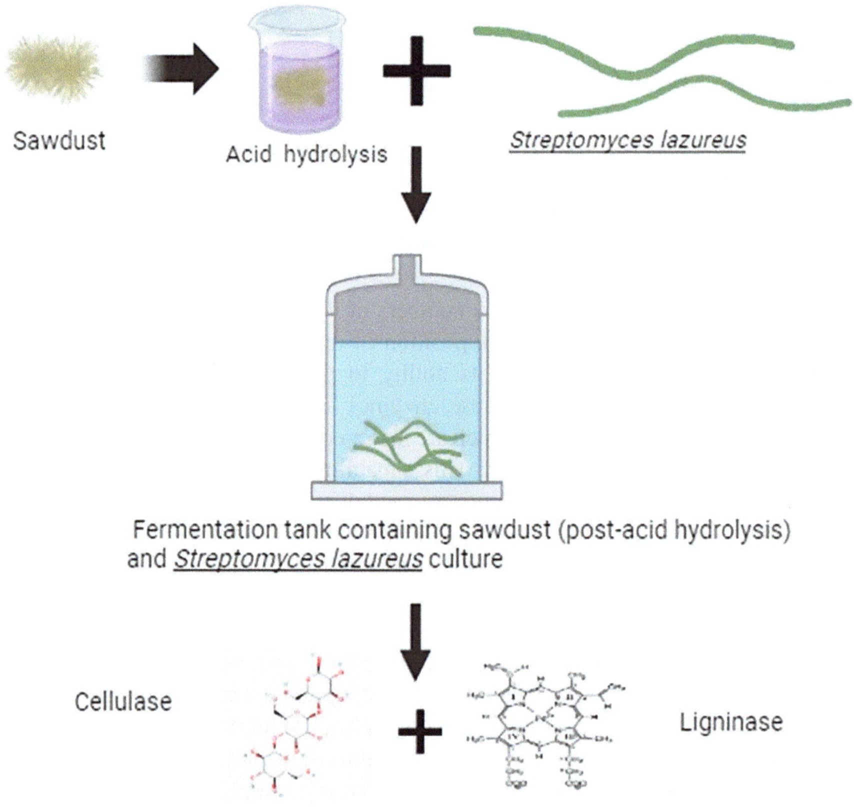

Fig. 14.7 Cellulase and ligninase for glucose production from sawdust

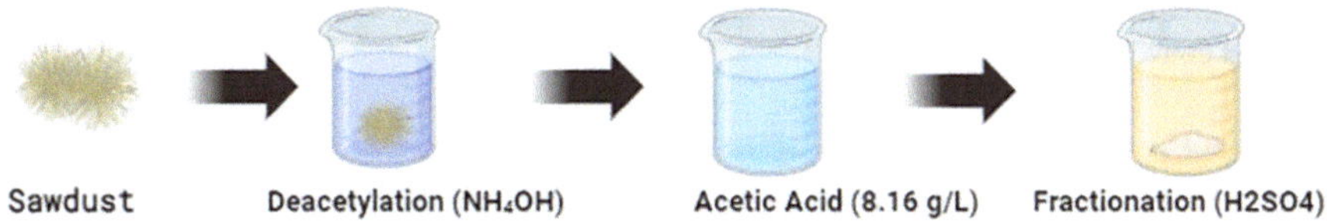

Fig. 14.8 Acetic acid formation from sawdust

sawdust undergoes fractionation with dilute sulfuric acid (H_2SO_4) at 130–150 °C for 10–80 min, breaking down hemicellulose into fermentable sugars (Behera et al. 2014). Acetic acid has diverse applications across various industries. Industrially, it serves as a chemical reagent in producing compounds like acetic anhydride, esters, and vinyl acetate monomer, and as a solvent for recrystallizing organic compounds. Medically, acetic acid acts as an antiseptic against various bacteria, aids in cervical cancer screening, treats infections, and lyses red blood cells for white blood cell examination (Murray et al. 2015). In the food industry, it is used in commercial pickling, condiments, and seasoning, and reacts with alkaline ingredients in baking. At home, diluted acetic acid is commonly used as vinegar for cleaning, laundry, and cooking (Fig. 14.8). In textiles, it is essential for dyeing and finishing processes, while in pharmaceuticals, it is used in medication production and as a disinfectant (Deshmukh and Manyar 2020).

14.6 Challenges and Prospects of Chemical Production and Biofuel Production from Sawdust

Securing a reliable sawdust supply is a significant challenge for biofuel production. Companies struggle with raw material procurement, especially in regions with limited forestry industries (Udokpoh and Nnaji 2023). Availability and quality fluctuations disrupt production efficiency and biofuel quality. Long-term supplier contracts, efficient collection and transportation systems, and alternative biomass sources can address this issue (Fig. 14.9). Partnerships with sawmills, furniture manufacturers, and construction companies ensure a steady supply of wood waste and sawdust (Adhikari and Ozarska 2018).

Environmental regulations add complexity, requiring investments in pollution control, waste treatment, and sustainable forestry practices. Non-compliance risks fines, legal issues, and operational shutdowns, impacting profitability. Companies must prioritize sustainability, implement environmental management systems, and optimize processes to reduce emissions and waste. Maintaining consistent biofuel pellet quality is crucial. Factors like moisture content, ash content, energy density, and durability vary with raw material quality and manufacturing processes (Wibowo et al. 2021). Rigorous quality control, advanced monitoring equipment, and skilled personnel are necessary. Regular testing and certifications enhance credibility and marketability. Collaboration with research institutions and industry associations fosters best practices and technology adoption. Lignocellulosic complexity in

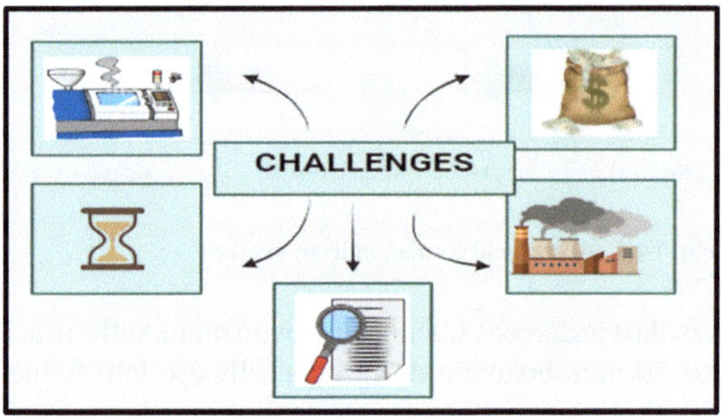

Fig. 14.9 Challenges in sawdust conversion

sawdust, comprising cellulose, hemicellulose, and lignin, complicates chemical production (Bajpai and Bajpai 2016). Effective lignin degradation is essential for converting sawdust into biofuels and chemicals. Innovations in catalytic and enzymatic methods can improve this process, enhancing biofuel production viability (Beig et al. 2021). Economic viability requires competing with low fossil fuel costs. This involves optimizing production rates and reducing costs. Energy-intensive pretreatment procedures and expensive enzymatic hydrolysis need technological advancements, process integration, and economies of scale. Despite challenges, sawdust offers significant prospects as a renewable, carbon-neutral biomass resource (Adegoke et al. 2022). Governments, industries, and communities must prioritize sawdust utilization for a sustainable future. Sawdust-derived fuels provide a reliable, locally available energy supply, enhancing energy independence and resilience (Silva et al. 2021). Ongoing research and development in gasification technology, briquetting processes, and combustion efficiency make conversion more cost-effective and efficient. The circular economy concept aligns with sawdust fuel production, minimizing waste and promoting sustainability (Adami and Schiavon 2021). Sawdust fuel production creates economic opportunities, job creation, local industry development, and revenue generation. Governments and investors must recognize its potential and provide support for growth, catalyzing economic prosperity and environmental stewardship (Fig. 14.10). Replacing fossil fuels with sawdust-based fuels reduces air pollution, lowers carbon emissions, and mitigates deforestation, contributing to cleaner air and combating climate change (Jeswani et al. 2020).

Beyond fuel, sawdust can produce biochar for soil fertility and carbon sequestration, and sawdust-based materials for manufacturing, fostering innovation and resource efficiency. Cross-sector collaborations maximize sawdust utilization (Elkhlifi et al. 2023). Supportive policies and incentives promote renewable energy and sustainable practices. Continued research and development in biomass energy are crucial for innovation, process efficiency, and product diversification (Fig. 14.11).

Engaging local communities enhances social acceptance and support for sawdust fuel projects. International collaboration on research and best practices accelerates sawdust fuel adoption globally. Sawdust fuel production enhances climate resilience by providing decentralized, adaptable energy, crucial for regions prone to

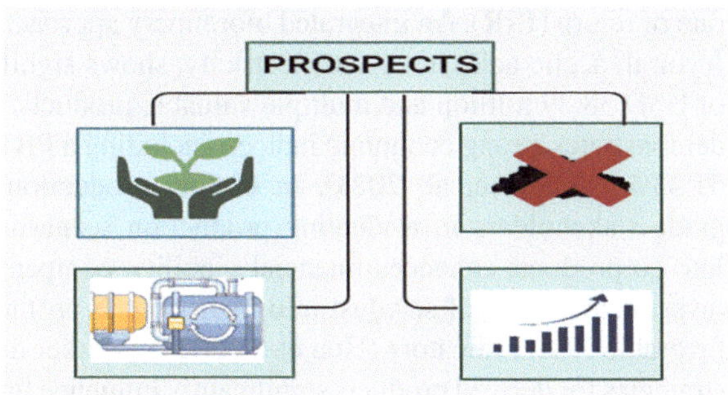

Fig. 14.10 Prospects of sawdust conversion

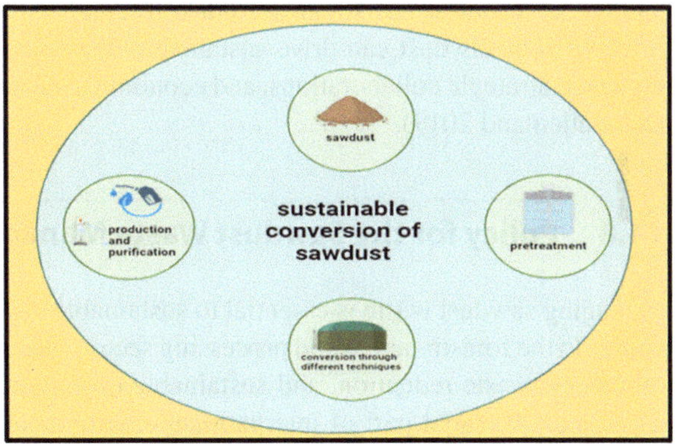

Fig. 14.11 Illustration showing the sustainability of sawdust conversion

climate-related disasters. Public awareness and education about sawdust fuel benefits drive adoption. Scaling up production through industrial partnerships, infrastructure investment, and market development accelerates commercialization, driving down costs, improving efficiency, and ensuring economic viability.

14.7 Feasibility and Economics of Chemical Production from Sawdust

Producing chemicals and biofuels from sawdust hinges on technological advances, process optimization, and overcoming the complexity of lignocellulosic materials. Enzymatic hydrolysis and pretreatment technologies, such as acid hydrolysis and steam explosion, improve the accessibility of cellulose and hemicellulose for subsequent processing (Sun et al. 2016). Hydrothermal carbonization (HTC) is a promising method, transforming sawdust into high-calorific, hydrophobic solid fuel. Studies show that industrial-scale HTC is economically viable with specific conditions, such as a reactor residence time of 1 h and a biomass-to-water ratio of 10% (Li et al. 2020). Economic viability in chemical production from sawdust depends on production costs, market demand, and profitability. Key economic indicators include cost analysis, payback time (PBT), net present value (NPV), and internal rate of return (IRR). An integrated biorefinery approach, producing torrefied pellets, furfural, acetic acid, heat, and electricity, shows significant promise, with an NPV of USD 38.29 million and multiple valuable products. Producing wood pellets also demonstrates strong economic indices, including a PBT of 2.49 years and an IRR of 51.33% (Vallejo et al. 2023). In biofuel production, techno-economic analyses guide stakeholders in evaluating production scenarios. Valorizing waste streams into co-products enhances financial viability, compensating for substantial capital investments. HTC of sawdust into hydrochar is profitable, with positive NPV and favorable ROE indicators (Hou et al. 2020). Market dynamics and optimal pricing strategies for derived products significantly influence financial outcomes. Embracing circular economy principles helps industries reduce their environmental footprint and unlock new revenue streams (Lüdeke-Freund et al. 2019). Thus, bioenergy production from sawdust can drive sustainable development through innovative technologies, strategic collaborations, and economic and environmental benefits (Scarlat and Dallemand 2019).

14.8 Policy for the Sawdust Waste Management Globally

Managing sawdust waste is essential to sustainable development, especially when it comes to the forestry and wood processing sectors. Because of its effects on resource efficiency, waste reduction, and sustainable development, sawdust waste management is an essential part of international environmental regulations (Đerčan et al. 2012). Promoting recycling, the production of bioenergy, and ecologically friendly disposal techniques are the main objectives of effective management strategies (Tsui and Wong 2019). The waste hierarchy is one of the main components of the policy; it places the highest priority on waste prevention, reuse, recycling, recovery, and disposal as a last choice. Businesses are encouraged to invest in sawdust recycling and bioenergy technologies through financial incentives such as grants, subsidies, tax benefits, feed-in tariffs, and carbon credits (Bowyer 2011). Permitting and compliance procedures guarantee that environmental regulations are followed, and regulatory frameworks impose stringent criteria for the processing, storing, and disposal of sawdust to prevent pollution. Research and development are supported through innovation grants and public–private partnerships to advance technologies for efficient sawdust utilization. Public awareness campaigns and training programs educate industries and the public on the benefits of proper sawdust management and its applications (Ellefson et al. 2010). Several policies around the world are good examples of sawdust waste management. The Waste Framework Directive and the Renewable Energy Directive in the European Union provide mandatory goals and guiding principles for the use of renewable energy sources and sustainable waste management (Udokpoh and Nnaji 2023). Frameworks for recovering wood waste and bioenergy research are provided under the Resource Conservation and Recovery Act and the Biomass Research and Development Act in the United States Sustainable wood waste management and energy production are supported by the Environmental Protection Act of Canada as well as regional initiatives like British Columbia's

Wood Waste-to-Energy Program (Yun et al. 2022). Australia's National Waste Policy encourages recycling and the production of bioenergy from organic waste, including sawdust, as do state programs like Victoria's Waste and Resource Recovery Policy. Japan's Basic Act on Establishing a Sound Material-Cycle Society and the Act on Promotion of Biomass Utilization encourage recycling and the use of biomass to reduce fossil fuel dependence (Begum 2016).

These international examples show different ways to handle sawdust waste, illustrating how policies may be implemented successfully and showing how international cooperation can advance sustainable practices. Hence, sawdust waste management rules that are effective are essential for encouraging sustainable behaviors, minimizing negative environmental effects, and improving resource efficiency in the wood processing sector.

14.9 Conclusions

The wood processing and furniture manufacturing industries produce significant sawdust, a lignocellulosic waste with potential for conversion into valuable chemicals and biofuels. Using advanced thermochemical and biochemical methods, this low-value by-product can be transformed into high-value products, promoting sustainable resource use and economic growth. Pyrolysis converts sawdust into bio-oils, rich in phenolic compounds, while anaerobic digestion produces biogas. Controlled pyrolysis yields biochar, useful for soil amendment and carbon sequestration, and direct combustion generates electricity. Acid hydrolysis extracts fermentable sugars from sawdust, producing bioethanol, and chemical activation creates high-performance activated carbons for purification and energy storage. Submerged fermentation yields glucose powder, and deacetylation extracts acetic acid.

Challenges include the recalcitrant nature of lignocellulosic components and the need for efficient pretreatment and conversion processes. Overcoming lignin degradation issues and establishing economic viability are critical. The benefits include reduced strain on food crops, waste diversion from landfills, decreased greenhouse gas emissions, and promotion of a circular bioeconomy. Sawdust valorization supports carbon sequestration, soil amendment, and environmental remediation. Through research, collaboration, policies, and investment, sawdust conversion can drive a sustainable bioeconomy, enhancing resource efficiency, energy security, and environmental sustainability.

References

Abd El Halim A, El Baroudy A (2014) Influence addition of fine sawdust on the physical properties of expansive soil in the middle Nile Delta. Egypt J Soil Sci Plant Nutr 14(2):483–490

Abou Rjeily M, Cazier F, Gennequin C, Randrianalisoa JH (2023) Detailed analysis of gas, char and bio-oil products of oak wood pyrolysis at different operating conditions. Waste Biomass Valoriz 14(1):325–343

Adami L, Schiavon M (2021) From circular economy to circular ecology: a review on the solution of environmental problems through circular waste management approaches. Sustain For 13(2):925

Adegoke KA, Adesina OO, Okon-Akan OA, Adegoke OR, Olabintan AB, Ajala OA, Bello OS (2022) Sawdust-biomass based materials for sequestration of organic and inorganic pollutants and potential for engineering applications. Curr Res Green Sustainable Chem 5:100274

Adhikari S, Ozarska B (2018) Minimizing environmental impacts of timber products through the production process "from sawmill to final products". Environ Syst Res 7(1):6

Ahmed A, Bakar MSA, Sukri RS, Hussain M, Farooq A, Moogi S, Park Y-K (2020) Sawdust pyrolysis from the furniture industry in an auger pyrolysis reactor system for biochar and bio-oil production. Energy Convers Manag 226:113502

Al-Dhabaan FA (2022) Sawdust biodegradation: cellulase and ligninase production via submerged fermentation and glucose production by Streptomyces lazureus. J Pure Appl Microbiol 16(4):2425–2438

Alizadeh P, Tabil LG, Mupondwa E, Li X, Cree D (2023) Technoeconomic feasibility of bioenergy production from wood sawdust. Energies 16(4):1914

Atiwesh G (2022) Lignin extraction, biodegradation and usage. Memorial University of Newfoundland, St. John's

Bajpai P, Bajpai P (2016) Pretreatment of lignocellulosic biomass. Springer, Cham

Bamidele Charles O, Mustapha Muhammad L, Kolawole Adisa O (2023) Utilization of sawdust composites in construction—a review

Begum S (2016) Assessment of alternative waste technologies for energy recovery from solid waste in Australia. CQUniversity, Rockhampton

Behera S, Arora R, Nandhagopal N, Kumar S (2014) Importance of chemical pretreatment for bioconversion of lignocellulosic biomass. Renew Sust Energ Rev 36:91–106

Beig B, Riaz M, Naqvi SR, Hassan M, Zheng Z, Karimi K, Chi NTL (2021) Current challenges and innovative developments in pretreatment of lignocellulosic residues for biofuel production: a review. Fuel 287:119670

Bowyer JL (2011) Potential impacts of climate and energy policy on forest sector industries: providing incentives for bio-energy while protecting established biomass-based industries: dovetail partners

Chan YH, Loh SK, Chin BLF, Yiin CL, How BS, Cheah KW, Lo SLY (2020) Fractionation and extraction of bio-oil for production of greener fuel and value-added chemicals: recent advances and future prospects. Chem Eng J 397:125406

Chukwuemeka IC, Ethel OC, Kalu AD (2020) Bio-production of ethanol and biomass from wood particles and sawdust using conventional yeast. GSC Biol Pharm Sci 11(1):069–077

Demirbas A (2009) Biofuels securing the planet's future energy needs. Energy Convers Manag 50(9):2239–2249

Đerčan B, Lukić T, Bubalo-Živković M, Đurđev B, Stojsavljević R, Pantelić M (2012) Possibility of efficient utilization of wood waste as a renewable energy resource in Serbia. Renew Sust Energ Rev 16(3):1516–1527

Deshmukh G, Manyar H (2020) Production pathways of acetic acid and its versatile applications in the food industry. In: Biomass. IntechOpen, London

Dhillon A, Sharma K, Rajulapati V, Goyal A (2017) Proteolytic enzymes. In: Current developments in biotechnology and bioengineering. Elsevier, Amsterdam, pp 149–173

Durán I, Álvarez-Gutiérrez N, Rubiera F, Pevida C (2018) Biogas purification by means of adsorption on pine sawdust-based activated carbon: impact of water vapor. Chem Eng J 353:197–207

Elhenawy Y, Fouad K, Bassyouni M, Al-Qabandi O, Majozi T (2024) Yield and energy outputs analysis of sawdust biomass pyrolysis. Energy Convers Manage X 22:100583

Elkhlifi Z, Iftikhar J, Sarraf M, Ali B, Saleem MH, Ibranshahib I, Torun Kayabasi E (2023) Potential role of biochar on capturing soil nutrients, carbon sequestration and managing environmental challenges: a review. Sustain For 15(3):2527

Ellefson PV, Kilgore MA, Skog KE, Risbrudt CD (2010) Wood utilization research and product development capacity in the United States: a review

Emenike EC, Iwuozor KO, Ighalo JO, Bamigbola JO, Omonayin EO, Ojo HT, Adeniyi AG (2024) Advancing the circular economy through the thermochemical conversion of waste to biochar: a review on sawdust waste-derived fuel. Biofuels 15(4):433–447

Gorman S, Scott E (2004) Chemical disinfectants, antiseptics and preservatives. In: Hugò and Russell's pharmaceutical microbiology. Wiley, Hoboken, pp 285–305

Hamad HN, Idrus S (2022) Recent developments in the application of bio-waste-derived adsorbents for the removal of methylene blue from wastewater: a review. Polymers 14(4):783

Hashemi B, Sarker S, Lamb JJ, Lien KM (2021) Yield improvements in anaerobic digestion of lignocellulosic feedstocks. J Clean Prod 288:125447

Hou P, Feng Y, Wang N, Petropoulos E, Li D, Yu S, Yang L (2020) Win-win: application of sawdust-derived hydrochar in low fertility soil improves rice yield and reduces greenhouse gas emissions from agricultural ecosystems. Sci Total Environ 748:142457

Hussain S, Reza M (2023) Environmental damage and global health: understanding the impacts and proposing mitigation strategies. J Big-Data Anal Cloud Comput 8(2):1–21

Issah A-A, Kabera T, Kemausuor F (2020) Biogas optimisation processes and effluent quality: a review. Biomass Bioenergy 133:105449

Jeswani HK, Chilvers A, Azapagic A (2020) Environmental sustainability of biofuels: a review. Proc R Soc A 476(2243):20200351

Kapoor R, Ghosh P, Kumar M, Vijay VK (2019) Evaluation of biogas upgrading technologies and future perspectives: a review. Environ Sci Pollut Res 26:11631–11661

Karuppiah T, Azariah VE (2019) Biomass pretreatment for enhancement of biogas production. Anaerob Dig 150:111509

Kumar AK, Sharma S (2017) Recent updates on different methods of pretreatment of lignocellulosic feedstocks: a review. Bioresour Bioprocess 4(1):7

Li J, Zhao P, Li T, Lei M, Yan W, Ge S (2020) Pyrolysis behavior of hydrochar from hydrothermal carbonization of pinewood sawdust. J Anal Appl Pyrolysis 146:104771

Lüdeke-Freund F, Gold S, Bocken NM (2019) A review and typology of circular economy business model patterns. J Ind Ecol 23(1):36–61

Mansur F, Faizal C, Samad N, Atnaw S, Sulaiman S (2020) Gasification performance of sawdust, pelletized sawdust and sub-bituminous coal in a downdraft gasifier. SN Applied Sciences 2:1–10

Megawati SWB, Sulistyo H, Hidayat M (2015) Sulfuric acid hydrolysis of various lignocellulosic materials and its mixture in ethanol production. Biofuels 6(5–6):331–340

Murray PR, Rosenthal KS, Pfaller MA (2015) Medical microbiology. Elsevier, Amsterdam

Nath HK (2015) Natural resources, the resource curse and sustainable development

Otulugbu K (2012) Production of ethanol from cellulose (sawdust)

Palange D (2009) Wood energy in the southeastern United States: a strategy for sustainable growth

Pitti A, Espinoza O, Smith R (2019) Marketing practices in the urban and reclaimed wood industries. BioProd Business 4:15–26

Rafiqul I, Sakinah AM (2012) Kinetic studies on acid hydrolysis of Meranti wood sawdust for xylose production. Chem Eng Sci 71:431–437

Rodriguez-Narvaez OM (2023) Activated carbon for environmental applications. In: Activated carbon: progress and applications. RSC, London, p 70

Rominiyi O, Adaramola B, Ikumapayi O, Oginni O, Akinola S (2017) Potential utilization of sawdust in energy, manufacturing and agricultural industry; waste to wealth. World J Eng Technol 5(3):526–539

Rusanen A, Lappalainen K, Kärkkäinen J, Tuuttila T, Mikola M, Lassi U (2019) Selective hemicellulose hydrolysis of scots pine sawdust. Biomass Convers Biorefinery 9:283–291

Salwan R, Sharma V (2020) Molecular and biotechnological aspects of secondary metabolites in actinobacteria. Microbiol Res 231:126374

Scarlat N, Dallemand J-F (2019) Future role of bioenergy. In: The role of bioenergy in the bioeconomy. Elsevier, Amsterdam, pp 435–547

Shrestha D, Gyawali G, Rajbhandari A (2018) Preparation and characterization of activated carbon from waste sawdust from saw mill. J Inst Sci Technol 22(2):103–108

Sichone K (2013) Pyrolysis of sawdust. University of Waikato, Hamilton

Silva J, Ferreira AC, Teixeira S, Martins L, Ferreira E, Teixeira JC (2021) Sawdust drying process in a large-scale pellets facility: an energy and exergy analysis. Clean Environ Syst 2:100037
Sun S, Sun S, Cao X, Sun R (2016) The role of pretreatment in improving the enzymatic hydrolysis of lignocellulosic materials. Bioresour Technol 199:49–58
Tran HM (2005) Quantifying the effects of sawdust application on soil chemical and physical properties and corn yield
Tsui T-H, Wong JW (2019) A critical review: emerging bioeconomy and waste-to-energy technologies for sustainable municipal solid waste management. Waste Disp Sustainable Energy 1:151–167
Tulashie SK, Akpari EEA, Appiah G, Adongo A, Andoh EK (2023) Acid hydrolysis of sawdust waste into bioethanol. Biomass Convers Biorefinery 13(7):5743–5756
Udokpoh U, Nnaji C (2023) Reuse of sawdust in developing countries in the light of sustainable development goals. Recent Progr Mater 5(1):1–33
Vallejo F, Díaz-Robles L, Germain I, Alejandro-Martin S, Ledesma B, Román S, Sabio E (2023) Economic and feasibility analysis of bioenergy production from sawdust via hydrothermal carbonisation for a circular economy. Chem Eng Trans 103:841–846
Vamvuka D (2011) Bio-oil, solid and gaseous biofuels from biomass pyrolysis processes—an overview. Int J Energy Res 35(10):835–862
Vieira Y, dos Santos JM, Georgin J, Oliveira ML, Pinto D, Dotto GL (2022) An overview of forest residues as promising low-cost adsorbents. Gondwana Res 110:393–420
Vu HP, Nguyen LN, Vu MT, Johir MAH, McLaughlan R, Nghiem LD (2020) A comprehensive review on the framework to valorise lignocellulosic biomass as biorefinery feedstocks. Sci Total Environ 743:140630
Wibowo S, Arief K, Waluyo T (2021) Characteristics of wood pellets from over-dry sawdust waste. IOP Conf Ser Earth Environ Sci 914:012069
Yun H, Wang H, Clift R, Bi X (2022) The role of torrefied wood pellets in the bio-economy: a case study from Western Canada. Biomass Bioenergy 163:106523

Conversion of Forestry Waste to Chemicals and Biofuels

15

G. Flora, Ashokkumar Veeramuthu, A. Lakshmi, Irudaya Antonat Sophia, V. P. Chandramughi, Agila Elango, and V. Vishnu Priya

Abstract

The transformation of forestry waste into chemical compounds and biofuels offers a sustainable solution for managing forest residues while producing valuable energy resources. Forest ecosystems, which cover approximately 31% of the Earth's land area, generate significant amounts of biomass residues from activities such as logging, thinning, pruning, and wood processing. These residues include branches, sawdust, bark, and leaf litter, which can be used as raw materials for bioenergy and biochemical production. This chapter discusses various pretreatment methods, including mechanical, chemical, biological, and hybrid techniques aimed at enhancing the usability of forest waste. Additionally, it examines thermochemical and biochemical conversion technologies, such as pyrolysis, gasification, torrefaction, fermentation, and anaerobic digestion for the production of biofuels like bioethanol, biodiesel, and bio-oil. The integration of these technologies transforms waste into renewable energy sources, reducing environmental impact while simultaneously enhancing economic benefits.

G. Flora
Department of Botany, St. Mary's College (Autonomous), Thoothukudi, Tamil Nadu, India

A. Veeramuthu (✉) · V. P. Chandramughi
Center for Waste Management and Renewable Energy, Saveetha Dental College and Hospitals, Saveetha Institute of Medical and Technical Sciences, Saveetha University, Chennai, Tamil Nadu, India

A. Lakshmi · I. A. Sophia
Department of Chemistry, St. Mary's College (Autonomous), Thoothukudi, Tamil Nadu, India

A. Elango · V. Vishnu Priya
Department of Biochemistry, Saveetha Dental College and Hospitals, Saveetha Institute of Medical & Technical Sciences, Saveetha University, Chennai, Tamil Nadu, India

A. Kumari et al. (eds.), *Valorization of Solid Wastes to Biofuels and Chemical Products for Sustainable World*, https://doi.org/10.1007/978-981-96-8594-3_15

Keywords

Forestry waste · Lignocellulosic biomass · Platform chemicals · Conversion technologies · Liquid biofuels

15.1 Introduction

Forest biodiversity covers the wide array of life forms present in the forest areas, including plants, animals, and microorganisms, along with their inherent diversity. It encompasses beyond the trees and includes the ecological roles played by these beasts within forest biota. Forest ecosystem plays a central role in the global biodiversity due to their high levels of collections compared to other types of ecosystems. They cover about 31% of the Earth's land area, with approximately half of this area remained undisturbed, and more than one-third of this is classified as primary forest, that is, naturally regenerated with native species and minimal human interference. Globally, forests span a total area of 4.06 billion hectares, which averages to roughly 5000 m^2/person (or about the size of a 50 × 100-m plot). However, the forest distribution is uneven across the globe. The majority of the world's forest covers are concentrated in Brazil, Russia, Canada, the United States, and China, which together account for more than half of global forest cover. India's total documented forest area is 767,419 km^2 (Forest Survey of India 2019). Here, forestry activities generate various forms of waste, including significant quantities of bark, wood chip, and brash. However, these materials are not considered waste if utilized for specific purpose such as energy production. Forest biomass residues present a promising opportunity in the field of new bio-based materials and biofuels in today's epoch of global climate change (Braghiroli and Passarini 2020). Thus, the necessity to investigate forest biodiversity and the use of forest biomass arises from the essential function trees serve in preserving global ecological equilibrium, fostering biodiversity and alleviating climate change. Given the escalating threats to forests from deforestation, habitat degradation, and climate change, it is imperative to comprehend their biodiversity and identify sustainable methods for utilizing biomass in the production of bio-based materials and biofuels to conserve ecosystems, mitigate carbon emission, and advance renewable energy alternative. This chapter explores the sources and chemical constituents of forest waste, as well as the conservation technologies used to convert them into valuable products such as biofuels and other biochemical derivatives.

15.2 Types and Sources of Forest Waste

Forest residues, including slender trees, branches, trimmings, and noncommercial wood that remains after forest cleaning, thinning, or the final falling of dead trees, are often considered forest wastes and are utilized as fuel without additional processing. These residues primarily come from three sources, which includes (1) slash from final downing of trees, (2) slash and small trees from pollarding, and (3) noncommercial wood. Figure 15.1 illustrates the various sources and types of forest waste in a structured and comprehensive manner.

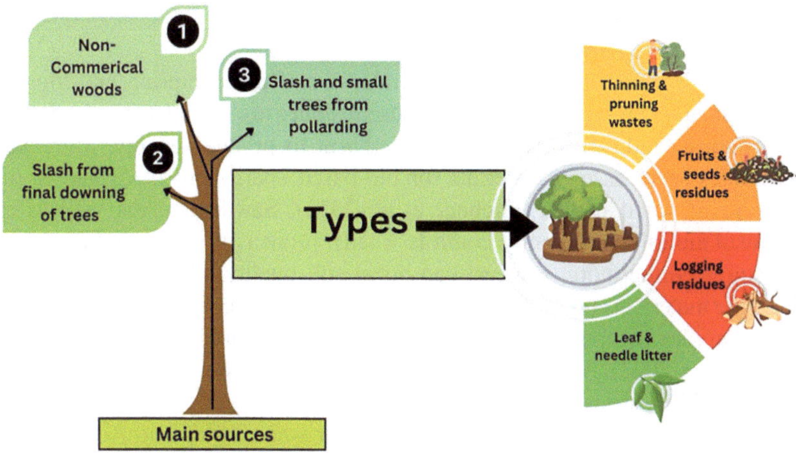

Fig. 15.1 Sources and types of forest waste

15.2.1 Logging Residues

Logging residues have a varied composition because they consist of different types of materials as they consist of round pieces of wood, twigs, bark, pinecones, and leaves or needles. These residues can be chopped into loose chips by using a chipper or crusher (Bozym et al. 2021). Forest biomass residue offers diverse opportunities in the development of novel biomaterials and efficient biofuels. The worldwide annual stock of biomass from agriculture and forestry is approximately 11.9 billion tons of dry mass, with forest contributing 39% in the global production. In the world's total wood biomass production, 23% is used as firewood, 8% is used as primary biomass as industrial roundwood, and another 8% is attributed to losses (FAO 2020).

15.2.2 Thinning and Pruning Wastes

Thinning is a more intense form of pruning that includes removing entire shoots, limbs, or branches at their point of origin. This technique is typically used to restore a plant by eliminating matured, weak, tough extensively grown branches. Appropriate thinning endorses the development of new shoot that is more probable to produce fruit and flowers. Pruning and thinning waste includes branches, twigs, leaves, and other plant debris. Thinning of trees in the forests leads to the higher groundwater recharge compared to clear-cut forests. This suggests that the groundwater recharge is strongly influenced by reduced vaporization in partly thinned forests than by the absolute removal of mature trees through clear-cutting (Schenk et al. 2020).

15.2.3 Bark, Sawdust, and Wood Chips

The wood residues include materials generated from manufacturing and logging practices that can be reused or recycled. These residues are often used as basic fuel in manufacturing industries, where they are burnt to produce energy. Moreover, wood residues serve as raw materials for bio-based energy processes and in the manufacture of wood pellets (Grebner et al. 2022). Sawdust serves well as an organic insulation that efficiently controls weeds, protects the soil, and preserves soil moisture. Furthermore, sawdust can boost soil quality when combined with other organic materials and forms a compost. Wood chips are rich in tannins, lignin, and other complex compounds that provide soil nutrients and increased absorption of water (Pandey 2022). A cheap and easily accessible by-product of forests and the pulp and wood industries is bark. It can be utilized in the production of furniture panels as well as a fuel source, mulch additive, or soil amendment. Additionally, bark can be used as a cheap and efficient sorbent material to remove contaminants from water, wastewater, as well as gas purification processes. Due to its porous nature, the bark of some conifer species can be used to monitor air pollution, particularly for the detection of heavy metals (Bozym et al. 2021).

15.2.4 Leaf and Needle Litter

Litter fall in terrestrial ecosystems shows an essential route for the returning of nutrients to the soil. In forests, leaf tissue will take 70% of the surface litter fall, while the rest consists of stems, small twigs, and propagative structures (Krishna and Mohan 2017). Prescott et al. reported that broad leaf litter decomposed more quickly than needle litter (Prescott et al. 2000). By raising the amounts of organic matter in the soil, leaf litter compost can help revegetate and rehabilitate disturbed soil . Replenishing the soil organic matter improves water retention volume, cation exchange capacity, and organic content, while reducing the soil bulk density (Solanki and Arora 2015).

15.3 Characterization of Forest Waste

Assessing the material goods and potential uses of forest waste ensures that it can be effectively reused, reducing the need for new resources. Characterizing the forest waste helps identify the opportunities to convert waste into valuable by-products, reducing the amount of waste sent to landfills. Adding composted forest waste to soil enhances its structure, fertility, and organic content. Forest waste can be used as insulation cover to protect soil from erosion and maintain soil moisture. Proper management of forest waste can subsidize carbon storage, which in turn helps mitigate climate change. Additionally, it results in the creation of cutting-edge goods and technologies including renewable energy sources and bio-based materials.

Exploring the waste produced from forest enhances our understanding of ecosystem dynamics, decomposition processes, and soil–plant interactions.

15.3.1 Physical Properties

Forest soils acquire their physical properties in the normal natural conditions because of long-term, stable vegetation. These properties are stable unless rehabilitated by human activities such as harvesting, cultivation, or forest fires. The physical nature of forest includes soil texture, structure, porosity, density, aeration, soil temperature, water retention capacity, and water movement. These properties significantly affect the soil fertility and productivity by changing the root penetration property, water absorption capacity, oxygen and gas levels, as well as water movement within the soil. Moreover, the physical properties of forest soils play an important role in determining the natural spreading of tree species, their development, and global forest biomass production. The size, distribution, and layer organization of the soil particles largely determine the soil's characteristics. Nurek et al. studied the elemental composition of forest residues comminuted from *Pinus sylvestris* (Nurek et al. 2019). Forest wood chips' bulk densities range from 110 kg/m^3 to 340 kg/m^3 depending on variables like composition, origin, and moisture content. Reports show that the carbon, nitrogen, sulfur, oxygen, and ash content from forest remains and wood chips forms homogeneous groups concerning their hydrogen content, and their energy value was found to be 15.78 ± 0.39 MJ/kg. The study also focuses on particle size distribution and its impact on the material's effectiveness in combustion and storage.

15.3.2 Chemical Composition

The chemical nature of forest waste contains many elements and nutrient contents that mainly impact its usability and environmental effects. The structural elements of plant biomass, such as cellulose, hemicellulose, and lignin, make up the majority of the lignocellulosic constituents found in forest debris. It usually has a high moisture content between 20% and 80% and contains a variety of extractives, including oils, phenolic compounds, terpenes, tannins, and volatile and non-volatile organic compounds that affect its energy potential and decomposition. Figure 15.2 presents the key chemical parameters present in forest waste. The forest waste is enriched with organic carbon, which is an important constituent for soil fertility and carbon cycling in forest biotas. Depending on the type of waste and stage of degradation, the percentage of carbon can change (Braghiroli and Passarini 2020; Khalili et al. 2016). Nitrogen is a very important element for plant growth and is present in varying amounts in forest waste. It is important for plant nutrient cycling in the ecosystem and soil fertility. High percentage of nitrogen can increase the decomposition speed and nutrient availability for plants. The percentage of nitrogen present in forest residues can be in the range from 0.1% to 6.5% (Aniszewska et al. 2017).

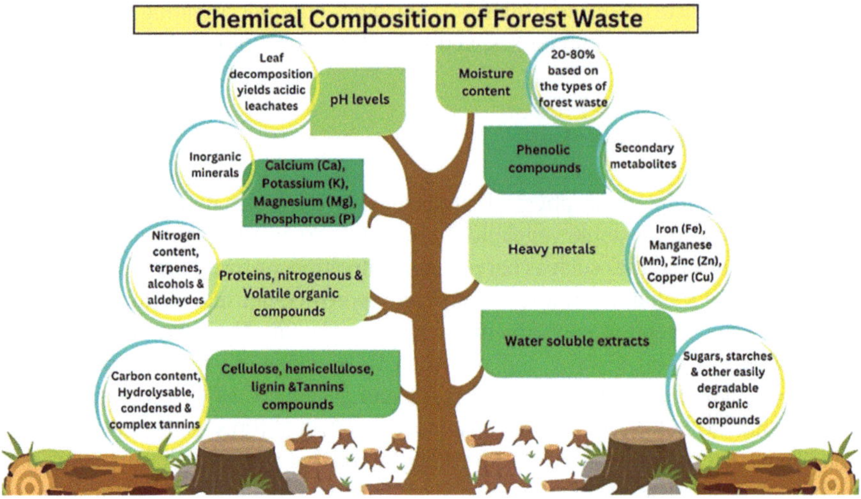

Fig. 15.2 Chemical parameters in forest waste

Sulfur is also an important nutrient present in forest residue, but in smaller amounts compared to carbon and nitrogen. The percentage of carbon and nitrogen is important for the synthesis of protein in plants and soil microbes. The percentage of sulfur in forest residue ranges from 0.01% to 0.05% (Bozym et al. 2021). Forest residue also contains oxygen as part of its organic compounds. This affects the decomposition process and the overall energy content of the forest residue. The amount of inorganic residue, which includes minerals like calcium, magnesium, potassium, and phosphorus, is left over after burning is indicated by the ash content. The ash content can affect the soil pH and the nutrient availability. The pH of forest waste can vary but is usually acidic to neutral. This affects the microbial action and nutrient availability in the soil (An and Park 2021).

15.3.3 Variability and Seasonal Influences

The chemical properties of forest residue vary notably based on seasonal influences, affecting its uses. Due to changes in the biological activity and plant composition with the seasonal effects, the percentage of nitrogen in forest residues is altered. The forest residues reported in spring and summer may have higher nitrogen percentage due to increased metabolic activities and nutrient uptake (Kumar et al. 2021). The percentage of sulfur in forest residues can also be varied by seasonal changes, such as fluctuating levels of atmospheric accumulation and plant uptake. For example, higher sulfur can be detected during periods with increased industrial action or specific weather that increases the sulfur accumulation (Bolat et al. 2015).

15.4 Pretreatment Technologies

Forest waste pretreatment technologies are crucial for improving its acceptability for a variety of uses, including soil texture, energy production, and value-added goods. These technologies help in the reduction of moisture content, increasing combustion efficiency and improving the accessibility of biomass components. The following is an overview of commonly used pretreatment technologies.

15.4.1 Mechanical Pretreatment

Chipping and shredding are the two process in which large pieces of forest waste would be broken down into smaller and more manageable sizes using chippers and shredders. This enlarges the surface area, making it easier for ensuing processing steps. The next process is grinding and milling that reduces the particle size of the forest waste that is particularly important for specific applications requiring uniform feedstock production (John et al. 2023).

15.4.2 Chemical Pretreatment

To improve the conversion of lignocellulosic biomass (LCB) like forest waste into value-added products, chemical pretreatment is commonly used. Since, LCB is highly recalcitrant various organic and inorganic chemical are applied to disrupt intra and interpolymer linkages in cellulose, hemicellulose and lignin. This breakdown enhances the release of biopolymeric constituents, facilitating more efficient biochemical conversion.

15.4.3 Acid Pretreatment

Acid pretreatment is an effective method to break down LCB by cleaving glucosidic bonds, solubilizing hemicellulose and partially disrupting lignin, thereby increasing the accessibility of cellulose. Both diluted and concentrated acids, such as ulfuric acid (H_2SO_4) or hydrochloric acid (HCl) are commonly employed to achieve this, making cellulose significantly more amenable for enzymatic hydrolysis (Mosier et al. 2004).

15.4.4 Alkaline Pretreatment

Alkaline pretreatment is considered a promising method for processing LCB due to its use of non-corrosive chemicals, non-polluting and operation under milder conditions than acid pretreatment. This method primarily targets lignin removal, enhancing the porosity and reactivity of the biomass. Alkaline agent such as sodium hydroxide (NaOH), potassium hydroxide (KOH), or ammonia (NH_3), effectively disrupt the LCB matrix, opening the structure and exposing cellulose components for effective biochemical conversion(Kim and Lee 2007). .

15.4.5 Biological Pretreatment

Biological pretreatment of forest waste comprises of using microbes such as fungi and bacteria to break down the complex LCB. By breaking down the lignin and hemicellulose, this procedure increases the amount of cellulose that is available for fermentation or enzymatic hydrolysis. *Phanerochaete chrysosporium*, *Trametes versicolor*, and *Pleurotus ostreatus* are among the fungi that have ligninolytic enzymes, such as lignin peroxidase, manganese peroxidase, and laccase, which are particularly effective at breaking down lignin (Salvachúa et al. 2011). Common bacteria such as *Bacillus*, *Cellulomonas*, and *Clostridium* species can be utilized to degrade lignin and hemicellulose (Kumar et al. 2008). Future studies should concentrate on biological pretreatment methods such as anaerobic co-digestion, metabolic engineering, and co-culture systems, which have been shown to promote lignocellulose hydrolysis and thus improve biogas production. Additionally, ruminants and xylophagous insects are examples of natural lignocellulose-degrading systems that may enhance the anaerobic digestion system (Gao et al. 2022).

15.4.6 Enzymatic Pretreatment

An effective method for converting LCB into fermentable sugars is the enzymatic pretreatment of forestry waste. Enzymes such as cellulase and hemicellulase have demonstrated high efficiency in enzymatic treatments, achieving hydrolysis rates of 87.14% for hemicellulose and 77.6% for cellulose, respectively. Enzymatic pretreatment is environmentally benign and circumvents the damaging and corrosive impacts of acid-based treatments (Bendaoud et al. 2023). In addition, commercial enzyme preparations containing cellulases, hemicellulases, and ligninases are utilized to break down the complex biomass structure (Alvira et al. 2010; Zheng et al. 2009).

15.4.7 Microbial Pretreatment

In this method, the natural breakdown of forest waste can be done by the activity of a diverse microbial community under controlled composting conditions (Bernal et al. 2009; Awasthi et al. 2015). Microbial pretreatment, particularly through the use of white-rot fungi, plays a critical role in enhancing the biomethanation of lignocellulosic substrates by degrading lignin and hemicellulose. Fungi such as *Auricularia auricula-judae* and *Ceriporiopsis subvermispora* have been shown to significantly improve biogas production by increasing cellulose digestibility and enhancing microbial accessibility to LCB. These fungi secrete essential enzymes like cellulases, hemicellulases, xylanases, and lignin-degrading enzymes, which contribute to the breakdown of complex lignocellulosic structures. Although many fungi produce laccase and manganese peroxidase, only a few are capable of secreting lignin peroxidase, which is crucial for effective lignin degradation during pretreatment (Mishra et al. 2018).

15.5 Combined and Novel Pretreatment Methods

15.5.1 Combined Physical and Chemical Pretreatments

To increase the efficiency of the breakdown of lignin and hemicellulose, the steam explosion method is combined with the addition of an acid, such as H_2SO_4 or an alkali, such as NaOH and KOH in the forest waste (Sun and Cheng 2002). This combined approach significantly enhances the breakdown of lignin and hemicellulose, making forest waste more suitable for subsequent processing and value-added applications.

15.5.2 Microwave-Assisted Chemical Pretreatment Method

Microwave-assisted chemical pretreatment is gaining attention for its high heating efficiency, rapid reaction rates, and operational simplicity. This technique uses microwave irradiation to heat chemical reagents such as acids or alkalis, significantly enhancing their interaction with biomass. By accelerating chemical reaction rates, microwave energy improves the disruption of the lignocellulosic matrix, leading to more effective breakdown of biomass components for subsequent conversion processes (Hu and Wen 2007; Chinnadurai and Muthukumarappan 2009).

15.6 Conversion Technologies

15.6.1 Thermochemical Conversion

15.6.1.1 Pyrolysis

Heat is a key component in thermochemical conversion processes used to transform forest waste into valuable compounds. Direct combustion of forest residue in the presence of oxygen generates a significant amount of heat and power. Forest residue is pyrolyzed at temperatures between 300 and 600 °C without oxygen for the production of biochar, syngas, and bio-oil (Van Loo and Koppejan 2012; McKendry 2002).

15.6.1.2 Gasification

Gasification is one of the most prominent thermochemical conversion processes used in both biomass to energy and waste to energy technologies. Originally applied to coal and natural gas, it has now gained widespread use for converting biomass into syngas, a valuable intermediate for biofuel synthesis. Compared to conventional methods like incineration, gasification offers significantly higher efficiency in electricity generation, making it a preferred alternative for sustainable energy recovery from organic waste and biomass feedstock. In particular, forest waste can be converted through gasification by heating it to temperature between 700–1500 °C in the presence of controlled amount of stream and oxygen, resulting in the production of syngas a mixture primarily composed of carbon monoxide, hydrogen and carbon dioxide (Basu 2010).

15.6.1.3 Torrefaction

Torrefaction enhances the quality of biomass, making it more suitable for energy production. It entails heating the biomass in a low-oxygen atmosphere to moderate temperatures (200–300 °C). It causes moisture and volatile organic substances to be eliminated. As a result of this process, the biomass has more energy and is simpler to transport, store, and use as fuel. Additionally, the outcome of torrefaction is a dry, brittle, and energy-dense substance called "torrefied biomass" or "bio-coal" (Van Der Stelt et al. 2011).

15.6.2 Biochemical Conversion

15.6.2.1 Fermentation

Primarily in biochemical conversion, which uses microbial fermentation to transform biomass into bioethanol. Cellulose and hemicellulose in the forest waste are broken down by enzymes into sugars that microbes like yeast may ferment (Limayem and Ricke 2012). The complex and resistant polymer lignin can be transformed into useful compounds using promising techniques provided by advanced biochemical processes. Lignin may be effectively converted into high-value compounds like cis, cis-muconic acid (ccMA), and pyrogallol by combining chemical depolymerization, plant metabolic engineering, and microbial synthetic pathway reconstruction. These procedures show how the difficulties of lignin valorization can be overcome by fusing synthetic microbial systems with plant-based metabolic pathways. These strategies open the door for environmentally friendly ways to use LCB in the manufacturing of bio-based products and specialty chemicals (Wu et al. 2017).

15.6.2.2 Anaerobic Digestion

This process produces biogas, which is mostly made up of CO_2 and CH_4, by using the anaerobic breakdown of forest waste. A variety of microorganisms, including hydrolytic, acidogenic, acetogenic, and methanogenic species, all of which carry out distinct metabolic functions, work in concert to drive the process. Complex organic polymers are broken down into simpler molecules by hydrolytic microorganisms and acidogenic bacteria then transform these simpler components into volatile fatty acids. The biogas-generating process is then completed when methanogens use these intermediates to generate CH_4 (Weiland 2009).

15.6.2.3 Integrated and Hybrid Approaches

Incorporating forest leftovers as organic matter into the soil is one way to combine ecological practices and waste management. By enhancing nutrient content, microbial activity, and soil structure, this improves soil fertility and supports the strength of the ecosystem as a whole. . Instead of being thrown away, forest trash can be put to profitable uses. For example, it can be transformed into bioenergy (such biofuels or electricity), compost (to improve soil), or utilized as raw material in the wood products industry (such as in creating particle boards, paper, or pulp), so generating money and minimizing environmental waste. Utilizing forest waste in an integrated and hybrid manner maximizes the value produced from forest wastes by combining various technologies and processes. A hybrid approach combining multicriteria decision methods and response surface methodology optimizes the selection of urban forest waste blends and torrefaction conditions. The optimum blend is determined based on feedstock properties and torrefaction is optimized to enhance energy content and combustion stability. The resulting torrefied optimum blend sho`ws improved calorific value, reduced ash and nitrogen content, also more stable combustion, making it a superior biofuel with lower nitric oxide emissions potential (Silveira et al. 2023).

15.7 Biofuels Production from Forest Waste

Biogas, liquid biofuels, including biodiesel, ethanol, methanol, and butanol, and solid biofuels, are all included in biomass energy. While biogas and liquid biofuels require certain conversion procedures to become useable fuels, solid biofuels can be burned directly to generate energy. Numerous techniques exist for turning forest biomass into fuels that can be utilized as a source of energy for houses, cars, and other applications. Direct combustion (burning) to generate heat, thermochemical conversion to generate solid, gaseous, and liquid fuels, as well as biochemical conversion to generate liquid fuels, are the three main methods for generating energy from biomass. To effectively transform LCB into useable energy, the process of producing biofuel from forest waste comprises several phases. In order to improve process efficiency and minimize particle size, it starts with the collecting of forest leftovers, including branches, sawdust, and bark. This is followed by sorting and physical processing like grinding. While biological pretreatment uses enzymes or microorganisms to simplify complicated polymers, chemical pretreatment uses acids or alkalis to break down lignin and hemicellulose linkages. While gasification turns biomass into syngas by reacting it with air, steam, or oxygen, direct liquefaction treats wet biomass with high pressure and high temperature to create bio-crude. Subcritical or supercritical water is used in hydrothermal liquefaction to produce bio-oil. Bioethanol is produced through the fermentation of cellulose and hemicellulose-derived sugars and bio-oil or syngas is converted into refined biofuels such as gasoline or diesel through catalytic upgrading. Biofuels are fractionated into their constituent parts, and purification guarantees that they fulfill quality requirements. Biochar is one example of a residual by-product that is used for energy recovery or as soil amendment (Song et al. 2020). While circular economy principles concentrate on resource recovery and sustainability, process integration integrates these processes for cost and energy efficiency, guaranteeing that the entire process reduces waste and environmental effect.

15.8 Liquid Biofuels

15.8.1 Bioethanol

Ethanol developed from plant-based sources is considered one of the cleanest liquid fuels (Akansu et al. 2017). Fermentation process is used to manufacture ethanol from biomass. Industrial production of ethanol utilizes sugarcane and starch-rich materials. Dissolving the lignocellulose into starch by pretreating with acid such as H_2SO_4 or an alkali such as NaOH, hydrolyzing the soluble starch to produce glucose, and fermenting the glucose to provide ethanol are the three main phases in the process. Crushing the biomass, adding water and yeast, further fermenting the combination in big tanks called fermenters are several ways to accomplish this process (Ho et al. 2012). For the production of ethanol, researchers utilize a variety of forest wastes, including pine needles, oil palm fronds, leaves, and sawdust from rubber wood.

15.8.2 Biodiesel

Biodiesel can be prepared from any natural lipids having a rich source of fatty acid. Tree species with a high oil content make good raw materials for the manufacturing of biodiesel. The transesterification technique used to make biodiesel has the potential to be a successful car fuel (Rajkumar and Kurinjimalar 2022). Another promising thermochemical valorization method for turning forest waste into biofuels at moderate temperatures (usually between 300 to 1300 °C) is pyrolysis. Gasification is another conversion process used to valorize forest biomass. Using the Fischer–Tropsch method, woody wastes and forest biomass can be converted into liquid biofuel, such as bio-oil, as an alternative to fossil diesel (Yu et al. 2021). This approach, along with liquefaction, could significantly reduce the overall environmental impact of the transportation sector.

15.8.3 Bio-Oil

Food waste, crop residues, municipal garbage, algae, agricultural waste biomass, and forest residues are all used to make bio-oil, a liquid biofuel (Awogbemi and Von Kallon 2023). The complex mixture of deoxygenated chemicals, including alcohols, ketones, carboxylic acids, esters, benzenoids, and aldehydes, is produced by thermochemical processes. For the past 30 years, pyrolysis has been the subject of extensive research as a method of producing bio-oils. The variations in the composition of the three components (cellulose, hemicellulose, and lignin) in the forest waste will manifest not only in the amount of bio-oil produced, but it is also anticipated that when various biomass types undergo pyrolysis, they would yield bio-oil with varying chemical compositions and quality (Tinwala et al. 2015). Bio-oil is a viscous, highly oxygenated, dark brown liquid with a significant water content that is polar. Its high energy density, low nitrogen, and the sulfur levels make it simple to handle, store, and transport. Among its notable uses are as a transportation fuel for internal combustion engines, particularly compression ignition engines, power-generating facilities (boiler and gas turbines), as a raw material for bio-based chemicals used in industry and as a chemical modifier for asphalt binder used in road construction. However, before the bio-oil can be used as a fuel for transportation, it must be refined. . Catalytic cracking, hydrodeoxygenation, decarboxylation, co-pyrolysis, hot vapor filtration, esterification, supercritical solvents, steam reforming, emulsification, and solvent addition are methods that can be used to upgrade the crude bio-oil by eliminating excess oxygen, water, and acids.

Utilizing bio-oil has increased compression ignition engines' braking power, brake-specific fuel consumption, and brake thermal efficiency under various loads and speeds. In addition, it has been demonstrated to produce less smoke, carbon monoxide, nitrogen oxides, and unburned hydrocarbons than traditional fossil fuels. Its oxygenated qualities, which increase fuel atomization and combustion efficiency, are the cause of this improved performance.

15.8.4 Gaseous Biofuels

Gaseous biofuels are mainly cost-effective as they require negligible capital and manual work, making them the most adopted fuel in rural communities. To ensure the renewability of gaseous biofuels, it is crucial to analyze their potential, address technological and economic limits, as well as alleviate the major environmental concerns (Mallikarjuna et al. 2021).

15.8.5 Biogas

Biogas is a sustainable energy transporter that may be used to generate heat or power as well as a transportation fuel. Any biomass waste can be utilized as a raw material for the anaerobic digestion of sewage sludge, a process that produces biogas. Anaerobic digestion is a method for turning forest waste into biogas. In the absence of oxygen, this mechanism breaks down organic matter using microbes. The biogas, a renewable energy source, can be used as fuel. Environmental changes can be significantly improved when biogas is used in place of traditional fossil fuels. In addition to the obvious environmental benefits of reducing land use and processing organic waste, producing one megajoule of electricity from biogas requires the consumption of 0.4 and 0.9 megajoules of fossil fuels. When waste biomass is utilized as source materials, biogas can reduce greenhouse gas emissions by 75–90% compared to fossil fuels (Liu and Yue 2011). Air pollution, eutrophication, acidification, and climate change can all be effectively addressed using biogas systems. Carefully planning and placing these systems is essential to maximizing their positive effects and minimizing their negative ones.

15.8.6 Syngas

Syngas, short for "synthesis gas," is a mixture of H_2, CO, and CO_2 that is mostly derived from waste biomass and may be used as a bridge to convert biomass into fuel. Syngas is mostly used in the production of diesel and methanol. Diesel fuel is made from syngas using the Fischer–Tropsch method, which is a sequence of chemical processes that turns CO and H_2 into liquid hydrocarbon. Simply extracting hydrogen from the gas and refining it for use in fuel cells allows syngas to be used directly to power hydrogen fuel cells (Capodaglio and Bolognesi 2019). By gasifying forest waste with hot blast furnace gas, it is possible to efficiently produce H_2-rich syngas while recycling 1.9 kg of CO_2 per kilogram of waste and attaining 74.75% energy efficiency. This process demonstrates excellent economic feasibility, with a net present value of $336.70 million and a payback period of 2.4 years. Additionally, it produces 1.67 kg of syngas for every kilogram of forest waste processed (Qin et al. 2024). It efficiently uses forest trash and provides the iron and steel sector with a low-carbon, sustainable alternative.

15.8.7 Solid Biofuels

These are derived mainly from plant biomass, animal waste, and municipal wastes for producing energy mainly for cooking, heating, and generating electricity. Humans have relied on wood, a solid biofuel, for heating, cooking, and other uses since the advent of fire. However, these solid biofuels are now derived from a variety of sources, such as agricultural waste, animal waste, and forest products (Bhattacharya et al. 2024).

15.8.8 Biochar

The carbonaceous compounds produced by the partial oxidation of wood and plants, excluding fossil fuels, are referred to as “biochar.” The carbon content of biochar made from biomass through thermal incineration in an oxygen-limited atmosphere is extremely high (Ngene et al. 2024). Forest residues can be used sustainably by producing biochar from them using techniques like hydrothermal carbonization and pyrolysis. It is a useful tool for tackling environmental issues because of its many uses, which include water treatment, carbon sequestration, and soil improvement. Depending on the kind of forest biomass utilized, biochar’s physicochemical characteristics change, impacting how well it works in various applications (Amalina et al. 2022).

15.8.9 Pellets and Briquettes

Briquettes of biochar are a sustainable and affordable alternative to fossil fuels in the industrial and energy sectors. Almost a renewable coal-like briquette is created by thermochemically converting biomass to biochar and then densifying the resulting material. According to recent studies, switching from coal to charcoal briquettes could result in a 40% reduction in CH_4 and CO_2 emissions. In order to transform low-density biochar (0.2–0.4 g/cm^3) into an energy product that resembles coal, densification is necessary. Careful consideration of binders, compaction techniques, and operation conditions is necessary to produce biochar briquettes of superior quality. An ideal binder loading is between 5 and 15%, depending on the pyrolysis temperature and feedstock. Briquettes of biochar made in these circumstances typically have durability ratings between around 70% and 90% (Huang 2018).

The leaves of forest trees are used to make forest leaves pellets, a type of biomass pellet. They are a sustainable and renewable energy source that may be applied to a number of tasks, such as industrial processes, power generation, and heating. Pellets made from forest leaves are usually made by a procedure known as pelleting (Huang 2018). The initial step in this process is gathering and drying the forest leaves. They are then ground into a powder after being crushed. Then, the powder is compacted into tiny, cylindrical pellets using heat. After cooling, the pellets are packaged for delivery. Compared to other forms of biomass pellets, forest leaf pellets provide several benefits. They generate a lot of heat per unit of volume because they have a high energy content.

15.9 Biochemicals Production

By 2030, the EU’s Bio-based Consortium wants to replace 30% of all chemicals produced with biochemicals obtained from biomass (Hassan et al. 2019). Organic acids such as citric, acetic, benzoic, lactic, and succinic acids, microbial enzymes such as amylase, cellulase, pectinase, xylanase, and mannanase are building blocks for bio-based polymers such as phenylpropanoids and polyhydroxyalkanoates and are among the biochemicals that are generated when lignocellulose is biorefined.

15.9.1 Platform Chemicals

Achieving high selectivity and yields when converting LCB into chemicals is still a challenge. Integrated biorefineries, which produce both bio-based products and energy (e.g., fuel and heat), are considered a superior approach for sustainable biomass valorization. To ensure economic viability and environmental benefits, expanding sustainable biorefineries requires market-competitive bioproducts. Biorefineries employ renewable biomass to produce platform chemicals such as ethanol, lactic acid, and succinic acid, as well as value-added products and polymers, many of which have been commercialized (Takkellapati et al. 2018).

15.9.2 Organic Acids

In the contemporary biotechnology sector, there is a need for organic acids, particularly those made by microbes. High yields of organic acids are covered, along with the latest developments in the worldwide organic acid market and sophisticated microbial processing technology (Panda et al. 2019). A renewable method for creating important organic acids, such as formic acid, glycolic acid, and sugar acids under oxidative conditions or lactic acid and levulinic acid under anaerobic conditions, is the catalytic conversion of cellulosic biomass, including forest waste. Complex reactions are involved in this process, and the development of multifunctional materials to improve manufacturing, the design of C–C coupling techniques, and the optimization of catalytic reactions are important breakthroughs (Yan et al. 2023). These initiatives offer a precise method for effectively and sustainably turning forest trash into high-value organic acids.

15.9.3 Alcohols

Alcohols made from biomass, such as crops, lignocellulosic wastes, and food waste, are known as bioalcohols. While second-generation bioalcohols use LCB, such as forestry and agricultural leftovers, first-generation bioalcohols are made from crops. The most promising and economical technique to generate second-generation biofuels is called consolidated bioprocessing (CBP), which combines microbial fermentation, enzymatic hydrolysis, and hydrolase synthesis in a single bioreactor (Jiasheng et al. 2023). Recent research emphasizes the use of both microbial monoculture and co-culture systems to produce ethanol and butanol from lignocellulose.

15.9.4 Intermediates and Specialty Chemicals

Converting building block biochemicals to secondary biochemicals can enhance the value of biorefinery operations. The novel bioproduct 5-(chloromethyl)furfural (CMF), for instance, is appropriate for conversion to intermediates. Biopolymers, biopharmaceuticals, and specialized biochemicals can be produced from CMF. Additionally, it enhances the value of cellulosic biorefineries, even though a sizable portion of biomass may still be transformed into high-volume, low-value biofuels. High-value bioproducts like phenols, quinones, dicarboxylic acids, carbon fiber, and bioplastics can be produced from lignocellulosic waste, such as leftover lignin from the manufacturing of cellulosic bioethanol. Furfural, 5-hydroxymethylfurfural, formic acid, and levulinic acid are among the biochemicals that can be created from lignin, an aromatic and low-value biomass fraction. Under mild circumstances, levulinic acid derived from cellulose is produced through catalytic hydrolysis. It is also possible to valorize lignocellulosic residues into sugars, which act as precursors for carboxylic acids, diols, and polyols. This helps produce biofuels and biochemicals like γ-valerolactone (GVL) in a sustainable manner (Budzianowski 2016). Although woody biomass, such as bark and branches, includes important extractives including oils, waxes, and tannins, their commercial usage is still restricted since extraction techniques are inefficient.

15.9.5 Sugar-Based Pharmaceuticals Compounds

Large-scale fermentation techniques can be used to create the complex natural compounds from readily available sugar-based feedstocks like forest waste (Chang and Keasling 2006). The production of these medications places a strong emphasis on working with farmers to meet regional needs for energy and bio-based components. This strategy seeks to improve the strain on biomass production systems by using biotechnological solutions that link underdeveloped areas with underutilized manufacturers (Kumar et al. 2022). Forest industry by-products like bark and peat contain bioactive compounds with pharmaceutical potential, including Rhaponticin from Norway spruce bark and peat distillate for nasal spray applications. The FORESTSPECS project showcased efficient extraction methods and highlighted their potential for medicines and other health-related uses (CORDIS 2024).

15.9.6 Flavors and Fragrances

Through the use of microbial engineering, which allows bacteria to manufacture chemicals like esters, terpenoids, aldehydes, and methyl ketones, fragrances that were previously collected from natural sources are now generated sustainably. By acting as an energy source for microbial fermentation, LCB lessens the need for conventional extraction techniques. This strategy provides a sustainable way to satisfy the rising demand for fragrances in soaps, cosmetics, and other goods worldwide (Carroll et al. 2016). Fragrances like essential oils, long oil, and turpentine are made from valuable terpenes found in forest waste, such as bark and wastes from forestry activities. Although storage drastically lowers the production and purity of these chemicals, which are produced from terpenes found naturally in wood and plant, they can be recovered through distillation. Since extended storage results in significant losses in volatile chemicals, extraction must be done as soon as possible after harvest to optimize production efficiency and maintain the aromatic qualities (Khan et al. 2014).

15.9.7 Bioplastics and Polymers

Bioplastics, such polyhydroxyalkanoates (PHAs), are environmentally benign polymers made from renewable resources that have qualities like low permeability and water resistance that are comparable to those of traditional plastics. PHAs are biocompatible, recyclable, and adaptable; their mechanical and thermal characteristics are determined by the molecular weight distribution, microstructure, and monomer makeup. Due to these qualities, PHAs hold promise for a range of industrial applications, where their microstructure can be altered to customize their features (Carvalheira et al. 2022). Biopolymers and bioplastics made from forest waste, like wood wastes and scraps, are becoming more popular as environmentally friendly substitutes for plastics made from fossil fuels. By adding wood waste as reinforcement, these materials can be utilized to create biocomposites with improved mechanical qualities, such as an increased Young's modulus (Ferrari et al. 2022). By turning waste into valuable bioplastics and biopolymers, valuing forest waste helps solve environmental issues while also advancing the circular economy. Figure 15.3 presents an overview of value-added products obtained from forest waste.

15.10 Advanced Technologies in Biofuels and Biochemicals Production

The generation of biofuels from biomass waste, agricultural residues, and microbiological materials is becoming more and more popular worldwide since it is environmentally benign and can help with problems like carbon emissions and garbage disposal. By accumulating biomass through photosynthesis, plants and algae are valuable renewable resources for the creation of biofuel (Ambaye et al. 2021). The previous section of this chapter reviewed conventional technologies for converting forest waste into valuable products such as biofuels, bio-oil, biochar, biochemicals, and flavors. In this section, we explore advanced conversion technologies currently being researched to maximize product yield from forest waste while minimizing costs.

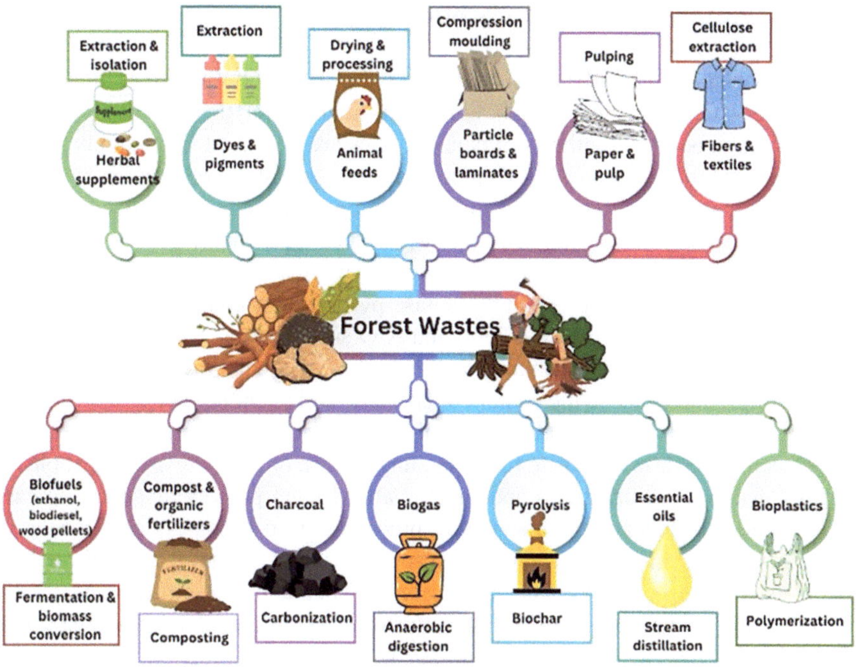

Fig. 15.3 Value-added products from forest waste

15.10.1 Genetic Engineering and Synthetic Biology

Optimizing the microorganisms used in these processes is the main obstacle to producing biofuels to meet the world's energy demands. By altering microbes, enzymes, and metabolic pathways, genetic engineering has become a potent technique for increasing the production of biofuel. This improves efficiency and makes it easier to convert biomass into biogas, biodiesel, and bioethanol. This development has significantly increased interest in advanced biofuels, spurring research aimed at enhancing their production and scalability (Ríos-Fránquez et al. 2021). Disrupting the pathways for butyrate, acetone, lactate, and acetate through gene deletion or knockdown and overexpression of solvent-production-related genes like SpoA and groESL is the main strategy for increasing butanol production. For improving screening efficiency in solvent fermentation, cutting-edge instruments such as TargeTron technology are being preferred over integrational plasmid technologies. It has been demonstrated that butanol production ratios can rise from 70% to 80–85% when the adc gene is disrupted, demonstrating how well it optimizes yields. To further increase the efficiency of butanol production, antisense RNA technology offers a flexible method for inhibiting genes such as ctfB without changing gene regulation (Rao et al. 2016).

15.10.2 Catalysis and Nanotechnology

Nano-biocatalysis has advanced because of the instability and short lifespan of enzymes, which are fundamental limitations in biocatalytic processes. Enzyme encapsulation and the synthesis of nanoparticles are improved by methods such as nano-entrapment and water-in-oil microemulsion schemes (Velvizhi et al. 2020). Using nanoparticles reconstituting apoenzymes on cofactor-functionalized nanomaterials for increased efficiency are examples of recent advances.

A promising method for transforming forestry wood waste, such as *Vernicia fordii*, into useful energy products is fast pyrolysis in conjunction with nano-catalysts. The nano-catalyst like FeO_3 and NiO significantly increase the yields of light hydrocarbons, acids, aromatic, and phenolics, with NiO promoting the breakdown of lignin to enhance the content of aromatic and phenolics (Li et al. 2019). By effectively recovering fuel and chemical feedstocks from forestry residues, this novel technique shows promise for expanding the use of bioenergy. Using sophisticated plasmonic nanoparticles and functionalized nano-photocatalysts to increase reactivity and efficiency, photocatalytic processes have become viable means of turning biomass into useful chemicals and fuels. Recent developments include methods for hydrogenating CO_2 into fuels like methanol and ethanol and photocatalytically oxidizing 5-hydroxymethylfurfural into high-value compounds like 2,5-furandicarboxylic acid (Katre et al. 2024).

15.10.3 Process Intensification and Optimization

Strategies for turning forest waste into value-added products that increase productivity, lower expenses, and limit waste include process intensification (PI) and optimization. The PI entails assessing methods to enhance a process by altering components or equipment. Technologies such as transesterification, which uses catalysts like acids, bases, or enzymes to transform oils into alkyl esters, are used in the manufacturing of biodiesel. High purity and the absence of soap production are two benefits of enzymatic catalysis; nevertheless, poor reaction rates and high costs are drawbacks that can be lessened by enzyme immobilization or heterogeneous catalysts made from biomass. Compared to chemical-based catalysts, biomass-derived catalysts made from waste materials like shells and plant biomass are more affordable and environmentally friendly. The PI technologies, like reactive distillation and microreactors, are being developed by researchers to increase the yield of biodiesel, shorten reaction times, and save production costs. The fatty acid alkyl ester profile of biodiesel, which is dependent on the oil composition used in manufacturing, affects its cetane number, viscosity, and heating value (Bhatia et al. 2021). Designing sustainable biomass-to-liquid processes for the production of biofuels and biochemicals is made possible by techniques like multiobjective optimization employing hybrid approaches and process simulators like Aspen Plus. These approaches concentrate on combining green chemistry concepts with economic, environmental, and safety criteria to develop creative and effective procedures that support the objectives of the circular economy (Ibarra-Gonzalez et al. 2021).

15.11 Environmental and Economic Impacts

Effectively converting biomass waste into innovative products can enhance economic efficiency, address environmental concern, and reduce reliance on petroleum resources. Malaysian forest waste, including as peat and wood leftovers, can be used responsibly for horticulture and biomass energy, lowering greenhouse gas emissions, and promoting the generation of renewable energy. Reduced environmental effects, long-term carbon storage, and carbon sequestration are all benefits of sustainably managed timber and wood-based goods. In the forestry sector, recycling wood products prolongs their lives and fosters environmental and economic sustainability (Mokhtar et al. 2022). . Conventional disposal methods for forestry and agricultural biowaste, such as landfilling and incineration, have adverse effects on the environment and the economy. These practices contribute to pollution and lead to the loss of valuable resources that could otherwise be utilized for sustainable applications like bioenergy or composting.It draws attention to how biowaste can be used to create composite materials, fertilizers, and renewable energy, which can lessen the need for fossil fuels and ease the strain on forests. Improvements in pretreatment, fermentation, and biorefinery technologies can maximize waste utilization and reduce secondary pollution, even in the face of obstacles like low water resistance in composites and the environmental issues of adhesives (Guo et al. 2024). To fully realize the economic and environmental benefits of forestry and agricultural biowaste as well as promote a circular bioeconomy, effective waste classification, technical innovation, and supportive legislation are necessary.

15.12 Challenges and Future Directions

There are several challenges and opportunities for future improvement in using forestry residue to produce chemicals and biofuels. A significant obstacle pertains to the gathering and conveyance of biomass from distant and scattered sites to processing facilities, posing logistical complexity and high costs. Efficient infrastructure is essential for the effective collection, storage, and transportation, necessitating substantial investments. Moreover, the diverse nature of forestry residue, encompassing bark, leaves, and wood, adds complexity to the preliminary treatments required for the conversion of this material into biofuels and chemicals. These pretreatment procedures can be energy-intensive and entail the utilization of hazardous substances, further complicating the process. Attaining elevated conversion efficiencies in methods like pyrolysis, gasification, and fermentation remains arduous, often demanding precise optimization of parameters and technologies. The management of the by-products generated from these conversion procedures is also vital to guarantee overall sustainability and economic feasibility. Economic hurdles encompass the considerable initial setup and continual operational expenses of biomass conversion facilities, alongside the necessity to compete with more affordable fossil fuels and traditional chemicals benefiting from established supply chains and economies of scale. Additionally, the environmental advantages of converting forestry residue must be counterbalanced against potential adverse effects, such as emissions from transportation and processing, as well as the utilization of nonrenewable resources in pretreatment. Prospective advancements in this domain encompass several critical domains. Technological progress is imperative, particularly in innovating more effective, cost-efficient, and environmentally sustainable pretreatment approaches, alongside refining conversion technologies to enhance yields and decrease production expenses. Integrated systems merging diverse conversion processes can boost efficiency and profitability by fully utilizing all biomass components. The establishment of biorefineries processing various biomass types to generate multiple outputs will bolster economic feasibility and resource utilization. Economic and policy backing are equally critical. The development and use of biomass conversion technology can be accelerated by government assistance in the form of grants, subsidies, and incentives. Establishing transparent regulatory frameworks governing the production, distribution, and utilization of biofuels also biochemicals can establish a steady market setting. Sustainability evaluations and environmental impact assessments, like life cycle analysis, are indispensable for ensuring the sustainable evolution of these technologies. Exploring the potential for carbon sequestration via biomass utilization can contribute to endeavors for mitigating climate change.

Collaboration and knowledge exchange will be pivotal in propelling this field forward. Partnerships between the public and private sectors can accelerate innovation and the implementation of novel technologies, while disseminating research discoveries, best practices, and technological progress through publications, conferences, and workshops can foster global advancement. Addressing these hurdles and pursuing these forthcoming paths will be indispensable for the effective and sustainable conversion of forestry residue into valuable chemicals and biofuels, ultimately aiding in the establishment of a circular bioeconomy and the global shift toward renewable energy sources.

15.13 Conclusion

This chapter emphasizes the vast potential of forest waste as a sustainable resource for energy, such as biofuels, biochemicals, and other high-value products, offering both economic and environmental advantages. Technologies such as anaerobic digestion, pyrolysis, gasification, and biorefining have enabled the conversion of forest waste into valuable outputs like organic acids, charcoal, bioalcohols, renewable energy sources, and specialty chemicals. In addition to providing financially feasible substitutes for fossil fuels and conventional industrial techniques, these inventions tackle environmental issues like greenhouse gas emissions, land use, and resource depletion. The use of forest trash can greatly support a circular bioeconomy and global sustainability goals by incorporating sustainable practices, streamlining production processes and promoting industry cooperation. Future research and development in this field should focus on optimizing conversion processes, improving the efficiency of biofuel production, and exploring new applications for forest residues.

References

Akansu SO, Tangöz S, Kahraman N, İlhak Mİ, Açıkgöz S (2017) Experimental study of gasoline-ethanol-hydrogen blends combustion in an SI engine. Int J Hydrog Energy 42(40):25781–25790. https://doi.org/10.1016/j.ijhydene.2017.07.014

Alvira P, Tomás-Pejó E, Ballesteros M, Negro MJ (2010) Pretreatment technologies for an efficient bioethanol production process based on enzymatic hydrolysis: a review. Bioresour Technol 101(13):4851–4861. https://doi.org/10.1016/j.biortech.2009.11.093

Amalina F, Abd Razak AS, Krishnan S, Sulaiman H, Zularisam AW, Nasrullah M (2022) Biochar production techniques utilizing biomass waste-derived materials and environmental applications—a review. J Hazard Mater Adv 7:100134. https://doi.org/10.1016/j.hazadv.2022.100134

Ambaye TG, Vaccari M, Bonilla-Petriciolet A, Prasad S, van Hullebusch ED, Rtimi S (2021) Emerging technologies for biofuel production: a critical review on recent progress, challenges and perspectives. J Environ Manag 290:112627. https://doi.org/10.1016/j.jenvman.2021.112627

An JY, Park BB (2021) Effects of wood ash and N fertilization on soil chemical properties and growth of Zelkova serrata across soil types. Sci Rep 11(1):14489. https://doi.org/10.1038/s41598-021-93805-5

Aniszewska M, Gendek A, Drożdżek M, Bożym M, Wojdalski J (2017) Physicochemical properties of seed extraction residues and their potential uses in energy production. Rocz Ochr Srodowiska 19:302–334

Awasthi MK, Pandey AK, Bundela PS, Khan J (2015) Co-composting of organic fraction of municipal solid waste mixed with different bulking waste: characterization of physicochemical parameters and microbial enzymatic dynamic. Bioresour Technol 182:200–207. https://doi.org/10.1016/j.biortech.2015.01.104

Awogbemi O, Von Kallon DV (2023) Achieving affordable and clean energy through conversion of waste plastic to liquid fuel. J Energy Inst 106:101154. https://doi.org/10.1016/j.joei.2022.101154

Basu P (2010) Biomass gasification and pyrolysis: practical design and theory. Academic, New York. https://archive.org/details/biomassgasificat0000basu

Bendaoud A, Belkhiri A, Maai M, Moubchir T, Hmamou A, Tlemcani S, Eloutassi N, Lahkimi A (2023) Simple and combined pretreatment of a mixture of forestry and aromatic-medicinal plant waste by chemical, physical, and enzymatic methods. J Ecol Eng 24(4):376–383. https://doi.org/10.12911/22998993/160094

Bernal M, Alburquerque J, Moral R (2009) Composting of animal manures and chemical criteria for compost maturity assessment. A review. Bioresour Technol 100(22):5444–5453. https://doi.org/10.1016/j.biortech.2008.11.027

Bhatia SK, Bhatia RK, Jeon J-M, Pugazhendhi A, Awasthi MK, Kumar D, Kumar G, Yoon J-J, Yang Y-H (2021) An overview on advancements in biobased transesterification methods for biodiesel production: oil resources, extraction, biocatalysts, and process intensification technologies. Fuel 285:119117. https://doi.org/10.1016/j.fuel.2020.119117

Bhattacharya T, Khan A, Ghosh T, Kim JT, Rhim J-W (2024) Advances and prospects for biochar utilization in food processing and packaging applications. Sustainable Mater Technol 39:e00831. https://doi.org/10.1016/j.susmat.2024.e00831

Bolat İ, Kara Ö, Tunay M (2015) Effects of seasonal changes on microbial biomass and respiration of forest floor and topsoil under Bornmullerian fir stand. Eurasian J For Sci 3(1):1–13. https://doi.org/10.31195/ejejfs.70190

Bozym M, Gendek A, Siemiatkowski G, Aniszewska M, Malaták J (2021) Assessment of the composition of forest waste in terms of its further use. Materials 14:973. https://doi.org/10.3390/ma14040973
Braghiroli FL, Passarini L (2020) Valorization of biomass residues from forest operations and wood manufacturing presents a wide range of sustainable and innovative possibilities. Curr For Rep 6(2):172–183. https://doi.org/10.1007/s40725-020-00112-9
Budzianowski WM (2016) High-value low-volume bioproducts coupled to bioenergies with potential to enhance business development of sustainable biorefineries. Renew Sust Energ Rev 70:793–804. https://doi.org/10.1016/j.rser.2016.11.260
Capodaglio AG, Bolognesi S (2019) Ecofuel feedstocks and their prospects. In: Elsevier eBooks. Elsevier, Amsterdam, pp 15–51. https://doi.org/10.1016/b978-0-08-102728-8.00002-4
Carroll AL, Desai SH, Atsumi S (2016) Microbial production of scent and flavor compounds. Curr Opin Biotechnol 37:8–15. https://doi.org/10.1016/j.copbio.2015.09.003
Carvalheira M, Marreiros BC, Reis M (2022) Acids (VFAs) and bioplastic (PHA) recovery. In: Elsevier eBooks. Elsevier, Amsterdam, pp 245–254. https://doi.org/10.1016/b978-0-323-90178-9.00016-0
Chang M, Keasling J (2006) Production of isoprenoid pharmaceuticals by engineered microbes. Nat Chem Biol 2:674–681. https://doi.org/10.1038/nchembio836
Chinnadurai NK, Muthukumarappan NK (2009) Combined effect of alkali soaking and extrusion conditions on fermentable sugar yields from switchgrass and prairie cord grass. 2009 Reno, Nevada, June 21 - June 24, 2009. https://doi.org/10.13031/2013.27201
CORDIS (2024) Wood bark and peat-based bioactive compounds, specialty chemicals, and remediation materials: from innovations to applications. New products from forest waste. https://cordis.europa.eu/article/id/91165-new-products-from-forest-waste
FAO (2020) Global forest resources assessment 2020: main report. FAO, Rome. https://doi.org/10.4060/ca9825en
Ferrari F, Striani R, Fico D, Alam MM, Greco A, Esposito Corcione C (2022) An overview on wood waste valorization as biopolymers and biocomposites: definition, classification, production, properties, and applications. Polymers 14(24):5519. https://doi.org/10.3390/polym14245519
Forest Survey of India (2019) India state of forest report 2019 (16th biennial assessment). Ministry of Environment, Forest and Climate Change. https://fsi.nic.in
Gao Z, Alshehri K, Li Y, Qian H, Sapsford D, Cleall P, Harbottle M (2022) Advances in biological techniques for sustainable lignocellulosic waste utilization in biogas production. Renew Sust Energ Rev 170:112995. https://doi.org/10.1016/j.rser.2022.112995
Grebner DL, Bettinger P, Siry JP, Boston K (2022) Forest products. In: Grebner DL, Bettinger P, Siry JP, Boston K (eds) Introduction to forestry and natural resources, 2nd edn. Academic, New York, pp 101–129. https://doi.org/10.1016/B978-0-12-819002-9.00004-3
Guo J, Zhang Y, Fang J, Ma Z, Li C, Yan M, Qiao N, Liu Y, Bian M (2024) Reduction and reuse of forestry and agricultural bio-waste through innovative green utilization approaches: a review. Forests 15(8):1372. https://doi.org/10.3390/f15081372
Hassan SS, Williams GA, Jaiswal AK (2019) Moving towards the second generation of lignocellulosic biorefineries in the EU: drivers, challenges, and opportunities. Renew Sust Energ Rev 101:590–599. https://doi.org/10.1016/j.rser.2018.11.041
Ho S, Huang S, Chen C, Hasunuma T, Kondo A, Chang J (2012) Bioethanol production using carbohydrate-rich microalgae biomass as feedstock. Bioresour Technol 135:191–198. https://doi.org/10.1016/j.biortech.2012.10.015
Hu Z, Wen Z (2007) Enhancing enzymatic digestibility of switchgrass by microwave-assisted alkali pretreatment. Biochem Eng J 38(3):369–378. https://doi.org/10.1016/j.bej.2007.08.001
Huang J (2018) Turn waste into treasure: make pellets with forestry residues. Geemco Energy from Biomass to Bioenergy https://www.biofuelmachines.com/application/make-biofuel-pellets-with-forestry-residues.html
Ibarra-Gonzalez P, Rong B-G, Segovia-Hernández JG, Sánchez-Ramírez E (2021) Multi-objective optimization methodology for process synthesis and intensification: gasification-based bio-

mass conversion into transportation fuels. Chem Eng Process Process Intensif 162:108327. https://doi.org/10.1016/j.cep.2021.108327

Jiasheng L, Zhang W, Lin CSK, Jiang Y, Xin F (2023) Biochemical production of bioalcohols. In: Luque R, Lin CSK, Wilson K, Chenyu D (eds) Handbook of biofuels production, 3rd edn. Woodhead Publishing, New Delhi, pp 231–252. https://doi.org/10.1016/B978-0-323-91193-1.00005-6

John N, Fathima PS, Harsha VS, Paul NM, Nisha P (2023) Physical conversion of biomass: dewatering, drying, size reduction, densification, and separation. In: Thomas S, Hosur M, Pasquini D, Jose Chirayil C (eds) Handbook of biomass. Springer, Singapore. https://doi.org/10.1007/978-981-19-6772-6_37-1

Katre S, Baghmare P, Giri AS (2024) Photocatalytic nanomaterials and their implications towards biomass conversion for renewable chemical and fuel production. Nanoscale Adv 6:5258–5284. https://doi.org/10.1039/D4NA00447G

Khalili S, Khoshandam B, Jahanshahi M (2016) A comparative study of CO_2 and CH_4 adsorption using activated carbon prepared from pine cone by phosphoric acid activation. Korean J Chem Eng 33(10):2943–2952. https://doi.org/10.1007/s11814-016-0138-y

Khan S, Krigstin S, Volpé S, Wetzel S (2014) Essential oil composition of forest biomass stored under industrial conditions in eastern Canada. Ind Crop Prod 56:35–42. https://doi.org/10.1016/j.indcrop.2014.02.026

Kim TH, Lee YY (2007) Pretreatment of corn Stover by soaking in aqueous ammonia at moderate temperatures. Appl Biochem Biotechnol 137(1–12):81–92. https://doi.org/10.1007/s12010-007-9041-7

Krishna MP, Mohan M (2017) Litter decomposition in forest ecosystems: a review. Energy Ecol Environ 2:236–249. https://doi.org/10.1007/s40974-017-0064-9

Kumar R, Singh S, Singh OV (2008) Bioconversion of lignocellulosic biomass: biochemical and molecular perspectives. J Ind Microbiol Biotechnol 35(5):377–391. https://doi.org/10.1007/s10295-008-0327-8

Kumar R, Sinha S, Ahirwar S, Rawat L (2021) Impact of seasonal variations on physico-chemical and exchangeable cations of forest soil in Doon Valley, Uttarakhand, India. Geol Ecol Landsc 7(3):258–268. https://doi.org/10.1080/24749508.2021.1952768

Kumar V, Bansal V, Madhavan A, Kumar M, Sindhu R, Awasthi MK, Saran S (2022) Active pharmaceutical ingredient (API) chemicals: a critical review of current biotechnological approaches. Bioengineered 13(2):4309–4327. https://doi.org/10.1080/21655979.2022.2031412

Li C, Yue X, Yang J, Yang Y, Gu H, Peng W (2019) Catalytic fast pyrolysis of forestry wood waste for bio-energy recovery using nano-catalysts. Energies 12:3972. https://doi.org/10.3390/en12203972

Limayem A, Ricke SC (2012) Lignocellulosic biomass for bioethanol production: current perspectives, potential issues and future prospects. Prog Energy Combust Sci 38(4):449–467. https://doi.org/10.1016/j.pecs.2012.03.002

Liu X, Yue Z-B (2011) Industrial biotechnology and commodity products. In: Comprehensive biotechnology, 2nd edn. Elsevier, Amsterdam

Mallikarjuna C, Dash RR, Bhunia P (2021) Sustainability of gaseous biofuels: potential uses technological constraints and environmental concerns. In: Pandey A, Tyagi RD, Varjani S (eds) Biomass, biofuels, biochemicals. Elsevier, Amsterdam, pp 247–268. https://doi.org/10.1016/B978-0-12-821878-5.00025-8

McKendry P (2002) Energy production from biomass (part 1): overview of biomass. Bioresour Technol 83(1):37–46. https://doi.org/10.1016/s0960-8524(01)00118-3

Mishra S, Singh PK, Dash S, Pattnaik R (2018) Microbial pretreatment of lignocellulosic biomass for enhanced biomethanation and waste management. 3 Biotech 8(11):458. https://doi.org/10.1007/s13205-018-1480-z

Mokhtar N, Razali SM, Sulaiman MS, Edin T, Wahab R (2022) Converting wood-related waste materials into other value-added products: a short review. IOP Conf Ser Earth Environ Sci 1053(1):012030. https://doi.org/10.1088/1755-1315/1053/1/012030

Mosier N, Wyman C, Dale B, Elander R, Lee Y, Holtzapple M, Ladisch M (2004) Features of promising technologies for pretreatment of lignocellulosic biomass. Bioresour Technol 96(6):673–686. https://doi.org/10.1016/j.biortech.2004.06.025

Ngene GI, Bouesso B, Martínez MG, Nzihou A (2024) A review on biochar briquetting: common practices and recommendations to enhance mechanical properties and environmental performances. J Clean Prod 469:143193. https://doi.org/10.1016/j.jclepro.2024.143193

Nurek T, Gendek A, Roman K (2019) Forest residues as a renewable source of energy: elemental composition and physical properties. Bioresources 14(1):6–20

Panda SK, Sahu L, Behera SK, Ray RC (2019) Research and production of organic acids and industrial potential. In: Bioprocessing for biomolecules production. Wiley, Hoboken, pp 195–209. https://doi.org/10.1002/9781119434436.ch9

Pandey S (2022) Wood waste utilization and associated product development from under-utilized low-quality wood and its prospects in Nepal. S N Appl Sci 4:168. https://doi.org/10.1007/s42452-022-05061-5

Prescott CE, Zabek LM, Staley CL, Kabzems R (2000) Decomposition of broadleaf and needle litter in forests of British Columbia: influences of litter type, forest type, and litter mixtures. Can J For Res 11:1742–1750. https://doi.org/10.1139/x00-097

Qin L, Zhu S, Qin W, Zhao B, Chen W, Asamoah EN, Han J (2024) H2-rich syngas generated from forestry waste gasification by recycling hot blast furnace gas. J Clean Prod 448:141671. https://doi.org/10.1016/j.jclepro.2024.141671

Rajkumar R, Kurinjimalar C (2022) Handbook of biofuels. Elsevier, Amsterdam. https://api.pageplace.de/preview/DT0400.9780128231333_A42706799/preview-9780128231333_A42706799.pdf

Rao A, Sathiavelu A, Mythili S (2016) Engineering, technology and techniques. Braz Arch Biol Technol 59:150612. https://doi.org/10.1590/1678-4324-2016150612

Ríos-Fránquez FJ, Alba-Fierro CA, Escamilla-Alvarado C (2021) Advances, challenges, and opportunities in genetic engineering for biofuels production. In: Ray RC (ed) Applied biotechnology reviews, sustainable biofuels. Academic, New York, pp 399–424. https://doi.org/10.1016/B978-0-12-820297-5.00014-1

Salvachúa D, Prieto A, López-Abelairas M, Lu-Chau T, Martínez ÁT, Martínez MJ (2011) Fungal pretreatment: an alternative in second-generation ethanol from wheat straw. Bioresour Technol 102(16):7500–7506. https://doi.org/10.1016/j.biortech.2011.05.027

Schenk ER, O'Donnell F, Springer AE, Stevens LE (2020) The impacts of tree stand thinning on groundwater recharge in aridland forests. Ecol Eng 145:105701. https://doi.org/10.1016/j.ecoleng.2019.105701

Silveira EA, Chaves BS, Macedo L, Ghesti GF, Evaristo RBW, Lamas GC, Luz SM, Protásio T, Rousset P (2023) A hybrid optimization approach towards energy recovery from torrefied waste blends. Renew Energy 212:151–165. https://doi.org/10.1016/j.renene.2023.05.053

Solanki R, Arora S (2015) Leaf litter dynamics in agroforestry system affecting microbial activity in saline soils. J Soil Water Conserv 14(4):332–339

Song C, Zhang C, Zhang S, Lin H, Kim Y, Ramakrishnan M, Du Y, Zhang Y, Zheng H, Barceló D (2020) Thermochemical liquefaction of agricultural and forestry wastes into biofuels and chemicals from circular economy perspectives. Sci Total Environ 749:141972. https://doi.org/10.1016/j.scitotenv.2020.141972

Sun Y, Cheng J (2002) Hydrolysis of lignocellulosic materials for ethanol production: a review. Bioresour Technol 83(1):1–11. https://doi.org/10.1016/s0960-8524(01)00212-7

Takkellapati S, Li T, Gonzalez MA (2018) An overview of biorefinery-derived platform chemicals from a cellulose and hemicellulose biorefinery. Clean Techn Environ Policy 20(7):1615–1630. https://doi.org/10.1007/s10098-018-1568-5

Tinwala F, Mohanty P, Parmar S, Patel A, Pant KK (2015) Intermediate pyrolysis of agro-industrial biomasses in bench-scale pyrolyser: product yields and its characterization. Bioresour Technol 188:258–264. https://doi.org/10.1016/j.biortech.2015.02.006

Van Der Stelt M, Gerhauser H, Kiel J, Ptasinski K (2011) Biomass upgrading by torrefaction for the production of biofuels: a review. Biomass Bioenergy. https://doi.org/10.1016/j.biombioe.2011.06.023

Van Loo S, Koppejan J (2012) The handbook of biomass combustion and co-firing. In: Routledge eBooks. Routledge, London. https://doi.org/10.4324/9781849773041

Velvizhi G, Ranjitha J, Vijayalakshmi S, Nikhil GN (2020) Advancements in nanobiocatalysis for bioenergy and biofuel production. In: Nanobiotechnology for sustainable bioenergy and biofuel production, 1st edn. CRC Press, Boca Raton, p 26. https://doi.org/10.1201/9780429023194

Weiland P (2009) Biogas production: current state and perspectives. Appl Microbiol Biotechnol 85(4):849–860. https://doi.org/10.1007/s00253-009-2246-7

Wu W, Dutta T, Varman AM, Eudes A, Manalansan B, Loqué D, Singh S (2017) Lignin valorization: two hybrid biochemical routes for the conversion of polymeric lignin into value-added chemicals. Sci Rep 7:8420. https://doi.org/10.1038/s41598-017-07895-1

Yan W, Guan Q, Jin F (2023) Catalytic conversion of cellulosic biomass to harvest high-valued organic acids. iScience 26(10):107933. https://doi.org/10.1016/j.isci.2023.107933

Yu Q, Wang Y, Van Le Q, Yang H, Hosseinzadeh-Bandbafha H, Yang Y, Sonne C, Tabatabaei M, Lam SS, Peng W (2021) An overview on the conversion of forest biomass into bioenergy updated. Front Energy Res 9:684234. https://doi.org/10.3389/fenrg.2021.684234

Zheng Y, Pan Z, Zhang R (2009) Overview of biomass pretreatment for cellulosic ethanol production. Int J Agric Biol Eng 2(3):51–68. https://doi.org/10.3965/j.issn.1934-6344.2009.03.051-068

Millet Waste as an Inexpensive Feedstock for Biofuel and Chemicals

16

Ravi K. Deshwal and Sujeet Pratap Singh

Abstract

With the increasing global demand for sustainable and eco-friendly energy sources, researchers and innovators have turned their attention to unconventional solutions. One such promising avenue is the use of millet waste to produce biofuel and chemicals, opening up a realm of possibilities for renewable energy production. Millet, a staple crop in many regions around the world, generates a significant amount of agricultural waste. However, this waste can now be transformed into a valuable resource for biofuel and chemical production. Through innovative biotechnological processes, scientists have unlocked the potential of millet waste, turning it into a renewable and cost-effective source of energy. The beauty of this breakthrough lies in its ability to address multiple challenges simultaneously. Millet is a drought-resistant cereal crop that is grown in many parts of the world, particularly in regions with arid or semi-arid climates. While the grain of millet is the primary edible part and used for food products, the plant produces a significant amount of agricultural residue or waste, including stalks, leaves, and husks. Utilizing millet waste, it can not only reduce agricultural waste and its environmental impact but also create an alternative to fossil fuels. Furthermore, this innovative approach promotes rural development and economic growth by providing new income streams for millet farmers. Millet waste can serve as a valuable and low-cost feedstock to produce biofuels and chemi-

R. K. Deshwal
Institute of Bioscience and Technology Sri Ramswaroop Memorial University, Lucknow, Uttar Pradesh, India

S. P. Singh (✉)
Amity Institute of Biotechnology, Amity University Lucknow Campus, Lucknow, Uttar Pradesh, India
e-mail: spsingh3@amity.edu

A. Kumari et al. (eds.), *Valorization of Solid Wastes to Biofuels and Chemical Products for Sustainable World*, https://doi.org/10.1007/978-981-96-8594-3_16

cals, contributing to both agricultural sustainability and renewable energy goals. In several studies, pearl millet has been used as a feedstock for fuel ethanol production in the presence of *Saccharomyces cerevisiae* acting as the fermenting microorganism. The pearl and other millets could be a promising feedstock for fuel ethanol production, especially in regions where it is challenging to grow corn and grain sorghum due to dry conditions. This highlights the versatility of pearl millets as a biofuel feedstock. Some studies show to recover cellulose from pearl millet biomass. To achieve this, various treatments are employed by different scientists. One study (Dresch et al., 2022) uses both chemical and physical treatments. The chemical treatments include alkali (using 4% NaOH) and acid (using $C_2H_2O_4$ 5%) extractions. The physical treatments involve vapor explosion and size reduction. Cellulose-rich fraction obtained from these treatments achieved an impressive removal rate of 100% for lignin and 99.86% for hemicellulose and high crystallinity index of 82.43%. For the extraction, 2- to 4-h alkaline extractions with 4% NaOH, followed by a 2-h extraction with $C_2H_2O_4$ 5% in the autoclave at a temperature of 125.6 °C and 1.4 bar. Various analytical techniques, including Fourier transform spectroscopy (FTIR-ATR), X-ray diffraction (XRD), and Raman spectroscopy, are also used to characterize the cellulosic fraction.

Keywords
Agriculture · Biofuel · Energy · Millet · Sustainable · Waste

16.1 Introduction

Millets are a broad category of cereals that include teff, sorghum (also known as large millet), fonio, pearl, proso, foxtail, barnyard, small, kodo, browntop, finger, and guinea millets. They are a vital food source for millions of people in Asia and sub-Saharan Africa. They contribute to ensuring food security in locations where they are culturally relevant and have a strong foundation in the culture and customs of indigenous peoples. The Food and Agriculture Organization (FAO) accepted India's proposal in 2018 to commemorate an International Year of Millets in 2023, and the United Nations General Assembly at its 75th session in March 2021 proclaimed 2023 to be the International Year of Millets (Patil et al. 2023).

16.1.1 The Importance of Alternate Feedstocks for Biofuel and Chemical Production

The importance of alternative feedstocks for biofuel and chemical production cannot be overemphasized in the context of modern sustainable development and resource management. Traditional fossil fuels and petrochemicals, which have long formed the backbone of our energy and chemical industries, are finite resources with serious environmental effects. As we confront increasing issues connected to climate change, resource shortages, and environmental deterioration, the need for more sustainable and eco-friendly options has become urgent (Shafique and Shafique 2021).

16.1.1.1 Reducing Dependency on Fossil Fuels

One of the major reasons for studying alternate feedstocks is to lessen our reliance on fossil fuels. The burning of fossil fuels for energy and the manufacturing of chemicals is a major contributor to greenhouse gas emissions, leading to climate change. Shifting toward biofuels and chemicals derived from renewable sources can dramatically cut carbon emissions and mitigate the consequences of global warming. Traditional energy sources are prone to geopolitical tensions and price changes. By diversifying our feedstock sources, we can boost energy security and reduce our vulnerability to supply interruptions. Locally supplied feedstocks, such as agricultural wastes or dedicated energy crops like millet waste, can improve energy self-sufficiency (Lakshm et al. 2023).

16.1.1.2 Utilizing Agricultural Residues to Promote Rural Economies and Better Environment

The usage of agricultural residues, such as millet trash, as feedstocks for biofuel and chemical manufacturing is particularly advantageous. These residues are commonly considered waste materials and are typically disposed of through burning or landfilling, leading to environmental difficulties. By turning them into useful biofuels and chemicals, we not only minimize waste but also add value to agricultural processes. The cultivation of energy crops and biofuel and chemical production facilities in rural regions can drive economic growth, create jobs, and provide extra income streams for farmers. This can help reinvigorate rural communities and lessen urbanization pressures. Alternative feedstocks often have a lesser environmental footprint compared to conventional sources. They can help reduce land and water use, minimize chemical inputs, and protect biodiversity. Furthermore, sustainable feedstock production strategies can contribute to soil health and ecosystem resilience (Koul et al. 2022).

16.1.1.3 Meeting Sustainable Development Goals

The exploitation of alternative feedstocks correlates with various Sustainable Development Goals (SDGs), including those relating to clean energy (SDG 7), responsible consumption and production (SDG 12), and climate action (SDG 13) (Chauhan et al. 2023). It represents a holistic approach to addressing many global concerns. The quest for alternative feedstocks for biofuel and chemical production is motivated by the pressing need to move to more sustainable and ecologically friendly sources of energy and chemicals. By embracing these options, we may lessen our environmental impact, boost energy security, and contribute to a more sustainable and resilient future for our world (Sieminski 2022).

16.1.2 Overview of Millet Waste as a Potential Feedstock

An examination of millet trash as a potential feedstock reveals a promising and underutilized resource in the domain of biofuel and chemical synthesis. Millets, a type of small-grain cereal crop, are predominantly farmed for human consumption, with their grains being a staple in diverse places worldwide. However, what often goes neglected is the significant amount of agricultural leftovers generated during millet production and processing. These leftovers, including millet straw, husks, and bran, show tremendous promise as important feedstocks for numerous industrial applications. Different properties of millet waste are as follows:

Abundance: Millet cultivation generates considerable volumes of waste materials, making them an abundant and consistent feedstock source. This abundance assures a consistent supply, vital for industrial activities.

Composition: Millet waste materials comprise a mix of cellulose, hemicellulose, and lignin, which are valuable components for biofuel and chemical synthesis. Cellulose and hemicellulose can be turned into sugars, whereas lignin can serve as a precursor for different compounds.

Low competition with food: Utilizing millet waste as a feedstock has the advantage of not competing with food production. This alleviates concerns about shifting food crops for nonfood applications, ensuring food security.

Environmental benefits: Millet waste usage contributes to waste reduction, limiting the environmental impact connected with agricultural leftovers' disposal, such as open-field burning.

Local sourcing: Millets are commonly farmed in places with restricted access to conventional feedstock sources. Harnessing millet trash for biofuels and chemicals can encourage localized, sustainable economic growth.

Biodiversity and soil health: Millet farming, when managed properly, can promote biodiversity and soil health. Using millet waste as a feedstock fosters environmentally sustainable farming techniques. In conclusion, millet waste offers a compelling prospect as a feedstock for biofuel and chemical manufacturing (Mohite et al. 2022).

Its availability, noncompetition with food crops, and environmental advantages make it an attractive candidate in the quest for sustainable alternatives to fossil-based resources. As research and technology continue to improve, millet waste's potential is positioned to play a crucial role in the transition toward a more sustainable and ecologically responsible bioeconomy.

16.2 Millete as a Versatile Crop

16.2.1 Millet Varieties and Cultivation Practices

Millet types and growth practices are crucial components of evaluating the potential of millet trash as a feedstock for biofuel and chemical manufacturing. Millets are a type of small-grain cereal crop recognized for their hardiness, adaptation to varied climates, and nutritional value. There are various millet types produced globally, with the most popular ones being pearl millet (*Pennisetum glaucum*), finger millet (*Eleusine coracana*), foxtail millet (*Setariaitalica*), and proso millet (*Panicum miliaceum*) (Gowri 2020). Every millet variety has its specific qualities and is cultivated utilizing different practices. Pearl millet, popularly known as "Bajra," is appropriate to parched and semi-dry conditions where water accessibility is confined. It is grown in India and Africa. Pearl millet needs planting during the stormy season, and it flourishes in hot, dry conditions. Finger millet, also called "Ragi," is generally grown in India and South Africa. It is noted for its surprising wholesome profile, especially its high calcium content. Development includes cultivating during the rainstorm season and is regularly intercropped with vegetables. Foxtail millet, also known as" Kangni" in India, is cultivated in a few parts of Asia and is noted for its fast turn of events. It is lenient toward dry spell conditions and might be developed in shifted soil types. Planting frequently happens during the stormy season. Proso millet also called as common millet and white millet is a warm-season crop that twists in calm settings with more limited developing seasons. It is regularly filled in segments of Europe, Asia, and North America. Techniques of cultivation vary considering local environment and soil attributes. Developing millets frequently requires spreading seeds straightforwardly onto arranged fields. Manageable cultivating systems incorporate yield turn, low culturing, and intercropping with vegetables to support soil fruitfulness (Ravikesavan et al. 2022). Understanding the variety of millet types and their resistance to fluctuated ecological conditions is imperative for utilizing millet waste as a feedstock. By expanding manageable millet production and streamlining rural strategies, we can keep a consistent stockpile of millet squander while advancing food security and soil well-being where millets are cultivated. This, in turn, contributes to the feasibility of millet waste as a useful resource for biofuel and chemical manufacturing (Fig. 16.1).

16.2.2 Nutritional Value and Traditional Applications of Millet

Millet, a category of small-grain cereal crops, is not only valuable for its potential as a feedstock for biofuel and chemical production but also for its outstanding nutritional content and long history of traditional applications. Here, we study the nutritional benefits and traditional uses of millet:

Nutritional value: Millet is famous for its impressive nutritional profile. Millet is a good source of critical nutrients, including carbs, proteins, dietary fiber, vitamins

Fig. 16.1 Different types of millets

(B-complex vitamins like niacin and folate), and minerals (such as magnesium, phosphorus, and iron). Millet contains antioxidants such as polyphenols, which help fight oxidative stress and reduce the risk of chronic diseases. People suffering from gluten sensitivity or celiac disease are advised to use millets in place of wheat. With low glycemic index, millets have a relatively low impact on blood sugar levels. This is a boon for people suffering from diabetes or want to control their sugar level in blood (Ratnavathi and Tonapi 2022).

Traditional uses: Millet has a long history of traditional applications in many cultures. Millet is a staple food in many parts of Asia and Africa. It was used to make many foods, including flatbreads, porridges, and fermented drinks such as "cod." Millet has been a reliable source of food, especially during times of food shortage or drought, as it can grow in adverse conditions. In some regions, millet is essential for rituals and cultural celebrations. It signifies prosperity, fertility, and communal bonding.

Livestock feed: Millet straw and residues left after grain harvest serve as valuable livestock feed, helping to sustain agriculture (Verma et al. 2021).

Medicinal uses: Traditional medicine systems have exploited millet for its possible health benefits, including easing digestion, supporting cardiovascular health, and supplying energy. In recent years, there has been a renaissance of interest in millet due to its nutritional value and tolerance to climate change. Beyond its traditional usage, millet is gaining prominence as a health-conscious food choice and an essential crop in sustainable agriculture. Understanding the nutritional richness and cultural significance of millet emphasizes its multidimensional value in our quest for healthy diets and sustainable agriculture (Paschapur et al. 2021).

16.3 Millete Waste Generation

16.3.1 Types of Millet Waste

Millet trash comprises diverse agricultural leftovers created during millet growth and processing. These waste products, often ignored, have considerable potential for different applications, including biofuel and chemical manufacture. Here are several main categories of millet waste:

Millet straw: Millet straw consists of the ground plant material that remains after the millet grains have been harvested. It includes the leaves, stems, and any remaining seed heads. Millet straw is fibrous in nature and contains an ample amount of cellulose, which makes it a suitable substrate to produce biofuels. Its lignin content is often lower than other agricultural wastes, which allows it to be easily converted into sugars for ethanol production. Millet straw can be processed to extract cellulose for the production of biofuels or as a source of lignocellulosic biomass in various chemical processes.

Millet husks: Millet husks are the protective outer layer of millet grains that is removed during milling process to produce edible millet grain. Millet hulls are rich in lignin and cellulose, which makes them an attractive source of lignocellulosic material for bio-based chemical production. They are often made of cellulose, hemicellulose, and lignin. Possible uses: Millet hulls can be used for bio-energy production using methods such as gasification, or as raw materials to produce bio-based compounds such as furfural or polymers (Gairola et al. 2022).

Millet bran: Millet bran is the outer layer of millet grains that is separated during milling to produce refined millet flour. Millet bran contains a lot of fiber, vitamins, and minerals. It is also a source of bioactive compounds, such as antioxidants and phytochemicals. Millet bran has applications in human and animal nutrition due to its nutritional worth. Additionally, it can be examined for its potential in bioactive chemical extraction for medicinal or functional food applications (Zheng et al. 2021).

Millet stems and roots: The entire millet plant, including stems and roots, might be considered waste after grain harvest. These portions of the millet plant may include lignocellulosic materials and other organic substances that can be exploited for various reasons. Millet stems and roots can be processed for cellulose extraction, utilized as mulch or organic matter in soil improvement, or investigated for other inventive applications (Fig. 16.2). Harnessing these numerous types of millet waste as feedstocks for biofuel and chemical production is a sustainable and environmentally sound method. It not only adds value to millet farming but also contributes to waste reduction and resource efficiency in agricultural and agro-industrial activities (Kumar et al. 2022a).

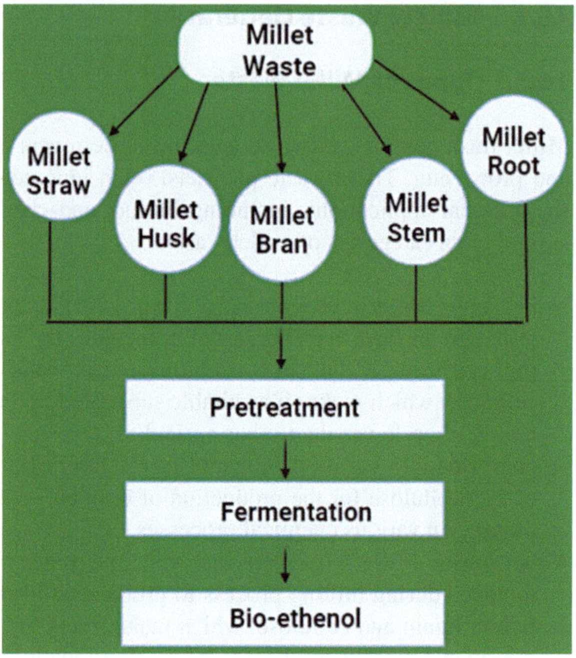

Fig. 16.2 Types of millet wastes

16.3.2 Quantity and Distribution of Millet Waste

The quantity and dispersion of millet waste are crucial criteria when considering its potential as a feedstock for various industrial applications. Millet trash is formed during the millet growing and processing stages, and its availability varies based on numerous factors. Geographic distribution: Millet is grown in numerous places globally, with India, Africa, China, and other nations in Asia being key producers. The distribution of millet trash largely correlates with these cultivation zones.

Crop output: The quantity of millet waste is closely related to crop output. Higher yields result in bigger volumes of waste materials. Factors such as climate, soil quality, and farming practices influence millet yields.

Millet variety: Different millet kinds produce varying quantities of trash. For example, pearl millet may yield more straw compared to finger millet, which has a relatively smaller straw component.

Harvesting and processing techniques: The efficiency of harvesting and processing processes might affect the quantity of waste generated. Traditional processes may result in more waste compared to current mechanical approaches.

Cultural and consumption practices: The demand for millet grains for human consumption affects the distribution of millet trash. In regions where millet is the main diet, there may be less waste available for alternate purposes.

Agricultural policies: Government policies, subsidies, and incentives can influence millet cultivation and, consequently, the quantity of millet trash. Support for millet as an energy or industrial crop can enhance waste availability. Understanding the quantity and distribution of millet waste is vital for assessing its suitability as a feedstock. Regions with large millet farming and waste output have better potential to gain from utilizing millet waste in biofuel and chemical manufacture. Additionally, measures to enhance millet cultivation and processing processes can help increase the quantity of waste available for sustainable and value-added applications, helping both agricultural and environmental sustainability (Devi and Kolli 2022).

16.4 Challenges in Millet Waste Management

16.4.1 Environmental Impact and Disposal Difficulties

The environmental effects and disposal concerns associated with millet waste are significant considerations when examining its potential as a feedstock for biofuel and chemical manufacturing. While millet waste has promise as a valuable resource, it is necessary to address its environmental implications and waste management challenges. In many locations, millet straw and other leftovers are disposed of through open burning, especially after the grain harvest. This practice contributes to air pollution by generating greenhouse gases, particulate matter, and hazardous pollutants, negatively damaging air quality and human health (Bhuvaneshwari et al. 2019). Another typical disposal method for millet waste is landfilling. When millet trash is transferred to landfills, it takes up scarce landfill space and can release methane, a strong greenhouse gas, as it decomposes anaerobically. Leaving millet remains in the field can improve soil health, prevent erosion, and promote soil fertility. Removing these leftovers without suitable management measures can lead to soil deterioration and diminished agricultural sustainability (Bimbraw 2019). Inadequate disposal of millet waste might result in water pollution as well. When a rainstorm washes residues into water bodies, it can contribute to nutrient runoff, potentially causing water pollution and eutrophication. To alleviate these environmental implications and disposal issues, sustainable waste management strategies should be advocated. These include bioenergy production: utilizing millet trash for biofuel and biogas production can lessen the environmental impact of open burning and contribute to renewable energy generation. Biological breakdown: Encouraging the spontaneous breakdown of millet residues in the field can promote soil health and reduce the need for disposal. Resource recovery: Exploring technology to extract useful components like cellulose from millet trash can add value to the waste and lessen its environmental footprint. Awareness and education: Raising awareness among farmers and stakeholders about sustainable waste management techniques is vital for decreasing the negative environmental repercussions of incorrect disposal (Devi and Kolli 2022). By addressing environmental issues and adopting sustainable waste management practices, millet waste can be exploited as a useful feedstock while limiting its ecological footprint, contributing to a more sustainable agricultural and industrial landscape.

16.4.2 Economic Strain on Millet Growers

The economic burden on millet farmers is a major element that needs addressed when contemplating the exploitation of millet trash as a feedstock for biofuel and chemical manufacturing. Millet cultivation is generally associated with particular economic concerns for farmers. Millets are frequently considered low-value crops compared to key staples like rice and wheat. As a result, millet growers may face diminished economic potential from their harvests. Many millet-growing regions have limited access to markets and encounter difficulty in selling their produce. This might lead to low pricing and limited income opportunities for farmers. Millet cultivation is predominantly rainfed, rendering it sensitive to weather swings and unpredictable rainfall patterns. Crop failure due to drought or other bad weather conditions might result in financial troubles for farmers. Millet cultivation sometimes lacks the technological assistance and infrastructure accessible to larger-scale crops. This can result in poorer yields and productivity, further reducing farmers' revenue. Despite being robust and drought-resistant, millet cultivation can nevertheless incur production costs, such as seeds, fertilizers, and labor. These charges can strain the budgets of small-scale millet producers. Millet farmers may not have access to technologies or information that would allow them to add value to their millet harvest, such as processing millet into value-added goods. The use of millet waste as a feedstock for biofuel and chemical manufacturing can potentially reduce some of these economic difficulties for millet growers. By establishing extra revenue streams from millet trash, farmers may diversify their income sources and increase their financial security. However, it is vital that the growth of millet waste-based companies emphasizes the equitable distribution of benefits to farmers and communities, ensuring that they receive fair recompense for their contributions to the supply chain. Additionally, support through agricultural extension services, market access improvements, and technology dissemination can help strengthen the overall economic sustainability of millet farming and its linked sectors (Dwivedi et al. 2023).

16.5 Millet Waste as a Feedstock for Biofuel Production

16.5.1 Pretreatment and Processing Procedures

Pretreatment and processing procedures are important stages in transforming millet waste into valuable biofuels and chemicals. Millet waste, constituted mostly of lignocellulosic elements like cellulose and hemicellulose, requires special treatment to make these components more accessible for downstream conversion processes. Mechanical operations like milling or grinding can reduce millet waste into smaller particles, increasing the surface area for future chemical treatments. Chemical pretreatment involves the use of chemicals or enzymes to break down the lignin and hemicellulose, making cellulose more accessible for enzymatic hydrolysis. This process is vital for the synthesis of sugars that can be fermented into biofuels.

Heat-based procedures, such as pyrolysis, can break the lignin structure and enable cellulose separation (Lamichhane et al. 2023). Thermochemical processes can also yield bio-oils and syngas, which have applications in the chemical industry. After pretreatment, enzymes like cellulase and hemicellulase are employed to convert cellulose and hemicellulose into fermentable sugars, which can be used for ethanol or biochemical synthesis. The acquired sugars are fermented by various microbes to make bioethanol, biobutanol, or other biofuels. Alternatively, they can be used as feedstock for chemical synthesis processes (Jamaldheen et al. 2019). For chemical manufacturing, millet waste may undergo extraction and purification methods to isolate certain constituents, such as organic acids, bio-based polymers, and other important chemicals. These pretreatment and processing processes are critical for increasing the yield and quality of biofuels and chemicals obtained from millet waste. Optimizing these phases is vital for establishing economic viability and environmental sustainability in millet waste utilization, ultimately helping to create a more sustainable bioeconomy.

16.5.2 Biofuel Kinds from Millet Waste

Millet waste can be a valuable source for manufacturing various biofuels, contributing to sustainability and reduced carbon emissions. Millet waste is rich in cellulose and hemicellulose, which can be enzymatically converted into fermentable sugars. These sugars are subsequently fermented by microorganisms like yeast to make bioethanol. Bioethanol is a clean-burning, renewable fuel that can be combined with gasoline or used independently. It minimizes greenhouse gas emissions and lessens our dependence on fossil fuels. Millet oil collected from millet bran or seeds can be turned into biodiesel by a process called transesterification. Biodiesel is a renewable, biodegradable alternative to traditional diesel fuel. It produces fewer particles and greenhouse gases, contributing to enhanced air quality and reduced environmental impact. Anaerobic digestion of millet waste can create biogas, principally methane and carbon dioxide. Biogas can be utilized as a sustainable energy source for electricity generation or as a clean cooking fuel. Thermochemical methods like pyrolysis may transform millet waste into pyrolysis oil, which can be further processed into biofuels or used as a source of heat or electricity. By utilizing the potential of millet waste for these biofuels, we can simultaneously address waste management concerns, promote sustainable agriculture, and lower our carbon footprint, aligning with worldwide efforts to battle climate change and shift to cleaner, renewable energy sources (Panpatte and Jhala 2019).

16.5.3 Advantages and Limitations

Advantages: Sustainability: Millet waste utilization supports sustainability by reducing agricultural waste and boosting resource efficiency. Carbon reduction: Millet-based biofuels contribute to greenhouse gas reduction as they generate lower carbon emissions compared to fossil fuels. Economic benefits: Millet waste valorization generates extra cash streams for farmers, promotes rural economies, and fosters employment development. Local sourcing: Millet is commonly farmed in places with restricted access to traditional feedstocks, supporting localized biofuel and chemical manufacturing. Soil health: Using millet waste leftovers can increase soil health, minimize erosion, and enhance agricultural sustainability. Limitations: Variable availability: Millet waste availability can be regionally uneven, impacting the scalability of biofuel and chemical production. Technological challenges: Conversion technologies may require significant upfront investment and continuous research and development. Ethical concerns: Balancing the use of millet for food and industrial reasons might create ethical problems concerning food security and equal distribution. Infrastructure and knowledge gaps: Some places may lack the essential infrastructure and knowledge for millet waste processing and usage. Competition with food: The exploitation of millet for non-food reasons may conflict with its use as a staple food source in particular places, needing careful analysis and planning.

Overall, the advantages of millet waste utilization seem promising for a sustainable bioeconomy, but addressing the limits requires a concentrated effort in research, technology development, and ethical considerations to achieve equitable and sustainable outcomes.

16.6 Millet Waste as a Feedstock for Chemicals

Millet waste serves as a great feedstock for chemical manufacturing, offering a sustainable alternative to petrochemicals. Its significant cellulose and hemicellulose content can be turned into bio-based compounds, including organic acids, bio-based polymers, and antioxidants. These compounds have applications in different industries, from pharmaceuticals and cosmetics to bioplastics and agriculture. Utilizing millet waste for chemical manufacture not only minimizes agricultural waste but also supports eco-friendly, renewable resources. However, obstacles such as efficient extraction methods and scaling up production must be overcome to fully utilize the potential of millet waste as a feedstock for chemicals, contributing to a greener and more sustainable chemical sector.

16.6.1 Chemical Extraction Procedures

Chemical extraction techniques are crucial processes for extracting useful compounds from millet waste, contributing to its application as a feedstock for chemicals. These procedures utilize numerous chemical reactions and separation methods to extract specific molecules of interest. Acid hydrolysis is a typical method for breaking down cellulose and hemicellulose in millet waste into soluble sugars. Dilute acids like sulfuric acid or hydrochloric acid are used to cleave the glycosidic linkages in these polysaccharides, releasing sugars that can be further transformed into other compounds, including bio-based organic acids. Alkaline hydrolysis utilizes strong bases like sodium hydroxide or potassium hydroxide to breakdown lignin and hemicellulose in millet waste. This technique can create phenolic compounds, which have applications in the chemical industry. Enzymatic methods utilize cellulase and hemicellulase enzymes to break down cellulose and hemicellulose into fermentable sugars. This method is more selective and environmentally favorable than acid hydrolysis. Solvent extraction is another method that includes employing organic solvents to dissolve and isolate certain components from industrial waste. It is widely applied for extracting lipids and bioactive substances, such as antioxidants or essential oils, from millet seeds or bran. Distillation is used to separate volatile components from millet waste extracts, particularly in the synthesis of bio-based products like biofuels and essential oils. These chemical extraction procedures are crucial for transforming millet waste into useful compounds with diverse industrial applications. Optimizing these processes, considering sustainability and environmental issues, is crucial to optimizing the efficiency and economic viability of millet waste utilization in the chemical sector (Jagadabhi et al. 2018).

16.6.2 Potential Compounds Generated from Millet Waste

Millet waste provides considerable potential for the synthesis of a wide range of useful compounds, contributing to sustainable and eco-friendly industrial applications. Acid hydrolysis of millet waste can provide organic acids such as acetic, formic, and levulinic acids, which find applications in the chemical, pharmaceutical, and food industries. The cellulose component of millet waste can be turned into bio-based polymers, giving sustainable alternatives to traditional plastics and packaging materials. Alkaline hydrolysis can yield phenolic compounds with antioxidant characteristics that are appropriate for use in the cosmetic and pharmaceutical industries (Fărcaş et al. 2021). Millet bran can be a source of essential oils, which are utilized in perfumes, aromatherapy, and the food sector. The sugars generated from millet waste can be fermented to produce biofuels like bioethanol and biobutanol, offering eco-friendly alternatives to fossil fuels. Millet waste can serve as a source for specialty chemicals, including furfural, which has applications in the production of resins, solvents, and polymers (Awogbemi and Von Kallon 2022). Certain components of millet waste, including bran, have bioactive chemicals with potential health advantages, making them appropriate for nutraceuticals and functional meals. Harnessing these potential compounds from millet trash not only adds value to agricultural residues but also corresponds with sustainable and ecologically favorable industrial practices. It promotes the shift toward a bio-based economy, lowering reliance on petrochemicals and contributing to a greener and more sustainable future (Fărcaş et al. 2022).

16.6.3 Industrial Uses

Millet waste-derived compounds find different industrial applications. Organic acids and bio-based polymers serve as eco-friendly options in the chemical and packaging industries. Phenolic chemicals from millet waste are beneficial for cosmetics and pharmaceuticals, while essential oils serve as perfumes and aromatherapy. Biofuels like bioethanol and biobutanol answer the energy sector's sustainability needs. Specialty compounds such as furfural contribute to resins, solvents, and polymers. Millet waste's nutraceutical components align with the health and wellness business, and its potential as a bio-based feedstock encourages a shift toward sustainable, environmentally conscious industrial operations, reducing dependence on traditional petrochemical sources (Gairola et al. 2022).

16.7 Economic Viability

16.7.1 Cost-Effectiveness of Millet Waste-Based Biofuel and Chemical Production

The cost-effectiveness of millet waste-based biofuel and chemical production depends on numerous factors, including feedstock availability, processing technologies, market demand, and government incentives. Here is an overview of essential considerations:

Feedstock availability: The abundance and consistent supply of millet waste influence cost-effectiveness. Regions with considerable millet farming can benefit from lower feedstock transportation costs. Processing technologies: Efficient and cost-effective pretreatment and conversion technologies are vital. Advances in enzyme technology, pyrolysis, and distillation can enhance process efficiency and lower production costs.

Economies of scale: Large-scale production facilities can generate economies of scale, driving down per-unit production costs. Co-location with millet processing facilities can help maximize resource efficiency.

Market demand: The demand for biofuels and bio-based chemicals affects pricing and profitability. A robust market for sustainable products can boost cost-effectiveness.

Government support: Government incentives, subsidies, and regulations favoring biofuels and renewable chemicals can dramatically impact cost-effectiveness. Tax subsidies, grants, and requirements might make millet waste-based production financially appealing.

Competition with food uses: Balancing millet waste usage with food production is crucial to prevent market distortions and ethical concerns, potentially compromising cost-effectiveness.

Research and innovation: Ongoing research and technical developments can cut production costs and enhance process efficiencies over time.

In summary, millet waste-based biofuel and chemical production can be cost-effective when feedstock availability, efficient processing technology, market demand, government assistance, and sustainable practices combine. Strategic planning, investment, and a supportive regulatory framework are important to unleashing the full economic potential of solid waste utilization, contributing to a more sustainable and eco-friendlier industrial sector (Haque et al. 2023).

16.7.2 Comparative Comparison with Other Feedstocks

A comparative analysis of millet waste as a feedstock for biofuel and chemical production against other feedstocks indicates numerous advantages and considerations:

1. Millet waste vs. maize stover: Millet waste has a reduced lignin concentration compared to maize stover, making it more suited to biofuel generation. However, corn stovers are more widely available in certain locations.
2. Millet waste vs. sugarcane bagasse: Both millet waste and sugarcane bagasse offer lignocellulosic feedstocks for biofuels; however, sugarcane bagasse has a more established industrial presence due to its link with sugarcane-based ethanol (Packiam et al. 2018).
3. Millet waste vs. algae: Algae may produce biofuels like biodiesel and biogas, but algae growth methods require major water resources and infrastructure, whereas millet waste does not.
4. Millet waste vs. jatropha: Jatropha seeds can be utilized for biodiesel generation, but they require proper cultivation and may compete with food crops, unlike millet waste.
5. Millet waste vs. woody biomass: Woody biomass, while abundant, frequently requires energy-intensive pretreatment due to its high lignin content, making millet waste an attractive option for its lower lignin levels.

Ultimately, the choice of feedstock depends on regional availability, sustainability, processing technology, and market demand. Millet waste's advantages lay in its lower lignin content and potential for use in regions where millet is a staple crop, presenting a promising alternative in the quest for sustainable biofuel and chemical production (Lalman et al. 2016).

16.8 Environmental Impact Assessment

16.8.1 Environmental Benefits of Millet Waste Utilization

The usage of millet waste has several substantial environmental benefits, coinciding with sustainability and ecological conservation goals: Millet waste use helps decrease agricultural waste that would otherwise be dumped or burned, contributing to cleaner and healthier habitats by decreasing air and soil pollution caused by open burning practices. Millet waste-based biofuels release fewer greenhouse gases compared to fossil fuels, contributing to climate change mitigation. Reduced carbon emissions translate into lower carbon footprints and reduced contributions to global warming. Leaving millet remains in the field instead of burning them can enhance soil health by boosting organic matter content, nutrient retention, and moisture retention. This approach mitigates soil erosion and promotes sustainable agriculture (Shahane and Shivay 2016). Sustainable millet farming practices, including waste utilization, can assist biodiversity conservation. By limiting the need for development in natural habitats, it helps safeguard ecosystems and their native species. The manufacturing of bio-based materials and chemicals from millet waste decreases the demand for traditional wood-based products, alleviating pressure on forests and protecting key ecosystems. Millet is drought-resistant and is typically cultivated in water-scarce places. By optimizing millet production and waste utilization, water resources can be conserved, and water-intensive crops replaced. Millet waste usage fosters sustainable agricultural techniques, such as no-tillage and crop rotation, contributing to long-term soil and environmental health.

In summary, millet waste use not only captures a valuable resource for biofuels and chemicals but also exemplifies a sustainable and eco-friendly approach that lowers waste, mitigates climate change, and supports environmental conservation. These environmental benefits align with worldwide initiatives to move to a more sustainable and greener economy (Devi et al. 2017).

16.8.2 Carbon Footprint Reduction and Sustainability

The exploitation of millet waste plays a key role in lowering the carbon footprint and furthering environmental goals. Millet waste-based biofuels generate much fewer greenhouse gases compared to fossil fuels. By employing renewable feedstock for bioethanol or biodiesel production, we cut carbon dioxide emissions, mitigating climate change and curbing the carbon footprint of transportation and industry (Kumar et al. 2022b). Millet waste utilization exhibits resource efficiency. Instead of abandoning agricultural wastes, we transform them into valuable biofuels and chemicals. This circular strategy eliminates waste, maximizes resource utilization, and supports sustainable agriculture. Sustainable millet farming techniques, driven by waste reuse, limit the need for expanding agricultural areas into natural habitats. This helps protect biodiversity, manage ecosystems, and preserve the carbon-absorbing capability of forests and wetlands. The move to millet waste-based biofuels fits with global initiatives to replace fossil fuels with renewable energy sources. It contributes to sustainable energy generation and decreases dependence on finite fossil fuel stocks. Millet waste integration into the soil improves its organic matter content, minimizes erosion, and enhances water retention. These approaches support soil health, reduce the need for synthetic fertilizers, and promote water conservation. Millet waste use exemplifies sustainability by minimizing carbon emissions, saving resources, maintaining natural ecosystems, and supporting renewable energy. It shows a cyclical and eco-friendly strategy that is important for tackling severe environmental concerns (Meena et al. 2022).

16.9 Technological Advancements

Recent advances and research in millet waste utilization have been important in harnessing the full potential of this plentiful and sustainable resource. These breakthroughs are driving advancements in agriculture, biofuels, and the chemical sector, supporting environmental sustainability and economic progress. Researchers are actively developing conversion technologies, including enzymatic hydrolysis and pyrolysis, to increase the efficiency of millet waste-to-biofuel and chemical processes (Tagade and Sawarkar 2023). These advances are raising yields and reducing energy usage, making millet waste utilization more economically viable. The research community is exploring new options for millet waste utilization by extracting a wider range of lucrative bioproducts. This includes bio-based polymers, organic acids, and bioactive chemicals like antioxidants, broadening the spectrum of potential uses. Genetic engineering and breeding procedures are being applied to generate millet varieties with higher biomass yields and biofuel-specific features. These activities aim to expand the total availability of millet waste as a feedstock (Rocha-Meneses et al. 2020). Research is also focusing on market development and product commercialization. Identifying and generating new markets for millet waste-derived goods, such as biofuels and specialty chemicals, is vital for scaling up usage. Researchers are undertaking extensive sustainability assessments to analyze the environmental and social implications of millet waste usage, ensuring that these procedures fit with sustainability goals (Kumar et al. 2022a).

These latest advancements and research activities are transforming the landscape of millet waste utilization, making it a vital participant in the shift to sustainable agriculture, renewable energy, and green chemistry. As these developments continue to evolve, millet waste is positioned to play a vital role in tackling global concerns linked to food security, climate change, and resource sustainability.

16.10 Case Studies

There were limited large-scale successful experiments specifically focused on millet waste utilization for biofuels and chemicals. However, there were some smaller-scale initiatives and research activities that demonstrated promising outcomes. Real-world examples of millet waste usage for biofuel and chemicals are sparse in comparison to more established feedstocks like corn or sugarcane. However, studies and pilot programs have shown the viability and potential of millet waste utilization. India's Ministry of Agriculture and Farmers Welfare began the "Millet Mission," promoting millet production and waste utilization. Millet waste is being examined for biofuel production to prevent agricultural residues such as parali burning and assist rural income development (Pandey and Bolia 2023). Research studies in countries like China and India have proved the conversion of millet straw into bioethanol using pretreatment and enzymatic hydrolysis. These efforts aim to produce economically feasible bioethanol from millet trash. Some research programs have concentrated on extracting important chemicals like levulinic acid and furfural from millet waste. These compounds have applications in the chemical and medicinal sectors (Rose 2022). While these examples are encouraging, greater study, development, and investment are needed to fully harness the potential of millet waste as a feedstock for biofuels and chemicals on a larger scale. The transition to more sustainable and diversified feedstock sources is a continuous process, and millet waste holds tremendous promise in this regard.

16.11 Challenges and Future Directions

16.11.1 Overcoming Barriers to Millet Waste Utilization

Overcoming constraints to millet waste utilization is vital for exploiting its full potential as a feedstock for biofuels and chemicals. More investment in research and development is needed to optimize millet waste conversion technologies, making them more efficient and cost-effective. Infrastructure development is required to handle millet waste efficiently and maintaining a constant supply chain. New markets and demand should be created for millet waste-derived products by promoting eco-friendly and sustainable alternatives. Implementation of new policies and incentives will encourage better waste usage, including tax credits, subsidies, and renewable energy standards. Sustainable farming practices should be promoted that enhance waste availability while reducing negative environmental impacts. Awareness and education among farmers and others should be increased about the benefits of millet waste utilization. Exploring opportunities to extract high-value chemicals and specialty goods from millet waste, will diversify its application. Foster collaboration between governments, industries, and research institutions to drive innovation and investment will enhance the prospects of millet waste utilization. By resolving these constraints, millet waste usage can become a sustainable and economically feasible alternative, contributing to cleaner energy production and less waste while boosting rural economies and sustainable agriculture (Kumar et al. 2022b).

According to agricultural and processed food products export development authority of Ministry of Commerce and Industry, Government of India, the world's largest producer of millet is India. In 2020, the two millets grown in India, pearl millet (Bajra) and sorghum (Jowar), would account for around 19% of global output. In 2020, the world's production of millets was made up of 40.51% pearl millet from India and 8.09% sorghum.

16.11.2 Emerging Patterns and Prospects

Emerging developments in millet waste utilization imply a hopeful future. Increasing awareness of sustainability is boosting interest in millet waste as a renewable resource for biofuels, chemicals, and materials. Advances in technology are also boosting process efficiency and diversifying product outputs. Collaborative initiatives between academia, business, and governments are encouraging research and investment in millet waste-based industries. Future potential includes expanding millet cultivation and trash collection, finding new markets for eco-friendly products, and addressing food–energy–ecosystem nexus concerns. With ongoing innovation, millet waste use is poised to contribute considerably to sustainable agriculture, renewable energy, and a greener chemical sector (Tagade and Sawarkar 2023).

16.12 Conclusion

A study on millet waste as an affordable feedstock for chemicals and biofuel highlights its promise as a sustainable substitute in the fields of chemical and energy production. A thorough review of the flexibility of millet waste as a crop and its potential benefits to the chemical and bioenergy sectors has been provided in response to the introduction's need for alternative feedstocks.

The comprehensive study of millet varieties, growing methods, and nutritional value demonstrates the crop's versatility and paves the way for a thorough investigation of millet waste production. The difficulties in managing millet waste, which include financial and environmental costs, highlight the need for long-term fixes. The chapters on millet waste as a feedstock for chemicals and biofuels give readers a comprehensive grasp of the advantages and disadvantages of the various chemical extraction processes, biofuel kinds, and pretreatment strategies. The evaluation of millet waste's economic feasibility and environmental impact highlights its potential in lowering costs and reducing carbon emissions.

The chapter is concluded with a forward-looking analysis of obstacles and future goals, case examples, and technology breakthroughs that present millet waste as a dynamic and attractive resource. This chapter essentially promotes the incorporation of millet waste into conventional industrial processes, emphasizing its ability to meet the increasing the needs for sustainable solutions in the manufacturing of chemicals and biofuels.

References

Awogbemi O, Von Kallon DV (2022) Valorization of agricultural wastes for biofuel applications. Heliyon 8(10):e11117. https://doi.org/10.1016/j.heliyon.2022.e11117

Bhuvaneshwari S, Hettiarachchi H, Meegoda JN (2019) Crop residue burning in India: policy challenges and potential solutions. Int J Environ Res Public Health 16(5):832. https://doi.org/10.3390/ijerph16050832

Bimbraw AS (2019) Generation and impact of crop residue and its management. Curr Agric Res J 7(3):304–309. https://doi.org/10.12944/carj.7.3.05

Chauhan D et al (2023) Millets: the sustainable ancient superfood for the modern world. J Drug Res Ayurved Sci 8(Suppl 1):S34–S41. https://doi.org/10.4103/jdras.jdras_248_23

Devi RU, Kolli B (2022) Crop waste management: perspectives on alternative uses in India. Cab Rev 2022:17022. https://doi.org/10.1079/cabireviews202217022

Devi S et al (2017) Crop residue recycling for economic and environmental sustainability: the case of India. Open Agric 2(1):486–494. https://doi.org/10.1515/opag-2017-0053
Dresch AP et al (2022) Different treatments of pearl millet for cellulose recovery: effects on lignocellulosic composition. Res Sq. https://doi.org/10.21203/rs.3.rs-1689771/v1
Dwivedi N, Rathore V, Sharma KK (2023) A review of millet crops for agricultural sustainability in India. Asian J Agric Ext Econ Sociol 41(10):216–224. https://doi.org/10.9734/ajaees/2023/v41i102162
Fărcaş AC et al (2021) Cereal processing by-products as rich sources of phenolic compounds and their potential bioactivities. Nutrients 13(11):3934. https://doi.org/10.3390/nu13113934
Fărcaş AC et al (2022) An update regarding the bioactive compound of cereal by-products: health benefits and potential applications. Nutrients 14(17):3470. https://doi.org/10.3390/nu14173470
Gairola S, Sinha S, Singh I (2022) Novel millet husk crop-residue based thermoplastic composites: waste to value creation. Ind Crop Prod 182:114891. https://doi.org/10.1016/j.indcrop.2022.114891
Gowri MU (2020) Millet scenario in India. Econ Aff 65(3):7. https://doi.org/10.46852/0424-2513.3.2020.7
Haque F, Fan C, Lee Y-Y (2023) From waste to value: addressing the relevance of waste recovery to agricultural sector in line with circular economy. J Clean Prod 415:137873. https://doi.org/10.1016/j.jclepro.2023.137873
Jagadabhi PS et al (2018) Physico-chemical, microbial and phytotoxicity evaluation of composts from sorghum, finger millet and soybean straws. Int J Recycl Org Waste Agric 8(3):279–293. https://doi.org/10.1007/s40093-018-0240-8
Jamaldheen SB, Thakur A, Moholkar VS (2019) Enzymatic hydrolysis of hemicellulose from pretreated finger millet (Eleusine coracana) straw by recombinant endo-1,4-β-xylanase and exo-1,4-β-xylosidase. Int J Biol Macromol 135:1098–1106. https://doi.org/10.1016/j.ijbiomac.2019.06.010
Koul B, Yakoob MY, Shah MP (2022) Agricultural waste management strategies for environmental sustainability. Environ Res 206:112285. https://doi.org/10.1016/j.envres.2021.112285
Kumar S et al (2022a) By-products from millet processing industry. In: Handbook of millets - processing, quality, and nutrition status. Springer, Cham, pp 301–314. https://doi.org/10.1007/978-981-16-7224-8_13
Kumar V et al (2022b) Emerging challenges for the agro-industrial food waste utilization: a review on food waste biorefinery. Bioresour Technol 362:127790. https://doi.org/10.1016/j.biortech.2022.127790
Lakshm PJ et al (2023) Sustainable biofuels from algae- process and methods adapted: a review. Int J Curr Res Rev 15(16):6–12. https://doi.org/10.31782/ijcrr.2023.151603
Lalman JA et al (2016) Biofuels production from renewable feedstocks. In: Green chemistry and sustainable technology. Springer, Cham, pp 193–220. https://doi.org/10.1007/978-3-662-53704-6_8
Lamichhane G et al (2023) Correction to: pretreatment of finger millet straw (Eleusine coracana) for enzymatic hydrolysis towards bioethanol production. Biomass Convers Biorefinery 13(8):7397. https://doi.org/10.1007/s13399-023-04315-5
Meena OP et al (2022) Energy-carbon footprint Vis-à-Vis system productivity and profitability of diversified crop rotations in semi-arid plains of north-West India. J Soil Sci Plant Nutr 22(2):2026–2041. https://doi.org/10.1007/s42729-022-00791-2
Mohite AS et al (2022) Recycling of major agriculture crop residues and its application in polymer industry: a review in the context of waste to energy nexus. Energy Nexus 7:100134. https://doi.org/10.1016/j.nexus.2022.100134
Packiam M et al (2018) Suitability of pearl millet as an alternate lignocellulosic feedstock for biofuel production in India. J Appl Environ Microbiol 6(2):51–58. https://doi.org/10.12691/jaem-6-2-4
Pandey AP, Bolia NB (2023) Millet value chain revolution for sustainability: a proposal for India. Socio Econ Plan Sci 87:101592. https://doi.org/10.1016/j.seps.2023.101592

Panpatte DG, Jhala YK (2019) Agricultural waste: a suitable source for biofuel production. In: Biofuel and biorefinery technologies. Springer, Cham, pp 337–355. https://doi.org/10.1007/978-3-030-14463-0_13

Paschapur AU et al (2021) Millets for life: a brief introduction. In: Springer eBooks. Springer, Cham, pp 1–32. https://doi.org/10.1007/978-981-16-0676-2_1

Patil PB et al (2023) Millets: empowering the society with nutrient-rich superfoods to achieve sustainable development goals. J Drug Res Ayurved Sci 8(Suppl 1):S100–S114. https://doi.org/10.4103/jdras.jdras_207_23

Ratnavathi CV, Tonapi VA (2022) Nutritional properties of millets: nutricereals with health benefits to reduce lifestyle diseases and malnutrition. In: Handbook of millets - processing, quality, and nutrition status. Springer, Cham, pp 129–149. https://doi.org/10.1007/978-981-16-7224-8_7

Ravikesavan R et al (2022) Millet cultivation: an overview. In: Handbook of millets - processing, quality, and nutrition status. Springer, Cham, pp 23–47. https://doi.org/10.1007/978-981-16-7224-8_2

Rocha-Meneses L et al (2020) Genetic modification of cereal plants: a strategy to enhance bioethanol yields from agricultural waste. Ind Crop Prod 150:112408. https://doi.org/10.1016/j.indcrop.2020.112408

Rose PK (2022) Bioconversion of agricultural residue into biofuel and high-value biochemicals: recent advancement. In: Energy, environment, and sustainability. Springer, Cham, pp 233–268. https://doi.org/10.1007/978-981-16-8682-5_9

Shafique T, Shafique J (2021) Scaling up sustainable biofuels for a low-carbon future. Sustainable Energy Invest. https://doi.org/10.5772/intechopen.92652

Shahane AA, Shivay YS (2016) Cereal residues - not a waste until we waste it: a review. Int J Bioresour Stress Manage 7(1):162–173. https://doi.org/10.23910/ijbsm/2016.7.1.1401b

Sieminski A (2022) Pathways to meeting climate goals with clean and affordable energy. Can For Policy J 28(3):321–327. https://doi.org/10.1080/11926422.2022.2107550

Tagade A, Sawarkar AN (2023) Valorization of millet agro-residues for bioenergy production through pyrolysis: recent inroads, technological bottlenecks, possible remedies, and future directions. Bioresour Technol 384:129335. https://doi.org/10.1016/j.biortech.2023.129335

Verma V et al (2021) Millet starch: current knowledge and emerging insights of structure, physiology, glycaemic attributes and uses. In: Springer eBooks. Springer, Cham, pp 121–142. https://doi.org/10.1007/978-981-16-0676-2_6

Zheng Y et al (2021) Effect of four modification methods on adsorption capacities and in vitro hypoglycemic properties of millet bran dietary fibre. Food Res Int 147:110565. https://doi.org/10.1016/j.foodres.2021.110565

Valorization of Pulse Waste to Biofuels and Chemicals

17

Chandra Jeet Singh, Manish Kumar, Mayuri Srivastava, and Vivek Jaiswal

Abstract

Food waste is a pressing global issue, with approximately 1.3 billion tons wasted annually, significantly impacting the environment, socioeconomic conditions, and food security. In developing countries like India, 22% of food grain output is lost due to inefficiencies in supply chains, storage, and transportation. The United Nations' Sustainable Development Goals (SDGs) aim to address these challenges by promoting sustainable practices to reduce food loss and waste. Food waste, rich in proteins, carbohydrates, and lipids, is highly biodegradable but often improperly disposed of, leading to environmental pollution through incineration and landfilling. Pulse waste, generated from crops like lentils, chickpeas, and beans, presents a valuable opportunity for valorization. Rich in lignocellulosic materials, proteins, and starch, pulse waste can be converted into biofuels (e.g., bioethanol, biogas, biodiesel), biochemicals, and bioplastics. Anaerobic digestion (AD) is a promising technology for managing food waste, producing biofuels like biomethane and biohydrogen, while also generating value-added products such as enzymes, bioplastics, and biochar. This review explores the composition of pulse waste, current disposal practices, and advanced conversion

C. J. Singh · V. Jaiswal
Department of Textile and Fibre Engineering, Indian Institute of Technology Delhi, New Delhi, India
e-mail: chandrajeettz@gmail.com

M. Kumar (✉)
Department of Plastic Technology, Harcourt Butler Technical University, Kanpur, Uttar Pradesh, India
e-mail: manish.kr@hbtu.ac.in

M. Srivastava
School of Interdisciplinary Research, Indian Institute of Technology Delhi, New Delhi, India

A. Kumari et al. (eds.), *Valorization of Solid Wastes to Biofuels and Chemical Products for Sustainable World*, https://doi.org/10.1007/978-981-96-8594-3_17

technologies, including biochemical and thermochemical processes. It highlights the potential of pulse waste as a sustainable feedstock for renewable energy and chemical production, aligning with circular economy principles. Future prospects for pulse waste valorization are promising, driven by technological advancements, sustainability goals, and the need to address global food waste challenges. By transforming waste into valuable resources, pulse waste valorization offers a pathway to enhance food security, reduce environmental impact, and promote economic growth.

Keywords

Agriculture residue · Biochemical production · Biomass · Waste management · Sustainable approach

17.1 Introduction

Food waste is a significant global issue, with an estimated 1.3 billion tons of food wasted each year (Leser 2013). In developing countries such as India, approximately 74 million tons of food are lost annually, which accounts for about 22% of the total food grain output (Nath et al. 2024). This wastage arises from various factors, including inefficient food supply chains, inadequate storage facilities, economic challenges, and delays in transportation. The consequences of food waste extend beyond environmental impacts; they also affect socioeconomic conditions and food security. In response to these challenges, the United Nations established the "Sustainable Development Goals" (SDGs) in 2015 to address food loss and waste while combating poverty. Implementing these goals can help nations mitigate food scarcity and enhance food security. Food waste (FW) typically originates from diverse sources such as kitchens, restaurants, agricultural fields, and processing plants, comprising nearly one-third of municipal waste. Decomposing food waste releases strong odors and is rich in proteins, carbohydrates, and lipids, making it more biodegradable than other organic wastes (Kannah et al. 2020). The biochemical composition of food waste varies based on its source and consumer preferences. Often, food waste is collected and disposed of at landfills or incinerated without proper pretreatment, leading to significant environmental issues (Nayak and Bhushan 2019). For example, incineration contributes to air pollution and respiratory problems due to harmful emissions. Similarly, landfilling generates toxic by-products such as leachate that can contaminate groundwater and emit corrosive gases like methane and hydrogen sulfide. Therefore, effective technologies for safe disposal are essential.

Among the various biomass resources available for sustainable practices, agricultural residues from pulses such as lentils, chickpeas, beans, and peas hold considerable potential for biofuel and biochemical production. Pulses are integral to global food systems; however, their processing generates significant waste in the form of husks and damaged grains. These residues are rich in carbohydrates and proteins, making them valuable feedstock for producing biofuels such as bioethanol and biogas as well as platform chemicals and bioplastics (Leser 2013). Pulses have a low food wastage footprint compared to other crops due to their shelf stability; they can be stored for extended periods without spoilage (Leser 2013). This characteristic reduces the likelihood of consumer food waste due to spoilage and make pulses a smart choice for households facing food insecurity. Furthermore, the valorization of pulse waste not only addresses the issue of food wastage but also contributes to renewable energy production and chemical synthesis.

Anaerobic digestion (AD) has been identified by a number of studies as the most promising method for the treatment of fuel waste (Ren et al. 2018; Zhang et al. 2020; Xu et al. 2018). Management or stabilization of FW is performed as part of the AD process. In addition to this, it produces biofuels that are lucrative and have a high thermal value. Using the AD method, it is possible to manufacture a wide variety of biofuels from FW, including biomethane (Kavitha et al. 2017), biohydrogen via dark fermentation (Yeshanew et al. 2016), biohythane via two stage AD process (Ghimire et al. 2017), and biodiesel via transesterification (Bizkarra et al. 2017). Recently, the biochemical composition of FW has drawn the attention of researchers and policymakers to recover different types of value-added products such as bioplastic (Sanchez-Vazquez et al. 2013), enzymes (Merrylin et al. 2020), and biochar (Alghashm et al. 2018). Figure 17.1 illustrates the bioconversion process of food waste and its resulting products. This review aims to provide an in-depth understanding of the current status of food waste generation, the properties of food waste, and the various pretreatment techniques employed for bioenergy recovery. Additionally, it discusses the different types of biofuels produced from food waste, as well as other value-added products that can be recovered from it (Kannah et al. 2020).

17.2 Pulse Waste: Composition and Current Disposal Practices

17.2.1 Types of Pulse Waste

Pulse waste is generated during the cultivation, harvesting, processing, and storage of pulse crops such as lentils, chickpeas, peas, and beans (Tassoni et al. 2020). This waste can be broadly categorized into the following types:

17.2.1.1 Agricultural Residues

Agricultural residues are the by-products and waste materials generated after the cultivation and harvesting of pulse crops. These residues, which include various plant components such as stems, stalks, and leaves, are produced in substantial quantities but are often underutilized despite their potential for valorization (Snapp et al. 2018; Keshavarz et al. 2020).

The residues from pulse crops are rich in organic matter and can be repurposed for a variety of applications, including biofuel production, biochemical processes, and animal feed (Bhuvaneshwari et al. 2019). Effective management of these residues is crucial as they can enhance soil fertility and contribute to sustainable

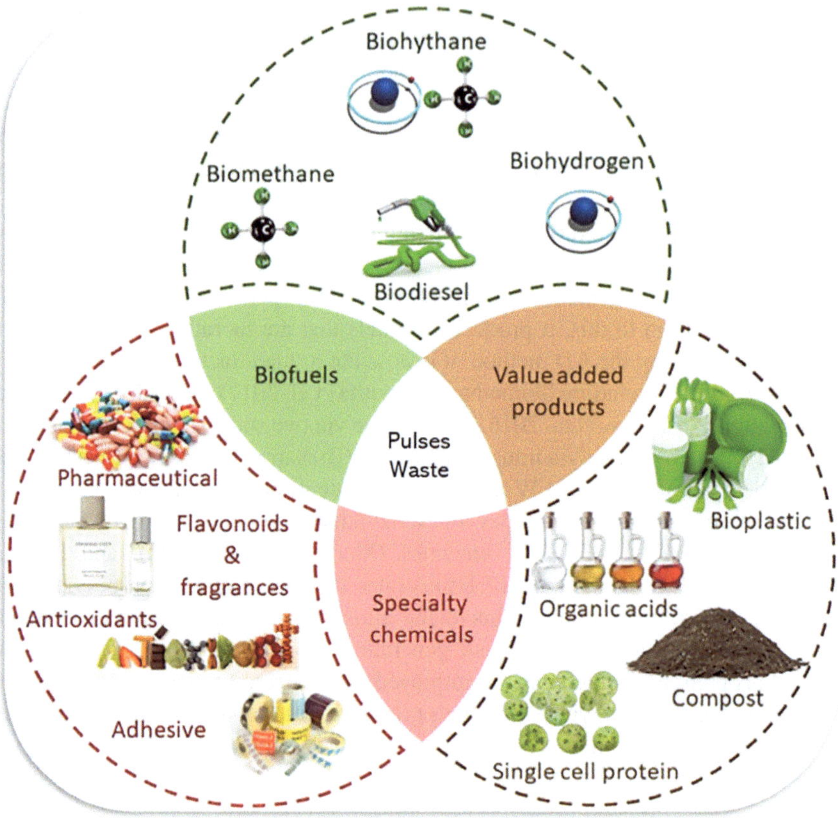

Fig. 17.1 Bioconversion of pulse waste and products. (Adapted with permission from Kannah et al. (2020))

agricultural practices. For instance, incorporating these residues back into the soil can improve its structure and nutrient content, thereby supporting crop productivity (Bhuvaneshwari et al. 2019). Despite their benefits, many agricultural practices still involve the burning of these residues, which poses environmental challenges such as air pollution and loss of valuable nutrients. Therefore, finding innovative ways to utilize pulse crop residues not only aligns with sustainability goals but also maximizes their economic value (Bhuvaneshwari et al. 2019; Gürdil et al. 2021; Shakuntala et al. 2024).

17.2.1.2 Processing By-Products

Processing by-products refer to the waste materials generated during the postharvest processing of pulse crops. These by-products are produced during cleaning, de-hulling, sorting, grading, and milling. Despite being regarded as waste, they are rich in organic compounds, such as cellulose, starch, protein, and fiber, which make them valuable for conversion into biofuels, biochemicals, and other high-value products. Husks, shells, and seed coats are generated during milling or dehulling (Selvamuthukumaran 2024).

17.2.1.3 Rejected Materials

Rejected materials are waste products generated during quality control, grading, and sorting of pulse crops. These materials primarily consist of pulse grains and related residues that fail to meet the standards required for human consumption or industrial use. Rejected materials are typically discarded, but they contain valuable organic components, such as starch, protein, and fiber, which can be repurposed for sustainable applications in biofuels, biochemicals, and other bioproducts. Damaged or broken grains and other residues from sorting and grading processes (Rathinam and Sani 2019).

17.2.2 Composition of Pulse Waste

Pulse waste generated during cultivation, processing, and consumption of pulses contains various organic and inorganic components. Its composition depends on the specific type of pulse, part of the plant (e.g., husks, stems, or seeds), and processing method. Pulse waste is generally rich in lignocellulosic material, starch, protein, fiber, and micronutrients, making it a valuable resource for valorization into biofuels, biochemicals, and other products. Pulse waste is rich in organic compounds, which are suitable for conversion into value-added products (Luzardo-Ocampo et al. 2020).

17.2.2.1 Lignocellulosic Components

Lignocellulosic components are the structural building blocks of plant biomass found in the cell walls of pulse residues, such as stems, husks, shells, and seed coats. These components primarily include cellulose, hemicellulose, and lignin, which provide rigidity and strength to plant materials. The abundance of lignocellulosic components in the pulse waste makes it a valuable feedstock for the production of biofuels, biochemicals, and biomaterials. Cellulose (30–50%), hemicellulose (15–25%), and lignin (15–30%) content (Diwan et al. 2020). All components of lignocellulosic biomass are capable of producing value-added products. Figure 17.2 displays these components of lignocellulosic biomass along with their products. Lignin is the principal recalcitrant component in lignocellulosic biomass due to its complex properties such as its chemical composition and molecular weight. It inhibits biological conversion by retarding the activity of enzymes as they bind nonproductively. Moreover, high lignin content prevents access of cellulase enzyme to cellulose. It hinders the fermentation process, consequently affecting the efficiency of bioethanol production. Therefore, delignification is a must to improve biological conversion (Lee et al. 2014). Lignin can also be directly burned to generate heat and power or can be treated thermochemically to produce low-grade fuels, phenolics, and adhesives. Cellulose and hemicellulose are polysaccharides; therefore, they are potential precursors of fermentable sugars such as glucose and xylose.

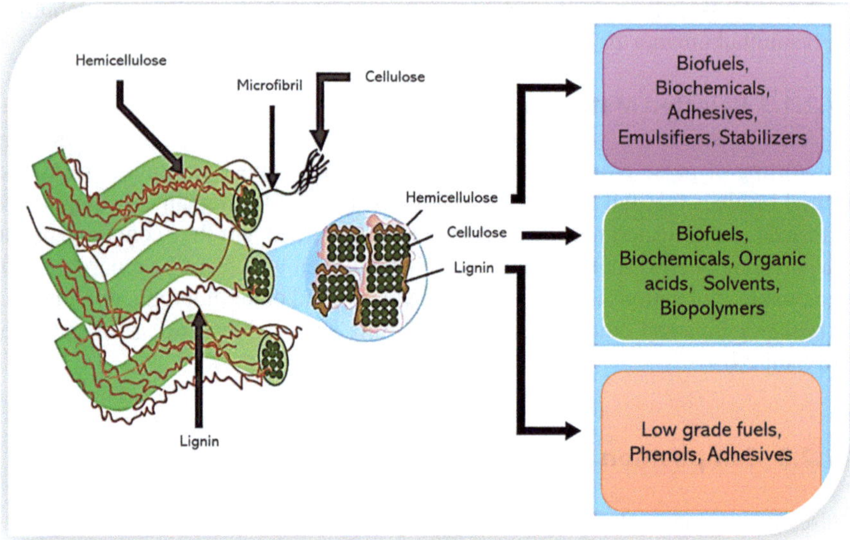

Fig. 17.2 Structure of lignocellulosic biomass and the products of its components (Lee et al. 2014)

Bioconversion of these reducing sugars produces bioethanol. Thermochemical conversion of cellulose and hemicellulose produces fuels such as bio-oil and gases along with certain chemicals and solvents.

17.2.2.2 Proteins

Proteins are a significant component of pulse waste, particularly in rejected grains, seed coats, broken seeds, and processing by-products, such as milling residues. Pulse proteins are highly valued because of their rich amino acid profile and functional properties, which make them suitable for diverse applications including animal feed, food additives, biofuels, and biochemicals. Sustainable utilization of protein-rich pulse waste is a key avenue for waste valorization and the development of bio-based industries. The residual proteins in rejected grains and shells can be harnessed for various applications (Yadav et al. 2024).

17.2.2.3 Starch and Sugars

Starch and sugars are valuable carbohydrate components present in pulse waste, particularly in rejected seeds, broken grains, and processed residues. These components are energy-dense and can be utilized as feedstocks for biofuels, biochemicals, and bio-based materials. The high starch content in many pulses and residual sugars in processing by-products makes this waste stream a promising resource for sustainable valorization. Presence of damaged grains provides a source for fermentation (Yadav et al. 2024).

17.2.3 Traditional Disposal Methods

Traditional disposal methods for pulse waste are often low-tech and driven by the need for simplicity and cost efficiency (Ankita et al. 2023). However, these methods typically fail to capitalize on the potential value of pulse waste and may lead to environmental challenges such as pollution and greenhouse gas emissions. Below are the common traditional practices for disposing of pulse waste, along with their associated challenges. Current practices for managing pulse waste include the following.

17.2.3.1 Burning

While the burning of pulse waste offers a short-term, low-cost solution to waste management, this practice has significant negative environmental, health, and economic consequences. The harmful effects on air quality, soil fertility, and public health highlight the need for alternative, sustainable disposal and valorization methods. Moving toward methods such as composting, biogas production, and biofuel generation would not only reduce the adverse impacts of burning, but also unlock the potential value of pulse waste, contributing to a more sustainable and circular economy. Burning contributes significantly to air pollution and leads to the loss of valuable organic materials that could otherwise be utilized productively (Ankita et al. 2023).

17.2.3.2 Landfilling

Landfilling is one of the most commonly employed methods of waste disposal worldwide, including pulsed waste generated from agricultural activities and processing. Despite their widespread use, landfilling has significant environmental, economic, and social implications. This method typically involves depositing pulse waste in designated landfill sites where it is compacted and covered with soil or other materials. Over time, waste decomposes anaerobically, potentially causing environmental damage and preventing the recovery of valuable resources. It occupies significant space and generates methane emissions (Ankita et al. 2023).

17.2.3.3 Animal Feed

The conversion of pulse waste into animal feed is a sustainable and effective method for valorizing agricultural by-products. Pulses such as lentils, chickpeas, peas, and beans generate a variety of residues during harvesting, threshing, and processing. These residues, including husks, stems, leaves, and broken seeds, contain valuable nutrients that can be used as animal feeds. By repurposing pulse waste for animal nutrition, this approach not only reduces waste, but also provides a cost-effective, nutrient-rich source of feed for livestock, poultry, and aquaculture. There is limited use due to variability in nutrient content and processing requirements (Ankita et al. 2023).

17.3 Potential for Valorization of Pulse Waste

Pulse waste is a great feedstock for the generation of biofuels and an excellent material for platform chemical synthesis due to its rich biological content. It is possible to turn waste materials from the processing of pulses such as lentils, chickpeas, and beans into useful resources using this opportunity (Yadav et al. 2024; Xia et al. 2016). By-products such as husks and shells can be recycled for a variety of useful applications, which is an alternative to throwing them away. For instance, they may be used as feed for animals, compost could be used to replenish soil, and either biofuels or electricity could be produced from them. Furthermore, certain components of the waste can be made into food ingredients, such as dietary fiber, by processing them (Xia et al. 2016). This approach promotes waste reduction, efficient use of resources, and the development of sustainable products and industries (Yadav et al. 2024; Ankita et al. 2023). Key valorization pathways include the following.

17.3.1 Different Types of Biofuel Production

The process of producing fuel from biological materials, such as plants and agricultural waste, is referred to as biofuel production. An example of this process is shown in Fig. 17.3. These fuels are renewable and serve as alternatives to traditional fossil fuels like petrol and diesel (Yadav et al. 2024; Ankita et al. 2023). Ethanol, which is derived from crops like maize or sugarcane, and biodiesel, which is created from vegetable or animal fats or recycled cooking oil, are examples of common biofuels. In order to produce useful fuel, the biological material is usually broken down into simpler molecules, like sugars, which are subsequently fermented or chemically

Fig. 17.3 Pulses feedstock for biofuel production. (**a**) Pea pods, (**b**) broad beans, (**c**) pigeon pea (stalk and pods), (**d**) soyabean hull, (**e**) soyabean stalk, (**f**) fava beans

treated. Biofuels are valued for being more environmentally friendly than fossil fuels as they produce fewer greenhouse gas emissions and reduce dependence on nonrenewable energy sources (Ankita et al. 2023).

17.3.1.1 Biomethane

Biomethane recovery from organic biomass is most successful by the anaerobic digestion process. Typically, anaerobic microorganism or methanogenic bacteria and temperature plays a major role in anaerobic digestion to convert the organic FW into biomethane (Ren et al. 2018; Kavitha et al. 2017). Macronutrients (protein, lipids, carbohydrate, and starch) present in the FW are hydrolyzed into simple monomers (amino acid, fat, and sugar) with the help of enzymes secreted by hydrolytic microbes. This in turn enhances the activity of methanogenic microbes to produce biogas (Zhang et al. 2020; Xu et al. 2018). The recovered biogas comprises merely 50–75% biomethane and remaining 25–50% of carbon dioxide. Anaerobic digestion process is broadly classified into two types, based on digester operating temperature, for example, (1) mesophilic and (2) thermophilic (Obulisamy et al. 2016). During mesophilic digestion (at 37 °C) and thermophilic digestion (at 55 °C), mesophilic bacteria and thermophilic bacteria play a major role in the conversion of organic waste into methane and carbon dioxide. In the AD process, a main by-product acetate was formed and maximum biomethane yield was suppressed. Hydrolysis is the rate-limiting step for biomethane production. On the other hand, the organic waste biomass gets stabilized and also reduces the environmental issues like ground water contamination, odor production, and avoid breeding of disease-causing flies. Biomethane is otherwise known as green energy from organic waste and acts as an alternative fuel to conventional fossil fuel. According to Yeshanew et al. (2016), 442 ± 8.6 mL CH_4/gVS was achieved through anaerobic digestion process and thermal pretreatment, resulting in 25% of COD solubilization.

17.3.1.2 Biohythane

Biohythane is the combination of biohydrogen and biomethane gas generated from FW through anaerobic fermentation. Biohythane consists of 10–15% of biohydrogen and 50–60% biomethane (Ghimire et al. 2017). In general, two-stage anaerobic fermentation was implemented to achieve effective yield of biohythane from FW (Ghimire et al. 2017). In two-stage anaerobic fermentation, at the first stage, hydrogen-consuming microorganism plays a major role in biohydrogen production at a shorter retention time. Then, in the second stage of fermentation process, methanogenic microorganism plays a major role in biomethane production at a longer retention time. It is an advanced technology to increase green economic growth worldwide. The chemical composition of biohythane will upturn the quality and market value compared to traditional fuel. Biohythane is well recognized as a high-commercial-value fuel that is used in combustion engines and motor vehicles (Sen et al. 2016). Similarly, Ding et al. (2017) have reported a yield of 43.0 mL H_2/gVS at the first stage and 511.6 mL CH_4/h at the second stage of anaerobic digestion. It has several advantages compared to conventional available fossil fuel such as (1) less amount of input energy is required to ignite combine and motor vehicles engines, (2) high-quality fuel than the conventional fuel, and (3), on the other hand, the biohythane has drawback in bioelectricity generation. For example, biohythane has a high market value than conventional fuel, and at present in the United States and India it has a high commercial value (Cavinato et al. 2016).

17.3.1.3 Bioethanol

The fermentation of biomass, which involves enzymatic breakdown by microorganisms at low pressure and temperature, is the main process that produces bioethanol (Chen and Fu 2016). Numerous investigations on bioethanol have been conducted using a range of feedstocks and modern technology (Jambo et al. 2016). Recent research has shed light on the potential for bioethanol generation in the future. Nigeria is investing a lot of money in the growth of its bioethanol sector. Current car biofuel plants in the states of Agasha, Guma, Bukaru, and Kupto, Benue, Gombe, and Benue, respectively, have the capacity to generate 1.8 million tons of feedstock and 59 MW of energy per year. Additionally, with 3.5 million tons of feedstock, Ebenebe, Anambra State's car biofuel facility, has a larger capacity (Ben-Iwo et al. 2016). Nigeria is investing a lot of money in the growth of its bioethanol sector. Current car biofuel plants in the states of Agasha, Guma, Bukaru, and Kupto, Benue, Gombe, and Benue, respectively, have the capacity to generate 1.8 million tons of feedstock and 59 MW of energy per year. Additionally, with 3.5 million tons of feedstock, Ebenebe, Anambra State's car biofuel facility, has a larger capacity (Adeleke et al. 2023). In Nigerian bioethanol operations, sorghum, sugarcane, and cassava are the primary feedstocks. However, as these feedstocks are important food crops, it is crucial to prevent food crises while ensuring energy security. Nigeria has enormous uncultivated areas that might be used for energy crops, thus methods for boosting capacity are needed to keep bioethanol production going (Adeleke et al. 2023). Bioethanol, a renewable fuel, can be produced from sugar and lignocellulosic fractions in pulse waste. The process involves

1. Pretreatment: chemical, enzymatic, or mechanical methods to release fermentable sugars from cellulose and hemicellulose.
2. Hydrolysis: enzymatic conversion of polysaccharides into monosaccharides.
3. Fermentation: microorganisms, such as *Saccharomyces cerevisiae*, ferment sugars into ethanol.
4. Distillation: recovery and purification of ethanol.

17.3.1.4 Biogas

Anaerobic digestion of organic materials, including food waste, agricultural wastes, and animal manure, produces biogas, a renewable energy source. In this process, these components are broken down by microorganisms without oxygen, resulting in a gas mixture that is mostly made up of carbon dioxide and methane. Because of its many uses, including heating, cooking, and electricity production, biogas is well known as a sustainable energy source. Additionally, by cutting greenhouse gas emissions, decreasing organic waste, and offering a substitute for fossil fuels, biogas production promotes environmental sustainability (Atelge et al. 2020). The by-product of this process, known as digestate, serves as a nutrient-rich fertilizer, further enhancing its value in agricultural systems (Chojnacka and Moustakas 2024).

Microorganisms break down organic matter during the complex biological process known as anaerobic digestion of pulse waste, which takes place in the absence of oxygen. Hydrolysis is the first important step in this process, where complex organic materials like proteins, fats, and carbohydrates are broken down into simpler compounds like sugars, fatty acids, and amino acids; acidogenesis is the next step, where acidogenic bacteria further ferment these simpler compounds into volatile fatty acids (VFAs), hydrogen (H_2), and carbon dioxide (CO_2); acetogenesis is the third step, where the VFAs and hydrogen are converted into acetic acid, along with additional hydrogen and carbon dioxide by acetogenic bacteria; and methanogenesis is the final critical step, where methanogenic archaea convert hydrogen and acetic acid into biogas, which is mainly made up of methane (CH_4) and carbon dioxide (CO_2). Sustainable waste management and energy production can be achieved by using the residual digestate as a nutrient-rich fertilizer and the generated biogas as a renewable energy source (Ramos-Suarez et al. 2021).

The anaerobic digestion of pulse waste produces biogas comprising methane (CH_4) and carbon dioxide (CO_2). This process occurs in the following stages.

- Hydrolysis: breakdown of complex polymers into simpler compounds.
- Acidogenesis: conversion of simple compounds into organic acids and alcohols.
- Methanogenesis: microbial production of methane.

17.3.1.5 Biohydrogen

Biohydrogen is a renewable and environmentally friendly energy source produced through biological processes using microorganisms such as bacteria and algae (Xu et al. 2022). It is primarily generated via methods like dark fermentation, photo fermentation, and microbial electrolysis, which utilize organic materials, including agricultural residues, wastewater, and other biomass, as feedstocks. Biohydrogen is considered a promising alternative to fossil fuels due to its high energy content and clean combustion, producing only water as a by-product (Saravanan et al. 2021). Additionally, its production supports waste management by utilizing organic waste streams and contributes to reducing greenhouse gas emissions (Sánchez et al. 2015). Global initiatives to mitigate climate change and create sustainable energy systems are in line with the development of biohydrogen technologies. A by-product of processing pulses like beans, chickpeas, and lentils, pulse waste has a lot of potential for producing biohydrogen. Pulse waste, which is high in organic matter and carbohydrates, is a great source of feedstock for microbial processes such as dark fermentation, in which bacteria break down organic compounds without oxygen to make hydrogen. In addition to providing a clean, renewable energy source, using pulse waste to produce biohydrogen offers a sustainable way to manage agricultural waste. This strategy lessens pollution in the environment, encourages the use of circular economy principles, and facilitates the switch to renewable energy sources (Saratale et al. 2019).

17.3.1.6 Biodiesel

Through a process known as transesterification, vegetable oils, animal fats, or recycled cooking oil can be converted into biodiesel, a renewable fuel (Kayode and Hart 2019). It is biodegradable, has less hazardous emissions, and burns cleaner than ordinary diesel. Biodiesel is an environmentally friendly substitute for conventional diesel fuel since it can frequently be utilized in diesel engines without requiring significant modifications. Biodiesel can be made by transesterifying pulse waste oils with alcohol. By converting the oils into glycerol and biodiesel (methyl esters), this process helps manage waste and provides a renewable energy source (Babadi et al. 2022). Although less common, pulsed waste oils can be trans-esterified to produce biodiesel. The process involves extraction of oils from pulse residues and transesterification with alcohols in the presence of a catalyst to produce biodiesel and glycerol (Ahmed et al. 2022).

17.3.2 Biochemical Production

Utilizing biological systems like microbes, enzymes, or plants, biochemical production is the process of creating valuable chemicals, fuels, and materials (Straathof 2014). In order to transform basic materials like sugars, biomass, or waste products into the needed biochemicals, this production is usually accomplished via fermentation, enzymatic reactions, or other biotechnological techniques (Kumar et al. 2016). Biofuels such as ethanol and biodiesel, organic acids, amino acids, and bioplastics are examples of biochemicals. Because biochemical production frequently employs renewable resources, generates less pollutants, and depends on environmentally friendly techniques, it is regarded as a sustainable alternative to traditional chemical manufacture. The area is essential for developing green chemistry and lessening the negative effects of industrial production on the environment. Biochemical production is the process of harnessing biological systems, such as microbes or enzymes, to create valuable chemicals, fuels, and materials. Biochemicals, such as biofuels, organic acids, and bioplastics, are produced from raw materials like sugars or biomass by procedures like fermentation. Compared to conventional chemical manufacturing, this process produces less pollutants and uses renewable resources, making it more environmentally friendly. Basic chemicals known as "platform chemicals" are utilized as building blocks to create a wide range of other goods, including fuels, polymers, and medications. In the chemical industry, they are indispensable and multipurpose (Parihar and Bhattacharya 2020). Waste from lignocellulosic pulses can be transformed into platform chemicals in the following ways:

- Furfural: derived from hemicellulose, used in resin production.
- Levulinic acid: a precursor for bio-based solvents and plastics.
- Lactic acid: fermented from sugars, essential for bioplastics.

17.3.2.1 Bioplastics

Bioplastics are plastics made from renewable natural materials like plants, starch, or vegetable oils, instead of fossil fuels like oil and gas. They are considered more environmentally friendly because they can be biodegradable or produced with less environmental impact (Singh and Verma 2020). Bioplastics are used in packaging, bottles, and various products, offering a sustainable alternative to traditional plastics. Pulse waste starch and proteins are also valuable for the production of bioplastics. For example: Starch-based plastics are created by polymerizing the starch extracted from rejected grains. Protein-based films derived from residual proteins for packaging applications. Pulse wastes are also utilized to produce organic polymers such as polyhyroxyalkanoates (PHA) and polyhydroxybutyrate (PHB), which are considered bioplastics and are completely biodegradable into carbon dioxide and water within a few months of burial (Phadke and Rawtani 2023). Microbes such as *Ralstoniaeutropha*, *E. coli.*, *Clostridia* sp., *Bacillus megaterium*, and natural isolates of *Actinobacillus*, *Azotobacter*, *Agrobacterium*, *Rhodobacter*, and *Sphaerotilius* utilize foodwaste as substrate and produce these polymers either as intracellular metabolite or as carbon reserves when there is limited availability of nitrogen and excess carbon (Nieto-López et al. 2013). PHA is produced in two steps. In the initial stage, the organic compounds are utilized by the biomass and are converted into volatile fatty acids (VFA). This step takes place under anaerobic condition. In the second step, the PHA-accumulating bacteria such as *Ralstonia eutropha* was cultivated to exploit short-chain fatty acids and polymerize these acids to polyhydroxyacetones under nutrient-limited conditions (Phadke and Rawtani 2023; Nieto-López et al. 2013).

17.3.2.2 Enzymes and Industrial Proteins

Enzymes and industrial proteins can be produced using agricultural waste as a source of raw material. Agricultural waste, such as crop residues, fruit peels, pulse agricultural waste (like lentil or chickpea by-products), or other plant by-products, contains valuable organic compounds that microorganisms can break down to produce enzymes or proteins. These enzymes and proteins have many industrial applications, such as in food processing, biofuels production, detergent manufacturing, and waste management (Verma et al. 2017).

By using agricultural waste, the production of enzymes and industrial proteins not only helps reduce waste but also offers a sustainable and cost-effective method for creating valuable bioproducts. This approach contributes to a circular economy by transforming waste into useful materials while minimizing environmental impact. Pulse waste proteins can be hydrolyzed to produce enzymes and amino acids for use in food, pharmaceutical, and industrial processes.

During valorization of FW to various value-added products, breaking down of complex carbohydrates into simple sugars is the rate-limiting step. FW generally contains a high amount of carbohydrates and should be hydrolyzed to monosaccharides by the saccharifying enzymes. Thus, the production of cellulases and amylases are of great interest (Teeri 1997). Enzymes such as pectinase, lipase, protease, and amylase are extracted from various types of FW using solid-state fermentation.

Wastes such as potato peel are fermented by *Bacillus* at 35 °C and pH 7.5 to generate about 676 U/mL of amylase. Microbes such as *Penicillium* sp., *Bacillus* sp., and *Aspergillus* sp. can ferment various FW such as apple pomace, orange peel, and citrus wastes and produce pectinase (Martinez Sabajanes et al. 2012). Based on the purification process of the foodwaste and its transportation cost, the recovered enzymes can have various industrial applications. The enzymes thus produced through fermentation can be incorporated into other biorefinery process to generate various platform chemicals and biofuels.

17.4 Conversion Technologies

Technologies such as fermentation, where microorganisms convert waste into bioethanol or biogas; transesterification, which turns pulse oils into biodiesel; gasification, which creates syngas for biofuels or chemicals; and anaerobic digestion, which produces biogas, are examples of conversion technologies for valorizing pulse waste into biofuels and chemicals. Furthermore, pulse waste is heated by pyrolysis to create charcoal and bio-oil. By converting agricultural waste into useful products, these technologies promote sustainability and lessen their negative effects on the environment. Pretreatment methods, biochemical procedures, and thermochemical processes are some of the crucial phases of conversion technologies for turning pulse waste into biofuels and chemicals. By combining these technologies, pulse waste may be efficiently converted into chemical goods and sustainable energy, which lowers waste and its impact on the environment (Arya et al. 2022).

17.4.1 Pretreatment Techniques

Efficient pretreatment is essential for breaking down complex biomass structures and enhancing the accessibility of fermentable components (Behera et al. 2014). Techniques include

- Physical: grinding, milling, and steam explosion.
- Chemical: acidic or alkaline hydrolysis to break down lignin and hemicellulose.
- Biological: enzymatic degradation using cellulases or microbial action.

17.4.2 Biochemical Processes

In biochemical processes, methods like fermentation and anaerobic digestion transform pulse waste into biofuels like bioethanol and biogas, while transesterification of pulse oils produces biodiesel (Ankita et al. 2023).

- Fermentation: microbial conversion of sugars into ethanol, organic acids, or other chemicals.

- Anaerobic digestion: multistage microbial processes for biogas production.

17.4.3 Thermochemical Processes

Biochar: The process of decomposing FW using high thermal temperature (between 400 and 900 °C) is known as pyrolysis (Premchand et al. 2023). During this process, a limited amount of oxygen that is availed in that environment has been utilized. Most probably, this process takes place under strict anaerobic condition. As a result, incomplete combustion of feedstock will generate a stable carbon-rich solid product such as biochar that is otherwise termed as black carbon (Robertson et al. 2012). It mainly comprises a high amount of carbon, material with low density, and is highly porous in nature (Kumar et al. 2016). Especially biochar is used as an excellent absorbent to remove organic and inorganic chemical compounds from wastewater. In some cases, biochar is also used as soil conditioner, which has a high water-holding capacity, to enhance the soil fertility and nutrient content for effective growth of plants and crop residues. Biochar has several advantages such as (1) reducing the concentration of biological oxygen demand, (2) total nitrogen, (3) nutrient leaching, (4) avoid emission of greenhouse gases, (5) less fertility requirement, (6) effective in soil–water management, and (7) increase soil fertility and plant productivity (Kumar et al. 2016). The recovery of value-added products from various food waste is as follows:

- Pyrolysis: heating the biomass in the absence of oxygen to produce bio-oil, syngas, and biochar.
- Gasification: the high-temperature conversion of biomass into syngas is used for fuel or chemical synthesis.
- Hydrothermal liquefaction: conversion of wet biomass into bio-crude oil at high temperature and pressure.

17.5 Specialty Chemicals and Its Components

Specialty chemicals generally indicated as functional compounds have a distinctive capability to be utilized as antioxidants, antimicrobials, flavor compounds, and colorants in the food, pharmaceutical, and cosmetics industry (Kawatra et al. 2023). Numerous studies have revealed that diverse functional compounds extracted from food waste such as phenolic compounds, bioactive peptides, carotenoid, and dietary fiber serve as a beneficial source of nutraceuticals, thereby augmenting human health. Phenolic compounds form a larger class of bioactive compounds that contain a polyphenol structure with a considerable number of hydroxyl groups in aromatic rings (Kawatra et al. 2023; Ho et al. 2012). Polyphenols are befitted to be used as antioxidants owing to their potentiality of scavenging free radicals and are further classified as flavonoids, phenolic acids, tannins (hydrolyzable and condensed), stilbenes, and lignans (Ho et al. 2012). Past studies have affirmed that a high volume of phenolic compounds are found in the peels, rinds, and seeds from fruits and vegetable waste (FVW). Resveratrol, a significant stilbene compound extracted from grape by-products using super-critical CO_2, has attained commercialization status in recent times (Zhou et al. 2021). Choi et al. (2016) revealed that potato peel extracts were found to be rich in phenolic acids (mainly chlorogenic acid and caffeic acid) than in the potato cortex.

17.6 Future Prospects on Pulses Waste Valorization

The future prospects for pulse waste valorization are increasingly promising, driven by advancements in technology, a growing emphasis on sustainability, and the need to address global food waste challenges. As awareness of the environmental impact of food waste rises, there is a concerted effort to transform pulse waste into valuable products such as biofuels, biochemicals, and bioactive compounds. Emerging technologies, including anaerobic digestion and fermentation, are being refined to optimize the conversion of pulse by-products into bioenergy and high-value chemicals. Additionally, thermochemical processes like pyrolysis and gasification are gaining traction, allowing for the efficient transformation of pulse waste into bio-oil and syngas, which can serve as renewable energy sources. The economic viability of these processes is bolstered by increasing demand for renewable energy and government incentives aimed at promoting sustainable practices. Furthermore, the circular economy model is becoming integral to valorization strategies, emphasizing resource recovery and minimizing waste generation. This approach not only enhances economic opportunities for farmers and food processors but also contributes to environmental sustainability by reducing landfill use and greenhouse gas emissions. However, challenges remain, including the need for further research to optimize valorization techniques and develop infrastructure for processing and distribution. As novel extraction methods for bioactive compounds from pulse waste continue to be explored, the potential for creating functional ingredients for food and pharmaceutical industries expands. Overall, the future of pulse waste valorization is characterized by innovation and sustainability, positioning it as a key player in addressing global energy needs while promoting environmental stewardship.

17.7 Conclusion

The valorization of pulse waste into biofuels and chemicals represents a promising avenue for sustainable development. By employing biochemical and thermochemical conversion techniques alongside extraction methods for valuable compounds, it is possible to transform agricultural by-products into economically viable resources. This not only addresses environmental challenges but also promotes a circular economy where waste is minimized, resources are reused, and new value is created from existing materials. Continued research and innovation in this field will be crucial for optimizing these processes and expanding their applications in the future.

References

Adeleke AA, Petrus N, Ayuba S, Yahya AM, Ikubanni PP, Okafor IS, Emmanuel SS, Olosho AI, Adesibikan AA (2023) Nigerian biomass for bioenergy applications: a review on the potential and challenges. J Renewable Mater 11(12):4123–4141

Ahmed M, Abdullah A, Patle DS, Shahadat M, Ahmad Z, Athar M, Aslam M, Vo DVN (2022) Feedstocks, catalysts, process variables and techniques for biodiesel production by one-pot extraction-transesterification: a review. Environ Chem Lett 20(1):335–378

Alghashm S, Qian S, Hua Y, Wu J, Zhang H, Chen W, Shen G (2018) Properties of biochar from anaerobically digested food waste and its potential use in phosphorus recovery and soil amendment. Sustain For 10(12):4692

Ankita, Choudhary S, Bakala HS, Sarao LK, Kaur S (2023) Pulses waste to biofuels. In: Agroindustrial waste for green fuel application. Springer, Singapore, pp 1–26

Arya PS, Yagnik SM, Rajput KN, Panchal RR, Raval VH (2022) Valorization of agro-food wastes: ease of concomitant-enzymes production with application in food and biofuel industries. Bioresour Technol 361:127738

Atelge MR, Krisa D, Kumar G, Eskicioglu C, Nguyen DD, Chang SW, Atabani AE, Al-Muhtaseb AH, Unalan S (2020) Biogas production from organic waste: recent progress and perspectives. Waste Biomass Valoriz 11:1019–1040

Babadi AA, Rahmati S, Fakhlaei R, Barati B, Wang S, Doherty W, Ostrikov KK (2022) Emerging technologies for biodiesel production: processes, challenges, and opportunities. Biomass Bioenergy 163:106521

Behera S, Arora R, Nandhagopal N, Kumar S (2014) Importance of chemical pretreatment for bioconversion of lignocellulosic biomass. Renew Sust Energ Rev 36:91–106

Ben-Iwo J, Manovic V, Longhurst P (2016) Biomass resources and biofuels potential for the production of transportation fuels in Nigeria. Renew Sust Energ Rev 63:172–192

Bhuvaneshwari S, Hettiarachchi H, Meegoda JN (2019) Crop residue burning in India: policy challenges and potential solutions. Int J Environ Res Public Health 16(5):832

Bizkarra K, Barrio VL, Arias PL, Cambra JF (2017) Biomass fast pyrolysis for hydrogen production from bio-oil. In: Hydrogen production technologies. Wiley, Hoboken, pp 305–362

Cavinato C, Bolzonella D, Pavan P, Cecchi F (2016) Two-phase anaerobic digestion of food wastes for hydrogen and methane production. In: Enriched methane: the first step towards the hydrogen economy. Springer, Cham, pp 75–90

Chen H, Fu X (2016) Industrial technologies for bioethanol production from lignocellulosic biomass. Renew Sust Energ Rev 57:468–478

Choi SH, Kozukue N, Kim HJ, Friedman M (2016) Analysis of protein amino acids, non-protein amino acids and metabolites, dietary protein, glucose, fructose, sucrose, phenolic, and flavonoid content and antioxidative properties of potato tubers, peels, and cortexes (pulps). J Food Compos Anal 50:77–87

Chojnacka K, Moustakas K (2024) Anaerobic digestate management for carbon neutrality and fertilizer use: a review of current practices and future opportunities. Biomass Bioenergy 180:106991

Ding B, Li Z, Qin Y (2017) Nitrogen loss from anaerobic ammonium oxidation coupled to iron (III) reduction in a riparian zone. Environ Pollut 231:379–386

Diwan B, Mukhopadhyay D, Gupta P (2020) Recent trends in biorefinery-based valorisation of lignocellulosic biomass. In: Biovalorisation of wastes to renewable chemicals and biofuels. Elsevier, Amsterdam, pp 219–242

Ghimire A, Kumar G, Sivagurunathan P, Shobana S, Saratale GD, Kim HW, Luongo V, Esposito G, Munoz R (2017) Bio-hythane production from microalgae biomass: key challenges and potential opportunities for algal bio-refineries. Bioresour Technol 241:525–536

Gürdil G, Mengstu M, Omer AK (2021) Utilization of agricultural wastes for sustainable development. Black Sea J Agric 4(4):146–152

Ho YL, Huang SS, Deng JS, Lin YH, Chang YS, Huang GJ (2012) In vitro antioxidant properties and total phenolic contents of wetland medicinal plants in Taiwan. Bot Stud 53:1
Jambo SA, Abdulla R, Azhar SHM, Marbawi H, Gansau JA, Ravindra P (2016) A review on third generation bioethanol feedstock. Renew Sust Energ Rev 65:756–769
Kannah RY, Merrylin J, Devi TP, Kavitha S, Sivashanmugam P, Kumar G, Banu JR (2020) Food waste valorization: biofuels and value added product recovery. Bioresour Technol Rep 11:100524
Kavitha S, Banu JR, Priya AA, Yeom IT (2017) Liquefaction of food waste and its impacts on anaerobic biodegradability, energy ratio and economic feasibility. Appl Energy 208:228–238
Kawatra A, Gupta S, Dhankhar R, Singh P, Gulati P (2023) Application of phytochemicals in therapeutic, food, flavor, and cosmetic industries. In: Phytochemical genomics: plant metabolomics and medicinal plant genomics. Springer, Singapore, pp 85–108
Kayode B, Hart A (2019) An overview of transesterification methods for producing biodiesel from waste vegetable oils. Biofuels 10(3):419–437
Keshavarz R, Didinger C, Duncan A, Thompson H (2020) Pulse crops and their key role as staple foods in healthful eating patterns–0.313 arrow. Pulse 970:491
Kumar R, Gupta SS, Katiyar S, Raman VK, Varigala SK, Pradeep T, Sharma A (2016) Carbon aerogels through organo-inorganic co-assembly and their application in water desalination by capacitive deionization. Carbon 99:375–383
Lee HV, Hamid SBA, Zain SK (2014) Conversion of lignocellulosic biomass to nanocellulose: structure and chemical process. Sci World J 2014(1):631013
Leser S (2013) The 2013 FAO report on dietary protein quality evaluation in human nutrition: recommendations and implications. Nutr Bull 38(4):421–428
Luzardo-Ocampo I, Cuellar-Nuñez ML, Oomah BD, Loarca-Piña G (2020) Pulse by-products. In: Food wastes and by-products: nutraceutical and health potential. Wiley, Hoboken, pp 59–92
Martinez Sabajanes M, Yáñez R, Alonso JL, Parajó JC (2012) Pectic oligosaccharides production from orange peel waste by enzymatic hydrolysis. Int J Food Sci Technol 47(4):747–754
Merrylin J, Kannah RY, Banu JR, Yeom IT (2020) Production of organic acids and enzymes/biocatalysts from food waste. In: Food waste to valuable resources. Academic, New York, pp 119–141
Nath B, Chen G, O'Sullivan CM, Zare D (2024) Research and technologies to reduce grain post-harvest losses: a review. Food Secur 13(12):1875
Nayak A, Bhushan B (2019) An overview of the recent trends on the waste valorization techniques for food wastes. J Environ Manag 233:352–370
Nieto-López I, Sanchez-Vazquez M, Bonilla-Cruz J (2013) A theoretical and experimental study. Macromol Symp 325(1):132–140
Obulisamy PK, May JSY, Rajasekar B (2016) Gradient packing bed bio-filter for landfill methane mitigation. Bioresour Technol 217:205–209
Parihar A, Bhattacharya S (2020) Cellulose fast pyrolysis for platform chemicals: assessment of potential targets and suitable reactor technology. Biofuels Bioprod Biorefin 14(2):446–468
Phadke G, Rawtani D (2023) Bioplastics as polymeric building blocks: paving the way for greener and cleaner environment. Eur Polym J 199:112453
Premchand P, Demichelis F, Chiaramonti D, Bensaid S, Fino D (2023) Study on the effects of carbon dioxide atmosphere on the production of biochar derived from slow pyrolysis of organic agro-urban waste. Waste Manag 172:308–319
Ramos-Suarez M, Zhang Y, Outram V (2021) Current perspectives on acidogenic fermentation to produce volatile fatty acids from waste. Rev Environ Sci Biotechnol 20(2):439–478
Rathinam NK, Sani R (2019) Biovalorisation of wastes to renewable chemicals and biofuels. Elsevier, Amsterdam
Ren Y, Yu M, Wu C, Wang Q, Gao M, Huang Q, Liu Y (2018) A comprehensive review on food waste anaerobic digestion: research updates and tendencies. Bioresour Technol 247:1069–1076
Robertson SJ, Rutherford PM, Lopez-Gutierrez JC, Massicotte HB (2012) Biochar enhances seedling growth and alters root symbioses and properties of sub-boreal forest soils. Can J Soil Sci 92(2):329–340

Sánchez A, Artola A, Font X, Gea T, Barrena R, Gabriel D, Sánchez-Monedero MÁ, Roig A, Cayuela ML, Mondini C (2015) Greenhouse gas emissions from organic waste composting. Environ Chem Lett 13:223–238
Sanchez-Vazquez SA, Hailes HC, Evans JRG (2013) Hydrophobic polymers from food waste: resources and synthesis. Polym Rev 53(4):627–694
Saratale GD, Saratale RG, Banu JR, Chang JS (2019) Biohydrogen production from renewable biomass resources. In: Biohydrogen. Elsevier, Amsterdam, pp 247–277
Saravanan A, Kumar PS, Khoo KS, Show PL, Carolin CF, Jackulin CF, Jeevanantham S, Karishma S, Show KY, Lee DJ, Chang JS (2021) Biohydrogen from organic wastes as a clean and environment-friendly energy source: production pathways, feedstock types, and future prospects. Bioresour Technol 342:126021
Selvamuthukumaran M (2024) Wealth out of food processing waste: ingredient recovery and valorization. CRC Press, Boca Raton
Sen B, Aravind J, Kanmani P, Lay CH (2016) State of the art and future concept of food waste fermentation to bioenergy. Renew Sust Energ Rev 53:547–557
Shakuntala VK, Kumari P, Khan SA, Meena BS, Shakuntala PK, Kumar V (2024) Integrated pest management approach in pulse crops for sustainability of farmers income. Indian J Agric Sci 92(4):531–536
Singh P, Verma R (2020) Bioplastics: a green approach toward sustainable environment. In: Environmental microbiology and biotechnology: volume 1: biovalorization of solid wastes and wastewater treatment. Springer, Cham, pp 35–53
Snapp S, Rahmanian M, Batello C (2018) Pulse crops for sustainable farms in sub-Saharan Africa. FAO, Rome
Straathof AJ (2014) Transformation of biomass into commodity chemicals using enzymes or cells. Chem Rev 114(3):1871–1908
Tassoni A, Tedeschi T, Zurlini C, Cigognini IM, Petrusan JI, Rodríguez Ó, Neri S, Celli A, Sisti L, Cinelli P, Signori F (2020) State-of-the-art production chains for peas, beans and chickpeas—valorization of agro-industrial residues and applications of derived extracts. Molecules 25(6):1383
Teeri TT (1997) Crystalline cellulose degradation: new insight into the function of cellobiohydrolases. Trends Biotechnol 15(5):160–167
Verma A, Singh H, Anwar S, Chattopadhyay A, Tiwari KK, Kaur S, Dhilon GS (2017) Microbial keratinases: industrial enzymes with waste management potential. Crit Rev Biotechnol 37(4):476–491
Xia H, Houghton JA, Clark JH, Matharu AS (2016) Potential utilization of unavoidable food supply chain wastes–valorization of pea vine wastes. ACS Sustain Chem Eng 4(11):6002–6009
Xu F, Li Y, Ge X, Yang L, Li Y (2018) Anaerobic digestion of food waste–challenges and opportunities. Bioresour Technol 247:1047–1058
Xu X, Zhou Q, Yu D (2022) The future of hydrogen energy: bio-hydrogen production technology. Int J Hydrog Energy 47(79):33677–33698
Yadav M, Sreerama YN, Meera MS (2024) Influence of protein on calcium bioaccessibility with phytic acid in milled by-product fractions of pulses. Waste Biomass Valoriz 15:6649–6664
Yeshanew MM, Frunzo L, Lens PN, Pirozzi F, Esposito G (2016) Mass loss controlled thermal pretreatment system to assess the effects of pretreatment temperature on organic matter solubilization and methane yield from food waste. Front Environ Sci 4:62
Zhang H, Liu G, Xue L, Zuo J, Chen T, Vuppaladadiyam A, Duan H (2020) Anaerobic digestion based waste-to-energy technologies can halve the climate impact of China's fast-growing food waste by 2040. J Clean Prod 277:123490
Zhou J, Gullon B, Wang M, Gullon P, Lorenzo JM, Barba FJ (2021) The application of supercritical fluids technology to recover healthy valuable compounds from marine and agricultural food processing by-products: a review. PRO 9(2):357

Challenges and Future Prospects of Conversion of Agricultural Solid Waste (ASW) to Biofuel

18

Pushpendra Sharma, Sagish Sagia, and Rajeev Kaushik

Abstract

Agricultural solid waste (ASW) is a byproduct of global food production, representing both an environmental challenge and a significant opportunity for renewable energy generation. This chapter examines the potential of ASW as a sustainable feedstock for biofuel production, including bioethanol, biodiesel, and biogas. It reviews existing conversion technologies, evaluates environmental benefits, and socio-economic benefits, and outlines conventional value-added applications. Key barriers such as technological constraints, economic viability, inconsistent feedstock supply, and inadequate infrastructure are discussed. The chapter underscores the need for policy support, advanced bioconversion techniques, integrated biorefineries, and circular economy approaches to overcome these challenges. By fostering strategic investments and collaborative efforts, ASW can be transformed from an environmental burden into a cornerstone of sustainable energy and agricultural waste management.

Keywords

Agricultural solid waste · Bioethanol · Circular economy · Bioconversion

18.1 Introduction

The exponential growth of the human population has led to a corresponding increase in agricultural production over the past 50 years, attributed to advancements like the green revolution and expanded arable land (Food and Agriculture Organization of the United Nations (FAO) 2017a; Food and Agriculture Organization of the United Nations (FAO) and Organization for Economic Co-operation and Development (OECD) 2019). Approximately 24 million tons of food are supplied globally by the

P. Sharma · S. Sagia · R. Kaushik (✉)
Division of Microbiology, ICAR-Indian Agricultural Research Institute, New Delhi, India

A. Kumari et al. (eds.), *Valorization of Solid Wastes to Biofuels and Chemical Products for Sustainable World*, https://doi.org/10.1007/978-981-96-8594-3_18

agricultural sector, although this brings health risks and environmental threats (Food and Agriculture Organization of the United Nations (FAO) 2017b). While agriculture is indispensable for global food security, it also contributes significantly to environmental degradation; for instance, around 21% of greenhouse gas emissions originate from agricultural activities. Consequently, there is a pressing need to enhance agricultural practices, especially in managing solid waste effectively (Food and Agriculture Organization of the United Nations (FAO) 2016).

Global leaders face the dual challenge of meeting the growing demand for food and energy due to population growth, expected to surpass 10 billion by 2050, while mitigating the adverse impacts of agriculture (European Commission Investing in Sustainable Development 2018; United Nations 2019). This necessitates a shift toward sustainable practices with reduced emissions, minimal waste, and decreased reliance on fossil fuels. Future projections indicate a rise in agricultural production to meet diverse needs, including industrial requirements and animal feed (Food and Agriculture Organization of the United Nations (FAO) 2017c). However, this expansion will inevitably lead to increased generation of agricultural solid waste, contingent on various factors like production methods and product types.

The agricultural sector, while essential for the global economy, is a significant source of solid waste, presenting both environmental and economic concerns (Bracco et al. 2018; European Commission 2015). Effective management of this waste offers numerous potential advantages, including mitigating greenhouse gas emissions, stimulating new markets, and facilitating the production of bioenergy and animal feed (McCormick and Kautto 2013; Scarlat et al. 2015). Given its far-reaching implications for health, environmental sustainability, and food security, prioritizing agricultural waste management is of paramount importance. The recycling of agricultural solid waste presents a multitude of advantages, notably the reduction of greenhouse gas emissions and a decreased dependence on fossil fuels. Moreover, it serves as a catalyst for the emergence of new environmentally friendly markets, generating employment opportunities, fostering bioenergy production, and enabling the transformation of agricultural solid waste into valuable cattle feed (McCormick and Kautto 2013; Scarlat et al. 2015). The significance of effectively managing agricultural solid wastes cannot be emphasized enough. These wastes originate from diverse sources, with pesticides like insecticides and herbicides being prominent contributors. Research suggests that completely stopping pesticide usage would lead to a substantial 42% decrease in global food production (Northerly 2019). Considering the severe environmental, animal welfare, and human health repercussions of mishandling agricultural solid waste, concerted efforts are imperative to tackle this pressing threat.

One of the primary drivers of agricultural solid waste mismanagement is a lack of awareness. Many individuals involved in generating this waste, including homeowners and farmers, are often uninformed about appropriate disposal practices (Wang et al. 2015a, b). This knowledge gap, coupled with potential logistical challenges and a limited understanding of associated health hazards, contributes to the estimated 7.5% annual increase in agricultural solid waste production. The consequences of this mismanagement are significant. In many developing countries, agricultural waste is often discarded haphazardly in public spaces, leading to a range of environmental problems. These include air and water pollution, soil contamination, and the release of harmful gases, smoke, and dust (Varma et al. 2015; Karak et al. 2015; Wang et al. 2016). Such practices can also disrupt aquatic ecosystems and contribute to soil erosion and sedimentation.

Furthermore, the use of pesticides to control pests that thrive in these mismanaged waste environments introduces additional risks. These chemicals can contaminate agricultural products, negatively impact human health, and disrupt the ecological balance (Varma et al. 2015). Ultimately, the mismanagement of agricultural solid waste poses a multifaceted threat to both human and environmental well-being.

18.2 Status of Agricultural Solid Waste

The global agricultural and food sectors are massive generators of waste, encompassing a wide range of materials such as crop residues, animal manure, packaging, and agrochemical containers (Cherubin et al. 2018). The sheer volume of this waste is staggering; global production of agricultural residues alone was estimated to be 3.8 billion metric tons of dry matter in 2018 (Cherubin et al. 2018). To illustrate the scale further, consider specific examples: approximately 377 million tons of corn stover are produced globally, while rice straw generation reaches an estimated 658 million tons. Cereal straw production amounts to around 1580 million tons, including barley and oat in Europe, corn and sorghum in the United States, and rice and wheat in China (Bhuyan et al. 2020). The European Union alone contributes a significant portion of this waste, generating up to 230 million tons of agro-industrial waste from its food and beverage industries, along with cereal and forestry residues (European Commission 2023). Worryingly, this already substantial amount of waste is projected to increase. With the global population expected to reach 9.3 billion by 2050, a corresponding rise in agro-industrial waste is anticipated (Fierascu et al. 2019). This highlights the urgent need for effective waste management strategies in the agricultural sector. The generation of agricultural residues in major countries during 2023–2024 is displayed in Fig. 18.1.

In terms of biofuel potential, China generates 649 MT of residues, largely from corn stalks and rice straw, concentrated in Jiangsu and Zhejiang provinces, generating 35 billion liters of potential biofuel (El-Sayed et al. 2023). The United States, with 581 MT of residues like corn stover and wheat straw, primarily in the Midwest, has the highest efficiency, yielding 50.0 billion liters (USDA 2023). Brazil's 195 MT of sugarcane bagasse provides 30.0 billion liters (EPE (Energy Research Company) 2023). Other contributors include Indonesia (20 billion liters) (Institute for Essential Services Reform (IESR) 2023), Russia (6.6 billion liters) (Food and Agriculture Organization (FAO) 2023), Pakistan (12 billion liters) (Pakistan Agricultural Research Council (PARC) 2023), Bangladesh (8.0 billion liters) (Renewable Energy Bangladesh 2023), and Nigeria (10.0 billion liters) (Nigerian Energy Commission 2023).

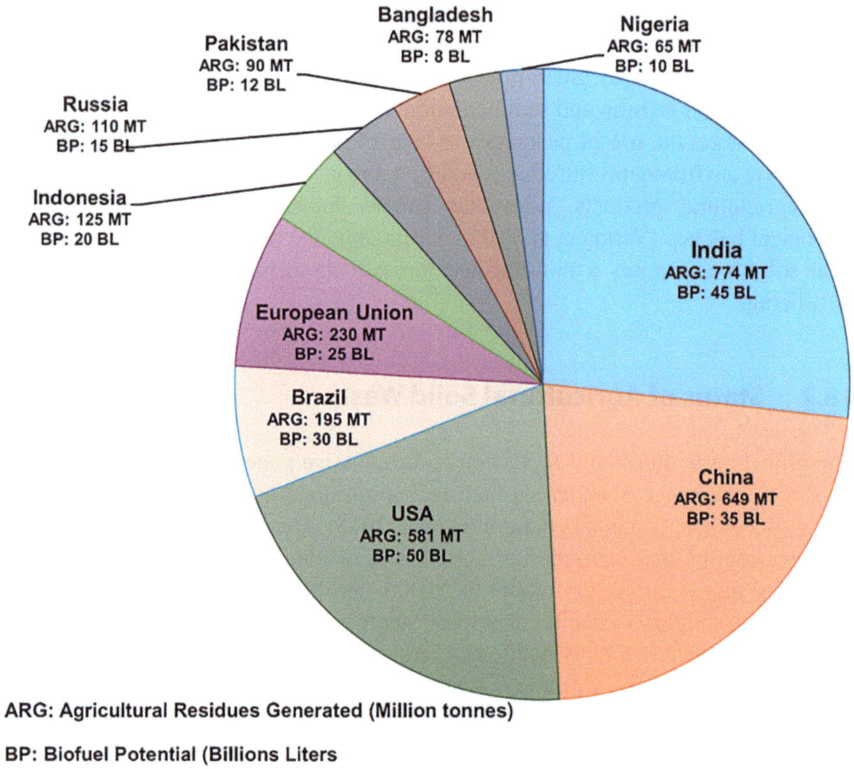

Fig. 18.1 Global agricultural residue generation and biofuel potential for 2023

India, with its diverse agricultural practices across various agro-climatic zones, generates a massive amount of crop residue annually, estimated at around 774 million tons. This includes rice husks, wheat straw, and sugarcane bagasse, primarily from Punjab, Haryana, and Uttar Pradesh (Dhingra and Sharma 2024). According to the Indian Ministry of New and Renewable Energy (MNRE), this waste represents a significant biofuel potential of 45.0 billion liters. However, improper disposal, such as the burning of an estimated 92 million tons of the 140 million ton surplus, leads to serious environmental problems. These include soil degradation, water pollution, greenhouse gas emissions, and habitat destruction, all of which negatively impact ecosystems and human health. As the world's second-largest producer of rice and wheat, both significant waste generators, India faces a particular challenge in managing this residue. Despite regulations aimed at curbing practices like crop burning, significant quantities continue to be burned each year, especially in Punjab and Haryana, contributing to air pollution and other environmental issues (MNRE 2021). While some agricultural waste is utilized beneficially as fertilizer or for energy production, the excess often overwhelms available disposal options, leading to pollution and groundwater contamination. This highlights the urgent need for comprehensive and sustainable agricultural waste management strategies in India.

18.3 Various Conventional Uses of Agricultural Solid Waste (ASW)

Agricultural solid waste (ASW), a by-product of agricultural practices, encompasses a diverse range of materials. These include

- Crop residues: This category comprises the remnants of plants after harvest, such as stems, leaves, husks, straw, stalks, and bagasse (the fibrous residue of sugarcane).
- Livestock waste: This includes manure and bedding materials from various livestock, including cows, pigs, poultry, and sheep. It contains organic matter, nutrients like nitrogen and phosphorus, and potentially harmful pathogens.
- Agro-industrial waste: This category encompasses a wide range of waste products from the agricultural industry, including packaging materials (plastic and paper), empty pesticide containers, discarded machinery parts, and processing residues from food and beverage industries (e.g., fruit pomace and brewery waste).

The composition of these residues, particularly crop residues, is primarily composed of complex carbohydrate polymers, including cellulose (40–50%), hemicellulose (25–35%), and lignin (15–20%).

Recognizing the potential of ASW, there is a growing emphasis on its effective utilization. By exploiting these renewable resources, particularly the renewable carbon present, we can contribute to sustainable industrial development. ASW can replace fossil resources in the production of various products, including chemicals, materials, polymers, fuels, and energy. This can be achieved through various approaches:

- Specialty product extraction: extracting valuable compounds like essential oils or bioactive substances.
- Conversion by green chemical or biotechnological processes: utilizing environmentally friendly methods to transform ASW into valuable products.
- Integrated biorefining: processing ASW to produce a range of products, including biofuels, chemicals, and materials.
- Industrial symbiosis: creating partnerships where the waste of one industry becomes the raw material for another.
- Cascade processing: utilizing ASW in a series of steps, maximizing resource utilization, and minimizing waste.
- On-site processing of seasonal waste streams: processing ASW directly at the farm or production site.

Beyond its potential for biofuel production, ASW can be utilized to produce a variety of value-added compounds, promoting sustainable practices and resource utilization. Here are some examples:

1. Composting: Organic fertilizer can be produced through the composting of agricultural waste, such as crop residues and livestock manure. This composting process yields a valuable soil amendment that enhances soil structure, boosts fertility, and improves water retention. Consequently, this promotes healthy plant growth and reduces the reliance on synthetic fertilizers, which can have detrimental environmental effects.
2. Animal bedding: Utilizing dry crop residues, straw, and other plant materials as bedding for livestock offers multiple benefits. This practice not only provides a comfortable and hygienic environment for the animals but also effectively absorbs moisture, contributing to improved animal health and welfare. Furthermore, the used bedding can be composted, facilitating nutrient recycling and minimizing waste. This approach supports sustainable livestock management by repurposing agricultural by-products and reducing environmental impact.
3. Soil amendment: Incorporating certain agricultural wastes, such as rice husks and sugarcane bagasse, as soil amendments can significantly enhance soil quality. These materials improve soil aeration, promoting better oxygen flow to plant roots. They also increase moisture retention, reducing the need for frequent irrigation. Additionally, they contribute to nutrient availability, ensuring essential elements are accessible to plants. These improvements benefit crop productivity and contribute to more sustainable agricultural practices.
4. Biodegradable packaging: Plant-based agricultural residues, like corn stalks and wheat straw, offer a promising avenue for producing biodegradable packaging materials. By utilizing these renewable resources, we can create eco-friendly alternatives to conventional plastics, which contribute significantly to environmental pollution and waste accumulation. This shift towards biodegradable packaging presents an opportunity to reduce our reliance on fossil fuels, mitigate plastic pollution, and promote a more sustainable approach to packaging and waste management.
5. Animal feed: Many agricultural by-products hold significant value as nutritious feed ingredients for livestock. For instance, citrus pulp, brewers' grains, and oilseed meals can be incorporated into animal diets to provide valuable sources of protein, fiber, and energy. Similarly, by-products like dried distillers' grains (a co-product of bioethanol production) and oilseed meals (such as soybean meal and canola meal) serve as excellent feed additives. These additives enrich animal diets with essential nutrients, promoting improved nutrition, growth, and overall health. This approach not only reduces waste but also enhances the sustainability of livestock production by utilizing valuable resources that might otherwise be discarded.
6. Biochar production: Biochar, a charcoal-like material rich in carbon, can be produced through the pyrolysis of agricultural residues. This process involves

heating the biomass in a low-oxygen environment. Biochar offers numerous benefits when applied to agricultural lands, such as

(a) *Enhanced soil fertility*: Biochar's porous structure improves soil aeration and water retention, creating a more hospitable environment for beneficial microbes and promoting nutrient availability.
(b) *Increased water retention*: The high surface area of biochar allows it to absorb and hold significant amounts of water, reducing the need for frequent irrigation and improving drought resilience.
(c) *Carbon sequestration*: Biochar is highly stable and can store carbon in the soil for long periods, effectively removing carbon dioxide from the atmosphere and mitigating climate change.

7. Mushroom cultivation: Agricultural wastes, such as sawdust, straw, and spent brewery grains, are utilized as substrates for mushroom cultivation. Mushroom farming not only transforms waste into nutritious food but also generates additional income for farmers.
8. Construction materials: Fibrous agricultural residues, like coconut coir and hemp stalks, are used in manufacturing eco-friendly construction materials such as biocomposites, insulation panels, and natural fibers for reinforced concrete.
9. Land reclamation: Certain types of agricultural waste, such as dredged sediments and plant residues, are utilized in land reclamation projects. They help stabilize soil, control erosion, and restore degraded landscapes, particularly in coastal and wetland areas.
10. Water filtration: Some agricultural wastes, such as coconut shells and activated carbon derived from agricultural residues, are used in water filtration systems. They remove contaminants and improve water quality for drinking and irrigation purposes.
11. Biochemicals: Lignocellulosic biomass from agricultural waste can be converted into biochemicals like organic acids, enzymes, and amino acids. These have diverse uses: organic acids act as preservatives and flavor enhancers in food; biochemicals can be used in pharmaceuticals and cosmetics; and bio-based solvents offer sustainable alternatives in industrial processes. This approach adds value to agricultural waste and promotes eco-friendly products.
12. Biocomposites: Fibrous agricultural residues, such as wheat straw, hemp fibers, and coconut coir, can be combined with biopolymers to produce biocomposites. Biocomposites are used in construction materials, automotive parts, packaging, and furniture, offering lightweight and eco-friendly alternatives.
13. Phytochemical extracts: Agricultural waste, such as fruit pomace, grape marc, and tea leaves, contains phytochemical compounds (e.g., polyphenols, antioxidants) with potential health benefits. Extracting and purifying these phytochemicals can lead to functional ingredients for dietary supplements, nutraceuticals, and food fortification.
14. Biological enzymes: Enzymes derived from microbial fermentation of agricultural waste or genetic engineering (e.g., cellulases, proteases, lipases) have

industrial applications in biofuel production, food processing, textile manufacturing, and waste treatment.

15. Natural dyes and pigments: Plant-based agricultural waste, such as fruit peels, leaves, and flowers, contains natural pigments and dyes (e.g., anthocyanins, carotenoids) used in textiles, cosmetics, and food coloring applications.
16. Essential oils and aromatics: Aromatic plants and herbal residues can be processed to extract essential oils, terpenes, and fragrances used in aromatherapy, perfumery, and personal care products.
17. Renewable energy co-generation: Integrated biorefinery concepts utilize agricultural waste for simultaneous production of biofuels, power generation (via combustion or gasification), and heat recovery, maximizing energy efficiency and resource utilization.

18.4 Potential of ASW for Biofuel

ASW holds significant potential for biofuel production, contributing to renewable energy sources and sustainable fuel alternatives. Here are some key aspects of its potential for biofuel production. As a feedstock, ASW consists of crop residues (straw, husks), forestry residues, and organic by-products (sugarcane bagasse, corn stover), provides abundant feedstock for biofuel production. This waste is generated seasonally, creating a continuous and renewable supply for bioenergy production. Some of the diversified biofuel described as

(a) Bioethanol: Bioethanol, a renewable fuel produced from the fermentation of crop residues and starchy crops such as corn and sugarcane, presents a viable alternative to traditional fossil fuels. Its ability to be blended with gasoline makes it a promising solution for reducing greenhouse gas emissions and decreasing reliance on non-renewable energy sources. While various sugar and starch-rich feedstocks can be used in its production, second-generation bioethanol derived from agricultural solid wastes (ASW) offers a particularly cost-effective approach. The hydrolysates from the ASW rich in hexose sugars and pentose sugars are fermented into ethanol under anerobic/aerobic conditions. Microorganisms commonly employed for the bioethanol production from lignocellulosic wastes include yeasts like *Saccharomyces cerevisiae*, *Saccharomyces pastorianus*, *Pichia stipitis*, *Pachysolen tannophilus*, *Candida shehatae* and bacteria like *Zymomonas mobilis*, *Thermoanaerobacterium saccharolyticum* and *Thermoanaerobacter ethanolicus* (Priyanka et al. 2019).

Global bioethanol production is expected to escalate by 20% attaining 130 billion liters (Samantaray et al. 2023). The United States is one of the largest producers of ethanol globally, with a production capacity exceeding 16 billion gallons per year. Ethanol production in the United States primarily relies on corn as a feedstock, accounting for over 60% of global ethanol production (Renewable Fuels Association (RFA) 2021). Brazil is another major ethanol producer, mainly from sugarcane. It accounts for around 25% of global ethanol production. The EU produces ethanol from various feedstocks, including grains, sugar beets, and wheat. It contributes approximately 5% of global ethanol production (European Commission 2021). Other countries like China, Canada, Thailand, and India also contribute to global ethanol production, collectively accounting for the remaining percentage (International Energy Agency 2021; Global Renewable Fuels Alliance (GRFA) 2021).

In recent years, India has witnessed a considerable surge in biofuel production, aligning with its commitment to renewable energy targets. Sugarcane molasses, a by-product of the sugar industry, serves as the primary raw material for ethanol production in the country. Promoting ethanol blending in petrol has been a key strategy to curb emissions and bolster energy security. Currently, India contributes approximately 3–4% to global ethanol production (International Energy Agency 2021). Furthermore, the Indian government has set an ambitious goal of achieving 20% ethanol blending in petrol by 2025–2026. To achieve this target, the government has implemented various policies and programs aimed at stimulating biofuel production and utilization. These initiatives include the Ethanol Blending Program (EBP) for petrol, incentives for biodiesel production, and the National Policy on Biofuels, which encourages investment in the bioenergy sector. Despite these efforts, the economic viability of bioethanol production in India faces several techno-economic hurdles. These challenges encompass efficient biomass handling, the delignification and pretreatment of biomass, the selection and availability of suitable fermentative microorganisms, and optimization of transportation processes. However, these challenges also present opportunities for innovation, research collaboration, and strategic leveraging of India's abundant agricultural and biomass resources to further enhance biofuel production.

(b) Biodiesel: Biodiesel is a renewable fuel that offers a cleaner-burning alternative to traditional petroleum diesel. It can be used in various applications, including transportation and heating, making it a versatile and sustainable energy source. Oils extracted from oilseed crops (e.g., soybean, canola) and waste cooking oils can be converted into biodiesel through transesterification. Oils synthesized and stored by oleaginous microorganisms like *Trichosporon cutaneum*, *Trichosporon fermentans*, *Rhodosporidium toruloides*, and *Lipomyces starkey* and micro-algae can also be transesterified into biodiesel. These oleaginous microorganisms and micro-algae are also able to utilize lignocellulosic hydrolysates thus facilitating economically feasible second and third generation bio-diesel production respectively (Chintagunta et al. 2021). The biomass generated after lipid extraction can also be used as manure for agriculture.

Globally, the EU is a significant producer of biodiesel, particularly vegetable oils such as rapeseed oil. It accounts for around 35% of global biodiesel production. The United States produces biodiesel primarily from soybean oil, accounting for approximately 20% of global biodiesel production. Indonesia and Malaysia are the countries that used palm oil to produce biodiesel, collectively contributing around 30% of global biodiesel production. While other countries like Argentina, Brazil, Canada, and Thailand also contribute to global biodiesel production, making up the remaining percentage. Biodiesel production in India utilizes a variety of feedstocks, including nonedible oils such as Jatropha, Mahua, and Karanja, as well as used cooking oil (UCO) and animal fats. These resources contribute to India's position in the global biodiesel market, with the country estimated to produce approximately 2–3% of the world's biodiesel supply (Global Renewable Fuels Alliance (GRFA) 2021).

(c) Biogas: Anaerobic digestion of organic waste, including animal manure, crop residues, and food waste, produces biogas composed of methane (60–70%) and carbon dioxide (28–30%). Biogas can be used for electricity generation, heating, and as a vehicle fuel (compressed natural gas or upgraded biogas). The biogas obtained from the anaerobic digestion of agricultural solid waste residues can be purified and compressed to produce CBG (compressed biogas). Unpurified biogas can be used as clean cooking fuel substituting LPG and for generation of electricity whereas CBG can be used in diesel engines to substitute diesel up to 80–100%. As of recent years, biogas production from ASW has been gaining attention globally as a renewable energy source. European countries like Germany, Denmark, and Sweden have established biogas plants that utilize ASW along with other organic waste streams to produce biogas for electricity generation, heat, and transportation fuels. Around 90% of the global biogas production comes from Europe, China, and the United States. Asia-Pacific countries such as China, India, and Thailand are increasingly investing in biogas infrastructure, utilizing ASW from agriculture and organic waste from urban areas to produce biogas for cooking, heating, and electricity. India also produces biomass-based fuels such as biogas from organic waste and bio-CNG from agricultural residues and municipal solid waste. These fuels contribute to a smaller percentage of the overall biofuel production in India (Raja and Iftikhar 2021; IEA 2020).

18.5 Environmental Benefits of Biofuels

Biofuels, especially those derived from biomass sources like crops, agricultural waste, and algae, offer a significant advantage over fossil fuels in terms of greenhouse gas emissions. Utilizing agricultural waste for biofuel production effectively reduces these emissions by diverting organic matter from landfills, where it would decompose and release methane, a potent greenhouse gas. Instead, this waste is transformed into renewable alternatives, contributing to a more sustainable energy cycle. This reduction is achieved because the carbon released during biofuel combustion is part of the ongoing natural carbon cycle, unlike fossil fuels, which release carbon that has been stored underground for millions of years (Solomon et al. 2007). Furthermore, biofuels are renewable energy sources that can be continuously produced from biomass feedstocks, offering a sustainable alternative to finite fossil fuel reserves. Implementing sustainable biofuel production practices, such as utilizing nonfood feedstocks and efficient land use strategies, further enhances their environmental benefits (Sims et al. 2010). In addition to reducing greenhouse gas

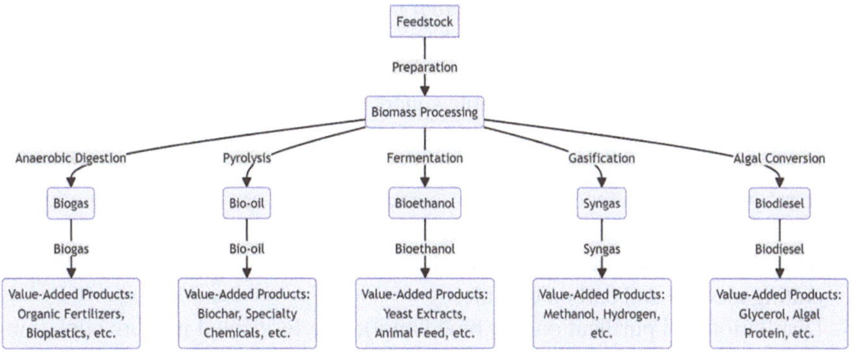

Fig. 18.2 Conversion of ASW to biofuel and value-added products

emissions, biofuels generally produce lower levels of pollutants like SO_2, particulate matter (PM), and nitrogen oxides (NOx) compared to conventional fossil fuels. This leads to improved air quality and reduced health risks associated with air pollution (Renewable Fuels Association (RFA) 2021). Moreover, utilizing biomass feedstocks for biofuel production can significantly reduce the amount of organic waste sent to landfills, mitigating methane emissions from decomposing organic matter. Agricultural residues and by-products can be effectively converted into biofuels, reducing waste and improving resource utilization (Demirbas 2009). Finally, biofuel production can stimulate economic growth in rural areas by supporting agriculture, bioenergy industries, and job creation. This can lead to sustainable rural development and reduced reliance on imported fossil fuels (OECD 2021).

18.6 Process for Biofuel Production

The conversion of agricultural solid waste into biofuel is a multistage process that varies depending on the specific feedstock used and the desired biofuel product. Here is a general overview of the typical steps involved in the biofuel production process from agricultural waste (Fig. 18.2):

1. Feedstock preparation: Agricultural solid waste, such as crop residues (e.g., straw, husks), forestry residues, or organic by-products (e.g., sugarcane bagasse, corn stover), is collected and prepared for further processing. This may include shredding, chopping, or drying the feedstock to improve handling and efficiency.
2. Pretreatment (if necessary): Some agricultural waste feedstocks require pretreatment to enhance accessibility to the biomass's cellulose and hemicellulose components. Pretreatment methods include physical processes (e.g., milling, grinding), chemical treatments (e.g., acid or alkali hydrolysis), or biological treatments (e.g., enzymatic hydrolysis).
3. Enzymatic hydrolysis (for cellulosic feedstocks): Cellulosic agricultural waste, such as crop residues and forestry waste, needs to be broken down into simpler

sugars before it can be used for biofuel production. This is done through a process called enzymatic hydrolysis, where enzymes like cellulases and hemicellulases are used to break down the complex carbohydrates in the waste (cellulose and hemicellulose) into fermentable sugars like glucose and xylose. This step is crucial in preparing the feedstock for the subsequent fermentation process.

4. Fermentation: Fermentable sugars obtained from enzymatic hydrolysis or starchy agricultural feedstocks (e.g., corn, sugarcane) are fermented by microorganisms (yeast or bacteria) to produce bioethanol or other fermentation-based biofuels. Yeast converts sugars into ethanol through anaerobic fermentation, generating alcohol and carbon dioxide.
5. Distillation and purification (for bioethanol): The fermented mixture containing ethanol is subjected to distillation to separate ethanol from water and other components. Distillation produces a high-concentration ethanol solution, which is further purified through dehydration processes (e.g., molecular sieves) to obtain fuel-grade ethanol.
6. Transesterification (for biodiesel): Oil-rich agricultural waste, such as oilseed crops (e.g., soybean, canola) or waste cooking oils, undergoes transesterification to produce biodiesel. Transesterification involves reacting vegetable oils or animal fats with alcohol (e.g., methanol or ethanol) and a catalyst (e.g., sodium hydroxide) to form biodiesel and glycerol.
7. Biogas production (for anaerobic digestion): Organic agricultural waste, including animal manure, crop residues, and food waste, is processed through anaerobic digestion to produce biogas. Anaerobic digestion involves microbial decomposition of organic matter in the absence of oxygen, producing methane-rich biogas, which can be used for electricity generation or upgraded for vehicle fuel use.
8. Upgrading and utilization: Biogas produced from anaerobic digestion can be upgraded (e.g., removing impurities like carbon dioxide) to produce biomethane, a renewable natural gas suitable for injection into the natural gas grid or as a vehicle fuel [compressed natural gas (CNG) or liquefied natural gas (LNG)].

18.7 Challenges in Conversion of ASW to Biofuel

Converting ASW into biofuel presents several challenges, ranging from feedstock availability and composition to technological and economic factors. Here are some key challenges with ASW conversion to biofuel:

1. Lack of infrastructure: Insufficient facilities for waste collection, recycling, and disposal in rural areas, leading to open burning or dumping.
2. Technological and process optimization: Limited access to cost-effective technologies for waste treatment and conversion, especially in developing countries. Developing cost-effective and scalable technologies for ASW conversion to biofuel remains a challenge. Process optimization, including pretreatment methods, enzyme cocktails, fermentation conditions, and product recovery

techniques, requires continuous research and development efforts (Kumar et al. 2009).

3. Awareness and education: Low awareness among farmers and stakeholders about the environmental impacts of improper waste management practices. Need for educational programs and training on sustainable waste management techniques.
4. Feedstock variability: Agricultural waste feedstocks can vary in composition, moisture content, and seasonal availability, posing challenges for consistent and efficient biofuel production processes (Alvira et al. 2010).
5. Pretreatment complexity: ASW often contains complex organic compounds such as lignocellulose, hemicellulose, and cellulose, which require specialized conversion processes for biofuel production. Preparing these feedstocks for breakdown into fermentable sugars or hydrocarbons involves advanced technologies like enzymatic hydrolysis often requires energy-intensive and costly pretreatment methods, thermochemical conversion, or microbial fermentation, limiting process scalability and economic viability (Mosier et al. 2005).
6. Enzyme costs: Enzymes used in biomass conversion processes, such as cellulases and hemicellulases, can be expensive, impacting the overall cost-effectiveness of biofuel production.
7. Fermentation efficiency: Fermentation of sugars from agricultural waste may face challenges related to microbial strain robustness, fermentation inhibitors, and product yield optimization.
8. Technological maturation: While biofuel conversion technologies hold great promise, some, particularly those focused on cellulosic bioethanol and advanced biofuels, are still in their early stages of commercial development. These technologies require further research and development to improve their efficiency, scalability, and economic viability.
9. Economic viability and market dynamics: The economic viability of ASW conversion to biofuel depends on factors such as feedstock availability, conversion efficiency, capital investment, operational costs, government policies, and market demand. Achieving cost-competitive biofuels from ASW requires innovative strategies, technological advancements, and supportive regulatory frameworks (Taherzadeh and Karimi 2008).
10. Environmental concerns: It is crucial to ensure that biofuel production from agricultural solid waste (ASW) is done sustainably. This means carefully considering the potential environmental impacts. Factors like land use changes, water consumption during processing, nutrient cycling, greenhouse gas emissions, and the potential for competition with food production or other land uses need to be thoroughly evaluated (Ragauskas et al. 2006). A balanced approach is essential to maximize the benefits of biofuels while minimizing any negative environmental consequences.

18.8 Future Prospects

The future prospects of converting ASW into biofuel are promising, driven by technological advancements, sustainability goals, and the need for renewable energy sources.

1. Advanced and conversion technologies: Continued research and innovation in biofuel production technologies, such as advanced pre-treatment methods, enzyme engineering, and microbial bioprocessing, will enhance process efficiency and cost-effectiveness. These technologies can efficiently convert diverse ASW feedstocks into biofuels, biochar, and valuable chemicals (Kumar et al. 2019).
2. Biological conversion and biorefineries integration: Biological conversion methods such as enzymatic hydrolysis, microbial fermentation, and anaerobic digestion offer prospects for converting ASW into biofuels like bioethanol, biogas, and biodiesel. Integrated biorefinery concepts, combining biofuel production with value-added co-products (e.g., biochemicals, bioplastics), can improve overall process economics and resource utilization (Jagtap et al. 2021).
3. Feedstock diversification: Exploring diverse feedstock sources, including agricultural residues, algae biomass, and waste streams from food processing industries, can expand feedstock availability and reduce competition with food production (Dijkman et al. 2019).
4. Circular economy approaches: Embracing circular economy principles, such as waste valorization, by-product utilization, and resource recycling, will contribute to sustainable biofuel supply chains and waste management strategies (Dijkman et al. 2019).
5. Policy support: Continued policy support, incentives, and regulatory frameworks that promote sustainable biofuel production from ASW are critical future prospects. Market expansion, international collaborations, and public–private partnerships can drive investments and innovation in ASW-based biofuel technologies (National Renewable Energy Laboratory (NREL) 2021).
6. Collaborative partnerships: Collaboration is key to advancing the bioenergy sector. By fostering partnerships between industry stakeholders, research institutions, government agencies, and NGOs, we can facilitate knowledge exchange, accelerate technology transfer, and build capacity. This collaborative approach can drive innovation, overcome challenges, and promote the wider adoption of sustainable bioenergy solutions.
7. Emerging technologies: These include synthetic biology, algae biofuels, and electrochemical conversion processes, offering promising avenues for diversifying biofuel production pathways and enhancing overall energy sustainability.
8. Technological integration and process optimization: Integration of multiple conversion technologies, process optimization, and scaling up pilot projects to commercial-scale operations are future prospects for enhancing the efficiency and cost-effectiveness of ASW conversion to biofuel (Gnansounou and Panichelli 2017).

References

Alvira P, Tomás-Pejó E, Ballesteros M, Negro MJ (2010) Pretreatment technologies for an efficient bioethanol production process based on enzymatic hydrolysis: a review. Bioresour Technol 101(13):4851–4861. https://doi.org/10.1016/j.biortech.2009.11.093

Bhuyan N, Narzari R, Gogoi L, Bordoloi N, Hiloidhari M, Palsaniya DR, Kataki R (2020) Valorization of agricultural wastes for multidimensional use. In: Current developments in biotechnology and bioengineering. Elsevier, Amsterdam, pp 41–78. https://doi.org/10.1016/B978-0-444-64309-4.00002-7

Bracco S, Calicioglu O, Juan GS, M., and Flammini, A. (2018) Assessing the contribution of bioeconomy to the total economy: a review of national frameworks. Sustain For 10(6):1698. https://doi.org/10.3390/su10061698

Cherubin MR, Oliveira DMDS, Feigl BJ, Pimentel LG, Lisboa IP, Gmach MR, Cerri CC (2018) Crop residue harvest for bioenergy production and its implications on soil functioning and plant growth: a review. Sci Agric 75:255–272. https://doi.org/10.1590/1678-992X-2016-0459

Chintagunta AD, Zuccaro G, Kumar M, Kumar SJ, Garlapati VK, Postemsky PD, Simal-Gandara J (2021) Biodiesel production from lignocellulosic biomass using oleaginous microbes: prospects for integrated biofuel production. Front Microbiol 2:658284. https://doi.org/10.3389/fmicb.2021.658284

Demirbas A (2009) Biofuels securing the planet's future energy needs. Energy Convers Manag 50(9):2239–2249. https://doi.org/10.1016/j.enconman.2009.05.010

Dhingra S, Sharma VK (2024) Crop residues based pellets market opportunities in India. J Biomed Res Environ Sci 5(4):403–410

Dijkman TJ, Gerlagh T, Kooroshy J (2019) Renewable energy and the circular economy. Science 363(6429):5479

El-Sayed SA, Ismail MA, Mostafa ME, Noseir EH, Khass TM (2023) Availability, sustainability and accessibility of agro crop residue production and solar radiation in Egypt for producing highly chemical products through pyrolysis processes. Clean Energy 7(3):654–670. https://doi.org/10.1093/ce/zkad016

EPE (Energy Research Company) (2023) Brazil's sugarcane bioenergy potential and residue utilization. https://www.epe.gov.br/en/

European Commission (2015) EIP-AGRI workshop opportunities for agriculture and forestry in the circular economy. Workshop Report 28-29 October 2015. Brussels, Belgium. https://ec.europa.eu/eip/agriculture/sites/agri-eip/files/eip-agri_ws_circular_economy_final_report_2015_en.pdf

European Commission (2021) Renewable Energy Directive (RED)

European Commission (2023) Biomass and agricultural residues in the European Union. https://ec.europa.eu/energy/topics/renewable-energy/biomass_en

European Commission Investing in Sustainable Development (2018) The EU at the forefront in implementing the Addis Ababa action agenda European Commission, Brussels, Belgium. https://doi.org/10.2841/874861

Fierascu RC, Fierascu I, Avramescu SM, Sieniawska E (2019) Recovery of natural antioxidants from agro-industrial side streams through advanced extraction techniques. Mol 24(23):4212

Food and Agriculture Organization (FAO) (2023) Agricultural residue utilization and biomass potential in Russia. https://www.fao.org/russian-federation

Food and Agriculture Organization of the United Nations (FAO) (2016) The state of food and agriculture. Climate Change, Agriculture and Food Security. http://www.fao.org/3/a-i6030e.pdf

Food and Agriculture Organization of the United Nations (FAO) (2017a) Strategic work of FAO for sustainable food and agriculture. http://www.fao.org/3/a-i6488e.pdf

Food and Agriculture Organization of the United Nations (FAO) (2017b) The state of food and agriculture: leveraging food systems for inclusive rural transformation. http://www.fao.org/3/a--i7658e.pdf

Food and Agriculture Organization of the United Nations (FAO) (2017c) The future of food and agriculture: trends and challenges. http://www.fao.org/3/a-i6881e.pdf

Food and Agriculture Organization of the United Nations (FAO) and Organization for Economic Co-operation and Development (OECD) (2019) Background notes on sustainable, productive and resilient agro-food systems: value chains, human capital, and the 2030 agenda. A report to the G20 agriculture deputies. https://www.oecd-ilibrary.org/docserver/dca82200-en.pdf

Global Renewable Fuels Alliance (GRFA) (2021) Global biofuels production data

Gnansounou E, Panichelli L (2017) Techno-economic analysis of biofuels: a wiki-based review. Biofuel Res J 4(1):498–510

IEA (2020) Outlook for biogas and biomethane: prospects for organic growth. IEA, Paris. https://www.iea.org/reports/outlook-for-biogas-and-biomethane-prospects-for-organic-growth

Institute for Essential Services Reform (IESR) (2023) Renewable energy and palm oil residue potential in Indonesia. https://iesr.or.id/en/

International Energy Agency (2021) Analysis and forecast to 2026. International Energy Agency, Paris

Jagtap SK, Bedekar SS, Bhatia AA, Patel RK, Pant AK, Ghosh S (2021) Advances in biorefineries: biomass and waste supply chain exploitation. Bioresour Technol 341:125807

Karak T, Sonar I, Nath JR, Paul RK, Das S, Boruah RK, Das K (2015) Struvite for composting of agricultural wastes with termite mound: utilizing the unutilized. Bioresour Technol 187:49–59. https://doi.org/10.1016/j.biortech.2015.03.070

Kumar P, Barrett DM, Delwiche MJ, Stroeve P (2009) Methods for pretreatment of lignocellulosic biomass for efficient hydrolysis and biofuel production. Ind Eng Chem Res 48(8):3713–3729. https://doi.org/10.1021/ie801542g

Kumar S, Bhattacharya S, Kumar A (2019) Advances in the development of thermochemical conversion technologies for lignocellulosic biomass. Renew Sust Energ Rev 103:111–130. https://doi.org/10.1016/j.rser.2018.12.036

McCormick K, Kautto N (2013) The bioeconomy in Europe: an overview. Sustain For 5(6):2589–2608. https://doi.org/10.3390/su5062589

Ministry of New and Renewable Energy (MNRE), Government of India (2021) National Policy on Biofuels

Mosier N, Wyman C, Dale B, Elander R, Lee YY, Holtzapple M, Ladisch M (2005) Features of promising technologies for pretreatment of lignocellulosic biomass. Bioresour Technol 96(6):673–686. https://doi.org/10.1016/j.biortech.2004.06.025

National Renewable Energy Laboratory (NREL) (2021) Bioenergy market report. U.S. Department of Energy, Washington

Nigerian Energy Commission (2023) Biomass residue potential for bioenergy in Nigeria. https://energy.gov.ng/

Northerly (2019) Top causes of agricultural waste, and how northerly works to combat them. https://northerly.ag/causesof-agricultural-waste

OECD (2021) Biofuels: linking support to performance. https://www.oecd-ilibrary.org/sites/89d2ac54-en/index.html?itemId=/content/component/89d2ac54-en

Pakistan Agricultural Research Council (PARC) (2023) Agricultural residue management in Pakistan. https://www.parc.gov.pk/

Priyanka M, Kumar D, Shankar U, Yadav A, Yadav K (2019) Agricultural waste management for bioethanol production. In: Biotechnology: concepts, methodologies, tools, and applications. IGI Global, Hershey, pp 492–524. https://doi.org/10.4018/978-1-5225-8903-7.ch019

Ragauskas AJ, Williams CK, Davison BH, Britovsek G, Cairney J, Eckert CA, Tschaplinski T (2006) The path forward for biofuels and biomaterials. Science 311(5760):484–489. https://doi.org/10.1126/science.1114736

Raja IA, Iftikhar AR (2021) Agriculture residue: a potential source for biogas production. Ann Agric Crop Sci 6(3):3. https://doi.org/10.1016/j.seta.2022.102843

Renewable Energy Bangladesh (2023) Jute and rice straw residue utilization in Bangladesh. https://www.renewableenergy-bd.org/

Renewable Fuels Association (RFA) (2021) Biofuels and the environment

Samantaray B, Mohapatra S, Mishra RR, Behera BC, Thatoi H (2023) Bioethanol production from agro-wastes: a comprehensive review with a focus on pretreatment, enzymatic hydrolysis, and fermentation. Int J Green Energy. https://doi.org/10.1080/15435075.2023.2253871

Scarlat N, Dallemand JF, Monforti-Ferrario F, Nita V (2015) The role of biomass and bioenergy in a future bioeconomy: policies and facts. Environ Dev 15:3–34. https://doi.org/10.1016/j.envdev.2015.03.006

Sims RE, Mabee W, Saddler JN, Taylor M (2010) An overview of second generation biofuel technologies. Bioresour Technol 101(6):1570–1580. https://doi.org/10.1016/j.biortech.2009.11.046

Solomon S, Qin D, Manning M, Chen Z, Marquis M, Averyt K, Miller H (2007) IPCC fourth assessment report (AR4). IPCC, Geneva, p 374

Taherzadeh MJ, Karimi K (2008) Pretreatment of lignocellulosic wastes to improve ethanol and biogas production: a review. Int J Mol Sci 9(9):1621–1651. https://doi.org/10.3390/ijms90911612

United Nations (2019) Programme performance report 2018. UN Environment Programme. https://wedocs.unep.org/bitstream/handle/20.500.11822/27734/PPR_2018_FINAL.pdf

USDA (2023) Bioenergy production and agricultural residues in the United States. U.S. Department of Agriculture. https://www.usda.gov/topics/energy/bioenergy

Varma VS, Yadav J, Das S, Kalamdhad AS (2015) Potential of waste carbide sludge addition on earthworm growth and organic matter degradation during vermicomposting of agricultural wastes. Ecol Eng 83:90–95. https://doi.org/10.1016/j.ecoleng.2015.05.050

Wang HY, Shen LIU, Zhai LM, Zhang JZ, Ren TZ, Fan BQ, Liu HB (2015a) Preparation and utilization of phosphate biofertilizers using agricultural waste. J Integr Agric 4(1):158–167. https://doi.org/10.1016/S2095-3119(14)60760-7

Wang S, Ru B, Dai G, Sun W, Qiu K, Zhou J (2015b) Pyrolysis mechanism study of minimally damaged hemicellulose polymers isolated from agricultural waste straw samples. Bioresour Technol 190:211–218. https://doi.org/10.1016/j.biortech.2015.04.098

Wang B, Dong F, Chen M, Zhu J, Tan J, Fu X, Chen S (2016) Advances in recycling and utilization of agricultural wastes in China: based on environmental risk, crucial pathways, influencing factors, policy mechanism. Procedia Environ Sci 31:12–17. https://doi.org/10.1016/j.proenv.2016.02.002

Part III

Industrial Waste to Biofuel and Chemical Products

Valorization of Petrochemical Industrial Waste

19

Naveen Kumar Karri, Rajasekhar Reddy Busigari,
Venkata Chandra Sekhar Palla,
and Hemanth Kumar Tanneru

Abstract

In this chapter, the petrochemical industrial wastes and their valorization techniques are discussed in a systematic manner. Wastes from the petroleum industry are classified into liquid and solid wastes. Liquid wastes are mainly categorized into oily sludge and waste lube oil. Solid wastes from the petrochemical industry are spent catalysts from the different units of refineries and the solid components of oily sludges. Valorization of liquid wastes can be achieved by recovering oil from the oily sludges and producing base oil from the waste lube oil respectively. Methods to recover the oil from oily sludges and base oil recovery from used lubricant are described. Valorization of solid wastes from petroleum industries can be done by recovering metals from the waste, using the solid waste as a precursor in other processes, and as an additive in cement and ceramics manufacturing, etc.

N. K. Karri · H. K. Tanneru (✉)
Department of Chemical Engineering, Indian Institute of Petroleum and Energy, Visakhapatnam, Andhra Pradesh, India,
e-mail: hemanth.che@iipe.ac.in

R. R. Busigari
Department of Fuel, Mineral and Metallurgical Engineering, Indian Institute of Technology (Indian School of Mines), Dhanbad, Jharkhand, India,

V. C. S. Palla
Material Resource Efficiency Division (MRED), CSIR-Indian Institute of Petroleum (IIP), Dehradun, Uttarakhand, India,

A. Kumari et al. (eds.), *Valorization of Solid Wastes to Biofuels and Chemical Products for Sustainable World*, https://doi.org/10.1007/978-981-96-8594-3_19

Keywords

Petrochemical waste · Oily sludge · Crude oil refining · Oil recovery technologies · Resource recovery

19.1 Introduction

Crude oil is the major source of world's energy requirement. Global crude oil production has seen an increment of 5.4% in 2022, which is greater than its 2021 growth (+1.6%). In 2010–2019, the average increase in world crude oil production was +1.3%/year (Enerdata 2023). Figure 19.1 shows the increasing oil production status of major oil-producing countries. As per the study conducted by Da Silva et al. (2012), the oily sludge is generated at the rate of 0.5% of global oil production.

The production, refining, and transportation of crude oil, and usage of some of its products generate the waste associated with the petrochemical industries (Hui et al. 2020). In this section, the status of the generation of these petroleum wastes, their compositions, and their potential as a feed stock to extract oil and produce other chemicals are discussed.

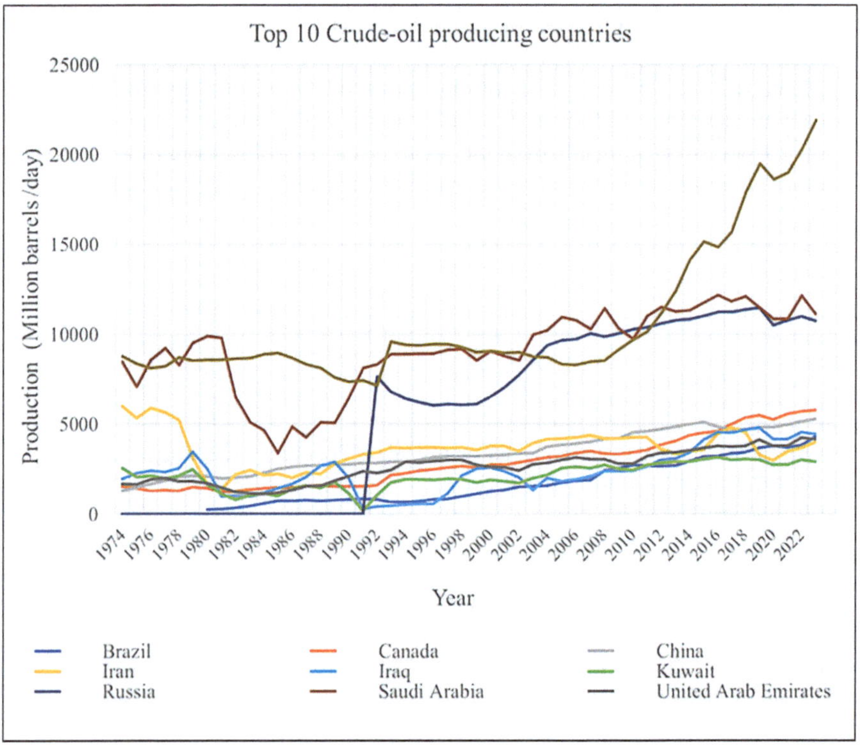

Fig. 19.1 Status of oil production in the top 10 crude oil-producing countries. (Source: EIA 2021)

19.2 Generation Status of Petrochemical Industrial Waste Worldwide and their Potential as Feedstock

19.2.1 Oily Sludges

As per the US Environmental Protection Agency (US EPA), the annual oily sludge produced from the refineries under this agency is estimated to be 30,000 tons/year. Similarly, China produces 3 million tons/year (Hasan et al. 2024). Baiji oil refineries in Iraq produce 3000–3500 m^3/year of oily sludge (Al-Doury 2019). It was reported that for every 500 tons of crude oil refined, 1 ton of oily sludge is produced (Mirghaffari 2018). For every 200–500 barrels of oil processed in a year, sludge of about 10,000 m^3 is generated (El Mahdi et al. 2017). Similarly, 50 tons/year oily sludge is produced from refinery, which refines the crude of 105,000 barrels/year. Thus, it is predicted that 0.3–0.5% of sludge is produced from every ton of crude oil refined. The Indian petroleum refinery produces 28,000 tons of sludge per year (Bhattacharyya and Shekdar 2003). It was reported that the optimistic, realistic, and pessimistic estimation of sludge production as a percentage of crude oil processed are 0.1, 0.5, and 1.5, respectively (Da Silva et al. 2012). The US EIA estimates the average oily sludge generation across the globe was 228.29 Mt/year in 2020. Therefore, the oily sludge is a potential feedstock to recover crude oil.

19.2.2 Waste Lubricating Oil

The lubricating oil is being used mainly in automobiles as well as in other industries to decrease friction in between machinery in machines/engines. To keep the engines working efficiently, lube oil is to be replaced at regular intervals. This produces large quantities of waste lube oil. The predicted generation of waste lube oil is several million tons annually. As per the reports, 89.64% waste oils are not collected annually in Algeria (Audibert 2006). Serbia produces more than 22,000 metric tons of base oil annually. Hence, the waste lube oil is a potential feedstock to recover base oil.

19.2.3 Solid Petroleum Waste

Typically, 11.81% of oily sludge is in liquid form and the remaining 88.19% are solids (Abbasi and Kamalan 2018a). Solid waste is generated mainly from the spent catalysts, metallic materials, plastic barrel, coke, wood, glass, cooling tower packing, and other material. The Petrochemical Especial Economic Zone (PEEZ) incorporates the Marun Petrochemical Complex (MPC) produces 6199 tons of waste per year. The solid waste generated from the MPC is around 80.1% and the remaining are liquid wastes (Abbasi and Kamalan 2018b). The average annual amount of total solid waste production from different units of gas and oil industries in the UAE is estimated at 4061 tons. In Asab fields, for every 1000 barrel of oil produced, 371 kg of solid waste, and for every 1000 m^3 of natural gas liquids (NGL) produced, 1588 kg of solid waste is generated (Elshorbagy and Alkamali 2005).

19.3 Classification of the Petrochemical Industrial Waste and their Variation

Petrochemical industrial waste is classified as liquid (oily sludge and waste lube oil) and the solid wastes. In this section, composition of each of the petrochemical wastes is described.

19.3.1 Oily Sludge

Oily sludge is an emulsion of water and oil along with solids, petroleum hydrocarbons (PHC), and different metals as solid components (Mazlova and Meshcheryakov 1999). The oily sludge is obtained from various operations such as crude oil exploration, crude oil gathering, and crude oil refining (see Fig. 19.2). More details on the subprocesses from which the oily sludge is generated in each of the above measures are available elsewhere (Hui et al. 2020). The oily sludge composition depends on the crude oil source, the steps of refining, and the reagents used in refining process. Total petroleum hydrocarbon (TPH) contents in oily sludge, water, and solids can range from 5–86.2%, 30–85%, and 5–46%, respectively. Oily sludge has a pH of between 6.5 and 7.5 (Ramaswamy et al. 2007; Tahhan et al. 2011). PHCs identified in oily sludge fall into four categories: aromatics, aliphatic, compounds combining nitrogen, sulfur, and oxygen (NSO), and asphaltenes (Mrayyan and Battikhi 2005). Aliphatic and aromatic hydrocarbons together contribute to nearly around 75% of PHCs in oily sludge (Kriipsalu et al. 2008). Oily sludge has a composition of alkanes, aromatics, and asphaltenes in the range of 40–52%, 28–31%, and 8–10%, respectively. Alkanes are the major constituents in the oily sludge (Mishra et al.

Fig. 19.2 Oil-sludge formation routes

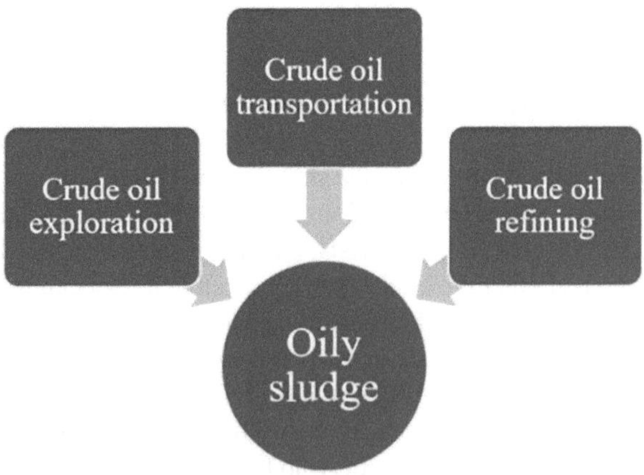

1999). The NSO fraction contains nitrogen, sulfur, and oxygen in the form of naphthenic acid, mercaptans, thiophenes, and pyridines. Around 7–22.4% by mass of oily sludge are resins. Heavy metals like Fe, Cu, or Ni and some other solid particles form a solid part of oily sludges (Jasmine and Mukherji 2015).

19.3.2 Waste Lube Oil

Waste lube oil (WLO) contains aromatic and aliphatic hydrocarbons, inorganic and organic chemical compounds, additives, and metals (Kim et al. 2003). The waste oil composition depends on additives/dilutants mixed with virgin oil. The additives interact with oil and produce corrosive acids and harmful substances after they were used (Jafari et al. 2015). Lead contamination occurs in WLOs when they are employed in vehicles that run on leaded gasoline (Cotton et al. 1977). The organic contaminants of WLO and their concentrations are given in Table 19.1. The physiochemical properties of fresh and WLO are available elsewhere (Mishra et al. 2021).

19.3.3 Solid Waste

Solid waste from the petroleum industries (refineries) is obtained from spent catalyst across several units of refinery such as hydroprocessing, fluidized catalytic cracking, and reforming (Akcil et al. 2015). Hydroprocessing catalysts contain metals, including V, Mo, W, Cr, Cu, Co, Ni, and Zn. Considering all of these factors, the treatment of spent refinery catalysts for metal production is generally feasible (Marafi et al. 2007). Hydrodesulphurization (HDS) is a method to remove sulfur from organic spent material in the presence of catalyst with Ni/Co on an aluminum oxide support (Akcil et al. 2015). Fluidized catalytic cracking converts the residues from atmospheric and vacuum distillates to light fractions in the presence of catalyst. After prolong exposer, spent catalyst is formed because of reduction in its activity. 20 metric tons of FCC catalyst will be used per day. It is prepared by using activated silica dioxide and aluminum oxide. The spent waste with approximately 250 kg of aluminum is disposed at the site (Al-Jabri et al. 2013). Many countries

Table 19.1 Organic compounds in waste lubricating oil with their concentrations

Organic contaminants		Approximate range (μg/l)
Aromatic hydrocarbons	Benzo pyrene	360–62,000
	Benzo anthracene	870–30,000
	Pyrene	1670–33,000
	Alkylbenzenes	900,000
	Naphthalene	440,000
Chlorinated hydrocarbons	Trichloroethanes	18–1800
	Trichloroethylene	18–2600
	Perchloroethylene	3–1300

Adapted from Cotton et al. (1977)

focus on the recovery of metal from spent catalyst using two types of metallurgical techniques, namely hydrometallurgy and pyrometallurgy (Marafi and Stanislaus 2003).

19.4 Valorization of Liquid Petrochemical Industrial Wastes

In this section, we focus on the oil recovery technologies from oily sludges and base-oil from waste lube oil. Solvent extraction, ultrasonic, surfactant enhancement oil recovery centrifugation, freeze/thaw, pyrolysis, froth flotation, and electrokinetic methods are technologies that are available for oil recovery from sludges. Few technologies, namely solvent extraction, centrifugation, pyrolysis, and ultrasonication technologies, are discussed here. The readers may refer to the following references for the details on other technologies (Hu et al. 2013).

19.4.1 Oil Recovery Technologies from Oily Sludge

Solvent Extraction

Solvent extraction is an extremely effective technique used to recover valuables from oily sludge. The process involves the use of solvents to separate and recover valuable components such as hydrocarbons, metals, and other organic compounds from the sludge shown in Fig. 19.3. One of the major advantages of solvent extraction is its ability to target specific compounds, permitting for a more efficient and selective recovery process. The solvent-rich phase containing the extracted compounds is then further processed to recover and purify the valuable components, while the remaining sludge can be treated or disposed of appropriately. Solvent extraction offers an environmentally friendly and cost-effective solution for both resource conservation and waste management.

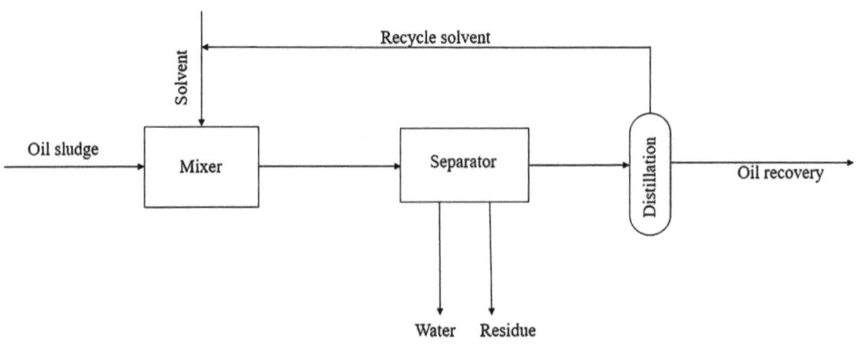

Fig. 19.3 Solvent extraction. (Adapted from Hu et al. 2013)

Centrifugation

Centrifugation is a generally used method for the upgradation of oily sludge to fuels and chemicals due to its efficiency in separating different components based on their densities. In this process, oily sludge is first mixed with a suitable solvent or reagent to enhance the extraction of valuable compounds. The slurry is then subjected to centrifugal forces, causing the denser solids and water phases to separate from the lighter hydrocarbons and organic compounds shown in Fig. 19.4.

This separation allows for the recovery of valuable components such as hydrocarbons, metals, and organic matter, which can be further processed into fuels, chemicals, and other useful products. Configurations to achieve optimal separation efficiency make centrifugation suitable for both small-scale laboratory experiments and large-scale industrial applications.

Sludge Pyrolysis

The breakdown of organic molecules in an inert atmosphere at temperatures between 500 and 1000 °C is known as pyrolysis. Pyrolysis differs from gasification, which turns organic molecules into a syngas with 20–40% oxygen or a flammable gas. Reduced molecular weight hydrocarbons are produced during pyrolysis in both condensable and noncondensable gases. Moreover, it leaves behind char, a solid residue (Fonts et al. 2012). Figure 19.5 is a schematic of pyrolysis of oily sludge.

19.4.1.1 Pilot Scale Study of Recovery of Oil from Oily Sludge

Ultrasonication

Ultrasonic waves enhance mass and heat transfer within the sludge, accelerating the pyrolysis process and increasing the yield of valuable products. This method offers advantages such as shorter reaction times, improvement in product quality and less energy consumption compared to conventional pyrolysis techniques. Ultrasonic irradiation can also be used in the catalytic conversion of oily sludge to chemicals and fuels. Ultrasonic waves enhance the dispersion of catalysts and improve contact

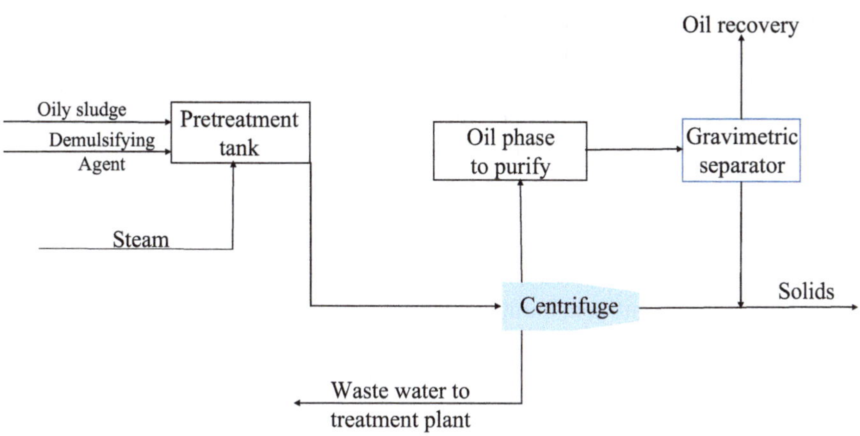

Fig. 19.4 Centrifugation. (Adapted from Hu et al. 2013)

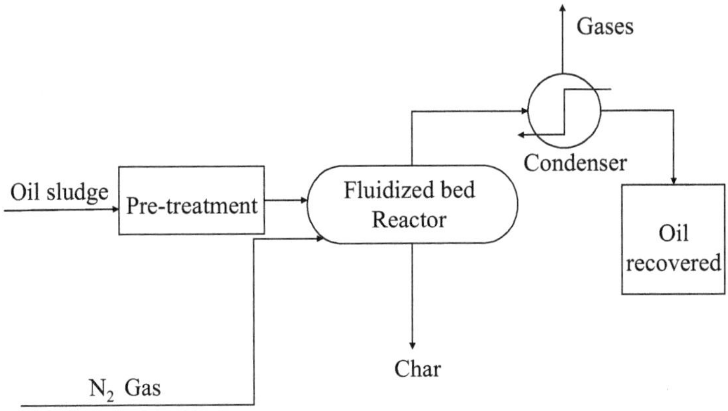

Fig. 19.5 Sludge pyrolysis. (Adapted from Hu et al. 2013)

between reactants, leading to enhanced catalytic activity and selectivity. This approach is particularly useful for producing high-quality fuels and chemicals with controlled compositions and properties. Furthermore, ultrasonic irradiation can be used in the oily sludge hydrothermal treatment, where high-pressure and high-temperature conditions are applied to promote the conversion of organic matter into biofuels and chemicals. Ultrasonic waves assist in breaking down biomass and enhancing the efficiency of hydrothermal reactions, resulting in increased yields of target products.

19.4.1.2 Challenges and Prospects of Valorization of Liquid Petroleum Industry Waste

There are many challenges encountered during the recovery of fuel from petroleum industrial waste. The technologies and the challenges involved in oil recovery are provided in Table 19.2.

19.4.1.3 Feasibility and Economics of Oil Recovery from Oily Sludge

Feasibility and economics of technology are assessed by the scale of application of the technology with the field application being more feasible, followed by pilot and laboratory scale. A comparison of oil recovery processes with their scale, efficiency, and advantages is provided in Table 19.3.

19.4.2 Technologies for Base oil Recovery from WLO

WLOs are disposed into soil, burnt, and discarded in water bodies, causing environmental damage. To protect environment, WLO is recycled using physical–chemical process for generating base oil (Carlos Oliveira Santos 2018). The processes to recover base oil from WLO are provided next.

Table 19.2 Challenges in different technologies to recovery of oil from oily sludge

Technology	Challenges
Solvent extraction	Organic solvents are expensive and cause secondary contamination
Centrifugation	Higher equipment maintenance, energy consumption and noise, and preconditioning is necessary
Surfactant	Cost is expensive, surfactants are hazardous and cause secondary pollutants, and not effective for all heavy metals treatment
Freezing/thawing	Excessive expense, low efficiency, and excessive energy usage
Pyrolysis	High energy consumption, high equipment maintenance costs, and unsuitability for sludges with large water content
Electro kinetic method	Limited processing capacity and complex equipment manufacturing process
Ultrasonic	Low handling capacity, high manufacturing and maintenance costs for equipment
Forth flotation	Excessive water use, inefficient, and unsuitable for sludge with a high viscosity

Adapted from Hu et al. (2013)

Table 19.3 Feasibility and economical for recovery of fuel from petroleum industry waste

Technology	Scale	Efficiency	Advantages
Solvent extraction	Field application	30–88	Simple, fast, efficient
Centrifugation	Field application	40–70	Convenient, fast, efficient, high yield
Surfactant	Pilot application	70–90	Simple, efficient, large handling capacity
Freezing/thawing	Laboratory application	50–70	Convenient, durable, suitable for areas with low temperatures
Pyrolysis	Field application	50–90	Efficient, fast, high-quality oil recovery
Electro kinetic method	Laboratory application	50–70	Fast and efficient, low-energy consumption
Ultrasonic	Pilot application	45–70	Quick, efficient, and environmentally friendly
Forth flotation	Laboratory application	55–70	Low-energy consumption and convenient

Adapted from Hu et al. (2013)

Sedimentation

It requires that spent oil remain in a cuneiform bottom tank for a long enough amount of time without stirring it. This process depends on the densities of oil, solid contaminants, and water; fluid oils separate from suspended particles more readily.

Filtration

In this process, WLO is passed through materials that capture solid particles. Adsorbent filters have been used a lot recently, with activated bauxite being one of the most common ones. Large quantities of water in WLO reduce filtration efficiency.

Centrifugation

Centrifugal purification is a mechanical method that uses high peripheral speed spinning to separate base oil from waste lube oils.

Degasification

Degasification unit is constructed using a high vacuum chamber with a ceramic or tiny metal ring insert to provide a huge surface area. During this procedure, any dissolved gasses, water, and solvents in the lubricant are removed.

Thermal Dehydration

In thermal dehydration water content in WLO is removed by means of a two-stage process. First, the oil is heated under normal pressure to the temperature between 160 °C and 200 °C. Oil flows continuously through a heat exchanger, moving some of the dehydrated oil to the second stage at the top of the column where it is cooled. Both the base oil and water portions are seperated.

Neutralization

This method aims to convert excess acid or base in the WLO into salts by neutralizing it. During this process, the WLOs pH will be brought to around 7. It entails mixing potash or caustic soda with the WLO. This treatment takes place in a bottom tank that has a reactor in it.

Hydro Finishing

This process includes hydrogenolysis in which unstable hydrocarbons are treated with hydrogen under intense pressure. Unstable hydrocarbons are saturable by using hydrogen.

Fractional Distillation

It separates the less volatiles present in the WLOs. Fractional distillation consists of a strip of distillation where steam is used to boil the lube oil and separates its less volatile lube fractions.

Acid Clay Process

Used oil is usually filtered and heated in the acid clay process to eliminate waste, solid particles, and debris. Sulfuric acid is then added, extracting metals, salts, acids, aromatics, asphaltenes, and other contaminants. Oil is recovered with acidic sludge as a by-product.

19.4.2.1 Challenges of WLO recycling and Impact on Human Health and Environment

The challenges to separate base oil from use WLO depends on the contaminants in the WLO. The effects of contaminants and the challenges posed by them are given in Table 19.4.

Table 19.4 Challenges posed by the oil contaminants in WLO for base oil recovery

Oil contaminants	Effects
Water	Leads steel or iron to rust Leads to the formation of water sludge, causes issues with foaming
Solid particles	Excessive wear, surface damage to the bearings, and inability to withstand metal fatigue Blockage of oil routes and clearances by sludge deposits
Sludge and lacquers	Hinders oil pump's functioning by making the valves sticky
Liquid contaminants	Engine residue dilutes LO and may reduce its viscosity Effects of oil pollution
Base oils derived from crude oil	Biodegradability, high carcinogenicity, and ecotoxicity Oil mist causes serious health issues related to heart, adrenal glands, liver, kidneys, and lungs
Oil mist	It has an adverse effect on the neurological and respiratory systems Additionally, it promotes skin cancer when absorbed by the skin

Adapted from Anisuzzaman et al. (2021)

Table 19.5 Feasibility and economics of processes for base oil recovery from WLO

Regenerative technology	Energy requirement	Economic costs	Recycling rate (%)
Acid/clay	Lower	Less	63
Distillation	Higher	Less	50
Solvent de-asphalting	Higher	More	70–65
TFE with hydro finishing	Higher	More	72
TFE with clay finishing	Higher	More	72
TFE with solvent finishing	Higher	More	72
TDA	Higher	More	74–77
Solvent extraction hydro finishing	Higher	More	74

Adapted from Jafari et al. (2015)

19.4.2.2 Feasibility and Economics of Processes to Recover Base Oil from WLO

The feasibility of several technologies for the separation of base oil from WLO is given in Table 19.5.

19.4.3 Technologies for Metal Recovery from Solid Petroleum Waste

Recovery of metals from solid waste in the petroleum refinery industries is done through ultrasonic hydrometallurgical leaching process. By promoting the rupture and refining of the solid's surface layer, the use of ultrasonic treatment might potentially improve the leaching process. The activation energy needed can be lowered by defects in the surface crystal structure, which would speed up leaching. Based on the complexity of the metals present in industrial solid wastes, an appropriate leaching agent is selected. As a result, using ultrasonic treatment can greatly increase the rate at which metals leach (Ran et al. 2024). Few types of leaching based on the leaching agent are provided next.

Acid Leaching

Involves the use of acidic solutions, typically sulfuric or hydrochloric acid, to dissolve and extract metals from ores. This method is most useful to remove heavy metals from crude oil or refinery waste. Valuable metals from spent catalysts can be recovered using acid leaching.

Alkali Leaching

It uses alkaline solutions, such as sodium hydroxide or ammonia, to leach metals from ores. Alkali leaching is useful in desulfurization processes to remove sulfur compounds from oil, improving fuel quality.

Oxidation Leaching

It utilizes oxidizing agents like oxygen, hydrogen peroxide, or chlorine to enhance the leaching of metals by converting them into a more soluble form. It is applied to treat refinery sludges and spent catalysts, making metals more extractable. It helps in breaking down complex organic compounds in oil, facilitating cleaner processing.

Bacterial Leaching

This method involves the use of bacteria to biologically convert insoluble metals into soluble forms, making them easier to extract. This process is used for bioremediation of oil-contaminated sites, aiding in the breakdown of hydrocarbons, which helps in the recovery of valuable metals from refinery waste using microorganisms.

19.4.3.1 Feasibility of Extracting Oil from Solid Waste

The types of leaching techniques and their advantages are indicated in Table 19.6.

19.4.3.2 Challenges in Metal Recovery from Solid Petroleum Wastes

Challenges in the different leaching techniques for the metal recovery from the solid petroleum wastes are given in Table 19.7.

Table 19.6 Feasibility of technologies for metal recovery from solid petroleum waste

Leaching process	Advantages
Acid leaching	Effectively dissolve metal ions, less expensive, and easier to operate
Alkaline leaching	Less expensive and easy to operate
Oxidation leaching	Improved controllability and stability Increased effectiveness of leaching
Bacterial leaching	Environmentally friendly and greater leaching efficiency

Adapted from Ran et al. (2024)

Table 19.7 Challenges in oil recovery from waste solid wastes

Leaching process	Challenges
Acid leaching	Large equipment needed; acidic waste liquid treatment is challenging
Alkaline leaching	Long leaching time, low efficiency
Oxidation leaching	Slow reaction rate and more operational expenses
Bacterial leaching	Longer leaching times, more complex operations, and more maintenance expenses
Electrolytic leaching	Energy intensive, difficult to control required current and voltage required for operation
Novel leaching	More complex functioning, highly expensive

Adapted from Ran et al. (2024)

19.4.4 Other Applications of Solid Waste from the Petroleum Industry

The utilization of spent catalysts from petroleum industries has many applications like production of concrete, super-plasticized mortars, fired brick, nonstructural blocks and floor pavers, hydraulic binders, building ceramics, clay-based bricks, Portland cement clinker, etc. (Asim et al. 2021).

Solid petroleum waste (SPW) is replacing natural clay up to 30 wt.% in the making of ceramics using clay such as roofing tiles and clay bricks (Souza et al. 2011). Concrete is an important material in construction. Cement acts like a binding agent in concrete. The cement industry is using solid wastes from petroleum industry to replace limestone and clay. Petroleum industries drilling cuttings solid wastes are used in the construction of roads, asphalt pavements, cement manufacturing, concrete, bricks, and also building ceramics (Bernardo et al. 2007).

The petroleum waste sludge is useful in sustainable construction like cement clinker production, mortar production, building blocks, ceramics, bricks, and tile fabrication. Petroleum sludges are also used in the road construction as organic binders, roadbed material, paving layer mixes, dust suppressor for unpaved road maintenance, bitumen supplement, and also asphalt production. For more details on the valorization of solid waste from petrochemical industry, refer to Jagaba et al. (2022).

The solid wastes from petroleum refineries like biochar and activated carbon are used as absorbent material in waste water treatment. Anaerobic digestion is a process of generating gas like methane, which is solid wastes are used as fuel. Solid wastes from the petroleum industry are used to manufacture metal catalysts, carbon clay composites, rubber tires, and levan.

19.5 Impact of Petroleum Industry Waste on the Environment

Heavy metals and some pathogens are frequently found in oily sludge, and these contaminants have the potential to damage the ecosystem (Barry et al. 2019). A high-temperature process is taking place in the petroleum industries that can

Table 19.8 Different petroleum wastes and their effects on the environment and humans

Solid waste	Compounds	Effects	References
Nitrogenous waste	Amine, pyridine, etc.	Respiratory problems	Wang et al. (2019)
Sulfur waste	Aliphatic and heterocyclic	Acid rain and corrosive	Tang et al. (2019)
Aromatic waste	Styrene, pyrene	Cancer	Haroni et al. (2019)
Heavy metals	Hexavalent chromium	Respiratory system and digestive track	Huang et al. (2015)
	Vanadium	Diarrhea and stomach cramps	
	Lead	Human brain cells	

eliminate pathogens, but the final products these processes still include heavy metals. During the pyrolysis process, sulfur and nitrogen compounds may move and alter, dispersing in the char or pyrolysis oil and syngas discharging into the atmosphere.

Numerous investigations state that oily sludge typically includes a wide range of metals, namely Li, Na, K, Ca, Ba, Cr, Cu, Ni, Zn, Pb, and V (Nansa et al. 2018; Huang et al. 2015). Heavy metals are not biodegradable, highly hazardous, and pollute soil over time. Hexavalent chromium may enter animals through the skin, mucosal membranes, respiratory system, and digestive tract. It also accumulates in the kidney, liver, and endocrine glands. Although vanadium alone is not very hazardous, its compounds are very harmful to both humans and animals. During petroleum process like pyrolysis, these chemicals undergo change, lowering the quality of the products like pyrolysis oil. Generally, nitrogen in the form of amine-N, pyridine-N, and pyrrole-N (Wang et al. 2019). Sulfur exists in aliphatic and heterocyclic forms. The latter produces H_2 S during pyrolysis, which pollute the environment (Tang et al. 2019). Polyaromatic hydrocarbons (PAHs) are very poisonous and highly carcinogenic (Haroni et al. 2019). Some compounds like fluorine, chrysene, pyrene, anthracene, and other poly hydrocarbons (PHC) are also carcinogenic when inhaled or come into direct contact with the skin (Haroni et al. 2019). The health effects and environmental damage of different wastes are provided in Table 19.8.

19.6 Policy for the Petrochemical Industrial Waste Management

Waste produced by humans is a huge environmental hazard on a global level. Landfilling, incineration, and recycling are a few methods in the waste management techniques used in many cities throughout the world. In this section, policies related to petroleum disposal wastes are discussed.

In general, wastes generated from the exploration and production of crude oil and natural gas are governed by state regulations and Subtitle D of the Resource Conservation and Recovery Act (RCRA). Every rule issued under the RCRA act

Table 19.9 Codes and their regulations by RCRA

S. No.	Code	Rule
1	K169	Sediments from petroleum refining operations in crude oil storage tanks
2	K170	Slurry from in-line filters and/or separating solids from petroleum refining processes, as well as clarified slurry oil storage tanks
3	K171	Spent hydrotreating catalyst from petroleum refining plants, including guard beds for desulfurizing inputs to other catalytic reactors
4	K172	Refined petroleum hydrocarbon catalyst waste, including guard beds for de-sulfurizing feedstock for more catalytic reactors

Modified from Kraus (2011)

must be reviewed in Section 2002(b). The Exploration and Production Activities (EPA) continues to work together with states and other organizations to find areas in demand for improvement and to face new challenges in order to make sure that waste from exploration, development, and production is handled in a way that protects the environment as well as public health (United States Environmental Protection Agency (EPA) 2020). The Bureau of Land Management (BLM) of the U.S. Department of Interior handles exploration and production activities on federal lands have certain guidelines and regulations. The EPA is determined that hazardous component emissions into the environment should not occur during the management of exploration and production wastes (Kraus 2011).

The RCRA act has mainly four codes that are signified as regulations; each code has its own rules as mentioned in Table 19.9.

19.7 Conclusions

The major petrochemical wastes were categorized as liquids (oily sludge (emulsion), lube-oil waste (liquid), and the solid wastes arising from the residue from oily sludge and spent catalysts. From the oily sludge, recovery of oil is of prime importance and several methods for oil recovery from the oily sludge are discussed. As the oily sludge composition changes depending upon the crude oil, operating procedures in refining, etc., there is no one method that fits for all kinds of oily sludges for recovery of oil from oily sludges. Waste lube oil can be processed to recover the base oil, and the relevant methods are discussed. The methods to recover the metals from spent catalysts and the application of solid waste of petroleum in construction industries are discussed.

References

Abbasi M, Kamalan HR (2018a) Quality and quantity of wastes generated in marun petrochemical complex and evacuating recovery potential. J Hydrosci Environ 1(3):1–8. https://doi.org/10.22111/JHE.2018.3841

Abbasi M, Kamalan HR (2018b) Waste management planning in amirkabir petrochemical complex. Environ Energy Econ Res 2(1):63–74. https://doi.org/10.22097/EEER.2018.96582.1011

Akcil A et al (2015) A review of metal recovery from spent petroleum catalysts and ash. Waste Manag 45:420–433. https://doi.org/10.1016/j.wasman.2015.07.007
Al-Doury MMI (2019) Treatment of oily sludge using solvent extraction. Pet Sci Technol 37(2):190–196. https://doi.org/10.1080/10916466.2018.1533859
Al-Jabri K et al (2013) Use of waste spent catalyst in construction. Constr Build Mater 39:77–81
Anisuzzaman SM, Jumaidi MH, Nasir NNM (2021) Used lubricating oil recovery process and treatment methods: a review. IOP Conf Ser Mater Sci Eng 1195(1):012031. https://doi.org/10.1088/1757-899x/1195/1/012031
Asim N et al (2021) Wastes from the petroleum industries as sustainable resource materials in construction sectors: opportunities, limitations, and directions. J Clean Prod 284:125459. https://doi.org/10.1016/j.jclepro.2020.125459
Audibert F (2006) 'Waste engine oils: rerefining and energy recovery', waste engine oils: rerefining and energy recovery. Elsevier, Amsterdam, pp 1–323. https://doi.org/10.1016/B9780444522023.X5018-8
Barry D et al (2019) Pyrolysis as an economical and ecological treatment option for municipal sewage sludge. Biomass Bioenergy 122:472–480. https://doi.org/10.1016/J.BIOMBIOE.2019.01.041
Bernardo G et al (2007) The use of oil well-derived drilling waste and electric arc furnace slag as alternative raw materials in clinker production. Resour Conserv Recycl 52(1):95–102. https://doi.org/10.1016/j.resconrec.2007.02.004
Bhattacharyya JK, Shekdar AV (2003) Treatment and disposal of refinery sludges: Indian scenario. Waste Manage Res 21(3):249–261. https://doi.org/10.1177/0734242X0302100309
Carlos Oliveira Santos J (2018) Recovery of used lubricating oils - a brief review. Progr Petrochem Sci 1(4):516. https://doi.org/10.31031/PPS.2018.01.000516
Cotton FO et al (1977) Analysis of 30 used motor oils. Hydrocarb Process 56(9):131–140
Da Silva LJ, Alves FC, De França FP (2012) A review of the technological solutions for the treatment of oily sludges from petroleum refineries. Waste Manage Res 30(10):1016–1030. https://doi.org/10.1177/0734242X12448517
EIA (2021) The 10 largest oil producers and share of total world oil production in 2021, Frequently Asked Questions (FAQs). http://www.eia.gov/tools/faqs/faq.cfm?id=74&t=11. Accessed 20 May 2024
El Mahdi AM, Aziz HA, Eqab ES (2017) Review on innovative techniques in oil sludge bioremediation. AIP Conf Proc 1892(1):040026. https://doi.org/10.1063/1.5005706
Elshorbagy W, Alkamali A (2005) Solid waste generation from oil and gas industries in United Arab Emirates. J Hazard Mater 120(1-3):89–99. https://doi.org/10.1016/j.jhazmat.2004.12.036
Enerdata (2023) Oil and energy trends. Wiley, Hoboken. https://doi.org/10.1111/oet.3_12989
Fonts I et al (2012) Sewage sludge pyrolysis for liquid production: a review. Renew Sust Energ Rev 16(5):2781–2805. https://doi.org/10.1016/j.rser.2012.02.070
Haroni NN et al (2019) The effect of oil sludge contamination on morphological and physiological characteristics of some tree species. Ecotoxicology 28(5):507–519. https://doi.org/10.1007/S10646-019-02034-0/METRICS
Hasan AMA et al (2024) Petroleum sludge formation and its treatment methodologies: a review. Environ Sci Pollut Res 31(6):8369–8386. https://doi.org/10.1007/s11356-023-31674-3
Hu G, Li J, Zeng G (2013) Recent development in the treatment of oily sludge from petroleum industry: a review. J Hazard Mater 261:470–490. https://doi.org/10.1016/J.JHAZMAT.2013.07.069
Huang Q et al (2015) Catalytic pyrolysis of petroleum sludge for production of hydrogen-enriched syngas. Int J Hydrog Energy 40(46):16077–16085. https://doi.org/10.1016/j.ijhydene.2015.09.139
Hui K et al (2020) Status and prospect of oil recovery from oily sludge: a review. Arab J Chem 13(8):6523–6543. https://doi.org/10.1016/j.arabjc.2020.06.009
Jafari A et al (2015) (2015) 'analysis and comparison of used lubricants, regenerative technologies in the world'. Resour Conserv Recycl 103:179–191. https://doi.org/10.1016/j.resconrec.2015.07.026

Jagaba AH et al (2022) Diverse sustainable materials for the treatment of petroleum sludge and remediation of contaminated sites: a review. Cleaner Waste Syst 2:100010. https://doi.org/10.1016/j.clwas.2022.100010

Jasmine J, Mukherji S (2015) Characterization of oily sludge from a refinery and biodegradability assessment using various hydrocarbon degrading strains and reconstituted consortia. J Environ Manag 149:118–125. https://doi.org/10.1016/j.jenvman.2014.10.007

Kim YS et al (2003) Tar-formation kinetics and adsorption characteristics of pyrolyzed waste lubricating oil. J Anal Appl Pyrolysis 70(1):19–33. https://doi.org/10.1016/S0165-2370(02)00072-4

Kraus RS (2011) Petroleum refining process. In: Encyclopaedia of occupational health & safety. International Labour Office, Geneva, pp 1–31

Kriipsalu M, Marques M, Maastik A (2008) Characterization of oily sludge from a wastewater treatment plant flocculation-flotation unit in a petroleum refinery and its treatment implications. J Mater Cycl Waste Manage 10(1):79–86. https://doi.org/10.1007/s10163-007-0188-7

Marafi M, Stanislaus A (2003) Options and processes for spent catalyst handling and utilization. J Hazard Mater 101(2):123–132. https://doi.org/10.1016/S0304-3894(03)00145-6

Marafi A, Hauser A, Stanislaus A (2007) Deactivation patterns of Mo/Al_2O_3, $Ni\text{-}Mo/Al_2O_3$ and $Ni\text{-}MoP/Al_2O_3$ catalysts in atmospheric residue hydrodesulphurization. Catal Today 125(3-4):192–202. https://doi.org/10.1016/j.cattod.2007.03.060

Mazlova EA, Meshcheryakov SV (1999) Ecological characteristics of oil sludges. Chem Technol Fuels Oils 35(1):49–53. https://doi.org/10.1007/BF02694263

Mirghaffari N (2018) Treatment and recycling of oily sludges produced in the petroleum industry. In: 2017 international conference on environmental impacts of the oil and gas industries: Kurdistan region of Iraq as a case study (EIOGI). IEEE, Piscataway, pp 1–2. https://doi.org/10.1109/eiogi.2017.8267638

Mishra S et al (1999) 'Field study: in situ bioremediation of oily sludge contaminated land using 'oilzapper. In: Hazardous and industrial wastes - proceedings of the mid-Atlantic industrial waste conference. Piscataway, IEEE, pp 177–186

Mishra A et al (2021) Pyrolysis of waste lubricating oil/waste motor oil to generate high-grade fuel oil: a comprehensive review. Renew Sust Energ Rev 150:111446. https://doi.org/10.1016/j.rser.2021.111446

Mrayyan B, Battikhi MN (2005) Biodegradation of total organic carbons (TOC) in Jordanian petroleum sludge. J Hazard Mater 120(1-3):127–134. https://doi.org/10.1016/j.jhazmat.2004.12.033

Nansa VM et al (2018) Analysing petroleum effluent samples for determination of cadmium using carbon nanotubes followed by atomic absorption spectrometry. Curr Nanosci 14(6):545–554. https://doi.org/10.2174/1573413714666180829115204

Ramaswamy B, Kar DD, De S (2007) A study on recovery of oil from sludge containing oil using froth flotation. J Environ Manag 85(1):150–154. https://doi.org/10.1016/j.jenvman.2006.08.009

Ran J et al (2024) Metal recovery from industrial solid waste by ultrasonic-assisted hydrometallurgical leaching: a review. Environ Chem Lett. https://doi.org/10.1007/S10311-024-01743-1

Souza AJ, Pinheiro BCA, Holanda JNF (2011) Valorization of solid petroleum waste as a potential raw material for clay-based ceramics. Waste Biomass Valoriz 2(4):381–388. https://doi.org/10.1007/s12649-011-9090-6

Tahhan RA et al (2011) Enhancing the biodegradation of total petroleum hydrocarbons in oily sludge by a modified bioaugmentation strategy. Int Biodeterior Biodegrad 65(1):130–134. https://doi.org/10.1016/j.ibiod.2010.09.007

Tang X, Wei X, Chen S (2019) Continuous pyrolysis technology for oily sludge treatment in the chain-slap conveyors. Sustain For 11(13):3614. https://doi.org/10.3390/SU11133614

United States Environmental Protection Agency (EPA) (2020) Management of oil and gas exploration and production waste | hazardous waste. https://www.epa.gov/hw/management-oil-and-gas-exploration-and-production-waste. Accessed 27 May 2024

Wang Y et al (2019) Nitrogen transformation during pyrolysis of oilfield sludge with high polymer content. Chemosphere 219:383–389. https://doi.org/10.1016/J.CHEMOSPHERE.2018.11.171

Conversion of Textile Waste to Biofuels and Chemicals

20

Ramon E. Vera, Keren A. Vivas, Ronald Marquez, Fernando Urdaneta, Ronalds Gonzalez, and Hasan Jameel

Abstract

The textile industry is a significant polluter, contributing to 10% of global emissions. Annually, over 100 million tons of textile waste are generated worldwide, and the linear model of make-use-waste is prevalent. Consequently, 85% of textile products end up in landfills or are incinerated. Natural fibers (e.g., cotton) and nondegradable synthetic fibers (e.g., polyester) are the main raw materials for textiles, with synthetic fibers accounting for over 60% of the share. Given the nonbiodegradability of synthetic fibers and the increasing textile landfilling, there is a pressing need to extend the end of life of textile products and adopt a circular economy approach. Natural fibers such as cotton, due to their cellulosic nature, present viable opportunities to transform textile waste into value-added biochemicals. Enzymatic hydrolysis offers a promising and sustainable conversion pathway for this purpose. Glucose, the main chemical resulting from the enzymatic hydrolysis of cellulosic materials such as cotton textile waste, serves as a crucial building block to produce biofuels and other value-added chemicals with potential worth of thousands of dollars per ton. This chapter comprises the fundamentals of enzymatic hydrolysis of textile wastes, pretreatments needed to overcome conversion challenges (e.g., the cellulosic structure itself, and the presence of dyes), processes for upcycling glucose into biofuels and other chemicals, and the recycling of synthetic fibers. Additionally, it explores the economic potential of utilizing textile waste as an advanced resource within the circular economy framework.

R. E. Vera · K. A. Vivas · R. Marquez · F. Urdaneta · R. Gonzalez · H. Jameel (✉)
Conversion Economics and Sustainability, Department of Forest Biomaterials, North Carolina State University, Raleigh, NC, USA
e-mail: jameel@ncsu.edu

A. Kumari et al. (eds.), *Valorization of Solid Wastes to Biofuels and Chemical Products for Sustainable World*, https://doi.org/10.1007/978-981-96-8594-3_20

Keywords

Textile industry · Circular economy · Pretreatment · Conversion

20.1 Introduction

The textile industry is one of the largest and most polluting industries globally, consuming large amounts of water, energy, and other primary resources (e.g., cotton, petroleum) to produce a single product like a T-shirt (Grover and Grover 2018). This is a significant environmental issue, given that textile products are nowadays produced rapidly and cheaply, encouraging consumers to indiscriminately buy and quickly dispose of them, so-called fast fashion (Chavan 2014; Grover and Grover 2018). Consequently, post-consumer textile waste is increasing at an alarming rate, resulting in over 100 million tons of textile waste being generated across the world annually (Bick et al. 2018; Pandit et al. 2018; United States Environmental Protection Agency 2021). More concerningly, 85% of this waste is either landfilled or incinerated, taking up valuable space and releasing greenhouse gases that contribute to climate change. Additionally, the degradation of textile waste in landfills leads to soil and water pollution due to leaching of toxic chemicals (e.g., dyes) (Pandit et al. 2018; Shen et al. 2010).

Recycling and upcycling textiles can help reduce waste and extend the lifespan of clothing items, thereby decreasing the amount of textile waste sent to landfills (Damayanti et al. 2021). However, there is currently a lack of effective and feasible recycling practices, as evidenced by the fact that only 15% of textile waste is recycled (Pandit et al. 2018; Shen et al. 2010).

After synthetic fibers, cotton is the second most used fiber for textile production (Fig. 2.1b). Cotton and other natural fiber fabrics account for 35% of the total textile waste globally (Sandin and Peters 2018). Different strategies have been developed to either repurpose or upcycle cotton textile waste. These cellulosic fibers can be hydrolyzed with enzymes to produce glucose, an important building block for biobased products (Sasaki et al. 2019; Werpy and Petersen 2004; Yousef et al. 2019).

The main component of natural fibers used in the textile industry is cellulose (Heinze and Rojas 2016). Cotton, in particular, is pure cellulose, and its enzymatic hydrolysis involves using enzymes to break down the polymeric structure of this cellulosic material into glucose, its monomeric component (Hafyan et al. 2020; Kapanji et al. 2019; Kohli et al. 2019; Kwan et al. 2018; Magalhães et al. 2019; Özüdoğru et al. 2019; Sadhukhan et al. 2016). Glucose can be further processed into a variety of biobased products (Hafyan et al. 2020; Sadhukhan et al. 2016; Schmidt et al. 2017). For example, glucose can be fermented into ethanol, which can be used as a biofuel, or used as a feedstock to produce chemicals such as lactic acid, which is utilized in the production of biodegradable plastics (Kohli et al. 2019; Kwan et al. 2018) (Fig. 1.1). Although converting cotton textile waste into glucose is an attractive concept, several challenges exist regarding the sustainable and feasible enzymatic conversion processes. Contaminants such as finishing chemicals, metals, and textile dyes must be addressed due to their potential inhibitory effects on enzyme performance (Buschle-Diller and Traore 1998; Zambrano et al. 2021). Therefore, pretreatments are required to manage these contaminants but also to modify fiber and cellulose morphology to improve enzyme effectiveness (typically cellulases) and achieve reasonable yields in the enzymatic conversion process of textiles (Daramola et al. 2020). Moreover, due to the high cost of enzymes, these pretreatment pathways must be capable of promoting effective enzyme and chemical usage while maintaining acceptable energy consumption.

Considering the dramatic increase in textile waste generation, it is clear that suitable and sustainable alternatives are needed to properly manage post-consumer textile wastes and implement a circular economy. Beyond biochemical production, enzymatic hydrolysis is an attractive pathway to separate, purify, and recycle synthetic fibers from cotton/polyester blends. In this context, the goal of this chapter is to highlight the importance of textile wastes as an emerging source to produce building blocks and platform biochemicals and biofuels. Converting cotton textiles into value-added chemicals is presented as a viable solution to address a significant part of the total textile waste generation.

This chapter is structured as follows:

1. Cotton structure and composition.
2. Enzymatic hydrolysis of cotton to produce glucose.
3. The need for pretreatments to increase conversion yield.
4. Mechanical refining as a feasible, sustainable, and low-risk pretreatment.
5. Dyes and their negative effect on textile conversion through enzymatic hydrolysis.
6. Factors affecting the economics of textile waste pretreatments intended for enzymatic hydrolysis.
7. Upcycling technologies after the enzymatic hydrolysis of textiles for value-added chemical production.
8. Challenges, opportunities, and future perspective.

20.2 Cellulose and Cotton Structure

Cellulose $(C_6H_{10}O_5)_n$ is one of the most abundant polymers in nature. Its primary source is the cell wall of wood and nonwood fibers, and it can be transformed into multiple products that are crucial today, including paper, packaging materials, and cotton garments. Cellulose is built by a linear chain of D-glucose linked by β-(1,4)-glycosidic bonds (Heinze and Rojas 2016; Zambrano et al. 2020a). The cellulose content can vary depending on the biomass source (Ang et al. 2020; Habibi et al. 2010; Nazhad et al. 1995). Among natural fibers, cotton is the biomass with the highest cellulose content. Additionally, the negligible lignin content in cotton favors high glucose yields (up to 90% in some cases) during enzymatic hydrolysis, which constitutes a significant advantage compared to other lignocellulosic feedstocks (Bajpai 2016; Singh and Satapathy 2018).

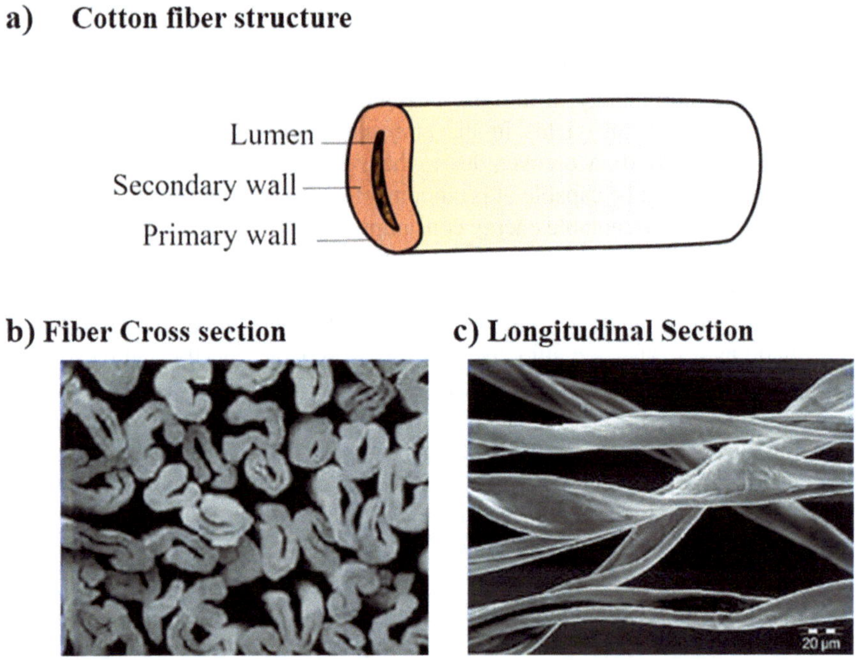

Fig. 20.1 (**a**) Schematic representation of the cotton fiber structure. The cuticle (not shown) is the most external wall covering the primary wall. Scanning microscopic views of cotton fibers: (**b**) cross-section and (**c**) longitudinal view. (Adapted from Vera et al. 2022)

20.2.1 Cotton Fiber Morphology

Cotton fibers have a four-layered cell wall that consists of a cuticle, a primary wall, a secondary wall, and a central core or lumen (Flint 1950), shown schematically in Fig. 20.1a. The width of cotton fiber varies between 12 and 20 µm (Shaker and Nawab 2020). Micrographs of cotton fibers can be seen in Fig. 20.1b, c. The average degree of polymerization of cotton ranges between 9,000 and 15,000, with an average crystallinity of 73%. These values are significantly higher than other lignocellulosic materials, such as wood pulp (D.P. = 600–1,500 and crystallinity = 35%) and viscose rayon (D.P. = 250–450 and crystallinity = 60%) (Heinze and Rojas 2016).

20.2.2 Cotton Composition

Cotton presents a cellulose content of approximately 90% or higher. The noncellulosic constituents of cotton include proteins (1.1–1.9%), pectic substances (0.7–1.2%), ash (0.7–1.6%), wax (0.4–1.0%), organic acids (0.5–1.0%), and other components such as amino acids and nitrogen-containing compounds, which are mainly located in the cuticle and the lumen (Heinze and Rojas 2016).

20.3 Enzymatic Hydrolysis of Cotton Textile Wastes

In cotton textile waste, cellulose is readily available for enzymatic hydrolysis due to its very low hemicellulose and lignin content (Czilik et al. 2002; Hafyan et al. 2020; Huh et al. 2006; Kapanji et al. 2019; Klement and Büchs 2013; Kohli et al. 2019; Kwan et al. 2018; Magalhães et al. 2019; Nieder-Heitmann et al. 2019; Özüdoğru et al. 2019; Sadhukhan et al. 2016; Schmidt et al. 2017). However, unless cotton textile fibers are pretreated and their morphology changed, the conversion of cellulose into glucose via enzymatic hydrolysis is slow and has a low yield. This is primarily due to the high degree of crystallinity, small pore size in the fiber cell wall, and the presence of dyes. These factors need to be addressed to increase the availability of cellulose chains to react with enzymes (Chandra et al. 2007; Galbe and Zacchi 2007; Singh and Satapathy 2018).

Enzymatic hydrolysis is performed using cellulase enzymes, which exhibit maximum activity at 50–60 °C and pH between 5.0–5.2 (Fockink et al. 2015; Giakoumakis et al. 2021; Hasanzadeh et al. 2018; Jeihanipour and Taherzadeh 2009; Wang et al. 2016). The enzymatic hydrolysis mechanism consists of three stages: (1) adsorption of cellulase onto the cellulose surface, (2) biodegradation of cellulose to sugars (mainly glucose), and (3) desorption of cellulase, as depicted in Fig. 20.2 (Kumar and Sharma 2017; Pu et al. 2013; Raspolli Galletti and Antonetti 2012; Singh and Satapathy 2018). Enzyme adsorption is the limiting stage on the kinetics of enzymatic hydrolysis considering that it is a function of (1) enzyme concentration, (2) available surface area of the substrate and its nature, (3) physical properties of the enzymes, and (4) the hydrolysis environment (i.e., pH, salt concentration, temperature) (Ghose and Bisaria 1979). The behavior follows a first-order kinetics:

$$\frac{dC_A}{dt} = k\left(C_A - C_S\right)$$

where C_A is the concentration of adsorbed enzyme, t is the reaction time, C_S is the saturation concentration of adsorbed enzyme, and k is the adsorption rate constant (Walker and Wilson 1991). To overcome the structural characteristics of cellulose (mainly the high crystallinity, amorphous regions, and refractory lignin-containing sites), engineered cellulases have been developed (Morjanoff and Gray 1987; Sun et al. 2015; Walker and Wilson 1991). Cellulases are multi-enzyme mixtures that integrate three different enzymes to complete the conversion of cellulose into glucose (Fig. 20.2). In the first conversion step, endoglucanases cleave individual bonds to yield oligosaccharides with degrees polymerization of 2–6, which are then converted by exoglucanases into small oligosaccharides (cellobiose) and finally hydrolyzed by β-glucosidases into glucose (Ghose and Bisaria 1979; Singhania et al. 2013; Zhang et al. 2018).

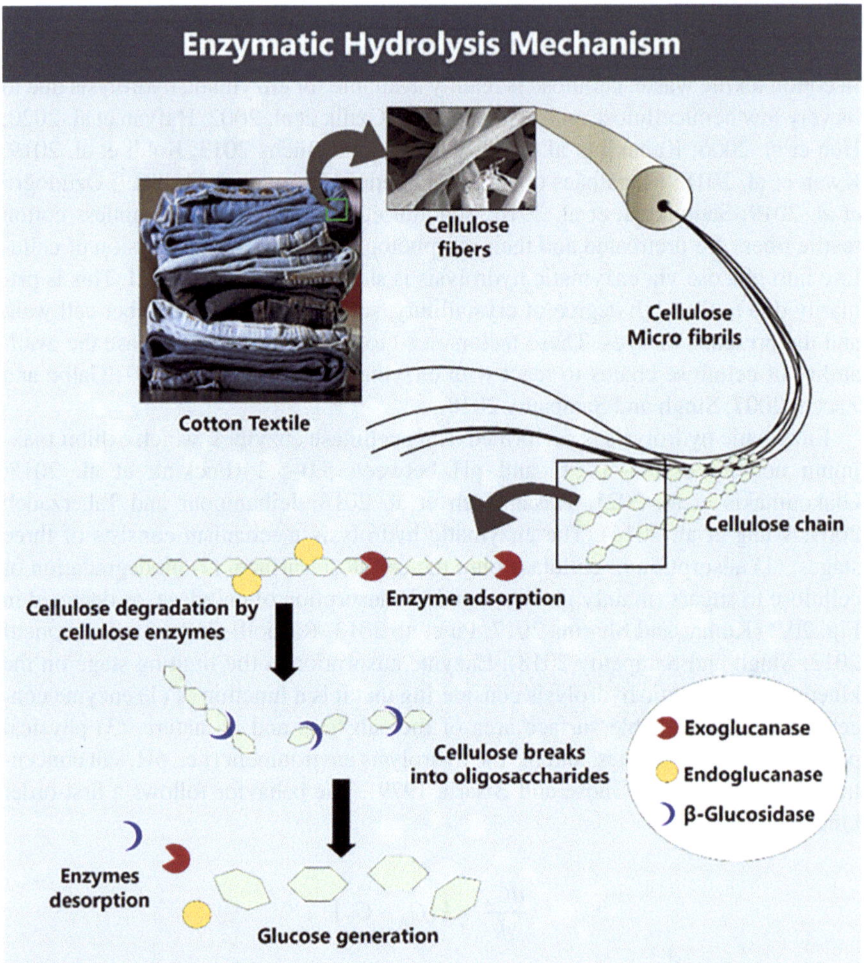

Fig. 20.2 Schematic representation of the mechanism of cellulose enzymatic hydrolysis. (Adapted from Vera et al. 2022)

20.4 The Need for Pretreatments to Improve Enzymatic Conversion of Textiles

Although enzymatic hydrolysis seems like an attractive and sustainable method to convert cotton textile waste, the nature of the cellulosic structure and the presence of contaminants like persistent dyes pose significant challenges to achieving feasible conversion to produce glucose. The surface area available in cotton waste materials to undergo enzymatic hydrolysis has been considered a key factor for increasing glucose yield (Cantero et al. 2019; Giakoumakis et al. 2021; Hasanzadeh et al. 2018). Specific surface area, cellulose crystallinity, and degree of polymerization have also been regarded as key physicochemical properties influencing glucose

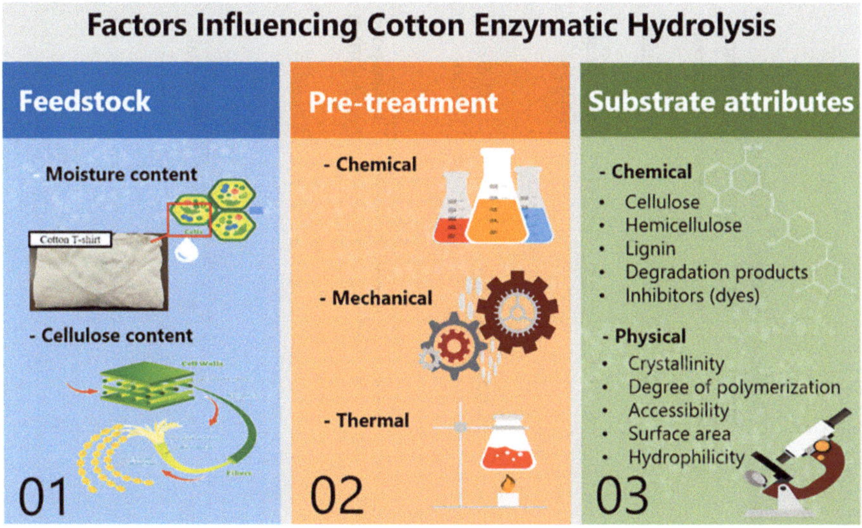

Fig. 20.3 Factors influencing enzymatic hydrolysis yield of cotton substrates. (Adapted from Vera et al. 2022)

yields (Fig. 20.3) (Shuhua et al. 2020). Moreover, the presence of dyes in cotton textiles has been reported to decrease the performance of enzymatic hydrolysis (Czilik et al. 2002). This variable, inherent to textile substrates, will be discussed in the following section.

Different pretreatments have been developed to increase enzymatic performance by modifying both the chemical and physical attributes of cellulose (Fig. 20.3). Chemical and thermal pretreatments have been the most researched pathways, while mechanical pretreatments such as mechanical refining have recently emerged as an attractive, low-risk, and feasible option. Table 20.1 shows that without pretreatment the conversion of cotton into glucose can be as low as 30%. On the other hand, pretreatments like alkaline extraction, total chlorine-free bleaching, mechanical refining, or steam explosion can significantly boost the yield to over 90% (Vera et al. 2022, 2023a).

20.5 Mechanical Refining as a Feasible, Sustainable, and Low-Risk Pretreatment

Mechanical refining has been a well-known, proven, and widely utilized pretreatment in the papermaking and biorefinery industry to produce biofuels from enzymatically hydrolyzed lignocellulosic biomass (Bajpai 2005; Baker and Pira International 2000; Chen et al. 2013; Corbett et al. 2018; de Assis et al. 2018; Jones et al. 2013; Vera et al. 2022, 2023a). This technique has been used for many years in the pulp and paper industry (Smook and Kocurek 1982), and it differs from

Table 20.1 Conditions for a variety of common pretreatments to upcycle cotton textile waste via enzymatic hydrolysis

Textile waste features	Type of pretreatment	Pretreatment method	Pretreatment condition	Enzyme charge/g of substrate	EH yield without pretreatment (%)	EH yield with pretreatment (%)	Chemical use[a]	Energy use[b]	Enzyme use[c]	References
Blue jeans/ cotton dyed	Chemical and mechanical	Shredded and ground + NaOH	NaOH 12%, 0 °C, 96 h	Cellulase (20 FPU/g) and β-glucosidase (30 IU/g)	24.1	99	+ + +	+ + + +	+ + + +	Jeihanipour and Taherzadeh (2009)
50/50 polyester/ cotton dyed	Dissolving and regeneration of cellulose	Milled + phosphoric acid	H_3PO_4 85%, 50 °C, 7 h	Cellulase (7.5 FPU/g) and cellobiase (15 CBU/g)	12	79	+ + + +	+ + +	+ + +	Shen et al. (2013)
40/60 polyester/ cotton dyed	Dissolving and regeneration of cellulose	Freezing alkali/urea soaking	NaOH 7%, urea 12%, −20 °C, 1 h	Cellulase (30 FPU/g) and β-glucosidase (60 IU/g)	46	91	+ + +	+ + +	+ + + +	Li et al. (2019)
50/50 polyester/ cotton dyed	Dissolving and regeneration of cellulose	Shredded and ground + N-methyl morpholine-N-oxide (NMMO)	NMMO 85%, 120 °C, 2 h	Cellulase (20 FPU/g) and β-glucosidase (30 IU/g)	16	90	+ + +	+ + +	+ + + +	Jeihanipour et al. (2010)
Cotton T-shirts (our previous work)	Mechanical	Mechanical refining	No chemicals 25 °C	Cellulase (6 FPU/g)	60	95	+	+	+	Vera et al. (2023a)

Cotton T-shirts dyed	Chemo-mechanical	Mechanical refining + TCF oxidation processes	2% NaOH, 1% H_2O_2, 3.2% O_3 70 °C, 1 h	Cellulase (6 FPU/g)	30	90	+ +	+ +	+	Vera et al. (2023b)

+++ high, ++ medium, + low

Adapted from Vera et al. (2023b)

[a] Refers to chemical use for pretreatments

[b] Refers to energy use during pretreatment

[c] Refers to enzyme use during the enzymatic hydrolysis stage

mechanical milling mainly due to the nature of the mechanical action on the fibers. Mechanical refining is used to modify fiber morphology and increase the available surface area through delamination and fibrillation without a significant decrease in fiber length (Zambrano et al. 2020a, b), whereas milling primarily reduces fiber length. Fibrillation is generally an external effect that involves peeling smaller fiber fragments away from the main fiber mass due to shearing forces. Delamination, on the other hand, is generally an internal effect arising from repeated compression and decompression of fibers as they are squeezed between the bars and grooves of the refiner plates (Corbett et al. 2020; de Assis et al. 2018). Figure 20.4 (top panel) shows schematically the forces involved in the mechanical refining process and their effect on fiber morphology.

It was only a few years ago that this technology was introduced to improve the enzymatic hydrolysis of textile waste to implement the biorefinery concept toward producing value-added chemicals. Vera et al. (2023a) were the first to show that mechanical refining on fiber properties maximized the reactivity of textile fibers toward enzymatic hydrolysis at lower energy consumption and, more importantly, with either low or no chemical usage (Fig. 20.4, middle panel). These findings indicate that a cotton conversion yield greater than 90% can be achieved using mechanical refining as a pretreatment (Fig. 20.4, lower panel). Moreover, compared to conventional pretreatment processes shown in Table 20.1, mechanical refining requires significantly lower enzyme concentrations (e.g., 5 FPU per gram O.D. fibers). This highlights the economic potential for scaling up this technology with reasonable capital expenditure and production costs of glucose.

20.6 Dyes and Their Negative Effect on Textile Conversion Through Enzymatic Hydrolysis

Dyes have been demonstrated to have a negative effect on the performance of enzymes in converting textiles into biochemicals (Vera et al. 2022, 2023b). Figure 20.5 shows that enzymatic hydrolysis (EH) is drastically affected by the presence of dyes, even when mechanical refining is utilized as a pretreatment. Figure 20.5a compares the EH of both undyed and black-dyed cotton at various levels of mechanical refining (2,500, 5,000, 10,000, and 20,000 PFI revs) and an enzyme charge 6 FPU/g O.D. cotton. The sugar yield improved with increasing mechanical refining intensity in both cases, but the conversion rate was significantly lower for dyed cotton. Thus, it is important to corroborate the inhibitory effect of dyes on cotton EH performance (Buschle-Diller and Traore 1998; Zambrano et al. 2021). Dyes confer a hydrophobic nature to the cotton surface, hindering enzyme accessibility to the cellulosic surface and negatively impacting their performance in converting cellulose into sugars (Zambrano et al. 2021).

The enzymatic conversion of cotton is significantly influenced by the nature of the dye, which is observed in Fig. 20.1b for refined and differently colored cotton textile waste. While certain dyes (e.g., grey, green, and blue) did not result in a

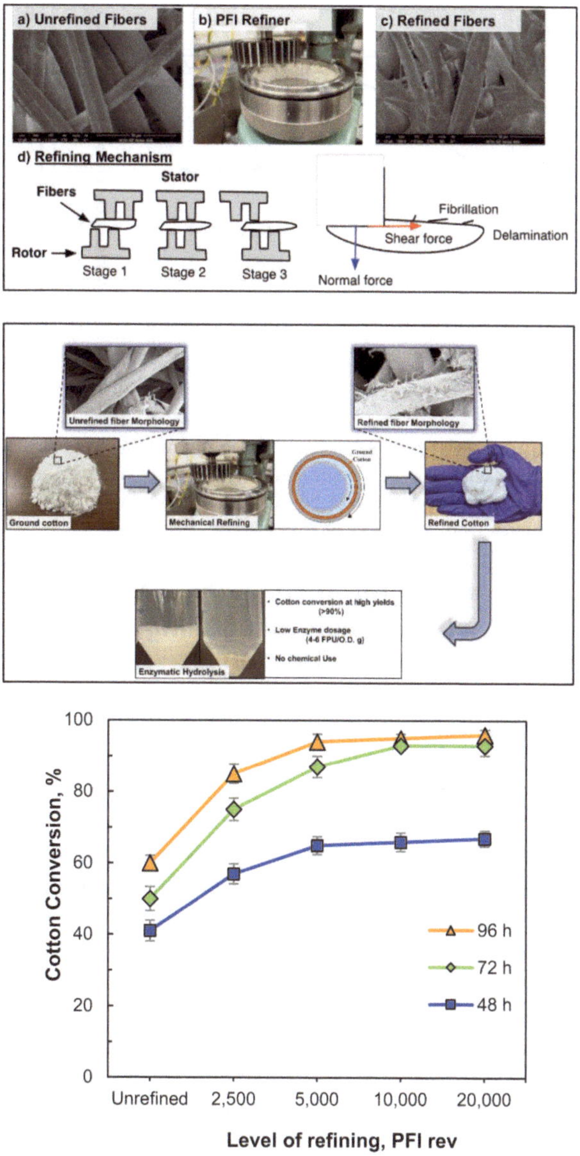

Fig. 20.4 (top panel) General mechanism of mechanical refining. (middle panel) Mechanical refining applied to textile waste. (lower panel) Effect of the mechanical refining on improving the enzymatic hydrolysis of cotton textile waste. (Adapted from Vera et al. 2022, 2023a)

significant decrease in cotton conversion, red and black dye reduced the conversion yield by up to 40% when compared to undyed cotton conversion. The substantial variation in the ability of dyes to inhibit EH could be directly linked to the dye's molecular structure and the mechanism by which they are attached to the cellulosic structure of cotton (Buschle-Diller and Traore 1998; Guo et al. 2016; Vera et al. 2022).

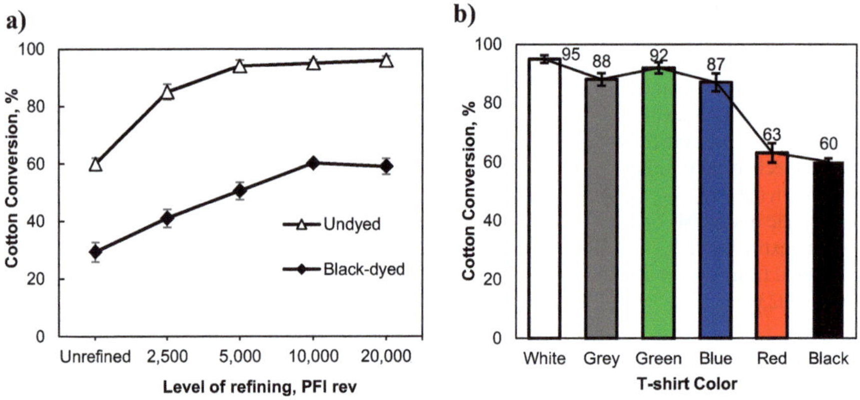

Fig. 20.5 (**a**) Effect of the mechanical refining and (**b**) dyes on enzymatic hydrolysis of dyed textile wastes. All enzymatic hydrolysis tests were assessed with an enzyme charge of 6FPU/g O.D. cotton during 96 h at 50 °C. Dyed textiles were mechanically refined at 10,000 PFI revs. (Adapted from Vera et al. 2023b)

20.7 Total Chlorine-Free Bleaching Combined with Mechanical Refining to Overcome Textile Dye's Inhibitory Effect on Enzymatic Hydrolysis

Environmentally friendly and total chlorine-free oxidation methods such as ozone and alkaline hydrogen peroxide have been combined with mechanical refining pretreatment to mitigate the low conversion of textile into glucose via enzymatic hydrolysis (Vera et al. 2023b). Recent research has focused on total chlorine-free (TCF) methods where the most commonly used chemicals are ozone (Z) and alkaline hydrogen peroxide (AHP) because of their capability to bleach and their low environmental impact because of the absence of chlorine-containing compounds (Javaid and Qazi 2019; Powar et al. 2020; Wang et al. 2020). The performance of oxidation (bleaching) processes will depend on several factors such as the nature of the dye, type, and charge of oxidation agents, time, pH, and temperature (Eren et al. 2016; Gias Uddin and Islam 2011; Li 2021; Powar et al. 2020; Wu et al. 2008). Recent research showed that the enzymatic conversion of black-dyed cotton, without oxidation, resulted in a glucose yield of only 60% compared to 95% for undyed cotton fibers. On the other hand, the inclusion of oxidation processes in the pretreatment stage resulted in a glucose yield of 90% via enzymatic hydrolysis at the expense of using low oxidation chemicals and low enzyme charges.

Regarding the effect of AHP to bleach and improve the enzymatic hydrolysis yield, it removes dye from cotton surface to improve EH significantly as shown in Fig. 20.6a. Additionally, the chemical charges and thermal conditions used in the AHP process to achieve an enzymatic cotton conversion close to 90% are much lower (e.g., 3% NaOH, 4% H_2O_2, and 90 °C) compared to conventional

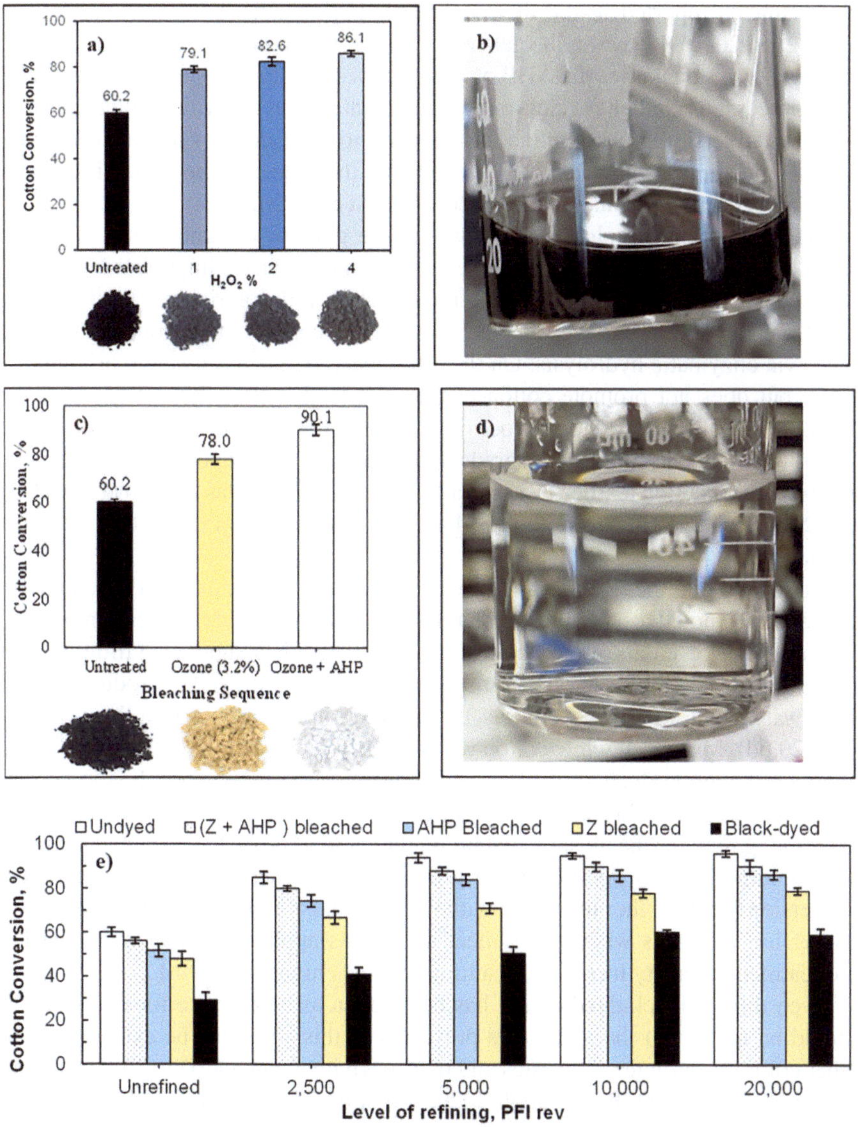

Fig. 20.6 (a) Effect of alkaline hydrogen peroxide charge on enzymatic hydrolysis of mechanically refined cotton (10,000 PFI revs), (**b**) colored glucose after enzymatic hydrolysis of cotton treated with AHP. The AHP process has the following base case: 4% H_2O_2 and 3%NaOH during 90 min at 90 °C and a solid concentration of 5wt%. (**c**) Effect of ozone bleaching followed by AHP extraction on enzymatic hydrolysis of mechanically refined cotton (10,000 PFI revs). (**d**) Glucose obtained after the ozone bleaching process. (e) Enzymatic hydrolysis of cotton textile waste at different levels of mechanical refining and bleaching processes. The ozone process has the following base case: 3.2% O_3, pH 3, and a solid concentration of 3%. Enzymatic hydrolysis was assessed by using CTec3 at 6FPU/g O.D. cotton during 96 h at 50 °C and a solid content of 5wt%. Enzymatic hydrolysis experiments were done by triplicate. (e) (Adapted from Vera et al. 2023b)

pretreatments reported in the literature (e.g., 12% NaOH, 85% H_3PO_4, 150 °C) (Vera et al. 2022). However, the generation of a colored glucose stream and colored wastewater residual after the AHP process poses significant sustainability and glucose processing challenges due to the detrimental effect that dye could have on glucose conversion processes like fermentation to produce the value-added chemicals (Turhan and Ozturkcan 2013). To mitigate this issue, further oxidation strategies like ozone treatments were considered (Vera et al. 2023b).

Transformation of colored textile waste into glucose is significantly improved when both AHP and ozone are applied, even though they have important limitations (Fig. 20.6c). The AHP process, while not effective at decolorizing, can extract the dye from the textile surface, thus achieving high cotton conversion (higher than 80%) via enzymatic hydrolysis. On the other hand, ozone has a potent brightening effect but does not promote cotton conversions higher than 78%. Therefore, the combination of both oxidation processes has been designed into a sequence where ozone was applied initially to degrade and break down the dye chemical structure and bleach the cotton textile waste. As a second stage, AHP was applied at low H_2O_2 and NaOH charge to extract the residual dye left on the cotton surface after the ozone stage.

Compared to either Z or AHP processes separately, the ozone + AHP oxidation sequence promoted a synergistic effect that resulted in a higher conversion of dyed cotton (~90%) even at lower chemical charges. There is a significant difference in the effluents generated after the entire oxidation process, with AHP producing dyed wastewater and ozone generating an undyed and nearly crystalline effluent (Fig. 20.6d). Moreover, white cotton would yield an undyed hydrolysate, leading to a more efficient production of value-added chemicals from glucose. It is important to emphasize that while bleaching is required to remove dye from textiles, mechanical refining remains the most crucial pretreatment to improve the enzymatic conversion of cotton textiles. As shown in Fig. 20.6e, even bleached material cannot have conversion yields higher than 60% without mechanical refining. In contrast, cotton conversion increases when mechanical refining is applied, particularly when the pretreatment intensity increases, resulting in yields higher than 90%.

Even though black-dyed cotton has been taken as a reference for this study, it should be noted that the treatments presented in this work can be extended to all other colored cotton textile wastes and dyes (Arivithamani and Dev 2017; Gül and Özcan-Yildirim 2009; Powar et al. 2020; Wu et al. 2008; Yousef et al. 2019). The ease of bleaching and process conditions are expected to change based on the nature of the dye, its chemical structure, molecular weight, chromophore groups, etc. (Vera et al. 2022). These observations highlight the potential for future environmental policies and regulations toward the development of dyes that are easier to bleach and degrade for optimum textile waste management. For example, the implementation of either natural dyes or dyes with low molecular weight seems to be a promising alternative for a more sustainable life cycle of textile products.

20.8 Factors Affecting the Economics of Textile Waste Pretreatments Intended for Enzymatic Hydrolysis

Enzymatic hydrolysis by itself has been considered an expensive pathway to produce chemicals from cellulosic feedstock. Factors such as the high cost of enzymes and increased capital expenditure, the latter caused by long residence times, considerably drive up the conversion cost (Abbati De Assis et al. 2018; Bajpai 2016; Gonzalez et al. 2011; Singh et al. 2015). Implementing a pretreatment stage helps mitigate this problem by increasing yields and reducing enzyme usage, making the process more economically attractive (Bhutto et al. 2017; Giakoumakis et al. 2021; Park et al. 2016). Moreover, enzymes represent one of the most critical components of the cost structure in the enzymatic hydrolysis process. In that sense, high yields may not necessarily compensate for the cost associated with a very high enzyme dosage, making these processes nonprofitable. This could be the case for some of the strategies described in Table 20.1 where high enzyme loads (30 FPU/g O.D. cotton) are implemented to get 90% glucose yields. Considering that cellulase enzyme prices can range between 2,000 and 5,000 USD per ton, an enzyme charge of 30 FPU/g could be worth between 470 and 780 USD/ton of glucose produced (Davis et al. 2013; de Assis et al. 2018; Gonzalez et al. 2011; Novozymes 2010, 2017; Ou et al. 2021). Hence, the conversion process would be unviable considering that the sugar price is approximately 400 USD per ton (Trading Economics 2022). In this context, the mechanical refining pretreatment effect on boosting the enzymatic hydrolysis yield up to 90% at low enzyme dosage looks promising in terms of getting a reasonable glucose production cost, also considering the low energy and low to no chemistry utilization (Vera et al. 2023a).

20.9 Upcycling Technologies After the Enzymatic Hydrolysis of Textiles for Value-Added Chemical Production

Cotton textile waste has been explored more recently as biomass to obtain ethanol and biobased building blocks (via enzymatic hydrolysis) that can be upcycled into value-added materials from glucose (e.g., levulinic acid, succinic acid, itaconic acid, and glutamic acid) (Cok et al. 2014; Humbird et al. 2011; Ortiz et al. 2022; Werpy and Petersen 2004). Table 20.2 presents a preliminary performance, technology/processes, and economic screening of selected value-added biobased chemicals obtained from glucose (Johnson et al. 2020). From an economic standpoint, it is evident that the production of glucose from textile waste through enzymatic hydrolysis processes would probably not be attractive due to the lower market price of such product (~500 USD/ton) (Parra-Ramírez et al. 2019). On the other hand, there might be an added value in the hundreds to thousands of dollars per ton for different types of glucose-derived chemicals (Table 20.2). For instance, levulinic, succinic, and itaconic acids are the first in the ranking based on their market price and yield. Levulinic acid can be transformed into building blocks (5-HMF, 1,4-butanediol esters, γ-valerolactone) to produce fuels, lubricants, and resins, among others (Kohli et al. 2019; Schmidt et al. 2017; Werpy and Petersen 2004). Succinic acid is a potential platform chemical to obtain petrochemical products such as 1,4-butanediol, tetrahydrofuran, γ-butyrolactone, and pyrrolidinone derivatives (Bozell and Petersen 2010; Cok et al. 2014; Delhomme et al. 2009). Lastly, itaconic acid can be converted into polymers with vinylic monomers, polyols, polyamines, etc. (De Carvalho et al. 2018; Okabe et al. 2009; Willke and Vorlop 2001). Table 20.2 shows common technologies and processes like fermentation and oxidation useful to convert glucose into tailored value-added chemicals. One of the simplest in the fermentation process is implemented to produce bioethanol, which is used as a biofuel with significantly lower environmental impact compared to fossil fuels.

20.10 Challenges, Opportunities, and Future Perspectives

Mechanical refining is presented as a highly efficient pretreatment method for converting cotton textile waste into glucose through enzymatic hydrolysis. This pretreatment eliminates the need for hazardous chemicals during this stage of the process, as reported in previous work (Corbett et al. 2018; de Assis et al. 2018; Jones et al. 2017), and more importantly, high conversion rates are achieved at low loads of enzymes. Thus, mechanical refining can potentially make the enzymatic management of textile wastes more eco-friendly (chemical-free) than conventional pretreatments (de Assis et al. 2018; Vera et al. 2022; Zambrano et al. 2020a, b). In addition, mechanical refining is a low-cost process widely used in the paper industry, and it can be easily implemented on a commercial scale (Chen et al. 2012; Koo et al. 2011; Zhu and Pan 2010). Since refiners are available in different sizes, equipment for textile conversion plants with different production rates could be easily designed. Mechanical refining is widely used in the pulp and paper and biorefinery industry, and can be considered a low-risk-proven technology (de Assis et al. 2018; Jones et al. 2013; Park et al. 2016). Other significant advantages of mechanical refining as a pretreatment method for enzymatic hydrolysis of cotton include the excellent conversion rates obtained (up to 97%) (de Assis et al. 2018; Sandin and Peters 2018). These high conversion rates can also facilitate the separation and recycling of synthetic fibers such as polyester.

Although recycling synthetic fibers can be promising, there is still uncertainty regarding their quality, particularly purity (e.g., residual cellulose or the presence of mixed synthetic fibers) and performance in desired applications (e.g., textiles). A better understanding of those aspects would enable more accurate pricing for recycled synthetic fibers from the enzymatic hydrolysis of cellulosic textile waste discussed herein.

Another challenge relates to the effect of polyester on the enzymatic hydrolysis performance. Existing literature has considered synthetic fibers as a noninhibitor of the enzyme's performance to convert cotton into glucose (Quartinello et al. 2018;

Table 20.2 Review and ranking of value-added biobased chemicals derived from glucose

Value-added chemical	Formula	Application	Process	Theoretical yield (g product g^{-1} glucose)	Yield (g product g^{-1} glucose)	Conditions (T, P, residence time)	Catalyst	Market price (2018) (USD/ton)	Approximate market volume (2018) (MMton/year)	References
Levulinic acid	$C_5H_8O_3$	Methyltetrahydrofuran (MeTHF) δ-aminolevulinic acid 1,4-butanediol esters γ-Valerolactone Na levulinates Valeric acid	Hydrogenation of 5 HMF from glucose	0.6	0.44	200 °C, 30 min, 50 bar N_2	Methanesulfonic or sulfuric acid 0.63M (6%) approximate	5,350–8,880	0.01	Hafyan et al. (2020), Sadhukhan et al. (2016), Schmidt et al. (2017)
Succinic acid	$C_4H_6O_4$	1,4-butanediol tetrahydrofuran γ-Butyrolactone Pyrrolidinone Maleic acid	Fermentation of glucose	0.7	0.73	30 °C, 2 h	*Saccharomyces cerevisiae*	2,500	0.04	Huh et al. (2006), Nieder-Heitmann et al. (2019)
Itaconic acid	$C_5H_6O_4$	Building block for biobased polymers: Styrene derivatives Polyacrylates Polyester Polyamides Diols, diamines	Fermentation of glucose	0.7	0.6	35 °C, 7 days	*A. terreus*	1,900	0.08	Klement and Büchs (2013), Magalhães et al. (2019)
Glutamic acid (amino acid)	$C_5H_9NO_4$	Polyesters Foods (neurotransmitter and flavor enhancer, nutrient) Plant growth	Fermentation of glucose	0.8	0.4	33 °C, 28.5 h	*Brevibacterium divaricatum*	3,625	1	Delaunay et al. (1999), Özüdoğru et al. (2019)

(continued)

Table 20.2 (continued)

Value-added chemical	Formula	Application	Process	Theoretical yield (g product g^{-1} glucose)	Yield (g product g^{-1} glucose)	Conditions (T, P, residence time)	Catalyst	Market price (2018) (USD/ton)	Approximate market volume (2018) (MMton/year)	References
Sorbitol	$C_6H_{14}O_6$	Sugar alcohol L-ascorbic acid Polyols	Reduction of glucose	1	0.56	50% glucose solution, 120 °C,70 bar H_2	5% Raney nickel catalyst	655	2.6	Kapanji et al. (2019)
Glucaric acid	$C_6H_{10}O_8$	Biobased adipic acid: Polyurethanes Non-phthalate Biodegradable Polyesters Nylon-6,6 fibers	Glucose oxidation with nitric acid	1.2	0.58	40 °C, 2 bar	Oxidation with nitric acid	2,910	2.6	Kapanji et al. (2019)
Lactic acid	$C_3H_6O_3$	Polylactic acid Lactate esters (green solvents)	Fermentation of glucose	0.5	0.72	37 °C, 36 h	*Lactobacillus casei*	1,874	1.4	Kohli et al. (2019), Kwan et al. (2018)

Adapted from Vera et al. (2022)

Subramanian et al. 2020). However, this assumption must be demonstrated. On the other hand, it is important to consider the enzymatic hydrolysis performance for converting other cellulosic fibers (e.g., rayon) and non-cellulosic ones (e.g., nylon). Moreover, although previous work shows that the glucose syrup obtained in this process is colorless (Vera et al. 2023a, b), the quality and reactivity of the glucose hydrolysate must be corroborated since this product will be used as a raw material for chemicals with higher prices (Table 20.2). Thus, it is important to evaluate whether dye molecule leftovers after bleaching and other impurities can negatively affect the conversion yield from glucose to downstream chemicals (e.g., bioethanol, furfural, or levulinic acid).

Implementating mechanical refining combined with a dye oxidation process (ozone + peroxide) may be a cost-effective pretreatment for the industrial-scale production of glucose from textile waste via enzymatic hydrolysis. These pretreatments have been proven to significantly boost the enzymatic hydrolysis of dyed textile wastes at low chemical and energy use, compared to EH process without these pretreatments (Vera et al. 2023a, b). On the other hand, the efficacy of textile bleaching coupled with the high yield in the enzymatic hydrolysis of cotton, generated by mechanical refining, opens the door to utilize textile waste containing synthetic fibers to further recycle them. Such synthetic fibers of glucose production.

The production of glucose syrup and further biochemicals from cotton textiles alongside the recovery of polyester is presented as a potential alternative to implement the circular economy in the highly pollutant textile industry. This would lead to reduced textile landfilling, currently estimated at more than 100 million ton/year (Vera et al. 2022). Moreover, glucose and recycled polyester are presented as potentially more sustainable sources to replace fossil-based chemicals or materials.

Understanding the sustainability of the conversion process in terms of CO_2 emissions is an important next step to compare this process with conventional ones. The conversion process considering mechanical refining as a pretreatment is expected to be more sustainable due to low chemical use, reduced energy requirements, and minimal enzyme demand (Gonzalez et al. 2023; Vera et al. 2022, 2023b; Vivas et al. 2024). The combined approach of enzymatic hydrolysis, mechanical refining, and dye oxidation emerges as a technologically and economically feasible pathway for converting cotton textile waste into glucose at an industrial scale.

References

Abbati De Assis C, Greca LG, Ago M, Balakshin MY, Jameel H, Gonzalez R, Rojas OJ (2018) Techno-economic assessment, scalability, and applications of aerosol lignin micro- and nanoparticles. ACS Sustainable Chem Eng 6(9):11853–11868. https://doi.org/10.1021/acssuschemeng.8b02151

Ang S, Haritos V, Batchelor W (2020) Cellulose nanofibers from recycled and virgin wood pulp: a comparative study of fiber development. Carbohydr Polym 234:115900. https://doi.org/10.1016/j.carbpol.2020.115900

Arivithamani N, Dev VRG (2017) Cationization of cotton for industrial scale salt-free reactive dyeing of garments. Clean Techn Environ Policy 19(9):2317–2326. https://doi.org/10.1007/s10098-017-1425-y

Bajpai P (2005) Technology developments in refining. Pira International, Leatherhead. https://www.academia.edu/8206083/Technology_Developments_in_Refining_Pratima_Bajpai_Pira_International_Ltd

Bajpai P (2016) Pretreatment of lignocellulosic biomass. Springer, Singapore, pp 17–70. https://doi.org/10.1007/978-981-10-0687-6_4

Baker C, Pira International (2000) Refining technology. Pira International, Leatherhead

Bhutto AW, Qureshi K, Harijan K, Abro R, Abbas T, Bazmi AA, Karim S, Yu G (2017) Insight into progress in pre-treatment of lignocellulosic biomass. Energy 122:724–745. https://doi.org/10.1016/j.energy.2017.01.005

Bick R, Halsey E, Ekenga CC (2018) The global environmental injustice of fast fashion. In: Environmental health: a global access science source, vol 17. BioMed Central Ltd, London. https://doi.org/10.1186/s12940-018-0433-7

Bozell JJ, Petersen GR (2010) Technology development for the production of biobased products from biorefinery carbohydrates—the US Department of Energy's "top 10" revisited. Green Chem 12(4):539–555. https://doi.org/10.1039/b922014c

Buschle-Diller G, Traore MK (1998) Influence of direct and reactive dyes on the enzymatic hydrolysis of cotton. Text Res J 68(3):185–192. https://doi.org/10.1177/004051759806800306

Cantero D, Jara R, Navarrete A, Pelaz L, Queiroz J, Rodríguez-Rojo S, Cocero MJ (2019) Pretreatment processes of biomass for biorefineries: current status and prospects. Annu Rev 10:289–310. https://doi.org/10.1146/annurev-chembioeng

Chandra RP, Bura R, Mabee WE, Berlin A, Pan X, Saddler JN (2007) Substrate pretreatment: the key to effective enzymatic hydrolysis of lignocellulosics? In: Biofuels, vol 108. Springer, Berlin, pp 67–93. https://doi.org/10.1007/10_2007_064

Chavan RB (2014) Environmental sustainability through textile recycling. https://doi.org/10.4172/2165-8064.S2-007

Chen X, Tao L, Shekiro J, Mohaghaghi A, Decker S, Wang W, Smith H, Park S, Himmel ME, Tucker M (2012) Improved ethanol yield and reduced minimum ethanol selling price (MESP) by modifying low severity dilute acid pretreatment with deacetylation and mechanical refining: 1 experimental. Biotechnol Biofuels 5(1):1–11. https://doi.org/10.1186/1754-6834-5-60

Chen X, Kuhn E, Wang W, Park S, Flanegan K, Trass O, Tenlep L, Tao L, Tucker M (2013) Comparison of different mechanical refining technologies on the enzymatic digestibility of low severity acid pretreated corn Stover. Bioresour Technol. https://doi.org/10.1016/j.biortech.2013.07.109

Cok B, Tsiropoulos I, Roes AL, Patel MK (2014) Succinic acid production derived from carbohydrates: an energy and greenhouse gas assessment of a platform chemical toward a bio-based economy. Biofuels Bioprod Biorefin 8(1):16–29. https://doi.org/10.1002/bbb.1427

Corbett DB, Venditti R, Jameel H, Park S (2018) Effect of mechanical refining energy on the enzymatic digestibility of lignocellulosic biomass. Ind Eng Chem Res 57(43):14648–14655. https://doi.org/10.1021/acs.iecr.8b02932

Corbett DB, Knoll C, Venditti R, Jameel H, Park S (2020) Fiber fractionation to understand the effect of mechanical refining on fiber structure and resulting enzymatic digestibility of biomass. Biotechnol Bioeng 117(4):924–932. https://doi.org/10.1002/bit.27258

Czilik M, Pászt É, Réczey I, Alt J, Rusznák I, Kárpáti É, Víg A (2002) Effects of reactive dyes on the enzymatic depolymerization of cellulose. Dyes Pigments 54(2):95–106. https://doi.org/10.1016/S0143-7208(02)00042-6

Damayanti D, Wulandari LA, Bagaskoro A, Rianjanu A, Wu HS (2021) Possibility routes for textile recycling technology. Polymers 13:3834. https://doi.org/10.3390/POLYM13213834

Daramola MO, Ayeni AO, Trends C (2020) Valorization of biomass to value-added commodities. Springer, Cham. https://doi.org/10.1007/978-3-030-38032-8

Davis R, Tao L, Tan ECD, Biddy MJ, Beckham GT, Scarlata C, Jacobson J, Cafferty K, Ross J, Lukas J, Knorr D, Schoen P (2013) Process design and economics for the conversion of lignocellulosic biomass to hydrocarbons: dilute-acid and enzymatic deconstruction of biomass to sugars and biological conversion of sugars to hydrocarbons. National Renewable Energy Laboratory, Applewood

De Assis T, Huang S, Driemeier CE, Donohoe BS, Kim C, Kim SH, Gonzalez R, Jameel H, Park S (2018) Toward an understanding of the increase in enzymatic hydrolysis by mechanical refining. Biotechnol Biofuels. https://doi.org/10.1186/s13068-018-1289-3

De Carvalho JC, Magalhães AI, Soccol CR (2018) Biobased itaconic acid market and research trends-is it really a promising chemical? Chem Today 36(4):56–58

Delaunay S, Uy D, Baucher MF, Engasser JM, Guyonvarch A, Goergen JL (1999) Importance of phosphoenolpyruvate carboxylase of Corynebacterium glutamicum during the temperature triggered glutamic acid fermentation. Metab Eng 1(4):334–343. https://doi.org/10.1006/mben.1999.0131

Delhomme C, Weuster-Botz D, Kühn FE (2009) Succinic acid from renewable resources as a C4 building-block chemical—a review of the catalytic possibilities in aqueous media. Green Chem 11(1):13–26. https://doi.org/10.1039/b810684c

Eren S, Ge B, Useyin H, Eren A (2016) Colour stripping of reactive-dyed cotton by ozone treatment. Wiley, Hoboken. https://doi.org/10.1111/cote.12240

Flint EA (1950) The structure and development of a cotton fiber. Biol Rev 25(4):414–434. https://doi.org/10.1111/j.1469-185X.1950.tb00767.x

Fockink DH, Maceno MAC, Ramos LP (2015) Production of cellulosic ethanol from cotton processing residues after pretreatment with dilute sodium hydroxide and enzymatic hydrolysis. Bioresour Technol 187:91–96. https://doi.org/10.1016/J.BIORTECH.2015.03.096

Galbe M, Zacchi G (2007) Pretreatment of lignocellulosic materials for efficient bioethanol production. Adv Biochem Eng Biotechnol 108:41–65. https://doi.org/10.1007/10_2007_070

Ghose TK, Bisaria VS (1979) Studies on the mechanism of enzymatic hydrolysis of cellulosic substances. Biotechnol Bioeng 21(1):131–146. https://doi.org/10.1002/bit.260210110

Giakoumakis G, Karnaouri A, Topakas E, Sidiras D (2021) Simulation and optimization of combined acid pretreatment and enzymatic saccharification of medical cotton waste. Biomass Convers Biorefinery 11:2. https://doi.org/10.1007/s13399-020-00694-1

Gias Uddin M, Islam MR (2011) Effects of reductive stripping of reactive dyes on the quality of cotton fabric. Fash Text 2:8. https://doi.org/10.1186/s40691-015-0032-y

Gonzalez R, Treasure T, Phillips R, Jameel H, Saloni D (2011) Economics of cellulosic ethanol production: green liquor pretreatment for softwood and hardwood, greenfield and repurpose scenarios. Bioresources 6(3):2551–2567

Gonzalez R, Vera RE, Zambrano F, Marquez R, Vivas K, Forfora N, Bedard J, Farrell M, Ankeny M, Jameel H (2023) Environmentally friendly oxidation pretreatments to produce sugar-based building blocks from dyed textile wastes via enzymatic hydrolysis. SSRN Electron J. https://doi.org/10.2139/SSRN.4385542

Grover T, Grover T (2018) Integrating sustainable strategies in fashion design by detox 2020 plan—case studies from different brands. In: Detox fashion. Springer, Cham, pp 45–84. https://doi.org/10.1007/978-981-10-4783-1_3

Gül Ş, Özcan-Yildirim Ö (2009) Degradation of reactive red 194 and reactive yellow 145 azo dyes by O_3 and H_2O_2/UV-C processes. Chem Eng J 155(3):684–690. https://doi.org/10.1016/j.cej.2009.08.029

Guo X, Chen L, Tang J, Jönsson LJ, Hong FF (2016) Production of bacterial nanocellulose and enzyme from [AMIM]cl-pretreated waste cotton fabrics: effects of dyes on enzymatic saccharification and nanocellulose production. J Chem Technol Biotechnol 91(5):1413–1421. https://doi.org/10.1002/jctb.4738

Habibi Y, Lucia LA, Rojas OJ (2010) Cellulose nanocrystals: chemistry, self-assembly, and applications. Chem Rev 110(6):3479–3500. https://doi.org/10.1021/cr900339w

Hafyan RH, Bhullar L, Putra ZA, Bilad MR, Wirzal MDH, Nordin NAHM (2020) Sustainability assessment of levulinic acid and succinic acid production from empty fruit bunch. IOP Conf Ser Mater Sci Eng 778(1):140. https://doi.org/10.1088/1757-899X/778/1/012140

Hasanzadeh E, Mirmohamadsadeghi S, Karimi K (2018) Enhancing energy production from waste textile by hydrolysis of synthetic parts. Fuel 218:41–48. https://doi.org/10.1016/J.FUEL.2018.01.035

Heinze T, Rojas OJ (2016) Cellulose chemistry and properties: fibers, nanocelluloses and advanced materials. In: Rojas OJ (ed) Advances in polymer science, vol 271. Springer, Cham. https://doi.org/10.1007/978-3-319-26015-0

Huh YS, Jun YS, Hong YK, Song H, Lee SY, Hong WH (2006) Effective purification of succinic acid from fermentation broth produced by Mannheimia succiniciproducens. Process Biochem 41(6):1461–1465. https://doi.org/10.1016/J.PROCBIO.2006.01.020

Humbird D, Davis R, Tao L, Kinchin C, Hsu D, Aden A, Schoen P, Lukas J, Olthof B, Worley M, Sexton D, Dudgeon D (2011) Process design and economics for biochemical conversion of lignocellulosic biomass to ethanol: dilute-acid pretreatment and enzymatic hydrolysis of corn Stover. National Renewable Energy Laboratory, Applewood. https://doi.org/10.2172/1107470

Javaid R, Qazi UY (2019) Catalytic oxidation process for the degradation of synthetic dyes: an overview. Int J Environ Res Public Health 16(11):1–27. https://doi.org/10.3390/ijerph16112066

Jeihanipour A, Taherzadeh MJ (2009) Ethanol production from cotton-based waste textiles. Bioresour Technol 100(2):1007–1010. https://doi.org/10.1016/J.BIORTECH.2008.07.020

Jeihanipour A, Karimi K, Niklasson C, Taherzadeh MJ (2010) A novel process for ethanol or biogas production from cellulose in blended-fibers waste textiles. Waste Manag 30(12):2504–2509. https://doi.org/10.1016/J.WASMAN.2010.06.026

Johnson S, Echeverria D, Venditti R, Jameel H, Yao Y (2020) Supply chain of waste cotton recycling and reuse: a review. AATCC J Res 7(1):19–31. https://doi.org/10.14504/ajr.7.S1.3

Jones BW, Venditti R, Park S, Jameel H, Koo B (2013) Enhancement in enzymatic hydrolysis by mechanical refining for pretreated hardwood lignocellulosics. Bioresour Technol 147:353–360. https://doi.org/10.1016/J.BIORTECH.2013.08.030

Jones BW, Venditti R, Park S, Jameel H (2017) Optimization of pilot scale mechanical disk refining for improvements in enzymatic digestibility of pretreated hardwood lignocellulosics. Bioresources 12(3):4567–4593. https://doi.org/10.15376/biores.12.3.4567-4593

Kapanji KK, Haigh KF, Görgens JF (2019) Techno-economic analysis of chemically catalysed lignocellulose biorefineries at a typical sugar mill: sorbitol or glucaric acid and electricity co-production. Bioresour Technol 289:121635. https://doi.org/10.1016/J.BIORTECH.2019.121635

Klement T, Büchs J (2013) Itaconic acid - a biotechnological process in change. Bioresour Technol 135:422–431. https://doi.org/10.1016/j.biortech.2012.11.141

Kohli K, Prajapati R, Sharma BK (2019) Bio-based chemicals from renewable biomass for integrated biorefineries. Energies 12(2):233. https://doi.org/10.3390/en12020233

Koo BW, Treasure TH, Jameel H, Phillips RB, Chang HM, Park S (2011) Reduction of enzyme dosage by oxygen delignification and mechanical refining for enzymatic hydrolysis of green liquor-pretreated hardwood. Appl Biochem Biotechnol 165(3–4):832–844. https://doi.org/10.1007/s12010-011-9301-4

Kumar AK, Sharma S (2017) Recent updates on different methods of pretreatment of lignocellulosic feedstocks: a review. Bioresour Bioprocess 4(1):7. https://doi.org/10.1186/s40643-017-0137-9

Kwan TH, Hu Y, Lin CSK (2018) Techno-economic analysis of a food waste valorisation process for lactic acid, lactide and poly(lactic acid) production. J Clean Prod 181:72–87. https://doi.org/10.1016/j.jclepro.2018.01.179

Li R (2021) Effects of various decolorizing agents on waste cotton fabric dyed with reactive dyes

Li X, Hu Y, Du C, Lin CSK (2019) Recovery of glucose and polyester from textile waste by enzymatic hydrolysis. Waste Biomass Valoriz 10(12):3763–3772. https://doi.org/10.1007/s12649-018-0483-7

Magalhães AI, de Carvalho JC, Thoms JF, Medina JDC, Soccol CR (2019) Techno-economic analysis of downstream processes in itaconic acid production from fermentation broth. J Clean Prod 206:336–348. https://doi.org/10.1016/J.JCLEPRO.2018.09.204

Morjanoff PJ, Gray PP (1987) Optimization of steam explosion as a method for increasing susceptibility of sugarcane bagasse to enzymatic saccharification. Biotechnol Bioeng 29(6):733–741. https://doi.org/10.1002/bit.260290610

Nazhad MM, Ramos LP, Paszner L, Saddler JN (1995) Structural constraints affecting the initial enzymatic hydrolysis of recycled paper. Enzym Microb Technol 17(1):68–74. https://doi.org/10.1016/0141-0229(94)00057-X

Nieder-Heitmann M, Haigh K, Görgens JF (2019) Process design and economic evaluation of integrated, multi-product biorefineries for the co-production of bio-energy, succinic acid, and polyhydroxybutyrate (PHB) from sugarcane bagasse and trash lignocelluloses. Biofuels Bioprod Biorefin 13(3):599–617. https://doi.org/10.1002/bbb.1972

Novozymes (2010) Application Sheet: Cellic® CTec2 and HTec2 - enzymes for hydrolysis of lignocellulosic materials

Novozymes (2017) Novozymes Cellic® CTec3 HS application sheet - secure your plant's lowest cost

Okabe M, Lies D, Kanamasa S, Park EY (2009) Biotechnological production of itaconic acid and its biosynthesis in aspergillus terreus. Appl Microbiol Biotechnol 84(4):597–606. https://doi.org/10.1007/s00253-009-2132-3

Ortiz MS, Alvarado JG, Zambrano F, Marquez R (2022) Surfactants produced from carbohydrate derivatives: a review of the biobased building blocks used in their synthesis. J Surfactant Deterg 25(2):147–183. https://doi.org/10.1002/jsde.12581

Ou L, Dou C, Yu JH, Kim H, Park YC, Park S, Kelley S, Lee EY (2021) Techno-economic analysis of sugar production from lignocellulosic biomass with utilization of hemicellulose and lignin for high-value co-products. Biofuels Bioprod Biorefin 15(2):404–415. https://doi.org/10.1002/bbb.2170

Özüdoğru HMR, Nieder-Heitmann M, Haigh KF, Görgens JF (2019) Techno-economic analysis of product biorefineries utilizing sugarcane lignocelluloses: xylitol, citric acid and glutamic acid scenarios annexed to sugar mills with electricity co-production. Ind Crop Prod 133:259–268. https://doi.org/10.1016/J.INDCROP.2019.03.015

Pandit P, Nadathur GT, Jose S (2018) Upcycled and low-cost sustainable business for value-added textiles and fashion. In: Circular economy in textiles and apparel: processing, manufacturing, and design. Elsevier, Amsterdam, pp 95–122. https://doi.org/10.1016/B978-0-08-102630-4.00005-4

Park J, Jones B, Koo B, Chen X, Tucker M, Yu JH, Pschorn T, Venditti R, Park S (2016) Use of mechanical refining to improve the production of low-cost sugars from lignocellulosic biomass. Bioresour Technol 199:59–67. https://doi.org/10.1016/J.BIORTECH.2015.08.059

Parra-Ramírez D, Martinez A, Cardona CA (2019) Lactic acid production from glucose and xylose using the lactogenic Escherichia coli strain JU15: experiments and techno-economic results. Bioresour Technol 273:86–92. https://doi.org/10.1016/j.biortech.2018.10.061

Powar AS, Perwuelz A, Behary N, Hoang L, Aussenac T (2020) Application of ozone treatment for the decolorization of the reactive-dyed fabrics in a pilot-scale process-optimization through response surface methodology. Sustain For 12(2):471. https://doi.org/10.3390/su12020471

Pu Y, Treasure T, Gonzalez R, Venditti RA, Jameel H (2013) Autohydrolysis pretreatment of mixed softwood to produce value prior to combustion. Bioenergy Res 6(3):1094–1103. https://doi.org/10.1007/s12155-013-9343-2

Quartinello F, Vecchiato S, Weinberger S, Kremenser K, Skopek L, Pellis A, Guebitz GM (2018) Highly selective enzymatic recovery of building blocks from wool-cotton-polyester textile waste blends. Polymers 10(10):1107. https://doi.org/10.3390/polym10101107

Raspolli Galletti AM, Antonetti C (2012) Biomass pretreatment: separation of cellulose, hemicellulose, and lignin—existing technologies and perspectives. Biorefinery. https://doi.org/10.1515/9783110260281.101

Sadhukhan J, Ng KS, Martinez-Hernandez E (2016) Novel integrated mechanical biological chemical treatment (MBCT) systems for the production of levulinic acid from fraction of munici-

pal solid waste: a comprehensive techno-economic analysis. Bioresour Technol 215:131–143. https://doi.org/10.1016/J.BIORTECH.2016.04.030
Sandin G, Peters GM (2018) Environmental impact of textile reuse and recycling—a review. J Clean Prod 184:353–365. https://doi.org/10.1016/J.JCLEPRO.2018.02.266
Sasaki C, Kiyokawa A, Asada C, Nakamura Y (2019) Glucose and valuable chemicals production from cotton waste using hydrothermal method. Waste Biomass Valoriz 10(3):599–607. https://doi.org/10.1007/s12649-017-0084-x
Schmidt LM, Mthembu LD, Reddy P, Deenadayalu N, Kaltschmitt M, Smirnova I (2017) Levulinic acid production integrated into a sugarcane bagasse based biorefinery using thermal-enzymatic pretreatment. Ind Crop Prod 99:172–178. https://doi.org/10.1016/j.indcrop.2017.02.010
Shaker K, Nawab Y (2020) Fibers for protective textiles. In: Topics in mining, metallurgy and materials engineering. Springer, Cham. https://doi.org/10.1007/978-3-030-49224-3_4
Shen L, Worrell E, Patel MK (2010) Environmental impact assessment of man-made cellulose fibres. Resour Conserv Recycl 55(2):260–274. https://doi.org/10.1016/j.resconrec.2010.10.001
Shen F, Xiao W, Lin L, Yang G, Zhang Y, Deng S (2013) Enzymatic saccharification coupling with polyester recovery from cotton-based waste textiles by phosphoric acid pretreatment. Bioresour Technol 130:248–255. https://doi.org/10.1016/J.BIORTECH.2012.12.025
Shuhua W, Xiaoying Y, Xiaogang C, Wensheng H, Mei N (2020) Recycling of cotton fibers separated from the waste blend fabric. J Nat Fib 17(4):520–531. https://doi.org/10.1080/15440478.2018.1503130
Singh YD, Satapathy KB (2018) Conversion of lignocellulosic biomass to bioethanol: an overview with a focus on pretreatment. Int J Eng Technol 15:17–43. https://doi.org/10.18052/www.scipress.com/ijet.15.17
Singh J, Suhag M, Dhaka A (2015) Augmented digestion of lignocellulose by steam explosion, acid and alkaline pretreatment methods: a review. Carbohydr Polym 117:624–631. https://doi.org/10.1016/j.carbpol.2014.10.012
Singhania RR, Patel AK, Sukumaran RK, Larroche C, Pandey A (2013) Role and significance of beta-glucosidases in the hydrolysis of cellulose for bioethanol production. Bioresour Technol 127:500–507
Smook GA, Kocurek MJ (1982) Handbook for pulp & paper technologists. Canadian Pulp and Paper Association, Montréal
Subramanian K, Chopra SS, Cakin E, Li X, Lin CSK (2020) Environmental life cycle assessment of textile bio-recycling—valorizing cotton-polyester textile waste to pet fiber and glucose syrup. Resour Conserv Recycl 161:104989. https://doi.org/10.1016/j.resconrec.2020.104989
Sun FF, Hong J, Hu J, Saddler JN, Fang X, Zhang Z, Shen S (2015) Accessory enzymes influence cellulase hydrolysis of the model substrate and the realistic lignocellulosic biomass. Enzym Microb Technol 79–80:42–48. https://doi.org/10.1016/j.enzmictec.2015.06.020
Trading Economics (2022) Sugar | 2022 data | 2023 forecast | 1912–2021 historical | Price | Quote | Chart. https://tradingeconomics.com/commodity/sugar
Turhan K, Ozturkcan SA (2013) Decolorization and degradation of reactive dye in aqueous solution by ozonation in a semi-batch bubble column reactor. Water Air Soil Pollut 224(1):1–13. https://doi.org/10.1007/s11270-012-1353-8
United States Environmental Protection Agency (2021) Textiles: material-specific data. https://www.epa.gov/Facts-and-Figures-about-Materials-Waste-and-Recycling/Textiles-Material-Specific-Data
Vera RE, Zambrano F, Suarez A, Pifano A, Marquez R, Farrell M, Ankeny M, Jameel H, Gonzalez R (2022) Transforming textile wastes into biobased building blocks via enzymatic hydrolysis: a review of key challenges and opportunities. Cleaner Circ Bioecon 3:100026
Vera RE, Suarez A, Zambrano F, Marquez R, Bedard J, Vivas KA, Pifano A, Farrell M, Ankeny M, Jameel H, Gonzalez R (2023a) Upcycling cotton textile waste into bio-based building blocks through an environmentally friendly and high-yield conversion process. Resour Conserv Recycl. https://doi.org/10.1016/j.resconrec.2022.106715
Vera RE, Zambrano F, Marquez R, Vivas KA, Forfora N, Bedard J, Farrell M, Ankeny M, Pal L, Jameel H, Gonzalez R (2023b) Environmentally friendly oxidation pretreatments to produce

sugar-based building blocks from dyed textile wastes via enzymatic hydrolysis. Chem Eng J 467:143321. https://doi.org/10.1016/j.cej.2023.143321

Walker LP, Wilson DB (1991) Enzymatic hydrolysis of cellulose: an overview. Bioresour Technol 36(1):3–14. https://doi.org/10.1016/0960-8524(91)90095-2

Wang M, Zhou D, Wang Y, Wei S, Yang W, Kuang M, Ma L, Fang D, Xu S, Du S (2016) Bioethanol production from cotton stalk: a comparative study of various pretreatments. Fuel 184:527–532. https://doi.org/10.1016/j.fuel.2016.07.061

Wang N, Tang P, Zhao C, Zhang Z, Sun G (2020) An environmentally friendly bleaching process for cotton fabrics: mechanism and application of UV/H_2O_2 system. Cellulose 27(2):1071–1083. https://doi.org/10.1007/s10570-019-02812-3

Werpy T, Petersen G (2004) Top value added chemicals from biomass volume I. NREL, Applewood. https://doi.org/10.2172/15008859

Willke T, Vorlop K-D (2001) Biotechnological production of itaconic acid. Appl Microbiol Biotechnol 56(3):289–295. https://doi.org/10.1007/s002530100685

Wu J, Doan H, Upreti S (2008) Decolorization of aqueous textile reactive dye by ozone. Chem Eng J 142(2):156–160. https://doi.org/10.1016/J.CEJ.2007.11.019

Yousef S, Tatariants M, Tichonovas M, Sarwar Z, Jonuškienė I, Kliucininkas L (2019) A new strategy for using textile waste as a sustainable source of recovered cotton. Resour Conserv Recycl 145:359–369. https://doi.org/10.1016/J.RESCONREC.2019.02.031

Zambrano F, Starkey H, Wang Y, Abbati De Assis C, Venditti R, Pal L, Jameel H, Hubbe MA, Rojas OJ, Gonzalez R (2020a) Using micro- and nanofibrillated cellulose as a means to reduce weight of paper products: a review. Bioresources 15(2):4553–4590

Zambrano F, Wang Y, Zwilling JD, Venditti R, Jameel H, Rojas O, Gonzalez R (2020b) Micro-and nanofibrillated cellulose from virgin and recycled fibers: a comparative study of its effects on the properties of hygiene tissue paper. Carbohydr Polym 254:117430. https://doi.org/10.1016/j.carbpol.2020.117430

Zambrano MC, Pawlak JJ, Daystar J, Ankeny M, Venditti RA (2021) Impact of dyes and finishes on the aquatic biodegradability of cotton textile fibers and microfibers released on laundering clothes: correlations between enzyme adsorption and activity and biodegradation rates. Mar Pollut Bull 165:112030. https://doi.org/10.1016/j.marpolbul.2021.112030

Zhang P, Chen M, Duan Y, Huang R, Su R, Qi W, Thielemans W, He Z (2018) Real-time adsorption of exo- and endoglucanases on cellulose: effect of pH, temperature, and inhibitors. Langmuir 34(45):13514–13522. https://doi.org/10.1021/acs.langmuir.8b02260

Zhu JY, Pan XJ (2010) Woody biomass pretreatment for cellulosic ethanol production: technology and energy consumption evaluation. Bioresour Technol 101(13):4992–5002. https://doi.org/10.1016/j.biortech.2009.11.007

Valorization of Pharmaceutical Wastes for the Production of Biofuels and Chemicals

21

Vaishali Chaudhary, Monika Prakash Rai, and Savita Chaurasia

Abstract

Pharmaceutical wastes pose a significant environmental challenge, majorly due to their complex chemical composition and persistent nature. Conventional disposal methods, like incineration or landfill, can lead to toxic emissions and chronic contamination of soil and water systems. In contrast, innovative and advanced conversion technologies are emerging as effective solutions to mitigate the environmental impact of pharmaceutical waste. Advanced treatment methods, including catalytic hydrothermal liquefaction, supercritical fluid extraction, and microbial degradation, offer promising ways to convert pharmaceutical waste into valuable products. These technologies ease the breakdown of hazardous compounds into simpler and less toxic compounds, more manageable forms, and allow the recovery of high-value chemicals and biofuels. For instance, catalytic hydrothermal liquefaction can convert pharmaceutical wastes into bio-oil and biogas, while microbial processes have the potential to produce bioethanol and biogas from waste. Furthermore, supercritical fluid extraction has proven effective in isolating valuable compounds from waste streams, which can be repurposed as high-quality chemicals or additives. Employing these advanced conversion technologies, it is possible to not only reduce the environmental footprint of pharmaceutical waste but also create sustainable, economically feasible

V. Chaudhary · M. P. Rai (✉)
Department of Biotechnology, Motilal Nehru National Institute of Technology Allahabad, Prayagraj, Uttar Pradesh, India, 211004
e-mail: mprai@mnnit.ac.in

S. Chaurasia (✉)
Biochemistry and Molecular Biology Program, Department of Chemistry, Bellarmine University, Louisville, KY 40205, USA
e-mail: schaurasia@bellarmine.edu

A. Kumari et al. (eds.), *Valorization of Solid Wastes to Biofuels and Chemical Products for Sustainable World*, https://doi.org/10.1007/978-981-96-8594-3_21

compounds that contribute to the circular economy. This approach aligns with global sustainability goals and provides a promising pathway for addressing the growing challenge of pharmaceutical waste management.

Keywords

Pharmaceutical wastes · Conversion technologies · Valuable chemicals · Biofuels · Sustainability · Environmental impact

21.1 Introduction

Pharmaceutical waste comprises expired, unused, or discarded medications and their by-products from various sources, including households, healthcare facilities, and pharmaceutical industries. The environment's health is seriously threatened by the inappropriate disposal of pharmaceutical waste (Kümmerer 2010). Prescription drugs can pollute soil, water, and even the air when they get into the environment, which can have a negative impact on ecosystems and public health. Research has demonstrated that pharmaceutical compounds, such as hormones, antibiotics, and other bioactive molecules, can linger in the environment and cause problems for aquatic life as well as the emergence of bacteria resistant to antibiotics (Kümmerer 2009a, b). A study reported that pharmaceutical pollutants in water bodies can impact aquatic organisms' reproductive systems, which feminizes them and causes reproductive failures (Arnold et al. 2014). Soil contaminated with pharmaceutical waste can have an impact on microbial communities, which can change soil fertility and nutrient cycling. The presence of drugs in minimal quantities in drinking water raises questions about possible long-term health impacts on people, such as antibiotic resistance and endocrine disruption (Fent et al. 2006, WHO 2022)

Waste management through sustainable practices plays an important role in mitigating the environmental impact of pharmaceutical waste. Traditional waste disposal methods, such as incineration and landfilling, are often inadequate and environmentally harmful. Where incineration can release toxic compounds, landfilling can result in leachate formation leading to the contamination of groundwater. Therefore, advanced and sustainable approaches to managing pharmaceutical waste are a necessity of the time for protecting environmental and public health. Recovery of pharmaceutical waste is one such method that turns waste materials into high-value products. A number of techniques, including chemical, biological, and thermal processes, can be used to achieve this. By producing value-added compounds like chemicals, biofuels, and energy, valorization not only limits the waste volume but also produces economic profits (Achillas et al. 2013). One efficient method of producing renewable energy and managing waste is the anaerobic digestion of pharmaceutical waste, which generates biogas. Pharmaceutical waste can be recycled into high-value chemicals and biofuels, which is an innovative and environmentally friendly approach to such waste management. For this, a number of processes can be used, including fermentation, gasification, and pyrolysis.

The introduction of cutting-edge technologies and integrated strategies that encompass resource recovery and waste treatment possesses the key to the pharmaceutical waste management. Biotechnological revolution, like the use of genetically modified bacteria, can enhance the efficacy of pharmaceutical waste transformation into chemicals and biofuels. The yields and purity of value-added products made from such waste can be upgraded by doing research on innovative catalysts and reactors (Clark and Deswarte 2014). To further encourage environmentally friendly waste management techniques, laws and regulations need to be strengthened. Guidelines for the appropriate handling and disposal of pharmaceutical waste should be developed in cooperation between the government and business sectors. Public awareness campaigns are also essential for educating the public about the environmental damage caused by pharmaceutical waste and also about the importance of sustainable disposal methods (Daughton 2003).

Waste should be seen as a potential resource rather than a burden in the circular economy that is being advocated. We can reduce our influence on the environment, preserve natural resources, and generate income by including pharmaceutical waste management into the circular economy framework. Healthcare, business, academia, and government stakeholders are among the multidisciplinary groups that must be involved in this (Stahel 2016).

21.2 Pharmaceutical Wastes: Characteristics and Types

Pharmaceutical waste comprises medications, vaccines, and sera of no longer uses and must be properly disposed of, as well as spilled, expired, unutilized, and contaminated pharmaceutical products (Freitas and Radis-Baptista 2021). Discarded pharmaceutical handling equipment, such as drug vials, gloves, masks, connecting tubing, and bottles or boxes with residues, is also included in this category (Pruss et al. 1999; Rajbongshi et al. 2016). From a pharmaceutical management perspective, pharmaceutical waste can be broadly classified into two categories: pharmaceutical waste from hospitals, healthcare, and research organizations and pharmaceutical waste from domestic households and the healthcare treatment industry, which includes expired or unopened drugs and vials (Vallini and Townend 2010).

Typically, pharmaceutical waste might consist of:

(a) Finished medications patients' abandoned prescription drugs.
(b) Waste items (such as vials, IV bags, tubing, and syringes) that contain extra medication.
(c) Trash that still contains traces of chemotherapeutic drugs.
(d) Medications in open containers that are not meant to be utilized.
(e) Containers for acute hazardous waste medicaments.
(f) Discarded medications, spill cleaning dispenses, contaminated clothing, and absorbents.
(g) Expired over-the-counter medications.
(h) Contaminated personal protective equipment (PPE).

21.2.1 Characteristics of Pharmaceutical Waste

Pharmaceutical waste has several key characteristics that make it a concern for disposal and environmental management (EPA 2022). Here's a breakdown of these characteristics:

(a) Persistence: A lot of pharmaceutical ingredients are made to work in the body for a long time. This trait translates to persistence in the environment, even when it is advantageous for treatment. Long after they are released, these persistent medications can still be harmful to living things (Dawood et al. 2023).
(b) Biological activity: Pharmaceuticals, in contrast to many other contaminants, have biological effects even at low quantities. Thus, even in minute quantities, they have the potential to impair ecological processes and cause harm to living things (Jaseem et al. 2017).
(c) Toxicity: Since they are made to specifically target biological functions within the body, many medicines are by nature harmful. If improperly disposed of, its intrinsic toxicity can affect aquatic life, humans, and animals (Jaseem et al. 2017).

21.2.2 Classification of Pharmaceutical Waste

Pharmaceutical waste is classified into three different categories as mentioned below (Sreekanth et al. 2014):

21.2.2.1 Hazardous Waste

The term "hazardous pharmaceutical waste" (HPW) describes used, expired, or contaminated drugs that, if improperly disposed of, could endanger public health or the environment. Federal EPA laws classify drugs into two categories: the "P list" and the "U list."

P-listed drugs (e.g., phentermine, epinephrine, physostigmine, nicotine, nitroglycerin, and warfarin >3%) are indicative of acute toxicity; the medicine and its container are deemed hazardous and should be disposed of in an RCRA-approved container.

U-listed materials, including lindane, phenol, chloral hydrate, and certain antineoplastic waste, are considered as hazardous. Pharmaceuticals that produce wastes show any of the following properties listed below (Sreekanth et al. 2014):

The following dangerous traits may be present in one or more of these medications.

Toxicity: The ability to cause injury when absorbed through the skin, breathed, or consumed (e.g., cytotoxic medications used in chemotherapy).

Ignitability: Easily ignited or flammable (such as cough syrups with alcohol).

Corrosivity: When strong acids or bases—like those found in some medications—come into contact with living tissues, they can cause harm or even destruction.

Reactivity: Under some circumstances, it may undergo uncontrollable reactions and release toxic gases or fumes (unusual for medications).

21.2.2.2 Nonhazardous Waste

Medication that has expired, been unused, or is left over but does not fit the requirements for hazardous waste as defined by environmental rules is referred to as nonhazardous pharmaceutical waste (NHPW). Even though NHPW isn't as dangerous as HPW, improper disposal can still impose negative effects on the environment and public well-being. The hazardous qualities (toxicity, ignitability, corrosivity, and reactivity) outlined by laws such as the Resource Conservation and Recovery Act (RCRA) in the United States are generally absent from nonhazardous petroleum waste (NHPW).

21.2.2.3 Controlled Substance Pharmaceutical Waste

Pharmaceutical waste containing restricted chemicals, such as stimulants, depressants, opioids, or hallucinogens, that may be abused, become dependent on, or divert attention, is referred to as controlled substance pharmaceutical waste (DEA 2020). Pharmaceutical waste involving prohibited substances includes, for example:

(a) Opioids include morphine, hydrocodone (Vicodin), and oxycodone (OxyContin) (DEA 2020).
(b) Amphetamine (Adderall) and methylphenidate (Ritalin) are stimulants (DEA 2020).
(c) Depressants: diazepam (Valium), alprazolam (Xanax) (DEA 2020).
(d) DEA, 2020 lists ketamine (Ketalar) and phencyclidine (PCP) as hallucinogens.

Apart from these major classes of pharmaceutical waste, a number of types are also there based on their properties and usage, listed in Table 21.1, along with their potential hazard.

21.3 Technologies for Conversion of Pharmaceutical Wastes

Because pharmaceutical waste is complicated and frequently poisonous, it presents serious risks to the environment and human health. When pharmaceutical waste is not properly disposed of, it can contaminate land, water, and air, endangering ecosystems and human health. But because of developments in waste conversion technologies, pharmaceutical waste may now be converted into useful chemicals and biofuels. Three categories comprise the majority of conversion technologies: chemical, biological, and thermal processes (Table 21.2).

Table 21.1 Varieties of pharmaceutical wastes and the associated hazards

Category of pharmaceutical waste	Examples	Potential hazards	References
Expired pharmaceuticals	Expired medications like antibiotics and painkillers	Environmental contamination and antimicrobial resistance	Nepal et al. (2020)
Contaminated pharmaceuticals	Medications that are mixed with bodily fluids or chemicals	Risk of infection and chemical burns	Vinaik et al. (2019)
Unusable pharmaceuticals	IV bags, syringes partially used vials	Environmental contamination and toxicity	Ruhoy and Daughton (2008)
Nonhazardous pharmaceutical waste	Vitamins, glucose solutions, saline solutions	Low risk but still required proper disposal	Nyaga et al. (2020)
Hazardous pharmaceutical waste	Chemotherapy drugs, mercury thermometers, and immunosuppressants	Toxicity, mutagenic effects, carcinogenic and teratogenic effects	Karri et al. (2021)
Controlled substances	Benzodiazepines and opioids	Addiction and poisoning	Gudin et al. (2013)
Cytotoxic waste	Chemotherapy drugs	Mutagenic, teratogenic, and carcinogenic effects	Biron et al. (2006)
Infectious waste	Vaccines containing live-attenuated viruses	Risk of infection	Makajic-Nikolic et al. (2016)
Sharps waste	Needles, scalpels, syringes	Risk of infection and needle-stick injuries	Bazie (2020)

21.3.1 Thermal Process

High temperatures are used in thermal processes to convert pharmaceutical waste from complex organic compounds into simpler molecules, neutralizing hazardous chemicals and minimizing waste volume. Methods including gasification, pyrolysis, and incineration are frequently used. According to Keller amd Heindel (2010) and Gogate and Pandit (2004), these procedures have the potential to significantly lower environmental pollution while recovering important by-products like energy and raw materials.

(a) Pyrolysis: The process of heat breakdown known as pyrolysis is carried out in an oxygen lacking environment. Temperatures between 300 and 700 °C are used to break down organic materials into bio-oil, charcoal, and syngas. Complex molecules are broken down into smaller, volatile chemicals in a pyrolysis reactor by applying high temperatures to pharmaceutical waste. These substances condense to produce bio-oil, which is abundant in phenolic and hydrocarbon chemicals. Pyrolysis not only yields bio-oil but also charcoal and

Table 21.2 List of technologies for the conversion of pharmaceutical wastes into valuable compounds

Technology	Description	Produced compounds	References
Thermal processes	Processes like pyrolysis, gasification that convert waste into chemicals and biofuels	Bio-oil, biochar, syngas	Kumar et al. (2009)
Biological processes	Employs microorganisms for the degradation and bioconversion waste	Organic acids, butanol, ethanol	Pandey and Szakacs (2005)
Chemical processes	Employs chemical reactions for the conversion of waste	Succinic acid, lactic acid, etc.	Clark and Deswarte (2014)
Catalytic conversion	Involves catalysts to enhance the conversion of waste into value-added products	Biofuels, fine chemicals	Luque et al. (2008)
Hydrothermal liquefaction	Conversion of wet biomass under high pressure and moderate temperature	Biocrude oil, biogasoline	Elliott (2007)
Anaerobic digestion	Biological process that converts organic waste into biogas and digestate	Biogas, digestate	Angelidaki and Ahring (1994)
Fermentation	Conversion of organic waste into biofuels using microorganisms	Bioethanol, butanol	Lynd et al. (2002)
Transesterification	Conversion of fats and oils into biodiesel using alcohol and a catalyst	Biodiesel, glycerol	Meher et al. (2006)

syngas. While syngas is a blend of hydrogen and carbon monoxide along with some other gases that can be utilized to generate energy, biochar is a carbon-rich material that can be added to soil (Singh et al. 2020; Patel et al. 2019).

(b) Gasification: Gasification is the process of partially oxidizing organic molecules at temperatures between 800 and 1200 °C. The result is syngas, which is a combination of methane, hydrogen, carbon dioxide, and carbon monoxide. Waste from pharmaceuticals is put into a gasifier, where it interacts with steam or oxygen in a controlled amount to produce syngas. To create useful compounds, this syngas can be further treated using the Fischer-Tropsch or methanol synthesis methods. Pharmaceutical waste syngas has been utilized to produce methanol and other alcohols, which are then employed as precursors for other kinds of industrial chemicals (Chen et al. 2021a; Smith et al. 2020).

(c) Hydrothermal liquefaction (HTL): HTL transforms wet organic waste into biocrude oil by operating in the presence of water at moderate temperatures (250–374 °C) and high pressures (10–25 MPa). This procedure involves heating pharmaceutical waste in an HTL reactor while it is combined with water. Heat and high pressure convert the waste to a liquid biocrude oil. Through catalytic hydrodeoxygenation, this biocrude oil can be improved to remove oxygenates and generate cleaner chemicals and fuels. Diesel-range alkanes and aromatics can be produced by upgrading biocrude oil from HTL of pharmaceutical waste (Wang et al. 2021; Zhang et al. 2020).

(d) Combustion: It is the process of burning pharmaceutical waste in the presence of extra oxygen releases energy and transforms it into carbon dioxide, water, and ash. In a combustion chamber, pharmaceutical waste is burned, generating heat energy that can be used to generate electricity. Emissions control technologies are integrated into advanced combustion systems to minimize the discharge of detrimental pollutants. Pharmaceutical waste can be burned in modern incinerators that are outfitted with filters and scrubbers to reduce the amount of harmful chemicals released into the air (Jones et al. 2019; Martinez et al. 2020).

21.3.2 Biological Process

The biological conversion of pharmaceutical waste into chemicals and biofuels involves several advanced methods and technologies. These approaches leverage the capabilities of microorganisms and enzymes to degrade complex pharmaceutical compounds into simpler, valuable products. The primary methods are described below:

(a) Microbial degradation

Metabolic engineering: The process of breaking down pharmaceutical waste into simpler molecules involves engineering microbes, such as fungi and bacteria, to perform certain tasks. According to Zhang et al. (2015), genetic engineering can improve the ability of these microbes to break down pharmaceutical contaminants into organic acids, alcohols, and other chemical precursors (Chen et al. 2021d).

Microbial consortia: Microorganisms in mixed cultures can destroy a variety of medicinal substances in a synergistic manner. Different microorganisms have complementary metabolic pathways that enhance each other's effectiveness in the breakdown and conversion processes (Aydin et al. 2019).

(b) Anaerobic digestion

Biogas production: Anaerobic digesters are used to treat pharmaceutical waste. Here, microorganisms break down the waste in the absence of oxygen to produce digestate and biogas, which includes carbon dioxide and methane. While the digestate can be utilized as a fertilizer with lots of nutrients, the biogas can be used as a renewable energy source (Mudhoo and Kumar 2013).

(c) Bioaugmentation

Tailored microbial communities: Recalcitrant pharmaceutical chemicals are broken down faster in waste treatment systems by adding particular strains or consortia of microorganisms. Pharmaceutical leftovers can be metabolized more effectively by these customized communities, producing intermediate molecules that can then be processed into industrial chemicals or biofuels (Verlicchi et al. 2012).

(d) Enzymatic treatment

Isolated enzymes: Pharmaceutical substances are broken down by specific enzymes called laccases and peroxidases (Jin et al. 2021). Recombinant DNA technology can be used to create these enzymes, which can then be used in waste treatment procedures to turn pharmaceutical waste into useful goods (Wang and Wang 2016).

(e) Fermentation

Biochemical conversion: In fermentation procedures, microorganisms are used to turn pharmaceutical waste into biofuels like butanol and ethanol. By choosing microorganisms with a high production of the desired products and a high resistance to medicinal substances, fermentation can be enhanced (Koutinas et al. 2014).

(f) Composting

Organic degradation: Composting pharmaceutical waste with organic matter allows the waste to break down into simpler organic molecules due to microbial action. Compost rich in nutrients and suitable for use as a soil conditioner can be produced by this procedure (Huang et al. 2013).

21.3.3 Chemical Process

Pharmaceutical waste is chemically converted into chemicals and biofuels by means of catalytic hydrothermal liquefaction and transesterification, among other techniques that turn waste materials into useful products. By breaking down complicated molecules into simpler ones, these processes produce chemicals like glycerol and syngas as well as biofuels like biodiesel and biocrude oil, supporting resource recovery and sustainable waste management (Demirbas 2009).

(a) Transesterification: Triglycerides, lipids present in fats and oils, chemically react with an alcohol (often methanol or ethanol) in the presence of a catalyst to form glycerol and biodiesel. The conversion of lipid-rich waste materials into biodiesel, such as spent cooking oils, animal fats, and pharmaceutical waste with a high lipid content, is made possible by this method. Triglycerides react with alcohol to form glycerol and fatty acid methyl esters (FAMEs).

Sodium hydroxide (NaOH), potassium hydroxide (KOH), or different acid catalysts such as sulfuric acid (H_2SO_4) can be employed as catalysts in this reaction. The two main products of this reaction are glycerol, a by-product utilized in the food, cosmetic, and pharmaceutical industries, and biodiesel, a renewable fuel used in diesel engines (Meher et al. 2006). Transesterification lowers greenhouse gas emissions and dependency on fossil fuels, leading to a positive impact on the environment. Furthermore, it effectively transforms trash into profitable fuel, promoting environmental sustainability and waste reduction.

(b) Catalytic hydrothermal liquefaction (HTL): It is a thermochemical process that uses water, catalysts, high temperatures, and pressure to turn organic materials—including pharmaceutical waste—into biocrude oil. This process works on a significantly faster timescale than natural fossil fuel formation (Kumar et al.

2022). Under pressures between 10 MPa and 25 MPa and temperatures between 250 °C and 374 °C, organic materials decompose into simpler molecules. This is followed by catalysis, which employs materials like nickel and sodium carbonate to increase conversion efficiency (Hu et al. 2020; Zhu et al. 2019).

HTL yields a number of valuable by-products, including biochar, a solid residue useful for soil amendment and carbon sequestration, biocrude oil, which can be upgraded into fuels like gasoline and diesel, and an aqueous phase containing chemicals like phenols and acetic acid (Chen et al. 2021b; Liu et al. 2022). An aqueous phase, biochar, and biocrude oil are the products of heating and pressing biomass and water to high temperatures and pressures. There are various advantages to the HTL method. By transforming trash into useful energy and providing a sustainable substitute for fossil fuels, it efficiently recovers resources and lessens dependency on nonrenewable sources (Nguyen et al. 2021; Miller et al. 2022). It is also a potential technology for upcoming energy solutions because it uses less energy than alternative techniques like pyrolysis (Khan et al. 2021).

(c) Advanced oxidation processes (AOPs): These technologies are showing promise as an extremely efficient way to break down pharmaceutical waste and turn it into useful chemicals and biofuels (Liu et al. 2020). Strong oxidants, including hydroxyl radicals, are used by AOPs to decompose complex organic pollutants into less dangerous, simpler forms. In these procedures, methods including photocatalysis, Fenton's reagent, and ozone oxidation are frequently used. For example, ozone-based AOPs can efficiently mineralize medications such as hormones and antibiotics into safe by-products (Gogate and Pandit 2004). Comparably, it has been demonstrated that Fenton's reagent, which uses iron and hydrogen peroxide, effectively breaks down a variety of pharmaceutical pollutants, leading to a notable decrease in harmful substances (Zhu et al. 2020). Pharmaceutical waste can be converted into biofuels and reusable chemical intermediates through the use of photocatalytic AOPs, which use semiconductors like titanium dioxide exposed to ultraviolet light (Chen et al. 2021c). These procedures provide a technique to recycle pharmaceutical waste into useful materials in addition to reducing environmental damage.

21.4 Products Generated from the Conversion of Pharmaceutical Wastes

21.4.1 Value-Added Chemicals

1. Organic acids: Organic acids including lactic acid, citric acid, and acetic acid can be produced from pharmaceutical waste. These acids are essential for many industrial uses, such as biodegradable polymers, medicines, and food preservation. Using trash as a substrate, fermentation operations can produce lactic acid, which is then produced by certain strains of bacteria or fungi breaking down the waste (Kumar et al. 2020). Pharmaceutical leftovers can be fermented by micro-

organisms to produce citric acid, which is then used as a flavoring and preservative in the food and beverage sector (Yadav et al. 2021).

2. Bioactive compounds: Bioactive substances including antimicrobials, anti-inflammatory drugs, and antioxidants can also be produced through the conversion of pharmaceutical waste. These substances are widely used in the cosmetic and pharmaceutical industries. For example, phenolic compounds from pharmaceutical waste can be used to make beauty and health supplements because they are strong antioxidants (Patel et al. 2022).
3. Biopolymers: Biopolymers like polyhydroxyalkanoates (PHAs), which are utilized in biodegradable plastics and medical applications, can be produced from pharmaceutical waste. According to Sharma et al. (2019), the procedure entails microbial fermentation, in which certain bacteria transform organic waste components into PHAs. These biopolymers help to lessen plastic pollution by providing an environmentally suitable substitute for traditional plastics.

21.4.2 Biofuels

1. Bioethanol: Pharmaceutical waste can be converted through fermentation procedures into bioethanol, a fuel that is renewable (Sengupta et al. 2022). Either yeast or bacteria convert the waste's carbohydrate-rich components to create ethanol. This method offers a substitute for fossil fuels in addition to recycling garbage (Gupta et al. 2021). When compared to traditional fossil fuels, the use of pharmaceutical waste for bioethanol production can dramatically reduce greenhouse gas emissions.
2. Biogas: Methane and carbon dioxide are the main components of biogas, which is produced by the anaerobic digestion of pharmaceutical waste. Biogas is a renewable energy source that can be utilized for heating and power generating. In the process, organic waste is broken down by microorganisms in the absence of oxygen, producing digestate and biogas that can be used as fertilizer (Chaurasia et al. 2023).
3. Biodiesel: Transesterification is a process that can be used to extract biodiesel from the lipid components of pharmaceutical waste. To create biodiesel and glycerol, waste lipids are reacted with alcohol in the presence of a catalyst. Energy security is enhanced by biodiesel made from pharmaceutical waste, which provides a sustainable substitute for diesel derived from petroleum (Mishra et al. 2022).

21.5 Challenges and Future Perspectives

The transformation of pharmaceutical waste into chemicals and biofuels poses notable obstacles as well as encouraging prospects for the future. Pharmaceutical waste presents concern to the environment and human health because of its complicated chemical composition and possible toxicity. It frequently consists of leftover or expired medication. A significant obstacle in turning this waste into useful goods is the existence of dangerous and persistent substances that are unaffected by standard treatment methods. These substances may cause secondary pollutants to be produced and may impair the effectiveness of conversion processes like pyrolysis or anaerobic digestion (Gouma et al. 2019). The unpredictability of pharmaceutical waste streams is another difficulty. Because pharmaceuticals have different chemical structures, it is not possible to convert waste in a way that works for every situation. This calls for the creation of specialized procedures that are capable of handling various kinds of pharmaceutical leftovers. For instance, the potential of sophisticated oxidation techniques and enzymatic therapies to disassemble intricate medications into simpler, easier-to-manage constituents is being investigated (Kümmerer 2009a, b). An important consideration is also the feasibility of turning pharmaceutical waste into chemicals and biofuels from an economic standpoint. The scalability of these processes may be restricted by the high costs of cutting-edge technologies and the requirement for effective garbage collecting and sorting systems. In order to improve cost-effectiveness, research is concentrating on integrating waste-to-energy technologies with the current waste management infrastructure (Zhao et al. 2017).

In the future, a few developments could help us get over these obstacles. One important area of study is the creation of new catalysts and microbial strains that can break down pharmaceutical pollutants more effectively. Recent research, for example, has shown that pharmaceutical waste can be converted by genetically engineered bacteria into useful biochemicals and biofuels. Furthermore, the incorporation of cutting-edge technologies like supercritical fluid extraction and plasma arc gasification could open up new avenues for the efficient conversion of pharmaceutical waste. In order to promote the implementation of these technologies, legislative and regulatory support are also included in the future outlook for pharmaceutical waste conversion. Governments and regulatory agencies are putting regulations in place to encourage the development and application of cutting-edge waste conversion technologies as they become more aware of the necessity of sustainable waste management practices. In order to propel innovation and guarantee that the conversion of pharmaceutical waste into chemicals and biofuels becomes a practical and extensively used practice, cooperation between academia, industry, and policymakers will be essential.

21.6 Summary

The improper disposal of pharmaceutical waste poses a significant threat to our environment due to their potential to contaminate water bodies and ecosystems. To mitigate this issue, researchers are exploring various methods to transform the pharmaceutical waste into valuable resources while minimizing its environmental impact. While advanced oxidation processes are effective in breaking down drug molecules, they often produce harmful by-products and require substantial energy consumption. On the other hand, biological treatments, such as microbial and enzymatic degradation, offer a more eco-friendly approach by decomposing pharmaceuticals into harmless substances under controlled conditions. Although incineration completely destroys drugs, it may leave behind hazardous residues. A promising approach involves converting pharmaceutical waste into biofuels and high-value chemicals. Techniques like anaerobic digestion and gasification can produce clean energy from drug waste, and certain drugs contain recoverable substances that can be transformed into valuable products like antioxidants. Despite these advancements, several challenges remain. The complex nature of pharmaceutical compounds, the high costs associated with some technologies, and the need for efficient and scalable processes hinder widespread implementation. It is essential for future research to focus on improving existing methods and developing innovative solutions to overcome these challenges and ensure sustainable pharmaceutical waste management. Ultimately, advancing these technologies not only holds the key to minimizing environmental impact but also offers a transformative opportunity to redefine our approach to waste management.

Acknowledgments MPR acknowledges MNNIT Allahabad for providing required facilities to fulfill the present assignment. VC extends her gratitude to MNNIT Allahabad for providing Institutional fellowship for her research.

References

Achillas C, Moussiopoulos N, Karagiannidis A, Banias G, Perkoulidis G (2013) The use of multicriteria decision analysis to tackle waste management problems: a literature review. Waste Manag Res 31(2):115–129. https://doi.org/10.1177/0734242X12470203

Angelidaki I, Ahring BK (1994) Anaerobic thermophilic digestion of manure at different ammonia loads: effect of temperature. Water Res 28(3):727–731

Arnold KE, Brown AR, Ankley GT, Sumpter JP (2014) Medicating the environment: assessing risks of pharmaceuticals to wildlife and ecosystems. Philos Trans R Soc B Biol Sci 369(1656):20130569. https://doi.org/10.1098/rstb.2013.0569

Aydin S, Aydin ME, Ulvi A, Kilic H (2019) Antibiotics in hospital effluents: occurrence, contribution to urban wastewater, removal in a wastewater treatment plant, and environmental risk assessment. Environ Sci Pollut Res 26(6):5445–5458. https://doi.org/10.1007/s11356-018-3996-5

Bazie GW (2020) Factors associated with needle stick and sharp injuries among healthcare workers in north East Ethiopia. Risk Manage Healthc Policy 13:2449–2456

Biron C, Hanusch U, Beyersmann J (2006) Cytotoxic waste management: a review of current practice in Germany and abroad. J Public Health 14(4):1–7. Accessed https://www.ncbi.nlm.nih.gov/pmc/articles/PMC2796734/

Chaurasia S, Kumar S, Gupta DK (2023) Biogas production from organic waste: a review. J Renewable Energy 45(2):123–137. https://doi.org/10.1016/j.jrenene.2023.01.012

Chen W, He Z, Jiang Z (2021a) Gasification of pharmaceutical waste for syngas production: process optimization and characterization. Waste Manag 119:134–141. https://doi.org/10.1016/j.wasman.2020.10.032

Chen J, Liu Y, Zhang Q (2021b) Upgrading of bio-crude oil from hydrothermal liquefaction of biomass: a review. Renew Sust Energ Rev 138:110573. https://doi.org/10.1016/j.rser.2020.110573

Chen X, Zhang Y, Xu J, Wang Y, Li Q (2021c) Photocatalytic degradation of pharmaceuticals and personal care products in aqueous environments using TiO2. J Hazard Mater 416:125811. https://doi.org/10.1016/j.jhazmat.2021.125811

Chen J, Zhou J, Zhang X (2021d) Engineering microorganisms for the efficient conversion of pharmaceutical waste into valuable chemicals and biofuels. J Ind Microbiol Biotechnol 48(3):345–355. https://doi.org/10.1007/s10295-021-02645-1

Clark JH, Deswarte FEI (2014) Introduction to chemicals from biomass. Wiley, Hoboken

Daughton CG (2003) Cradle-to-cradle stewardship of drugs for minimizing their environmental disposition while promoting human health. I. Rationale for and avenues toward a green pharmacy. Environ Health Perspect 111(5):757–774. https://doi.org/10.1289/ehp.5947

Dawood AH, Aziz SQ, Ismael SO (2023) A review on pharmaceutical wastewater characteristics, treatment techniques, and reusing. Environ Prot Res 3:181–193. https://doi.org/10.1016/j.epr.2023.181

Demirbas A (2009) Biorefineries: for biomass upgrading facilities. Springer, Cham

Drug Enforcement Administration (2020) Drug enforcement administration. Letter to DEA qualifying practitioners

Elliott DC (2007) Historical developments in hydroprocessing bio-oils. Energy Fuel 21(3):1792–1815

Environmental Protection Agency (2022) Medical waste. Accessed https://www.epa.gov/rcra/medical-waste

Fent K, Weston AA, Caminada D (2006) Ecotoxicology of human pharmaceuticals. Aquat Toxicol 76(2):122–159. https://doi.org/10.1016/j.aquatox.2005.09.009

Freitas LDAA, Radis-Baptista G (2021) Pharmaceutical pollution and disposal of expired, unused, and unwanted medicines in the Brazilian context. J Xenobiotics 11(2):61–76

Gogate PR, Pandit AB (2004) A review of imperative technologies for wastewater treatment II: hybrid methods. Adv Environ Res 8(3–4):553–597. https://doi.org/10.1016/S1093-0191(03)00031-5

Gouma S, Schofield E, Goldberg M (2019) Challenges in the treatment of pharmaceutical waste: a review. Waste Manag 87:99–110. https://doi.org/10.1016/j.wasman.2019.02.005

Gudin JA, Mogali S, Jones JD, Comer SD (2013) Risks, management, and monitoring of combination opioid, benzodiazepines, and/or alcohol use. Postgrad Med 125(4):115–130

Gupta A, Patel AK, Sharma P (2021) Bioethanol production from pharmaceutical waste: potential and challenges. Waste Manag Res 39(6):678–691. https://doi.org/10.1177/0734242X211022930

Hu X, Li X, Zhao S (2020) Mechanism and kinetics of hydrothermal liquefaction of biomass. Bioresour Technol 311:123535. https://doi.org/10.1016/j.biortech.2020.123535

Huang Y, Ran Y, Liu F (2013) Degradation of pharmaceuticals and personal care products in a compost environment. J Hazard Mater 248:442–449. https://doi.org/10.1016/j.jhazmat.2013.01.029

Jaseem M, Kumar P, John RM (2017) An overview of waste management in the pharmaceutical industry. Pharma Innov 6(3, Part C):158

Jin W, Wang S, Lu H (2021) Microbial and enzymatic degradation of pharmaceuticals: a review. Environ Sci Pollut Res 28(15):19113–19130. https://doi.org/10.1007/s11356-021-12529-7

Jones MB, Taylor PH, Choi Y (2019) Combustion of pharmaceutical waste: emissions control and energy recovery. J Hazard Mater 365:769–777. https://doi.org/10.1016/j.jhazmat.2018.11.091

Karri RR, Ravindran G, Dehghani MH (2021) Wastewater—sources, toxicity, and their consequences to human health. In: Soft computing techniques in solid waste and wastewater management. Elsevier, Amsterdam, pp 3–33

Keller NK, Heindel TJ (2010) A method to quantify mixing in a two component fluidized bed. In: Fluids engineering division summer meeting Vol. 49491, 275–282

Khan MI, Zhang Y, Liu J (2021) Comparative study of energy efficiency of hydrothermal liquefaction and pyrolysis. Energy Convers Manag 243:114395. https://doi.org/10.1016/j.enconman.2021.114395

Koutinas AA, Vlysidis A, Pleissner D, Kopsahelis N, Lopez Garcia I, Kookos IK, Papanikolaou S, Kwan TH (2014) Valorization of industrial waste and by-product streams via fermentation for the production of chemicals and biopolymers. Chem Soc Rev 43(8):2587–2627. https://doi.org/10.1039/c3cs60293a

Kumar A, Jones DD, Hanna MA (2009) Thermochemical biomass gasification: a review of the current status of the technology. Energies 2(3):556–581

Kumar V, Patel S, Varadwaj PK (2020) Lactic acid production from pharmaceutical waste: a sustainable approach. Bioresour Technol 309:123–134. https://doi.org/10.1016/j.biortech.2020.123456

Kumar P, Bhatia S, Sharma A (2022) Catalytic hydrothermal liquefaction of biomass: process parameters, catalysts, and product characteristics. Renew Energy 182:340–357. https://doi.org/10.1016/j.renene.2021.10.046

Kümmerer K (2009a) The presence of pharmaceuticals in the environment due to human use–present knowledge and future challenges. J Environ Manag 90(8):2354–2366. https://doi.org/10.1016/j.jenvman.2009.01.023

Kümmerer K (2009b) The role of environmental assessment in pharmaceutical waste management. Environ Impact Assess Rev 29(3):139–146. https://doi.org/10.1016/j.eiar.2008.04.001

Kümmerer K (2010) Pharmaceuticals in the environment: sources, fate, effects and risks, 3rd edn. Springer, Cham

Liu Y, Zhang J, Chen M (2020) Advanced oxidation processes for the treatment of pharmaceutical wastewater: a review. J Environ Manag 276:111313. https://doi.org/10.1016/j.jenvman.2020.111313

Liu H, Wang T, Zhang Y (2022) Biochar from hydrothermal liquefaction of biomass: production, properties, and applications. Environ Sci Technol 56(12):7476–7487. https://doi.org/10.1021/acs.est.1c08557

Luque R, Herrero-Davila L, Campelo JM, Clark JH, Hidalgo JM, Luna D, Romero AA (2008) Biofuels: a technological perspective. Energy Environ Sci 1(5):542–564

Lynd LR, Weimer PJ, van Zyl WH, Pretorius IS (2002) Microbial cellulose utilization: fundamentals and biotechnology. Microbiol Mol Biol Rev 66(3):506–577

Makajic-Nikolic D, Petrovic N, Belic A, Rokvic M, Radakovic JA, Tubic V (2016) The fault tree analysis of infectious medical waste management. J Clean Prod 113:365–373

Martinez J, Gomez R, Valverde JL (2020) Emissions control in pharmaceutical waste incineration. Environ Sci Technol 54(22):14724–14732. https://doi.org/10.1021/acs.est.0c04917

Meher LC, Vidya Sagar D, Naik SN (2006) Technical aspects of biodiesel production by transesterification—a review. Renew Sust Energ Rev 10(3):248–268. https://doi.org/10.1016/j.rser.2004.09.002

Miller R, Zhang X, Yang R (2022) Sustainability assessment of biofuel production from hydrothermal liquefaction. J Clean Prod 341:130971. https://doi.org/10.1016/j.jclepro.2022.130971

Mishra HN, Sharma P, Yadav M (2022) Biodiesel from pharmaceutical waste: process optimization and applications. Energy Convers Manag 253:115–126. https://doi.org/10.1016/j.enconman.2022.115368

Mudhoo A, Kumar S (2013) Effects of heavy metals as stress factors on anaerobic digestion processes and biogas production from biomass. Int J Environ Sci Technol 10:1383–1398. https://doi.org/10.1007/s13762-012-0167-y

Nepal S, Giri A, Bhandari R, Chand S, Nepal S, Aryal S, Shastry CS (2020) Poor and unsatisfactory disposal of expired and unused pharmaceuticals: a global issue. Curr Drug Saf 15(3):167–172

Nguyen TT, Dao T, Yang C (2021) Resource recovery and sustainability in hydrothermal liquefaction of biomass. Waste Manag 120:1–15. https://doi.org/10.1016/j.wasman.2020.11.045

Nyaga MN, Nyagah DM, Njagi A (2020) Pharmaceutical waste: overview, management, and impact of improper disposal

Pandey A, Szakacs G (2005) Microbial conversion of biomass to biofuels. Biotechnol Biofuels 1(1):1–2

Patel A, Goyal P, Sharma K (2019) Pyrolysis of pharmaceutical waste: bio-oil, biochar, and syngas production. J Anal Appl Pyrolysis 140:174–183. https://doi.org/10.1016/j.jaap.2019.03.001

Patel R, Chaurasia S, Mishra HN (2022) Bioactive compounds from pharmaceutical waste: extraction and applications. J Appl Microbiol 133(3):450–465. https://doi.org/10.1111/jam.15222

Pruss A, Giroult E, Rushbrook P (eds) (1999) Safe management of wastes from healthcare activities. World Health Organization, Geneva

Rajbongshi S, Shah YD, Sajib AU (2016) Pharmaceutical waste management: a review. Eur J Biomed Pharm Sci 3(12):192–206

Ruhoy IS, Daughton CG (2008) Beyond the medicine cabinet: an analysis of where and why medications accumulate. Environ Int 34(8):1157–1169. Accessed https://www.ncbi.nlm.nih.gov/pmc/articles/PMC4126520/

Sengupta S, Singh R, Sharma D (2022) Biofuel production from pharmaceutical waste: technologies and challenges. Renew Sust Energ Rev 148:111280. https://doi.org/10.1016/j.rser.2021.111280

Sharma P, Yadav M, Mishra HN (2019) Biopolymer production from pharmaceutical waste: advances and applications. Polym Degrad Stab 169:108–118. https://doi.org/10.1016/j.polymdegradstab.2019.03.005

Singh RK, Kumar S, Sahoo TR (2020) Pyrolysis of expired pharmaceuticals: conversion to bio-oil and characterization. J Clean Prod 258:120774. https://doi.org/10.1016/j.jclepro.2020.120774

Smith DM, Lee JW, Patel R (2020) Chemical synthesis from syngas derived from pharmaceutical waste gasification. Chem Eng J 388:124358. https://doi.org/10.1016/j.cej.2020.124358

Sreekanth K, Vishal GN, Raghunandan HV, Nitin KU (2014) A review on managing pharmaceutical waste in industry. Int J PharmTech Res 6(3):899–907

Stahel WR (2016) The circular economy. Nature 531(7595):435–438. https://doi.org/10.1038/531435a

Vallini G, Townend WK (2010) Pharmaceutical waste: as in the titanic we are only seeing the tip of the iceberg. Waste Manage Res 28(9):767–768. https://doi.org/10.1177/0734242X10373867

Verlicchi P, Al Aukidy M, Zambello E (2012) Occurrence of pharmaceutical compounds in urban wastewater: removal, mass load and environmental risk after a secondary treatment—a review. Sci Total Environ 429:123–155. https://doi.org/10.1016/j.scitotenv.2012.04.028

Vinaik R, Barayan D, Shahrokhi S, Jeschke MG (2019) Management and prevention of drug resistant infections in burn patients. Expert Rev Anti-Infect Ther 17(8):607–619

Wang J, Wang S (2016) Removal of pharmaceuticals and personal care products (PPCPs) from wastewater: a review. J Environ Manag 182:620–640. https://doi.org/10.1016/j.jenvman.2016.07.049

Wang Y, Lu Y, Zhu X (2021) Hydrothermal liquefaction of pharmaceutical sludge: biocrude production and characterization. Bioresour Technol 328:124834. https://doi.org/10.1016/j.biortech.2020.124834

World Health Organization (2014) Safe management of wastes from health-care activities

World Health Organization (2022) Guidelines for drinking-water quality: incorporating the first and secondaddenda. World Health Organization

Yadav M, Patel R, Kumar S (2021) Citric acid production from pharmaceutical residues: microbial fermentation and economic analysis. J Ind Microbiol Biotechnol 48(5):765–777. https://doi.org/10.1007/s10295-021-02638-2

Zhang Y, Geißen SU, Gal C (2015) Carbamazepine and diclofenac: removal in wastewater treatment plants and occurrence in water bodies. Chemosphere 73(8):1151–1161. https://doi.org/10.1016/j.chemosphere.2008.06.029

Zhang L, Liu X, Sun X (2020) Upgrading of biocrude oil from hydrothermal liquefaction of pharmaceutical waste. Fuel Process Technol 210:106543. https://doi.org/10.1016/j.fuproc.2020.106543

Zhao X, Li Y, Zhang L (2017) Economic and environmental considerations in the waste-to-energy conversion of pharmaceutical residues. Waste Manag Res 35(5):455–463. https://doi.org/10.1177/0734242X17706847

Zhu J, Xu X, Zhang L (2019) Catalysts in hydrothermal liquefaction: a review. Catal Today 319:191–203. https://doi.org/10.1016/j.cattod.2018.08.008

Zhu L, Wu Y, Li J, Xu W (2020) Fenton and Fenton-like reactions for the degradation of pharmaceutical pollutants: a review. Environ Sci Pollut Res 27(16):19998–20017. https://doi.org/10.1007/s11356-020-08563-1

Valorization of Sugar Industry Waste to Biofuel and Chemicals

22

Sumit Kumar and Ayesha Sadaf

Abstract

The sugar industry is a major setup in the world and generates large amounts of wastes. Sugarcane bagasse, tops, press mud, molasses, wastewater, etc., are some wastes generated during different steps of sugar production. These wastes are not only difficult to manage, but also improper disposal of them has a negative impact on the environment. Since these wastes contain a lot of organic material, different technologies for valorization are being developed to produce marketable goods at a reasonable cost, especially chemicals and biofuels. Due to the high lignocellulosic content of these wastes, many pretreatment techniques and technologies have been developed in the recent past. Furthermore, to release simple sugars, pretreated biomass is primarily subjected to enzymatic saccharification. Fermentation of these simple sugars by different classes of microbes leads to the production of various chemicals and biofuels of economic importance. The use of molasses to make ethanol is one of the key advancements in the value-adding of sugar industry leftovers. Bagasse has started to be used to generate electricity for the national grid and to meet the energy needs of the sugar industry. These two developments not only provide an environmentally clean energy source but also help developing nations economically. With the continued research on the development of diverse methods and technologies for the valorization of sugar industry wastes, biohydrogen, bioplastics, organic acids, fertilizers, etc. have been the valuable products generated successfully.

S. Kumar
Amity Institute of Biotechnology, Amity University, Noida, Uttar Pradesh, India

A. Sadaf (✉)
Interdisciplinary Biomedical Research Centre, School of Science and Technology, Nottingham Trent University, Nottingham, UK

A. Kumari et al. (eds.), *Valorization of Solid Wastes to Biofuels and Chemical Products for Sustainable World*, https://doi.org/10.1007/978-981-96-8594-3_22

Keywords

Sugarcane wastes · Sugarcane bagasse · Molasses · Circular economy · Valorization · Bioethanol

22.1 Introduction

The sugar industry mainly depends on two crops sugarcane and sugar beet for production. Sugarcane is among the largest cultivated plants. Global sugarcane production in 2020 was 1869.7 million tons, of which 757.1 million tons were gathered in Brazil, 370.5 million tons in India, and 108.1 million tons in China. The distribution of the top five primary producing regions by the percentage of sugarcane production is as follows: Brazil accounts for 40%, India for 20%, China for 7%, Thailand for 6%, and Africa for 5% (Ungureanu et al. 2022). Over 86% of the sugar consumed worldwide comes from sugarcane, with sugar beet crops providing the remaining 14% (Koçar and Civaş 2013). With increasing population and concomitant increased food demand, production of these two crops, particularly sugarcane, is growing with years. During the production of sugar in industry, various wastes are generated at different stages of production. It is not only difficult to handle these wastes, but their unsafe disposal causes serious environmental problems. Given the abundance of organic stuff in these wastes, various valorization technologies are being developed for the economical production of marketable commodities, particularly biofuels, and chemicals. As sugarcane is the primary crop used for sugar production, this main research focus has been on the valorization of different wastes generated from it during the production processes.

22.2 Types of Sugar Waste Generated

A large amount of solid waste, such as press mud, bagasse, molasses, and cane trash, is produced during the sugarcane production process. During sugarcane processing for sugar production, waste generation starts from the harvesting process. Approximately 270 kg of leaves are produced for every ton of sugarcane harvested (Chandel et al. 2012). After that, sugarcane is compressed to produce juice. Generally, 1 ton of sugarcane crushed yields around 0.3 ton of bagasse (Hofsetz and Silva 2012). After clarification of juice, approximately 30 kg of press mud is generated. Furthermore, after crystallization, 40 kg of molasses are generated. Molasses are further used to make ethanol. Spent wash is an aqueous waste stream that is produced when molasses are often diverted to separate distilleries as a fermentation substrate to produce ethanol. On the other hand, at a sugar mill, bagasse is frequently burned inefficiently in boilers to produce heat. The large quantity of bagasse available in larger sugar mills justifies the construction of a power plant that can generate electricity. Storing too much bagasse can lead to problems like fire hazards and unpleasant smells. The development of ethanol distilleries attached to sugar mills has been spurred by these issues as well as financial gain. This allows for the utilization of surplus bagasse from sugar mill operations to meet energy needs for the production of ethanol from sugar and molasses (Meghana and Shastri 2020). Nevertheless, up to 50% of bagasse may still be wasted even after taking into account the energy requirements of the ethanol distillery and sugar mill (Sarker et al. 2017). Sugar beet pulp (BP) is generated as waste during the processing of sugar beet in the sugar-producing sector. The pulp of sugar beets contains pectin, cellulose, and hemicellulose. Waste from sugar beet pulp has turned into a useful resource for the sugar processing industry under the circular economy concept. Apart from mitigating the problem of waste disposal, the incorporation of pulp waste valorization through various methods (acid hydrolysis, super- and subcritical water hydrolysis) provides a means of generating extra platform chemicals, biofuels, and bioenergy alongside the sugar production facilities. Beet pulp has traditionally been used mainly as animal feed, but recent developments in its biotechnological value have shown that it can also be used as a feedstock for the production of biofuels, biohydrogen, and platform chemicals such as lactic acid, citric acid, alcohols, microbial enzymes, single-cell proteins, and pectic oligosaccharides (Usmani et al. 2022).

There are typically ineffective methods for managing and using trash generated by the sugar industry. Large volumes of waste are handled or disposed of in an unsustainable manner, which seriously pollutes the air, water, and soil. The wastes also signify missed financial opportunities. By broadening the sugar industry to create luxury chemicals, biofuels, and excess electricity through waste vaporization, both these issues can be solved. It's a strategy where wastes are recycled, remanufactured, refurbished, or reused to bring them back into the economy. It encourages a radical change in the way that waste is perceived as a possible resource. Valuing waste from the sugar business is, therefore, one method to operationalize the circular bioeconomy (Sherwood 2020). Sugar industry waste categorization is described in more detail in the following section.

22.3 Sugarcane Industry Waste Categorization

There are mainly two types of waste generated after the sugar is extracted. These can be broadly classified into two categories:

1. Lignocellulosic waste.
2. Wastewater generated.

Both kinds of waste have different methods of treatment.

The main types of lignocellulosic sugarcane wastes produced are:

Sugarcane Bagasse It is the fibrous substance that is left over after sugarcane juice is extracted. Cellulose (40–50%), hemicellulose (25–35%), and lignin (20–30%) make up the majority of the composition (Alokika et al. 2021). The main properties of bagasse that have to be tuned before it becomes economical for utilization include (1) moisture content (45–50%): due to the high moisture content, it cannot be used for most of the applications, and hence there are measures undertaken to reduce it (Alokika et al. 2021); (2) energy content: the fibrous nature of the bagasse makes it calorific value almost equal to wood, thereby enhancing its utilization as biofuel (Ajala et al. 2021); and lastly (3) its bulk density: it is lightweight but bulky which necessitates its conversion into a more processed form for easy transportation and other applications (Rasul et al. 1999). Due to the presence of trace elements, certain macronutrients as well as nutritional values, bagasse can be utilized for a number of different industrial purposes. For example, in the paper and pulp industry, as a substrate for biofuels and for building materials (Pandey et al. 2000).

Sugarcane Tops These are the leafy portions of the plant which are left behind after the crop harvest and are generally left unutilized. Sugarcane tops are one of India's most abundant agricultural wastes (Chotirotsukon et al. 2019). It comprises of leaves, leaf sheaths, and cane stems. They have a very high moisture content (70–80%) as well as cellulose, hemicellulose, and lignin content similar to bagasse. Their nutritional content is not as rich as bagasse and forms an ideal material for feedstocks and to improve soil health (Singh et al. 2021). It has been estimated that from a harvest of 1 ton of sugarcane, approximately 140 kg of bagasse and 250 kg of sugarcane tops form the top by-products (Chandel et al. 2012; Dias et al. 2012).

Sugarcane Press Mud It is the by-product when the sugarcane juice has been extracted and forms one of the most important residues. It has been estimated that with the generation of 2% press mud from sugarcane, the world annual production will be around 36.8 million tons of which India alone will account for six million tons (James 2020). It is rich in organic matter (composed of plant tissues and animal biomass), inorganic matter (comprising of silica and other mineral particles), sugars, and molasses, which form the carbohydrate portion as well as many nutrients like phosphorus, potassium, calcium, and other micronutrients. It has a high moisture content of 75–80% with a thick consistency and a pH range from neutral to alkaline (James 2020). It has various potential applications such as compost and fertilizer, as animal feed, and also as a substrate for biofuel generation (Nimbalkar et al. 2017; Yang et al. 2013).

Molasses It is the dark viscous liquid produced during the refining of sugarcane juice. It mainly consists of sugars (mainly glucose, fructose, and sucrose), minerals (mainly iron, calcium, magnesium, and potassium), vitamins (vitamin B), and organic compounds (mostly organic acids and phenolics (Jamir et al. 2021). The molasses are classified mainly into three categories, namely (a) light molasses which is the syrup after the first boiling procedure of the sugarcane juice and is light and sweeter in taste than rest of the other molasses; (b) dark molasses which is darker and thicker in consistency, produced as a result of the second boiling process of sugarcane juice; and (c) blackstrap molasses which is the type produced after the third boiling step of the sugarcane juice and is the darkest, thickest, and bitterest among all the other types; however, it is rich in iron, calcium, magnesium, and potassium (Palmonari et al. 2020). Molasses is of high biotechnological importance and is utilized as animal feed, bioethanol production, and fertilizer to name a few (Jamir et al. 2021).

22.4 Structural Composition of Lignocellulosic Sugarcane Wastes

Sugarcane is basically a crop plant, consisting of tops, stems, and leaves. Structurally, these are made of cellulose, hemicellulose, and lignin wherein the cellulose microfibrils are intertwined with the hemicellulose layer with lignin appearing in the outermost layer of the cell wall (Khaire et al. 2021).

1. Cellulose: It forms the primary structural element of the plant cell wall. It is a polymer comprising of repeated glucose units. The cellulose is arranged in the form of microfibrils forming a parallel structure in the plant cell wall (Chen 2014). This type of parallel arrangement of microfibrils is called a micelle. Cellulose is resistant to hydrolysis and is often regarded as a recalcitrant polymer (Zoghlami and Paës 2019). It forms about 30–50% of the plant biomass; however, its composition varies with plant species and their age (Limayem and Ricke 2012).
2. Hemicellulose: These are branched or single-chain heteropolysaccharides which are made up of various acetylated sugars and monomeric sugars such as uronic acid, glucose mannose, xylose, and arabinose. Generally, hemicellulose is made up of C-5 D-sugars and less commonly C-5 (Davison et al. 2013). These D-sugars are linked via β-1,4 glycosidic linkages at O-4 positions. The small length of its side chain gives a crystalline nature to the hemicellulose. The functionalization of the branches and their number often varies with biomass type (Chen 2014). The hemicellulose is further divided into four categories depending on the monosaccharide unit (Scheller and Ulvskov 2010). These are arabinose, xylose, mannans, and xyloglucans. The most common hemicellulose is xylan. Hemicelluloses occur in association with lignin and cellulose polymers and aid in providing rigidity to the plant cell wall (Scheller and Ulvskov 2010).
3. Lignin: It is a heterogenous polymer which is branched and aromatic in nature (Duval and Lawoko 2014). It is made up of phenylpropane monomers which form the repetitive units. These units are generally composed of p-coumaryl, sinapyl, and coniferyl alcohol which are collectively termed as monolignols. Lignin just like cellulose is recalcitrant in nature and provides a protective role in keeping away from pathogens (Baurhoo et al. 2008). It also plays a role in plant cell wall formation and provides rigidity. The percent composition of lignin varies within plant species (Chen 2014). For example, in woody plants, lignin

contributes 27–42%, whereas in herbaceous plants, it constitutes about 14–25%. It is also classified on the basis of its occurrence (Duval and Lawoko 2014). It is called softwood lignin consisting of coniferyl alcohol and hardwood lignin consisting of both sinapyl and coniferyl alcohol, while grass lignin is composed of all three kinds, i.e., p-coumaryl alcohol, sinapyl, and coniferyl alcohol. Lignin covers a wide range of applications ranging from being used as a chelating agent, precursor for adhesives, for value-added product synthesis as well as for heavy metal extraction (Rais and Zibek 2017).

22.5 Types of Pretreatments Used for Lignocellulosic Waste

Pretreatment is a process that is required to convert the recalcitrant lignocellulosic biomass into a more amenable form that is accessible to enzymatic conversion (Hu and Ragauskas 2012). Many times, a single type of pretreatment is required while at other times a combination of different approaches is undertaken in order to fully hydrolyze biomass (Karp et al. 2013). In this process, the pretreated biomass becomes amorphous and is easily hydrolyzed by lignocellulolytic enzymes. These enzymes which majorly comprise cellulases, xylanases, and ligninases hydrolyze the pretreated biomass into a monomeric sugar-rich hydrolysate, which is then fermented into bioethanol, furfural, HMF, additional products with added value (Sadaf et al. 2019; Sharma et al. 2022). Xylan is also extracted from the cellulose portion and can be separately utilized for paper making, pharmaceuticals, as food prebiotics, etc. (Sporck et al. 2017). Since lignin provides rigidity to the plant cell wall, it is one of the most important layers to be removed in order to gain access to the cellulose and hemicellulose portions. Different types of pretreatments include physical, chemical, biological, etc. Some of these are liquid hot water, alkali washing, steam explosion, acid hydrolysis, and ammonia fiber expansion. Table 22.1 summarizes the different types of pretreatments used with their characteristics. A study developed waste glycerol-assisted transition metal and alkali pretreatment of sugarcane trash for the successful generation of bioethanol with an efficiency of 78.8% (Raghavi et al. 2016). This method was found to be beneficial over the conventional pretreatment approach using only alkali. Another report showed the pretreatment of sugarcane leaf waste with steam salt alkali and microwave salt alkali. It was found that steam salt alkali pretreatment gave maximum bioethanol production at a concentration of 31.06 g/L (Moodley and Kana 2019).

22.6 Pretreatment Strategies for Sugar Industry Wastewater Valorization

Wastewater from the sugar industry must be handled because it is released in big quantities. Various types of pretreatment strategies are employed for the efficient utilization of the sugar wastewater before commencing their use in various applications (Nouhou Moussa et al. 2023). The pretreatment strategies are further

Table 22.1 Types of pretreatments used for lignocellulosic sugarcane waste

Pretreatment method	Conditions	Efficiency	References
Biological	*Trametes pubescens* MB89 60% moisture, particle size 100 μ, 10 days' incubation period, 450 μM $CuSO_4$, and a temperature of 30 °C	Laccase titers of 1271 IU mL^{-1} *Neobacillus sedimentimangrovi* UE25 produced 4.574 IU mL^{-1} of endoglucanase and 1.812 IU mL^{-1} of β-glucosidase	Abbas et al. (2024)
Alkali	1% NaOH at a solid-to-liquid ratio of 1:15 for 60 min, respectively	Reducing sugar yield obtained 443.52 mg/g of sugarcane bagasse. Lignin yield and solid yield were 82.3 and 54.6, respectively	Ruan et al. (2024)
Ultrasound-combined alkaline pretreatment	3% NaOH at 188 W/ml ultrasonic power density for 60 min	Glucose released was 487 g/kg of biomass	Sitotaw et al. (2024)
$C_6H_8O_7$ Citric acid	Loading of 25 g biomass, liquid–solid 25 mL/g citric acid 10% (w/w)	Percent composition of cellulose, hemicellulose, and lignin after pretreatment were 34.85, 15.16, and 23.2%, respectively	Gomes et al. (2020)
$NaOH^+$ formaldehyde	0.7% w/v NaOH to reach 12% (w/v) solid content at 80 °C for 2 h	Percent composition of cellulose, hemicellulose, and lignin after pretreatment were 90.63, 73.29, and 45.03%, respectively	Jin et al. (2020)
Ionic liquid	1-Butyl-3-methylimidazolium chloride [Bmim] Cl + HCl 0,5% as a catalyst, 130 °C for 120 min	Xylose was removed by 75.2%, whereas lignin and glucan got concentrated by 97.3 and 24.3%, respectively	Trinh et al. (2019)
Alkaline	0.5M Na_2CO_3 at 140 °C for 80 min	Lignin was removed by 87%. Cellulose and xylose concentrated by 18.6 and 21.39%, respectively	Nasirpour and Mousavi (2018)
Acid	H_2SO_4 in solution 1:30 (w/v) at 100 °C for 60 min	Percent composition of cellulose, hemicellulose and lignin after pretreatment were 66, 0.06, and 0.36%, respectively	Bernier-Oviedo et al. (2018)
Organosolv	Ethanol and water in the ratio of 60:40 + $FeCl_3$ at 160 °C for 60 min	Lignin and xylose were removed by 20 and 100%, respectively. Glucan got concentrated by 83.98%	Zhang et al. (2018)
Alkaline	0.3M/L NaOH at 60 °C for 20 min	Lignin was removed by 30.45%. Cellulose and hemicellulose concentrated by 10.14 and 4.7%, respectively	Terán-Hilares et al. (2016)
Direct acid	1% H_2SO_4 at 140 °C, for 60 min	Xylose and arabinoxylan removal by 85.9 and 100%. Glucan and lignin concentrated by 47 and 22%, respectively	Li et al. (2017)

categorized depending on the type of waste generated. The following methods are generally used to treat the effluent wastewater from the sugar industry. These are as follows:

1. Physical pretreatment

   It consists of processes like (a) screening and filtration, (b) sedimentation, (c) centrifugation, and (d) thermal treatment (Patil et al. 2023). Primary methods to remove solids include screening, sedimentation, and skimming. Once the solids have been separated, the other type of treatment, particularly thermal one becomes important. A report discusses the use of thermocatalytic treatment for treating wastewater (Sahu 2017). The technique utilizes metal salts as catalysts such as magnesium oxide (MnO), copper oxide (CuO), zinc oxide (ZnO), etc. in combination with high temperature for complete decomposition of the pollutants. The pH and temperature are thus important parameters to be considered under this process. It was found that by using copper oxide, a COD (chemical oxygen demand) of 84.2% and 89.6% of color could be removed at a loading of 5 kg/m^3 biomass. The optimum pH and temperature for this process were 5 and 85 °C, respectively. The treated sludge can therefore be used as a fuel owing to its high calorific value.
2. Biological treatment

   These use microbes for the digestion of pollutants into an environmentally acceptable form. These are typically of two types: (a) aerobic and (b) anaerobic (Kushwaha 2015). The aerobic procedures require a large amount of oxygen although they lead to complete degradation. They include majorly activated sludge, trickling filters, and aerated lagoons or a combination of each of these (Fito et al. 2018; Kocaturk and Erguder 2015). Due to its high energy consumption as well as inefficiency in performing at higher COD levels, it becomes less amenable. Anaerobic processes on the other hand take place in the absence of any oxygen while producing CO_2 and CH_4 as the by-products. These processes consume less energy than the aerobic processes. Over the years, many types of anaerobic reactors have been generated, for example, an anaerobic batch reactor (ABR), anaerobic fixed bed reactor (AFR), and upflow anaerobic fixed (UAFB) reactor (Jaria et al. 2019; Wang et al. 2017).
3. Physicochemical treatment

   The main methods included in this category are coagulation or flocculation. The suspended colloids are generally removed by this process. Coagulation is the process in which the positively charged metal ion causes charge destabilization leading to neutralization (Duan and Gregory 2003). The flocculation process works on the collision between these destabilized particles which are brought closer by hydraulic shear forces (Ramli et al. 2022). These are generally carried out by sedimentation, filtration, and disinfection and are completed by chlorination procedure. The most commonly used coagulants include aluminum sulfate ($Al_2(SO_4)_3$), ferric sulfate (III) ($Fe_2(SO_4)_3$), $AlCl_3$, and ferric chloride ($FeCl_3$) (Sher et al. 2013). Some studies have also started using activated charcoal for the coagulation process wherein the pollutant is trapped in the pores of activated charcoal and is later decomposed by treatment at high temperature (Fahim et al. 2006; Panhwar et al. 2022).

4. Electrochemical/electrocoagulation treatment

   This method is generally much faster and more convenient than conventional physicochemical treatment methods. The redox reaction takes place on the anode and cathode which further leads to mineralization of the pollutants. Major factors which influence this process include the pH of the solution, electrode gap, intensity of the current as well as the treatment gap (Yadav et al. 2021). More commonly, the electrodes are made up of aluminum due to their electron transfer efficiency, ease of availability, and an electrode potential of −1.66 V (Choi et al. 2017). A study successfully used aluminum electrodes for treating sugar industry waste and got 78% COD and color removal by 70% at pH 6.0 (Sahu et al. 2015). Due to the shortcomings of aluminum electrodes like hydrogen formation and spontaneous redox reactions, they may alter the oxidation potential and affect the overall efficiency of the process. Hence, to overcome this, iron and copper are also used due to their low cost, durability, and good conductivity (Moreno et al. 2009; Sahu 2019). In a study, by using an iron electrode, 84% and 86% removal of COD and color, respectively, were obtained. In another study, the use of copper electrodes led to a COD and color removal of 92 and 95%, respectively (Sahu and Chaudhari 2015). Apart from the above-stated electrochemical processes, there are other alterations such as indirect oxidation electrolysis, combination electrodes made of iron and aluminium, etc. Although most of these processes are best suited for hazardous waste management, it is still advised to opt for a biological treatment procedure after following an electrochemical or oxidation process (Yadav et al. 2021).

22.7 Valorization Techniques Used and Types of Chemicals and By-Products Formed/Industrial Significance

One of the main developments in the value-adding of waste from the sugar industry is the use of molasses to produce ethanol. This resulted in the setting up of an ethanol distillery plant in combination with the sugar industry. Molasses has been widely used as a viable substrate in yeast fermentation to produce high-value biochemicals and liquid biofuels (Saxena et al. 2009). Later, bagasse utilization started to fulfill the energy needs of the sugar industry. Still, bagasse was left out and was used for electricity generation in sugar mills connected to ethanol distilleries (derived from molasses) in Brazil, India, Australia, Mauritius, and Colombia. Nowadays, facilities are generating more electricity than they require, and that extra energy is being sent to the national grid. In countries like Brazil, cogeneration actually yields more profits than the production of ethanol and sugar (Eggleston and Lima 2015). Cogeneration accounted for 19.3% of Brazil's renewable energy supply (Hofsetz and Silva 2012). Through cogeneration, Cuba was able to generate 600 MW in the operational sugar mills (Alonso-Pippo et al. 2008). On the other hand, according to To et al. (2018), Mauritius's portion of the country's electricity generation in 2015 was 17%.

Following biomass devolatilization, researchers have shown that turning sugarcane waste into biochar is an effective way to capture energy. There is scientific interest in the transformation of biomass into products with added value, such as biofuel, biogas, and charcoal. Numerous studies have been conducted to investigate the various pyrolysis methods that can be used to turn sugarcane bagasse (SCB) into char materials, highlighting the material's potential for a wide range of uses. Thermochemical techniques such as pyrolysis are commonly used to produce biochar (Zafeer et al. 2024). The produced biochar finds widespread application in different sectors. The main advantage of biochar is that it can effectively eliminate both organic and inorganic pollutants from soil and water systems. It has been used for malachite green dye removal (Vyavahare et al. 2018). The use of sugarcane bagasse biochar as a sustainable and affordable substitute for CO_2 adsorption has become popular (Lefebvre et al. 2021). Sugarcane bagasse-based biochar has also been extensively researched for use in composites, building materials, and soil amendments (Zafeer et al. 2024).

Sugarcane by-products can be used as fertilizer and have an improved effect on soil characteristics and crop productivity. By adding sugar industry by-products to the soil, such as bagasse and press mud, crop quality and yield are increased, and the chemical, physical, and biological characteristics of the soil are enhanced. By-products from the sugarcane industry are used in the soil to increase its organic matter during crop cultivation and to lower the needed dosage of fertilizers. About 32,464, 28,077, 14,038, 3434, 393, 1030, and 240 tons (t) of N, P, K, Fe, Zn, Mn, and Cu, respectively, can be obtained if 100% of the press mud is recycled through agriculture. This helps to reduce the need for pricey chemical fertilizers. It can also be packaged and sold with commercial fertilizer specifically designed for a given cropping system, and together with inorganic chemical fertilizers, it can be utilized. This lessens the issue of storing the by-products of the sugarcane industry (Dotaniya et al. 2016).

The production of biohydrogen from wastewater and solid waste from agriculture and agro-industrial sources is thought to be very beneficial because these materials are readily available, reasonably priced, and biodegradable. Various research studies have been carried out in this direction. Hydrogen gas production has also been demonstrated from sugar industry wastes (Urbaniec and Bakker 2015). Pretreatment sugarcane bagasse (SCB) in mildly alkaline conditions was utilized for the production of biohydrogen utilizing a new coculture of *Clostridium thermocellum* and *Thermoanaerobacterium aotearoense*. The ideal pretreatment parameters were determined to be 3% NaOH and a 25:1 liquid-to-solid ratio for 3 h at 80 °C. At 55 °C and with 40 g/L pretreated SCB, a final hydrogen production of 50.05 ± 1.51 mmol/L was attained. For the bioconversion of SCB to biohydrogen, the established coculture system offers a novel unified bioprocessing technique (Cheng and Zhu 2013). Investigations were conducted into the effects of pretreatment techniques (alkaline, thermal, microwave, thermal-alkaline, and microwave-alkaline) on the synthesis of biohydrogen from sugar beet pulp during dark fermentation. In comparison to other reactors, those that underwent alkaline, microwave-alkaline, and thermal-alkaline pretreatments produced a notable solubilization of beet pulp. A maximum yield of 115.6 mL H_2/g COD was observed in the reactor containing alkaline-pretreated beet pulp during the biohydrogen generation process (Ozkan et al. 2011).

Due to their similar physicochemical properties to synthetic plastics and their ability to biodegrade and be compatible with biological systems, polyhydroxyalkanoates (PHAs) present an appealing alternative to traditional plastics. PHA has been produced utilizing wastes from sugar processing industries. Molasses has been the primary sugar industry waste for PHA production, and various research has been conducted for its production (Saratale et al. 2021). Using sugarcane molasses, *Azotobacter vinelandii* was initially investigated for PHB production, yielding a maximum DCW of 11.14 g/L and PHB production of 7.8 g/L (Page 1992). Vinasse and sugarcane molasses have been used for PHA production using *Cupriavidus necator.* The molasses underwent prior hydrolysis using either acid (HCl or H_2SO_4) or enzymatic methods (invertase or baker's yeast), and the growth kinetics of *C. necator* were subsequently compared. Under optimized conditions, 1.7 g/L of PHB and 56% of product accumulation were achieved (Dalsasso et al. 2019).

Sugar waste-derived materials are used as adsorbents in water treatment for pollutant removal. Various types of research have been reported for heavy metals, dyes, and other miscellaneous pollutant removal (Anastopoulos et al. 2017). It was looked into whether it would be possible to use sugarcane bagasse (SCB) and beet pulp (BP), two inexpensive natural wastes, to adsorb Mn (II) from groundwater and eradicate or lessen the chance of it happening. The percentages of Mn (II) removed under ideal circumstances with SCB and BP were 62.5 and 86.36%, respectively. This technique has been effectively used to lower the danger of Mn(II) in groundwater samples (Ahmed et al. 2015).

Sugar industry wastes have also been utilized for biosurfactant production for varied industrial and environmental applications (Sundaram et al. 2024). Molasses is the most commonly used substrate due to its high vitamin content. Several recent reviews have looked at various elements of biosurfactant types and how they work in a variety of bioremediation applications, such as oil removal, heavy metal extraction, dye degradation, and hydrocarbon breakdown (Femina et al. 2023). The enhanced surface tension and emulsification capabilities of the biosurfactants generated from molasses feedstock suggest their possible application in bioremediation projects (Begum et al. 2023).

Sugarcane bagasse is unquestionably composed of lignocellulosic biomass (LCB), which embeds around 60% of carbohydrates, giving it a renewable supply of sugars that ferment. If the biochemical approach is used to produce the sugar platform, then the tremendous potential of SCB can be unlocked. Sugars are used as a feedstock for the fermentative synthesis of a number of chemicals and renewable fuels, which are thought to be important factors accelerating industrialization. In this direction, many of the platform chemicals, particularly organic acids, have been produced from SCB. Lactic acid (LA) has a multibillion-dollar industry and a wide range of applications, making it one of the top platform chemicals created from biomass, according to the US Department of Energy. At the moment, microbial fermentation (~90%) dominates the industrial production of LA and primarily uses starchy or sugar-rich food feedstocks. LA production using SCB as feedstock offers

Table 22.2 Commercial products from the sugarcane industry waste

Type of sugarcane waste	Commercial product	References
Sugarcane bagasse	Epoxy	Marichelvam et al. (2021)
Sugarcane bagasse	Polyethylene (polythene, PE)	Chen et al. (2020)
Sugarcane bagasse	Orthophthalic polyester resin	Candido et al. (2017)
Brazilian sugarcane ethanol	Biobased polyethylene	Mohsenzadeh et al. (2017)
Filter mud and vinasse	Methane biogas	Leme and Seabra (2017)
Sugarcane bagasse and sugarcane tops	Bioelectricity	Arshad and Ahmed (2016), Silalertruksa et al. (2015)
Sugarcane molasse	2,3-Butanediol using *Enterobacter cloacae* CGMCC 6053	Dai et al. (2015)
Sugarcane bagasse and sugarcane tops and molasses	Bioethanol	Sindhu et al. (2011)

a low-cost production route. Many techniques have been designed for LA production from SCB in the fields of saccharification, fermentation, and pretreatment that bridge the gap between lab-scale and industrial LA production (Agrawal and Kumar 2023; Yankov 2022). One-pot simultaneous saccharification and cofermentation (SSCF) was designed as a unique technique. Sugars produced by saccharification of ionic liquid-pretreated feedstocks using nanoimmobilized cellulase were fermented with the help of *Lactobacillus brevis*. Cottonseed cake, wheat straw, and sugarcane bagasse were shown to generate lactic acid at concentrations of 0.22, 0.49, and 0.52 g/g, respectively. Together with the ionic liquid-tolerant *L. brevis*, a cellulolytic, reusable nanoimmobilized enzyme, the use of renewable feedstocks offers a comprehensive plan for future biorefineries that generate bioproducts in a sustainable manner (Grewal and Khare 2018). Table 22.2 summarizes some of the major commercial products developed from the sugarcane industry waste. In order to create a circular economy, research on the value-adding of sugar industry waste has been conducted in order to produce biofuels and other chemicals on a greater scale.

22.8 Conclusion

The production of sugar has increased along with the population, and with it, the associated waste. If disposed of unsafely, these wastes pose serious environmental problems. An effective strategy has been the valorization of these wastes to generate useful commodities. Continued research has been done on the pretreatment, saccharification, and fermentation of these wastes to generate biofuels and chemicals. Production of bioethanol and electricity has been a significant step in this direction. Biohydrogen, bioplastics, butanediol, organic acids, fertilizers, etc. are some of the other valuable products. Continued research has been carried out for the valorization of sugar industry waste to synthesize biofuels and varied chemicals on a larger scale and cost effective manner to establish a circular economy.

Acknowledgments The financial support by Amity University Uttar Pradesh, Noida, is gratefully acknowledged.

References

Abbas M, Ejaz U, Sohail M, Alanazi AK (2024) Application of *Trametes pubescens* for dye removal and biological pretreatment of sugarcane bagasse. Biofuels Bioprod Biorefin 18:453–463. https://doi.org/10.1002/bbb.2591

Agrawal D, Kumar V (2023) Recent progress on sugarcane-bagasse based lactic acid production: technical advancements, potential and limitations. Ind Crop Prod. https://doi.org/10.1016/j.indcrop.2022.116132

Ahmed SA, El-Roudi AM, Salem AAA (2015) Removal of Mn(II) from ground water by solid wastes of sugar industry. J Environ Sci Technol 8:338–351. https://doi.org/10.3923/jest.2015.338.351

Ajala EO, Ighalo JO, Ajala MA, Adeniyi AG, Ayanshola AM (2021) Sugarcane bagasse: a biomass sufficiently applied for improving global energy, environment and economic sustainability. Bioresour Bioprocess 8:87. https://doi.org/10.1186/s40643-021-00440-z

Alokika A, Kumar A, Kumar V, Singh B (2021) Cellulosic and hemicellulosic fractions of sugarcane bagasse: potential, challenges and future perspective. Int J Biol Macromol 169:564–582. https://doi.org/10.1016/j.ijbiomac.2020.12.175

Alonso-Pippo W, Luengo CA, Koehlinger J, Garzone P, Cornacchia G (2008) Sugarcane energy use: the Cuban case. Energy Policy 36:2163–2181. https://doi.org/10.1016/j.enpol.2008.02.025

Anastopoulos I, Bhatnagar A, Hameed BH, Ok YS, Omirou M (2017) A review on waste-derived adsorbents from sugar industry for pollutant removal in water and wastewater. J Mol Liq. https://doi.org/10.1016/j.molliq.2017.05.063

Arshad M, Ahmed S (2016) Cogeneration through bagasse: a renewable strategy to meet the future energy needs. Renew Sust Energ Rev 54:732–737. https://doi.org/10.1016/j.rser.2015.10.145

Baurhoo B, Ruiz-Feria CA, Zhao X (2008) Purified lignin: nutritional and health impacts on farm animals—a review. Anim Feed Sci Technol 144:175–184. https://doi.org/10.1016/j.anifeedsci.2007.10.016

Begum W, Saha B, Mandal U (2023) A comprehensive review on production of bio-surfactants by bio-degradation of waste carbohydrate feedstocks: an approach towards sustainable development. RSC Adv. https://doi.org/10.1039/d3ra05051c

Bernier-Oviedo DJ, Rincón-Moreno JA, Solanilla-Duqué JF, Muñoz-Hernández JA, Váquiro-Herrera HA (2018) Comparison of two pretreatments methods to produce second-generation bioethanol resulting from sugarcane bagasse. Ind Crop Prod 122:414–421. https://doi.org/10.1016/j.indcrop.2018.06.012

Candido VS, Silva ACR, Simonassi NT, Luz FS, Monteiro SN (2017) Toughness of polyester matrix composites reinforced with sugarcane bagasse fibers evaluated by Charpy impact tests. J Mater Res Technol 6:334–338. https://doi.org/10.1016/j.jmrt.2017.06.001

Chandel AK, da Silva SS, Carvalho W, Singh OV (2012) Sugarcane bagasse and leaves: foreseeable biomass of biofuel and bio-products. J Chem Technol Biotechnol 87:11–20. https://doi.org/10.1002/jctb.2742

Chen H (2014) Chemical composition and structure of natural lignocellulose. In: Biotechnology of lignocellulose. Springer, Dordrecht, pp 25–71. https://doi.org/10.1007/978-94-007-6898-7_2

Chen B, Luo Z, Chen H, Chen C, Cai D, Qin P, Cao H, Tan T (2020) Wood plastic composites from the waste lignocellulosic biomass fibers of bio-fuels processes: a comparative study on mechanical properties and weathering effects. Waste Biomass Valoriz 11:1701–1710. https://doi.org/10.1007/s12649-018-0413-8

Cheng J, Zhu M (2013) A novel anaerobic co-culture system for bio-hydrogen production from sugarcane bagasse. Bioresour Technol 144:623–631. https://doi.org/10.1016/j.biortech.2013.07.018

Choi S, Go H, Lee G, Tak Y (2017) Electrochemical properties of an aluminum anode in an ionic liquid electrolyte for rechargeable aluminum-ion batteries. Phys Chem Chem Phys 19:8653–8656. https://doi.org/10.1039/C6CP08776K

Chotirotsukon C, Raita M, Champreda V, Laosiripojana N (2019) Fractionation of sugarcane trash by oxalic-acid catalyzed glycerol-based organosolv followed by mild solvent delignification. Ind Crop Prod 141:111753. https://doi.org/10.1016/j.indcrop.2019.111753

Dai JY, Zhao P, Cheng X-L, Xiu Z-L (2015) Enhanced production of 2,3-butanediol from sugarcane molasses. Appl Biochem Biotechnol 175:3014–3024. https://doi.org/10.1007/s12010-015-1481-x

Dalsasso RR, Pavan FA, Bordignon SE, de Aragão GMF, Poletto P (2019) Polyhydroxybutyrate (PHB) production by Cupriavidus necator from sugarcane vinasse and molasses as mixed substrate. Process Biochem 85:12–18. https://doi.org/10.1016/j.procbio.2019.07.007

Davison BH, Parks J, Davis MF, Donohoe BS (2013) Plant cell walls: basics of structure, chemistry, accessibility and the influence on conversion. In: Aqueous pretreatment of plant biomass for biological and chemical conversion to fuels and chemicals. Wiley, Hoboken, pp 23–38. https://doi.org/10.1002/9780470975831.ch3

Dias MOS, Junqueira TL, Cavalett O, Cunha MP, Jesus CDF, Rossell CEV, Maciel Filho R, Bonomi A (2012) Integrated versus stand-alone second generation ethanol production from sugarcane bagasse and trash. Bioresour Technol 103:152–161. https://doi.org/10.1016/j.biortech.2011.09.120

Dotaniya ML, Datta SC, Biswas DR, Dotaniya CK, Meena BL, Rajendiran S, Regar KL, Lata M (2016) Use of sugarcane industrial by-products for improving sugarcane productivity and soil health. Int J Recycl Org Waste Agric. https://doi.org/10.1007/s40093-016-0132-8

Duan J, Gregory J (2003) Coagulation by hydrolysing metal salts. Adv Colloid Interf Sci 100:475–502. https://doi.org/10.1016/S0001-8686(02)00067-2

Duval A, Lawoko M (2014) A review on lignin-based polymeric, micro- and nano-structured materials. React Funct Polym 85:78–96. https://doi.org/10.1016/j.reactfunctpolym.2014.09.017

Eggleston G, Lima I (2015) Sustainability issues and opportunities in the sugar and sugar-bioproduct industries. Sustain For. https://doi.org/10.3390/su70912209

Fahim N, Barsoum B, Eid A, Khalil M (2006) Removal of chromium(III) from tannery wastewater using activated carbon from sugar industrial waste. J Hazard Mater 136:303–309. https://doi.org/10.1016/j.jhazmat.2005.12.014

Femina CC, Kumar PS, Mohanakrishna G, Hemavathy RV, Rangasamy G, Aminabhavi TM (2023) Sustainable production of biosurfactants via valorisation of industrial wastes as alternate feedstocks. Chemosphere 312:137326. https://doi.org/10.1016/j.chemosphere.2022.137326

Fito J, Tefera N, Kloos H, Van Hulle SWH (2018) Anaerobic treatment of blended sugar industry and ethanol distillery wastewater through biphasic high rate reactor. J Environ Sci Health A 53:676–685. https://doi.org/10.1080/10934529.2018.1438826

Gomes MG, Gurgel LVA, Baffi MA, Pasquini D (2020) Pretreatment of sugarcane bagasse using citric acid and its use in enzymatic hydrolysis. Renew Energy 157:332–341. https://doi.org/10.1016/j.renene.2020.05.002

Grewal J, Khare SK (2018) One-pot bioprocess for lactic acid production from lignocellulosic agro-wastes by using ionic liquid stable lactobacillus brevis. Bioresour Technol 251:268–273. https://doi.org/10.1016/j.biortech.2017.12.056

Hofsetz K, Silva MA (2012) Brazilian sugarcane bagasse: energy and non-energy consumption. Biomass Bioenergy 46:564–573. https://doi.org/10.1016/j.biombioe.2012.06.038

Hu F, Ragauskas A (2012) Pretreatment and lignocellulosic chemistry. Bioenergy Res 5:1043–1066. https://doi.org/10.1007/s12155-012-9208-0

James J (2020) Sugarcane press mud modification of expansive soil stabilized at optimum lime content: strength, mineralogy and microstructural investigation. J Rock Mech Geotech Eng 12:395–402. https://doi.org/10.1016/j.jrmge.2019.10.005

Jamir L, Kumar V, Kaur J, Kumar S, Singh H (2021) Composition, valorization and therapeutical potential of molasses: a critical review. Environ Technol Rev 10:131–142. https://doi.org/10.1080/21622515.2021.1892203

Jaria G, Calisto V, Silva CP, Gil MV, Otero M, Esteves VI (2019) Fixed-bed performance of a waste-derived granular activated carbon for the removal of micropollutants from municipal wastewater. Sci Total Environ 683:699–708. https://doi.org/10.1016/j.scitotenv.2019.05.198

Jin Y, Shi Z, Xu G, Yang H, Yang J (2020) A stepwise pretreatment of sugarcane bagasse by alkaline and hydroxymethyl reagent for bioethanol production. Ind Crop Prod 145:112136. https://doi.org/10.1016/j.indcrop.2020.112136

Karp SG, Woiciechowski AL, Soccol VT, Soccol CR (2013) Pretreatment strategies for delignification of sugarcane bagasse: a review. Braz Arch Biol Technol 56:679–689. https://doi.org/10.1590/S1516-89132013000400019

Khaire KC, Moholkar VS, Goyal A (2021) Bioconversion of sugarcane tops to bioethanol and other value added products: an overview. Mater Sci Energy Technol 4:54–68. https://doi.org/10.1016/j.mset.2020.12.004

Koçar G, Civaş N (2013) An overview of biofuels from energy crops: current status and future prospects. Renew Sust Energ Rev. https://doi.org/10.1016/j.rser.2013.08.022

Kocaturk I, Erguder TH (2015) Investigation of the use of aerobic granules for the treatment of sugar beet processing wastewater. Environ Technol 36:2577–2587. https://doi.org/10.1080/09593330.2015.1039070

Kushwaha JP (2015) A review on sugar industry wastewater: sources, treatment technologies, and reuse. Desalination Water Treat 53:309–318. https://doi.org/10.1080/19443994.2013.838526

Lefebvre D, Williams A, Kirk GJ, Meersmans J, Sohi S, Goglio P, Smith P (2021) An anticipatory life cycle assessment of the use of biochar from sugarcane residues as a greenhouse gas removal technology. J Clean Prod 312:127764. https://doi.org/10.1016/j.jclepro.2021.127764

Leme RM, Seabra JEA (2017) Technical-economic assessment of different biogas upgrading routes from vinasse anaerobic digestion in the Brazilian bioethanol industry. Energy 119:754–766. https://doi.org/10.1016/j.energy.2016.11.029

Li H, Xiong L, Chen X, Wang C, Qi G, Huang C, Luo M, Chen X (2017) Enhanced enzymatic hydrolysis and acetone-butanol-ethanol fermentation of sugarcane bagasse by combined diluted acid with oxidate ammonolysis pretreatment. Bioresour Technol 228:257–263. https://doi.org/10.1016/j.biortech.2016.12.119

Limayem A, Ricke SC (2012) Lignocellulosic biomass for bioethanol production: current perspectives, potential issues and future prospects. Prog Energy Combust Sci 38:449–467. https://doi.org/10.1016/j.pecs.2012.03.002

Marichelvam MK, Manimaran P, Verma A, Sanjay MR, Siengchin S, Kandakodeeswaran K, Geetha M (2021) A novel palm sheath and sugarcane bagasse fiber based hybrid composites for automotive applications: an experimental approach. Polym Compos 42:512–521. https://doi.org/10.1002/pc.25843

Meghana M, Shastri Y (2020) Sustainable valorization of sugar industry waste: status, opportunities, and challenges. Bioresour Technol. https://doi.org/10.1016/j.biortech.2020.122929

Mohsenzadeh A, Zamani A, Taherzadeh MJ (2017) Bioethylene production from ethanol: a review and techno-economical evaluation. ChemBioEng Rev 4:75–91. https://doi.org/10.1002/cben.201600025

Moodley P, Kana EBG (2019) Bioethanol production from sugarcane leaf waste: effect of various optimized pretreatments and fermentation conditions on process kinetics. Biotechnol Rep 22:e00329. https://doi.org/10.1016/j.btre.2019.e00329

Moreno C, Cocke DL, Gomes JAG, Morkovsky P, Parga JR, Peterson E, Garcia C (2009) Electrochemical reactions for electrocoagulation using iron electrodes. Ind Eng Chem Res 48:2275–2282. https://doi.org/10.1021/ie8013007

Nasirpour N, Mousavi SM (2018) RSM based optimization of PEG assisted ionic liquid pretreatment of sugarcane bagasse for enhanced bioethanol production: effect of process parameters. Biomass Bioenergy 116:89–98. https://doi.org/10.1016/j.biombioe.2018.06.008

Nimbalkar PR, Khedkar MA, Gaikwad SG, Chavan PV, Bankar SB (2017) New insight into sugarcane industry waste utilization (press mud) for cleaner biobutanol production by using C. Acetobutylicum NRRL B-527. Appl Biochem Biotechnol 183:1008–1025. https://doi.org/10.1007/s12010-017-2479-3

Nouhou Moussa AW, Sawadogo B, Konate Y, Sidibe DS, Heran M (2023) Critical state of the art of sugarcane industry wastewater treatment technologies and perspectives for sustainability. Membranes 13:709. https://doi.org/10.3390/membranes13080709

Ozkan L, Erguder TH, Demirer GN (2011) Effects of pretreatment methods on solubilization of beet-pulp and bio-hydrogen production yield. Int J Hydrog Energy 36:382–389. https://doi.org/10.1016/j.ijhydene.2010.10.006

Page WJ (1992) Suitability of commercial beet molasses fractions as substrates from polyhydroxyalkanoate production by *Azotobacter vinelandii* UWD. Biotechnol Lett 14:385–390. https://doi.org/10.1007/BF01021252

Palmonari A, Cavallini D, Sniffen CJ, Fernandes L, Holder P, Fagioli L, Fusaro I, Biagi G, Formigoni A, Mammi L (2020) Short communication: characterization of molasses chemical composition. J Dairy Sci 103:6244–6249. https://doi.org/10.3168/jds.2019-17644

Pandey A, Soccol CR, Nigam P, Soccol VT (2000) Biotechnological potential of agro-industrial residues. I: sugarcane bagasse. Bioresour Technol 74:69–80. https://doi.org/10.1016/S0960-8524(99)00142-X

Panhwar A, Faryal K, Kandhro A, Brohi N, Ahmed K, Iqbal M (2022) Treatment of sugarcane industry waste water by combination of chemical coagulation and activated carbon. Pak J Sci Ind Res Ser A Phys Sci 65:59–63. https://doi.org/10.52763/PJSIR.PHYS.SCI.65.1.2022.59.63

Patil PM, Ingavale RR, Matkar AR, Hwang S, Gurav R, Dhanavade MJ (2023) An innovative and effective industrial wastewater treatments: a brief history and present scenario. In: Advanced and innovative approaches of environmental biotechnology in industrial wastewater treatment. Springer, Singapore, pp 191–219. https://doi.org/10.1007/978-981-99-2598-8_10

Raghavi S, Sindhu R, Binod P, Gnansounou E, Pandey A (2016) Development of a novel sequential pretreatment strategy for the production of bioethanol from sugarcane trash. Bioresour Technol 199:202–210. https://doi.org/10.1016/j.biortech.2015.08.062

Rais D, Zibek S (2017) Biotechnological and biochemical utilization of lignin. Adv Biochem Eng Biotechnol 166:469–518. https://doi.org/10.1007/10_2017_6

Ramli SF, Aziz HA, Omar FM, Yusoff MS, Halim H, Kamaruddin MA, Ariffin KS, Hung Y-T (2022) Influence of particle size and zeta potential in treating highly coloured old landfill leachate by tin tetrachloride and rubber seed. Int J Environ Res Public Health 19:3016. https://doi.org/10.3390/ijerph19053016

Rasul MG, Rudolph V, Carsky M (1999) Physical properties of bagasse. Fuel 78:905–910. https://doi.org/10.1016/S0016-2361(99)00011-3

Ruan L, Wu H, Wu S, Zhou L, Wu S, Shang C (2024) Optimizing the conditions of pretreatment and enzymatic hydrolysis of sugarcane bagasse for bioethanol production. ACS Omega 9:29566–29575. https://doi.org/10.1021/acsomega.4c02485

Sadaf A, Fatima SW, Khare SK (2019) Sporotrichum thermophile xylanases and their biotechnological applications. In: Fungi in extreme environments: ecological role and biotechnological significance. Springer, Cham, pp 307–328. https://doi.org/10.1007/978-3-030-19030-9_16

Sahu O (2017) Catalytic thermal pre-treatments of sugar industry wastewater with metal oxides: thermal treatment. Exp Thermal Fluid Sci 85:379–387. https://doi.org/10.1016/j.expthermflusci.2017.03.022

Sahu O (2019) Electro-oxidation and chemical oxidation treatment of sugar industry wastewater with ferrous material: an investigation of physicochemical characteristic of sludge. S Afr J Chem Eng 28:26–38. https://doi.org/10.1016/j.sajce.2019.01.004

Sahu OP, Chaudhari PK (2015) Electrochemical treatment of sugar industry wastewater: COD and color removal. J Electroanal Chem 739:122–129. https://doi.org/10.1016/j.jelechem.2014.11.037
Sahu OP, Gupta V, Chaudhari PK, Srivastava VC (2015) Electrochemical treatment of actual sugar industry wastewater using aluminum electrode. Int J Environ Sci Technol 12:3519–3530. https://doi.org/10.1007/s13762-015-0774-5
Saratale RG, Cho SK, Saratale GD, Kadam AA, Ghodake GS, Kumar M, Bharagava RN, Kumar G, Kim DS, Mulla SI, Shin HS (2021) A comprehensive overview and recent advances on polyhydroxyalkanoates (PHA) production using various organic waste streams. Bioresour Technol. https://doi.org/10.1016/j.biortech.2021.124685
Sarker TC, Azam SMGG, Bonanomi G (2017) Recent advances in sugarcane industry solid by-products valorization. Waste Biomass Valoriz. https://doi.org/10.1007/s12649-016-9665-3
Saxena RC, Adhikari DK, Goyal HB (2009) Biomass-based energy fuel through biochemical routes: a review. Renew Sust Energ Rev. https://doi.org/10.1016/j.rser.2007.07.011
Scheller HV, Ulvskov P (2010) Hemicelluloses. Annu Rev Plant Biol 61:263–289. https://doi.org/10.1146/annurev-arplant-042809-112315
Sharma S, Tsai M-L, Sharma V, Sun P-P, Nargotra P, Bajaj BK, Chen C-W, Dong C-D (2022) Environment friendly pretreatment approaches for the bioconversion of lignocellulosic biomass into biofuels and value-added products. Environments 10:6. https://doi.org/10.3390/environments10010006
Sher F, Malik A, Liu H (2013) Industrial polymer effluent treatment by chemical coagulation and flocculation. J Environ Chem Eng 1:684–689. https://doi.org/10.1016/j.jece.2013.07.003
Sherwood J (2020) The significance of biomass in a circular economy. Bioresour Technol. https://doi.org/10.1016/j.biortech.2020.122755
Silalertruksa T, Gheewala SH, Pongpat P (2015) Sustainability assessment of sugarcane biorefinery and molasses ethanol production in Thailand using eco-efficiency indicator. Appl Energy 160:603–609. https://doi.org/10.1016/j.apenergy.2015.08.087
Sindhu R, Kuttiraja M, Binod P, Janu KU, Sukumaran RK, Pandey A (2011) Dilute acid pretreatment and enzymatic saccharification of sugarcane tops for bioethanol production. Bioresour Technol 102:10915–10921. https://doi.org/10.1016/j.biortech.2011.09.066
Singh SP, Jawaid M, Chandrasekar M, Senthilkumar K, Yadav B, Saba N, Siengchin S (2021) Sugarcane wastes into commercial products: processing methods, production optimization and challenges. J Clean Prod 328:129453. https://doi.org/10.1016/j.jclepro.2021.129453
Sitotaw YW, Habtu NG, Gebreyohannes AY, Van Gerven T (2024) Effect of alkaline combined with ultrasonic pretreatment to enhance enzymatic saccharification of sugar cane bagasse and Enset fiber (*Ensete ventricosum*). Biomass Convers Biorefinery 14:3897–3915. https://doi.org/10.1007/s13399-022-03021-y
Sporck D, Reinoso FAM, Rencoret J, Gutiérrez A, del Rio JC, Ferraz A, Milagres AMF (2017) Xylan extraction from pretreated sugarcane bagasse using alkaline and enzymatic approaches. Biotechnol Biofuels 10:296. https://doi.org/10.1186/s13068-017-0981-z
Sundaram T, Govindarajan RK, Vinayagam S, Krishnan V, Nagarajan S, Gnanasekaran GR, Baek KH, Rajamani Sekar SK (2024) Advancements in biosurfactant production using agro-industrial waste for industrial and environmental applications. Front Microbiol. https://doi.org/10.3389/fmicb.2024.1357302
Terán-Hilares R, Reséndiz AL, Martínez RT, Silva SS, Santos JC (2016) Successive pretreatment and enzymatic saccharification of sugarcane bagasse in a packed bed flow-through column reactor aiming to support biorefineries. Bioresour Technol 203:42–49. https://doi.org/10.1016/j.biortech.2015.12.026
To LS, Seebaluck V, Leach M (2018) Future energy transitions for bagasse cogeneration: lessons from multi-level and policy innovations in Mauritius. Energy Res Soc Sci 35:68–77. https://doi.org/10.1016/j.erss.2017.10.051
Trinh LTP, Lee Y-J, Park CS, Bae H-J (2019) Aqueous acidified ionic liquid pretreatment for bioethanol production and concentration of produced ethanol by pervaporation. J Ind Eng Chem 69:57–65. https://doi.org/10.1016/j.jiec.2018.09.008

Ungureanu N, Vlăduț V, Biriș SȘ (2022) Sustainable valorization of waste and by-products from sugarcane processing. Sustain For. https://doi.org/10.3390/su141711089

Urbaniec K, Bakker RR (2015) Biomass residues as raw material for dark hydrogen fermentation - a review. Int J Hydrog Energy. https://doi.org/10.1016/j.ijhydene.2015.01.073

Usmani Z, Sharma M, Diwan D, Tripathi M, Whale E, Jayakody LN, Moreau B, Thakur VK, Tuohy M, Gupta VK (2022) Valorization of sugar beet pulp to value-added products: a review. Bioresour Technol. https://doi.org/10.1016/j.biortech.2021.126580

Vyavahare GD, Gurav RG, Jadhav PP, Patil RR, Aware CB, Jadhav JP (2018) Response surface methodology optimization for sorption of malachite green dye on sugarcane bagasse biochar and evaluating the residual dye for phyto and cytogenotoxicity. Chemosphere 194:306–315. https://doi.org/10.1016/j.chemosphere.2017.11.180

Wang D, Ma W, Han H, Li K, Hao X (2017) Enhanced treatment of Fischer-Tropsch (F-T) wastewater by novel anaerobic biofilm system with scrap zero valent iron (SZVI) assisted. Biochem Eng J 117:66–76. https://doi.org/10.1016/j.bej.2016.09.012

Yadav M, Jyoti, Yadav RK, Gole VL (2021) Sugar industry wastewater treatment: current practices and advances. In: Microbial ecology of wastewater treatment plants. Elsevier, Amsterdam, pp 151–174. https://doi.org/10.1016/B978-0-12-822503-5.00017-5

Yang S-D, Liu J-X, Wu J, Tan H-W, Li Y-R (2013) Effects of vinasse and press mud application on the biological properties of soils and productivity of sugarcane. Sugar Tech 15:152–158. https://doi.org/10.1007/s12355-012-0200-y

Yankov D (2022) Fermentative lactic acid production from lignocellulosic feedstocks: from source to purified product. Front Chem. https://doi.org/10.3389/fchem.2022.823005

Zafeer MK, Menezes RA, Venkatachalam H, Bhat KS (2024) Sugarcane bagasse-based biochar and its potential applications: a review. Emerg Mater. https://doi.org/10.1007/s42247-023-00603-y

Zhang H, Fan M, Li X, Zhang A, Xie J (2018) Enhancing enzymatic hydrolysis of sugarcane bagasse by ferric chloride catalyzed organosolv pretreatment and tween 80. Bioresour Technol 258:295–301. https://doi.org/10.1016/j.biortech.2018.03.004

Zoghlami A, Paës G (2019) Lignocellulosic biomass: understanding recalcitrance and predicting hydrolysis. Front Chem 7:874. https://doi.org/10.3389/fchem.2019.00874

Utilization of Dairy Industry Waste for Biofuels and Chemicals

23

Priya Shukla, Pradeep Srivastava, and Abha Mishra

Abstract

Despite producing diversified dairy products, the dairy business produces excessive waste and by-products. As these wastes have high chemical oxygen demand levels and abundances of nutrients (fat, lactose, and protein), other possible industrial utilization must be found. Considering the definition of the efficient use of dairy wastewater in manufacturing different microbiological products such as enzymes, biofuel, polymer, biomass production, organic acid, etc., this work critically assesses the fermentation-engineering issues. To improve microbial cell activities, techno-economic studies have been performed on the viability of associated processes and the homogenization of specific microbiological products in biorefineries. Overall, it appears promising for business strategies to include dairy wastes in the specified microbial processes. The production of biogas at the facility will have a payback period (PBP) and a positive net present value (NPV) in the span of a decade, mainly when dairy wastes are digested with cow dung and/or other substrates.

Keywords

Dairy industry · Resource recovery · Fermentation · Bioconversions · Bioproducts

P. Shukla · P. Srivastava · A. Mishra (✉)
School of Biochemical Engineering, Indian Institute of Technology (Banaras Hindu University), Varanasi, Uttar Pradesh, India
e-mail: abham.bce@itbhu.ac.in

A. Kumari et al. (eds.), *Valorization of Solid Wastes to Biofuels and Chemical Products for Sustainable World*, https://doi.org/10.1007/978-981-96-8594-3_23

23.1 Introduction

Technological advances greatly impact the advancement of a country. Of all the manufacturing industries, the food industry has the greatest influence on global pollution concerns due to its excessive water consumption and substantial wastewater creation for units (Ahmad et al. 2019). The development of innovative technologies to increase dairy products merchandise manufacturing has led to a notable expansion in the dairy industry in the previous couple of years. The dairy business's growth has resulted in an immense quantity of waste being created. An estimated 6–10 l of waste are produced while processing 1 l of dairy milk (Gramegna et al. 2020). To keep both humans and the surroundings safe, effluent from dairy farms needs to be handled correctly. Consequently, prior to dumping garbage into waterways, businesses ought to comply with the regulations. Nonetheless, there are plenty of techniques, both chemical-based and physical in nature, for purifying runoff from dairy pollutants (Vialkova et al. 2019).

Dairy effluent contains a variety of colloidal particles. These suspended particles are capable of being eliminated utilizing a variety of standard methodologies, including flocculation, flotation, ion exchange, coagulation, solvent extraction, and membrane filtering. The most appropriate techniques among them are flocculation and coagulation, which include the inclusion of artificial coagulants including chitosan, Moringa oleifera, common bean, and fava bean (Tripathi et al. 2021). Wastewater can be recycled using membrane separation (Pasotti et al. 2020). However, during membrane separation, the proteinaceous components adhere to the membranes, giving off an unpleasant odor. Prior to membrane separation, the application of coagulants and absorbents may offer a remedy to this issue. Recovering protein, lactose, and water reusability may be possible with the use of modern technology like nanofiltration and reverse osmosis. These new technologies are more reliable, adaptable, clean, and energy-efficient than traditional ones (Roufou et al. 2021).

Cheese whey, pasteurized milk, and a variety of dairy industry by-products have been evaluated as inexpensive raw materials owing to having plentiful nutrients and as a potential replacement substrate for microbiological manufacturing methods (enzymes, biopolymers, organic acids, biomass, bioenergy, etc.). It is possible to contribute to four different "Global Sustainable Goals" by analyzing these wastes in microbiological processes. The generation of certain commodities with additional value (including lactic acid, biomass, and nisin) can be incorporated into objectives for sustainable production and consumption. Other products like bioethanol, biogas, and biohydrogen can be produced in a way that makes clean energy inexpensive. As part of the objective of sustainable industrialization, bringing ingenious, cutting-edge technologies to manufacturing sectors might assist with solving socioeconomic and sustainability challenges and create additional employment opportunities. By using the wastes produced during microbial processes, it helps to achieve the goal of reducing marine pollution concurrently with metabolite production. The objective of zero waste will also be furthered by the bioprocesses' valuation of these by-products.

There are several value-added items that can be made from dairy waste, such as biodiesel (Kavitha et al. 2019) bioplastic (Tripathi et al. 2021), ethanol (Zheng et al. 2022), functional beverages (Lawton et al. 2021), single-cell protein (Suman et al. 2017), organic fertilizers (Alharbi et al. 2021), and microbial fuel cell (Ahmad et al. 2019). Among the many beneficial products, polyhydroxyalkanoates, or PHAs, are highly sought-after and can be made with dairy waste, including milk whey. Lactose, lipids, and proteins are all present in milk whey. As a result, PHAs can be produced by fermenting it (Tripathi et al. 2021). Biodiesel can also be produced from dairy waste. Dairy industries provide the scum from their waste products for the manufacture. The procedure involves transesterifying dairy scum using calcium oxide (CaO) derived from eggshells. In a study, *Lactobacillus bulgaricus* was utilized to manufacture D-lactic acid from cheese whey powder (Liu et al. 2018).

The review paper that is being presented here examines the subsequent fields present state-of-the-art: (1) composition of dairy waste; (2) methodologies for elimination of dairy waste; (3) merchandise with an additional appeal made from dairy waste.

23.2 Waste and By-Products of Dairy Industry

The dairy sector's core business is processing raw milk to create a variety of products, including pasteurized and sterilized milk and processed dairy products (Ahmad et al. 2019; Rivas et al. 2010). The two distinct groups of dairy sector side-streams screened throughout handling are dairy company wastes that are not identifiable as food (contaminated, spoiled, or expired dairy products) and milk industry by-products that comply with consumption requirements (including ice cream, whey, butter, cheese, and milk residues). While throwaways and by-products can be employed in microbiological procedures including anaerobic digestion and fermentation, it seems more prudent to employ by-products to produce food-grade materials like biomass or lactic acid (Fig. 23.1). Throughout product manufacture and packing, the initial wastewater generated all over the process is released, particularly while sterilizing bottles, tanks, jars, and other equipment (Yonar et al. 2018). Dairy goods, particularly yogurt, cream, and butter, produce wasteful consumption due to use-by dates or inadequate manufacturing, shipping, and preservation methods (Tostivint et al. 2017).

The Food and Agriculture Organization (FAO) estimates that losses or waste account for almost one-third of the food generated for consumption by humans. It has been estimated that, for a variety of reasons, about 18% of milk output is squandered (Chang et al. 2018). Throughout the cheese-making procedure, enormous amounts of whey (9–10 L for 1 kg of cheese manufacture) are also released, besides this copious quantity of milk. The greenish-yellow substance known as whey fluctuates according to the kind of milk (sheep, buffalo, goat, cow, and other mammals) and the methodologies used to make the cheese (resulting in acidic, sweet, and salty whey) (Blaschek et al. 2007). Whey could also be ultrafiltered to generate

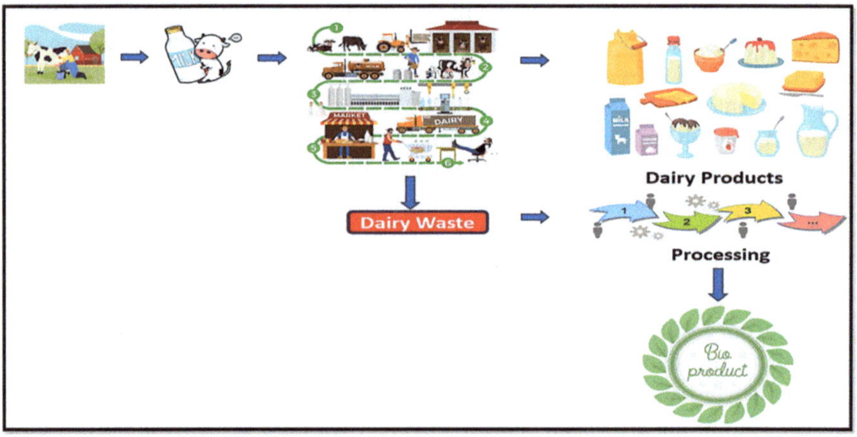

Fig. 23.1 Diagrammatic representation of the process used to turn dairy waste into biological products

miscellaneous by-products (whey protein, lactose, and powder) that can be utilized for a variety of applications, mostly as dietary supplements (Atra et al. 2005).

23.3 Composition of Dairy Waste

Soluble proteins, dissolved solids, D-lactose, salts, sulfates, chlorides, and lipids make up dairy effluents. There is a significant quantity of organic content in dairy wastewater. As a result, it has a high biological and chemical oxygen demand. It is also quite high in nitrogen and phosphate. In turn, it might end up in eutrophication, which could jeopardize aquatic ecosystems if introduced into the water column unregulated. Grease and oil, which are fatty pollutants, cause a light coating on the outermost layer of water bodies that block the circulation through which air circulates. Dairy effluent also constitutes of casein, inorganic salts, lactose, phosphorus, potassium, and nitrogen (Ahmad et al. 2019). Whey is a significant unwanted product of the dairy business, generated along with casein and cheese. It has soluble vitamins, carbohydrates, and fats (Arefin et al. 2020). Another undesirable outcome of dairy processing is the sedimentary residue from refined butter created while manufacturing butter or clarified ghee. It is mostly comprised of proteins and lipids. The pH, which varies from slightly acidic to highly basic in dairy runoff, suggests the type and amount of cleaning agents required for sterilization. The effluent from dairy farms is typically between 17 and 25 °C in temperature, and it smells bad (Mishra et al. 2021; Li et al. 2021). The range of chemical and biological oxygen demand, phosphorus, and total suspended particles in the untreated effluents has been observed to be between 1–3 kg per metric ton, 1–4 kg per metric ton, and 10–100 mg/L and 100–1000 mg/L, respectively (Ahmad et al. 2019).

23.4 Conventional Methods for Elimination of Waste and by-Products from the Dairy Sector

Processing dairy to make a variety of goods, including cheese, yogurt, butter, and so on, results in waste streams with complicated compositions. The formulation of dairy wastewater is influenced by the manufacturing of the milk and each dairy good, which varies according to the animal source. Dairy sector by-products have a high organic content; therefore, disposing of them directly can have negative environmental effects (Asunis et al. 2020). An extensively used technique for treating streams is in-plant chemical, mechanical, biological (aerobic and anaerobic), and physicochemical treatments (Kolev 2017). Dairy waste that has been mechanically treated has its suspended particles removed. This procedure is not a therapeutic procedure, but rather a pretreatment. Chemical treatment, which involves computing flocculants including Fenton reagents (H_2O_2 and $FeSO_4$), is a pretreatment technique utilized for disposing of dairy processing waste of colloids and soluble contaminants (Vlyssides et al. 2012). In addition, biological processes have been used more often in treatment systems because the physicochemical method can need expensive equipment and is ineffective for wastes containing significant levels of soluble carbon dioxide (Fig. 23.2) (Kolev 2017).

Microorganisms cultivated on organic materials in either anaerobic or aerobic environments form the basis of biological treatment. Techniques commonly employed in aerobic processes include sequencing batch reactors, membrane bioreactors, biofilm reactors, and rotating biological contactors (Swain et al. 2018;

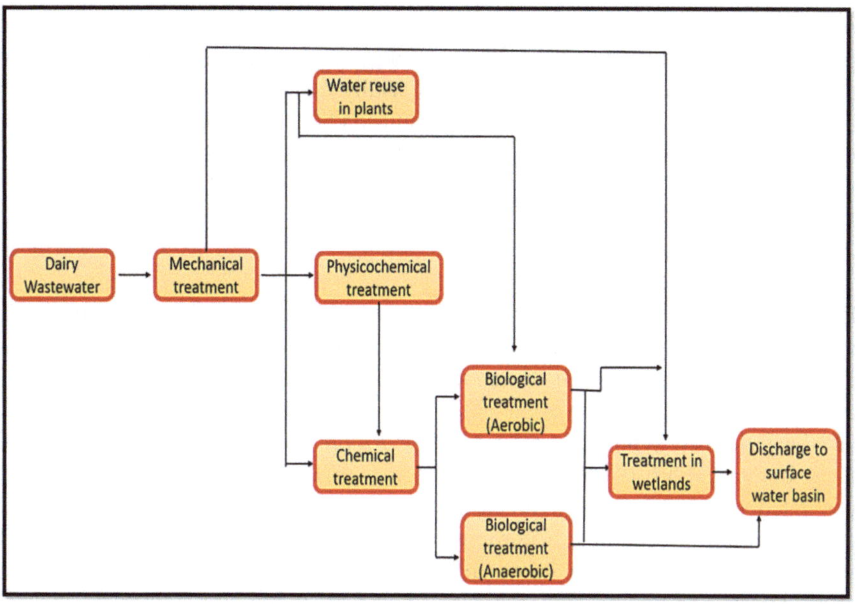

Fig. 23.2 Techniques for treating dairy wastewater

Collivignarelli et al. 2019). Wetland remediation continues to be one of the aerobic strategies used in practice. Dairy wastewater can be released into the water basin after being treated in aerobic ponds for 5 days (Britz et al. 2004). Nevertheless, this technology presents many drawbacks, including contaminating water supplies and creating a bottleneck due to the enormous surface area requirement. Furthermore, aerobic systems were unable to fully breakdown fat and protein, two components found in dairy waste (Omil et al. 2003). As a different option, two thermophilic waste treatment techniques using continuous-stirred tank reactors—aerobic and two anaerobic, or just anaerobic digestion—were disclosed (Arvanitoyannis and Giakoundis 2006). Even though anaerobic and aerobic system combinations are utilized to deal with dairy waste, these multistep procedures can be reasonably priced (Demirel et al. 2005).

23.5 Microbiological Processes

One of the major global generators of trash is the food sector, a significant portion of which is attributed to the dairy sector. The most noticeable waste product from the dairy business is whey, which is removed from milk and endangers both ecosystems and human well-being (Chandra et al. 2018). The development of various technologies, including drying, reverse osmosis, ion exchange, nanofiltration, hydrolysis, electrodialysis, and ultrafiltration, can purify dairy waste and associated by-products and use them in other sectors (Ostojić et al. 2005; Ryan and Walsh 2016). Nonetheless, biological techniques such as anaerobic digestion and microbial fermentation may be chosen since they can yield more value molecules and be more sustainable than conventional techniques (Chandra et al. 2018). In the current period, bioconversion of dairy goods including whey, milk, and others into innovative products is an indispensable option. Dairy industry by-products, although highly polluting, are a great source of nutrients and lactose, which promotes microbiological development in order to produce bioproducts (Fig. 23.3).

23.5.1 Biofuels

High chemical oxygen demand by-products from the dairy sector may be utilized as a feedstock for anaerobic digestion, a procedure which generates biogas (Kozłowski et al. 2019). The bacteria that cause the hydrolysis are *Bacillus*, *Acetivibrio*, *Vibrio*, and *Peptococcus*. In the anaerobic digestion system, several methanogenic archaea and bacteria also synthesize biogas, while *Propionibacterium*, *Clostridium*, and *Butyribacterium* are producers of volatile-free acids (Wainaina et al. 2019). The generated biogas could be utilized as thermal energy to generate steam in the cheese-making procedures, according to the circular economy theory. If not, it might be utilized to provide the wastewater treatment unit with electricity (Comino et al. 2009). Additionally, the anaerobic digesters' liquid and solid parts include beneficial nutrients that can be applied as soil fertilizers (Kavacik and Topaloglu

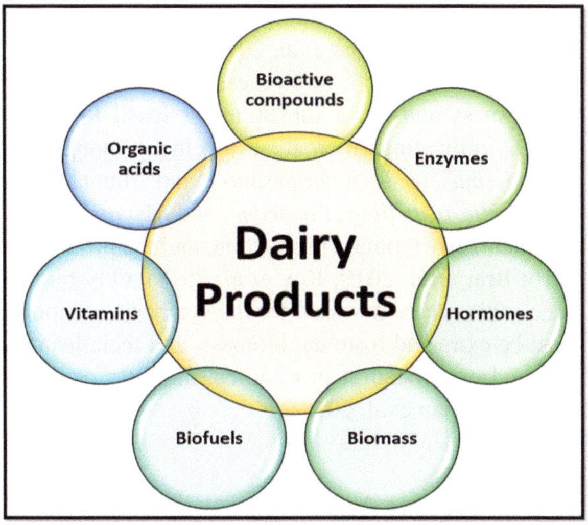

Fig. 23.3 Promising bioproducts made from dairy industry by-products employing microbial processes

2010). As a result, the anaerobic digestion method can reduce wastewater pollution caused by the dairy sector and save energy (Asunis et al. 2020).

Apart from biomethane, burning fossil fuels can also be replaced by biohydrogen. Since its combustion releases only water into the atmosphere and produces no greenhouse emissions, it is a clean energy source (Das 2009). However, carbon dioxide, the primary greenhouse gas, is produced throughout the standard hydrogen production process (Show et al. 2012). Anaerobic fermentation may be a good option in this situation instead of producing hydrogen from wastes that include a lot of organic material. Microbial electrohydrogenesis cells (MECs) represent an additional approach for the regulated synthesis of biohydrogen from cheese whey. Since low pH levels in this system can impair exoelectrogen function, pH management is necessary. Additionally, anaerobic digestion effluent is a preferable option for the microbial electrohydrogenesis process when it comes to the cheese whey's valorization (Rivera et al. 2017).

Currently, it is recommended to utilize inexpensive feedstocks to produce bioethanol, and spoiled milk or lactose-containing whey provides a benefit over lignocellulosic feedstocks because it doesn't require any particular pretreatment (Okamoto et al. 2019). Microbes which can naturally digest lactose or the enzyme β-galactosidase can be utilized to perform a chemical hydrolysis in order to produce bioethanol from dairy effluents rich in lactose (Guimarães et al. 2010). Nevertheless, the efficiency of producing ethanol is decreased by the diauxic proliferation of microbes in the presence of both galactose and glucose following hydrolysis (Panesar and Kennedy 2012). The most prevalent yeasts that can directly ferment lactose into ethanol are *Kluyveromyces marxianus* and *K. lactis* (Yamahata et al. 2020). Bioethanol generation from whey powder, whey, and expired milk is determined by several factors, including aeration rate, strain type, cell immobilization, substrate concentration, initial pH, temperature, and type of fermentation systems (Asunis et al. 2020; Sar et al. 2019).

Methyl ethyl esters, or biodiesel, are derived from plant, microalgal, or oily fungal biomass and are a substitute for fossil fuels (Mathew et al. 2021). In dairy wastes, cultivating microalgae (such as *Anabaena*, *Chlorella*, *Chlamydomonas*, *Scenedesmus*, and *Acutodesmus*) and fungal (such as *Thamnidium elegans*, *Mortierella isabellina*, *Fusarium*, and *Mucor* sp.) species can promote wastewater treatment, oily biomass production, and the production of bioenergy (Chokshi et al. 2016; Brar et al. 2019; Roy et al. 2018). Oily biomass that contains palmitic and oleic acids can be assessed for the synthesis of biodiesel in two stages: first, the oil must be extracted from the biomass, and then, using acid-catalyzed transesterification and esterification procedures, the fatty acid methyl esters (FAMEs) can be obtained (Chan et al. 2018).

23.5.2 Organic Acids

Organic acids are beneficial microbiological compounds that have an acidic characteristic. Lactic, acetic, succinic, propionic, oxalic, citric, valeric, and capric acids are the most prevalent organic acids (Louasté and Eloutassi 2020). These acids, often referred to as carboxylic acids, are rapidly becoming more and more in demand worldwide since they are valued as vital raw materials in many different industries. Thus, researchers are interested in discovering cheap, readily available, and renewable feedstock for high-yield organic acid synthesis, such as dairy wastes.

In the conventional process of producing acetic acid (vinegar), carbohydrates are fermented to ethanol, which is subsequently converted to acetic acid with the help of acetic acid bacteria. Nevertheless, because it requires a lot of energy, the aerobic phase of this process is not cheap. Therefore, it was suggested in previous decades to use *Clostridium* to synthesize acetic acid using whey employing an anaerobic homo-fermentative method. Though the estimated yield of acetate concentration in the above method was about 3%, in practice, the acetate output was just under 1% (Yang et al. 1992).

Natural flavoring and preservative lactic acid is utilized in the food, chemical, pharmaceutical, and textile sectors (Trakarnpaiboon et al. 2017). Waste products from the dairy business, including whey and skim milk, may be used as an appropriate medium for LAB culture. Numerous studies have demonstrated that when bacterial strains (*Streptococcus thermophiles*, *Lactobacillus rhamnosus*, *L. paracasei* subsp. *tolerans*, *L. casei*, and *L. lactis*) are cultivated on whey medium, the $L^{(+)}$ isomer of lactic acid rather than the $D^{(-)}$ isomer of lactic acid was synthesized (Pescuma et al. 2015; Ghasemi et al. 2017). Probiotics are made from lactic acid bacteria, which are naturally found in milk from cows, goats, sheep, and camels (Fguiri et al. 2016). To do this, dairy products such as expired milk and cheese whey have been utilized to generate lactic acid. But in order to encourage the proliferation of bacteria and maximize lactic acid yields, cheese whey must be enriched with nitrogen (peptones, yeast extract, corn steep, or molasses) and minerals (ammonium sulfate, potassium phosphate, magnesium sulfate, or manganese chloride). This is because the formulation of cheese whey is inadequate for the production of lactic acid (Panesar and Kennedy 2012).

Citrus fruits naturally contain citric acid, a typical organic acid utilized in the culinary, beverage, and pharmaceutical sectors. Furthermore, *Yarrowia lipolytica*, a well-known yeast species, possesses a unique capacity for synthesis and medium secretion. *Aspergillus niger* is a filamentous fungus that is commonly utilized to produce citric acid (Panesar and Kennedy 2012). Citric acid has many uses; hence, new, inexpensive process technologies are needed to increase production output. In general, high sugar raw materials are required for the fabrication of citric acid with great yields. Numerous low-cost waste materials, including date seeds, starch hydrolysates, molasses, apple pomace, and grape must, are used to produce citric acid due to their high carbohydrate content. The problems with garbage disposal and the industry's dependence on other manufacturers will be resolved by the advancements in the citric acid sector. Thus, the citric acid business will benefit both environmentally and financially (Sawant et al. 2018).

23.5.3 Enzyme

The statement that enzymes form the basis of biotechnological operations is not hyperbole. They have been used both directly and indirectly throughout human history. Enzyme production on an industrial scale began in the 1960s and expanded quickly. The majority of industrial enzymes are made by potentially genetically engineered microbes (Singhania et al. 2010). The synthesis of enzymes has been extensively studied, although it is still frequently an expensive process. The choice of substrates can have a big impact on production costs and enzyme production. Dairy sector by-products are among the inexpensive resources for the manufacture of enzymes (Ryan and Walsh 2016).

According to Oliveira et al. (2011), *Kluyveromyces* and *Aspergillus* species have historically shown an overwhelming capacity for producing β-galactosidase, or lactase. Although this enzyme can be produced by few microbes (*E. coli*, *Streptococcus*, *Leuconostoc*, *Lactobacillus*), *Bacillus* sp. has produced the enzyme that is commercially available (Oliveira et al. 2011). Plenty of investigations have been carried out regarding the possible application of whey and other dairy by-products to produce enzymes (Bosso et al. 2019). The β-galactosidase production rate can be affected by the starting pH and the incubation temperature of whey (Alves et al. 2019). The intracellular nature of lactase production is another disadvantage, which drives up the cost of downstream processing. Numerous studies were carried out in order to identify lactase purification techniques that are easy to use, practical, and reasonably priced (Bansal et al. 2008).

Dairy waste can also help in the fabrication of the α-amylase, an additional commercialized enzyme. It can be made from agro-dairy wastes such as cheese whey, wheat bran, hazelnut oil cake, and soybean cake by fermentation (Selen and Saban 2017). Furthermore, *Anoxybacillus beppuensis* may also manufacture a thermally stable amylase from dairy and agricultural wastes, which are a combination of sugarcane bagasse and whey (Jabeen et al. 2019). To the contrary, dairy wastes combined with leftovers from the fruit sector, including orange peels, might also be utilized to produce α-amylase (Uygut and Tanyildizi 2018).

23.5.4 Bioactive Compounds

In recent times, the significance of bioactive chemicals for human health has grown. As a result, a lot of work has gone into coming up with creative and affordable methods for producing them in big quantities (Kaur et al. 2020). Dairy products like whey can be used to produce healthy goods by microbial fermentation that are comparable to milk's bioactive peptides, which are used for the manufacture of functional and nutraceutical foods (Yadav et al. 2015).

Kluyveromyces marxianus was the only strain of bacteria and yeast that demonstrated a strong capacity to produce β-lactoglobulin from whey. Furthermore, whey has been transformed by *L. helveticus* and *S. cerevisiae* into a strong antihypertensive peptide. According to (Didelot et al. 2006), a coculture of *Lactobacillus paracasei* and *Candida parapsilosis* yielded a peptide (bioactive) that was resistant to pepsin and trypsin in vitro and comprised of tryptophan-leucine-alanine-histamine-lysine. This peptide had an ACE inhibitory function. Furthermore, the strain *of Pseudomonas taetrolens* is capable of oxidizing lactose to lactobionic acid (LBA), a bioactive substance. This polyhydroxy acid derivative of gluconic acid has been used in the medical, beauty, and food industries. It possesses acidulant, humectant, antioxidant, and anti-ageing qualities. The chemical processes required to create LBA are very costly, and side reactions could result in the development of dangerous by-products. Therefore, research is focused on the biodegradable prefabrication of LBA using reasonably priced raw materials. Cow scrota and cheese whey are examples of primal materials that are suitable as substrates for the microbial fermentation process that produces LBA (De Giorgi et al. 2018).

Whey can also be used to make antimicrobial peptides like bacteriocins. These substances might work well against either Gram-positive or Gram-negative bacteria or both. Two typical examples of bacteriocins generated in the whey-containing medium are nisin and pediocin. According to Ryan and Walsh (2016), these bacteriocins show promise as antibiotics for treating illnesses that are resistant to drugs. According to Panesar and Kennedy (2012), *Pediococcus acidilactici* NRRL B-5627 strain and *Lactococcus lactis* subsp. *lactis* CECT 539 may produce bacteriocin on diluted whey as a suitable substrate.

23.5.5 Single-Cell Protein

Several research studies have concentrated on microbial fermentation for the synthesis of single-cell proteins (SCPs). It could be fed to animals or consumed by humans since dairy wastes make suitable bioprocessing substrates (Ryan and Walsh

2016). *K. fragilis*, a yeast which utilizes lactose and is designated as generally regarded as safe (GRAS), was employed in the majority of research (Ghaly et al. 2005). Three yeasts were utilized in the industrial manufacture of SCP from deproteinized whey: *Candida pintolopesii*, and *K. marxianus* var. *marxianus* (Panesar and Kennedy 2012).

Through the culture of *Neurospora intermedia* and *Aspergillus oryzae* strains, the possible utilization of a variety of dairy wastes (such as milk, cheese whey, cream, and yogurt) as a substitute to the microbial feed utilized for fungal biomass synthesis was examined. *N. intermedia* and *A. oryzae* were reported to produce 7 g/L and 11 g/L of fungal biomass, respectively, from expired milk (Thunuguntla et al. 2018).

23.5.6 Biopolymers

For the manufacture of biopolymers, dairy waste is a suitable substrate (Tripathi et al. 2021). Biopolymers are useful in a multitude of applications. They are regarded as suitable substitutes for oil-based polymers and can have emulsifying, gel formation, biocatalyst, anticancer, and other properties (Zikmanis et al. 2020). According to Gironi and Piemonte (2011), certain biopolymers, like PLA and polyhydroxyalkanoates (PHAs), have mechanical and plastic qualities that are similar to those of oil-based plastics. Furthermore, they can be used for food packaging, health care applications, insulation, or disposable clothing and tableware (Ryan and Walsh 2016).

In a similar vein, exopolysaccharides (EPS) can be produced through bioprocessing dairy wastes. These extracellular biopolymeric materials are frequently utilized in food products to enhance the texture of the dish (Iliev et al. 2001). Zikmanis et al. (2020) have reported that whey and its derivatives can yield several bacterial extracellular polymer suspensions (EPS), such as dextran, alginate, xanthan, pullulan, curdlan, cellulose, or hyaluronic acid.

23.6 Conclusion

Due to their high lactose, protein, and fat levels, dairy sector by-products and wastes are considered a significant contaminant. These wastes have high COD values, and no appropriate treatment plan has been identified for them. The possibility of using dairy wastes as a carbon source in microbiological production methods has been thoroughly investigated (organic acids, biofuels, polymers, enzymes, and biomass), as well as for their suitability as a substrate. Dairy wastes can also be combined with other microbial feeds (like potato stem or cow dung) and incorporated into anaerobic digestion processes. Depending on the results of a techno-economic analysis, these codigestion procedures may prove profitable for the facilities involved.

References

Ahmad T, Aadil RM, Ahmed H et al (2019) Treatment and utilization of dairy industrial waste: a review. Trends Food Sci Technol 88:361–372

Alharbi S, Majrashi A, Ghoneim AM et al (2021) A new method to recycle dairy waste for the nutrition of wheat plants. Agronomy 11(5):840

Alves ÉP, Morioka LRI, Suguimoto HH (2019) Comparison of bioethanol and beta-galactosidase production by Kluyveromyces and saccharomyces strains grown in cheese whey. Int J Dairy Technol 72(3):409–415

Arefin MA, Nabi MN, McIntosh S (2020) Harnessing energy from Australian dairy waste: utilizing five methodologies. Biofuels Bioprod Biorefin 14(6):1180–1196

Arvanitoyannis IS, Giakoundis A (2006) Current strategies for dairy waste management: a review. Crit Rev Food Sci Nutr 46(5):379–390

Asunis F, De Gioannis G, Dessì P et al (2020) The dairy biorefinery: integrating treatment processes for cheese whey valorisation. J Environ Manag 276:111240

Atra R, Vatai G, Bekassy-Molnar E et al (2005) Investigation of ultra-and nanofiltration for utilization of whey protein and lactose. J Food Eng 67(3):325–332

Bansal S, Oberoi HS, Dhillon GS et al (2008) Production of β-galactosidase by Kluyveromyces marxianus MTCC 1388 using whey and effect of four different methods of enzyme extraction on β-galactosidase activity. Indian J Microbiol 48:337–341

Blaschek K, Wendorff W, Rankin S (2007) Survey of salty and sweet whey composition from various cheese plants in Wisconsin. J Dairy Sci 90(4):2029–2034

Bosso A, Setti ACI, Tomal AB et al (2019) Substrate consumption and beta-galactosidase production by Saccharomyces fragilis IZ 275 grown in cheese whey as a function of cell growth rate. Biocatal Agric Biotechnol 21:101335

Brar A, Kumar M, Pareek N (2019) Comparative appraisal of biomass production, remediation, and bioenergy generation potential of microalgae in dairy wastewater. Front Microbiol 10:678

Britz TJ, van Schalkwyk C, Hung Y-T (2004) Treatment of dairy processing wastewaters. In: Handbook of industrial and hazardous wastes treatment. CRC Press, Boca Raton, pp 673–705

Chan LG, Cohen JL, Ozturk G et al (2018) Bioconversion of cheese whey permeate into fungal oil by Mucor circinelloides. J Biol Eng 12(1):1–14

Chandra R, Castillo-Zacarias C, Delgado P et al (2018) A biorefinery approach for dairy wastewater treatment and product recovery towards establishing a biorefinery complexity index. J Clean Prod 183:1184–1196

Chang I, Im J, Chung M-K et al (2018) Bovine casein as a new soil strengthening binder from diary wastes. Constr Build Mater 160:1–9

Chokshi K, Pancha I, Ghosh A et al (2016) Microalgal biomass generation by phycoremediation of dairy industry wastewater: an integrated approach towards sustainable biofuel production. Bioresour Technol 221:455–460

Collivignarelli MC, Bertanza G, Abbà A et al (2019) Wastewater treatment by means of thermophilic aerobic membrane reactors: respirometric tests and numerical models for the determination of stoichiometric/kinetic parameters. Environ Technol 40(2):182–191

Comino E, Rosso M, Riggio V (2009) Development of a pilot scale anaerobic digester for biogas production from cow manure and whey mix. Bioresour Technol 100(21):5072–5078

Das D (2009) Advances in biohydrogen production processes: an approach towards commercialization. Int J Hydrog Energy 34(17):7349–7357

De Giorgi S, Raddadi N, Fabbri A et al (2018) Potential use of ricotta cheese whey for the production of lactobionic acid by pseudomonas taetrolens strains. New Biotechnol 42:71–76

Demirel B, Yenigun O, Onay TT (2005) Anaerobic treatment of dairy wastewaters: a review. Process Biochem 40(8):2583–2595

Didelot S, Bordenave-Juchereau S, Rosenfeld E et al (2006) Preparation of angiotensin-I-converting enzyme inhibitory hydrolysates from unsupplemented caprine whey fermentation by various cheese microflora. Int Dairy J 16(9):976–983

Fguiri I, Ziadi M, Atigui M et al (2016) Isolation and characterisation of lactic acid bacteria strains from raw camel milk for potential use in the production of fermented Tunisian dairy products. Int J Dairy Technol 69(1):103–113
Ghaly A, Kamal M, Correia L (2005) Kinetic modelling of continuous submerged fermentation of cheese whey for single cell protein production. Bioresour Technol 96(10):1143–1152
Ghasemi M, Ahmad A, Jafary T et al (2017) Assessment of immobilized cell reactor and microbial fuel cell for simultaneous cheese whey treatment and lactic acid/electricity production. Int J Hydrog Energy 42(14):9107–9115
Gironi F, Piemonte V (2011) Bioplastics and petroleum-based plastics: strengths and weaknesses. Energy Sour Part A 33(21):1949–1959
Gramegna G, Scortica A, Scafati V et al (2020) Exploring the potential of microalgae in the recycling of dairy wastes. Bioresour Technol Rep 12:100604
Guimarães PM, Teixeira JA, Domingues L (2010) Fermentation of lactose to bio-ethanol by yeasts as part of integrated solutions for the valorisation of cheese whey. Biotechnol Adv 28(3):375–384
Iliev I, Radoilska E, Ivanova I et al (2001) Biosynthesis of exopolysaccharides by two strains of lactobacillus bulgaricus in whey-based media. Mededelingen 66(3):511–516
Jabeen F, Hussain A, Younis T et al (2019) Isolation of thermophilic Anoxybacillus beppuensis JF84 and production of thermostable amylase utilizing agro–dairy wastes. Environ Prog Sustain Energy 38(2):417–423
Kaur J, Kumar V, Sharma K et al (2020) Opioid peptides: an overview of functional significance. Int J Pept Res Ther 26:33–41
Kavacik B, Topaloglu B (2010) Biogas production from co-digestion of a mixture of cheese whey and dairy manure. Biomass Bioenergy 34(9):1321–1329
Kavitha V, Geetha V, Jacqueline PJ (2019) Production of biodiesel from dairy waste scum using eggshell waste. Process Saf Environ Prot 125:279–287
Kolev SA (2017) General characteristics and treatment possibilities of dairy wastewater–a review. Food Technol Biotechnol 55(1):14–28
Kozłowski K, Pietrzykowski M, Czekała W et al (2019) Energetic and economic analysis of biogas plant with using the dairy industry waste. Energy 183:1023–1031
Lawton MR, Dana L, Alcaine SD (2021) Lactose utilization by Brettanomyces claussenii expands potential for valorization of dairy by-products to functional beverages through fermentation. Current opinion in food. Science 42:93–101
Li Y, Qi C, Zhang Y et al (2021) Anaerobic digestion of agricultural wastes from liquid to solid state: performance and environ-economic comparison. Bioresour Technol 332:125080
Liu P, Zheng Z, Xu Q et al (2018) Valorization of dairy waste for enhanced D-lactic acid production at low cost. Process Biochem 71:18–22
Louasté B, Eloutassi N (2020) Succinic acid production from whey and lactose by Actinobacillus succinogenes 130Z in batch fermentation. Biotechnol Rep 27:e00481
Mathew GM, Raina D, Narisetty V et al (2021) Recent advances in biodiesel production: challenges and solutions. Sci Total Environ 794:148751
Mishra A, Kumar M, Bolan NS et al (2021) Multidimensional approaches of biogas production and up-gradation: opportunities and challenges. Bioresour Technol 338:125514
Okamoto K, Nakagawa S, Kanawaku R et al (2019) Ethanol production from cheese whey and expired milk by the brown rot fungus Neolentinus lepideus. Fermentation 5(2):49
Oliveira C, Guimarães PM, Domingues L (2011) Recombinant microbial systems for improved β-galactosidase production and biotechnological applications. Biotechnol Adv 29(6):600–609
Omil F, Garrido JM, Arrojo B et al (2003) Anaerobic filter reactor performance for the treatment of complex dairy wastewater at industrial scale. Water Res 37(17):4099–4108
Ostojić S, Pavlović M, Živić M et al (2005) Processing of whey from dairy industry waste. Environ Chem Lett 3(1):29–32
Panesar PS, Kennedy JF (2012) Biotechnological approaches for the value addition of whey. Crit Rev Biotechnol 32(4):327–348

Pasotti L, De Marchi D, Casanova M et al (2020) Engineering endogenous fermentative routes in ethanologenic Escherichia coli W for bioethanol production from concentrated whey permeate. New Biotechnol 57:55–66

Pescuma M, de Valdez GF, Mozzi F (2015) Whey-derived valuable products obtained by microbial fermentation. Appl Microbiol Biotechnol 99:6183–6196

Rivas J, Prazeres AR, Carvalho F et al (2010) Treatment of cheese whey wastewater: combined coagulation– flocculation and aerobic biodegradation. J Agric Food Chem 58(13):7871–7877

Rivera I, Bakonyi P, Cuautle-Marín MA et al (2017) Evaluation of various cheese whey treatment scenarios in single-chamber microbial electrolysis cells for improved biohydrogen production. Chemosphere 174:253–259

Roufou S, Griffin S, Katsini L et al (2021) The (potential) impact of seasonality and climate change on the physicochemical and microbial properties of dairy waste and its management. Trends Food Sci Technol 116:1–10

Roy M, Kumar R, Ramteke A et al (2018) Identification of lipase producing fungus isolated from dairy waste contaminated soil and optimization of culture conditions for lipase production by the isolated fungus. J Microbiol Biotechnol Food Sci 8(1):698

Ryan MP, Walsh G (2016) The biotechnological potential of whey. Rev Environ Sci Biotechnol 15:479–498

Sar T, Stark BC, Akbas MY (2019) Bioethanol production from whey powder by immobilized E. coli expressing Vitreoscilla hemoglobin: optimization of sugar concentration and inoculum size. Biofuels 12:1103–1108

Sawant O, Mahale S, Ramchandran V et al (2018) Fungal citric acid production using waste materials: a mini-review. J Microbiol Biotechnol Food Sci 8(2):821–828

Selen V, Saban TM (2017) Optimization of low-cost medium composition for the production of α-amylase by bacillus amyloliquefaciens using a semi-solid substrate. Curr Phys Chem 7(4):305–312

Show K, Lee D, Tay J et al (2012) Biohydrogen production: current perspectives and the way forward. Int J Hydrog Energy 37(20):15616–15631

Singhania RR, Patel AK, Pandey A (2010) The industrial production of enzymes. Ind Biotechnol 12:207–225

Suman G, Nupur M, Anuradha S et al (2017) Characterization of dairy waste and its utilisation as substrate for production of single cell protein. J Biotechnol Biochem 3:73–78

Swain AK, Sahoo A, Jena HM et al (2018) Industrial wastewater treatment by aerobic inverse fluidized bed biofilm reactors (AIFBBRs): a review. J Water Process Eng 23:61–74

Thunuguntla R, Mahboubi A, Ferreira JA et al (2018) Integration of membrane bioreactors with edible filamentous fungi for valorization of expired milk. Sustain For 10(6):1940

Tostivint C, de Veron S, Jan O et al (2017) Measuring food waste in a dairy supply chain in Pakistan. J Clean Prod 145:221–231

Trakarnpaiboon S, Srisuk N, Piyachomkwan K et al (2017) L-lactic acid production from liquefied cassava starch by thermotolerant Rhizopus microsporus: characterization and optimization. Process Biochem 63:26–34

Tripathi AD, Paul V, Agarwal A et al (2021) Production of polyhydroxyalkanoates using dairy processing waste–a review. Bioresour Technol 326:124735

Uygut MA, Tanyildizi MŞ (2018) Optimization of alpha-amylase production by bacillus amyloliquefaciens grown on orange peels. Iran J Sci Technol Trans A Sci 42:443–449

Vialkova E, Sidorenko O, Glushchenko E (2019) Qualitative composition and local pretreatment of dairy wastewaters. IOP Conf Ser Mater Sci Eng 687:066049

Vlyssides AG, Tsimas ES, Barampouti EMP et al (2012) Anaerobic digestion of cheese dairy wastewater following chemical oxidation. Biosyst Eng 113(3):253–258

Wainaina S, Lukitawesa, Kumar Awasthi M et al (2019) Bioengineering of anaerobic digestion for volatile fatty acids, hydrogen or methane production: a critical review. Bioengineered 10(1):437–458

Yadav JSS, Yan S, Pilli S et al (2015) Cheese whey: a potential resource to transform into bioprotein, functional/nutritional proteins and bioactive peptides. Biotechnol Adv 33(6):756–774

Yamahata N, Toyotake Y, Kunieda S et al (2020) Application of multiple sensory evaluations to produce fermented beverages made from sole whey using Kluyveromyces marxianus. Int J Food Sci Technol 55(4):1698–1704

Yang S-T, Tang I-C, Zhu H (1992) A novel fermentation process for calcium magnesium acetate (CMA) production from cheese whey. Appl Biochem Biotechnol 34:569–583

Yonar T, Sivrioğlu Ö, Özengin N (2018) Physico-chemical treatment of dairy industry wastewaters: a review. In: Technological approaches for novel applications in dairy processing. IntechOpen, London, p 179

Zheng L, Xu Y, Geng H et al (2022) Enhancing short-term ethanol-type fermentation of waste activated sludge by adding saccharomycetes and the implications for bioenergy and resource recovery. J Environ Sci 113:179–189

Zikmanis P, Kolesovs S, Semjonovs P (2020) Production of biodegradable microbial polymers from whey. Bioresour Bioprocess 7(1):1–15

Valorization of Leather Industry Waste to Biofuel and Chemicals

24

Vishal Kumar Sandhwar, Abhishek Kumar Lal,
Shivendu Saxena, Diksha Saxena, Siddharth Pandey,
and Alok Tiwari

Abstract

The leather industry, a cornerstone of global manufacturing, faces significant environmental challenges due to the substantial waste it generates. Traditional disposal methods exacerbate pollution and resource depletion, necessitating sustainable solutions. Valorization of leather industry waste offers a promising approach, transforming discarded materials into valuable biofuels and chemicals through processes such as anaerobic digestion, pyrolysis, and transesterification. These techniques not only minimize waste but also produce renewable energy sources like biogas and bio-oil. Advances in technology enhance process efficiency and broaden the range of valuable products, including collagen for pharmaceuticals and bioplastics from leather fibers. Despite challenges, such as technological scalability and environmental impact assessment, ongoing research and investment promise a sustainable future where leather waste contributes to a circular economy and mitigates environmental impact.

Keywords

Leather industry waste · Valorization · Biofuels · Chemicals · Sustainability

V. K. Sandhwar · S. Saxena · D. Saxena · S. Pandey · A. Tiwari
Department of Chemical Engineering, Parul Institute of Technology, Parul University, Vadodara, Gujarat, India

A. K. Lal (✉)
Department of Leather Technology, School of Chemical Technology, HBTU, Kanpur, Uttar Pradesh, India
e-mail: abhisheklt@hbtu.ac.in

A. Kumari et al. (eds.), *Valorization of Solid Wastes to Biofuels and Chemical Products for Sustainable World*, https://doi.org/10.1007/978-981-96-8594-3_24

24.1 Introduction

The leather industry, among the oldest and most influential sectors globally, plays a crucial role in the economy by producing a vast array of products, from footwear and apparel to automotive upholstery and accessories. However, the sector is also a major benefactor to environmental pollution due to the substantial waste it generates. Traditional waste management practices, such as landfilling and incineration, exacerbate environmental degradation and resource depletion, necessitating innovative and sustainable solutions. This discussion focuses on the valorization of leather industry waste, a process that transforms waste materials into valuable products like biofuels and chemicals, thereby addressing environmental concerns and promoting economic benefits. Valorization involves various methods to convert leather industry waste into biofuels and chemicals. Anaerobic digestion, pyrolysis, and transesterification are commonly used to produce biofuels, while processes like collagen extraction, hydrolysis, and the production of composite materials yield valuable chemicals and bioproducts. By adopting these methods, the leather industry can align with circular economy philosophies, elevating sustainability and resource efficiency. For example, organic waste breaks down in the absence of oxygen in anaerobic digestion to produce renewable energy sources and biogas. Pyrolysis thermally decomposes organic material to generate bio-oil, which can be further purified into biofuels, while transesterification converts fats and oils from leather waste into biodiesel. These processes not only minimize the volume of waste but also produce renewable energy, promoting a greener and more sustainable industry (Kandasamy et al. 2020).

Recent advancements in technology and growing environmental awareness have led to important progress in the valorization of waste in the leather industry. Innovations in biotechnological processes and chemical engineering have improved the efficiency and feasibility of converting waste into biofuels and chemicals. Enhanced anaerobic digestion techniques have increased biogas yields, while advancements in pyrolysis have optimized bio-oil production. The development of biocomposites and biopolymers from leather waste has opened new possibilities for sustainable material science, creating environmentally friendly alternatives to conventional plastics. Additionally, governments and industry stakeholders are increasingly investing in research and development to scale these technologies and integrate them into mainstream waste management practices. For instance, the global market for biofuels has been growing at an annual rate of 5.5%, indicating a strong demand for sustainable energy solutions. This collaborative effort aims to create a sustainable and economically viable leather industry that minimizes environmental impact (Kandasamy et al. 2020; Rathinam and Sani 2019).

Despite these promising advancements, several challenges remain in the valorization of leather industry waste. One major challenge is the requirement for more economical and scalable technologies capable of managing the diverse composition of leather waste. The complexity and variability of waste materials require tailored approaches to ensure efficient conversion processes. Moreover, comprehensive studies on the long-term environmental impacts and economic viability of these valorization methods are lacking. Understanding the full life cycle impacts of these processes is crucial for developing sustainable solutions. Future research is also required to explore the potential of evolving technologies like enzymatic hydrolysis and advanced gasification techniques, which could offer more effective and eco-friendly substitutes to existing methods. Addressing these challenges through inter-disciplinary research and collaboration can pave the way for more sustainable and cost-effective solutions for the management of waste in leather sector (Rathinam and Sani 2019).

24.1.1 Generation Status of Leather Industry Waste around the World

The leather industry is a significant contributor to global manufacturing, producing essential goods such as footwear, clothing, upholstery, and accessories. However, the production processes generate a substantial amount of waste, posing environmental and economic challenges. Leather industry waste can be broadly categorized into solid, liquid, and gaseous emissions, each with distinct characteristics and impacts.

24.1.1.1 Solid Waste

Solid waste from the leather industry primarily includes:

- Animal hides and skin trimmings: Generated during the pretanning processes.
- Fleshings: Residual animal tissue removed during fleshing.
- Shavings and splits: Produced during the trimming and finishing stages.
- Tanned leather scraps: Resulting from cutting and shaping finished products.

Globally, it is projected that the leather industry produced more than 20 million tons of solid waste annually. Regions with large-scale leather production, such as Asia, Europe, and South America, are the primary contributors to this waste stream. India and China, as major leather exporters, produce significant quantities of both rawhide and finished leather waste.

The tanning sector globally generates a considerable volume of effluents and solid waste. Environmental concerns related to waste discharge and the growing cost of landfills have become increasingly urgent issues. The management of these wastes is complex, with limited effective utilization methods currently available. Huge amounts of solid waste are generated through various steps of leather processing, including splitting, fleshing, trimming, and shaving, as well as sludge from wastewater treatment plants. These collectively contribute to the growing volume of waste. On average, from 1 ton of rawhide, approximately 0.8 tons of solid waste is produced, with only 0.2 tons converted into usable products. Worldwide, leather industries generate about 600,000 tons of solid waste annually (Cotance 2012;

Yorgancioglu et al. 2020). Rao et al. provide an overview of the variety and amount of solid waste generated during the processing of 1 ton of raw hides or skins (Rao et al. 2004).

24.1.1.2 Liquid Waste

Liquid waste, predominantly in the form of wastewater, originates from:

- Soaking, liming, and deliming processes: High in organic matter, sulfides, and ammonia.
- Tanning and dyeing: Containing chromium salts, dyes, and other heavy metals.

Leather tanning processes generate approximately 30–40 l of wastewater per kilogram of rawhide processed. The effluents are often laden with pollutants that can lead to severe water contamination if untreated.

24.1.1.3 Gaseous Emissions

The industry also produces various gases, including volatile organic compounds (VOCs) and ammonia, through tanning and finishing. These emissions contribute to air pollution and are hazardous to workers and nearby communities.

24.1.1.4 Environmental Impact of Waste Generated from Leather Industry

The improper management of leather industry waste leads to significant environmental consequences: Untreated tannery wastewater is often disposed of on land as a containment measure. However, this practice severely impacts soil quality. Soils that directly absorb such waste or are irrigated with contaminated groundwater lose their productivity due to the accumulation of toxic substances. Tanneries are notorious for generating malodorous emissions. The rehydration of salted hides and skins releases volatile fatty acids and amino acids due to biological decomposition in the presence of water. Additionally, hydrogen sulfide, a toxic gas produced during liming, deliming, and tanning processes, significantly contributes to air pollution within tanneries. The venting of these malodorous substances into the atmosphere and their subsequent dispersion over long distances exacerbate the issue.

- Hydrogen sulfide (H_2S): At concentrations of 20 ppm (30 mg/m^3) in ambient air, H_2S is lethal to humans.
- Ammonia (NH_3): Released during deliming operations, ammonia is pungent and odorous. The maximum permissible level of ammonia in air is 50 mg/m^3.
- Phenolics: Monohydric, dihydric, and trihydric phenolics are emitted during post-tanning and finishing operations. The permissible level of phenol in water is 0.0102 g/L, and it must not exceed 0.0003 g/L in drinking water.

Moreover. the wastewater discharged from tanneries contains harmful levels of sulfide, ammonia, phenol, and chromium, which are toxic to freshwater fish and other aquatic organisms. These pollutants pose a significant threat to biodiversity and the health of aquatic ecosystems (Mariappan 1997; Kanagaraj et al. 2006).

24.2 Sources and Composition of Leather Industry Waste

Leather industry waste is produced at various steps of production, each contributing to the overall waste stream. Initially, during the preparation phase, raw hides and skins undergo processes such as soaking, liming, fleshing, and unhairing. This phase is essential for removing extraneous materials such as hair, flesh, and dirt, resulting in significant solid waste primarily composed of organic materials. Approximately 50–60% of the weight of raw hides and skins is discarded as waste during this stage. This waste includes trimmings, fleshings, and hair, which, if not managed properly, can decompose and produce methane, a potent greenhouse gas. The preparation stage also generates wastewater containing high levels of organic matter, suspended solids, and chemicals used in the soaking and liming processes. Proper management and disposal of waste are crucial to minimize environmental pollution (Chojnacka et al. 2021).

The tanning process, which transforms raw hides into stable and durable leather, is another major source of waste. Chrome tanning, the most prevalent method, uses chromium salts to stabilize the collagen fibers in the hides. This process generates both solid and liquid waste streams, including shavings, trimmings, and sludge, as well as effluents rich in chemicals and organic matter. Up to 20–25% of the total weight of raw hides is lost as solid waste during tanning. The solid waste contains a mix of organic materials (collagen, fats, and proteins) and inorganic substances (chromium and other tanning agents). Chromium, in particular, poses significant environmental and health risks if not properly treated, as it can oxidize from the less toxic chromium (III) to the more toxic chromium (VI). Liquid waste from tanning contains high levels of chromium, sulfates, and other chemicals, necessitating advanced treatment technologies such as adsorption and chemical precipitation to remove contaminants. Beyond the tanning stage, finishing processes like dyeing and coating further contribute to waste generation, adding dyes, pigments, and finishing agents to the waste stream. The composition of leather industry waste is diverse, encompassing organic matter, chemicals, and heavy metals, each requiring specialized management and valorization strategies to mitigate environmental impact and unlock potential value (Chojnacka et al. 2021).

24.3 Biofuel Production from™ Leather Waste

S. No.	Process	Description	Type of biofuel produced	Yield	Additional products	References
1	Anaerobic digestion	Biological process by microbes breaking down organic waste in absence of oxygen.	Biogas (methane, CO_2)	200–300 m^3 methane per ton of organic waste	Digestate (used as fertilizer)	Amdouni et al. (2021)
2	Anaerobic digestion	Anaerobic digestion of chrome-tanned leather waste	Biogas	Biogas yield (162.2 mL g^{-1}) and methane fraction (73.7%)	–	Priebe et al. (2016)
3	Pyrolysis	Thermochemical decomposition of organic material at high temperatures in absence of oxygen	Bio-oil, syngas	40–60% bio-oil yield of dry feedstock	Biochar (used as soil amendment)	Appala et al. (2024)
4	Pyrolysis	Pyrolysis of tannery fleshing wastes	Bio-oil, syngas	60% bio-oil, 35.9 wt% syngas	Biochar (biofertilizer or low-cost dye adsorbent)	Amdouni et al. (2021)
5	Transesterification	Chemical reaction converting fats and oils into biodiesel using an alcohol and a catalyst	Biodiesel	80–90% of feedstock weight	Glycerol (used in various industries)	Mozhiarasi et al. (2021)
6	Transesterification	Biodiesel production from tannery fleshing waste	Biodiesel	12.05–26.7% yield	–	Dagne et al. (2019)
7	Transesterification	Biodiesel production from tannery sludge	Biodiesel	89.7% biodiesel yield	–	Kumar and Thakur (2020)
8	Gasification	Partial oxidation of organic waste at high temperatures to produce syngas	Syngas	Variable, depending on feedstock and process	Ash (potential use as construction material)	Hari et al. (2023)
9	Gasification	Gasification of leather waste at both laboratory and industrial scales	Producer gas	Producer gas with a calorific value of 4.1–6.5 MJ/m^3	–	Dudyński et al. (2021)

24.3.1 Anaerobic Digestion

A biological process called anaerobic digestion makes it easier for organic waste-including leftovers of the leather industry to break down without oxygen. This process occurs naturally and involves a series of microbial-mediated stages within sealed digesters. Leather waste, rich in collagen, fats, and proteins, undergoes sequential transformations facilitated by anaerobic bacteria:

Hydrolysis is the first step in the process, which breaks down complex chemical molecules like proteins and fats into simpler ones like fatty acids and amino acids. Through the process of acidogenesis, these substances are subsequently transformed into volatile fatty acids, alcohols, and gases such as carbon dioxide and hydrogen. Volatile fatty acids undergo further metabolism in the following stage of acetogenesis, producing hydrogen, carbon dioxide, and acetic acid. Finally, methanogenesis results in the production of methane (CH_4) and additional carbon dioxide from acetic acid and hydrogen. The output of anaerobic digestion includes biogas, typically composed of 50–70% CH_4 and 30–40% CO_2, along with trace amounts of other gases. Leather waste can yield approximately 200–300 cubic meters of methane per ton of organic material processed. Biogas has diverse applications, including electricity generation, heating, and as a vehicle fuel. The remaining material after digestion, known as digestate, is nutrient-rich and can be used as an organic fertilizer. This not only reduces the volume of organic waste but also enhances soil fertility, minimizing the need for chemical fertilizers and supporting sustainable agricultural practices. Anaerobic digestion thus serves as a dual-purpose solution, managing organic waste effectively while contributing to renewable energy production and environmental sustainability (Amdouni et al. 2021). The production of biogas from the anaerobic digestion of waste from chrome-tanned leather was examined by Priebe et al. (2016). Through the production of clean fuel (biogas) from tannery waste, this study explores ways to improve the tanning industry's environmental performance. In order to ascertain whether pretreatment or mixing with other materials is necessary prior to anaerobic decomposition, it evaluates the biodegradability of collagenic substrates, and whether or not they have been tanned. The production of biogas from four substrates-soybean meal, hydrolyzed collagen, hide powder, and wet-blue leather shavings-inoculated with three biological sludges was investigated in bioreactor studies. Higher chromium concentrations in substrates decreased methane and biogas output, according to the results. For chrome leather shavings, the highest biogas yield (0.1622 L/g) and methane proportion (73.7%) were recorded in 3–36 days. Slaughterhouse and tannery sludges performed better due to preadaptation to collagen-based substrates. Mixing leather substrates with readily degradable materials improved biogas production by 74–181% by destabilizing chrome–collagen complexes and enhancing substrate degradability (Priebe et al. 2016).

24.3.2 Pyrolysis

Pyrolysis is a thermochemical process that breaks down organic materials, such as leather waste, at elevated temperatures (typically 300–700 °C) in the absence of oxygen. This process can be tailored into two main methods: fast pyrolysis and slow pyrolysis:

Fast pyrolysis involves rapidly heating leather waste to around 500 °C, maximizing the yield of bio-oil, which constitutes approximately 40–60% of the feedstock's dry weight. This bio-oil is a valuable liquid fuel that can be further refined into transportation fuels or used directly in heating applications (Appala et al. 2024). Alongside bio-oil, fast pyrolysis also produces syngas, a mixture of hydrogen, carbon monoxide, methane, and other gases, which can serve as a versatile energy source for various industrial processes. In contrast, slow pyrolysis employs lower heating rates and longer residence times, resulting in a higher yield of biochar a solid residue rich in carbon. Biochar enhances soil fertility by improving water retention, nutrient availability, and microbial activity, making it an effective soil amendment and carbon sequestration tool. The applications of pyrolysis are manifold: bio-oil can substitute conventional fossil fuels in boilers, engines, and industrial processes, reducing greenhouse gas emissions. Syngas is used for electricity generation or further processed into synthetic fuels, supporting energy security and sustainability. Biochar contributes to sustainable agriculture by enhancing soil health and mitigating climate change through carbon sequestration. Pyrolysis thus offers a comprehensive approach to valorizing leather waste, converting it into valuable biofuels and soil enhancers while promoting resource efficiency and environmental stewardship (Bhatt et al. 2021). Amdouni et al. (2021) had explored the pyrolysis of tannery fleshing wastes (TFW) for biofuel and biochar production. At optimal conditions (600 °C, 15 °C/min), the highest bio-oil yield (60 wt%) with excellent fuel properties (HHV ~39.36 MJ/kg) was achieved, suitable for liquid fuel applications. The resulting char has potential as a biofertilizer or low-cost dye adsorbent. Syngas yield (35.9 wt%) at 700 °C showed potential as a gaseous biofuel. This approach reduces TFW management costs, generates valuable products, and promotes sustainable practices for the tannery industry.

24.3.3 Transesterification

A chemical process called transesterification turns the fats and oils found in leather waste into glycerol and biodiesel. In this reaction, fats and oils are combined with an alcohol, usually methanol or ethanol, while a catalyst, usually sodium or potassium hydroxide, is present. Triglycerides are esterified to produce biodiesel methyl or ethyl esters of fatty acids and glycerol as part of the transesterification process. Biodiesel, which makes up 80–90% of the feedstock weight, is appropriate for use in diesel engines that have not been modified since it shares characteristics with regular diesel fuel, such as high cetane numbers and low sulfur levels. The transesterification by-product glycerol is used in the food processing, cosmetics, and pharmaceutical industries, increasing the process' economic feasibility. Transesterification contributes significantly to sustainable energy solutions by reducing dependency on fossil fuels and mitigating greenhouse gas emissions from transport sectors. Biodiesel derived from leather waste supports circular economy practices by utilizing renewable resources and reducing environmental impact. The versatility of glycerol expands its industrial applications, adding value to the overall biodiesel production process. Transesterification exemplifies a practical approach to valorizing leather industry waste, transforming it into renewable fuels and beneficial by-products while promoting environmental stewardship and economic growth (Mozhiarasi et al. 2021). A study explores biodiesel production from tannery fleshing waste. Delimed, dried, and chopped waste was extracted for oil, yielding 12.05–26.7% depending on the source. Using a KOH-catalyzed transesterification process, biodiesel conversion exceeded 96% under optimized conditions: 6:1 methanol-to-oil molar ratio, 1% catalyst, and 60 °C for 1 h. The biodiesel produced has low emissions and performance similar to diesel fuel, offering a sustainable solution to manage costly and troublesome tannery waste disposal while addressing the rising demand for diesel alternatives (Dagne et al. 2019). Another study evaluated the effect of different solvents on lipid extraction efficiency and biodiesel production from tannery sludge. The chloroform-methanol method yielded the highest lipid extraction efficiency of 98.5%, corresponding to a lipid yield of 22.8% (w/v). Optimized in situ transesterification conditions—such as 3% catalyst concentration, a sludge-to-methanol ratio of 1:25 (w/v), 60 °C reaction temperature, and 12-h reaction time—achieved an 89.7% biodiesel yield. The FAME profile, analyzed using FT-IR, NMR, and GC-MS, confirmed a purity of over 97.12%. The physicochemical properties of 5% biodiesel blends met ASTM standards. The study concluded that biodiesel production from tannery sludge is techno-economically viable due to its high lipid yield and low-cost feedstock (Kumar and Thakur 2020).

24.3.4 Gasification

Gasification is a thermochemical process which transforms organic waste, such as leather waste, into syngas through partial oxidation at high temperatures (700–1500 °C) and controlled supply of oxygen or steam. This process offers flexibility in handling diverse feedstocks and produces a mixture of gases:

Organic materials undergo partial oxidation within a gasifier, yielding syngas a combination of carbon monoxide, hydrogen, methane, and other hydrocarbons depending on the specific feedstock and operational conditions. Syngas serves as a versatile energy source for electricity generation, industrial processes, and chemical synthesis, contributing to energy security and sustainability. The residual solid material, known as ash, contains minerals and carbon and can be utilized in construction materials, such as concrete or road surfaces, reducing waste and promoting resource efficiency. Gasification technology encompasses various approaches, including fixed-bed, fluidized-bed, and plasma gasification, each offering distinct advantages in terms of efficiency and product quality. Fixed-bed gasifiers are robust and simpler to operate but may produce higher tar content, whereas fluidized-bed gasifiers offer better mixing and heat transfer, resulting in cleaner syngas production. Plasma gasification employs a plasma arc to achieve extremely high temperatures, facilitating clean syngas production with minimal emissions and ash residues. Gasification of leather waste exemplifies an innovative method for the management of waste and renewable energy production, transforming organic residues into valuable energy resources and construction materials while addressing environmental challenges and supporting sustainable development goals (Hari et al. 2023). The gasification of leather waste at both laboratory and industrial scales converts 1000 kg/h of waste into syngas with a calorific value of 4.1–6.5 MJ/m^3 for steam generation. The process produces ash with up to 55% chromium (III) oxide for chromium recovery and shows low Cr (VI) content, demonstrating environmental safety. Flue gas purification using $Ca(OH)_2$ effectively removes SO_2 and HCl. Compared with incineration, gasification proves more efficient, producing valuable syngas and safer ash, supporting a closed chromium recovery cycle in the leather industry (Dudyński et al. 2021).

The leather industry generates a considerable amount of fat-rich liquid and solid waste during the proceedings of hides and skins. If not managed properly, these wastes can lead to significant challenges in the ecosystem. Solid waste originates from pretanning processes, including trimming, prefleshing, fleshing, and shaving. Prefleshing involves the removal of flesh to facilitate the penetration of chemicals, resulting in waste rich in fat and protein (İşler et al. 2010). The high-fat content of leather industry waste makes it a valuable but underutilized resource, as there are currently limited methods for its recovery. One promising approach is to valorize these wastes and use them as feed generation of biodiesel due to their high-fat content (Colak et al. 2005; Özgünay et al. 2007; Alptekin et al. 2012). By converting these wastes into biodiesel, the environmental impact of leather industry waste can be mitigated, and valuable products can be generated, especially in countries with a well-developed leather industry, such as Turkey (Alptekin et al. 2012). Several studies have explored the potential of leather waste-derived fat for biodiesel production. For example, İşler et al. (2010) looked into the usage of fleshing waste as a biodiesel oil source. In their investigation, fleshing fat with an acid value of 0.28 mg potassium hydroxide (KOH)/g was converted to biodiesel by transesterification with methanol and an alkali catalyst. The generated biodiesel met the EN 14214 fuel property criteria, with the exception of cold filter plugging point (CFPP) and oxidative stability. Colak et al. (2005) retrieved fat from prefleshing sheep skins using boiling with water and high-speed grinding. Using methanol and potassium methoxide (30% in methanol) as a catalyst, they achieved a high ester yield of up to 95%. Similarly, Alptekin et al. (2012) utilized fat from fleshing waste with an acid value of 24.30 mg KOH/g (equivalent to a free fatty acid level of approximately 12.15%) to generate methyl esters. Due to the high free fatty acid content, a pretreatment reaction using sulfuric acid and methanol was performed to reduce the acid value to below 1%. The subsequent transesterification reaction with an alkaline catalyst produced biodiesel that complied with standards, except for CFPP, ester, and sulfur content. Additionally, Özgünay et al. (2007) evaluated the engine performance and emissions of biodiesel derived from prefleshing waste in comparison to petroleum diesel. Their findings indicated that biodiesel significantly reduced hydrocarbon and particulate matter emissions, though its heating value was slightly lower than that of petroleum diesel. They concluded that waste by-products from the leather industry could be used to generate an environmentally friendly fuel, suitable for certain diesel engines without major modifications, providing both economic and ecological advantages. These studies collectively highlight the potential of leather industry waste as a sustainable feedstock for biodiesel production, offering a pathway to decrease pollution and produce value-added products.

24.4 Extraction Techniques for Collagen and Gelatin

Leather waste, a by-product of the leather industry primarily composed of animal hides, contains collagen as its main protein component. Collagen extraction from leather waste involves several techniques aimed at separating and purifying this valuable protein. One of the commonly employed methods is enzymatic extraction. Enzymes, such as proteases, are used to break down the collagen fibers into smaller peptide fragments and amino acids. Typical yields range from 50% to 70% of collagen content, depending on enzyme efficiency and treatment conditions (e.g., pH 6.5–7.5, 30–40 °C, and 12–24 h). Another method involves acidic extraction, where the leather waste is treated with acids such as hydrochloric acid or acetic acid. Acidic conditions disrupt the collagen cross-links and solubilize collagen and gelatin into the solution. This method is effective but requires careful control to prevent excessive degradation and ensure the desired quality of extracted proteins (Šimat 2021).

- Because of their various properties such as anti-ageing, moisturizing, and wound-healing, collagen and gelatin are used in skin care products in the cosmetic industry. These proteins enhance skin elasticity and texture, making them valuable ingredients in creams, lotions, and serums aimed at rejuvenating and repairing damaged skin. Typical formulations contain 0.3–5% collagen/gelatin for skin elasticity and rejuvenation (Yorgancioglu et al. 2020).
- The pharmaceutical sector also benefits from collagen and gelatin derived from leather waste. Collagen is used in medical exercises such as wound dressings, scaffolds for tissue engineering, and drug delivery systems because of its biocompatibility and capacity to stimulate tissue regeneration. Gelatin finds application in encapsulating drugs and supplements, ensuring controlled release and improved bioavailability (Reddy et al. 2021).
- In the food industry, gelatin serves as a versatile ingredient in different food products, such as desserts, confectionery, and dairy products. Its thickening, gelling, and stabilizing properties make gelatin a preferred additive in food formulations, contributing to texture enhancement and shelf-life extension (Yorgancioglu et al. 2020).

S. No.	Extraction method	Description	Yield and efficiency	Applications	References
1	Enzymatic extraction	Utilizes specific enzymes (e.g., collagenases, proteases) to break down collagen and gelatin into peptides and amino acids	High yield (50–70% of collagen) with minimal denaturation of proteins	Food industry (gelatin production), biomedical applications (tissue engineering, drug delivery)	Yorgancioglu et al. (2020)
2	Acidic extraction	Uses acids like hydrochloric acid (HCl) to dissolve collagen and gelatin, followed by neutralization and purification steps	Efficient in breaking down proteins, yields vary (40–60% of collagen), requires careful pH control to avoid protein degradation	Industrial applications (cosmetics, pharmaceuticals—Adhesives, capsules)	Yorgancioglu et al. (2020)
3	Alkaline extraction	Involves treatment with bases such as sodium hydroxide (NaOH) to solubilize collagen and gelatin, followed by neutralization and purification steps	Effective for large-scale extraction, yields vary (30–50% of collagen), may alter protein structure	Biomedical research (studies on collagen structure), tissue engineering (scaffolds)	Reddy et al. (2021)
4	Enzyme-assisted extraction	Combination of enzymes and mild acids to enhance breakdown of collagen and gelatin, optimizing yield and protein preservation	Yield range (40–60% of collagen), maintains protein integrity compared to harsher methods	Nutraceuticals (collagen supplements), wound care products (dressings)	Reddy et al. (2021)
5	Supercritical fluid extraction	Uses supercritical CO_2 as a solvent to extract collagen and gelatin, offering high selectivity and minimal environmental impact	Yield range (20–40% of collagen), solvent is recyclable and nontoxic	Green chemistry (environmentally friendly extraction), pharmaceutical applications (highly purified collagen)	Reddy et al. (2021)

24.5 Hydrolysis for Amino Acids and Peptides

Hydrolysis of leather waste involves breaking down the proteinaceous components (primarily collagen and gelatin) into smaller peptides and amino acids. This process is essential for recovering valuable bioactive compounds that can be utilized in various industrial applications. Because it is efficient and selective in cleaving peptide bonds without the need for harsh chemical treatments, enzymatic hydrolysis is the method of choice. Proteolytic enzymes, such as collagenases and proteases, selectively degrade collagen and gelatin into peptides of varying lengths and free amino acids. Chemical hydrolysis, on the other hand, utilizes acids or alkalis to break down proteins into soluble fragments. Acidic hydrolysis involves treating the leather waste with acids like hydrochloric acid or sulfuric acid under controlled conditions of temperature and pH. Alkaline hydrolysis uses bases such as sodium hydroxide or potassium hydroxide to achieve similar results. While effective, chemical hydrolysis may lead to protein denaturation and require extensive purification steps to obtain high-quality peptides and amino acids (Selvaraj et al. 2019).

- The amino acids and peptides obtained from hydrolyzed leather waste have significant applications in agriculture and animal nutrition. In animal feed formulations, these bioactive compounds serve as essential amino acid supplements that improve protein content and nutrient balance. Amino acids namely lysine, methionine, and tryptophan are vital for animal growth, development, and overall health. Hydrolyzed peptides can also act as natural growth promoters, enhancing feed efficiency and livestock performance (Selvaraj et al. 2019; Chatterjee et al. 2023).
- In the fertilizer industry, peptides derived from leather waste hydrolysis are utilized as biostimulants and soil amendments. These peptides contain nitrogen, phosphorus, and potassium (NPK), along with micronutrients essential for plant growth. By enhancing nutrient uptake and soil fertility, peptide-based fertilizers contribute to increased crop yield and quality. Moreover, peptides improve soil structure and water retention capacity, promoting sustainable agricultural practices and reducing dependency on synthetic fertilizers (Chojnacka et al. 2024).

24.6 Production of Bioplastics and Biocomposites

Leather waste can be a valued source for manufacturing bioplastics, which are biodegradable polymers derived from renewable biomass sources. Several types of bioplastics can be synthesized using components extracted from leather waste, including:

- Polyhydroxyalkanoates (PHA): PHA bioplastics are produced by bacteria through fermentation of organic substrates, such as sugars derived from leather waste hydrolysates. These bioplastics exhibit properties similar to conventional plastics but are fully biodegradable under natural conditions, offering a sustainable alternative for packaging and disposable products (Vinayak et al. 2022).

- Polylactic acid (PLA): PLA is a bioplastic derived from fermented plant starch, such as corn or sugarcane. By incorporating amino acids and peptides from leather waste, PLA production can be enhanced to improve mechanical strength, thermal stability, and biodegradability. PLA bioplastics are used in a variety range of applications such as packaging of food, consumer goods, and medical appliances (Gupta et al. 2022).

24.7 Incorporation of Leather Waste Fibers

Leather waste fibers can be effectively incorporated into bioplastics and biocomposites to enhance their mechanical properties and sustainability profile. These fibers act as reinforcing agents, improving tensile strength, flexural modulus, and impact resistance of the resulting materials. The incorporation of leather waste fibers also reduces the overall plastic content in biocomposites, making them lighter and more cost-effective compared to conventional materials. Advanced processing techniques, such as extrusion, injection molding, or compression molding, are employed to ensure uniform dispersion of leather waste fibers within the bioplastic matrix. This integration enhances the structural integrity and durability of biocomposites, making them suitable for demanding applications in the automotive, construction, and consumer goods industries (Mikula et al. 2023; Romaní et al. 2020).

Bioplastics and biocomposites derived from leather waste find diverse applications across various sectors:

- Packaging: Bioplastics offer a sustainable substitute for packaging materials, reducing their adverse environmental consequences and meeting consumer demand for more environmentally friendly solutions. Leather waste-derived bioplastics can be used in food packaging, agricultural films, and disposable containers, providing effective barrier properties and ensuring product integrity (Romaní et al. 2020).
- Automotive industry: Biocomposites containing leather waste fibers are utilized in automotive interior components, such as door panels, dashboard covers, and seat upholstery. These materials offer lightweight construction, improved acoustic insulation, and enhanced thermal stability, contributing to fuel efficiency and vehicle performance (Soyer et al. 2023).
- Construction sector: Bioplastics are increasingly used in construction materials, including flooring, insulation, and structural components. By incorporating leather waste fibers, biocomposites enhance the mechanical strength, fire resistance, and durability of building materials, supporting sustainable building practices and reducing carbon footprint (Arora et al. 2021).

The mechanical properties of bioplastics and biocomposites, for example, tensile strength, modulus of elasticity, and impact resistance are tailored to meet specific application requirements. Through continuous research and development efforts, advancements in processing technologies and material formulations are driving the adoption of leather waste-derived bioplastics in mainstream industrial applications (Mikula et al. 2023; Arora et al. 2021.

24.8 Environmental Benefits

- Waste reduction: The leather industry globally generates approximately 2.5 million tons of solid waste annually, including trimmings, fleshings, and shavings. These by-products traditionally end up in landfills or are incinerated, contributing to environmental pollution and resource wastage. By adopting valorization technologies such as anaerobic digestion, pyrolysis, and gasification, a significant portion of this waste can be diverted from landfills (Sreekala et al. 2022).
- Methane mitigation: Landfills are one of the primary sources of methane emissions, a potent greenhouse gas that has a much higher potential to cause global warming than carbon dioxide (CO_2). Anaerobic digestion of organic leather waste produces biogas, primarily CO_2 and methane (CH_4), from biodegradable components. Methane, which can be used as a renewable energy source to reduce greenhouse gas emissions and mitigate the consequences of climate change, is produced by anaerobic digestion. For instance, compared to landfilling, anaerobic digestion can save about 30–40 tons of CO_2 equivalent emissions from the conversion of 1 ton of organic waste (Ufitikirezi et al. 2024).

24.9 Resource Conservation and Renewable Energy Production

- Biochar production: Pyrolysis of leather waste not only produces biofuels but also biochar, a stable form of carbon-rich material. Biochar improves soil fertility and water retention while sequestering carbon dioxide from the atmosphere. Approximately 1 ton of biochar can sequester around 3 tons of CO_2 equivalent over its lifetime, making it a valuable tool in climate change mitigation strategies (Sodhi et al. 2022).
- Biogas and biofuel production: Anaerobic digestion and pyrolysis processes can yield significant amounts of biogas and biofuels. For instance, 1 ton of leather waste can potentially produce 200–300 m^3 of biogas containing 50–70% methane. This biogas can be utilized for the generation of electricity, heating, or as a fuel in automobiles. Similarly, pyrolysis of leather waste can yield approximately 300–500 l of biodiesel or bio-oil per ton of feedstock, contributing to the generation of renewable energy production and minimizing the dependency on nonrenewable energy sources (Sodhi et al. 2022).

24.10 Economic Viability and Market Dynamics

As the global economy shifts towards sustainability, the economic viability of leather waste valorization will continue to improve, driven by market dynamics and supportive policies.

- Market demand for biofuels and green chemicals: The global market for biofuels and green chemicals is anticipated to rise significantly in the future. Market research indicates that the global biofuels market is anticipated to reach $218.7 billion by 2027, with a compound annual growth rate (CAGR) of 8.3% during the period from 2020 to 2027. Similarly, the demand for green chemicals, driven by consumer preference for sustainable products and stringent environmental regulations, will bolster the economic prospects of leather waste valorization.
- Cost reduction through scale and innovation: As technologies mature and economies of scale are realized, the cost of producing biofuels and chemicals from leather waste will decrease. Innovations in process efficiency, waste pretreatment, and product upgrading will further reduce production costs, making these biobased products more competitive with their fossil-derived counterparts.
- Investment and financing: Increasing investments in renewable energy and circular economy initiatives will provide the necessary capital to scale up leather waste valorization projects. Government incentives, green bonds, and private equity investments focused on sustainability will play a crucial role in financing these ventures.

24.11 Conclusion and Future Perspectives

The valorization of leather industry waste into biofuels and chemicals represents a pivotal opportunity to address the environmental and economic challenges associated with this sector. By leveraging the organic richness of leather waste, including collagen, fats, and proteins, it is possible to produce renewable energy sources such as biogas, bio-oil, and biodiesel through innovative processes like anaerobic digestion, pyrolysis, and transesterification. These technologies not only reduce dependency on fossil fuels but also significantly lower greenhouse gas emissions. Moreover, the recovery of valuable by-products like collagen, gelatin, amino acids, and peptides supports various industries, promoting a circular economy and sustainable resource utilization. Incorporating leather waste into bioplastics and biocomposites offers a promising alternative to conventional plastics, addressing the growing need for biodegradable materials and reducing plastic pollution. Collectively, these approaches mitigate environmental pollution, reduce landfill burdens, and contribute to carbon emission reductions, aligning with global sustainability goals.

Looking forward, the future of leather waste valorization lies in advancing technologies and developing integrated systems for resource recovery.

- Advanced anaerobic digestion: Future developments in high-rate digesters, pretreatment methods, and biogas upgrading technologies will enhance efficiency and expand applications for biomethane as a clean energy source.

- Pyrolysis and gasification innovations: Emerging catalytic and plasma-based processes will improve conversion efficiencies and product quality, while enhanced biochar production techniques will add value through applications in soil enhancement and carbon sequestration.
- Integrated biorefineries: The evolution of biorefineries capable of combining multiple conversion pathways will optimize resource recovery, enabling the simultaneous production of biofuels, chemicals, and materials.

Economic incentives, regulatory frameworks, and continued investment in research and development will play critical roles in accelerating these advancements. The transition toward a circular and sustainable leather industry will not only minimize environmental impacts but also unlock new economic opportunities, ensuring a resilient and sustainable future.

References

Alptekin E, Canakci M, Sanli H (2012) Evaluation of leather industry wastes as a feedstock for biodiesel production. Fuel 95:214–220

Amdouni S, Trabelsi ABH, Elasmi AM, Chagtmi R, Haddad K, Jamaaoui F, Khedhira H, Chérif C (2021) Tannery fleshing wastes conversion into high value-added biofuels and biochars using pyrolysis process. Fuel 294:120423

Appala VG, Pandhare NN, Bajpai S, Nikhil GN (2024) Valorization of leather industry solid waste for sustainability through a biorefinery approach. In: From waste to wealth. Springer, Singapore, pp 1119–1137

Arora K, Kaur P, Kumar P, Singh A, Patel SKS, Li X, Yang YH, Bhatia SK, Kulshrestha S (2021) Valorization of wastewater resources into biofuel and value-added products using microalgal system. Front Energy Res 9:646571

Bhatt M, Chakinala AG, Joshi JB, Sharma A, Pant KK, Shah K, Sharma A (2021) Valorization of solid waste using advanced thermo-chemical process: a review. J Environ Chem Eng 9(4):105434

Chatterjee S, Das A, Paul D, Chakraborty S, Choudhury P (2023) Utilization of fleshing waste of leather processing for the growth of zygomycetes: a new substrate for economical production of bio-polymer chitosan. J Environ Manag 343:118141

Chojnacka K, Skrzypczak D, Mikula K, Witek-Krowiak A, Izydorczyk G, Kuligowski K, Bandrow P, Kułażyński M (2021) Progress in sustainable technologies of leather wastes valorization as solutions for the circular economy. J Clean Prod 313:127902

Chojnacka K, Mikula K, Skrzypczak D, Izydorczyk G, Gorazda K, Kulczycka J, Kominko H, Moustakas K, Witek-Krowiak A (2024) Practical aspects of biowastes conversion to fertilizers. Biomass Convers Biorefinery 14(2):1515–1533

Colak SE, Zengin GÖ, Ozgunay H, Sarikahya H, Sari O, Yuceer L (2005) Utilization of leather industry prefleshings in biodiesel production. J Am Leather Chem Assoc 100(4):137–141

Cotance S (2012) Environmental report, European Leather Industry

Dagne H, Karthikeyan R, Feleke S (2019) Waste to energy: response surface methodology for optimization of biodiesel production from leather fleshing waste. J Energy 2019(1):7329269

Dudyński M, Dudyński K, Kluska J, Ochnio M, Kazimierski P, Kardaś D (2021) Gasification of leather waste for energy production: laboratory scale and industrial tests. Int J Energy Res 45(13):18540–18553

Gupta S, Nadda AK, Gupta A, Singh J, Mulla SI, Sharma S (2022) Transforming wastes into high value-added products: an introduction. In: Biopolymers: recent updates, challenges and opportunities. Springer, Cham, pp 1–18

Hari S, Harshan PB, Punniavan S, Sudharsan MS, Ramadoss G (2023) Valorization of industrial wastes. In: Valorization of wastes for sustainable development. Elsevier, Amsterdam, pp 371–394

İşler A, Sundu S, Tüter M, Karaosmanoğlu F (2010) Transesterification reaction of the fat originated from solid waste of the leather industry. Waste Manag 30(12):2631–2635

Kanagaraj J, Velappan KC, Babu NK, Sadulla S (2006) Solid wastes generation in the leather industry and its utilization for cleaner environment-A review. J Sci Ind Res 65:541–548

Kandasamy R, Venkatesan SK, Uddin MI, Ganesan S (2020) Anaerobic biovalorization of leather industry solid waste and production of high value-added biomolecules and biofuels. In: Biovalorisation of wastes to renewable chemicals and biofuels. Elsevier, Amsterdam, pp 3–25

Kumar V, Thakur IS (2020) Extraction of lipids and production of biodiesel from secondary tannery sludge by in situ transesterification. Bioresour Technol Rep 11:100446

Mariappan M (1997) Environmental protection initiatives in Indian tanneries—a perspective view. In: Proceedings of leather research industry get-together. CLRI, Chennai, pp 150–165

Mikula K, Konieczka M, Taf R, Skrzypczak D, Izydorczyk G, Moustakas K, Kułażyński M, Chojnacka K, Witek-Krowiak A (2023) Tannery waste as a renewable source of nitrogen for production of multicomponent fertilizers with biostimulating properties. Environ Sci Pollut Res 30(4):8759–8777

Mozhiarasi V, Krishna BB, Nagabalaji V, Srinivasan SV, Bhaskar T, Suthanthararajan R (2021) Leather industry waste based biorefinery. In: Waste biorefinery. Elsevier, Amsterdam, pp 267–304

Özgünay H, Çolak S, Zengin G, Sari Ö, Sarikahya H, Yüceer L (2007) Performance and emission study of biodiesel from leather industry pre-fleshings. Waste Manag 27(12):1897–1901

Priebe GPS, Kipper E, Gusmão AL, Marcilio NR, Gutterres M (2016) Anaerobic digestion of chrome-tanned leather waste for biogas production. J Clean Prod 129:410–416

Rao RJ, Thanikaivelan P, Sreeram KJ, Unni Nair B (2004) Tanning studies with basic chromium sulfate prepared using chrome shavings as a reductant: a call for 'wealth from waste' approach to the tanning industry. J Am Leather Chem Assoc 99:170–176

Rathinam NK, Sani R (2019) Biovalorisation of wastes to renewable chemicals and biofuels. Elsevier, Amsterdam

Reddy CC, Khilji IA, Gupta A, Bhuyar P, Mahmood S, Al-Japairai KAS, Chua GK (2021) Valorization of keratin waste biomass and its potential applications. J Water Process Eng 40:101707

Romaní A, Rocha CM, Michelin M, Domingues L, Teixeira JA (2020) Valorization of lignocellulosic-based wastes. In: Current developments in biotechnology and bioengineering. Elsevier, Amsterdam, pp 383–410

Selvaraj S, Jeevan V, Jonnalagadda RR, Fathima NN (2019) Conversion of tannery solid waste to sound absorbing nanofibrous materials: a road to sustainability. J Clean Prod 213:375–383

Šimat V (2021) Valorization of seafood processing by-products. In: Valorization of agri-food wastes and by-products. Academic, New York, pp 515–536

Sodhi AS, Sharma N, Bhatia S, Verma A, Soni S, Batra N (2022) Insights on sustainable approaches for production and applications of value added products. Chemosphere 286:131623

Soyer Ş, Gürlek G, Kılıç E (2023) Valorization of leather industry waste in polyurethane composites for reduced flammability. J Mater Cycl Waste Manag 25(1):314–323

Sreekala AGV, Ismail MHB, Nathan VK (2022) Biotechnological interventions in food waste treatment for obtaining value-added compounds to combat pollution. Environ Sci Pollut Res 29(42):62755–62784

Ufitikirezi JDDM, Filip M, Ghorbani M, Zoubek T, Olšan P, Bumbálek R, Strob M, Bartoš P, Umurungi SN, Murindangabo YT, Heřmánek A (2024) Agricultural waste valorization: exploring environmentally friendly approaches to bioenergy conversion. Sustain For 16(9):3617

Vinayak A, Sharma S, Singh GB (2022) Biopolymers from industrial waste. In: Biopolymers: recent updates, challenges and opportunities. Springer, Cham, pp 129–149

Yorgancioglu A, Başaran B, Sancakli A (2020) Industry wastes and by-products: hydrolyzed collagen and collagen. In: Waste in textile and leather sectors. IntechOpen, London, p 131

Pulp and Paper Industry Waste as Low-Cost Materials to Produce Biofuel and Chemicals

25

Manasvi Vashisht, Surbhi Sharma, Aryan Prajapati, Smriti Shrivastava, Abha Kumari, and Minsik Kim

Abstract

The paper and pulp industry is foundation of global economy. It is essential for producing a diverse array of products which includes printing, packaging, and hygiene materials. These industries fundamentally alter wood into pulp, thereafter producing paper and paperboard products through the processes of mechanical and chemical pulping, bleaching, and paper manufacturing. Pulp and paper industry faces critical challenge which is the management of wastewater practices. This chapter highlights tactical emphasis on decreasing freshwater usage through recycling and reuse of treated wastewater. This chapter also investigates perspective of transforming this waste into high-valued biomaterials using integrated and conventional biorefinery technologies.

Keywords

Paper industry waste · Black liquor · Industrial waste management · Biofuels · Biochemicals

M. Vashisht · S. Sharma · A. Prajapati · S. Shrivastava · A. Kumari
Amity Institute of Biotechnology, Amity University, Noida, Uttar Pradesh, India

M. Kim (✉)
Department of Biological Engineering, Inha University, Incheon, Republic of Korea
e-mail: minsik.kim@inha.ac.kr

A. Kumari et al. (eds.), *Valorization of Solid Wastes to Biofuels and Chemical Products for Sustainable World*, https://doi.org/10.1007/978-981-96-8594-3_25

25.1 Introduction

Utilization of paper and pulp has long been a barometer of prevailing evolution and industrial advancement, reflecting a level of civilization across different eras and regions (Lwako et al. 2013). Paper and pulp industries are essential for producing a wide range of products, including packaging, printing, and hygiene products. Alongside, these industries also give rise to a significant amount of waste which is particularly in form of effluents. These effluents include liquid, solid, and air emissions which arise from paper manufacturing processes. Managing these wastes builds consequential ecological challenges due to their compound and intoxicated composition. Impact of the industry on the ecosystem and human health coupled with the sheer volume of waste generated, underscores the need for effective waste management and recycling strategies and techniques. This matter is particularly pronounced in developing countries, where panoramic qualitative and quantitative evaluation of biomass waste from these mills remains conspicuously absent. Moreover, except for a small portion of waste paper that is recycled, most of the waste turns out to be disposed of in landfills or is incinerated without any commercial benefit (Gupta and Gupta, 2019; Bajpai 2015).

Environmental sustainability has emerged as a pivotal facet of the global development agenda, driven by desperate need to mark challenges such as resource depletion, climate change, global warming, and environmental degradation (McKinnon 2010; Bajpai 2015). Wastewater from these industries is released from different stages, which includes washing of raw materials, cooked pulp, and bleached pulp (Gupta and Shukla 2020). Solid waste is produced from primary and secondary sludge and lime sludge from chemical recovery (Đurđević et al. 2019). Pulp and paper business generates waste at several points in the process, from washing of raw wood sources through pulping and bleaching procedures and, lastly, the chemical recovery phases. This waste consists of inorganic chemicals, organic acids, lignin, chlorinated compounds, and organic debris and alarms the need for recycling of materials. Anaerobic digestion is one technology that may turn organic waste into biogas, which is a sustainable energy source. Likewise, processing of lignocellulosic biomass can yield bioethanol and other biofuels, providing a blessing to human life. According to Landrigan and Fuller (2012), "Waste to Wealth" strategy recognizes that efficient use of waste can have a substantial positive impact on economic growth and product diversification in addition to protecting the environment. Increased acquisition of biorefineries which can efficiently convert waste into valuable products such as biomaterials, bioethanol, and biochemicals, is required in the future. This chapter highlights the status of waste generation in the pulp and paper industry and evaluates environmental catastrophe linked with this and explores sustainable recycling methods.

25.2 Generation Status of Waste in Pulp and Paper Industry

The pulp and paper industry is considered among the largest generators of industrial waste, producing substantial amounts of both liquid and solid wastes throughout its various production processes. Solid waste in this industry is mainly produced from paper processing and it is wastewater treatment (Roy et al. 2022). An approximate amount of 70% primary sludge and 30% secondary sludge is generated from 40 to 50 kg of dry sludge obtained by the production of around 1 ton paper. The amount of dry sludge varies greatly, for example, it may be around 20% in a newsprint mill and, at the same time, be 40% in a tissue mill (Gottumukkala et al. 2016).

First step is wood reparation, the debarking and chipping of wood procedure with significant amounts of bark, wood chips, and sawdust. Debarking can generate up to 50 kg of bark per ton of processed wood (Chahal and Ciolkosz 2019). Wastewater from this stage contains organic matter and wood debris. Second stage is pulping, which is further divided into three types, that is, mechanical pulping, chemical pulping, and semichemical pulping.

Next stage is bleaching which generates effluent containing chlorinated organic compounds (AOX), residual bleaching agents (e.g., chlorine oxide, hydrogen peroxide), and various dissolved organic materials. Bleaching can produce approximately 15–20 m^3 of effluent per ton of paper. Further is paper making, this stage produces effluent with fine fibers, fillers, coating materials, and various chemicals used in paper formation. Effluents from paper-making process typically range from 5 to 10 m^3/ton of paper produced (Pokhrel and Viraraghavan 2004). Paper making is followed by chemical recovery, where lime sludge, green liquor dregs, and boiler ash are primary solid wastes. Chemical recovery systems, such as those in kraft mills, generate these solid wastes alongside any liquid residues. The recovery process can produce several tons of lime sludge per day, depending on the scale of operations (Bajpai 2015).

There is a critical need for improved waste management and environmental practices within the global paper industry. As environmental regulations become sustainable and uncompromising, it becomes more critical to responsibility, and the industry must establish and adopt more sustainable practices to lessen its environmental footprint. An overview of the various kinds and quantities of waste generated by the pulp and paper industry worldwide along with the major companies involved is given in Table 25.1.

25.3 Composition of Waste from Paper and Pulp Industries

Waste produced by the paper and pulp industries varies in composition and volume due to the diverse processes and products involved in these functions. These pulp and paper mills experience variation due to factors which demand fluctuations, raw material availability, technological advancements, and environmental regulations.

Table 25.1 Overview of various kinds and quantities of waste generated by the pulp and paper industry

S. No.	Country	Type of waste	Wastage production (metric million tons)	Companies involved	References
1	The United States of America	Sludge, air emissions	46.66	International Paper, Georgia Pacific	He and Wang (2019)
2	Japan	Solid, wastewater effluent, air emissions	64.3	Oji holdings, Nippon paper industries	Solanki and Pitroda (2012)
3	China	Sludge	85.87	Nine dragon paper, Lee & man Paper	Hryb (2015)
4	Germany	Solid waste, sludge, fly ash	22.7	Stora Enso, Smurfit kappa	Morlok et al. (2017)
5	Canada	Solid residue, sludge	12.1	Domtar, resolute Forest products	Kourkoumpas et al. (2015)
6	South Korea	Sludge, chemical residue, wood processing	11.5	Hansol paper, Moorim paper	Ganguly et al. (2018)
7	Finland	Black liquor, sludge, fly ash	11.3	UPM, Stora Enso, Metsa group	Kuparinen et al. (2022)
8	Brazil	Wood and bark residue	10.2	Suzano, Klabin	Worku et al. (2023)
9	Indonesia	Solid waste, wastewater effluent, chemical pulp emissions	10.0	APRIL group, Asia Pulp & Paper	Lee et al. (2022)

Changes in consumer preferences and global economic conditions also play a worthy role. They are majorly classified as solid, liquid, and gaseous waste based on state and hazardous and nonhazardous based on environmental impact (Bajpai 2015). Solid waste primarily comprises of bark and wood residues, which are generated during the debarking and chipping processes. The type of wood used (softwood and hardwood) can affect the volume and the type of bark and wood residues.

Liquid waste, composed of black liquor, is a by-product of the kraft process and contains lignin, hemicellulose, and inorganic chemicals. These specific pulping processes used (kraft, sulfide, and mechanical) determine the types of liquor produced (Pola et al. 2022). Secondly, white liquor is used in the pulping process, containing chemicals like sodium hydroxide and sodium sulfide. Also, the green liquor is the intermediate product in the chemical recovery process. Efficiency of the recovery process can influence the volume of liquid waste (Sewsynker- Sukai et al. 2020). Air emissions include the particulate matter consisting of fly ash from the combustion process. The volatile organic compounds (VOCs) are emitted during various stages of paper production and sulfur compounds such as sulfur dioxide (SO_2) are used for the kraft pulping process. Nitrogen oxides emitted from combustion processes and carbon dioxide from burning fossil fuels and biomass. Effectiveness of air pollution control equipment, for example, scrubbers and electrostatic precipitators can vary affecting emission levels. Hazardous waste comprises spent chemicals that are used in bleaching. Solvents are used for cleaning and other manufacturing processes. Environmentally friendly and sustainable practices can be adopted by the sector with the support of strict regulatory frameworks and innovations in green techniques and technologies (de Azevedo et al. 2019).

25.4 Biofuel from Pulp and Paper Industry

Pulp and paper sector produces a substantial quantity of waste in the form of ligno-cellulosic materials, including wood residues, pulp rejects, and spent liquor that can be effectively employed for production of biofuels through diverse processes (Branco et al. 2018). Biofuels, derived mainly from by-products and residues produced during pulp and paper manufacturing, capitalize on the substantial biomass resources inherent in the pulp and paper industry. Transforming these resources into biofuels presents environmentally friendly process and creates economic prospects while aiding in the reduction in dependence on fossil fuels (Ubando et al. 2020).

Several production methods including cellulosic ethanol production and biogas generation together with biomass gasification and pyrolysis and biomass to liquid conversion among others allow the manufacturing of biofuel. Scientists transform ethanol using saccharification techniques to convert lignocellulosic materials including wood and paper waste into sugars before using fermentation to produce ethanol which is known as cellulosic ethanol production. Finally the fermentation and enzymatic hydrolysis processes convert both cellulose and hemicellulose into bioethanol (Mäki et al. 2021). Systematic biogas production requires the anaerobic decomposition of existing organic waste products. Karlsson et al. (2011) explained microbial decomposition of organic matter under oxygen-free conditions creates renewable biofuel, which largely consists of methane. The thermochemical transformation of biomass through gasification generates another route to create biofuel. The syngas can be further processed to produce biofuels like synthetic natural gas (SNG) or liquid biofuels. Gasification is applicable to diverse biomass feedstocks, including wood residues (Chhiti and Kemiha 2013). Pyrolysis is a thermal decomposition process transforming biomass into bio-oil, biochar, and gases (Das et al. 2016a). Bio-oil obtained from pyrolysis can be upgraded to generate liquid biofuels, and wood and paper waste prove to be suitable feedstocks for this process (Pütün 2016). Biomass-to-liquid (BTL) technology involves converting solid biomass into liquid biofuels, such as biodiesel or synthetic diesel. Synthetic gas produced from biomass gasification can be converted into liquid fuels through Fischer-Tropsch synthesis (Swain et al. 2011). Advanced biofuel production focuses on deriving biojet fuel and renewable diesel from lignocellulosic materials through advanced conversion processes like hydrothermal liquefaction or upgrading bio-oil (Lahijani et al. 2022).

Production of lignin-based biofuel involves isolation of lignin from pulping liquors, followed by a refinement process to yield a form suitable for use as a fuel (Kocaturk et al. 2023). This biofuel exhibits versatility in applications, including its utilization for heat and power generation, thereby serving as a potential substitute or supplement to conventional fossil fuels (Kaniapan et al. 2021; Poveda-Giraldo et al. 2021). Moreover, the residual cellulose fibers, postextraction of lignin, present an opportunity for the generation of cellulosic biofuels, such as bioethanol. Advanced methodologies, such as enzymatic hydrolysis and fermentation, are employed to break down cellulose into sugars, subsequently undergoing fermentation processes to produce ethanol (Devi et al. 2021; Haile et al. 2021).

Utilization of pulp and paper industry waste for biofuel production aligns with the overarching concept of a biobased circular economy. Through the conversion of waste materials into valuable biofuels, the industry can mitigate its environmental impact, optimize resource utilization, and contribute to the advancement of a sustainable energy landscape. Additionally, the adoption of biofuels enables pulp and paper companies to adhere to increasingly stringent environmental regulations while fostering innovation and diversification within the sector (Del Rio et al. 2022; Mohanty et al. 2022).

25.5 Challenges and Prospects of Biofuel Production from Pulp and Paper Industry

The global emphasis on sustainable and renewable energy sources has led to increased attention on biofuels as a viable alternative to traditional fossil fuels (Mahapatra et al. 2021). Pulp and paper industry, being a significant contributor to biomass production, holds substantial promise for biofuel generation (Nandal et al. 2021). However, effective utilization of this potential is accompanied by inherent challenges and opportunities.

Key considerations impacting biofuel production include the accessibility and characteristics of biomass feedstock for effective biofuel production (Ambaye et al. 2021; Mäki et al. 2021; Ethaib et al. 2020). The refinement and enhancement of these techniques for extensive production pose challenges and demand significant investments in research and development endeavors. The economic feasibility of producing biofuels from residues persists as a notable hurdle (Holland and Shapira 2023). Biofuel production encounters regulatory impediments associated with land utilization, environmental consequences, and adherence to global sustainability benchmarks (Johnson 2022; Del Rio et al. 2022; Mäki et al. 2021).

Despite persistent challenges, outlook for biofuel production derived from residues of the pulp and paper industry presents resolutions to environmental, economic, and energy security considerations as biofuels originating from residues will diminish greenhouse gas emissions in contrast to conventional fossil fuels (Chakraborty et al. 2019; Poudyal et al. 2017; Nigam and Singh 2011; Dixit et al. 2023).

25.6 Challenges and Prospects of Chemical Products Production from Pulp and Paper Industry Waste

Various types of waste from paper and pulp operations include rejects containing fiber lumps alongside staples and metals and glass materials and sizing agents along with deinking sludge made from short fibers combined with inks and additional additives and primary sludge produced from flotation systems transporting fillers and fines and secondary sludge which emerges from biological wastewater treatment clarifiers containing high microbial contents. The primary sludge consists of organic alongside inorganic substances and contains appreciable quantities of both inorganic compounds and cellulose-rich fibers. Secondary sludge stands out as a difficult-to-manage form of waste because its production rate originates from multiple stages of water processing combined with high microbial cell density (Haq et al. 2016; Varjani and Upasani 2017; Bajpai 2015; Hu et al. 2017).

Due to the rising need for energy, the paper and pulp industries have worked hard to raise their energy sufficiency (Bayr et al. 2013). One innovative concept for environmental sustainability and a reduction in potentially harmful effluents is the production of resources from waste from the pulp sector, such as clinker preparation, green energy, and sorbent development.

Commercialization of the anaerobic dark fermentation method for energy recovery and methane production and biological digestion is one of the greatest methods for reducing pulp mill waste (Singh et al. 2010). Owing to its massive production, lignin is often very resistant. Extensive chemical and biological pretreatment is required for lignin to be converted into usable sugars, flocculants, fuels, battery plates, adsorbents, dispersants, dust suppressants, etc. (Sewsynker-Sukai et al. 2020).

25.6.1 Treatment Techniques Used for Waste Treatment

Raising pH of effluents is a commonly used treatment technique for removing heavy metals and harmful compounds from wastewater. Ion exchange is the second approach that is employed the most (Jaria et al. 2017). Ion exchange serves as a metal and poisonous agent concentrator in this approach. However, because of its versatility in design and usage, the sorbent made from the wastewater of the paper industry is one of the most appealing and effective methods for the removal of heavy metals from wastewater. Numerous oil spills occur during the exploration, extraction, refining, transportation, and storage phases of the oil industry.

By use of technological innovation in green energy generation from waste, a large amount of pulp and paper industry waste can be diverted in different industries for better utilization and to remove the load on the environment. Valuable usage of this waste includes energy generation with the help of other integrated industries (Table 25.2).

Table 25.2 Components of waste materials as well as the value-added goods produced by methods of valorization and their benefits

S. No.	Waste material	Production stage	Constituents	Need for valorization	Treatment method	Resource manufactured	Advantages	References
1	Chlorine-free elemental bleaching wastewater	Kraft pulping	Adsorbable organic halogen and organic nutrients	Chemicals and organic material from wood are quite polluting	Anaerobic sludge bed reactors with upflow	Biogas	Potential for full-scale implementation	Liang et al. (2021)
2	Wastewater and coconut husk	Wastewater from paper/pulp industry and coconut husk	Metal ions and residues high in lignin	Wastewater is polluted, and coconut husk is not easily degradable	Supercritical gasification and pyrolysis	Fuel gas and hydrogen	Effort to generate fuel gas and reduce waste	Gupta and Shukla (2020)
3	Red liquor	Acidic sulfite cooking	Polysaccharide, lignosulphonate, and lignin	Energy-intensive condensation and separation are necessary for pretreatment	Copolymerization of acrylic acid and red liquor	Hydrogels	Approximately 80% water retention ratio and no red liquor treatment needed	Meng et al. (2019)
4	Flue gas and waste heat	Soda boiler of the factory	CO_2, NO, SO_2	Greenhouse gases	Oleaginous microalgae cultivation	Biomass production	Energy generation and sequestration of CO_2	Wang et al. (2024)

S. No.	Waste material	Production stage	Constituents	Need for valorization	Treatment method	Resource manufactured	Advantages	References
5	Pulp and paper solid waste	Pulping process	High BOD and COD content due to carbon, minerals, and moisture	Greenhouse gas emissions, contaminated groundwater, and provision of food for insects and microbes	Submerged and solid-state fermentation	Fumaric acid	Sustainable environmental practices, waste valuation, and safe disposal	Das et al. (2016b)
6	Waste liquor from monosodium glutamate and paper sludge	Primary and secondary sludge	Minerals, carbon dioxide, organic nitrogen compounds, volatile solids	Hard to dispose of and a breeding habitat for pests and diseases	Anaerobic codigestion	Methane gas	Generating energy, using waste materials sustainably, and breaking down large molecules	Lin et al. (2011)
7	Biomass fly ash and alkaline effluent ion	Biomass boilers and fluidized bed combustors	A-quartz, calcite, and mica group minerals	High energy intensity, greenhouse gas emissions, and dust pollution in cement industries	Integrating the use of wastes	Mortars and binders made of geopolymer	Affordable, easy-to-use, and replicable product	Saeli et al. (2019)

25.7 Chemical Products Production from Pulp and Paper Industry Waste

Around 70 million tons of lignin is generated worldwide through pulping processes (Dessbesell et al. 2020; Bajwa et al. 2019). Black liquor separated by washing of pulp is made up of organic polymers such as lignin, polysaccharides, low-molar-mass resin compounds as well as inorganic materials like soluble salt ions (Viel et al. 2020). The lignin and hemicellulose in black liquor can be utilized for conversion into valuable biomaterials (Vakkilainen and Välimäki 2009). Black liquor hemicellulose contains xylan (Yang et al. 2019; Liu et al. 2012), which can be biorefined to produce lactic acid. Xylan is transformed to low-molecular-weight carbohydrates through enzyme or acid hydrolysis which further form lactic acid via fermentation or alkali oxidation. Lactic acid produced from biomass can be used as a renewable chemical building component for bioplastic production (Fernández-Rodríguez et al. 2019; Chen 2012). Polylactic acid (PLA), a biodegradable polyester, derived from renewable feedstock is synthesized by polymerization of xylan lactic acid refinery, which can be used as a precursor for bioplastic production (Gordobil et al. 2015; Singhvi et al. 2019).

Polyhydroxyalkanoates (PHA) are biobased, biodegradable polymers, also referred to as "green plastic" (Koller et al. 2013) and hence being a sustainable replacement of traditional plastics. They are mostly made up of monomeric units called hydroxyalkanoates, with 3-hydroxybutyrate (HB) and 3-hydroxyvalerate (HV). Several bacteria and archaea produce PHA for internal carbon and energy storage (Laycock et al. 2013). Pure or mixed microbial cultures (MMC) or microbial population of unknown composition such as activated sludge from wastewater treatment plants produces PHA (Jiang et al. 2012; Al-Battashi et al. 2019).

Shi et al. (2017) found good results using kraft lignin as a carbon source for PHA synthesis by *Cupriavidus basilensis* B-8. Also, lignin is an important biomass and a promising substitute precursor for carbon fibers production. Carbon fiber possesses high strength-to-weight ratio, low density, and good mechanical properties, which makes it suitable for its utilization as a load-bearing element in lightweight high-performance composites (Haile et al. 2021). Lignin obtained from kraft pulp has high carbon content (60–65%) and is readily available, hence useful for carbon fiber production (Olsson et al. 2017; Zhang et al. 2020).

With progressive growth of pulp and paper mills, waste biomass combustion releases huge amount of fly ash, which is primarily composed of minerals like iron oxide, silica, and alumina and metal oxides present in fly ash such as sodium, calcium, magnesium, potassium, sulfur, and titanium (Na_2O, CaO, MgO, K_2O, SO_3, and TiO_2) in varying amounts. These metal oxides are polymerized by geopolymerization to produce biopolymers with a variety of applications. Polymerization method produces 3-D aluminosilicates and similar materials known as geopolymers (Cherian and Siddiqua 2019; Sjöström and Westermark 1999).

25.8 Policies for Waste Management in the Pulp and Paper Industry

To protect the environment, there are governmental and nongovernmental policies for wastewater management for sheltering the environment and human health while promoting sustainable measures. These policies often are found under the laws like Clean Waste Act, which sets the regulatory standards by permitting processes and enforcement measures for ensuring that wastewater discharge meets the acceptable quality meter and may not harm human health, the biodiversity, and the ecosystem. Additionally, the FSC and WWF-like organizations advocate for sustainable practices and transparency within this sector. By motivating the adoption of water conservation practices, recycling techniques and technologies, and cleaner production practices, these policies take the initiative of working toward reducing waste generation and minimizing the industry's ecological impact as governed in references such as the Clean Water Act and the reports generated by FSC and WWF (Gupta et al. 2019). Standardizing regulatory standards, permission and compliance, best management practices, certification programs like Forest Stewardship Council (FSC) and the Sustainable Forestry Initiative (SFI), and awareness through NGO advocacies and management are major areas in policy management.

References

Al-Battashi HS, Annamalai N, Sivakumar N, Al-Bahry S, Tripathi BN, Nguyen QD, Gupta VK (2019) Lignocellulosic biomass (LCB): a potential alternative biorefinery feedstock for polyhydroxyalkanoates production. Rev Environ Sci Biotechnol 18:183–205

Ambaye TG, Vaccari M, Bonilla-Petriciolet A, Prasad S, van Hullebusch ED, Rtimi S (2021) Emerging technologies for biofuel production: a critical review on recent progress, challenges and perspectives. J Environ Manag 290:112627

Bajpai P (2015) Management of pulp and paper mill waste, vol 431. Springer, Cham

Bajwa DS, Pourhashem G, Ullah AH, Bajwa SG (2019) A concise review of current lignin production, applications, products and their environmental impact. Ind Crop Prod 139:111526

Bayr S, Kaparaju P, Rintala J (2013) Screening pretreatment methods to enhance thermophilic anaerobic digestion of pulp and paper mill wastewater treatment secondary sludge. Chem Eng J 223:479–486

Branco RH, Serafim LS, Xavier AM (2018) Second generation bioethanol production: on the use of pulp and paper industry wastes as feedstock. Fermentation 5(1):4

Chahal A, Ciolkosz D (2019) A review of wood-bark adhesion: methods and mechanics of debarking for woody biomass. Wood Fiber Sci 51(3):1–12

Chakraborty D, Dahiya S, Amulya K, Srivastav V, Mohan SV (2019) Valorization of paper and pulp waste: opportunities and prospects of biorefinery. In: Industrial and municipal sludge. Butterworth-Heinemann, Oxford, pp 623–656

Chen K (2012) Bio-renewable fibers extracted from lignin/polylactide (PLA) blend. Carbon 68:159–166

Cherian C, Siddiqua S (2019) Pulp and paper mill fly ash: a review. Sustain For 11(16):4394

Chhiti Y, Kemiha M (2013) Thermal conversion of biomass, pyrolysis and gasification. Int J Eng Sci 2(3):75–85

Das O, Sarmah AK, Bhattacharyya D (2016a) Biocomposites from waste derived biochars: Mechanical, thermal, chemical, and morphological properties. Waste management, 49:560–570.

Das RK, Brar SK, Verma M (2016b) Potential use of pulp and paper solid waste for the bio-production of fumaric acid through submerged and solid state fermentation. Journal of Cleaner Production, 112:4435–4444.

de Azevedo ARG, Alexandre J, Pessanha LSP, Manhães RST, de Brito J, Marvila MT (2019) Characterizing the paper industry sludge for environmentally-safe disposal. Waste Manag 95:43–52

Del Rio DDF, Sovacool BK, Griffiths S, Bazilian M, Kim J, Foley AM, Rooney D (2022) Decarbonizing the pulp and paper industry: a critical and systematic review of sociotechnical developments and policy options. Renew Sust Energ Rev 167:112706

Dessbesell L, Paleologou M, Leitch M, Pulkki R, Xu CC (2020) Global lignin supply overview and Kraft lignin potential as an alternative for petroleum-based polymers. Renew Sust Energ Rev 123:109768

Devi A, Singh A, Bajar S, Pant D, Din ZU (2021) Ethanol from lignocellulosic biomass: an in-depth analysis of pre-treatment methods, fermentation approaches and detoxification processes. J Environ Chem Eng 9(5):105798

Dixit Y, Yadav P, Sharma AK, Pandey P, Kuila A (2023) Multiplex genome editing to construct cellulase engineered Saccharomyces cerevisiae for ethanol production from cellulosic biomass. Renew Sust Energ Rev 187:113772

Đurđević D, Blecich P, Jurić Ž (2019) Energy recovery from sewage sludge: the case study of Croatia. Energies 12(10):1927

Ethaib S, Omar R, Mazlina MS, Radiah AD, Zubaidi SL (2020) Toward sustainable processes of pretreatment technologies of lignocellulosic biomass for enzymatic production of biofuels and chemicals: a review. Bioresources 15(4):10063

Fernández-Rodríguez J, Erdocia X, Hernández-Ramos F, Alriols MG, Labidi J (2019) Lignin separation and fractionation by ultrafiltration. In: Separation of functional molecules in food by membrane technology. Academic, New York, pp 229–265

Ganguly R, Midya S, Chakraborty S (2018) Antioxidant and anticancer roles of a novel strain of bacillus anthracis isolated from vermicompost prepared from paper mill sludge. Biomed Res Int 2018(1):1–10

Gordobil O, Delucis R, Egüés I, Labidi J (2015) Kraft lignin as filler in PLA to improve ductility and thermal properties. Ind Crop Prod 72:46–53

Gottumukkala LD, Haigh K, Collard FX, Van Rensburg E, Görgens J (2016) Opportunities and prospects of biorefinery-based valorisation of pulp and paper sludge. Bioresour Technol 215:37–49

Gupta A, Gupta R (2019) Treatment and recycling of wastewater from pulp and paper mill. In: Advances in biological treatment of industrial waste water and their recycling for a sustainable future. Springer, Cham

Gupta GK, Shukla P (2020) Insights into the resources generation from pulp and paper industry wastes: challenges, perspectives and innovations. Bioresour Technol 297:122496

Gupta GK, Liu H, Shukla P (2019) Pulp and paper industry–based pollutants, their health hazards and environmental risks. Curr Opin Environ Sci Health 12:48–56

Haile A, Gelebo GG, Tesfaye T, Mengie W, Mebrate MA, Abuhay A, Limeneh DY (2021) Pulp and paper mill wastes: utilizations and prospects for high value-added biomaterials. Bioresour Bioprocess 8:1–22

Haq I et al (2016) Evaluation of the phytotoxic and genotoxic potential of pulp and paper mill effluent using Vigna radiata and Allium cepa. Adv Biol 2016:3

He B, Wang G (2019) Is ceramsite the last straw for sewage sludge disposal: a review of sewage sludge disposal by producing ceramsite in China. Water Sci Technol 80(1):1–10

Holland C, Shapira P (2023) Building the bioeconomy: a targeted assessment approach to identifying biobased technologies, challenges and opportunities in the UK. bioRxiv, 2023–08

Hryb W (2015) Sorting tests of unsorted municipal solid waste from Germany for a selected opto-pneumatic sorting machine. Pol J Environ Stud 24(1):99–105

Hu S, She X, Wei X, Hu B, Hu C, Qian Y, Chen Z (2017). Surplus sludge treatment in two sludge treatment beds under subtropical condition in China. International Biodeterioration & Biodegradation, 119:377–386.

Jaria G et al (2017) Sludge from paper mill effluent treatment as raw material to produce carbon adsorbents: an alternative waste management strategy. J Environ Manag 188:203–211

Jiang Y, Marang L, Tamis J, van Loosdrecht MC, Dijkman H, Kleerebezem R (2012) Waste to resource: converting paper mill wastewater to bioplastic. Water Res 46(17):5517–5530

Johnson A (2022) The roundtable on sustainable palm oil (RSPO) and transnational hybrid governance in Ecuador's palm oil industry. World Dev 149:105710

Kaniapan S, Hassan S, Ya H, Patma Nesan K, Azeem M (2021) The utilisation of palm oil and oil palm residues and the related challenges as a sustainable alternative in biofuel, bioenergy, and transportation sector: a review. Sustain For 13(6):3110

Karlsson A, Truong XB, Gustavsson J, Svensson BH, Nilsson F, Ejlertsson J (2011) Anaerobic treatment of activated sludge from Swedish pulp and paper mills–biogas production potential and limitations. Environ Technol 32(14):1559–1571

Kocaturk E, Salan T, Ozcelik O, Alma MH, Candan Z (2023) Recent advances in lignin-based biofuel production. Energies 16(8):3382

Koller M, Salerno A, Braunegg G (2013) Polyhydroxyalkanoates: basics, production and applications of microbial biopolyesters. In: Bio-based plastics: materials and applications. Wiley, Hoboken, pp 137–170

Kourkoumpas D, Karellas S, Kouloumoundras S, Koufodimos G, Grammelis P, Kakaras E (2015) Comparison of waste-to-energy processes by means of life cycle analysis principles regarding the global warming potential impact: applied case studies in Greece, France, and Germany. Waste Biomass Valoriz 6(4):605–621

Kuparinen K, Lipiäinen S, Vakkilainen E, Laukkanen T (2022) Effect of biomass-based carbon capture on the sustainability and economics of pulp and paper production in the Nordic mills. Environ Dev Sustain 25(4):648–668

Lahijani P, Mohammadi M, Mohamed AR, Ismail F, Lee KT, Amini G (2022) Upgrading biomass-derived pyrolysis bio-oil to bio-jet fuel through catalytic cracking and hydrodeoxygenation: a review of recent progress. Energy Convers Manag 268:115956

Landrigan PJ, Fuller R (2012) Environmental pollution: an enormous and invisible burden on health systems in low-and middle-income counties. World Hosp Health Serv 50(4):35

Laycock B, Halley P, Pratt S, Werker A, Lant P (2013) The chemomechanical properties of microbial polyhydroxyalkanoates. Prog Polym Sci 38(3–4):536–583

Lee CL, Chin K, H'ng P, Hafizuddin MS, Khoo PS (2022) Valorisation of underutilized grass fibre (stem) as a potential material for paper production. Polymers 14(10):1–15

Liang J, Mai W, Wang J, Li X, Su M, Du J, Wei Y (2021) Performance and microbial communities of a novel integrated industrial-scale pulp and paper wastewater treatment plant. Journal of Cleaner Production, 278:123896.

Liu YX, Sun B, Zhu Y, Li TZ (2012) Hemicelluloses separation from black liquor. Adv Mater Res 54:589–593

Lin Y, Wang D, Li Q, Xiao M (2011) Mesophilic batch anaerobic co-digestion of pulp and paper sludge and monosodium glutamate waste liquor for methane production in a bench-scale digester. Bioresource technology, 102(4):3673–3678.

Lwako MKO, Byaruhanga JK, Baptist KJ (2013) A review on pulp manufacture from non wood plant materials. Int J Chem Eng Appl 4(3):144–148

Mahapatra S, Kumar D, Singh B, Sachan PK (2021) Biofuels and their sources of production: a review on cleaner sustainable alternative against conventional fuel, in the framework of the food and energy nexus. Energy Nexus 4:100036

Mäki E, Saastamoinen H, Melin K, Matschegg D, Pihkola H (2021) Drivers and barriers in retrofitting pulp and paper industry with bioenergy for more efficient production of liquid, solid and gaseous biofuels: a review. Biomass Bioenergy 148:106036

McKinnon A (2010) Environmental sustainability: a new priority for logistics managers. In: Green logistics: improving the environmental sustainability of logistics. Kogan Page, New York, pp 3–30

Meng Y, Liu X, Li C, Liu H, Cheng Y, Lu J, Wang H (2019) Super-swelling lignin-based biopolymer hydrogels for soil water retention from paper industry waste. International journal of biological macromolecules, 135:815–820

Mohanty AK, Wu F, Mincheva R, Hakkarainen M, Raquez JM, Mielewski DF, Misra M (2022) Sustainable polymers. Nat Rev Methods Primers 2(1):46

Morlok J, Schoenberger H, Styles D, Gálvez-Martos J-L, Zeschmar-Lahl B (2017) The impact of pay-as-you-throw schemes on municipal solid waste management: the exemplar case of the county of Aschaffenburg, Germany. Resources 6:8

Nandal P, Arora A, Virmani S (2021) An appraisal on valorization of lignin: a byproduct from biorefineries and paper industries. Biomass Bioenergy 155:106295

Nigam PS, Singh A (2011) Production of liquid biofuels from renewable resources. Prog Energy Combust Sci 37(1):52–68

Olsson C, Sjoholm E, Reimanna A (2017) Carbon fibers from precursors produced by dry-jet wet-spinning of Kraft lignin blended with Kraft pulps. Holzforschung 71:275–283

Pokhrel D, Viraraghavan T (2004) Treatment of pulp and paper mill wastewater—a review. Science of the total environment, 333(1–3):37–58.

Pola L, Collado S, Oulego P, Díaz M (2022) Kraft black liquor as a renewable source of value-added chemicals. Chem Eng J 448:137728

Poudyal NC, Joshi O, Taylor AM, Hodges DG (2017) Prospects of wood-based energy alternatives in revitalizing the economy impacted by decline in the pulp and paper industry. For Prod J 67(7–8):427–434

Poveda-Giraldo JA, Solarte-Toro JC, Alzate CAC (2021) The potential use of lignin as a platform product in biorefineries: a review. Renew Sust Energ Rev 138:110688

Pütün E (2016) Pyrolysis: a sustainable way from biomass to biofuels and biochar. Cambridge University Press, Cambridge, p 239

Roy S, Muthoka RM, Park JH, Kim HC, Kim J (2022) Waste paper: a potential source for cellulose nanofiber and bio-nanocomposite applications. In: Re-use and recycling of materials. River Publishers, Kolkata, pp 327–344

Saeli M, Tobaldi DM, Seabra MP, Labrincha JA (2019) Mix design and mechanical performance of geopolymeric binders and mortars using biomass fly ash and alkaline effluent from paper-pulp industry. Journal of Cleaner Production, 208:1188–1197

Sewsynker-Sukai Y, David AN, Kana EG (2020) Recent developments in the application of Kraft pulping alkaline chemicals for lignocellulosic pretreatment: potential beneficiation of green liquor dregs waste. Bioresour Technol 306:123225

Shi Y, Yan X, Li Q, Wang X, Xie S, Chai L, Yuan J (2017) Directed bioconversion of Kraft lignin to polyhydroxyalkanoate by Cupriavidus basilensis B-8 without any pretreatment. Process Biochem 52:238–242

Singh A, Agrawal M, Marshall FM (2010) The role of organic vs. inorganic fertilizers in reducing phytoavailability of heavy metals in a wastewater-irrigated area. Ecol Eng 36(12):1733–1740

Singhvi MS, Zinjarde SS, Gokhale DV (2019) Polylactic acid: synthesis and biomedical applications. J Appl Microbiol 127(6):1612–1626

Sjöström E, Westermark U (1999) Chemical composition of wood and pulps: basic constituents and their distribution. In: Analytical methods in wood chemistry, pulping, and papermaking. Springer, Berlin, pp 1–19

Solanki J, Pitroda J (2012) Study of modulus of elasticity of concrete with partial replacement of cement by hypo sludge waste from paper industry. Global J Res Anal 2(3):40–41

Swain PK, Das LM, Naik SN (2011) Biomass to liquid: a prospective challenge to research and development in 21st century. Renew Sust Energ Rev 15(9):4917–4933

Ubando AT, Felix CB, Chen WH (2020) Biorefineries in circular bioeconomy: a comprehensive review. Bioresour Technol 299:122585

Vakkilainen E, Välimäki E (2009) Effect of lignin separation to black liquor and recovery boiler operation. Proceedings of the TAPPI's

Varjani SJ, Upasani VN (2017) A new look on factors affecting microbial degradation of petroleum hydrocarbon pollutants. Int Biodeterior Biodegradation 120:71–83

Viel M, Collet F, Lanos C (2020) Effect of compaction on multi-physical properties of hemp-black liquor composites. J Mater Res Technol 9(2):2487–2494

Wang Q, Lyu X, Liu Q, Liu X, Zhou W (2024) Recycle of soda waste sludge as the limestone material for Portland cement: Chemical composition, physical properties and hydration mechanism. Construction and Building Materials, 422:135790.

Worku LA, Bachheti A, Bachheti RK, Reis CER, Chandel A (2023) Agricultural residues as raw materials for pulp and paper production: overview and applications on membrane fabrication. Membranes 13(1):1–15

Yang J, Ching YC, Chuah CH (2019) Applications of lignocellulosic fibers and lignin in bioplastics: a review. Polymers 11(5):751

Zhang Q, Khan MU, Lin X, Yi W, Lei H (2020) Green-composites produced from waste residue in pulp and paper industry: a sustainable way to manage industrial wastes. J Clean Prod 262:121251

Challenges and Future Prospects of Conversion of Industrial Waste to Biofuel and Chemicals

26

Jagadeesh Chandra Bose and Anirudh Sharma

Abstract

The utilization of industrial waste to produce biofuels and chemicals is a novel and auspicious strategy for tackling environmental and economic issues. With the progression of technology, it is reasonable to anticipate the development of conversion procedures that are both more efficient and cost-effective. Ongoing research in the fields of catalyst development, biotechnology, and bioprocess engineering will be of paramount importance in enhancing the efficacy and functionality of these technologies. The notion of a circular economy, in which trash is regarded as a valuable resource, is increasingly gaining momentum. The conversion of industrial waste is in accordance with this method, hence facilitating the establishment of symbiotic relationships among different industries. With the escalating environmental concerns, industries may encounter mounting pressure to mitigate their carbon emissions. The conversion of industrial waste has the potential to assist companies in achieving their sustainability objectives and adhering to regulatory requirements. The increasing market share of biofuels and sustainable chemicals, driven by environmental concerns and legislative incentives, is expected to lead to a rise in demand for technology that facilitates the conversion of industrial waste. The exploration of a broader spectrum of feedstock materials, encompassing waste plastics, agricultural wastes, and other sources, has the potential to yield a more consistent and reliable supply of raw materials for conversion operations. The growing public consciousness regarding environmental concerns may lead to a rise in support and demand for sustain-

J. C. Bose (✉)
University Institute of Biotechnology, Chandigarh University, Mohali, Punjab, India

A. Sharma
Department of Biotechnology, Jaypee Institute of Information Technology, Noida, Uttar Pradesh, India

A. Kumari et al. (eds.), *Valorization of Solid Wastes to Biofuels and Chemical Products for Sustainable World*, https://doi.org/10.1007/978-981-96-8594-3_26

able waste management strategies, such as the utilization of garbage to produce biofuels and chemicals. The process of transforming industrial waste into biofuels and chemicals entails a set of obstacles and potential opportunities. The composition of industrial waste streams can exhibit substantial variability, hence posing difficulties in the development of conversion procedures that are both consistent and efficient. This book chapter aims at the conversion of industrial waste into biofuels and chemicals typically necessitating energy-intensive procedures and specialized equipment, resulting in significant implementation and operational costs. Handling and processing industrial waste products may necessitate compliance with regulatory barriers and acquisition of environmental permits, hence introducing intricacy and financial implications to such endeavors. The primary objective is to mitigate the environmental impact; however, it is important to acknowledge that the conversion process has the potential to produce emissions and trash, which might possibly undermine the environmental advantages if not effectively controlled. Recycling, landfilling, and incineration are well-established techniques for managing waste. Persuading enterprises to allocate resources toward the adoption of novel waste-to-biofuel technologies might present a formidable task.

Keywords

Industrial waste · Biodiesel · Bioethanol · Biogas · Waste management

26.1 Introduction

The processes of waste collection, transportation, processing, recycling, and disposal are all included in waste management. For the sake of public health, ecology, and community well-being, effective waste management is essential. The waste can be divided into recyclables, organic waste, and nonrecyclables at the point of origin for both individuals and companies. Garbage direcly collected from residences and commercial buildings. Conventional waste disposal techniques entail burying rubbish in specific locations. Burning waste to produce energy and reduce volume is known as incineration. Nonetheless, air pollution may result from this technique. The recycling and disposal of electronic waste require specialized techniques due to the potentially dangerous nature of some electronic components. Working together between nations overcomes problems with transboundary waste management and advances environmentally friendly behaviors. Industrial waste management is the organized collection, treatment, and disposal of garbage from industrial processes in a way that does not harm the environment and follows the rules (Nazemi et al. 2014). It involves using combined and long-lasting waste management methods to fix the environmental problems caused by throwing away waste in the wrong way (Hoveidi et al. 2013). There are many parts to industrial waste management, including managing hazardous waste, solid waste, and the growth of waste management trends in industrial zones (Kanagamani et al. 2021). Industrial waste is the collective term used to describe the waste materials that are generated because of various industrial operations. This encompasses a diverse array of substances that are produced throughout the course of manufacturing, mining, agriculture, and other industrial processes. The content, amount, and hazardousness of this waste can exhibit substantial variations, contingent upon the industry and specific operations in question. The management of industrial waste is of utmost importance for sustainable development due to the environmental, economic, and public health concerns it presents. The proper handling of industrial garbage is very important because it can pollute and damage the environment (Enes and Kipöz 2019). To manage industrial waste effectively, environmental laws and plans for industrial development must work together. This is so that solid waste does not hurt the environment and industrial growth does not stop (Nazemi et al. 2014). It also includes using economic, organizational, and technical methods to get the most out of processing, getting rid of, and getting secondary materials and energy resources from waste. Industrial waste management also includes picking the right technologies, capacities, sites, garbage suppliers, and heat consumers for waste facilities (Nixon et al. 2013). Managing industrial waste in emerging countries is harder because of laws that are not followed, pollution, and where the country is located. However, these problems need to be fixed if we want to lessen the damage that industry waste does to the environment in these areas. To sum up, industrial waste management is a complete way of dealing with the waste that comes from industrial processes in a way that is safe for the environment and follows the rules. Managing solid and hazardous waste and balancing industry growth with environmental rules are some of the things that are included. Taking on the problems that come with managing industrial waste is important for protecting the environment and promoting long-term industry growth.

26.2 Need for Industrial Waste Management

The management of industrial waste holds substantial worldwide implications due to the global scope of industrial activity. The implementation of global initiatives is crucial for effectively addressing the transnational difficulties related to the disposal of industrial waste and pollution. This academic inquiry focuses on the examination and application of methodologies aimed at efficiently managing and mitigating adverse consequences. The notion of waste reduction pertains to the implementation of strategies aimed at diminishing the quantity of waste produced in diverse processes and activities. The techniques that are designed to mitigate the development of waste at its source are implemented.

The adoption of sustainable industrial practices, efficient waste management systems, and the implementation of the concepts of circular economy are required in order to minimize the negative consequences on public health and the ecology. It is absolutely necessary to have effective hazardous waste management, biomedical waste management, and the reuse of industrial waste materials in order to reduce the negative effects that industrial activities have on the environment, safeguard public health, and maintain a healthy ecosystem.

26.3 The Concept of Recycling and Reuse

Recycling is the process of recovering and repurposing materials, such as glass, metal, plastic, and paper, to create new products. Composting is the process of breaking down organic waste into a nutrient-rich compost to improve soil. Advanced recycling technologies: Creating and putting into practice new technology to recycle materials more effectively and efficiently. Anaerobic digestion and gasification are two examples of waste-to-energy technologies that are being investigated.

Recycling and finding new uses for waste from industrial processes are necessary in order to protect human health and ecological systems. The idea of recycling and reusing industrial waste materials is consistent with the ideals of a sustainable circular economy, which aims to minimize the negative effects on the environment and preserve public health. Recycling of industrial waste, which includes e-waste, plastic waste, and hazardous garbage, plays an important part in lowering the burden placed on landfills and mitigating environmental contamination (Yu et al. 2009; Al-Ghabban et al. 2018). Some examples of industrial waste include hazardous waste, plastic waste, and e-waste. Additionally, the reuse of industrial waste materials, such as agri-food by-products and waste from the processing of kaolin, contributes to the conservation of resources, the savings of energy, and the reduction of production costs, while also providing alternative raw materials for a variety of industrial sectors (Gołębiewska et al. 2022).

According to Taufan and Purwanto (2018), the importance of recycling and reuse procedures in industrial sectors cannot be overstated. The handling of poisonous and hazardous waste materials in the food industry requires adequate utilization methods to maintain environmental and public health safety. In addition, the recycling of waste electrical and electronic equipment (WEEE) in China and the assessment of medical hazardous waste generation rates in Duhok Governorate highlight the necessity for integrated e-waste management and the development of alternative disposal and management methods to comply with international environmental regulations (Yu et al. 2009).

Recycling and reuse of waste from industrial processes have not only direct but also indirect effects on environmental justice and public health. According to Mohai and Saha (2007), the existence of racial disparities in the distribution of hazardous waste highlights the necessity of equitable waste management techniques as a means of shielding disadvantaged communities from the effects of environmental dangers. Furthermore, the recycling of the urban solid waste in mangrove forests in the Niger Delta highlights the necessity for proper waste management methods to minimize the degradation of the environment and preserve public health (Numbere 2019). Recycling and reuse of industrial waste are crucial for fostering a sustainable circular economy, limiting environmental pollution, and protecting public health. These goals can all be accomplished through recycling and reuse of industrial waste. In addition to addressing environmental justice concerns and ensuring fair waste management methods, these procedures help conserve resources, save energy, and cut production costs. In addition, they contribute to the reduction of overall production costs.

26.4 Numerous Types of Industrial Wastes that Can Be Converted into Biofuels

Several kinds of industrial wastes are capable of being converted into biofuels through a variety of procedures, contributing to the generation of sustainable energy as well as the management of waste: crop residues such as maize stover, wheat straw, and rice husks are examples of biomass by-products; sugarcane-processing by-products and waste products; waste products resulting from the processing of wood in the forestry and manufacturing industries; components of domestic trash can be broken down biologically, such as discarded food and yard debris; materials that can be recycled and utilized in the production of biofuels; waste produced during the production of food, including fruit peels, vegetable trimmings, and other by-products of the processing. Specific industries, like the textile or dairy industries, can encourage the growth of algae for the manufacture of biofuel. Used cooking oils from restaurants and individual homes can be recycled into a useful product called biodiesel. Waste or by-products from the cheese manufacturing process in the dairy industry are suitable for use in the generation of biogas or biofuel. It is possible to create biofuels out of waste items generated during the brewing process, including discarded grains and other by-products. Residues from the pulp and paper industry, such as black liquor, which is a by-product of the process of creating paper and can be converted into biofuels, are one example. Some of the waste products from oil extraction procedures, such as soybean or canola cake, can be used in the manufacturing of biofuel. Textile manufacturing of various types of waste, including cotton waste, and the processing of textiles result in the generation of waste that can be utilized in the production of biofuels.

The conversion of various types of industrial waste into biofuels presents a promising avenue for sustainable energy production and waste management. Agricultural waste, including crop residues and dung, has been identified as a potential source for biofuel production, contributing to the utilization of organic waste for energy generation (Yevich and Logan 2003). Additionally, waste biorefinery processes have been explored as a solution for converting waste materials into biofuels, highlighting the potential for global sustainability through circular economy initiatives (Leong et al. 2021). Next-generation biofuels, such as biomethane from waste and synthetic biofuels obtained via gasification of biomass, have garnered attention as sustainable alternatives for energy generation (Zinoviev et al. 2010).

Furthermore, the utilization of biofuel residues and waste materials has been investigated for sustainable bioenergy production. Processes such as anaerobic digestion for biogas production, as well as catalytic and enzymatic processes for liquid biofuels, have demonstrated the potential for converting waste into valuable energy resources (Sheehan et al. 2018). The synthesis and production of biofuels, including ethanol, methane, and biodiesel, from the fermentation of citrus waste exemplify the potential for bioenergy generation from organic waste streams (Mahato et al. 2021). Moreover, the role of agricultural and agro-industrial waste in boosting bioethanol production has been explored, emphasizing the potential for utilizing waste materials as feedstock for biofuel production (Elyamny et al. 2022).

Lignocellulosic biomass, including forest product waste and switchgrass, has been recognized as a potential feedstock for next-generation biofuels, emphasizing the importance of sustainable waste utilization for energy applications (Gopalakrishnan et al. 2013).

The potential for utilizing waste materials for biofuel production extends to microbial work on biowaste resources, with the aim of developing biofuels based on cellulosic biomass and lipid-producing microbes (Pranaw et al. 2022). Furthermore, the utilization of waste materials for biofuel production has implications for sustainable aviation biofuels, providing opportunities for the development of renewable energy sources in the aviation industry (Barbosa Cortez et al. 2015). The production of biofuels from waste materials, such as bioethanol and biogas, has been explored as a sustainable approach to addressing energy needs and waste management (Mahato et al. 2021). Additionally, the conversion of waste fish oil into biofuel has demonstrated the potential for utilizing unconventional waste streams for energy generation (Mrad et al. 2012).

The conversion of industrial waste into biofuels offers a sustainable approach to energy production and waste management. The utilization of various waste materials, including agricultural residues, industrial by-products, and organic waste, presents opportunities for developing renewable energy sources and promoting circular bioeconomy initiatives.

26.5 Various Technologies for Conversion into Biofuels

The process of converting organic waste into biogas (methane) and digestate is known as anaerobic digestion. During fermentation, sugars or other organic materials are converted into ethanol or other types of biofuels. Pyrolysis is the process of producing bio-oil, charcoal, and syngas from the decomposition of biomass that is stimulated by heat. The process of converting biomass into syngas, which can then be further processed into biofuels, is referred to as gasification. The production of biodiesel involves transesterification, in which fats or oils are converted into biodiesel. Industrial wastes can be converted into biofuels through the utilization of these conversion technologies, which not only provides a source of energy that is renewable and sustainable but also helps to address issues pertaining to waste management. The exact features of the waste feedstock and the desired biofuel output are often taken into consideration when deciding the technique of conversion to use.

26.6 Various Types of Industrial Wastes Can be Effectively Converted into Chemicals

It is possible to effectively transform a variety of different types of industrial waste into useful chemicals, which can both contribute to environmentally responsible waste management and the manufacturing of high-value goods. For instance, food waste has been recognized as a possible feedstock for the synthesis of chemicals and liquid fuels, which offers a second generation of waste valorization techniques. These techniques are referred to as "waste valorization techniques." In addition, the exploitation of flotation wastes of copper slag as raw material in the production of cement illustrates the potential for turning waste from industrial processes into important construction materials. Furthermore, the transformation of calcium hydroxide that has been polluted with gases that have evolved from the smokestacks of the glass sector into calcium hydroxyapatite is an example of the utilization of chemical wastes in order to decrease pollution and add economic benefits.

In addition, agro- and forest-industrial wastes have been recognized as plentiful and low-cost sources of carbohydrates and phenolic compounds. These wastes have the potential to be transformed into biofuels, biomaterials, and chemicals with a high added value through the use of hydrothermal treatments. Plasma gasification of refuse-derived fuel in a single-stage system utilizing multiple gasifying agents has been investigated for the generation of syngas. Syngas can be valorized for both energy and chemical end-purposes if it is produced in this manner. In addition, the exploitation of filters made from natural sources for the removal of toxins from wastewater is a demonstration of the potential for the utilization of chemical waste in environmental applications.

Furthermore, the potential for waste valorization and the creation of valuable chemicals is highlighted by the conversion of biomass to biofuels, carbohydrates, monosaccharides, or other chemicals such as organic acids and phenolic compounds through the processes of biorefinery.

There are many other different forms of industrial wastes that are capable of being properly transformed into chemicals that have applications that are useful for human use. Methanol is a versatile chemical that is utilized in a variety of industrial processes. It can be produced by converting methane that has been obtained through anaerobic digestion or from landfills. Hydrolysis and fermentation are two examples of the processes that can be used to convert agricultural by-products, such as maize stover, wheat straw, and rice husks, into biobased compounds. Through a procedure known as transesterification, used cooking oil can be transformed into biodiesel. It is possible to cultivate algae to create oils that can then be transformed into biofuels. Additionally, the biomass that is produced from algae can be utilized in order to extract a variety of biochemicals. Through a combination of enzymatic hydrolysis and fermentation, bagasse, which is a by-product of the processing of sugarcane, can be utilized to make cellulosic ethanol. In the process of producing paper, a by-product called black liquor is produced. This black liquor can be treated to recover compounds like lignin and hemicellulose.

The effective transformation of industrial waste into chemicals frequently calls for the application of cutting-edge technologies and novel processing methods. The objective is to establish a circular economy in which discarded materials are treated as valuable resources, so fostering sustainability and minimizing adverse effects on the environment. The waste stream and the chemical products that are needed should guide the selection of the appropriate conversion technique. In addition, ensuring that these conversion procedures are sustainable is dependent upon complying with any applicable regulations and considering any relevant environmental factors. The growth of activities associated with circular economies has the potential to further improve the exploitation of trash from industrial processes as valuable resources.

26.7 Various Types of Biofuels Generated by Industrial Waste

The feedstock used and the manufacturing method together can be used to categorize the many kinds of biofuels that are produced from the waste products of industrial processes. The following are some examples of frequent types:

26.7.1 Biodiesel

Waste cooking oil, animal fats, or vegetable oils can all be used as feedstock. Transesterification of triglycerides in the feedstock with alcohol (often methanol or ethanol) is the process that is used during production (Figs. 26.1 and 26.2).

26.7.2 Bioethanol

Feedstock can be agricultural wastes (such as maize stover or wheat straw), waste wood, or crops specifically grown for their energy content. The fermentation of any sugars or starches presents in the feedstock, followed by distillation, makes up the production process.

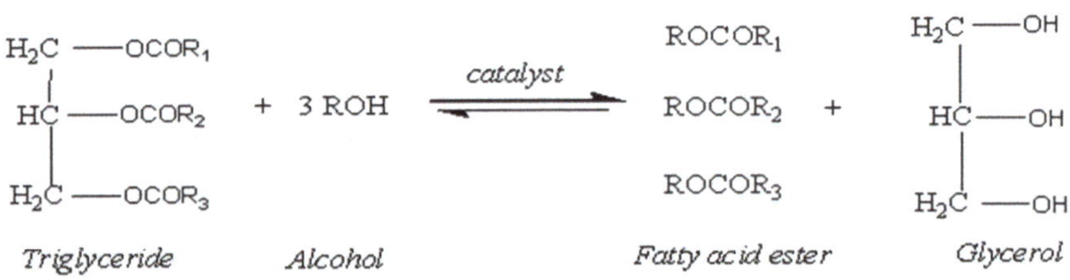

Fig. 26.1 Transesterification reaction to produce biodiesel

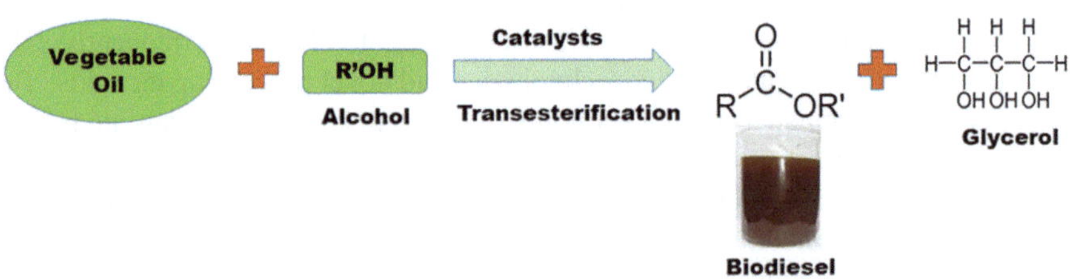

Fig. 26.2 Schematic representation of generation of biodiesel from vegetable oils

26.7.3 Biogas

Organic wastes, such as agricultural wastes, food waste, and organic waste from industrial processes, are all considered feedstock. Methane and carbon dioxide are the by-products of a production process that involves the anaerobic digestion of organic sources by microorganisms.

Syngas, also known as synthetic gas: Feedstock can come from a wide variety of biomass sources, such as discarded wood, agricultural wastes, or organic waste from industrial processes. The process begins with the gasification of biomass, which results in the production of a combination of carbon monoxide and hydrogen.

26.7.4 Hydrogen (H_2)

Organic waste, such as biomass or wastewater, is referred to as feedstock. Producing hydrogen can be accomplished by several processes, including steam methane reforming (SMR) and the gasification of biomass. Fermentation or other biochemical processes may be used during the production process in order to produce platform chemicals such as organic acids, alcohols, or other useful molecules.

26.7.5 Oil Created Via Pyrolysis

Materials that are used as feedstock include biomass and organic waste products. Pyrolysis, which is a thermal degradation process, is utilized as the production method for converting biomass into bio-oil.

Methane hydrates are known as organic waste and residues that are high in methane are examples of feedstock. The formation of methane hydrates, which are crystalline structures that contain methane, from organic waste as part of the production process.

26.7.6 Biofuels in Aviation

Feedstock comes from a wide variety of sources, including agricultural by-products and trash from manufacturing. Utilizing cutting-edge methods like hydroprocessing or Fischer-Tropsch synthesis to generate aviation-grade biofuels.

26.7.7 Synthetic Fuels Derived from Biomass

The production method utilizes cutting-edge procedures, such as the Fischer-Tropsch synthesis or other cutting-edge thermochemical techniques, to make synthetic fuels.

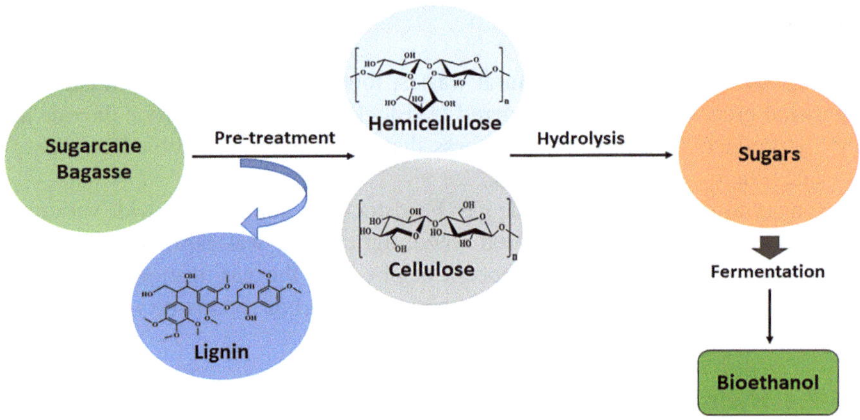

Fig. 26.3 Schematic representation of bioethanol production from lignocellulosic biomass

26.7.8 Biofuels Made from Lignocellulosic Material

Feedstock might be agricultural wastes, trash from forestry operations, or crops specifically grown for their energy content. Conversion of lignocellulosic biomass into biofuels such as cellulosic ethanol or other advanced biofuels is the process that underlies the production of these fuels (Fig. 26.3).

26.7.9 Fuels for Biojet

A wide variety of waste resources, such as municipal solid trash and agricultural wastes, are referred to as feedstock. The production process involves the utilization of cutting-edge processing techniques to generate aviation-grade biofuels.

The qualities of the waste feedstock, the ultimate use that is intended, and the technology that is used to convert the waste all play a role in determining the type of biofuel that should be used. The use of waste materials in the manufacture of biofuel not only creates a source of renewable energy but also contributes to the management of waste and creates a more environmentally friendly approach to the generation of energy.

26.8 Best Industrial Waste Materials Used for the Production of Biofuels

According to the references that have been offered, a number of different types of industrial waste have been recognized as potential feedstocks for the generation of biofuels. These waste products have the potential to be used in the manufacturing of sustainable biofuels, which would contribute to the valorization of waste and the development of renewable energy. The following types of industrial waste are particularly significant for the generation of biofuels due to their compatibility with the process:

1. Agro-industrial waste: Certain types of agro-industrial waste, including as rice straw, wheat straw, and corn stover, have been identified as desirable lignocellulosic feedstocks for the manufacture of biofuel. These waste products have an advantageous conversion factor for the manufacturing of biofuel, which positions them as a potentially useful resource for the development of sustainable energy (Elyamny et al. 2022). According to Zinoviev et al. (2010), these waste materials, which include trash from forestry and by-products of companies based on biomaterials, present prospects for the manufacture of sustainable biofuels and the valorization of waste.
2. Food waste: The use of food waste as a resource in the manufacturing of biofuels has been recognized as having significant potential value. According to Ki Lin et al. (2013), the usage of food waste as a raw material for the production of biofuel presents the possibility of transforming trash into clean energy and valuable chemicals, so contributing to the valorization of waste and the development of sustainable bioenergy.
3. Cocoa waste from industry: Researchers have investigated the possibility of using cocoa waste from industry to produce solid, liquid, and gaseous forms of biofuel. The exploitation of cocoa industrial waste as a direct combustion source reveals its potential for biofuel production. This strategy provides a sustainable method for waste valorization and energy generation.
4. Used cooking oil: Kitchen waste, and waste cooking oil in particular, has been discovered as a biodiesel raw material with a high yield. This indicates that there are good development prospects for the manufacturing of biofuels. According to Geng et al. (2021), the use of used cooking oil as a raw material in the manufacturing of biodiesel can result in both financial gains and the generation of environmentally friendly energy.
5. Waste bread: Enzymatically hydrolyzed waste bread has been investigated for its use as a potential sugar substitute in the production of wheat bread. The use of stale bread as a sugar substitute presents prospects for waste valorization as well as the manufacture of biofuels and biomaterials, so contributing to the sustainable exploitation of resources (Rosa-Sibakov et al. 2022).
6. Waste fats and oils: Waste fats and oils obtained from traditional industrial industries, such as leather fat, waste anchovy fish oil, and waste frying cottonseed oil, have been examined to see whether or not they are suitable for use as raw materials in the creation of biofuels. According to Altun and Rodrguez-Fernǎndez, the exploitation of waste fats and oils presents an opportunity for the manufacture of a sustainable biofuel as well as the valorization of waste.

26.9 Major Challenges in the Conversion of Industrial Waste into Biofuels

While the conversion of industrial waste into biofuels holds great promise for sustainable energy production and waste management, several challenges need to be addressed for the process to be effective and economically viable.

26.9.1 Diverse Feedstock Composition

Industrial waste streams vary widely in composition, making it challenging to develop standardized conversion processes. Different waste materials require specific technologies for efficient biofuel production. Industrial waste streams are diverse and vary widely in composition, making it challenging to develop standardized conversion processes. Different feedstocks require specific technologies, and adapting processes to diverse waste types can be complex.

The impact of diverse feedstock composition on biofuel generation is a critical aspect of bioenergy production. The selection of feedstock significantly influences the efficiency, sustainability, and economic viability of biofuel production. The diverse composition of feedstock materials plays a crucial role in determining the overall performance and environmental impact of biofuel generation. The following references provide insights into the impact of diverse feedstock composition on biofuel generation:

Jeswani et al. (2020) emphasize the significance of feedstock composition in biofuel production and its impact on biodiversity. The review highlights that the choice of feedstock and the scale of production can have implications for biodiversity and environmental sustainability. This underscores the importance of considering diverse feedstock compositions to minimize adverse effects on ecosystems.

Chandrasekar et al. (2021) discuss the interactions between feedstocks and biofuel production processes. The study emphasizes that environmental factors can have detrimental effects on saccharification and fermentation processes, indicating that diverse feedstock compositions may influence the efficiency of biofuel production processes.

Zhu (2015) provides insights into the types of feedstocks used for first- and second-generation biofuels. The review highlights the shift from edible feedstocks to waste and dedicated lignocellulosic materials for second-generation biofuels, indicating the importance of diverse feedstock compositions in sustainable biofuel production.

Rajarapu et al. (2015) focus on the metatranscriptomic profiles of termites fed on second-generation lignocellulosic feedstocks. The study underscores the potential of lignocellulosic feedstocks as alternatives to first-generation biofuels derived from grain starches and sugars, highlighting the impact of diverse feedstock compositions on biofuel innovation.

Cheah et al. (2020) discuss the pretreatment methods for lignocellulosic biofuels production. The study emphasizes that lignocellulosic materials are the most abundant feedstock for second-generation biofuels, indicating the importance of diverse feedstock compositions in advancing biofuel production technologies.

These references collectively highlight the importance of diverse feedstock compositions in biofuel generation. The selection of feedstock materials significantly influences the environmental, economic, and technological aspects of biofuel production, emphasizing the need to consider a wide range of feedstock compositions for sustainable and efficient bioenergy generation.

26.9.2 Feedstock Availability and Seasonality

The consistent and reliable availability of certain industrial waste feedstocks, such as crop residues or specific manufacturing by-products, can be influenced by seasonal variations and fluctuations in industrial production.

The production of biofuel is substantially affected by the availability of feedstock as well as the seasonality of feedstock, which in turn has an effect on the economics, sustainability, and logistics of the supply chain for bioenergy. The references offer some illumination into the impact of seasonality and the availability of feedstock on the production of biofuel.

Xu and Li (2017) stress the fluctuation in the availability of raw materials for the manufacture of biofuel. They also emphasize the necessity of stable feedstock availability in order to sustain consistent biofuel production.

In conclusion, the availability of feedstock and the seasonality of feedstock have a substantial influence on biofuel production. Because of this, it is essential to have a steady and diverse availability of feedstock in order to enable consistent and sustainable biofuel production.

26.9.3 Economic Viability

The economic viability of converting industrial waste into biofuels is often hindered by high capital costs, operational expenses, and competition with low-cost fossil fuels. Achieving cost-competitive biofuels is crucial for widespread adoption. The economic feasibility of industrial waste-to-biofuel conversion is a critical hurdle. High capital costs, operational expenses, and competition with low-cost fossil fuels can make it difficult for biofuels derived from industrial waste to be economically competitive.

Biofuels produced on a large scale present a number of obstacles that must be surmounted prior to commercialization. The following sources offer valuable perspectives on the difficulties associated with expanding the production of biofuels. Farrokh et al. (2019) underscore the critical concern of expanding the biomass production capacity of photosynthetic microorganisms to facilitate the manufacturing of biofuels on a large scale. The authors specifically highlight the obstacles that arise when attempting to achieve this. Cheah et al. (2020) highlight the significant obstacle that prevents the widespread implementation of lignocellulosic feedstock into biofuels: the exorbitant expense of pretreatment. This underscores the economic complexities that are inherent in the process of expanding biofuel production. Azimov et al. (2021) shed light on the obstacles that hinder the commercialization of advanced microbial biofuel candidates by describing the technological and commercial complexities involved in expanding the production of such fuels. Sur et al. (2020) highlight the difficulties that arise when attempting to increase the commercial-scale production capacity of bioethanol, noting that scaling up has become the greatest obstacle in the bioethanol manufacturing process.

26.10 Prospects of Industrial Waste Conversion into Biofuels and Other Chemicals

The possibilities for sustainable energy generation, waste management, and resource exploitation are all enhanced by the conversion of industrial waste into biofuels and other chemicals. Woodard (2006) covers the treatment of industrial waste, which has the potential to contribute to the creation of biofuels and other valuable chemicals. The author emphasizes the need proper waste management methods by pointing out that this treatment can potentially contribute to the manufacturing of these chemicals. Blaschek et al. (2010) investigate the possibility of producing biofuels from the waste and by-products of agricultural production, drawing attention to the opportunities for the generation of sustainable biofuels from agricultural waste streams. Bhatia et al. (2021) explore the future challenges and bioeconomic prospects of exploiting lignocellulosic waste biomass for biohydrogen generation. They place particular emphasis on the potential for reducing energy crises and boosting the creation of green energy. Jayathilakan et al. (2011) examine the utilization of byproducts and waste materials from the production of meat, poultry, and fish. They focus on the possibilities of recycling raw materials and converting them into useful products with greater value. Karmee and Ki Lin (2014) explore the process of converting food waste into biofuel. They place an emphasis on the future potential of industrial production of biodiesel and bioethanol from food waste as a solution to the problems of improper waste disposal and a lack of available energy. Ki Lin et al. (2013) emphasize the different applications of food supply chain waste for the manufacture of biofuel and chemicals, drawing attention to the potential of food waste as a useful resource for the production of chemicals, materials, and fuels. This is done by highlighting the potential of food waste as a valuable resource. Elyamny et al. (2022) examine the unparalleled catalysis of the sequential pretreatment of sustainable agricultural and agro-industrial wastes in the context of improving bioethanol production. They place an emphasis on the potential for the generation of clean energy as well as sustainable development. In their study of the developing circular economy, Pranaw et al. (2022) places a strong emphasis on the notion of "waste to wealth," which highlights the possibility of transforming waste materials into valuable industrial products. Kichonge and Kivevele examine the practicability of using nonedible oilseed plants and agricultural wastes as feedstock to produce biofuels. They place an emphasis on the prospects for the expansion of bioenergy in

Africa. The potential for converting industrial waste into biofuels and other chemicals in the future is both encouraging and congruent with the objectives of sustainability, circular economy, and the elimination of environmental problems.

References

Al-Ghabban MM, Mizzouri NS, Mahmood FR, Hassan HH, Abdulrahman KI (2018) Assessment of waste generation rate of medical hazardous in duhok governorate (proposal of alternative disposal and management methods). Acad J Nawroz Univ 7(4):139

Azimov U, Okoro O, Hernandez HH (2021) Recent progress and trends in the development of microbial biofuels from solid waste—a review. Energies 14(19):6011

Barbosa Cortez LA, Baccaro Nigro FE, Horta Nogueira LA, Nassar AM, Cantarella H, Ferraz Moraes MA, Baldassin R (2015) Perspectives for sustainable aviation biofuels in Brazil. Int J Aerospace Eng

Bhatia L, Sarangi PK, Singh AK, Prakash A, Shadangi KP (2021) Lignocellulosic waste biomass for biohydrogen production: future challenges and bio-economic perspectives. Biofuels Bioprod Biorefin

Blaschek HP, Ezeji TC, Scheffran J (2010) Biofuels from agricultural wastes and byproducts. Wiley, Hoboken

Chandrasekar M, Joshi L, Krieg K, Chipkar S, Burke E, Debrauske DJ, Ong RG (2021) A high solids field-to-fuel research pipeline to identify interactions between feedstocks and biofuel production. Biotechnol Biofuels

Cheah WY, Sankaran R, Show PL, Baizura Ibrahim TN, Chew KW, Culaba AB, Chang JS (2020) Pretreatment methods for lignocellulosic biofuels production: current advances, challenges and future prospects. Biofuel Res J

Elyamny S, Hamdy A, Ali RM, Hamad H (2022) Role of combined Na_2HPO_4 and $ZnCl_2$ in the unprecedented catalysis of the sequential pretreatment of sustainable agricultural and agro-industrial wastes in boosting bioethanol production. Int J Mol Sci

Enes E, Kipöz Ş (2019) Türkiye Moda Endüstrisinin Kesim Atık Problemi Ve Atık Yönetim Stratejileri. Tekstil Mühendis

Farrokh P, Sheikhpour M, Kasaeian A, Asadi H, Bavandi R (2019) Cyanobacteria as an eco-friendly resource for biofuel production: a critical review. Biotechnol Prog

Geng N, Fu Q, Sun Y (2021) Stochastic programming of sustainable waste cooking oil for biodiesel supply chain under uncertainty. J Adv Transp

Gołębiewska E, Kalinowska M, Yildiz G (2022) Sustainable use of apple pomace (AP) in different industrial sectors. Materials

Gopalakrishnan K, Ceylan H, Kim S (2013) Renewable biomass-derived lignin in transportation infrastructure strengthening applications. Int J Sustain Eng

Hoveidi H, Pari MA, Pazoki M, Koulaeian T (2013) Industrial waste management with application of RIAM environmental assessment: a case study on Toos industrial state, Mashhad. Iran J Energy Environ

Jayathilakan K, Sultana K, Radhakrishna K, Bawa AS (2011) Utilization of byproducts and waste materials from meat, poultry and fish processing industries: a review. J Food Sci Technol

Jeswani HK, Chilvers A, Azapagic A (2020) Environmental sustainability of biofuels: a review. Proc R Soc Math Phys Eng Sci

Kanagamani K, Geethamani P, Narmatha M (2021) Hazardous waste management. IntechOpen, London

Karmee SK, Ki Lin CS (2014) Valorisation of food waste to biofuel: current trends and technological challenges. Sustainable Chem Process

Ki Lin CS, Pfaltzgraff LA, Herrero-Davila L, Mubofu EB, Solhy A, Clark JH, Luque R (2013) Food waste as a valuable resource for the production of chemicals, materials and fuels. Current situation and global perspective. Energy Environ Sci
Leong HY, Chang CK, Khoo KS, Chew KW, Chia SR, Lim JW, Show PL (2021) Waste biorefinery towards a sustainable circular bioeconomy: a solution to global issues. Biotechnol Biofuels
Mahato N, Sharma K, Sinha M, Dhyani A, Pathak B, Jang H, Cho S (2021) Biotransformation of citrus waste-I: production of biofuel and valuable compounds by fermentation. Processes
Mohai P, Saha R (2007) Racial inequality in the distribution of hazardous waste: a national-level reassessment. Soc Probl
Mrad N, Aloui F, Tazerout M (2012) Performance and emissions of diesel engine using bio-fuel derived from waste fish oil
Nazemi S, Roudbari A, Yaghmaeian K (2014) Design and implementation of integrated solid wastes management pattern in industrial zones, case study of Shahroud, Iran. J Environ Health Sci Eng
Nixon J, Wright DG, Dey PK, Ghosh SK, Davies PA (2013) A comparative assessment of waste incinerators in the UK. Waste Manag
Numbere AO (2019) Municipal solid waste disposal in mangrove forest: environmental implication and management strategies in the Niger Delta, Nigeria
Pranaw K, Drewniak L, Nain L, Singh S (2022) Editorial: waste to wealth: a sustainable circular bioeconomy approach. Front Bioeng Biotechnol
Rajarapu SP, Shreve J, Bhide KP, Thimmapuram J, Scharf ME (2015) Metatranscriptomic profiles of eastern subterranean termites, reticulitermes flavipes (kollar) fed on second generation feedstocks. BMC Genomics
Rosa-Sibakov N, Sorsamäki L, Immonen M, Nihtilä H, Maina NH, Siika-aho M, Nordlund E (2022) Functionality and economic feasibility of enzymatically hydrolyzed waste bread as a sugar replacer in wheat bread making. J Food Process Preserv
Sheehan N, Plante L, Martinez E, Murray KE, Bier P, Ouellette C (2018) Sustainable bioenergy from biofuel residues and waste. Water Environ Res
Sur S, Dave V, Prakesh A, Sharma P (2020) Expansion and scale up of technology for ethanol production based on the concept of biorefinery. J Food Process Eng
Taufan HS, Purwanto P (2018) The management of toxic and hazardous waste materials in the food industry. E3s Web Conf
Woodard F (2006) Industrial waste treatment handbook. Elsevier, Amsterdam
Xu J, Li M (2017) Innovative technological paradigm-based approach towards biofuel feedstock. Energy Convers Manag
Yevich R, Logan JA (2003) An assessment of biofuel use and burning of agricultural waste in the developing world. Glob Biogeochem Cycles
Yu J, Mei J, Williams E (2009) Waste electrical and electronic equipment recycling in China: practices and strategies
Zhu L (2015) Microalgal culture strategies for biofuel production: a review. Biofuels Bioprod Bioref
Zinoviev S, Müller-Langer F, Das P, Bertero NM, Fornasiero P, Kaltschmitt M, Miertus S (2010) Next-generation biofuels: survey of emerging technologies and sustainability issues. ChemSusChem

Part IV

Life Cycle Assessment (LCA) of valorisation of solid waste

Life Cycle Assessment of Valorization of Municipal Solid Waste to Biofuels

27

Manoj Kumar, Alisha Verma, Zeeshan Hussain Hashmi, Renu Yadav, and Zia Zehra Zaidi

Abstract

Because of sustainability, municipal solid waste management (MSWM) has changed from direct disposal to recycling and resource recovery. The MSWM structure's environmental impact can be evaluated using the life cycle assessment (LCA) approach. By combining information from waste management procedures, transportation, waste collection, and product usage, a LCA technique is applied. Functional unit definition, system boundaries, complementary system design, waste management, and partial energy consumption were all covered in the analytical framework. The purpose the study is to assess the effects of the MSWM system in Nagpur, India, fewer than four distinct situations for instance, using an LCA tool, landfilling and composting (S1), land filling and anaerobic digestion (AD) and composting (S2), material recovery facility (MRF) and composting and land filling (S3), and MRF, AD, and composting and land filling (S4). In one situation that was considered, the implications of the recycling rate of precious resources were further assessed using a sensitivity study. The

M. Kumar (✉)
Department of Mechanical Engineering, ABES Engineering College, Ghaziabad, Uttar Pradesh, India

A. Verma
Symbiosis Law School, Pune, Maharashtra, India

Z. H. Hashmi
Amity Law School, Amity University, Jaipur, Rajasthan, India

R. Yadav
NorthCap University, Gurugram, Haryana, India

Z. Z. Zaidi
GNIOT Institute of Management Studies (GIMS), Greater Noida, Uttar Pradesh, India

A. Kumari et al. (eds.), *Valorization of Solid Wastes to Biofuels and Chemical Products for Sustainable World*, https://doi.org/10.1007/978-981-96-8594-3_27

CML-1A impact characterization approach and the Gabi 8.5.0.79 model were used to compare the scenarios. It was discovered that S2 had the least negative effects on the possible categories of photochemical ozone formation, eutrophication, human toxicity, and global warming. According to the sensitivity analysis, changes in recycling rates and overall environmental impacts are inversely related.

Keywords

LCD · Landfill · Biofuels · Anaerobic digestion · MSWM

27.1 Introduction

In many nations, municipal solid waste (MSW) is a complicated issue that affects both the populace's welfare and the environment's quality (Charkhestani and Yousefi Kebria 2022; Ghazali et al. 2021). Previously, the change from linear to circular economic management in waste management programs worldwide was driven by the amount of uncontrolled garbage production that was connected with population expansion and its heterogenous character (Kurniawan et al. 2022). A strategy known as sustainability aims to strike a balance between social, economic, and environmental factors. In the context of municipal solid waste (MSW) creation, sustainability emphasizes the necessity of reducing trash generation, managing garbage in an eco-friendly way, and promoting efficient production cycles (Bakan et al. 2022; Weekes et al. 2021).

Landfills pose a threat to soil and water systems in addition to air pollution. Groundwater and nearby ecosystems are contaminated by leachate, the liquid residue of waste decomposition, which frequently contains organic pollutants, heavy metals, and hazardous chemicals (Vaverková 2019). Furthermore, landfills are becoming known to be sources of microplastics that leak into the environment and may have an impact on human health and biodiversity (Wojnowska-Baryła et al. 2022).

The advanced thermal incineration process can employ raw waste as a feedstock. Incineration procedures have been carried out at 850 °C with air present, converting waste into carbon dioxide, water, and nonflammable compounds with solid residue (bottom ash) (Slorach et al. 2020). Because it allows waste management facilities to generate both heat and power, incineration is a widely popular technique in Europe. Incineration processes primarily comprise two parts (Adipah and Kwame 2019): (a) burning the trash to produce flue gas, and (b) cleaning the flue gas before it is released into the atmosphere.

Waste management includes the transportation, recycling, collection, processing, and disposal of waste items as well as their monitoring (Das et al. 2019). It is projected that the amount of MSW generated worldwide in 2007 was 2 billion tons, up from 0.49 billion tons in 1997, with an estimated yearly growth rate of 2–3% in emerging nations. Local governments are always in charge of managing MSW.

A tool called LCA can assess the environmental aspects and possible environmental repercussions of a good or service that is provided to the public. According to ISO, the general framework for LCA comprises four primary phases: definition of the purpose and scope, analysis of the life cycle inventory (LCI), evaluation of the life cycle impact (LCIA), and interpretation. Quantifying the environmental characteristics of the system under study, such as GHG emissions, particle emissions, etc., as well as any potential environmental repercussions, such as climate change, human health, etc., would result from following these processes. LCA can also be applied to waste management to assist in evaluating various situations and treatments in order to determine and choose the best waste management approach (Schrijvers et al. 2020).

27.2 Objective of Study

To provide the optimal waste management strategy for Nagpur city based on the lowest emissions and maybe extrapolate the findings of the present research to other areas as well, the study compared the environmental effects of three other proposed MSWM scenarios and current waste management practices from a life cycle perspective. Gate to grave is the life cycle period or scope taken into consideration. It entails transporting waste to facilities for treatment (anaerobic digestion [AD], composting, MRF), after which the remnants are dumped in a landfill.

27.3 Literature Review

27.3.1 The Source of Municipal Solid Waste

MSW is a type of garbage that is collected by a municipality in a specific area and primarily consists of domestic waste, occasionally with the inclusion of business waste. They come in solid or semisolid form, and industrial hazardous wastes are typically not included. By using energy, particularly the burning of fossil fuels involved in the transportation, use, and disposal of the product or substance that ends up as garbage, MSW can contribute to greenhouse gas emissions. Emissions of CH_4 from landfills where garbage is dumped, as well as CO_2 and N_2O from burning waste, are other major roots of greenhouse gas emissions that cause global warming (Anshassi et al. 2021). The effects of various wastes and waste management techniques on energy use also vary. Paper product recycling and source reduction, for instance, can boost forest carbon sequestration, lower energy consumption, and reduce emissions from combustion and landfills (Hockstad and Hanel 2018).

27.3.2 Landfill

The physical facilities used to dispose of solid waste residues and solid waste in the earth's surface soils are referred to as "landfills" (Chandrappa and Das 2024). Despite being quite old, landfills are still one of the most widely utilized systems for managing MSW and advanced sanitary landfills. They can also collect landfill gas and use it to produce power or biofuel. The landfill site is the source of landfill gas. There are typically five stages involved in the production of biogas in a landfill: initial adjustment, transition, acid phase, methane fermentation, and maturity phase (Cossu and Stegmann 2018; Paruti and Santhaveeranagoud 2019).

27.3.3 Anaerobic Digestion (AD)

One of the best approaches for managing organic waste, AD has a greater potential for resource recovery. With the aid of bacteria, the biological process of AD occurs in the absence of oxygen. The anaerobic process creates compost and biogas (Lin et al. 2019). After 2–3 weeks, biogas is created, which is composed of carbon dioxide (CO_2) and methane (between 55 and 70%). Depending on its nutrient level, compost can be used as organic fertilizer.

27.3.4 MSWM-LDA

The 1990s saw a significant increase in the use of MSW LCA to assess environmental impacts (Christensen et al. 2020). The International Organization for Standardization (ISO) published guidelines that make it easier. The LCA of MSW used a variety of techniques and methodologies when the assessment criterion was established. For MSW and its consequences for environmental effects, LCA was employed as a tool. Later on, it was seen to be a helpful tool for comparing two or more alternative production systems by assessing their ecological sustainability and environmental implications (Wang et al. 2020).

27.4 Methodology

27.4.1 Study Region and Composition of MSW

The 10 administrative wards of Nagpur, India, are included in the current study. With an estimated 24.06 million people living in the 227.4 km^2 total area under investigation, 1200 metric tons of MSW are produced daily, with a typical day to day MSW age pace of 0.50 kg/capita. 60.10% of the MSW delivered in the city is natural, 16.50% is plastic, 10.20% is paper, 5.50% is cardboard, 1.10% is every one of material, inerts, and thermocol, 2.80% is wood, 0.50% is glass, and 0.20% is metal. As per the organization, organic treatment techniques like fertilizing the AD would be the most reasonable for the district in light of the fact that most of the waste delivered in Nagpur is natural in nature.

27.4.2 Life Cycle Assessment (LCA)

A product, movement, or administration can be assessed at each phase of its life cycle, from the securing of unrefined components through handling, fabricating, dissemination, use, conceivable reuse or recycling, and management of waste, as indicated by studies. LCA, as characterized by ISO 14,040 and ISO 14044, is an ecological administration instrument (Zarea et al. 2019). The four stages of the LCA technique—inventory analysis, impact assessment and interpretation, aim and scope definition, and inventory analysis—were particularly incorporated into the current study. LCA is now a widely used tool that, depending on how it is applied, can provide a more thorough and equitable environmental assessment when comparing waste management systems (Edwards et al. 2018). In waste management, LCA is used to evaluate the potential environmental impacts of the waste life cycle, from generation to disposal.

27.4.3 Situations

There are four particular situations, including the pattern situation for Nagpur's MSWM rehearses, as shown in Table 27.1. In Table 27.1, COM is for compositing, AD is for anaerobic digestion, MRF is for material recovery facility, and Lf is for landfill.

27.4.4 LCI

An inventory of natural output and input is delivered by this part, which focusses on the ID and estimation of framework related ecological activities. The Municipal Corporation office of Nagpur (India), on location research, recently distributed writing, and the Gabi 8.5.0.79 data set gave the informational index utilized in this

Table 27.1 An explanation of the scenarios utilized in this study

Scenario	Description
S4 MRRF_COM_ AD_LF	70% of biodegradable waste to composite +30% of recyclables to MRF +45.5% landfill residual waste
S3 MRF_AD_LF	70% of biodegradable waste to anaerobic digestion +30% of recyclables to MRF +45.5% landfill residual waste
S2 MRF_COM_LF	70% of biodegradable waste to composite +30% of recyclables to MRF +45.5% landfill residual waste
S1 COM_LF	73% MSW to landfill +27% MSW to composting

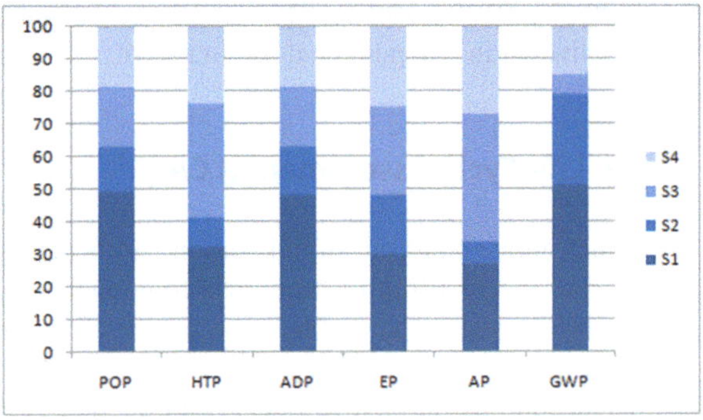

Fig. 27.1 LCA description of the impact categories under consideration

review. The framework limit can be isolated into two frameworks: the foundation framework, which incorporates the creation of power and mineral fertilizer, as well as the diesel and power necessities for the frontal area framework, and the closer view framework, which incorporates outflows connected with different MSW treatment frameworks thought about in the review, like MRF, COM, AD, and LF.

27.4.5 Impact Evaluation of the Life Cycle and Sensitivity Analysis

The impact assessment was conducted in the current study utilizing the LCA software Gabi 8.5.0.79 and the CML-1A technique and to estimate the environmental burden and benefits. EP (in PO_4 3-eq), GWP (in kg CO_2 eq), POCP (in C_2H_4 eq), ADP (in MJ), AP (in kg SO_2 eq), and HTP (in DCB eq.) are the six effect classes that were analyzed.

Sensitivity analysis determines which parameters are sensitive to changes in order to determine whether a slight change in one or more of them would cause the impact by environmental category(s) to rise. Here, the adjustment of the recycling rate fills in as the information boundary for the sensitivity examination (Khandelwal et al. 2019).

POCP, GWP, HTP, EP, AP, and ADP are the six impact categories that are mentioned in the results. The outcomes for each of the four situations under consideration are displayed in Fig. 27.1. Furthermore, the sensitivity analysis took into account the shift in the recycling rate. This section presents a thorough analysis of the six damage classifications.

27.4.6 Potential for Global Warming

Temperature increases brought on by GHG emissions, including CO_2, CH_4, N_2O, and CFCs, are the causes of global warming. The breakdown of a MSW biodegradable part produces biogenic CO_2, while the consuming of materials of nonbiodegradable like plastic, material, and calfskin produces fossil CO_2. By and large,

"MSW treatment creates both biogenic and fossil CO_2." In any case, as per studies, biogenic CO_2 makes no difference and doesn't cause an unnatural weather change. Since there is no waste isolation, no LFG control framework, and high fossil CO_2 and CH_4 discharges, the pattern situation (COM_LF) has the most elevated GHG outflows (1368.77 kg CO_2 eq). However, by implementing MRF, these emissions can be significantly decreased. With GWP of 834.67 kg CO_2 equivalent, scenario 2 had the lowest.

27.4.7 Potential for Acidification

The emanation of acidifying synthetic substances, like NO_x and SO_x, causes fermentation; the fermentation impact of NOx is somewhat more noteworthy than that of SO_x. Due to the huge SO_x and NO_x discharges during the energy age from biogas, situation 3 has the most noteworthy acidifying influence (7.65 E-01 kg SO_2 eq.). The pattern situation (1.83 E-01 kg SO_2 eq.), situation MRF_COM_AD_LF (5.34 E-01 kg SO_2 eq.), and situation MRF_COM_LF (1.88 E-01 kg SO_2 eq.) have the least effect on acidifying.

27.4.8 Potential for AD

It is observed that situation 1 (COM_LF) has the maximum ability for AD (146.47 MJ). Since the majority of the waste in the COM_LF scenario is delivered to landfills, this outcome was anticipated. Other options include sending a significant amount to energy and resource recovery facilities, which stop natural resources from being extracted.

27.4.9 Potential for Eutrophication

The discharge of ammonia, nitrogen oxide, and phosphate causes eutrophication. Because of the significant arrival of nitrogen oxides from the promotion interaction, situation 3's nitrogen oxide had the biggest eutrophication influence (8.6 E-02 kg PO_4^{3-} eq.).

27.4.10 Potential Toxicity to Humans

PM, SO_x, NO_x, and heavy metal emissions are among the pollutants that are harmful to humans. Because of the significant SO_x and NO_x emissions, the AD situation (situation: MRF_AD_LF) has an environmental cost of 9.79 E-01 kg DCB eq. Since composting has some positive environmental effects when paired with AD (S4), the overall impact of this scenario is comparatively lower than that of the AD scenario alone (7.48 E-01 kg DCB eq.). Of all the possible outcomes, S2 has the minimum influence (5.37 E-01 kg DCB eq.).

27.4.11 Potential for Photochemical Ozone Production

In the presence of sunshine, a variety of chemical interactions involving NO_x and VOCs produce ozone at lower atmospheric levels. Asthma and other respiratory conditions can be brought on by low ozone levels. The baseline scenario, in which NO_x emissions and fuel use are the main contributors, has the highest potential for photochemical oxidation (4.12 E-01 kg C_2H_4 eq.).

27.4.12 Analysis of Sensitivity

This part makes sense of the sensitivity investigation's discoveries in view of the MRF cycle's recycling rate increase, which goes from 30% to 100%. This exploration likewise showed that the term MSW extent of organic treatment (fertilizer and anaerobic absorption) will remain at 48.1%, the MSW extent for the landfill will change from 45.5% to 15.5% as the recycling extents of range of MSW from 15% to 100%. The discoveries of the sensitivity examination of the natural effect coming about because of the variety in recycling rate across all situations are displayed in Fig. 27.2.

As per studies, there is a generally straight contrarily proportionate relationship among the adjustment of recycling rate and the natural advantage (Khandelwal et al. 2019). This linear relationship came to the conclusion that a rise in recycling efficiency would result in a rise in overall environmental benefits through the use of MRF and landfill composting. Resource recycling has several advantages, demonstrating that the MRF will play a significant role in enhancing the MSWM system's future environmental benefits.

27.5 Conclusion

For Nagpur municipality, the LCA technique has been used to assess municipal solid waste management and its impact on energy efficiency and overall environmental effects. The usage of alternative fuels may have a greater influence on the environment than other sensitivity analysis considerations. The system may need to be expanded in order to comprehend other systems, such as the social and economic systems, as the results do not offer any economic recommendations. Nagpur's Municipal Solid Waste Management (MSWM) system manages waste through segregation, collection, transportation, processing, and disposal, aiming for sustainability with technology and partnerships. As per the ends in light of Gwp, EP, HTP, and POCP, the best strategy is to join recycling, fertilizing the soil the biodegradable part, and discarding the losses in a landfill. Sensitivity concentrate likewise showed

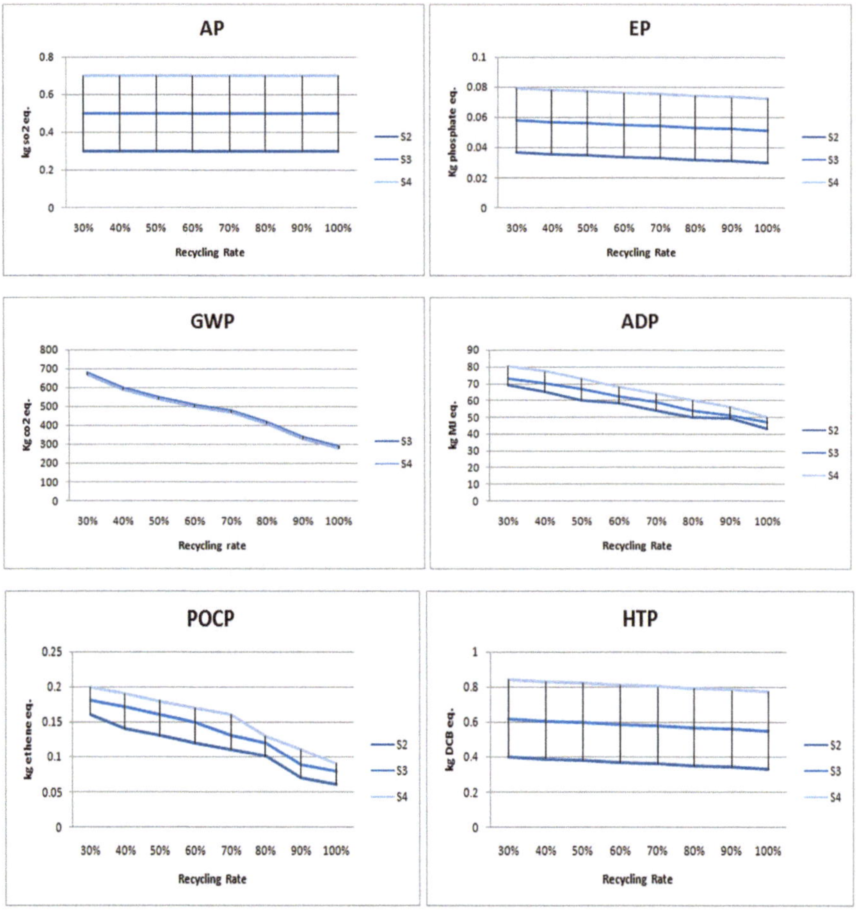

Fig. 27.2 Evaluation of the sensitivity of the suggested situations to variations in the recycling rate

that raising the reusing rate essentially reduces natural impacts. For extra examination and extreme direction, social (respiratory issues, and so on), economic (diesel costs, and so on), and mental (strange commotions, and so on) contemplations should likewise be considered to comprehend the impacts of MSWM completely.

References

Adipah S, Kwame ON (2019) A novel introduction of municipal solid waste management. J Environ Sci Publ Health 3(2):147–157

Anshassi M, Preuss B, Townsend TG (2021) Moving beyond recycling: examining steps for local government to integrate sustainable materials management. J Air Waste Manage Assoc 71(8):1039–1052

Bakan B, Bernet N, Bouchez T, Boutrou R, Choubert JM, Dabert P, Tremier A (2022) Circular economy applied to organic residues and wastewater: research challenges. Waste Biomass Valoriz 13(2):1267–1276
Chandrappa R, Das DB (2024) Solid waste management: principles and practice. Springer, Cham
Charkhestani A, Yousefi Kebria D (2022) Laboratory analysis to determine the accurate characteristics of urban food waste. Glob J Environ Sci Manage 8(2):225–236
Christensen TH, Damgaard A, Levis J, Zhao Y, Björklund A, Arena U, Bisinella V (2020) Application of LCA modelling in integrated waste management. Waste Manag 118:313–322
Cossu R, Stegmann R (2018) Solid waste landfilling: concepts, processes, technology. Elsevier, Amsterdam
Das S, Lee SH, Kumar P, Kim KH, Lee SS, Bhattacharya SS (2019) Solid waste management: scope and the challenge of sustainability. J Clean Prod 228:658–678
Edwards J, Othman M, Crossin E, Burn S (2018) Life cycle assessment to compare the environmental impact of seven contemporary food waste management systems. Bioresour Technol 248:156–173
Ghazali A, Tjakraatmadja JH, Pratiwi EYD (2021) Resident-based learning model for sustainable resident participation in municipal solid waste management program. Glob J Environ Sci Manage 7(4):599–624
Hockstad L, Hanel L (2018) Inventory of US greenhouse gas emissions and sinks (no. cdiac: EPA-EMISSIONS). Environmental system science data infrastructure for a virtual ecosystem (ESS-DIVE) (United States)
Khandelwal H, Dhar H, Thalla AK, Kumar S (2019) Application of life cycle assessment in municipal solid waste management: a worldwide critical review. J Clean Prod 209:630–654
Kurniawan TA, Othman MHD, Hwang GH, Gikas P (2022) Unlocking digital technologies for waste recycling in industry 4.0 era: a transformation towards a digitalization-based circular economy in Indonesia. J Clean Prod 357:131911
Lin L, Xu F, Ge X, Li Y (2019) Biological treatment of organic materials for energy and nutrients production—anaerobic digestion and composting. In: Advances in bioenergy, vol 4. Elsevier, Amsterdam, pp 121–181
Paruti B, Santhaveeranagoud B (2019) Characterisation of leachate in solid waste landfill site: a case study of S. Bingipur in Bangalore. Ann Fac Eng Hunedoara 17(1):205–208
Schrijvers D, Loubet P, Sonnemann G (2020) Archetypes of goal and scope definitions for consistent allocation in LCA. Sustain For 12(14):5587
Slorach PC, Jeswani HK, Cuéllar-Franca R, Azapagic A (2020) Assessing the economic and environmental sustainability of household food waste management in the UK: current situation and future scenarios. Sci Total Environ 710:135580
Vaverková MD (2019) Landfill impacts on the environment. Geosciences 9(10):431
Wang Z, Lv J, Gu F, Yang J, Guo J (2020) Environmental and economic performance of an integrated municipal solid waste treatment: a Chinese case study. Sci Total Environ 709:136096
Weekes JG, Musa Wasil JC, Malave Llamas K, Morales Agrinzoni C (2021) Solid waste management system for small Island developing states. Glob J Environ Sci Manage 7(2):259–272
Wojnowska-Baryła I, Bernat K, Zaborowska M (2022) Plastic waste degradation in landfill conditions: the problem with microplastics, and their direct and indirect environmental effects. Int J Environ Res Public Health 19(20):13223
Zarea MA, Moazed H, Ahmadmoazzam M, Malekghasemi S, Jaafarzadeh N (2019) Life cycle assessment for municipal solid waste management: a case study from Ahvaz. Iran Environ Monitor Assess 191:1–13

Life Cycle Assessment for Valorization of Municipal Solid Waste to Chemicals

28

Dibyarupa Pal, Sudipto Sarkar, and Rishi Kumar Singh

Abstract

Human actions result in the production of waste products, which are frequently discarded due to their perceived lack of value. Still, with the right kind of management, a lot of these recyclables can become something useful, like raw materials for new products or fuel for power plants. One of the biggest issues we have now is waste management. Finding solutions to the ever-increasing trash problem is a top priority for businesses, individuals, and state legislators. To determine the ecological effects of recycling trash into useful chemicals, the life cycle assessment (LCA) of municipal solid waste (MSW) to chemicals is an essential instrument. From garbage collection and transportation to grading and conversion technologies, chemical manufacturing, and finally, disposal, the full process of life cycle is examined in detail. Thermochemical procedures like gasification and pyrolysis and biochemical techniques like anaerobic digestion and fermentation are essential steps. Energy consumption, resource depletion, air and water pollution, toxicity, and the potential for global warming are some of the environmental impact areas that are evaluated in the LCA. In order to understand the pros and cons of MSW-to-chemicals operations, LCA takes a look at these aspects. Based on the conversion method, making chemicals from trash can either require a lot of energy or result in a lot of emissions, but it can also recover

D. Pal (✉)
Department of Biosciences, JIS University, Agarpara, Kolkata, West Bengal, India
e-mail: dibyarupa.pal@gmail.com

S. Sarkar
Department of Agricultural Engineering, Assam University, Silchar, Assam, India

R. K. Singh
Department of Biosciences, IMS Ghaziabad, University Courses Campus, Ghaziabad, Uttar Pradesh, India

A. Kumari et al. (eds.), *Valorization of Solid Wastes to Biofuels and Chemical Products for Sustainable World*, https://doi.org/10.1007/978-981-96-8594-3_28

important resources and decrease landfill use. Process optimization and improved decision-making are made possible by the LCA's identification of environmental hotspots. Promoting sustainable waste management practices and improving the circular economy through the conversion of garbage into renewable chemicals is crucial for minimizing environmental impact and dependency on fossil fuels. The full potential of MSW for contributing to chemical production while reducing environmental consequences can be achieved through LCA, opening the door to more efficient and environmentally friendly waste-to-chemicals technologies.

Keywords

Life cycle assessment · Municipal solid waste · Sustainable waste management · Waste-to-chemicals

28.1 Introduction

Since every manufactured goods has a "life", improving its environmental performance is a top priority in this age of heightened environmental consciousness. A product's life cycle begins during its design or creation and concludes during its end-of-life operations, which include collecting or sorting, recycling, disposal, and potential reuse. Environmental consequences are a consequence of resource consumption, substance emissions into the surrounding environment, and other interactions in the environment that occur throughout a product's lifetime (Belboom and Léonard 2016; Khandelwal et al. 2019). Environmental managers can use the life cycle assessment (LCA) to find out how harmful a product, process, or service could be over time. In terms of energy, materials used in production, and waste released into the environment, it compiles a product inventory (Benetto et al. 2004; Ghosh et al. 2016). During compilation, LCA considers all potential consequences associated with the identified discharges. Decisions affecting human and environmental health can be better understood with its help. From the extraction of raw materials through the final phases of disposal or recycling, the environmental implications of a product can be systematically evaluated using LCA (Fruergaard et al. 2010; Gupta et al. 2015).

Anything and everything that people throw out from their houses or companies that is no longer needed is considered municipal solid waste (MSW). Things like food scraps, paper, glass, plastic, textiles, wood, metal, leather, and sewage from septic tanks are all considered rubbish or garbage. As a result of rapid urbanization, India's MSW production is projected to skyrocket from its current 62 million tons/year to as much as 165 million tons by 2030 (TERI 2023). CPCB (2022) states that the massive amounts of MSW produced daily, along with environmental and aesthetic issues, have made MSW management a major issue in India. The physical properties of solid waste can vary significantly. There are five types of municipal waste: residential, commercial, construction and demolition, industrial, and agricultural (Pires et al. 2011; Ghosh et al. 2019). Each type is characterized by its unique combination of physical, chemical, and biological properties. LCA is a useful tool for determining the ecological footprint of various MSW management techniques, such as collection, treatment, reuse and recycling, and disposal (Nessi et al. 2021).

28.2 Management of Municipal Solid Waste (MSW)

Landfills and incinerators receive hazardous materials when people discard of things that contain harmful substances. Toxic chemicals are released into the environment by subsurface leachate, wastewater runoff, emissions into the air, or hazardous trash from these disposal sites (Eriksson et al. 2005; Narayana 2009). Ecological systems, animals, and human health are all at risk when harmful substances are discharged into the environment. The metals that escape from SLFs include numerous neurotoxins, carcinogens for humans (like lead, cadmium, arsenic, and beryllium), recognized reproductive toxins for humans (like lead and mercury), and acutely toxic elements for aquatic life (like copper and zinc). Reducing the harmful components of solid waste can be achieved through three main methods (Morrissey and Browne 2004; Pappu et al. 2007). The first is to lessen the amount of hazardous trash that ends up in landfills by enhancing current waste management procedures and recycling more. The majority of people use this method. The limitations in the long run outweigh its short-term impacts. In the second, we see an effort to alter the physical properties of goods utilized in both private and public spheres. Although this strategy will have greater effects in the long run, short-term results will be disappointing. The third one is an effort to alter manufacturing procedures in order to cut down on harmful materials. This strategy is long-term, but it has the potential to be quite effective. The transformation of MSW into chemicals has great potential because of the wide range of products it can supply, including polymers, fuels, and medicines (Sharholy et al. 2008; Rigamonti et al. 2009). In addition to contributing to a circular economy, the process of converting MSW into chemicals helps reduce trash accumulation and provides alternative resources for industrial production. Converting MSW into chemicals is an extensive process that includes evaluating the technical steps, financial viability, environmental advantages, and problems.

28.3 Composition of Municipal Solid Waste

Municipal solid waste comprises the nonhazardous waste collected from domestic, institutional, commercial, and industrial areas. The wastes needs to be collected from the sources and transported to the sorting or disposal areas, which includes both garbage and trash. Usually the garbage includes all biodegradable waste, whereas trash includes nonbiodegradable but recyclable waste, which requires special collection due to their volumes (Table 28.1). Knowing what makes up MSW is

Table 28.1 Sources of solid waste generation and their composition

Source of solid waste	Component of solid waste
Household/ residential	Food wastes, vegetable peels, paper, plastics cardboard, textiles, mats, glass, broken metals pieces, electronic waste, and household hazardous wastes
Commercial	Paper, cardboard, plastics, wood, food wastes, glass, metals, special wastes, hazardous wastes
Institutional	Paper, cardboard, plastics, wood, food wastes, glass, metals, special wastes, hazardous wastes

crucial before delving into the technology that turns it into chemicals. Disposal methods, socioeconomic level, and geographic location all influence waste composition. Components of municipal solid waste (MSW) typically include organic matter (30–50%), plastics (10–20%), paper and cardboard (15–25%), metals (5–10%), glass (5–10%), and inert substances and inorganic minerals (Rathi 2006). In order to ascertain whether or not these materials are suitable for chemical conversion, it is essential to understand their chemical composition. Consider this: organic materials are perfect for biobased chemical production because of their abundance of carbon, hydrogen, and oxygen. On the other hand, plastics include hydrocarbons, which can be transformed into useful chemical substrates (Ryu 2010). Hydrocarbons, hydrogen gas, and biogas, are some of the industrially significant compounds that can be produced by utilizing the chemical composition of trash. Additionally, innovative materials such as biofuels and biodegradable polymers can be created through waste-to-chemicals processes (Parthan et al. 2012). Through the creation of new revenue streams from trash, these procedures lessen the reliance on fossil fuels, decrease emissions of greenhouse gases, and aid in sustainable development.

28.4 Recovery Opportunities of Chemicals from MSW

Resource recovery, less landfills use, and the manufacture of useful chemicals and fuels are just a few of the many advantages of converting MSW to chemicals. Factors such as waste type, conversion technique, and energy sources employed determine the environmental implications of these operations, which in turn involve complex systems. From garbage collection to the finished chemical products, an LCA may assist evaluate the entire environmental effect of these operations, including resource consumption, energy usage, emissions, and more (Sonnemann et al. 2004). A life cycle assessment (LCA) can help optimize procedures to reduce their environmental impact by identifying critical hotspots, highlighting trade-offs, and offering a comprehensive review of the full life cycle (Finnveden et al. 2005). The environmental impact is increased throughout the recovery life cycle of chemicals made from municipal solid waste, which is achieved through numerous stages.

28.4.1 Waste Collection, Segregation, and Transfer

Global interest in source reduction—sometimes called waste prevention—has been on the rise in recent years. The primary objective of a source reduction is to eliminate the need to manage material by minimizing its generation, which in turn reduces its amount and toxicity. The elimination or simplification of the need to handle garbage is what sets source reduction apart from recycling and yard waste collection, two additional types of solid waste management. Reduced pollution and the associated costs of solid waste management and disposal are two of the most important environmental benefits of source reduction, which also has economic benefits (Astrup et al. 2009). Furthermore, the waste composition can alter as a result of source reduction initiatives. Reduced or eliminated toxicity in product manufacturing and packaging, improved product longevity, reusability, and repair ability; shifting consumer preferences toward products with reduced residual toxicity and waste; and shifting consumption and material use patterns toward ones that produce less waste are the main tenets of source reduction (Agarwal et al. 2005). Reducing pollution and the expense of solid waste handling and elimination are two of the main environmental and economic benefits of source reduction. Furthermore, the solid waste composition can alter as a result of source reduction initiatives.

Everything from collecting materials to processing and disposing of them adds to the overall cost of a solid waste management system. Several methods exist for lowering the expense of MSW management through source reduction, the most important of which are a decrease in the amount of garbage that needs management, an avoidance of purchasing costs, and the collection of revenues from the resale of products. Reusing, repairing, and refurbishing items can often be more cost-effective than throwing them away (Chand et al. 2019). There are long-term financial benefits to selecting reusable products as well. Performing routine maintenance and repairs extends the life of an item, reducing the frequency with which it needs to be disposed of and replaced. Electronic mail and news are two examples of modern technology that promote paperless communication and allow for more effective information flow. Electronic data transfer allows for the limited use of paper to only be utilized when a hard copy is required or desired (Rajkumar et al. 2010). Reduced product packaging, less production waste, and increased reutilization of scrap materials are all ways in which industries can save money. Cutting down on packing weight and volume also helps with shipping and advertising expenses. More effective material utilization and lower purchasing and disposal costs are the outcomes of reducing harmful and wasteful industrial processes.

Waste and toxicity reduction has several positive effects on the environment, including less energy and pollutants caused by materials not being reprocessed, less material transferred to landfills and waste combustion facilities, and less demand for natural resources. The energy required to gather and dispose of residual trash after a product's useful life has finished, as well as to mine, transport, process, manufacture, and transport raw materials, results in the emission of greenhouse gases such nitrous oxide, carbon dioxide, and methane (Singh et al. 2011). Cutting down trees for paper, putting trash in landfills, and burning trash all contribute to higher levels of greenhouse gas emissions. It is well acknowledged that there is a great deal of potential to decrease emissions of greenhouse gases by reducing MSW at the source. The precise consequences of releasing massive amounts of greenhouse gases into the atmosphere without any sort of regulation are yet unknown. Nevertheless, it's quite probable that MSW-related activities add to the problem of climate change. There is evidence that initiatives like beverage container deposit and refund schemes can significantly cut down on litter and boost recycling rates to over 80% in most areas. It is unclear how some actions regarding waste management may affect the environment. One way to reduce water usage is to switch from disposable to washable cutlery, cups, utensils, and towels. There may be negative effects on both water supply and wastewater treatment operations due to an increase in biological and dispersed solid material in effluent caused by increased water use. More energy may be required to run repaired and reused older, less efficient electronics and appliances. Energy efficiency is one potential upside of new technology, but the displacement of existing products and an increase in the waste management load are two potential downsides. LCA can be used to forecast the environmental impact of a product or substance due to the complexity of doing so. Using the following steps: selecting and extracting raw materials; transporting and processing those materials; making products from those materials; using those products; and finally, the product's fate at the end of its life cycle is considered in the LCA (Barlaz 2004). By incorporating LCA into MSW systems, we can look at how decisions about MSW affect environmental systems as a whole, rather than only at the process level.

The development of efficient programs to reduce sources relies heavily on planning. It is critical that community source reduction planners have a clear understanding of their goals, the amount of reduction they hope to accomplish, and the metrics by which they will evaluate their success before they begin crafting targeted programmatic efforts. An official policy for reducing sources, well-defined objectives, and practical methods for measuring progress should all be part of any municipal solid waste plan. The efficacy of such programs cannot be assessed without these metrics. Creating a foundation to back a source reduction plan is also an important part of implementing one. For a program to be successful, it needs strong, autonomous leadership, sufficient resources, and qualified employees (Joshi and Ahmed 2016). Making items or their packaging more reusable, using lighter materials, or reducing the amount of harmful compounds used; reusing and recycling means putting previously used goods, supplies, or packaging to new uses; this includes things like dual-sided printing, refillable bottles, and reusable pallets. Cutting down on product or packaging waste; delaying disposal by extending the life of products or materials, for example by performing routine maintenance or opting to repair an item; putting a product in packaging that keeps it fresh for longer; composting on-site or finding other solutions to dispose of organic waste.

Quantities of municipal solid waste (MSW) can be estimated or measured. Solid waste toxicity, on the other hand, is harder to quantify. Household garbage has always included harmful elements, but the percentage of synthetically produced poisonous compounds has grown significantly since the mid-century, when synthetic materials started to replace many conventional ones (Khajuria et al. 2010). Chemicals like naphthalene and toluene, chlorinated hydrocarbons like perchloroethylene and trichloroethylene, aromatic compounds like naphthalene and toluene, pesticides and other biocides, and old motor oil are among the hazardous components found in solid waste. Heavy metals like lead, cadmium, nickel, and mercury are also present.

Some of these harmful substances get up in MSW streams as by-products of industrial or residential activities. An example of this would be the used oil from petrol stations. Some hazardous materials are harmful by-products that should not be disposed of after only a fraction of the product has been consumed. One example is waste paints. Unfortunately, the majority of harmful substances end up in landfills as by-products of businesses' discarded goods. As an illustration, the annual sales of dry cell batteries in the India exceed 3 billion. Among these are zinc–carbon, alkaline, and lithium-based cells, with zinc–carbon being the most common. While many of these metals aren't harmful on their own, dry cell batteries are thrown out after their useful life has elapsed, which can be anywhere from a few hours to a few months. The degradation of a battery is an inevitable consequence of its disposal in a landfill. Metals can seep into the earth and groundwater as a result of this degradation. The combustion process releases gases from the incinerator, and some metals fall out of the batteries as bottom ash. Filters used in incinerators can gather fly ash, which is a by-product of metal combustion. Once again, landfills pose a threat to the contamination of the ground and groundwater when disposing of bottom ash and fly ash.

28.4.2 Processing and Conversion of MSW

Separating recyclables and removing pollutants that could hinder chemical conversion processes are two goals of sorting and pretreatment, which MSW goes through after collection. For the chemical conversion process to be more efficient and of higher quality, this step is crucial. Sorting and preparation for specialized processing methods is done for different fractions of trash, including plastics, paper, and organic materials. Automated technology is used in mechanical sorting to separate various waste types according to their physical attributes, including size, density, and magnetic. Screening, magnetic separation, and shredding are common procedures. Mechanical sorting lessens the processing load by recovering valuable items like metals and polymers. To be ready for conversion operations, MSW may sometimes need to go through further pretreatment such drying, grinding, or pelletizing. To make thermochemical conversion more efficient, organic waste may need to be dried to lower moisture content, while plastic garbage can be shred to increase efficiency in gasification and pyrolysis.

There are a number of environmentally and economically conscious technologies that can be employed to transform garbage into useful chemical products. Biochemical and thermochemical conversion are the most frequent types of conversion; however, other methods like plasma gasification are also becoming increasingly popular. Using heat, waste materials can be converted into simpler chemical compounds in a process known as thermochemical conversion. Gasification is a process that involves heating waste in an oxygen-poor atmosphere to create syngas, which is a combination of hydrogen and carbon monoxide. Methanol, ethanol, and hydrogen are among the compounds that can be refined from syngas. By subjecting trash to high temperatures in an oxygen-free environment, pyrolysis can separate it into three distinct phases: bio-oil, syngas, and char. When it comes to recycling plastic, pyrolysis is the way to go for producing chemical feedstock like hydrocarbons. Incineration is a method of producing heat and power by completely burning waste materials. Optimizing incineration with energy recovery can yield syngas and other chemical by-products, while it is not a direct chemical conversion process. Environmental consequences, including energy consumption, emissions, and waste generation, are unique to each thermochemical process. When compared to conventional incineration, gasification and pyrolysis often result in lower emissions; however, they may necessitate more intricate energy and infrastructure systems.

The process of biochemical conversion involves the breakdown of organic waste into chemicals by biological means. Anaerobic digestion is a process where bacteria decompose organic waste without the presence of oxygen. The result is digestate and biogas, which are composed of methane and carbon dioxide. Chemicals such as methanol and hydrogen can be made from biogas, or it can be refined into renewable natural gas (RNG). Waste sugars and starches can be transformed into alcohols, acids, and gases by fermentation, which makes use of microbes. One example is the fermentation of lignocellulosic materials found in organic waste and paper to create biobased compounds like lactic acid and ethanol. Because they use lower temperatures and generate less emissions, biochemical conversion processes are generally better for the environment than thermochemical ones (Cleary 2009). But they can only handle certain kinds of trash, mostly organic stuff, and you might have to pretreat it or use more energy than usual to use them. There are new ways to recycle municipal solid waste (MSW) into chemicals, made possible by emerging technologies. Plasma gasification is a process that separates garbage into its component elements by utilizing plasma torches to produce extraordinarily high temperatures. This method can handle challenging waste streams, including electronic and toxic trash, and produces cleaner syngas compared to conventional gasification. Catalytic processes enhance the efficiency of waste-to-chemicals operations by using catalysts to speed up chemical reactions. Catalytic reforming can transform syngas into compounds like methanol, and catalytic cracking can transform plastic waste into high-quality hydrocarbons. Though they can be more difficult and costly to adopt, emerging technologies have the potential to enhance efficiency and decrease emissions, two major environmental concerns. Considerations such as MSW composition, product needs, and existing infrastructure all play a role in determining the best technology to use.

28.5 Conversion to Chemicals

Intermediate products like syngas or bio-oil can be made from waste and then used to make valuable compounds. MSW can be used to make a number of important chemical types. The conversion process and waste type utilized determine the particular chemicals that are produced. Biogas, ethanol, and biodiesel are biofuels that can be produced from MSW due to the high concentration of organic materials in this waste. Both transportation and power generation can benefit from these biofuels. Many different types of products rely on platform chemicals as their foundation, such as plastics, textiles, and pharmaceuticals (Grosso et al. 2010). Platform chemicals such as aromatics, syngas, and methanol are produced from municipal solid waste. Many different sectors rely on specialty chemicals, such as the pharmaceutical, agricultural, and electronics industries. Organic acids, amino acids, and pigments are some examples of specialized chemicals that can be made from MSW.

Chemicals made from MSW are then supplied to consumers or businesses for usage in a variety of contexts. There are several potential applications for these substances, such as fuels, inputs into the pharmaceutical industry, and feedstock for the manufacturing of polymers and chemicals. The exact uses of the chemicals and the ways of transportation and distribution determine the environmental impact of this stage. One way to lessen our impact on the environment is to switch from fossil fuels to renewable fuels made from trash. As the life cycle comes to a close, the last step is to treat and dispose of any leftover waste or by-products from the MSW to chemicals conversion. Landfilling or incineration may be necessary for certain waste streams, such as nonrecyclable residues or industrial by-products; however, recycling and reusing are viable options for other waste streams (Agarwal et al. 2005). In order to make MSW-to-chemicals processes more environmentally friendly, it is important to reduce waste and make the most of by-products. There are technologies that can assist lessen the environmental impact of waste disposal. For example, plasma gasification can produce inert slag. Biochemical processes can generate digestate, which can be used as fertilizer.

28.6 Life Cycle Impact

An MSW-to-chemicals process' overall sustainability is evaluated in a life cycle evaluation by taking into account different environmental effect categories. A process' global warming potential is the amount of greenhouse gases it could release into the atmosphere, including carbon dioxide (CO_2), methane (CH_4), and nitrous oxide (N_2O). By removing organic matter from landfills, where anaerobic decomposition produces methane, the conversion of municipal solid waste to chemicals can mitigate emissions of greenhouse gases. Further reductions in emissions can be achieved by the manufacture of renewable chemicals and fuels from waste, which can replace products derived from fossil fuels. The total GWP is affected by variables such as energy inputs, and the type of waste, and process efficiency; nevertheless, gasification and pyrolysis are two processes that may still generate CO_2 and other pollutants while converting waste (Patil and Shekdar 2001). Throughout the MSW-to-chemicals process life cycle, natural resources such as water, fossil fuels, and minerals are consumed. This is known as resource depletion. These techniques help conserve resources by reducing the requirement for fresh raw materials by turning trash into valuable compounds. One way to lessen reliance on petrochemical feedstock is to turn organic waste into biobased chemicals. Another way is to recover metals from waste streams, which means less mining is required. The efficiency of the technique and the affordability of sources of clean energy determine whether the energy and materials needed for garbage collection, transport, and processing contribute to resource reduction or not. To determine whether MSW-to-chemicals processes are sustainable, one must look at their energy consumption. Converting garbage into chemicals has environmental benefits, but if the energy to do so is expensive and originates from fossil fuels, those benefits can be nullified. Chemical production from waste is an example of an activity that involves both direct and indirect energy use, which life cycle assessment (LCA) measures. Gasification and pyrolysis are thermochemical processes that typically need more energy than biochemical processes. However, these processes have the potential to reclaim energy in the manner of syngas or bio-oil, which can be utilized to offset part of the energy inputs. Among the most important environmental issues with waste-to-chemicals processes are air emissions, which include particulates, organic volatile compounds (VOCs), and nitrogen oxides (NO_x). Filtration and scrubbing technologies are necessary for controlling gaseous emissions from thermochemical processes and incineration/gasification, two methods that burn waste and generate air pollutants. Additionally, it is important to think about water pollutants and the possibility of water contamination. Wastewater from fermentation and leachate from garbage sorting are two examples of the liquid waste streams that can be produced by various processes. To lessen the ecological impact on aquatic ecosystems and lessen water pollution, it is crucial to manage these waste streams properly. The conversion of MSW can result in the production of chemicals that pose risks to human and environmental health, including dioxins, toxic heavy metals, and other dangerous substances. LCA evaluates the effects of chemical manufacturing and disposal on communities, employees, and the environment, in addition to the possibility of harmful emissions during waste processing. By decomposing harmful substances or encasing them in innocuous by-products, processes such as plasma gasification and advanced catalytic conversion aim to reduce harmful emissions. But, improper management of contaminants in the waste stream, such as industrial or electronic waste, can still cause problems.

28.7 Limitations with MSW-to-Chemicals Life Cycle Assessment

Because of the variety and complexity of the waste flows, methods, and environmental impacts involved, performing an LCA for MSW-to-chemicals operations is no easy task. Obtaining consistent information for MSW-to-chemicals processes can be challenging owing to the variability of waste flows, conversion methods, and geographic considerations. However, accurate and thorough data are required for completing a reliable LCA. The findings of a life cycle assessment (LCA) can be drastically altered by changing the waste's composition, collecting techniques, or energy sources. To make sure the LCA results are comparable and meaningful, the system parameters and functional unit must be defined. Results can also be impacted by the functional unit chosen (e.g., a single KG of chemicals created vs. one ton of garbage processed). Multiple products, including chemicals, fuels, and energy, are commonly generated by MSW-to-chemicals processes. Because various allocation approaches (such as economic or mass-based) can provide different outcomes, it can be difficult to allocate the environmental burdens among these items. The unpredictability of the data, suppositions, and models utilized to conduct LCA makes it an inherently uncertain process. To reduce uncertainty, it is helpful to do a sensitivity analysis to determine which parameters are most influential on the LCA results. One way to test how well LCA results hold up is to play around with different assumptions on things like the nature of waste, energy efficiency, and emission variables.

28.8 Economic Concerns

The economic viability of MSW to chemicals conversion depends on several factors. The current waste management system and the area in question determine the MSW cost. Depending on the facility's size and the process used, the costs of MSW to chemicals methods of conversion can vary. Market demand as well as manufacturing costs determine the value of chemicals produced from MSW. Making the conversion of municipal solid waste (MSW) to chemicals economically appealing can be achieved through government programs like subsidies and tax incentives. The number of waste management facilities, including those that convert garbage into chemicals, recover usable materials, and others, is increasing at a rapid pace. Recycling is encouraged by the majority of states through various legislation and incentives. Additionally, composting and trash minimization are also supported by many governments. As we stray from the conventional wisdom of using packer trucks to collect trash and sending it to the local dump, the waste management systems we use become increasingly intricate. The standards for funding the new initiatives are becoming more difficult as waste management technology continues to advance. The conventional municipal solid waste (MSW) landfill isn't the sole facility that needs funding; increased capital expenditures, which typically involve borrowing money, are also required. Recycling, composting, burning, and landfilling are all components of a comprehensive waste management system that some towns opt to adopt. System finance, as opposed to financing individual facilities, is typically the result of the necessity for multiple sites to form an integrated system. A shift toward service privatization has also occurred as a consequence of the increasing difficulty of integrated waste management. Rather than try to handle things on their own, municipalities will typically hire private companies that focus on certain areas. The funding of solid waste management infrastructure and systems is also influenced by this aim for limited public involvement. The population to be served can sometimes have a significant impact on the decision of recycling method. If a small town or village doesn't produce enough recyclables to warrant the high cost of a central processing facility, then no amount of recycling will be accepted. Although larger population bases have more recycling operations options available to them, this does not necessarily translate into a major economic benefit for major towns or regional waste agencies. Calculated net expenditures for recycling are comparable across methods, despite substantial variation in collection expenses, treatment or consolidated costs, and net income.

28.9 Conclusions

One of the most important ways to measure the environmental impact of chemical processing operations and find ways to make them more efficient is to conduct a life cycle assessment. The benefits and drawbacks of different methods and waste streams can be better understood with the use of life cycle assessment (LCA), which examines the entire process from garbage collection to chemical manufacturing and disposal. Reducing waste and recovering resources are two of the many environmental benefits of chemically treating municipal solid waste (MSW). However, there are also concerns about the consumption of energy, emissions, and resource usage. A more environmentally friendly and circular economy can only be realized through the full utilization of waste-to-chemicals processes, which in turn depend on continuing technological advancements, governmental backing, and the incorporation of renewable energy sources. To conclude, in order to steer decision-makers, industry, and legislators toward more environmentally friendly chemical manufacturing and waste management systems, it is crucial to perform an exhaustive LCA for MSW-to-chemicals processes. We can save municipal solid trash for the next generation by making these procedures more efficient and fixing the problems they cause to the environment.

References

Abdel-Shafy HI, Mansour MS (2018) Solid waste issue: sources, composition, disposal, recycling, and valorization. Egypt J Pet 27(4):1275–1290

Agarwal R, Singhmar A, Kulshrestha M, Mittal AK (2005) Municipal solid waste recycling and associated markets in Delhi, India. Resour Conserv Recycl 44(1):73–90

Astrup T, Fruergaard T, Christensen TH (2009) Life cycle assessment of recycling of source-segregated organic household waste. Waste Manag 29(4):1312–1323

Barlaz MA (2004) Critical review of forest products decomposition in municipal solid waste landfills. Waste Manag 24(4):321–333

Belboom S, Léonard A (2016) Does the feedstock influence the environmental burden of municipal solid waste incineration? Waste Manag 48:184–194

Benetto E, Tiruta-Barna L, Perrodin Y, Rousseaux P (2004) LCA assessment of alternative municipal solid waste treatment scenarios in a French context. Int J Life Cycle Assess 9(6):387–393

Chand P, Pathak A, Saroj DP (2019) Current status of municipal solid waste management in India. Glob J Environ Sci Manage 5(1):83–94
Cherubini F, Bargigli S, Ulgiati S (2008) Life cycle assessment (LCA) of waste management strategies: landfilling, sorting plant and incineration. Energy 34(12):2116–2123
Cleary J (2009) Life cycle assessments of municipal solid waste management systems: a comparative analysis of selected peer-reviewed literature. Environ Int 35(8):1256–1266
CPCB (2022) Annual report of CPCB related to waste generation, segregation and processing (2020–21), New Delhi
Eriksson O, Reich MC, Frostell B, Björklund A, Assefa G, Sundqvist JO, Granath J, Baky A, Thyselius L (2005) Municipal solid waste management from a systems perspective. J Clean Prod 13(3):241–252
Finnveden G, Johansson J, Lind P, Moberg Å (2005) Life cycle assessment of energy from solid waste - part 1: general methodology and results. J Clean Prod 13(3):213–229
Fruergaard T, Astrup T, Ekvall T (2010) Energy use and recovery in waste management and implications for accounting of greenhouse gases and global warming contributions. Waste Manag Res 28(9):776–784
Ghosh P, Samadder SR, Bose S (2016) A review on life cycle assessment of solid waste management system: current trends and future challenges. Int J Environ Sci Technol 13(7):1739–1764
Ghosh A, Chakraborty S, Das D (2019) A comparative life cycle assessment of different municipal solid waste treatment technologies in India. Environ Eng Sci 36(8):1034–1045
Grosso M, Motta A, Rigamonti L (2010) Efficiency of energy recovery from waste incineration, in the light of the new waste framework directive. Waste Manag 30(7):1238–1243
Gupta N, Yadav KK, Kumar V (2015) A review on current status of municipal solid waste management in India. J Environ Sci 37:206–217
Joshi R, Ahmed S (2016) Status and challenges of municipal solid waste management in India: a review. Cogent Environ Sci 2(1):1139434
Khajuria A, Yamamoto Y, Morioka T (2010) Estimation of municipal solid waste generation and landfill area in Asian developing countries. J Environ Biol 31(5):649–654
Khandelwal H, Dhar H, Kumar S (2019) Life cycle assessment of municipal solid waste management options for India. Bioresour Technol 288:121515
Morrissey AJ, Browne J (2004) Waste management models and their application to sustainable waste management. Waste Manag 24(3):297–308
Narayana T (2009) Municipal solid waste management in India: from waste disposal to recovery of resources? Waste Manag 29(3):1163–1166
Nessi S, Rigamonti L, Grosso M (2021) Life cycle assessment of waste management systems: evolution over the last decades in the environmental profiles, causes and challenges. Environ Sci Technol 55(6):3432–3443
Pappu A, Saxena M, Asolekar SR (2007) Solid waste generation in India and their recycling potential in building materials. Build Environ 42(6):2311–2320
Parthan S, Milke MW, Wilson DC, Cocks JH (2012) Cost estimation for solid waste management in industrializing regions—precedents, problems, and prospects. Waste Manag 32(3):584–594
Patil AD, Shekdar AV (2001) Health-care waste management in India. J Environ Manag 63(2):211–220
Pires A, Martinho G, Chang NB (2011) Solid waste management in European countries: a review of systems analysis techniques. J Environ Manag 92(4):1033–1050
Rajkumar R, Meenambal T, Saravanan S, Rajasimman M (2010) Municipal solid waste management in Salem city, Tamil Nadu, India. J Eng Sci Technol 5(2):134–144
Rathi S (2006) Alternative approaches for better municipal solid waste management in Mumbai. India Waste Manage 26(10):1192–1200
Rigamonti L, Grosso M, Sunseri MC (2009) Influence of assumptions about selection and recycling efficiencies on the LCA of integrated waste management systems. Int J Life Cycle Assess 14(5):411–419
Ryu C (2010) Potential of municipal solid waste for renewable energy production and reduction of greenhouse gas emissions in South Korea. J Air Waste Manage Assoc 60(2):176–183

Sharholy M, Ahmad K, Mahmood G, Trivedi RC (2008) Municipal solid waste management in Indian cities—a review. Waste Manag 28(2):459–467

Singh RP, Tyagi VV, Allen T, Ibrahim MH, Kothari R (2011) An overview for exploring the possibilities of energy generation from municipal solid waste (MSW) in Indian scenario. Renew Sust Energ Rev 15(9):4797–4808

Sonnemann G, Castells F, Schuhmacher M (2004) Integrated life-cycle and risk assessment for industrial processes. CRC Press, Boca Raton

TERI (2023) TERI energy & environment data diary and yearbook (2022–23), New Delhi

Life Cycle Assessment for the Valorization of Agricultural Solid Waste to Biofuels

29

Anurag Deepak

Abstract

The agricultural system has long been a cornerstone of human civilization, providing food security and supporting economic growth worldwide. With an increasing population, global food demands for human consumption will need to be increased by 70% before 2050. Besides fulfilling food requirements, the agricultural sector produces waste, and the improper disposal of such waste adversely affects the environment. Climate change and dwindling fossil fuel supplies are driving interest in using agricultural solid waste (ASW) to lower greenhouse gas emissions (GHGs) and get more value from resources. This study explores the application of life cycle assessment (LCA) as a decision-making tool to examine the environmental feasibility of converting ASW into valuable biofuels. The LCA methodology allows us to comprehensively assess the entire life cycle of ASW valorization, i.e., from waste generation to the production, utilization, and disposal of biofuels. Furthermore, the analysis sheds light on the trade-offs and environmental hotspots of different valorization pathways, aiding in the identification of sustainable practices. By harnessing the potential of ASW as a resource, the decision-makers and stakeholders can decide on an environment-friendly scenario, which ultimately paves the way for a more sustainable and circular agricultural system.

A. Deepak (✉)
Department of Mechanical Engineering, Motilal Nehru National Institute of Technology Allahabad, Prayagraj, Uttar Pradesh, India
e-mail: anudeepak@mnnit.ac.in

A. Kumari et al. (eds.), *Valorization of Solid Wastes to Biofuels and Chemical Products for Sustainable World*, https://doi.org/10.1007/978-981-96-8594-3_29

Keywords

Agricultural waste · LCA · Biofuels · Environmental impact · Circular economy · Valorization

29.1 Introduction

Our planet grapples with a dual challenge: growing populations demand more food, leading to a surge in agricultural production, while simultaneously generating vast quantities of ASW. Agricultural activities generate a significant amount of waste, encompassing organic materials like crop residues and animal manure, alongside inorganic elements like agricultural plastics and chemical containers (Mengqi et al. 2023). Globally, agriculture generates an astonishing 23.7 million tons of food daily. The surge in production comes at a cost to the environment, jeopardizing human health and the long-term viability of ecosystems. With agriculture contributing a hefty 21% of greenhouse gas emissions, a shift towards more sustainable practices is critical (Duque-Acevedo et al. 2020).

The ASW presents a unique opportunity to promote environmental sustainability and economic prosperity within the agricultural sector. This vast resource stream holds immense potential for valorization into valuable products (Chilakamarry et al. 2022). The effective waste management strategies unlock the potential of ASW for biofuel production, generation of organic fertilizers for improved soil health, and development of bio-based materials (Bala et al. 2023). The shift towards a circular bioeconomy can contribute to reduced reliance on fossil fuels, minimize environmental pollution, and enhance overall agricultural productivity (Dey et al. 2021).

The increasing dependence on fossil fuels intensified the exploration of sustainable alternatives. ASW, a significant byproduct of modern farming practices, emerges as a promising source for biofuel generation like ethanol, hydrogen, biodiesel, etc. Utilization of ASW for biofuel production offers a multitude of benefits, promoting resource efficiency and fostering rural development (Awogbemi and Kallon 2022). With innovative techniques for ASW-to-biofuel conversion, we have a chance to unlock a more sustainable and secure energy future (Chilakamarry et al. 2021). The research across agricultural waste streams highlights the potential of biofuel generation. Taghizadeh-Alisaraei et al. (2017) and Sadeek et al. (2020) demonstrate the successful conversion of citrus fruit waste into bioethanol and waste cooking oil into biodiesel. Furthermore, multiple conversion processes have yielded promising results. Sharma and Dubey (2020) explored the use of hydrothermal and carbonization for converting food waste to bioethanol. Biological approaches offer an eco-friendly alternative. Shanmugam et al. (2018) identified specific microorganisms as effective agents for waste valorization into biofuels. These studies showcase the economic and environmental benefits of transforming ASW into valuable biofuels.

The valorization of ASW through biofuel generation presents a favourable solution for waste management and renewable energy, but it is crucial to evaluate its environmental impact across its entire lifecycle. The generation of greenhouse gases and pollutants from the conversion processes requires careful monitoring and mitigation strategies. The land-use changes to cultivate biofuel feedstocks could indirectly impact biodiversity and soil health. Additionally, efficient conversion processes can minimize resource consumption and emissions (Alhazmi and Loy 2021).

LCA serves as a powerful tool for comprehensively evaluating the environmental impact of valorizing ASW. This holistic approach considers the environmental burden throughout the entire life cycle, encompassing resource extraction to final product use (Sumiyati et al. 2024). LCA provides insights into potential environmental hotspots, areas with significant environmental impact. For short-term decision-making, LCA can guide the selection of the most environmentally friendly conversion processes by pinpointing those with lower energy consumption, reduced emissions, and minimal resource use. Furthermore, LCA informs long-term strategic planning by evaluating the potential long-term consequences of large-scale biofuel production from ASW (Nayal et al. 2016). By proactively identifying these potential challenges, LCA empowers stakeholders to implement mitigation strategies and ensure the long-term environmental sustainability of ASW (Wang et al. 2016).

This research delves into this critical topic by performing a comprehensive LCA of different ASW-to-biofuel conversion pathways. The research investigates the environmental footprints of various approaches, providing valuable insights for policymakers, industries, and researchers striving towards sustainable waste management and renewable energy production. Through this in-depth analysis, we hope to pave the way for a future where ASW is no longer a burden but a valuable resource, contributing to a cleaner and more sustainable world.

29.2 Life Cycle Assessment (LCA)

LCA emerged as a dominant tool for comprehensively evaluating the environmental impact of a product, process, or service. It transcends the limitations of traditional single-stage assessments by considering the entire life cycle (Finnveden et al. 2009). This holistic approach facilitates an informed understanding of environmental burdens and facilitates the identification of areas of improvement (Prado et al. 2017).

The LCA methodology is organized into four phases respectively. These phases provide a structured framework for administering a rigorous LCA study ISO 14040 (ISO 14044 2006). Figure 29.1 represents the conceptual framework for the LCA methodology.

29.2.1 Phase I: Goal and Scope Definition

This initial phase establishes the driving force behind the LCA study. This entails a comparative evaluation of alternative product designs or the guidance of

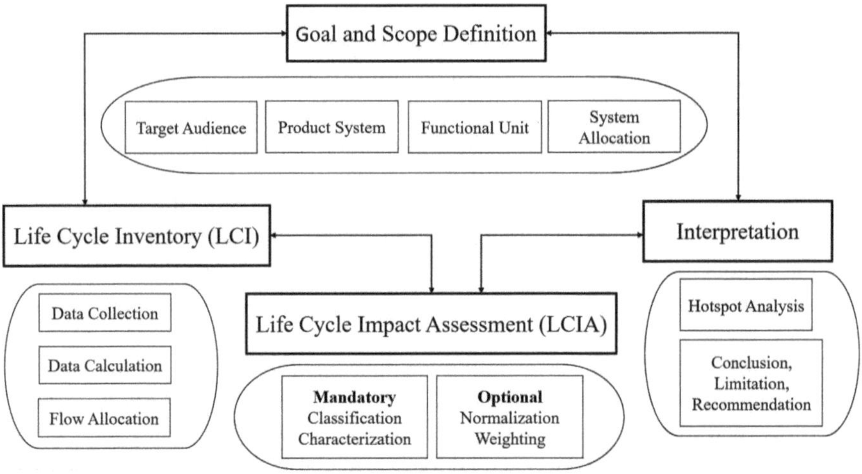

Fig. 29.1 Representation of conceptual framework for LCA methodology

eco-friendly service development. Furthermore, identifying the target audience is equally crucial (Finnveden et al. 2009). The LCA results are intended to inform policymakers, guide consumer choices, or drive sustainability initiatives. Finally, a meticulous system definition establishes the boundaries of the product system under investigation. This entails a comprehensive identification of all unit processes involved all over the life cycle (Laurent et al. 2014). The clear definition of these elements lays the foundation for a robust and targeted analysis.

29.2.2 Phase II: Life Cycle Inventory (LCI)

The LCI stage gathers detailed information on everything entering and leaving the product system at each step of its life span. The source of data can be industry databases, LCI databases, or through direct measurements from production facilities (Deepak et al. 2022). This phase involves creating a model for each unit process to quantify its environmental burdens. This model analyzes the raw data to uncover the environmental footprint of each step in the process (Laurent et al. 2014).

29.2.3 Phase III: Life Cycle Impact Assessment (LCIA)

The LCIA phase translates the ecological burdens identified in the previous step into understandable and comparable results. This two-step process begins with classification. In this, environmental burdens are categorized into relevant impact categories. This initial sorting helps to organize the diverse environmental effects associated with the system under study (Jolliet et al. 2004). Further, characterization factors are applied. These elements standardize the inventory findings within each area of impact by applying a consistent unit of measurement. This crucial step allows for the comparison of different types of environmental burdens. Depending on the specific goal and scope, optional normalization and weighting steps can be employed Verones et al. (2017).

29.2.4 Phase IV: Interpretation

The LCA culminates in the interpretation phase, where the analysis of results occurs within the defined goal and scope. This stage entails key environmental impacts, identifying hotspots that contribute most significantly to the environment, and exploring potential trade-offs between different impact categories. Subsequently, recommendations for improvement are formulated based on the findings (Laurent et al. 2014). Finally, the interpretation phase acknowledges any limitations inherent to the study, such as data gaps or uncertainties. It serves as a springboard for future research directions aimed at refining the LCA methodology and bolstering its effectiveness in driving environmental progress (Laurent et al. 2020).

29.3 LCA Contribution to Sustainable Agricultural Practices

ASW, traditionally a disposal challenge, emerges as a viable resource for biofuels. By emphasizing recycling, composting, and designing products for disassembly and reuse, the circular bioeconomy aims to minimize waste and pollution while fostering a more sustainable relationship between nature and our economic activity. LCA becomes a critical tool in this scenario. LCA can be used to determine how sustainable it is to create biofuels from ASW by considering its environmental effects throughout its life cycle. This allows researchers to identify areas for improvement, minimizing environmental impact and ensuring the overall viability of this circular bioeconomy approach. The present research focuses on five significant biofuels, i.e., biobutanol, biodiesel, bioethanol, biohydrogen, and biomethanol for analyzing environmental sustainability. Tables 29.1, 29.2, 29.3, 29.4, and 29.5 represent the major LCA studies for the biofuel generation from ASW.

Table 29.1 Extracts of the major LCA studies for biomethanol/methanol production from ASW

S. No.	References	Functional unit (FU)	LCIA method	Critical observations	Key findings
1.	Xiao et al. (2009)	1000 kg of biomethanol	"Research Center for eco-Environmental Sciences, Chinese Academy of Sciences"	• Bioethanol production from rice straw is less energy efficient (42.9%) compared to coal-based methanol (48.8%) • Bioethanol production has a higher unit investment cost (7880 RMB/t) compared to coal-based methanol (5186 RMB/t) due to the smaller scale of biomethanol plants • Bioethanol becomes more competitive as the price of fossil fuels increases	• The research found that converting rice straw into methanol yielded 0.308 kg of methanol per kilogram of rice straw • The cost to produce this biomethanol was 2347 RMB/ton, with the largest cost component being the raw material (rice straw) at 1046 RMB/ton
2.	Renó et al. (2011)	1 kg of produced methanol	CML 2001 method	• The study investigates how making methanol from sugarcane bagasse affects the environment • Biofuel production can have different environmental effects, even when made from the same starting ingredient	• Sugarcane bagasse emerges as a promising biofuel feedstock with a lower environmental footprint • The amount of energy produced from feedstock for methanol is promising, making it a competitive alternative to fossil fuel sources like natural gas • The study found that methanol from sugarcane bagasse has lower greenhouse gas emissions and other pollutants compared to fossil-based methanol
3.	Yadav et al. (2020)	1 ton of biomethanol	ReCiPe 20,165 (world-H) midpoint method	• This research investigates how a new method for producing methanol from wood affects the environment compared to the traditional method • Novel process modifications include pre-treatment, syngas cleaning, and gasification • The study suggests that optimizing the pre-treatment stage and carefully selecting catalysts and electricity sources can minimize environmental impacts in methanol production from wood	• The new method caused less harm to the environment in terms of global warming and harm to sea life compared to the traditional method • The pre-treatment stage of the new process was less environmentally friendly • Changing the catalyst from Na_2CO_3 to K_2CO_3 significantly increased the GWP

4.	Fózer et al. (2021)	1 ton of produced biomethanol	IMPACT2002+ v2.14	• The study emphasizes the importance of integrating variable clean energy sources to enhance the environmental performance of biomethanol production • It highlights the robustness and feasibility of hydrothermal gasification (HTG) for converting biomass into value-added fuels • The use of artificial neural networks (ANNs) is presented as an effective tool for optimizing and predicting the outcomes of complex chemical processes like HTG	• The study investigates a method for lowering GHGs in biomethanol production. It involves enriching the syngas with hydrogen derived from various renewable electricity sources • This approach to biomethanol synthesis, hydrothermal gasification, is attractive due to its flexibility in synthesis gas composition, ability to integrate heat effectively, and potential to reduce GHGs

Table 29.2 Extracts of the major LCA studies for biohydrogen/hydrogen production from ASW

S. No.	References	Functional unit (FU)	LCIA method	Critical observations	Key findings
1.	Wulf and Kaltschmitt (2013)	1 MJ hydrogen (at 15 °C, 1.5 MPa)	CO_2-eq, SO_2-eq, cumulative energy demand	• This research assessed the life cycle environmental impact of various biomass-to-hydrogen production pathways in Germany • Different biomass feedstocks like wood, biogas from fermentation, and glycerol (biodiesel byproduct) were analyzed • Allocation methods were used to account for co-products and byproducts in some processes	• Gasification of woody biomass offers the most promising option for environmentally sustainable biohydrogen production in Germany • Optimizing fertilizer use during bio-waste and energy crop production could potentially lower the environmental impact of biomethane reforming • Electrolysis using electricity generated from straw combustion had significant emissions due to the inefficiencies of the process and the use of fertilizer during straw cultivation
2.	Moreno and Dufour (2013)	1 Nm^3 of hydrogen (99.99% purity)	CML 2001 method	• The study finds that gasification of forestry residues is a more environmentally friendly way to produce hydrogen from biomass compared to using pruning residues • Recovering and utilizing non-converted methane from the process reduces CO_2 emissions and energy consumption • Forestry residues (pine and eucalyptus) are preferable over pruning residues (vine and almond) due to lower environmental impact during growth (no fertilizers or pesticides needed) and higher yield of gas during gasification	• Forestry residues, especially eucalyptus, have the lowest CO_2 emissions and energy requirements • Utilizing non-converted methane can lead to a negative CO_2 balance for forestry residues • Allocating the environmental impact of cultivation to fruits reduces energy consumption and other environmental impacts for pruning residues but significantly increases CO_2 emissions

3.	Reaño (2020)	1 kg of biohydrogen (27 °C, 99.99% purity)	CML 2001 method	• The research promotes dark fermentation using rice husk as a sustainable biohydrogen production method because it is eco-friendly and highly efficient • The study technology tackles environmental concerns by turning ASW into a source of clean energy for heating and powering buildings • Rice husks are a sustainable fuel source for co-generation plants, resulting in a process with minimal carbon released	• Dark fermentation requires the most freshwater, but using wastewater from breweries can help reduce the amount of freshwater needed • The key factors for improvement in dark fermentation include increased efficiency of rice husk CHP for electricity generation, improved hydrogen yield, and higher cellulose content in rice husk • Utilizing rice husk for CHP significantly reduces emissions compared to fossil fuel-based options
4.	Ramírez-Díaz et al. (2023)	1 kg of H_2 (99.99% purity)	ReCiPe-2016 v1.1. method	• Bio-hydrogen production from sugarcane vinasse has the potential to be a sustainable process • Utilizing excess steam from the sugarcane mill and integrating the H_2 production process could lead to a significant decrease in environmental impact • Optimizing energy usage throughout the process is crucial for improving sustainability	• Electricity and steam consumption are the main factors impacting environmental categories • With catalytic reforming and an H_2 purification system, the production of H_2 resulted in more GHGs • The production cost of H_2 falls within the range reported for other technologies
5.	Zheng et al. (2024)	1 kg H_2 (99.9% wt)	Cumulative energy demand (CED) v 1.11 and IPCC 2013 GWP 100a (v 1.03)	• Assessed the environmental impact of a biorefinery process (FW-H2) that utilizes food waste (FW) • Compared to steam methane reforming, FW-H_2 offers a cleaner approach to hydrogen production • Different allocation methods used in LCA can significantly affect the environmental impact results	• The FW-H2 process generates 10.1 kg CO_2 eq. of GHGs and requires 104 MJ of non-sustainable energy per kg of hydrogen • Sensitivity analysis showed that reducing energy consumption for gas compression by 20% could significantly decrease environmental impact • Economic allocation and system expansion allocation methods resulted in similar environmental impact values

Table 29.3 Extracts of the major LCA studies for biobutanol/butanol production from ASW

S. No.	References	Functional unit (FU)	LCIA method	Critical observations	Key findings
1.	Pereira et al. (2015)	1 kg of butanol, US$ earned, and km run	CML 2 baseline 2000 v2.05 method	• Bio-based butanol production from sugarcane likely exhibits lower environmental impact across various categories like global warming potential and ozone depletion • Using genetically modified strains with higher butanol yield could potentially reduce environmental impact • Integrating butanol production in bio-refineries might increase revenue compared to scenarios without butanol, potentially offsetting environmental costs	• The agricultural stage contributes significantly to the environmental impact due to factors like fertilizer application, vinasse disposal, and transportation • Butanol as a drop-in fuel for vehicles might offer environmental advantages over gasoline when considering the entire life cycle
2.	Brito and Martins (2017)	1 kg of n-butanol	IMPACT 2002+	• The potential of n-butanol from wheat straw acts as a renewable fuel source that aligns with circular economy principles • Using corn as a substrate raises sustainability concerns due to competition with food supply	• Among the substrates tested, corn resulted in the highest environmental impact during acetone, butanol, and ethanol (ABE) fermentation, whereas wheat straw exhibited the lowest • Focusing on the economic value of products in the ABE process can make it seem more environmentally damaging, even though these byproducts are also produced • Sensitivity analysis shows environmental impact is highly affected by the amount of gases produced during fermentation
3.	Marzocchella et al. (2022)	1 MJ of biobutanol	Environmental footprint (EF) 2.0 method	• The study evaluates a novel process for producing biobutanol from potato peel waste • Two distillation scenarios are investigated: a reduced one recovering only butanol, and a complete one recovering butanol, acetone, and ethanol • Fermentation is the biggest contributor to variable operating costs, though less significant environmentally	• Pre-treatment is the biggest environmental impact contributor due to citric acid production • The total production cost is high for both scenarios, with nearly 90% of variable operating costs • Biobutanol has a higher total production cost and environmental score than gasoline and E85, but significantly lower climate change impact

Table 29.4 Extracts of the major LCA studies for biodiesel production from ASW

S. No.	References	Functional unit (FU)	LCIA method	Critical observations	Key findings
1.	Yee et al. (2009)	GigaJoule/ton crude palm oil (tCPO)/year and ton CO_2/ton biodiesel/year	GHG emissions	• The study suggests that palm biodiesel has lower greenhouse gas (GHG) emissions compared to petroleum diesel • Making palm biodiesel creates more energy than it uses over its entire lifespan	• Palm biodiesel production utilizes energy from various sources like fertilizers, steam, electricity, and fossil fuels • Palm biodiesel production utilizes energy from various sources like fertilizers, steam, electricity, and fossil fuels • Produced alongside palm biodiesel, glycerol boasts its substantial energy content
2.	Pradhan et al. (2011)	1 L of biodiesel	Fossil energy ratio (FER)	• The study finds that soybean biodiesel production is becoming more renewable due to improvements in agricultural practices and processing technologies • Soybean yields and energy efficiency in processing facilities have improved significantly, which led to a substantial reduction in the fossil energy required to produce biodiesel • Improvements in soybean production stem from factors like increased yields, the adoption of eco-friendly processing plants, and the employment of genetically engineered soybeans which require less pesticide	• Achieved substantial reductions in energy consumption for both soybean crushing (58%) and transesterification (33%) per unit of biodiesel produced • The overall reduction in energy input for biodiesel production was 42%

(continued)

Table 29.4 (continued)

S. No.	References	Functional unit (FU)	LCIA method	Critical observations	Key findings
3.	Chung et al. (2019)	1000 kg of biodiesel produced per day	CML 2001 and Eco-indicator 99	• Biodiesel production using "waste cooking oil" (WCO) and a recycled eggshell catalyst has a smaller environmental footprint compared to jatropha oil and a KOH catalyst • Waste cooking oil as feedstock reduces environmental impact compared to Jatropha oil because it avoids the agricultural phase • Waste chicken eggshell-derived catalyst is a more sustainable option than KOH due to avoiding chemicals, complex purification, and neutralization processes	• The movement of raw materials has a substantial negative impact on the environment • Equipment operation is the primary source of the impact • Transesterification of WCO has the greatest environmental impact due to high electricity consumption for equipment, purification, and centrifugation
4.	Aghbashlo et al. (2020)	1 MJ of biodiesel	IMPACT 2002+ method	• The downside impact of biodiesel produced from WCO using an ultrasound-assisted transesterification process is significantly lower • The study found that varying the amount of methanol, the temperature of the methanolysis process, and the length of the reaction significantly impacted all of the environmental categories • Methanol content plays the biggest role. In comparison, methanolysis temperature has a much smaller effect	• Optimizing process parameters can significantly reduce the downside impact of biodiesel production • Biodiesel production from WCO has the potential to be more sustainable than using mineral diesel • The amount of electricity used and the phosphoric acid used to neutralize the crude glycerol both have a significant impact on the environment and human health • Methanol consumption significantly affects resource consumption
5.	Al-Mawali et al. (2021)	1000 kg of biodiesel production	CML-IA baseline V3.06 and ReCiPe 2016 endpoint V1.04	• The research converted waste date seed oil into biodiesel using a novel magnetic catalyst • The study identified optimal reaction conditions for maximizing biodiesel yield (90%) using the magnetic catalyst • Date seed oil exhibits good characteristics for biodiesel production, including suitable viscosity, fatty acid profile, and stability	• Optimized reaction conditions achieved a 90% conversion rate for FFAs to biodiesel at 55 °C, 47 min, and 1.5 wt% catalyst loading • Preparing and reusing catalysts had the greatest environmental burden among the studied processes • Carbon emissions were estimated at 1.11 kg CO_2 eq/kg of biodiesel

Table 29.5 Extracts of the major LCA studies for bioethanol production from ASW

S. No.	References	Functional unit (FU)	LCIA method	Critical observations	Key findings
1.	Cherubini and Ulgiati (2010)	477 kilo-tons$_{dry}$/ annum	CML 2 baseline 2000 V. 2.03	• Biorefineries using agricultural residues can significantly reduce GHG emissions (around 50%) and fossil energy demand (over 80%) compared to fossil fuel alternatives • Biorefineries that utilize ASW offer a potential solution for reducing our reliance on fossil fuels and combating climate change • Sustainable practices for residue harvest and land management are crucial to minimize negative environmental impacts	• Biorefinery using residues can save 1.8–3 tons CO_2-eq./ ha of land used for corn and wheat production • Land-use change effects, fertilizer production, and raw material supply significantly influence the final GHG balance • Quantifying soil carbon losses from residue management practices remains imprecise
2.	Vaskan et al. (2018)	1 ton of palm empty fruit bunches (EFB)	ReCiPe method	• Converting palm empty fruit bunches (EFB) into biofuels and co-products (ethanol, C5 syrup, electricity) offers a more sustainable solution than conventional options (fossil fuels, 1G biofuels) • The economic feasibility of the EFB biorefinery depends on several factors, including market prices, production costs, and capital investments • The study compares two approaches for processing EFB	• Both scenarios significantly reduce climate change impact (84% and 73%) and fossil fuel depletion (83% and 76%) compared to traditional methods • Both scenarios are currently economically unfavourable with negative profitability evaluation protocol (PEP) values • EFB OF's economic advantage comes from the higher market value of fuel ethanol compared to C5 syrup
3.	Liu et al. (2020)	1 km distance driven by a multi-fuel car	Midpoint and endpoint impact categories	• Lignocellulosic ethanol production from agricultural waste is a promising alternative for a more sustainable approach • The environmental benefits are proportional to the ethanol blending ratio	• E10 has a 6% lower fossil depletion potential and 1–10% lower greenhouse gas emissions than gasoline • Growing crops for fuel and converting them into ethanol significantly increases the risk of ozone depletion, acid rain, and nutrient pollution in water, especially for E10 and E85 fuel

(continued)

Table 29.5 (continued)

S. No.	References	Functional unit (FU)	LCIA method	Critical observations	Key findings
4.	Safaripour et al. (2023)	1 MJ of bioethanol	TRACI 2.1 method	• The environmental impact of bioethanol from wheat straw stems largely from agricultural practices along with the energy required for irrigation • Bioethanol conversion process generates surplus electricity, offsetting some environmental burdens like ozone depletion and fossil fuel depletion • Compared to gasoline, wheat straw bioethanol offers a significant reduction in greenhouse gas (GHG) emissions (60% reduction)	• The net CO_2 emission for 1 MJ of bioethanol is 37.3 g CO_2 eq, which is lower than gasoline (93 g CO_2 eq/MJ) • Wheat straw cultivation contributes most to global warming, acidification, eutrophication, and human health impacts • Uncertainty analysis for fertilizer application in the study region showed high confidence in the model for farm-related emissions
5.	Kovacs et al. (2022)	1 kg of bioethanol and 1 kg of pellets/ briquettes	ReCiPe midpoint and endpoint H	• This research supports policymakers and stakeholders in the viticulture sector to implement circular economy practices and promote sustainability • Energy consumption across most stages impacts the different stages of bioethanol production	• Bioethanol production has a higher environmental impact than pellets/briquette production across most studied categories • Delignification is the stage with the highest impact on bioethanol production, followed by SSF (simultaneous saccharification and fermentation) and autohydrolysis • Transportation of vine shoots has the lowest impact in both processes
6.	Correia et al. (2024)	1 MJ of ethanol	Environmental footprint (EF) 3.0 method	• Making ethanol from eucalyptus leftovers is cheaper than using corn stalks as it costs less to move and process them • Burning corn Stover for energy has a smaller EF in several ways compared to using eucalyptus leftovers • Key factors influencing environmental impact include enzyme production and energy sources	• The approach of combining on-site enzyme production with renewable energy sources leads to a substantial decrease in environmental impact • There is a gap between the production cost of bioethanol and its market price • By optimizing processes like heat recovery and in-house enzyme production, ethanol producers can bring down the minimum ethanol selling price
7.	Yin et al. (2024)	1 mg bioethanol	Selected impact categories	• Corn cobs are the most promising feedstock for bioethanol production based on energy efficiency and environmental impact • Straw (corn and wheat) can be a viable option when used in combination with corn cobs for comprehensive benefits • Bioethanol production from all three feedstocks reduces greenhouse gas (GHG) emissions compared to gasoline (48–54% reduction)	• Corn cobs achieve a net energy ratio of 1.30 and a net energy balance of 6902 MJ/mg bioethanol, making them a very promising feedstock • The allocation method used to distribute environmental burdens and energy use across co-products (lignin, xylose) significantly affects the final results • Extracting value from lignin offers more economic and environmental benefits than simply burning it as a fuel source

29.4 Conclusions

This research has comprehensively explored the environmental analysis of converting ASW into valuable biofuels. By analyzing the environmental footprints of various ASW-to-biofuel conversion pathways, this study provides valuable insights for policymakers, industries, and researchers striving towards resource recovery and clean energy production. The key findings of the LCA studies on different biofuels derived from ASW reveal:

Potential for greenhouse gas (GHG) reduction: Most biofuels derived from ASW offer significant reductions in GHG emissions compared to their fossil fuel counterparts.

Importance of feedstock selection: The environmental impact of biofuel production is significantly influenced by the type of ASW feedstock used.

Optimization opportunities: The studies identified various opportunities for process optimization to minimize environmental impact.

In conclusion, the valorization of ASW for biofuel production presents a promising solution for a zero-waste approach and sustainable energy generation. Through LCA, we can identify environmental hotspots in biofuel production. This approach allows for the development of more sustainable practices. By implementing these insights, stakeholders can contribute to a more circular bioeconomy and a cleaner future. Further research should explore combining ASW valorization with other waste strategies, investigate advanced conversion technologies for better yields and lower impact, and assess the economic feasibility and life cycle costs of large-scale ASW-to-biofuel production. By addressing these research directions, we can accelerate the transition towards a more sustainable future where ASW becomes a valuable resource, contributing to clean energy production and a circular bioeconomy.

Declaration The author declares no competing interests.

References

Aghbashlo M, Tabatabaei M, Amid S, Hosseinzadeh-Bandbafha H, Khoshnevisan B, Kianian G (2020) Life cycle assessment analysis of an ultrasound-assisted system converting waste cooking oil into biodiesel. Renew Energy 151:1352–1364. https://doi.org/10.1016/j.renene.2019.11.144

Alhazmi H, Loy ACM (2021) A review on environmental assessment of conversion of agriculture waste to bio-energy via different thermochemical routes: current and future trends. In: Bioresource technology reports, vol 14. Elsevier, Amsterdam. https://doi.org/10.1016/j.biteb.2021.100682

Al-Mawali KS, Osman AI, Al-Muhtaseb AH, Mehta N, Jamil F, Mjalli F, Vakili-Nezhaad GR, Rooney DW (2021) Life cycle assessment of biodiesel production utilising waste date seed oil and a novel magnetic catalyst: a circular bioeconomy approach. Renew Energy 170:832–846. https://doi.org/10.1016/j.renene.2021.02.027

Awogbemi O, Kallon DV (2022) Valorization of agricultural wastes for biofuel applications. Heliyon 8(10):e11117. https://doi.org/10.1016/j.heliyon.2022.e11117
Bala S, Garg D, Sridhar K, Inbaraj BS, Singh R, Kamma S, Tripathi M, Sharma M (2023) Transformation of agro-waste into value-added bioproducts and bioactive compounds: micro/nano formulations and application in the agri-food-pharma sector. Bioengineering 10(2):10020152. https://doi.org/10.3390/bioengineering10020152
Brito M, Martins F (2017) Life cycle assessment of butanol production. Fuel 208:476–482. https://doi.org/10.1016/j.fuel.2017.07.050
Cherubini F, Ulgiati S (2010) Crop residues as raw materials for biorefinery systems - a LCA case study. Appl Energy 87(1):47–57. https://doi.org/10.1016/j.apenergy.2009.08.024
Chilakamarry CR, Mimi Sakinah AM, Zularisam AW, Pandey A, Vo DVN (2021) Technological perspectives for utilisation of waste glycerol for the production of biofuels: a review. Environ Technol Innov 24:101902. https://doi.org/10.1016/j.eti.2021.101902
Chilakamarry CR, Mimi Sakinah AM, Zularisam AW, Sirohi R, Khilji IA, Ahmad N, Pandey A (2022) Advances in solid-state fermentation for bioconversion of agricultural wastes to value-added products: opportunities and challenges. Bioresour Technol 343:126065. https://doi.org/10.1016/j.biortech.2021.126065
Chung ZL, Tan YH, Chan YS, Kansedo J, Mubarak NM, Ghasemi M, Abdullah MO (2019) Life cycle assessment of waste cooking oil for biodiesel production using waste chicken eggshell derived CaO as catalyst via transesterification. Biocatal Agric Biotechnol 21:101317. https://doi.org/10.1016/j.bcab.2019.101317
Correia B, Matos HA, Lopes TF, Marques S, Gírio F (2024) Sustainability assessment of 2G bioethanol production from residual lignocellulosic biomass. PRO 12(5):987. https://doi.org/10.3390/pr12050987
Deepak A, Sharma V, Kumar D (2022) Life cycle assessment of biomedical waste management for reduced environmental impacts. J Clean Prod 349:131376. https://doi.org/10.1016/J.JCLEPRO.2022.131376
Dey T, Bhattacharjee T, Nag P, Ritika, Ghati A, Kuila A (2021) Valorization of agro-waste into value added products for sustainable development. Bioresour Technol Rep 16:100834. https://doi.org/10.1016/j.biteb.2021.100834
Duque-Acevedo M, Belmonte-Ureña LJ, Cortés-García FJ, Camacho-Ferre F (2020) Agricultural waste: review of the evolution, approaches and perspectives on alternative uses. Glob Ecol Conserv 22:e00902. https://doi.org/10.1016/j.gecco.2020.e00902
Finnveden G, Hauschild MZ, Ekvall T, Guinée J, Heijungs R, Hellweg S, Koehler A, Pennington D, Suh S (2009) Recent developments in life cycle assessment. J Environ Manag 91(1):1–21. https://doi.org/10.1016/j.jenvman.2009.06.018
Fózer D, Tóth AJ, Varbanov PS, Klemeš JJ, Mizsey P (2021) Sustainability assessment of biomethanol production via hydrothermal gasification supported by artificial neural network. J Clean Prod 318:128606. https://doi.org/10.1016/j.jclepro.2021.128606
ISO 14044 (2006) ISO 14044: 2006 Environmental management - life cycle assessment - requirements and guidelines. https://www.iso.org/standard/38498.html
Jolliet O, Miiller-Wenk R, Bare J, Brent A, Goedkoop M, Heijungs R, Itsubo N, Pefia C, Pennington D, Potting J, Rebitzer G, Stewart M, Udo De Haes H, Weidema B (2004) The LCIA midpoint-damage framework of the UNEP/SETAC life cycle initiative. Int J Life Cycle Assess 9(6):394–404. https://doi.org/10.1007/BF02979083
Kovacs E, Hoaghia MA, Senila L, Scurtu DA, Varaticeanu C, Roman C, Dumitras DE (2022) Life cycle assessment of biofuels production processes in viticulture in the context of circular economy. Agronomy 12(6):1320. https://doi.org/10.3390/agronomy12061320
Laurent A, Clavreul J, Bernstad A, Bakas I, Niero M, Gentil E, Christensen TH, Hauschild MZ (2014) Review of LCA studies of solid waste management systems—part II: methodological guidance for a better practice. Waste Manag 34(3):589–606. https://doi.org/10.1016/J.WASMAN.2013.12.004

Laurent A, Weidema BP, Bare J, Liao X, Maia de Souza D, Pizzol M, Sala S, Schreiber H, Thonemann N, Verones F (2020) Methodological review and detailed guidance for the life cycle interpretation phase. J Ind Ecol 24(5):986–1003. https://doi.org/10.1111/jiec.13012

Liu F, Short MD, Alvarez-Gaitan JP, Guo X, Duan J, Saint C, Chen G, Hou L (2020) Environmental life cycle assessment of lignocellulosic ethanol-blended fuels: a case study. J Clean Prod 245:118933. https://doi.org/10.1016/j.jclepro.2019.118933

Marzocchella A, Chirone R, Paulillo A, Lettieri P, Salatino P (2022) Sustainability assessment of biotechnological processes: LCA and LCC of second-generation biobutanol production. In: Teodosiu C, Fiore S, Hospido A (eds) Assessing progress towards sustainability frameworks, tools and case studies. Elsevier, Amsterdam

Mengqi Z, Shi A, Ajmal M, Ye L, Awais M (2023) Comprehensive review on agricultural waste utilization and high-temperature fermentation and composting. Biomass Convers Biorefinery 13(7):5445–5468. https://doi.org/10.1007/s13399-021-01438-5

Moreno J, Dufour J (2013) Life cycle assessment of hydrogen production from biomass gasification. Evaluation of different Spanish feedstocks. Int J Hydrog Energy 38(18):7616–7622. https://doi.org/10.1016/j.ijhydene.2012.11.076

Nayal FS, Mammadov A, Ciliz N (2016) Environmental assessment of energy generation from agricultural and farm waste through anaerobic digestion. J Environ Manag 184:389–399. https://doi.org/10.1016/j.jenvman.2016.09.058

Pereira LG, Chagas MF, Dias MOS, Cavalett O, Bonomi A (2015) Life cycle assessment of butanol production in sugarcane biorefineries in Brazil. J Clean Prod 96:557–568. https://doi.org/10.1016/j.jclepro.2014.01.059

Pradhan A, Shrestha DS, McAloon A, Yee W, Haas M, Duffield JA, Shapouri H (2011) Energy life-cycle assessment of soybean biodiesel. Trans ASABE 54(3):1031–1039

Prado V, Wender BA, Seager TP (2017) Interpretation of comparative LCAs: external normalization and a method of mutual differences. Int J Life Cycle Assess 22(12):2018–2029. https://doi.org/10.1007/s11367-017-1281-3

Ramírez-Díaz RC, Prato-Garcia D, Vasquez-Medrano R (2023) How sustainable is the biohydrogen produced from sugarcane vinasse? An approach based on life cycle assessment. Biomass Convers Biorefinery 13(16):14755–14775. https://doi.org/10.1007/s13399-022-03242-1

Reaño RL (2020) Assessment of environmental impact and energy performance of rice husk utilization in various biohydrogen production pathways. Bioresour Technol 299:122590. https://doi.org/10.1016/j.biortech.2019.122590

Renó MLG, Lora EES, Palacio JCE, Venturini OJ, Buchgeister J, Almazan O (2011) A LCA (life cycle assessment) of the methanol production from sugarcane bagasse. Energy 36(6):3716–3726. https://doi.org/10.1016/j.energy.2010.12.010

Sadeek SA, Mohammed EA, Shaban M, Abou Kana MTH, Negm NA (2020) Synthesis, characterization and catalytic performances of activated carbon-doped transition metals during biofuel production from waste cooking oils. J Mol Liq 306:112749. https://doi.org/10.1016/j.molliq.2020.112749

Safaripour M, Ghanbari A, Seyedabadi E, Pourhashem G (2023) Investigation of environmental impacts of bioethanol production from wheat straw in Kermanshah, Iran. Biomass Convers Biorefinery 13(7):5931–5941. https://doi.org/10.1007/s13399-021-01676-7

Shanmugam S, Hari A, Ulaganathan P, Yang F, Krishnaswamy S, Wu YR (2018) Potential of biohydrogen generation using the delignified lignocellulosic biomass by a newly identified thermostable laccase from Trichoderma asperellum strain BPLMBT1. Int J Hydrog Energy 43(7):3618–3628. https://doi.org/10.1016/J.IJHYDENE.2018.01.016

Sharma HB, Dubey BK (2020) Co-hydrothermal carbonization of food waste with yard waste for solid biofuel production: hydrochar characterization and its pelletization. Waste Manag 118:521–533. https://doi.org/10.1016/J.WASMAN.2020.09.009

Sumiyati S, Samadikun BP, Widiyanti A, Budihardjo MA, Al Qadar S, Puspita AS (2024) Life cycle assessment of agricultural waste recycling for sustainable environmental impact. Global J Environ Sci Manage 10(2):907–938. https://doi.org/10.22034/gjesm.2024.02.30

Taghizadeh-Alisaraei A, Hosseini SH, Ghobadian B, Motevali A (2017) Biofuel production from citrus wastes: a feasibility study in Iran. Renew Sust Energ Rev 69:1100–1112. https://doi.org/10.1016/J.RSER.2016.09.102

Vaskan P, Pachón ER, Gnansounou E (2018) Techno-economic and life-cycle assessments of biorefineries based on palm empty fruit bunches in Brazil. J Clean Prod 172:3655–3668. https://doi.org/10.1016/j.jclepro.2017.07.218

Verones F, Bare J, Bulle C, Frischknecht R, Hauschild M, Hellweg S, Henderson A, Jolliet O, Laurent A, Liao X, Lindner JP, Maia de Souza D, Michelsen O, Patouillard L, Pfister S, Posthuma L, Prado V, Ridoutt B, Rosenbaum RK, Fantke P (2017) LCIA framework and cross-cutting issues guidance within the UNEP-SETAC life cycle initiative. J Clean Prod 161:957–967. https://doi.org/10.1016/J.JCLEPRO.2017.05.206

Wang QL, Li W, Gao X, Li SJ (2016) Life cycle assessment on biogas production from straw and its sensitivity analysis. Bioresour Technol 201:208–214. https://doi.org/10.1016/j.biortech.2015.11.025

Wulf C, Kaltschmitt M (2013) Life cycle assessment of biohydrogen production as a transportation fuel in Germany. Bioresour Technol 150:466–475. https://doi.org/10.1016/j.biortech.2013.08.127

Xiao J, Shen L, Zhang Y, Gu J (2009) Integrated analysis of energy, economic, and environmental performance of biomethanol from rice straw in China. Ind Eng Chem Res 48(22):9999–10007. https://doi.org/10.1021/ie900680d

Yadav P, Athanassiadis D, Yacout DMM, Tysklind M, Upadhyayula VKK (2020) Environmental impact and environmental cost assessment of methanol production from wood biomass. Environ Pollut 265:114990. https://doi.org/10.1016/j.envpol.2020.114990

Yee KF, Tan KT, Abdullah AZ, Lee KT (2009) Life cycle assessment of palm biodiesel: revealing facts and benefits for sustainability. Appl Energy 86(suppl. 1):S189–S196. https://doi.org/10.1016/j.apenergy.2009.04.014

Yin T, Huhe T, Li X, Wang Q, Lei T, Zhou Z (2024) Research on life cycle assessment and performance comparison of bioethanol production from various biomass feedstocks. Sustain For 16(5):1788. https://doi.org/10.3390/su16051788

Zheng X, Wang J, Huang J, Xu X, Tang J, Hou P, Han W, Li H (2024) Environmental impact assessment of a combined bioprocess for hydrogen production from food waste. Waste Manag 173:152–159. https://doi.org/10.1016/j.wasman.2023.11.009

Life Cycle Assessment for the Valorization of Agricultural Solid Waste to Chemicals

30

Rishi Kumar Singh, Sanghdeep Gautam, Akhilesh Kumar Singh, and Sujit Pratap Singh

Abstract

The valorization of agricultural solid waste into chemicals represents a transformative approach towards achieving sustainable development. This chapter delves into the application of life cycle assessment (LCA) stands as a crucial methodology for comprehensively assessing the environmental repercussions associated with converting agricultural waste into valuable chemicals. The strategy of valorization addresses two major issues: the reduction of waste volumes and the mitigation of greenhouse gas emissions. This is particularly relevant in the context of increasing industrial activities and population growth, which exacerbate challenges related to waste management.

Agricultural solid waste, such as crop residues, animal manure, and food processing by-products, is often utilized as low-grade animal feed or is disposed of through methods like burning or landfilling. These conventional disposal methods have significant environmental drawbacks, including increased greenhouse gas emissions and soil degradation. Valorization, in contrast, offers a more sustainable approach by transforming these wastes into high-value products such as

R. K. Singh (✉) · S. Gautam
Department of Biosciences, IMS Ghaziabad, University Courses Campus, Ghaziabad, Uttar Pradesh, India
e-mail: rishikumar@imsuc.ac.in; sanghdeep.gautam@imsuc.ac.in

A. K. Singh
Department of Biotechnology, School of Life Sciences, Mahatma Gandhi Central University, Motihari, Bihar, India

S. P. Singh
Amity Institute of Biotechnology, Amity University, Lucknow, Uttar Pradesh, India
e-mail: spsingh3@amity.edu

A. Kumari et al. (eds.), *Valorization of Solid Wastes to Biofuels and Chemical Products for Sustainable World*, https://doi.org/10.1007/978-981-96-8594-3_30

biofuels, fertilizers, and chemicals. This chapter highlights the role of LCA in assessing these processes, from waste collection to product distribution.

A detailed examination of various valorization methods, including fermentation, transesterification, and anaerobic digestion, reveals their potential to convert agricultural residues into bioethanol, biodiesel, and biogas (Singh and Gu 2010). Each method has unique advantages and challenges, particularly concerning energy efficiency, resource utilization, and environmental impact. LCA provides a comprehensive framework for understanding these impacts by evaluating factors such as global warming potential, energy efficiency, and resource consumption.

Case studies, such as the conversion of rice husk into chemicals like silica, bio-oil, and furfural, underscore the practical application of LCA in assessing the sustainability of valorization processes. These studies illustrate how valorization can reduce greenhouse gas emissions and enhance resource efficiency compared to traditional waste management practices. The chapter also discusses the integration of LCA with techno-economic analysis to offer all-inclusive view of the environmental and economic paybacks of valorization.

In summary, the valorization of agricultural solid waste presents a promising strategy for enhancing sustainability, reducing environmental impacts, and supporting economic development. Future research should focus on optimizing these processes and further integrating LCA to refine and advance agricultural waste valorization strategies, aligning with global sustainability goals and fostering a circular bioeconomy.

Keywords

Agricultural waste valorization · Life cycle assessment (LCA) · Sustainable development · Biofuels · Circular economy · Environmental impact · Waste management · Biorefinery processes

30.1 Introduction

The valorization of agricultural solid waste into chemicals is an innovative strategy that significantly contributes to sustainable development. This approach not only helps in reducing the volume of waste but also mitigates greenhouse gas emissions, while providing alternative sources of valuable chemicals. The importance of such strategies has been increasingly recognized in the context of global sustainability goals. However, to ensure that the processes involved in agricultural waste valorization are genuinely beneficial from an environmental standpoint, a rigorous evaluation is required. Life cycle assessment (LCA) stands as a crucial methodology for comprehensively assessing tool, for analysis of the environmental impacts associated with converting agricultural solid waste into chemicals. This chapter explores the application of LCA in this context, offering an in-depth review of methodologies, case studies, and the potential environmental benefits of these processes.

The rapid increase in industrial activities and population growth has led to significant challenges related to waste management, which have become a focal point for many researchers (Ghosh et al. 2019). As industries have expanded, there have been several consequential developments, such as enhanced employment opportunities, changes in land use, and increased international trade (Singh et al. 2018; Kumar and Patel 2020). Agro-industries, in particular, generate substantial quantities of residues each year. For instance, cereal processing produces large amounts of husks, while coffee production leads to the generation of coffee pulp waste, and fruit and vegetable processing results in peels and other organic by-products (Roy et al. 2019). The growth of these industries necessitates considerable energy resources (Rai et al. 2021). The escalating demand for energy from traditional sources has driven efforts to explore alternative methods of energy generation (Sharma et al. 2022; Patel et al. 2022).

In rural areas, one effective approach for energy production is the deployment of agricultural waste. Bioenergy, a renewable form of energy, can be derived from organic agricultural residues (Gupta et al. 2016; Roy et al. 2018; Das et al. 2020). These residues can be converted into various upgraded products, such as electricity, bio-based fuels, and biogas (Mehta et al. 2021). Different organic substrates own distinctive characteristics that make them suitable for producing a range of products. Although organic waste can pose environmental risks, it also offers opportunities for creating valuable products like animal feed (Salemdeeb et al. 2017), biofertilizers, and biofuels (Sharma et al. 2019; Thakur et al. 2021). Rich in essential nutrients like minerals, proteins, and sugars, these materials can also foster microbial growth, which further contributes to their utilization.

The improper disposal of untreated agricultural residues can introduce contaminants into the ecosystem, adversely affecting both environmental and human well-being (Mishra 2016; Banerjee et al. 2017; Joshi et al. 2017). Many of these residues remain unprocessed and are often disposed of through methods like burning or landfilling, resulting in environmental concerns including increased greenhouse gas emissions and higher temperatures. Thus, there is an urgent need for sustainable waste management strategies. Adopting a circular economy approach—integrating the principles of biorefineries and the 3Rs (reduce, reuse, recycle)—could offer a pathway for more effective utilization of agricultural raw materials (Chakraborty et al. 2020). Developing a circular economy in agriculture provides a promising strategy to tackle these issues, although further refinement is needed to fully realize its environmental benefits.

Research on the agricultural circular economy is still evolving. Life cycle assessment (LCA) serves as a useful tool for assessing environmental impacts across different stages of this economy (Das and Singh 2020). This approach includes utilizing advanced technologies to minimize waste within a closed-cycling system (Khan and Bhatia 2021). The circular economy's development requires technologies that harmonize economic development, industrial growth, ecosystem preservation, and efficient resource use.

This chapter focuses on the potential of agro-industrial by-products in the production of bio-based products through economically viable and environmentally friendly biorefinery processes. Various biorefinery strategies are currently under development. This work also provides a comprehensive overview of the possibilities for valorizing agro-industrial waste to produce renewable energy products. It also discusses different methods for converting agricultural waste into valuable products. Furthermore, it examines the role of life cycle assessment (LCA) in promoting a circular bioeconomy in agriculture, offering key insights into future research areas aimed at optimizing the use of biogas and its derivatives (Angelidaki and Ahring 2000).

30.2 Overview of Agricultural Solid Waste Valorization

Agricultural solid waste, encompassing residual materials from agricultural activities, comprising crop residues (such as straw, husks, and stalks), animal manure, and waste generated during food processing, constitutes a substantial portion of the global waste stream. The quantity of such waste is particularly high in regions with intensive agricultural activities, such as India, China, and other agrarian economies. Traditionally, these materials have been managed through practices like field decomposition, burning, or their use as low-value animal feed. However, these methods have significant environmental drawbacks.

30.2.1 Environmental Challenges

Greenhouse Gas Emissions A significant environmental concerns associated with agricultural waste is its contribution to greenhouse gas emissions. When agricultural waste is left to decompose in fields or is burned, it can release substantial emissions of greenhouse gases (GHGs), notably methane (CH_4) and carbon dioxide (CO_2). Methane is produced during anaerobic decomposition in manure storage systems or under wet conditions in fields. The burning of crop residues—an unfortunately common practice in many developing countries—releases CO_2, particulate matter, and other pollutants, exacerbating air quality issues and contributing to climate change (Munasinghe and Khanal 2010).

Soil Degradation Residue removal practices from fields for burning or other uses can also lead to soil degradation. Crop residues are essential for maintaining soil health; they contribute organic matter, enhance water retention, and help prevent erosion. Continuous removal of these residues can deplete the soil of essential nutrients, thereby reducing its fertility over time.

30.2.2 Valorization of Agricultural Solid Waste

Valorization refers to the process of converting agricultural waste into higher-value products, thereby transforming what was once considered waste into a resource. This approach aligns with the principles of a circular economy, where waste materials are repurposed through reuse, recycling, or conversion into valuable products. This minimizes reliance on new raw resources and mitigates environmental impacts.

30.2.3 Biofuel Production Through Fermentation and Transesterification

Agricultural waste can be transformed into various chemicals and biofuels using methods such as anaerobic digestion, pyrolysis, fermentation, and transesterification. For instance, crop residues can be converted into bioethanol, a renewable fuel that serves as an alternative to fossil fuels for transportation. Similarly, animal manure can be processed into biogas, providing a sustainable energy (Kwiatkowski et al. 2006).

The production of biofuels is a vital field of research and development, especially in the pursuit of sustainable energy solutions. Two widely utilized methods in biofuel production are fermentation and transesterification, which employ different techniques to yield valuable products, thereby reducing dependence on fossil fuels.

30.2.3.1 Fermentation Process for Bioethanol Production

Fermentation is a commonly applied biochemical process for generating bioethanol, a popular type of biofuel. This method utilizes microorganisms to convert organic materials like starch, cellulose, or sugars into ethanol (Alam et al. 2019). Initially, the biomass undergoes mechanical processing to crush it, followed by the application of specific enzymes that break down complex starches into simpler sugars (Sharma et al. 2018). These sugars are then fermented by yeast strains, producing bioethanol as a result.

Ethanol produced via fermentation undergoes purification through distillation, an energy-demanding process. Importantly, the solid byproducts of fermentation, including animal feed and residual materials such as bagasse from sugarcane, possess diverse utilization potential. For example, bagasse is frequently employed as fuel in gasification processes, which enhances sustainability by ensuring that all parts of the biomass are utilized effectively (Singh et al. 2020).

30.2.3.2 Transesterification Process for Biodiesel Production

Transesterification is another key process used in the biofuel industry, specifically for producing biodiesel. This chemical process for biodiesel synthesis involves the transesterification of fats or oils with an alcohol, typically methanol or ethanol, in the presence of a catalyst like sodium or potassium hydroxide (Demirbas 2009). This reaction yields biodiesel and glycerol, the latter being a valuable by-product that has applications in various sectors, including pharmaceuticals and cosmetics.

Biodiesel produced through transesterification can be directly used in orthodox diesel engines, serving as an immediate substitute for traditional fossil-based diesel. The use of waste vegetable oils and animal fats as feedstocks for biodiesel generation not only provides a renewable energy source but also offers a method for repurposing waste materials, further contributing to the sustainability of the process (Rashid et al. 2008).

Both fermentation and transesterification play essential roles in the production of biofuels like bioethanol and biodiesel, respectively. These methods not only provide renewable alternatives to fossil fuels but also promote the efficient use of organic materials and by-products, supporting a more sustainable energy future.

Fertilizers The nutrient-rich nature of agricultural waste makes it an ideal raw material for producing organic fertilizers. For instance, composting animal manure or crop residues can yield nutrient-rich compost, which can contribute to improved soil health and fertility, reducing the reliance on synthetic fertilizers.

Anaerobic Digestion for Agricultural Waste Valorization: This is a biological process where organic matter, such as agricultural and food waste, decomposes in the absence of oxygen, yielding valuable byproducts (Pandey et al. 2019). The process involves four key stages: methanogenesis, acidogenesis, hydrolysis, and acetogenesis. These stages collectively contribute to the breakdown of the organic material into simpler compounds, ultimately leading to the production of biogas, which primarily consists of methane.

Temperature, pH levels, total solids concentration, retention time, and volatile solids content are crucial parameters that significantly impact the efficiency and effectiveness of anaerobic digestion (Cremonez et al. 2021). One of the significant benefits of anaerobic digestion is its cost-effectiveness and simplicity, as well as the fact that it does not require an external power source. This makes it an appealing method for the valorization of agronomic waste, especially damp materials with high organic content.

Currently, anaerobic digestion is widely used to convert agricultural waste into biogas, which can serve as a direct substitute for natural gas in various applications (Macias-Corral et al. 2008). The process operates within a temperature range of 30–65 °C, where microorganisms break down the biodegradable biomass in an oxygen-free environment. A notable product of this process is biogas, which can be upgraded to biomethane, providing a renewable alternative to conventional natural gas (Cherubini 2010). Furthermore, this process generates significantly fewer gaseous pollutant emissions compared to other waste treatment methods.

However, a major challenge in utilizing agricultural waste as a substrate for pretreatment is often necessary for successful anaerobic digestion. Agricultural wastes often require extensive pre-treatment to prepare them for digestion, adding complexity and potential costs to the process. Despite this, the benefits of anaerobic digestion, such as its environmental advantages and ability to produce renewable energy, make it a valuable method for waste valorization.

30.2.4 Anaerobic Digestion for Agricultural Waste Valorization

Anaerobic digestion is a critical biological process used to decompose organic materials like agricultural and food waste in an oxygen-free environment, producing valuable by-products. This process involves four main stages: methanogenesis, acidogenesis, hydrolysis, and acetogenesis, which work together to break down organic material into simpler compounds, ultimately leading to the generation of biogas, primarily methane.

The efficiency of anaerobic digestion is influenced by various factors, including temperature, pH, total solids, retention time, and volatile solids. One of the key advantages of anaerobic digestion is its cost-effectiveness and operational simplicity, as it does not require external power sources. This makes it an attractive option for valorizing agricultural waste, especially wet materials with high organic content.

Anaerobic digestion is widely used to convert agricultural waste into biogas, which can directly replace natural gas in various applications. The process typically operates at temperatures ranging from 30 to 65 °C, where microorganisms decompose the biodegradable biomass in an oxygen-free environment. A significant product of this process is biogas, which can be upgraded to biomethane, serving as a renewable alternative to conventional natural gas. Additionally, this method produces fewer gaseous pollutants compared to other waste treatment processes (Weiland 2010).

A significant hurdle in utilizing agricultural waste for anaerobic digestion is the often extensive pre-treatment required, increasing process complexity and potentially escalating costs. Despite these challenges, the environmental benefits and the production of renewable energy make anaerobic digestion a valuable method for waste valorization.

30.2.5 Agricultural Waste in India

India, with its vast agricultural sector, generates significant quantities of agricultural waste, including crop residues (such as rice straw, sugarcane bagasse, and wheat husks), animal waste (like manure), and waste from food processing industries. These wastes, if not managed properly, contribute to severe environmental issues like air and water pollution. India annually generates over 500 million tons of crop residues. A significant portion of this biomass is unfortunately burned in fields, particularly in northern regions, contributing significantly to severe air pollution (Singh et al. 2010).

Valorization in India: The valorization of agricultural waste in India has gained attention as a viable solution to these environmental challenges. Efforts include converting residues into biofuels, compost, and various chemicals. For instance, sugarcane bagasse, a byproduct of the sugar industry, is widely utilized for bioenergy production. Rice husk, another abundant residue, is processed into silica, which has numerous industrial applications.

30.2.6 Contribution to the Circular Economy

The valorization of agricultural solid waste is a crucial component of the circular economy, which emphasizes the sustainable use of resources. By transforming waste into valuable products, we can mitigate the environmental impact of waste disposal, diminish GHG emissions, and foster sustainable agricultural practices. Moreover, this approach provides economic benefits by creating new revenue streams for farmers and industries involved in waste processing (Chynoweth et al. 2001).

30.3 Life Cycle Assessment (LCA) Methodology

Life cycle assessment (LCA) is a systematic approach used to evaluate the environmental impacts of products, processes, or systems over their entire life cycle—from raw material extraction through production, use, and disposal.

30.3.1 Goal and Scope Definition

The initial stage of any LCA involves defining the assessment's goals and scope, including specifying the purpose of the study, the functional unit (e.g., 1 kg of chemical product), system boundaries, and any assumptions or limitations. For agricultural waste valorization, the system boundaries might encompass waste collection, transportation, pre-treatment, conversion processes, and product distribution.

30.3.2 Inventory Analysis

The inventory analysis phase involves compiling data on all relevant inputs (e.g., energy, water, raw materials) and outputs (e.g., emissions, waste) associated with the valorization process. This data collection can be complex due to the diversity of agricultural waste types and conversion technologies.

30.3.3 Impact Assessment

During the impact assessment phase, inventory data is translated into potential environmental consequences, such as global warming potential (GWP), eutrophication, acidification, and resource depletion. Various impact categories can be analyzed depending on the study's goal.

30.3.4 Interpretation

The final phase of LCA is interpretation, where results are analyzed to draw conclusions and make recommendations. This step often involves sensitivity analysis to understand the influence of key assumptions or parameters on the overall environmental impact.

30.4 Application of LCA in Agricultural Waste Valorization

30.4.1 Case Study: Conversion of Rice Husk to Chemicals

Rice husk, a byproduct of rice milling, is one of the most abundant agricultural residues globally, with significant quantities generated in major rice-producing countries such as India, China, and Southeast Asia. Traditionally, rice husk has been either disposed of through open burning, which contributes to air pollution and greenhouse gas emissions, or used as a low-value fuel in rural areas. However, recent advancements in chemical engineering and material science have paved the way for the valorization of rice husk into valuable chemicals and materials, transforming it from waste to resource.

30.4.1.1 Potential Products from Rice Husk Valorization

1. Silica extraction: One of the most valuable components of rice husk is silica (SiO_2), which constitutes about 20% of the husk's weight. Silica extracted from rice husk ash has high purity and is used in various industries, including electronics, construction, and agriculture. The extraction process typically involves the combustion of rice husk to produce ash, followed by chemical treatments to isolate silica. The high surface area and porous structure of silica derived from rice husk make it particularly suitable for applications in the production of advanced materials such as nanocomposites, adsorbents, and catalysts.

   Rice Husk → Carbonization → Silica-Rich Ash → Acid Leaching → Washing and Drying → Calcination → High-Purity Silica
2. Bio-oil production: Through pyrolysis, rice husk can be thermochemically converted into bio-oil condensate, a complex mixture of organic compounds that can serve as a renewable source of chemicals and fuels. The pyrolysis process involves heating the rice husk in the absence of oxygen, breaking down the biomass into solid biochar, liquid bio-oil, and gaseous products. Bio-oil can be further refined to produce valuable chemicals such as phenols, acids, and aldehydes, which are precursors for a wide range of industrial products. Moreover, bio-oil has potential as a sustainable alternative to fossil-based fuels in energy generation.

   Rice Husk → Pyrolysis → Vaporization → Condensation → Bio-Oil Collection → Refining
3. Furfural production: Rice husk is also a promising feedstock for the production of furfural, an organic compound derived from the hemicellulose portion of biomass. Furfural is a versatile chemical used as a solvent in the refining of lubricat-

ing oils, as well as a precursor for the synthesis of furan-based chemicals, resins, and biofuels. The production of furfural from rice husk typically involves acid hydrolysis, where the hemicellulose is broken down into sugars, which are then dehydrated to form furfural. This process not only adds value to the rice husk but also aligns with the principles of green chemistry by utilizing renewable biomass.

Rice Husk → Acid Hydrolysis → Pentose Sugars (Xylose) → Dehydration → Furfural Formation → Distillation → Condensation → Crude Furfural → Purification → High-Purity Furfural.

30.4.1.2 Environmental and Economic Assessment Through LCA

The conversion of rice husk into chemicals through these processes has been extensively studied using life cycle assessment (LCA) to evaluate their environmental impacts. LCA studies typically compare the environmental footprint of rice husk valorization to conventional waste management practices such as open burning or landfilling, as well as to the production of chemicals from fossil-based feedstocks.

1. Global warming potential (GWP): LCA studies have shown that converting rice husk into chemicals significantly decrease the global warming potential (GWP) compared to traditional practices. For instance, silica extraction from rice husk ash produces substantially lower CO_2 emissions than conventional silica production methods, which require high energy inputs and raw materials such as quartz. Similarly, the pyrolysis of rice husk to produce bio-oil and biochar can result in a net reduction in GHG emissions, as the biochar can sequester carbon in soils, and bio-oil can replace fossil fuels (Lehmann and Joseph 2015).
2. Energy efficiency: The energy efficiency of rice husk valorization processes is another critical parameter assessed in LCA studies. While the energy input required for pyrolysis or chemical extraction processes can be high, the overall energy balance is often favourable when considering the energy content of the bio-oil or the value of the chemicals produced. In the case of silica production, the energy savings from using rice husk as a raw material can be significant compared to traditional methods.
3. Resource utilization: LCA also considers resource utilization, particularly the use of water, chemicals, and other inputs in the valorization processes. For example, the acid hydrolysis process for furfural production requires careful management of acid and water inputs to minimize environmental impacts. However, the use of rice husk, a readily available and renewable resource, enhances the sustainability of these processes by reducing the reliance on non-renewable resources.
4. Economic viability: Beyond environmental impacts, LCA studies often integrate techno-economic analysis (TEA) to assess the economic viability of rice husk valorization. The production of high-purity silica, bio-oil, and furfural from rice husk has shown promising economic potential, particularly in regions where rice husk is abundant and disposal is a challenge. The revenue generated from these value-added products can offset the costs of processing, making rice husk valorization an attractive option for both small-scale and industrial operations.

30.4.1.3 Comparative Analysis with Conventional Feedstocks

Comparing rice husk valorization with conventional fossil-based feedstocks reveals several advantages. For example, the production of silica from rice husk not only consumes less energy but also reduces environmental pollution associated with mining and processing quartz. Similarly, bio-oil derived from rice husk has a lower environmental impact than conventional petroleum refining, as it avoids the extraction and combustion of fossil fuels. These comparative advantages underscore the potential of rice husk valorization as a sustainable alternative in the chemical and materials industries.

30.4.1.4 Challenges and Future Prospects

Despite its potential, the valorization of rice husk into chemicals faces several challenges. The variability in rice husk composition, influenced by factors such as rice variety, cultivation practices, and milling processes, can affect the efficiency and yield of the valorization processes. Additionally, the scaling up of these technologies from laboratory to industrial scale requires further research and development, particularly in optimizing process conditions and reducing costs.

Future research should focus on improving the efficiency of rice husk valorization processes, exploring new applications for rice husk-derived products, and integrating these processes into existing agricultural and industrial systems. Moreover, combining LCA with economic assessments will be crucial in guiding the development of sustainable and economically viable valorization pathways for rice husk and other agricultural residues.

The conversion of rice husk into valuable chemicals such as silica, bio-oil, and furfural represents a promising strategy for waste valorization, contributing to both environmental sustainability and economic development. Through the application of LCA, these processes can be optimized to minimalize environmental impacts and make the most of resource efficiency. As the demand for sustainable alternatives to fossil-based chemicals grows, the valorization of rice husk will likely play a gradually important role in the global circular economy.

30.4.2 Comparative LCA: Agricultural Waste Vs. Conventional Feedstocks

LCA can also compare the environmental impacts of producing chemicals from agricultural waste versus conventional fossil-based feedstocks. Such comparisons are crucial for highlighting the environmental benefits and trade-offs of transitioning to bio-based production systems.

30.4.3 Challenges and Considerations

Conducting LCA for agricultural waste valorization presents unique challenges, such as data variability, the heterogeneity of waste streams, and the multifunctionality of agricultural systems. Moreover, the geographic specificity of agricultural practices can significantly influence LCA outcomes.

30.5 Environmental and Economic Benefits

Valorizing agricultural solid waste into chemicals provides a range of environmental and economic benefits, addressing critical challenges related to waste management, resource use, and rural development.

30.5.1 Reduction in Greenhouse Gas Emissions

Agricultural waste is frequently disposed of in landfills or charred in open fields, both practices that contribute significantly to greenhouse gas emissions. Landfills generate methane, a greenhouse gas that is 25 times more effective than CO_2 over a 100-year period (Intergovernmental Panel on Climate Change 2014). Open burning, on the other hand, releases CO_2 and particulate matter, contributing to air pollution and climate change (Gadde et al. 2009). By converting agricultural waste into valuable chemicals through valorization, these waste materials are diverted from these harmful practices, leading to a substantial reduction in methane and CO_2 emissions. Studies show that waste valorization can cut emissions by up to 30–50%, making it a critical strategy for climate change mitigation (Mohan et al. 2016).

30.5.2 Resource Efficiency and Circular Economy

Valorizing agricultural waste supports the principles of a circular economy, which accentuates the repeated use of resources by generating closed-loop systems. Using waste as a raw material reduces dependence on virgin resources, such as fossil fuels or mined materials, thus minimizing the environmental impact of resource extraction and processing (Ellen MacArthur Foundation 2017). According to Balan et al. (2018), agricultural waste contains valuable organic compounds that can be converted into bio-based chemicals, fertilizers, and energy sources. This transformation not only promotes resource efficiency but also creates new market opportunities for renewable products, reducing overall material waste and advancing sustainable production processes.

30.5.3 Economic Viability and Rural Development

The economic benefits of valorizing agricultural waste extend to rural communities and local economies. By generating new value-added products from discarded materials, farmers and rural entrepreneurs can develop new revenue streams and reduce economic dependency on traditional agricultural practices (Pandey et al.

2019). This diversification helps to enhance economic resilience, especially in regions where agriculture is subject to market volatility or climate impacts. Furthermore, waste valorization processes can stimulate local job creation in rural areas, including roles in waste collection, processing, and management, thereby supporting sustainable development goals (Food and Agriculture Organization of the United Nations 2020). The ability to generate income from agricultural by-products can increase the financial stability of small-scale farmers, contributing to rural development and poverty reduction (Galanakis 2021).

30.6 Future Perspectives and Research Directions

The field of agricultural waste valorization is rapidly evolving, with ongoing research aimed at improving process efficiencies, developing new valorization pathways, and optimizing the environmental performance of these systems. Integrating LCA with TEA in future research will provide a valuable tool for assessing the environmental and economic viability of agricultural waste valorization strategies.

30.7 Conclusion

Valorizing agricultural solid waste offers a sustainable solution to many pressing environmental issues by turning waste materials into valuable resources. Processes such as bioenergy production, anaerobic digestion, and bioconversion are effective strategies that contribute to reducing waste volumes, lowering greenhouse gas emissions, and decreasing reliance on fossil fuels. By converting agricultural residues like crop straw, husks, and animal manure into biofuels, fertilizers, and biochemicals, these approaches promote a circular economy where waste is not merely discarded but reimagined as a resource.

Among these valorization methods, anaerobic digestion plays a significant role, not only by producing biogas that serves as a renewable energy source but also by generating digestate that can be used as a biofertilizer, enhancing soil health and reducing the need for synthetic fertilizers. Similarly, bioenergy production, including the generation of bioethanol and biodiesel, provides a sustainable alternative to petroleum-based fuels, thereby supporting energy security and mitigating climate change. Bioconversion processes, which transform waste into higher-value chemicals, are also gaining traction, particularly in developing countries where agricultural waste is abundant and often underutilized.

The application of life cycle assessment (LCA) is fundamental to maximizing the environmental benefits of these waste valorization processes. LCA provides a holistic approach to evaluate the environmental impacts associated with the entire life cycle of valorization, from the collection of waste to the final utilization of products. It enables stakeholders to quantify energy use, emissions, and resource depletion, thereby identifying hotspots where improvements can be made. For instance, LCA can reveal the environmental trade-offs involved in different conversion technologies, such as the balance between energy input and greenhouse gas emissions in anaerobic digestion versus bioethanol production. Moreover, it can inform policy decisions by highlighting the most sustainable pathways for agricultural waste management, aligning economic growth with ecological integrity.

However, while LCA is an invaluable tool, its effectiveness is often limited by the availability and quality of data. The current methodologies must be refined to capture the unique characteristics of diverse agricultural systems, especially in regions with limited data infrastructure.

Additionally, integrating social and economic dimensions into LCA could offer a more comprehensive evaluation framework. This would help address potential social impacts, such as the effects on local communities, job creation, and rural development. It would also incorporate economic considerations, such as the cost-effectiveness and market viability of various valorization technologies. As such, LCA could evolve into a more multidimensional tool that informs not only environmental policy but also socioeconomic planning in the context of agricultural waste management.

Ultimately, as global populations grow and pressures on agricultural systems intensify, the valorization of agricultural solid waste will become increasingly important. A greater understanding of the environmental implications of these processes through refined LCA methodologies will be crucial in ensuring that waste valorization contributes meaningfully to global sustainability goals. Strengthening collaboration between researchers, policymakers, and industry stakeholders will be essential to advancing these efforts, promoting innovation, and achieving a more circular and sustainable agricultural economy (Muench and Guenther 2013).

Conflicts of Interest The authors declare no conflicts of interest. The funders had no role in the design of the study; in the collection, analysis, or interpretation of data; in the writing of the manuscript; or in the decision to publish the results.

References

Alam F, Mobin S, Chowdhury H (2019) Third generation biofuel from algae. Proc Eng 105:763–768

Angelidaki I, Ahring BK (2000) Methods for increasing the biogas potential from the recalcitrant organic matter contained in manure. Water Sci Technol 41(3):189–194

Balan V, Chiaramonti D, Kumar S (2018) Review of US and EU initiatives toward development, demonstration, and commercialization of lignocellulosic biofuels. Biofuels Bioprod Biorefin 12(3):590–609

Banerjee S, Chakraborty R, Ghosh S (2017) Impact of agricultural waste disposal on environmental health: a comprehensive review. Environ Res 153:74–84

Chakraborty P, Jha M, Singh A (2020) Circular economy and waste management: a systematic review on sustainable agricultural practices. J Clean Prod 258:120933

Cherubini F (2010) The biorefinery concept: using biomass instead of fossil fuels to produce energy and chemicals. Energy Convers Manag 51(7):1412–1421

Chynoweth DP, Owens JM, Legrand R (2001) Renewable methane from anaerobic digestion of biomass. Renew Energy 22(1–2):1–8

Das S, Singh R (2020) Life cycle assessment in the agricultural circular economy: an approach for sustainable development. Renew Sust Energ Rev 119:109575

Das A, Chatterjee N, Roy S (2020) Bioenergy from agricultural waste: opportunities and challenges. Renew Energy 147:2097–2110

Cremonez PA, Teleken JG, Meier TRW, Alves HJ (2021) Two-Stage anaerobic digestion in agroindustrial waste treatment: A review. J Environ Manage 281:111854

Demirbas A (2009) Biodiesel from waste cooking oil via base-catalytic and supercritical methanol transesterification. Energy Convers Manag 50(4):923–927

Ellen MacArthur Foundation (2017) Circular economy in detail. https://ellenmacarthurfoundation.org

Food and Agriculture Organization of the United Nations (2020) The state of food and agriculture 2020: overcoming water challenges in agriculture

Gadde B, Bonnet S, Menke C, Garivait S (2009) Air pollutant emissions from rice straw open field burning in India, Thailand, and The Philippines. Environ Pollut 157(5):1554–1558

Galanakis CM (2021) The food waste valorization handbook. Academic, New York

Ghosh P, Mehta A, Sharma V (2019) Emerging challenges in waste management due to industrial growth: a review. Waste Manag 85:108–116

Gupta M, Thakur P, Roy S (2016) Potential of agricultural residues for bioenergy production. Biofuels 7(1):23–36

Intergovernmental Panel on Climate Change (2014) Climate change 2014: synthesis report

Joshi R, Patel A, Shah K (2017) Impact of untreated agricultural waste on environmental and human health. Environ Monit Assess 189:411

Khan M, Bhatia R (2021) Advanced technologies for waste minimization and circular economy development in agriculture. J Sustain Agric 15(2):89–103

Kumar R, Patel N (2020) The effect of industrialization on global trade and employment: a sectoral analysis. J Ind Econ 68(2):277–295

Kwiatkowski JR, McAloon AJ, Taylor F, Johnston DB (2006) Modeling the process and costs of fuel ethanol production by the corn dry-grind process. Ind Crop Prod 23(3):288–296

Lehmann J, Joseph S (eds) (2015) Biochar for environmental management: science, technology, and implementation. Routledge, London

Macias-Corral M, Samani Z, Hanson A, Smith G, Funk P, Yu H, Longworth J (2008) Anaerobic digestion of municipal solid waste and agricultural waste and the effect of co-digestion with dairy cow manure. Bioresour Technol 99(17):8288–8293

Mehta D, Sharma A, Jain P (2021) Valorization of agricultural waste for biofuel production: a sustainable approach. Biomass Bioenergy 147:105960

Mishra V (2016) Contaminants in the ecosystem from agricultural residue disposal: an assessment. Ecotoxicol Environ Saf 132:356–364

Mohan SV, Nikhil GN, Chiranjeevi P, Reddy CN, Rohit MV, Kumar AN, Sarkar O (2016) Waste biorefinery models towards sustainable circular bioeconomy: critical review and future perspectives. Bioresour Technol 215:2–12

Muench S, Guenther E (2013) A systematic review of bioenergy life cycle assessments. Appl Energy 112:257–273

Munasinghe PC, Khanal SK (2010) Biomass-derived syngas fermentation into biofuels: opportunities and challenges. Bioresour Technol 101(13):5013–5022

Pandey A, Negi S, Binod P, Larroche C (2019) Current developments in biotechnology and bioengineering: production of biofuels and chemicals from sustainable resources. Elsevier, Amsterdam

Patel S, Singh T, Sharma A (2022) Exploring alternative energy harvesting techniques in response to rising global demand. Renew Sust Energ Rev 153:111734

Rai P, Kumar V, Sharma S (2021) Energy requirements for developing industrial economies: implications for policy and practice. Energy Policy 150:112170

Rashid U, Anwar F, Moser BR, Knothe G (2008) Moringa oleifera oil: a possible source of biodiesel. Bioresour Technol 99(17):8175–8179

Roy S, Ghosh P, Mehta A (2018) Renewable bio-based energy from agricultural residues: technologies and perspectives. J Renewable Energy 128:23–34

Roy S, Thakur P, Das A (2019) Residues from agro-industries: types and potential for resource recovery. Resour Conserv Recycl 148:105–115

Salemdeeb R, Zu Ermgassen EKHJ, Kim MH, Balmford A, Al-Tabbaa A (2017) Environmental and health impacts of using food waste as animal feed: a comparative analysis of food waste management options. J Clean Prod 140:871–880

Sharma A, Ghosh P, Verma S (2018) Production of bioethanol from agricultural waste: an overview. Renew Sust Energ Rev 90:877–892

Sharma V, Ghosh P, Mehta D (2019) Organic waste valorization for value-added products: a review. J Clean Prod 228:784–802

Sharma N, Singh P, Patel A (2022) Innovative approaches in alternative energy production and sustainability. J Energy Res 246:19–35

Singh J, Gu S (2010) Biomass conversion to energy in India—a critique. Renew Sust Energ Rev 14(5):1367–1378

Singh A, Pant D, Korres NE, Nizami AS, Prasad S, Murphy JD (2010) Key issues in life cycle assessment of ethanol production from lignocellulosic biomass: challenges and perspectives. Bioresour Technol 101(13):5003–5012

Singh P, Patel A, Shah K (2018) Global trade dynamics and industrial growth: trends and implications. Int J Trade Glob Mark 11(3):231–245

Singh V, Pandey J, Sharma S (2020) Sustainable utilization of sugarcane bagasse: an overview. Environ Prog Sustain Energy 39(5):13349

Thakur P, Gupta M, Kumar R (2021) Agro-waste management and its impact on sustainable agriculture. Agric Environ 56(1):33–47

Weiland P (2010) Biogas production: current state and perspectives. Appl Microbiol Biotechnol 85(4):849–860

Life Cycle Assessment (LCA) Investigations on Valorisation of Waste to Biofuels

31

Anshul Nigam, Dev Pramanick, Aditya Singh, and Swechchha Mishra

Abstract

Energy requirements and solid waste poses major challenge to sustainable development. However, valorisation of solid waste to fuels and chemicals holds the key to cleaner environment. Although, valorisation studies at laboratory level seems to be promising but its actual evaluation in terms of environment benefit can only be evaluated by LCA as it involves multiple inputs. As far as fuels are concerned, we can conclude that biogas generation from waste is most efficient on LCA, while biohydrogen is in terms of calorific value, but very few LCA studies have been conducted on it. As far as bioethanol and biodiesel are concerned, it can be made both using feed stocks and waste. Bioethanol has been proven efficient on LCA from feed but its production on waste has been proven only at pilot scale level while biodiesel seems to be nearing being successful biofuel based on LCA data. On the basis of our survey, we recommend solid waste valorisation to biogas as the best waste-to-energy proposition.

Keywords

Biohydrogen · Cellulosic ethanol · Biogas · Biodiesel · Life cycle assessment

A. Nigam (✉) · D. Pramanick · A. Singh · S. Mishra
Department of Biotechnology, Kanpur Institute of Technology, Kanpur, Uttar Pradesh, India
e-mail: anshul.nigam@kit.ac.in

A. Kumari et al. (eds.), *Valorization of Solid Wastes to Biofuels and Chemical Products for Sustainable World*, https://doi.org/10.1007/978-981-96-8594-3_31

31.1 Introduction

Landfill produces gaseous mix of several gases, few of them are mentioned in Table 31.1.

Three processes involved in gas production from waste include bacterial decomposition, volatilisation and chemical reaction (ATSDR n.d.). According to World Bank data, most of the countries dump their 39–54% of their waste in landfills (The World Bank Group n.d.). The lower percentage is managed by incineration (22%) or composting/recycling (36%). Both landfill and incineration are not environmentally friendly, as they generate obnoxious gases, toxic percolates while accelerating climate change (Abubakar et al. 2022). Methane-related global warming is substantial and is responsible for 30% rise in global temperature and has 12 years of life in atmosphere (IEA n.d.). The other issue with methane is that in presence of intense sunlight of the higher atmosphere, methane forms chemicals that destroy ozone (Lovett 2002). Moreover, solid waste leads to contamination of water through leaching (Lin et al. 2012). Therefore, conversion of waste to fuels/chemicals not only provides energy but also limit damage to environment as these fuels are neutral towards CO_2 emissions and will be helpful in achieving Agenda 2030 of sustainable development (Bjelić et al. 2024). Systematic review by Mulya et al. (2022) suggests that LCA studies on solid waste are more focused on energy, with a preference for anaerobic digestion.

In this chapter, we are discussing LCA studies on four major biofuels biohydrogen, biogas, bioethanol and biodiesel along with chemicals derived solid waste as it provides big picture. LCA investigations not only discuss the limitations and caveats related to upstream production process but also concomitantly evaluate after-usage effects of end products over environment. This cradle-to-grave approach (Kandasamy et al. 2020) helps in identifying all parameters needed to be addressed prior to real-world application.

Table 31.1 Data adapted from Agency for Toxic Substances and Disease Registry

Gas	Fraction by volume
Methane, CH_4	0.45–0.60
Carbon dioxide, CO_2	0.40–0.60
Nitrogen, N_2	0.02–0.05
Ammonia, NH_3	0.001–0.01
Oxygen, O_2	0.001–0.01
Sulphides, gaseous mix of H_2S, $(CH_3)_2S$, mercaptans	0–0.01
Hydrogen, H_2	0–0.002
Carbon monoxide, CO	0–0.002
Organic compounds other than methane	0.0001–0.006

31.2 LCA Studies on Biohydrogen

Increasing concerns about environmental sustainability due to usage of fossil fuels, there is a need for an alternative fuel such as hydrogen. As a fuel, it is the cleanest and only produces water vapour during energy generation and releases a significant amount of energy per unit weight, which is around 120–142 MJ/kg when compared to the next best fuel, methane, whose energy per unit weight is around 50–55 MJ/kg. The energy equivalence by weight between hydrogen and petrol is 1:2.8 (Alternative Fuels Data Center n.d.). The hydrogen production currently involves steam reformation of natural gas and electrolysis of water. Hydrogen production from biomass, and hence referred to as biohydrogen, provides a sustainable and eco-friendly alternative to traditional approaches. Hydrogen produced from biomass is renewable, reduces the reliance on fossil fuels and helps in waste management.

Depending on the raw material and nature of process for its production, hydrogen is categorised into black (coal), brown (lignite), grey (natural gas), blue (from natural gas capturing its carbon) and green (it is produced by electrolysis process with power generated by renewables).

Biohydrogen is defined as the hydrogen derived from living microorganisms via biological processes like fermentation and photolysis or gasification of biomass. There are very limited LCA investigations, and dimensions of research involved include the nature of substrates, the biochemistry involved in bioprocess and processes involving power supplies from renewable sources.

Since considerable energy inputs are required for the biohydrogen production, Barghash et al. (2022) used solar energy inputs to reduce carbon footprint. In another investigation, energy inputs made in generating biohydrogen from biogas enhanced carbon footprint and proved it to be unsustainable (Cvetković et al. 2021). In a similar investigation, biohydrogen was produced from gasification of biomass only scores well on LCA parameters if gasification energy is drawn from biomass (Reaño and Halog 2020). Based on above LCA analysis, we can conclude that inputs from renewable source play essential role in achieving lower carbon footprint irrespective of nature of the process.

Direct production of hydrogen by lignocellulosic biomass through pyrolysis, gasification and liquefaction has been extensively reviewed by Jara-Cobos et al. (2023), and they have concluded that the technological innovations can make these methods commercially viable and environmentally sustainable.

The third set of LCA studies involving biohydrogen production focuses on variation in the fermentation process. In fermentation, there are three methods to produce biohydrogen, which include: (1) dark fermentation (DF), (2) photo-fermentation (PF) and (3) integrated fermentation (IF), combining both.

DF is a promising and cheaper way to produce biohydrogen. In addition, dark fermentation can use complex biomass waste as raw material and can also produce by-products like volatile fatty acids (VFAs) that have economic benefits. Compared with hydrogen generation via bio-photolysis, PF and DF can produce hydrogen at a

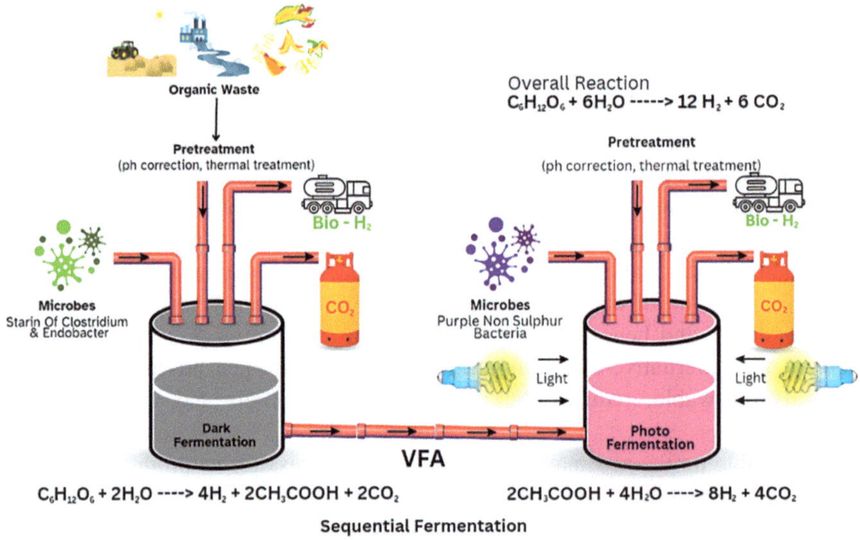

Fig. 31.1 Set up of sequential reactor combining photo-fermentation and dark fermentation

higher rate (Ahmad et al. 2023) and thus can be easily scaled up. Balachandar et al. (2019) were successful in achieving scale-up from lab to pilot level.

Photo-fermentation involves the use of photosynthetic bacteria to produce hydrogen by utilising light as an energy source. The effluents from dark fermentation (such as VFAs and alcohols) can be used for hydrogen production (Ahmed et al. 2021) in photo-fermentation to achieve a higher H_2 yield. This has led to integrated approaches combining both DF and PF in the same or sequential reactors (Fig. 31.1). The LCA studies on biohydrogen involving metabolic pathways clearly reveal that although these technologies are environmentally friendly, the milestone of achieving economies of scale is still beyond vision.

31.3 LCA Studies on Biogas

Biogas is the mixture primarily composed of CH_4 and CO_2, it is the simplest way to prevent release of methane into the atmosphere through waste, concomitantly acting as a source of fuel and organic manure. The process of biogas production, also known as anaerobic digestion, breaks down biopolymers via: (1) hydrolysis, (2) acidogenesis, (3) acetogenesis and (4) methanogenesis (Nwokolo et al. 2020), involving breaking down of complex biopolymers to simple molecules (Fig. 31.2). We are excluding discussion on these steps while maintaining our prime focus on LCA studies related to biogas production. In nearly all LCA investigations, irrespective of parameters studied, biogas production is found to be both environmentally and economically sustainable (Aziz and Hanafiah 2019; Mezzullo et al. 2012; Mertins and Wawer 2022; Kalita et al. 2023; Vu et al. 2015). It is important to note

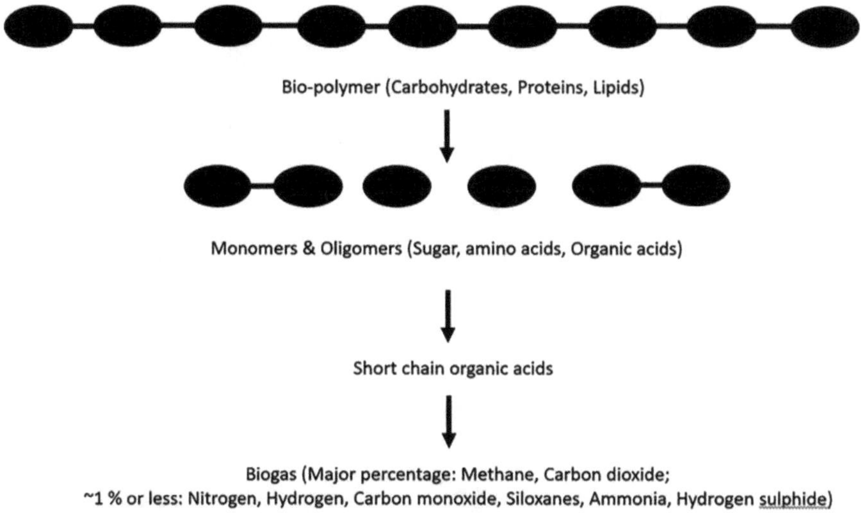

Fig. 31.2 Biogas formation involving the breakdown of complex biopolymers. Each step involves a variety of bacterial consortia in action

that manure produced as byproducts reduces burden of inputs, while ammonia generation, transportation and water consumption add to it. It is important to note that it is also helpful in reducing waste volume (Ghosh 2016). Still, its input requirements are the least compared to other biofuel technologies (Yousuf et al. 2020). Moreover, dry biogas technologies that reduce water requirements for anaerobic digestion are underway, consolidating its position as a sustainable fuel (Abu et al. 2023; Kanadevia Inova n.d.; Renergon Biogas 2025). Most importantly, manures generated through anaerobic digestion not only reduce carbon burden from fossil fuels but also benefit environment by reducing eutrophication, which is the most neglected parameter in many LCA investigations.

31.4 LCA Studies on Biodiesel

The biodiesel is primarily classified into first-generation or second-generation biofuel depending upon the substrate used for its production. The first-generation biodiesel has been proven to be effective both economically and environmentally (Yee et al. 2009), although they cannot escape food vs. fuel debate (Moodley 2021) as they are derived largely from edible oil. Apart from this, LCA studies related to it do not consider the ecological damage caused by enhancing oil crop plantations, although they consider inputs on agriculture (Arguelles-Arguelles et al. 2021). Although mentioning of 1 G biodiesel is beyond the scope of this chapter, we have included it because it is produced commercially. However, conversion of used cooking oil to biodiesel is best exemplified as valorisation of waste to fuel, which, if used for edible purpose, can lead to wide range of health problems and cause

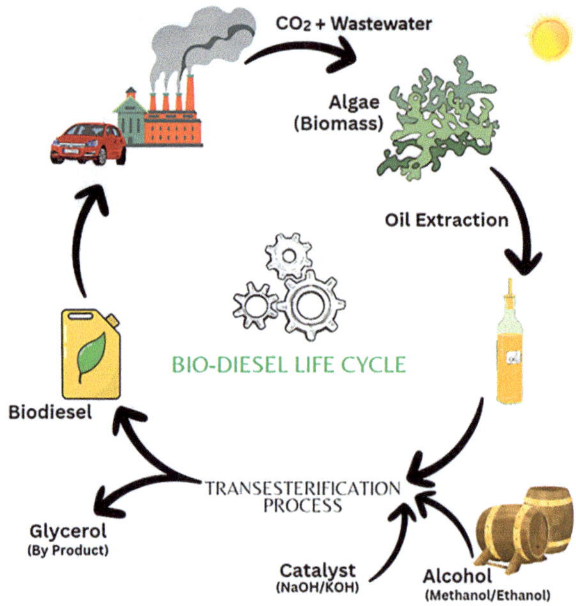

Fig. 31.3 Best possible scenario for second generation biodiesel

environmental issues if disposed (Adhikesavan et al. 2022). The major carbon-contributing steps in converting used cooking oil to biodiesel include: (1) transportation (Hatzisymeon et al. 2019) and (2) transesterification (Peiró et al. 2009; Zhao et al. 2021). Transesterification method is the first choice to convert the waste cooking oils to biodiesel, as it is comparatively low on environmental burden owing to lower temperature requirements (Hosseinzadeh-Bandbafha et al. 2022). The LCA investigations on second-generation biodiesel have attempted to achieve economic viability by enhancing the production of algal biomass using waste carbon dioxide from industry (Beal et al. 2015; Adesanya et al. 2014; Amer et al. 2011). This approach reduces carbon footprint effectively and results in costing in the range of $1–2/l, bringing second-generation biodiesel close to reality. Sasongko et al. (2018) reported reduction in cost of algal biomass generation for biofuels using wastewater along with the usage of waste CO_2 in their investigation (Fig. 31.3).

31.5 LCA Studies on Bioethanol

Like biodiesel, bioethanol is also represented as first generation and second generation, based on raw material used. The first-generation bioethanol is obtained from corn and sugarcane or molasses (by product of sugar industry) globally. Since first-generation ethanol is involved in food vs. fuel debate (Angili et al. 2021) creating interest for the second-generation bioethanol, which is derived from lignocellulosic biomass (Gupta et al. 2013) and does not require agriculture inputs.

The lignocellulosic biomass is primarily made of cellulose, hemicelluloses and lignin. Its productivity has potential to be sustainable alternative to fossil fuel

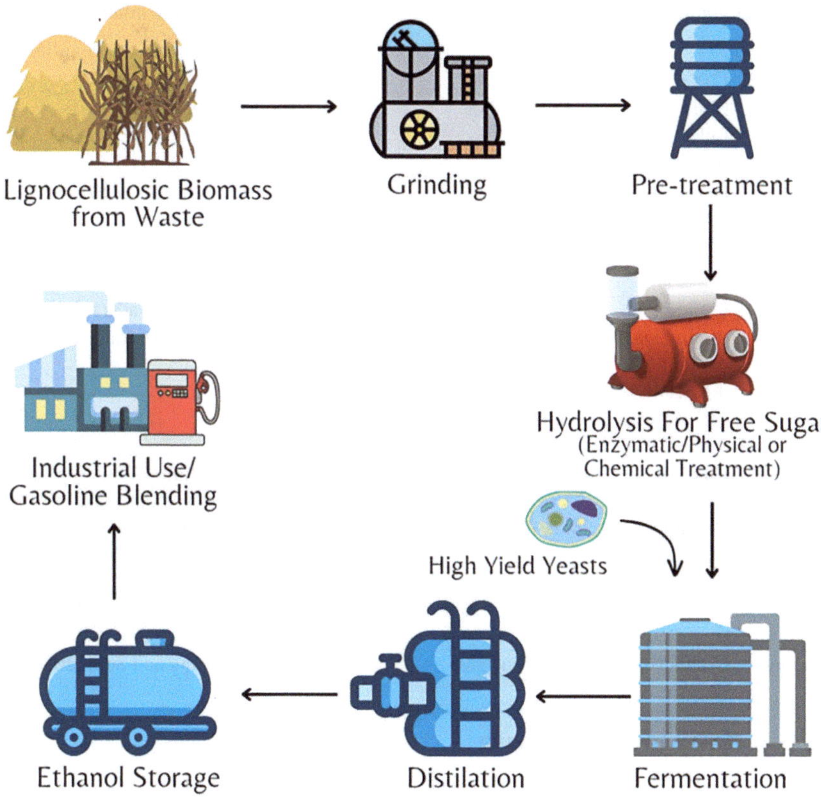

Fig. 31.4 Life cycle of second-generation ethanol from lignocellulosic biomass. The release of free sugar is key to making cellulosic ethanol a reality

(Zoghlami and Paës 2019). On comparison of both first- and second-generation bioethanol on LCA parameters, it can be concluded that the latter has the least environmental impact as it does not involve agriculture inputs (Shakelly et al. 2023; Junqueira et al. 2017). It is beyond doubt that the major hiccup in making second-generation bioethanol a reality is the release of free sugar from lignocellulosic biomass (Fig. 31.4), but operational pilot plants create positive outlook.

31.6 Brief Summary of Industrial Waste and Their Valorisation Methods

Industrial waste generation dates back with industrial revolution, but its recalcitrant effects on the environment were recognised nearly a century later due to its toxic and persistent nature. In this section, we have discussed the valorisation of industrial waste from various sectors (Table 31.2).

Table 31.2 Industries, process, waste type and quantity released into the environment

Industry	Process	Waste type	Global waste generation (in million tons)
Dairy	Cattle feeding, milk processing	Dung, whey	16 (Liu and Chen 2008)
Leather	Dehairing, tanning		8.5 (Kandasamy et al. 2020)
Paper	Bleaching	Saw dust, black liquor	100 (Ma et al. 2021)
Textile		Dyes, pigments, fibres	92 (Tang 2023)
Health care waste	During production, expired medication	Antibiotic fermentation residue and sludge, expired medication includes tablets, syrups, injectables, ointment, etc.	1470 (Kenny and Priyadarshini 2021), antibiotic residue: 10 (Yang et al. 2020)
Food	Fruit beverages, warehouse waste	Peels, household waste	66 from retails (https://www.epa.gov/facts-and-figures-about-materials-waste-and-recycling/food-material-specific-data), 25–40 food and beverage (Kandemir et al. 2022), food waste estimated: 931 (UNEP Food Waste Index Report 2021 n.d.)
Energy	Combustion	Ash, flue gas	3500–3700 (Wang and Song 2020)

31.6.1 Dairy Industry Waste

The environmental degradation brought about by the dairy industry primarily includes wastewater generation, eutrophication and greenhouse gas emission (Havlikova et al. 2008). The sustainable way of generating energy from dairy waste involves anaerobic digestion in form of methane (Adesra et al. 2021; Kalita et al. 2023). In some investigations, whey from dairy waste was used for production of bioethanol (Shen et al. 2019; Liu et al. 2016), although we could not find LCA investigations related to it. Similarly, lipids were extracted from dairy waste sludge and were converted to biodiesel (Balasubramanian et al. 2018; Abreu-Jauregui et al. 2023) and the production of biohydrogen from dairy waste was also attempted (Hellal et al. 2022; Xing et al. 2009). Overall, anaerobic digestion seems to be the most sustainable option.

31.6.2 Leather Industry Waste

Europe alone produces 10^8 kg of leather waste, which is disposed of either through landfilling or incineration (European Commission 2013). Through incineration, it can be directly converted into energy as it possesses sufficiently high energy values

(Bahillo et al. 2004), although the presence of chromium remains an issue. Pyrolysis of leather waste generates combustible gases and bio-oil for energy generation, along with char (Guan et al. 2019; Yilmaz et al. 2006). The other methods of energy production from leather waste involve anaerobic digestion for biogas production (Priebe et al. 2016; Moktadir and Rahman 2022). The gasification of leather waste has the most promising advantage, as it aids in the recovery of chromium (Dudyński et al. 2021), concomitantly limiting its release into the environment. Also, fats from leather waste have been converted to biodiesel through the transesterification process (Devarajan et al. 2022).

31.6.3 Pulp and Paper Industry Waste

The pulp and paper industry is a huge generator of waste, both liquid and solid. The liquid waste is primarily in the form of wastewater through processes like washing and bleaching. The solid waste is recovered through screening and, along with scrap, forms major component of landfills. The pulp and paper waste that can be valorised to fuel includes pulp liquor, wood waste, sludge waste and scraps. The major routes for valorisation of pulp and paper waste to energy are (1) incineration, (2) biogas, (3) gasification and (4) biofuel production (Haile et al. 2021). Among all the routes, incineration is most applied as it does not involve any kind of processing, although it causes the release of pollutants (Amândio et al. 2022). However, few investigations suggest that, on an LCA scale, the environmentally friendly approach is anaerobic digestion (Mohammadi et al. 2019a, b; Gievers et al. 2022).

31.6.4 Textile Industry Waste

Textiles are responsible for polluting a fifth of freshwater resources (European Parliament n.d.), and the waste involves scraps, trimmings and unused materials. Those materials that cannot be recycled can be converted into value-added chemicals and fuels. The conversion methods into fuel include thermal methods (Xu et al. 2020), anaerobic digestion and bioethanol production (Hasanzadeh et al. 2018). The LCA of the above processes was done against damage caused to the environment by landfilling methods of disposal and scored better than the latter's sustainability.

31.6.5 Flue Gas to Waste

Irrespective of the nature of end product or service, nearly all heavy industries generate flue gas. The carbon from this gas can be captured for generation of algal biomass and subsequently converted into biodiesel. The LCA investigation on second-generation biodiesel has attempted to achieve economic viability by enhancing production of algal biomass using waste carbon dioxide from industry (Beal et al. 2015; Adesanya et al. 2014; Amer et al. 2011). This approach reduces carbon footprint effectively and, costing in the range of $1–2/l, brings second generation biodiesel close to reality. Sasongko et al. (2018) reported a reduction in the cost of algal biomass generation for biofuels using wastewater along with usage of waste CO_2 in their investigation (Fig. 31.3).

31.6.6 Pharma Waste

Pharmaceuticals represent the major group of emerging contaminants in the environment (Samal et al. 2022), and antibiotics are the second largest group of pharmaceuticals in wastewater (Mora-Gamboa et al. 2022), beside other drugs. Pharmaceutical waste leads to considerable ecological damage by influencing sex determination in aquatic animals (Mgharbel et al. 2023). The major approaches involving pharma waste to energy are gasification, pyrolysis and incineration.

31.7 Conclusion

We have compared four major biofuels on LCA parameters that include biohydrogen, biogas, biodiesel and bioethanol. Among them, biohydrogen cannot be used in real-time as its usage technology is expensive at present, as well as its production methods from waste are far from being economical. Biogas cannot be used directly as liquid transport fuel at mass level, although its processing requirements are minimal, while biodiesel and bioethanol can be blended with existing transport fuels. Moreover, the first generation of biodiesel and bioethanol is being successful in achieving nearly carbon neutrality but cause ecological damage in their production, which is not accounted in LCA investigations we surveyed, though it has been published in news reports. We have also mentioned sector-wise waste generation so that it may be clear to policy makers where to focus the most (Table 31.2).

Considering ecological advantage and ease to implement, we recommend solid waste valorisation to biogas as best biofuel option, although it cannot be applied to all types of industrial waste like cosmetics, industrial glue, paint, and pharma, as they generate plethora of obnoxious aromatic compounds and are components of MSW (Li et al. 2019). It is important to note that only biogas has reached commercial-level scale-up in waste-to-energy scenarios, and this process further consolidates our recommendation with TOE values in the range of other biofuels (Table 31.3).

Table 31.3 Ton(s) of oil equivalent (TOE), by convention, is a measure of fuel energy in terms of crude oil. It is based on the normalisation that one ton of crude stores 41.868 GJ of energy

Fuel	TOE
Diesel	1.01
Petrol/ gasoline	1.05
Biodiesel	0.86
Bioethanol	0.64
Coal	0.7–0.75
Natural gas	0.9
Biogas	0.45–0.72

The above data is obtained from references (European Commission n.d.; https://enerteam.org/conversion-to-toe.html; Energy Conservation Research and Development Center n.d.). For biogas it is calculated from natural gas values considering methane concentration in the range of 50–80% (Chen et al. 2015)

References

Abreu-Jaureguí C, Reynel-Ávila H, Bonilla-Petriciolet A (2023) Biodiesel production from wastewater scum of dairy industry: lipid extraction studies and reaction routes. Fuel 342:127868. https://doi.org/10.1016/j.fuel.2023.127868

Abu R, Aziz M, Hassan CHC, Noor ZZ, Jalil RA (2023) A comparative life cycle assessment of dry and wet anaerobic digestion technologies for food waste management. J Kejuruteraan 35(2):317–349. https://doi.org/10.17576/jkukm-2023-35(2)-05

Abubakar IR, Maniruzzaman KM, Dano UL, AlShihri FS, AlShammari MS, Ahmed SMS, Al-Gehlani WAG, Alrawaf TI (2022) Environmental sustainability impacts of solid waste management practices in the global south. Int J Environ Res Public Health 19(19):12717. https://doi.org/10.3390/ijerph191912717

Adesanya VO, Cadena E, Scott SA, Smith AG (2014) Life cycle assessment on microalgal biodiesel production using a hybrid cultivation system. Bioresour Technol 163:343–355. https://doi.org/10.1016/j.biortech.2014.04.051

Adesra A, Srivastava VK, Varjani S (2021) Valorization of dairy wastes: integrative approaches for value added products. Indian J Microbiol 61(3):270–278. https://doi.org/10.1007/s12088-021-00943-5

Adhikesavan C, Ganesh D, Augustin VC (2022) Effect of quality of waste cooking oil on the properties of biodiesel, engine performance and emissions. Cleaner Chem Eng 4:100070. https://doi.org/10.1016/j.clce.2022.100070

Ahmad A, Hasan SW, Show PL, Banat F (2023) Biohydrogen production through dark fermentation: recent trends and advances in transition to a circular bioeconomy. Int J Hydrog Energy 52:335–357. https://doi.org/10.1016/j.ijhydene.2023.05.161

Ahmed SF, Rafa N, Mofijur M, Badruddin IA, Inayat A, Ali MS, Farrok O, Khan TMY (2021) Biohydrogen production from biomass sources: metabolic pathways and economic analysis. Front Energy Res 9:753878. https://doi.org/10.3389/fenrg.2021.753878

Alternative Fuels Data Center (n.d.) Hydrogen basics. https://afdc.energy.gov/fuels/hydrogen_basics.html#:~:text=The%20energy%20in%202.2%20pounds,driving%20range%20of%20conventional%20vehicles

Amândio MST, Pereira JM, Rocha JMS, Serafim LS, Xavier AMRB (2022) Getting value from pulp and paper industry wastes: on the way to sustainability and circular economy. Energies 15(11):4105. https://doi.org/10.3390/en15114105

Amer L, Adhikari B, Pellegrino J (2011) Technoeconomic analysis of five microalgae-to-biofuels processes of varying complexity. Bioresour Technol 102(20):9350–9359. https://doi.org/10.1016/j.biortech.2011.08.010

Angili TS, Grzesik K, Rödl A, Kaltschmitt M (2021) Life cycle assessment of bioethanol production: a review of feedstock, technology and methodology. Energies 14(10):2939. https://doi.org/10.3390/en14102939

Arguelles-Arguelles A, Amezcua-Allieri MA, Ramírez-Verduzco LF (2021) Life cycle assessment of green diesel production by hydrodeoxygenation of palm oil. Front Energy Res 9:690725. https://doi.org/10.3389/fenrg.2021.690725

ATSDR (n.d.) Landfill gas primer - chapter 2: landfill gas basics. https://www.atsdr.cdc.gov/hac/landfill/html/ch2.html#t2_1

Aziz NIHA, Hanafiah MM (2019) Life cycle analysis of biogas production from anaerobic digestion of palm oil mill effluent. Renew Energy 145:847–857. https://doi.org/10.1016/j.renene.2019.06.084

Bahillo A, Armesto L, Cabanillas A, Otero J (2004) Thermal valorization of footwear leather wastes in bubbling fluidized bed combustion. Waste Manag 24(9):935–944. https://doi.org/10.1016/j.wasman.2004.07.006

Balachandar G, Varanasi JL, Singh V, Singh H, Das D (2019) Biological hydrogen production via dark fermentation: a holistic approach from lab-scale to pilot-scale. Int J Hydrog Energy 45(8):5202–5215. https://doi.org/10.1016/j.ijhydene.2019.09.006

Balasubramanian R, Sircar A, Sivakumar P, Anbarasu K (2018) Production of biodiesel from dairy wastewater sludge: a laboratory and pilot scale study. Egypt J Pet 27(4):939–943. https://doi.org/10.1016/j.ejpe.2018.02.002

Barghash H, AlRashdi Z, Okedu K, Desmond P (2022) Life-cycle assessment study for biohydrogen gas production from sewage treatment plants using solar PVs. Energies 15(21):8056. https://doi.org/10.3390/en15218056

Beal CM, Gerber LN, Sills DL, Huntley ME, Machesky SC, Walsh MJ, Tester JW, Archibald I, Granados J, Greene CH (2015) Algal biofuel production for fuels and feed in a 100-ha facility: a comprehensive techno-economic analysis and life cycle assessment. Algal Res 10:266–279. https://doi.org/10.1016/j.algal.2015.04.017

Bjelić D, Markić DN, Prokić D, Malinović BN, Panić AA (2024) "Waste to energy" as a driver towards a sustainable and circular energy future for the Balkan countries. Energy Sustain Soc 14(1):3. https://doi.org/10.1186/s13705-023-00435-y

Chen XY, Vinh-Thang H, Ramirez AA, Rodrigue D, Kaliaguine S (2015) Membrane gas separation technologies for biogas upgrading. RSC Adv 5(31):24399–24448. https://doi.org/10.1039/c5ra00666j

Cvetković SM, Radoičić TK, Kijevčanin M, Novaković JG (2021) Life cycle energy assessment of biohydrogen production via biogas steam reforming: case study of biogas plant on a farm in Serbia. Int J Hydrog Energy 46(27):14130–14137. https://doi.org/10.1016/j.ijhydene.2021.01.181

Devarajan Y, Jayabal R, Munuswamy DB, Ganesan S, Varuvel EG (2022) Biofuel from leather waste fat to lower diesel engine emissions: valuable solution for lowering fossil fuel usage and perception on waste management. Process Saf Environ Prot 165:374–379. https://doi.org/10.1016/j.psep.2022.07.001

Dudyński M, Dudyński K, Kluska J, Ochnio M, Kazimierski P, Kardaś D (2021) Gasification of leather waste for energy production: laboratory scale and industrial tests. Int J Energy Res 45(13):18540–18553. https://doi.org/10.1002/er.6966

Energy Conservation Research and Development Center (n.d.) Conversion to TOE. https://enerteam.org/conversion-to-toe.html
European Commission (2013) Announcement: server inaccessibility. https://webgate.ec.europa.eu/
European Commission (n.d.) European Commission. https://ec.europa.eu/
European Parliament (n.d.)The impact of textile production and waste on the environment (infographics). https://www.europarl.europa.eu/topics/en/article/20201208STO93327/the-impact-of-textile-production-and-waste-on-the-environment-infographics
Ghosh SK (2016) Sustainable SWM in developing countries focusing on faster growing economies, India and China. Procedia Environ Sci 35:176–184. https://doi.org/10.1016/j.proenv.2016.07.073
Gievers F, Walz M, Loewe K, Bienert C, Loewen A (2022) Anaerobic co-digestion of paper sludge: feasibility of additional methane generation in mechanical–biological treatment plants. Waste Manag 144:502–512. https://doi.org/10.1016/j.wasman.2022.04.016
Guan Y, Liu C, Peng Q, Zaman F, Zhang H, Jin Z, Wang A, Wang W, Huang Y (2019) Pyrolysis kinetics behavior of solid leather wastes. Waste Manag 100:122–127. https://doi.org/10.1016/j.wasman.2019.09.005
Gupta D, Singh A, Sharma A, Nigam A (2013) Processing of biofuels. In: Babu V, Thapliyal A, Patel GK (eds) Biofuels production. Wiley, Hoboken. https://doi.org/10.1002/9781118835913.ch3
Haile A, Gelebo GG, Tesfaye T, Mengie W, Mebrate MA, Abuhay A, Limeneh DY (2021) Pulp and paper mill wastes: utilizations and prospects for high value-added biomaterials. Bioresour Bioprocess 8(1):35. https://doi.org/10.1186/s40643-021-00385-3
Hasanzadeh E, Mirmohamadsadeghi S, Karimi K (2018) Enhancing energy production from waste textile by hydrolysis of synthetic parts. Fuel 218:41–48. https://doi.org/10.1016/j.fuel.2018.01.035
Havlikova M, Kroeze C, Huijbregts M (2008) Environmental and health impact by dairy cattle livestock and manure management in The Czech Republic. Sci Total Environ 396(2–3):121–131. https://doi.org/10.1016/j.scitotenv.2008.02.027
Hellal MS, Abou-Taleb EM, Rashad AM, Hassan GK (2022) Boosting biohydrogen production from dairy wastewater via sludge immobilized beads incorporated with polyaniline nanoparticles. Biomass Bioenergy 162:106499. https://doi.org/10.1016/j.biombioe.2022.106499
Hosseinzadeh-Bandbafha H, Nizami A, Kalogirou SA, Gupta VK, Park Y, Fallahi A, Sulaiman A, Ranjbari M, Rahnama H, Aghbashlo M, Peng W, Tabatabaei M (2022) Environmental life cycle assessment of biodiesel production from waste cooking oil: a systematic review. Renew Sust Energ Rev 161:112411. https://doi.org/10.1016/j.rser.2022.112411
IEA (n.d.) Industry–sustainable recovery–analysis. https://www.iea.org/reports/sustainable-recovery/industry
Jara-Cobos L, Abril-González M, Pinos-Vélez V (2023) Production of hydrogen from lignocellulosic biomass: a review of technologies. Catalysts 13(4):766. https://doi.org/10.3390/catal13040766
Junqueira TL, Chagas MF, Gouveia VLR, Rezende MCAF, Watanabe MDB, Jesus CDF, Cavalett O, Milanez AY, Bonomi A (2017) Techno-economic analysis and climate change impacts of sugarcane biorefineries considering different time horizons. Biotechnol Biofuels 10(1):50. https://doi.org/10.1186/s13068-017-0722-3
Kalita S, Ohlsson JA, Potter HK, Nordberg Å, Sandgren M, Hansson P (2023) Energy performance of compressed biomethane gas production from co-digestion of Salix and dairy manure: factoring differences between Salix varieties. Biotechnol Biofuels Bioprod 16(1):165. https://doi.org/10.1186/s13068-023-02412-1
Kanadevia Inova (n.d.) About Kanadevia Inova group. Kanadevia Inova. https://hz-inova.com/
Kandasamy R, Venkatesan SK, Uddin MI, Ganesan S (2020) Anaerobic biovalorization of leather industry solid waste and production of high value-added biomolecules and biofuels. In: Elsevier eBooks. Elsevier, Amsterdam, pp 3–25. https://doi.org/10.1016/b978-0-12-817951-2.00001-8
Kandemir K, Piskin E, Xiao J, Tomas M, Capanoglu E (2022) Fruit juice industry wastes as a source of bioactives. J Agric Food Chem 70(23):6805–6832. https://doi.org/10.1021/acs.jafc.2c00756

Kenny C, Priyadarshini A (2021) Review of current healthcare waste management methods and their effect on global health. Healthcare 9(3):284. https://doi.org/10.3390/healthcare9030284

Li Y, Alaimo CP, Kim M, Kado NY, Peppers J, Xue J, Wan C, Green PG, Zhang R, Jenkins BM, Vogel CFA, Wuertz S, Young TM, Kleeman MJ (2019) Composition and toxicity of biogas produced from different feedstocks in California. Environ Sci Technol 53(19):11569–11579. https://doi.org/10.1021/acs.est.9b03003

Lin CSK, Pfaltzgraff LA, Herrero-Davila L, Mubofu EB, Abderrahim S, Clark JH, Koutinas AA, Kopsahelis N, Stamatelatou K, Dickson F, Thankappan S, Mohamed Z, Brocklesby R, Luque R (2012) Food waste as a valuable resource for the production of chemicals, materials and fuels. Current situation and global perspective. Energy Environ Sci 6(2):426. https://doi.org/10.1039/c2ee23440h

Liu Y, Chen J (2008) Phosphorus cycle. In: Elsevier eBooks. Elsevier, Amsterdam, pp 2715–2724. https://doi.org/10.1016/b978-008045405-4.00754-0

Liu J, Dantoft SH, Würtz A, Jensen PR, Solem C (2016) A novel cell factory for efficient production of ethanol from dairy waste. Biotechnol Biofuels 9(1):33. https://doi.org/10.1186/s13068-016-0448-7

Lovett RA (2002) Global warming tied to ozone recovery. Greenhouse gases, effects of warming may prolong ozone woes

Ma Z, Yang Y, Chen W, Wang P, Wang C, Zhang C, Gan J (2021) Material flow patterns of the global waste paper trade and potential impacts of China's import ban. Environ Sci Technol 55(13):8492–8501. https://doi.org/10.1021/acs.est.1c00642

Mertins A, Wawer T (2022) How to use biogas? A systematic review of biogas utilization pathways and business models. Bioresour Bioprocess 9(1):59. https://doi.org/10.1186/s40643-022-00545-z

Mezzullo WG, McManus MC, Hammond GP (2012) Life cycle assessment of a small-scale anaerobic digestion plant from cattle waste. Appl Energy 102:657–664. https://doi.org/10.1016/j.apenergy.2012.08.008

Mgharbel M, Halawy L, Milane A, Zeaiter J, Saad W (2023) Pyrolysis of pharmaceuticals as a novel means of disposal and material recovery from waste for a circular economy. J Anal Appl Pyrolysis 172:106014. https://doi.org/10.1016/j.jaap.2023.106014

Mohammadi A, Sandberg M, Venkatesh G, Eskandari S, Granström K (2019a) Anaerobic digestion of pulp and paper mill sludge: life cycle assessment of alternative uses of biogas. Conference: EGU general assembly 2019 Vienna, Austria

Mohammadi A, Sandberg M, Venkatesh G, Eskandari S, Dalgaard T, Joseph S, Granström K (2019b) Environmental performance of end-of-life handling alternatives for paper-and-pulp-mill sludge: using digestate as a source of energy or for biochar production. Energy 182:594–605. https://doi.org/10.1016/j.energy.2019.06.065

Moktadir MA, Rahman MM (2022) Energy production from leather solid wastes by anaerobic digestion: a critical review. Renew Sust Energ Rev 161:112378. https://doi.org/10.1016/j.rser.2022.112378

Moodley P (2021) Sustainable biofuels: opportunities and challenges. In: Elsevier eBooks. Elsevier, Amsterdam, pp 1–20. https://doi.org/10.1016/b978-0-12-820297-5.00003-7

Mora-Gamboa MPC, Rincón-Gamboa SM, Ardila-Leal LD, Poutou-Piñales RA, Pedroza-Rodríguez AM, Quevedo-Hidalgo BE (2022) Impact of antibiotics as waste, physical, chemical, and enzymatical degradation: use of laccases. Molecules 27(14):4436. https://doi.org/10.3390/molecules27144436

Mulya KS, Zhou J, Phuang ZX, Laner D, Woon KS (2022) A systematic review of life cycle assessment of solid waste management: methodological trends and prospects. Sci Total Environ 831:154903. https://doi.org/10.1016/j.scitotenv.2022.154903

Nwokolo N, Mukumba P, Obileke K, Enebe M (2020) Waste to energy: a focus on the impact of substrate type in biogas production. PRO 8(10):1224. https://doi.org/10.3390/pr8101224

Peiró LT, Lombardi L, Méndez GV, Durany XGI (2009) Life cycle assessment (LCA) and exergetic life cycle assessment (ELCA) of the production of biodiesel from used cooking oil (UCO). Energy 35(2):889–893. https://doi.org/10.1016/j.energy.2009.07.013

Priebe G, Kipper E, Gusmão A, Marcilio N, Gutterres M (2016) Anaerobic digestion of chrome-tanned leather waste for biogas production. J Clean Prod 129:410–416. https://doi.org/10.1016/j.jclepro.2016.04.038

Reaño RL, Halog A (2020) Analysis of carbon footprint and energy performance of biohydrogen production through gasification of different waste agricultural biomass from The Philippines. Biomass Convers Biorefinery 13(10):8685–8699. https://doi.org/10.1007/s13399-020-01151-9

Renergon Biogas (2025) Biogasanlage und biogas Technologie zur Trockenvergärung. Renergon - Biogasanlagen. https://renergon-biogas.com/

Samal K, Mahapatra S, Ali MH (2022) Pharmaceutical wastewater as emerging contaminants (EC): treatment technologies, impact on environment and human health. Energy Nexus 6:100076. https://doi.org/10.1016/j.nexus.2022.100076

Sasongko NA, Noguchi R, Ito J, Demura M, Ichikawa S, Nakajima M, Watanabe MM (2018) Engineering study of a pilot scale process plant for microalgae-oil production utilizing municipal wastewater and flue gases: Fukushima pilot plant. Energies 11(7):1693. https://doi.org/10.3390/en11071693

Shakelly N, Pérez-Cardona JR, Deng S, Maani T, Li Z, Sutherland JW (2023) Comparative life cycle assessment of bioethanol production from different generations of biomass and waste feedstocks. Proc CIRP 116:630–635. https://doi.org/10.1016/j.procir.2023.02.106

Shen J, Chen J, Jensen PR, Solem C (2019) Development of a novel, robust and cost-efficient process for valorizing dairy waste exemplified by ethanol production. Microb Cell Factories 18(1):51. https://doi.org/10.1186/s12934-019-1091-3

Tang KHD (2023) State of the art in textile waste management: a review. Text 3(4):454–467. https://doi.org/10.3390/textiles3040027

The World Bank Group (n.d.) Trends in solid waste management. https://datatopics.worldbank.org/what-a-waste/trends_in_solid_waste_management.html#:~:text=Upper%2Dmiddle%2Dincome%20countries%20have,and%2022%20percent%20to%20incineration

UNEP Food Waste Index Report 2021 (n.d.) UNEP - UN Environment Programme. https://www.unep.org/resources/report/unep-food-waste-index-report-2021

Vu TKV, Vu DQ, Jensen LS, Sommer SG, Bruun S (2015) Life cycle assessment of biogas production in small-scale household digesters in Vietnam. Asian Australas J Anim Sci 28(5):716–729. https://doi.org/10.5713/ajas.14.0683

Wang X, Song C (2020) Carbon capture from flue gas and the atmosphere: a perspective. Front Energy Res 8:560849. https://doi.org/10.3389/fenrg.2020.560849

Xing Y, Li Z, Fan Y, Hou H (2009) Biohydrogen production from dairy manures with acidification pretreatment by anaerobic fermentation. Environ Sci Pollut Res 17(2):392–399. https://doi.org/10.1007/s11356-009-0187-4

Xu Z, Qi R, Xiong M, Zhang D, Gu H, Chen W (2020) Conversion of cotton textile waste to clean solid fuel via surfactant-assisted hydrothermal carbonization: mechanisms and combustion behaviors. Bioresour Technol 321:124450. https://doi.org/10.1016/j.biortech.2020.124450

Yang J, Hong C, Li Z, Xing Y, Zhao X (2020) Study on hydrothermal liquefaction of antibiotic residues for bio-oil in ethanol-water system. Waste Manag 120:164–174. https://doi.org/10.1016/j.wasman.2020.11.026

Yee KF, Tan KT, Abdullah AZ, Lee KT (2009) Life cycle assessment of palm biodiesel: revealing facts and benefits for sustainability. Appl Energy 86:S189–S196. https://doi.org/10.1016/j.apenergy.2009.04.014

Yilmaz O, Kantarli IC, Yuksel M, Saglam M, Yanik J (2006) Conversion of leather wastes to useful products. Resour Conserv Recycl 49(4):436–448. https://doi.org/10.1016/j.resconrec.2006.05.006

Yousuf A, Karim A, Islam MA, Karim SU, Khan MMR, Faizal CKM (2020) Dry fermenters for biogas production. In: Elsevier eBooks. Elsevier, Amsterdam, pp 75–87. https://doi.org/10.1016/b978-0-12-821264-6.00006-1

Zhao Y, Wang C, Zhang L, Chang Y, Hao Y (2021) Converting waste cooking oil to biodiesel in China: environmental impacts and economic feasibility. Renew Sust Energ Rev 140:110661. https://doi.org/10.1016/j.rser.2020.110661

Zoghlami A, Paës G (2019) Lignocellulosic biomass: understanding recalcitrance and predicting hydrolysis. Front Chem 7:874. https://doi.org/10.3389/fchem.2019.00874

Hatzisymeon M, Kamenopoulos S, Tsoutsos T (2019) Risk assessment of the life-cycle of the used cooking oil-to-biodiesel supply chain. J Clean Prod 217:836–843. https://doi.org/10.1016/j.jclepro.2019.01.088

Part V

Contribution of Circular Bioeconomy

Contribution of Municipal Solid Waste to a Circular Economy

32

Prasann Kumar and Padmanabh Dwivedi

Abstract

This chapter explores the significant role of municipal solid waste (MSW) in contributing to the circular economy paradigm. With the burgeoning global population and urbanisation, managing MSW has become a pressing concern. The traditional linear "take-make-dispose" approach has led to environmental degradation and resource depletion. However, by integrating MSW management into the circular economy framework, waste can be transformed into valuable resources through reduction, reuse, recycling, and recovery. It examines various strategies and initiatives for optimising MSW management within the circular economy context, highlighting its potential benefits in resource conservation, waste reduction, energy recovery, and economic growth. Moreover, it discusses the challenges and opportunities associated with implementing circular economy principles in MSW management, emphasising the need for policy support, technological innovation, and, most importantly, public awareness. It underscores the importance of reimagining MSW as a valuable resource rather than a disposable burden in advancing the circular economy agenda, empowering the audience and making them feel responsible for the change.

P. Kumar
Departemnt of Agronomy, School of Agriculture, Lovely Professional University, Phagwara, Punjab, India

P. Dwivedi (✉)
Department of Plant Physiology, Institute of Agricultural Sciences, Banaras Hindu University, Varanasi, Uttar Pradesh, India

A. Kumari et al. (eds.), *Valorization of Solid Wastes to Biofuels and Chemical Products for Sustainable World*, https://doi.org/10.1007/978-981-96-8594-3_32

Keywords

Agriculture · Economic · Growth · Waste · Circular · Energy · Policy · Resource · Population

32.1 Introduction

In urban settings, metropolitan solid waste (MSW) refers to a diverse range of waste materials produced by residential, commercial, institutional, and non-industrial entities. The category encompasses commonplace objects, such as packaging, food remnants, paper, plastics, textiles, glass, metals, and various other materials that have been disposed of (Rondón Toro et al. 2022), Gronba-Chyła et al. (2024), Saravanakumar et al. (2024). The composition of municipal solid waste (MSW) varies based on consumer behaviour, economic development, cultural practices, and waste management infrastructure. Managing MSW is challenging due to its vast volume, diverse components, and potential environmental impact. The challenges at hand have been further intensified by the rapid urbanisation, population growth, and shifting consumption patterns, resulting in heightened strain on current waste management systems. The issue of environmental pollution, public health risks, and resource depletion is exacerbated in numerous regions due to insufficient infrastructure, limited financial resources, and the absence of effective policies. Throughout history, municipal solid waste (MSW) management has adhered to a linear framework called the "take-make-dispose" model. This approach involves extracting resources, their subsequent processing into products, consumption, and eventual disposal as waste. The linear system has attributed the depletion of natural resources, pollution of air, water, and soil, and accumulation of waste in landfills, incinerators, and oceans. Furthermore, it has fostered a societal inclination towards inefficiency and unsustainable consumption patterns, presenting enduring threats to ecological systems and human welfare. Unlike the linear economy model, the circular economy promotes a holistic and regenerative approach to resource management and economic development. Rooted in sustainability principles, this concept aims to decouple economic growth from resource consumption while minimising waste generation and maximising resource efficiency throughout a product's lifecycle (Tsai et al. 2020; Bertanza et al. 2021; Munir et al. 2021; Mazzanti et al. 2022; Ambaye et al. 2023; Pajura 2024). The significance of eco-design, product longevity, and material efficiency in reducing waste generation and environmental impact is underscored by designing out waste and pollution. By integrating durability, reparability, and recyclability into product design, businesses can enhance product longevity and reduce dependence on raw materials. The circular economy focuses on maintaining the value of products and materials by promoting reuse, repair, refurbishment, and remanufacturing instead of discarding them after use. Strategies such as product leasing, shared usage platforms, and reverse logistics help keep materials in circulation, reducing the demand for new resources and minimising waste. The circular economy seeks to reduce waste, improve resource efficiency, rejuvenate natural ecosystems, and reinstate ecological services. To achieve this objective, embracing regenerative agricultural methodologies, advocating for preserving biodiversity, and establishing circular water management systems that emulate natural processes is imperative. Through adopting these principles, the circular economy provides a framework for transforming linear value chains into closed-loop systems, where waste is recognised as a valuable resource rather than a disposable burden. The successful transition necessitates the cooperation of various entities involved in the value chain, such as governmental bodies, enterprises, consumers, and civil society organisations. This collaboration is crucial for stimulating innovation, enacting policy changes, and promoting sustainable consumption and production patterns. The multifaceted and crucial role of MSW in contributing to the circular economy is essential for attaining sustainable development goals. Given its substantial contribution to material resources and energy, MSW poses challenges and opportunities in adopting a circular economy framework. To begin with, MSW is a substantial repository of materials with the potential for recycling, organic substances, and energy-dense resources. These resources can be effectively recovered, reused, or repurposed within the circular economy framework. Municipalities can harness the inherent value of MSW and decrease reliance on new resources by adopting sophisticated waste sorting technologies, enhancing recycling infrastructure, and advocating for waste-to-energy conversion processes. Finally, incorporating MSW management within the circular economy framework can effectively address environmental pollution, preserve finite natural resources, and diminish greenhouse gas emissions. Municipalities can produce valuable soil amendments and biogas by diverting organic waste from landfills and employing composting or anaerobic digestion techniques while reducing methane emissions. Similarly, by promoting closed-loop recycling systems for materials like plastics, metals, and paper, managing MSW can help conserve resources and save energy, thus reducing the environmental impact of production and consumption. Moreover, the circular economy strategy presents economic prospects for municipalities, businesses, and communities by establishing fresh markets for recycled materials, promoting advancements in waste management technologies, and creating environmentally friendly employment opportunities in recycling, remanufacturing, and renewable energy industries (Allevi et al. 2021; Bakalár et al. 2021; Tong et al. 2021; Magazzino and Falcone 2022; Di Foggia and Beccarello 2023; Ersahin et al. 2023; Rebehy et al. 2023; Vines et al. 2023; Chenavaz and Dimitrov 2024; Munir et al. 2024). Municipalities can promote local economic growth, strengthen their ability to withstand global market fluctuations and establish a more inclusive and sustainable economy for future generations by utilising the economic opportunities presented by MSW. Nevertheless, to fully harness the potential of MSW in supporting the circular economy, policymakers, industry stakeholders, and the general public must collaborate and address a range of obstacles, including insufficient infrastructure, financial limitations, and social impediments. Municipalities can expedite the shift towards a more resource-efficient, resilient, and equitable society by enacting supportive policies, offering incentives for sustainable waste management practices, and enhancing public awareness regarding the advantages of the circular economy. Efficiently handling MSW is crucial for promoting the circular economy agenda and attaining a sustainable future for everyone. Municipalities can effectively address environmental impacts, generate economic opportunities, and enhance

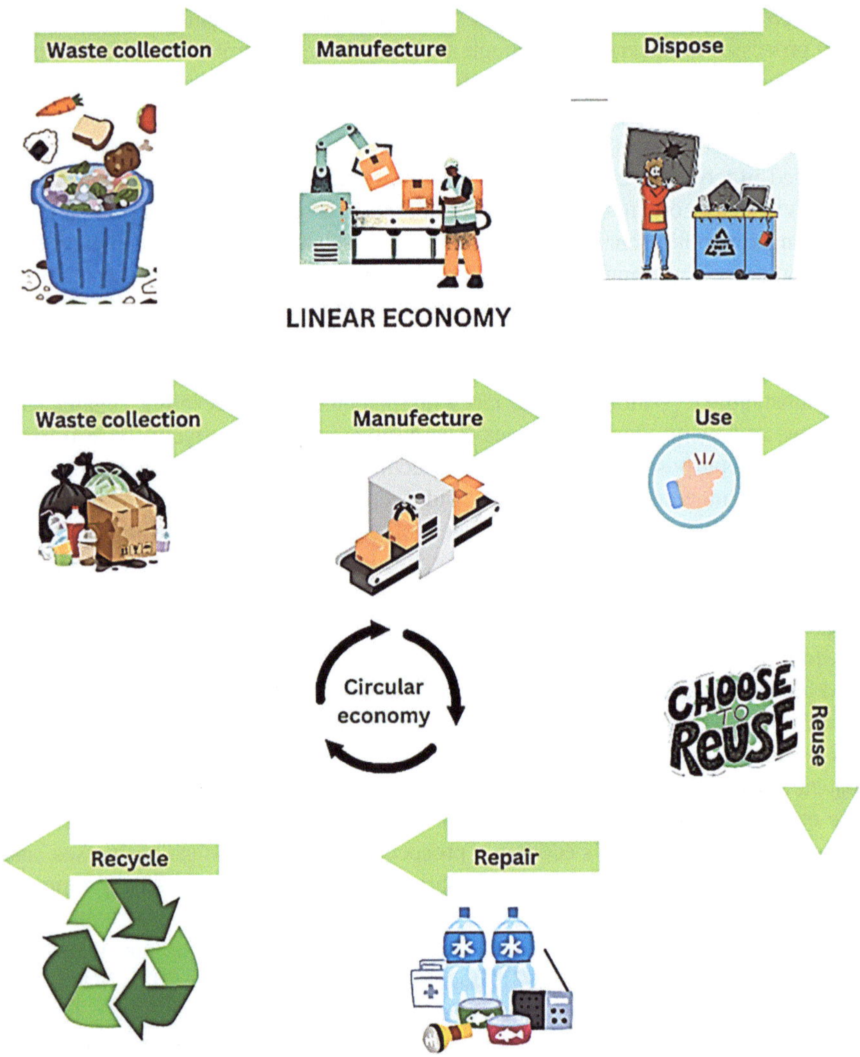

Fig. 32.1 Municipal solid waste. (Source: created by authors)

social well-being by acknowledging waste as a valuable resource and implementing circular economy principles in MSW management. This approach facilitates the transition towards a more sustainable and regenerative economy (Fig. 32.1).

32.2 The Limitations and Challenges of the Linear Economy Model

The linear economy, often referred to as the "take-make-dispose" model, follows a traditional approach where resources move in a one-way flow from extraction and production to consumption and eventual disposal. This linear system involves extracting raw materials from the environment, their subsequent processing into products, their consumption by end-users, and ultimately, their disposal as waste after reaching the end of their useful life. The linear economy functions based on the fundamental principle of perpetual growth and consumption, propelled by the objective of generating profit and fostering economic expansion. The first stage of the linear economy involves extracting natural resources from the environment, including minerals, fossil fuels, timber, water, and other essential materials. Subsequently, these unprocessed materials undergo various manufacturing procedures, including refining, milling, and assembly, to convert them into finished products. The manufacturing stage commonly encompasses substantial energy consumption, extraction of resources, and generation of waste, resulting in environmental degradation and the exhaustion of limited resources. Once produced, commodities are disseminated to consumers via retail channels and utilised for diverse objectives, encompassing individual consumption, industrial utilisation, and infrastructure advancement. Elevated levels of resource consumption, material throughput, and waste generation characterise the consumption phase of the linear economy. Consumer culture, advertising, and planned obsolescence drive the demand for new products, resulting in a continuous cycle of production and consumption (Yaashikaa et al. 2020; Awasthi et al. 2022; Hoang et al. 2022; Marques and Teixeira 2022; Paes et al. 2024; Zhao et al. 2024). When products conclude their life cycle, they are typically disposed off as waste and deposited in landfills, incinerators, or natural habitats. In the linear economy, the disposal phase leads to the build up of waste, contamination of air, water, and soil, and deterioration of ecological systems. Many products lack the necessary features for effortless disassembly or recycling, posing difficulties in retrieving valuable materials and resources from the waste stream. Consequently, substantial amounts of valuable resources are wasted through landfilling or incineration, worsening the problem of limited resources and environmental contamination. The linear economy depends on the ongoing extraction of limited resources, resulting in the depletion of resources and degradation of the environment. As natural resources become scarcer and extraction costs continue to rise, the linear economy becomes increasingly unsustainable. This model generates substantial waste at every stage of the production-consumption-disposal process, much of which ends up in landfills, incinerators, or marine ecosystems, leading to pollution and threatening both the environment and human health. The linear approach is a major contributor to environmental degradation, including deforestation, habitat destruction, air and water contamination, and climate change. Activities such as transportation, manufacturing, and waste disposal release greenhouse gases, reduce biodiversity, and disrupt ecosystems, jeopardising their stability and resilience. Furthermore, the linear economy is economically inefficient, as it fails to maximise the full value of resources throughout their lifecycle. Rooted in a "take-make-dispose" mentality, this model prioritises short-term gains over long-term sustainability. As a result, businesses and governments bear the financial burden of environmental restoration, waste management, and resource depletion, ultimately undermining economic growth and societal well-being.

The perpetuation of unsustainable consumption patterns and resource disparities in the linear economy exacerbates social inequality. The linear model emphasises economic growth and consumption, potentially resulting in disparities in the distribution of resources, opportunities, and wealth, compromising social equity and justice. Environmental pollution, waste incineration, and resource extraction impact marginalised communities more, intensifying social injustices and inequities. The linear economy is predicated upon the perpetual extraction and consumption of limited resources, resulting in the depletion of resources, degradation of the environment, and eventual collapse of ecological systems. With the growth of populations and the expansion of economies, there is an escalating demand for resources, which exerts additional pressure on ecosystems and natural habitats. The linear economy produces substantial waste throughout the production-consumption-disposal cycle, producing air, water, and soil pollution. The accumulation of discarded products in landfills, the accumulation of plastics in oceans, and the release of toxic pollutants into the atmosphere by incinerators pose significant threats to ecological systems and human well-being (Rada and Cioca 2017; Yong et al. 2021; Yadav et al. 2022; Chen and Yu 2023; Derdera and Ogato 2023; Ahmad et al. 2024; Hemali et al. 2024; Lederer and Schuch 2024; Nguyen et al. 2024; Saluja et al. 2024). The economic inefficiency of the linear economy is evident as it fails to adequately capture the comprehensive value of resources and materials across their entire life cycle. A significant portion of the economic worth of resources is forfeited due to inefficiency and contamination, compromising economic advancement and societal welfare. Moreover, the linear model is based on a mindset of "take-make-dispose," which places greater importance on immediate profits rather than long-term viability and adaptability, resulting in economic instability and susceptibility. The perpetuation of unsustainable consumption patterns and resource disparities in the linear economy exacerbates social inequality. Environmental pollution, waste incineration, and resource extraction impact marginalised communities more, intensifying social injustices and inequities. Furthermore, it can be argued that the linear model places a higher emphasis on economic growth and consumption, potentially resulting in inequities in the distribution of resources, opportunities, and wealth. The linear economy is a major driver of environmental degradation, contributing to deforestation, habitat destruction, air and water pollution, and climate change. Activities such as transportation, manufacturing, and waste disposal release greenhouse gases, reduce biodiversity, and disrupt ecosystems, threatening their long-term stability. As the impacts of climate change intensify, transitioning to a more sustainable and regenerative economic model becomes increasingly urgent.

The environmental and economic consequences of the linear economy are significant and unsustainable. With global population growth and economic expansion placing greater pressure on natural resources, ecosystems face escalating degradation, depletion, and socio-economic disparities. In response to these challenges, there is a growing recognition of the need to adopt a circular economy model that emphasises resource efficiency, waste reduction, and sustainable production and consumption practices.

The linear economy relies on the continuous extraction of finite resources—such as minerals, fossil fuels, and biomass—to meet rising consumer demands. However, this unsustainable rate of resource depletion leads to habitat destruction, biodiversity loss, and ecosystem collapse. The consequences of overexploitation, including deforestation, overfishing, and land degradation, weaken ecological systems and threaten essential services such as clean air, water, and food supply.

Furthermore, waste generation is a persistent issue throughout the linear production-consumption-disposal cycle, contributing to rising pollution levels in air, water, and soil. The build up of discarded materials in landfills, plastic accumulation in oceans, and the release of toxic pollutants from incineration pose severe risks to both ecosystems and human health. Additionally, manufacturing and product use contribute to greenhouse gas emissions, acidification, and the spread of hazardous chemicals, further accelerating environmental decline and climate change.

The linear economy plays a substantial role in exacerbating climate change, as it is responsible for a considerable proportion of worldwide greenhouse gas emissions (Abad et al. 2019; Taleb and Al Farooque 2021; Gómez Montoya et al. 2024; Kurniawan et al. 2024; Papamichael et al. 2024; Pomoni et al. 2024). The emission of carbon dioxide (CO_2), methane (CH_4), and other greenhouse gases into the atmosphere as a result of the extraction, processing, and combustion of fossil fuels for energy generation, transportation, and industrial activities leads to the retention of heat and subsequent global temperature increase. Climate change exerts extensive effects on ecological systems, meteorological patterns, and occurrences of natural calamities, presenting hazards to agriculture, water supplies, and human habitations. The linear economy is susceptible to resource scarcity and price volatility due to its dependence on finite resources. With the growing demand for resources and the depletion of reserves, the costs of extracting and producing them increase, resulting in elevated prices for goods and services. In addition, the scarcity of resources can cause disruptions in supply chains, diminish competitiveness, and weaken economic growth, especially in industries that heavily rely on resources such as mining, agriculture, and manufacturing. The linear economy incurs substantial expenses related to waste management, encompassing the gathering, conveyance, processing, and elimination of waste. Municipalities, businesses, and governments bear the financial responsibility for waste management, leading to potential strain on public budgets and the diversion of resources from critical sectors like healthcare, education, and infrastructure development. Furthermore, the environmental and social consequences of waste disposal, including pollution, health effects, and loss of biodiversity, are frequently not adequately considered, resulting in market failures and ineffective allocation of resources. The "take-make-dispose" mentality inherent in the linear economy leads to the deprivation of economic prospects linked to recycling, reuse, and resource recovery. Instead of being reintegrated into the economy through recycling, remanufacturing, or composting, valuable materials and resources are lost to landfilling or incineration. This situation signifies a failure to capitalise on the potential for job creation, innovation stimulation, and value generation from waste streams. Furthermore, the linear economy's framework possesses inherent unsustainability and susceptibility to external disturbances, including resource scarcities, price fluctuations, and regulation alterations. These factors can potentially disrupt supply chains and undermine the economy's stability (Poyyamozhi et al. 2024; Sikder et al. 2024; Singh et al. 2024; Zhang et al. 2024). Adopting a circular economy is imperative to advance resource efficiency and optimise the resource value across its entire life cycle. Business enterprises have the potential to mitigate waste generation, preserve natural resources, and diminish environmental consequences through the strategic development of products that prioritise durability, reparability, and recyclability. Furthermore, by implementing circular business models such as product-as-a-service, sharing platforms, and circular supply chains, organisations can optimise resource utilisation, strengthen their ability to withstand resource scarcity, and establish a competitive edge within the marketplace. The circular economy approach significantly emphasises reducing waste and recovering resources, mitigating the environmental and economic burdens associated with waste management (Rada and Cioca 2017; Yadav et al. 2022; Derdera and Ogato 2023; Liu et al. 2023; He et al. 2024; Kumareswaran et al. 2024; Lederer and Schuch 2024; Sandhi and Rosenlund 2024; Taouahria 2024). Municipalities can generate economic opportunities for recycling, remanufacturing, and composting industries by adopting advanced waste sorting technologies, enhancing recycling infrastructure, and advocating for circular product design. This will help divert valuable materials away from landfills and incinerators. Furthermore, by promoting sustainable consumption practices, such as the repair, repurposing, and sharing of products, the circular economy has the potential to reduce waste production and foster a culture of responsible resource management. The circular economy enhances economic resilience by separating economic growth from resource consumption while simultaneously fostering the development of renewable and diversified economic systems. The circular economy has the potential to bolster the resilience of supply chains, mitigate risks related to resource scarcity and price volatility, and foster innovation, job creation, and inclusive economic development through the closure of material loops, reduction of reliance on virgin resources, and promotion of local production and consumption. Furthermore, the circular economy has the potential to effectively tackle systemic issues such as waste management, pollution, and climate change while simultaneously advancing social equity and environmental justice through the cultivation of collaboration among various stakeholders throughout the value chain. These stakeholders encompass governments, businesses, consumers, and civil society organisations. Adopting a circular economy is crucial to effectively tackle the environmental and economic ramifications associated with the linear model and attain sustainable development objectives. The circular economy presents a comprehensive and regenerative approach to economic activity encompassing resource efficiency, waste reduction, and financial resilience. This approach can potentially enhance human well-being, safeguard the environment, and secure prosperity for future generations. However, transitioning to a circular economy requires concerted efforts from policymakers, industry stakeholders, and the public to overcome barriers such as vested interests, institutional inertia, and cultural norms. By embracing the principles of the circular economy and collaborating to implement innovative solutions, we can create a more sustainable and resilient future for all (Table 32.1 and Fig. 32.2).

32.3 Municipal Solid Waste Management: Challenges and Current Practices

The management of municipal solid waste presents numerous challenges on a global scale, which are further complicated by various socio-economic contexts and factors such as population growth and urbanisation patterns. The challenges of managing MSW arise from its diverse composition, encompassing various materials, including organic waste, plastics, metals, glass, and hazardous substances. With the rapid growth of urban populations, the amount of MSW produced increases significantly, placing a significant strain on the current waste management infrastructure. The challenges of managing municipal solid waste MSW are exacerbated in numerous regions due to inadequate waste collection systems, insufficient landfill capacity, and limited recycling facilities. These factors contribute to environmental pollution, public health hazards, and social inequities. An essential obstacle in managing MSW is the immense quantity of waste produced daily by urban areas. The rapid urbanisation and population growth increase consumption, leading to higher waste generation rates. Furthermore, the alteration of consumption patterns, exemplified by the widespread use of disposable plastics and electronic devices, plays a role in the increasing quantity of MSW, placing additional pressure on waste management infrastructure and intensifying environmental consequences. One notable obstacle is the diverse composition of municipal solid waste (MSW), which introduces sorting, recycling, and disposal complexities. Municipal solid waste comprises various organic, recyclable, and non-recyclable components. This requires the use of advanced waste sorting technologies and separation facilities (Bakalár et al. 2021; Awasthi et al. 2022; Magazzino and Falcone 2022; Mazzanti et al. 2022; Vines et al. 2023; Di Foggia and Beccarello 2023). Nevertheless, numerous municipalities lack the necessary infrastructure and resources to segregate and process MSW efficiently, resulting in elevated contamination levels and lower recycling rates. Moreover, MSW disposal presents environmental and health hazards, especially in areas with insufficient landfill infrastructure. Inadequately administered landfills have the potential to introduce leachate and hazardous chemicals into soil, groundwater, and surface water, thereby posing risks to both ecological systems and public well-being. In addition, the practice of openly disposing of waste in informal settlements and unregulated dumpsites amplifies pollution levels and heightens the potential for disease transmission, thereby disproportionately impacting marginalised communities. Municipalities utilise various MSW management techniques to address these challenges, such as landfilling, incineration, composting, and recycling. Despite its environmental drawbacks and limited capacity, landfilling continues to be the predominant method of MSW disposal on a global scale. Despite being promoted as a means to decrease waste volume and recover energy, incineration gives rise to apprehensions regarding air pollution, greenhouse gas emissions, and the disposal of toxic ash. Composting is a viable and environmentally friendly method for managing organic waste, effectively diverting it from landfills while generating valuable soil amendments. Nevertheless, composting necessitates substantial amounts of space, time, and resources, which restricts its ability to be

Table 32.1 From trash to treasure: how municipal solid waste fuels the circular economy

Aspect	Description
Linear economy model	The linear economy model is a traditional production and consumption approach characterised by the "take-make-dispose" cycle. It involves extracting raw materials, manufacturing products, consuming them, and disposing of them as waste
Pitfalls	
1. Resource depletion	– Extraction of finite resources without adequate consideration for sustainability leads to depletion of natural resources
	– Overreliance on non-renewable resources contributes to environmental degradation and threatens future availability
2. Pollution	– Production processes generate pollution and waste, including greenhouse gas emissions, water pollution, and air pollution
	– Improper waste disposal leads to land, water, and air contamination, impacting ecosystems and human health
3. Energy consumption	– High energy consumption throughout the linear economy model contributes to climate change and exacerbates environmental issues
	– Fossil fuel dependency for energy generation intensifies greenhouse gas emissions, driving global warming
4. Economic inefficiency	– Linear economy practices often result in inefficiencies, such as underutilisation of resources, high production costs, and reliance on linear supply chains
	– Lack of resource circulation limits economic opportunities for recycling, remanufacturing, and reuse, hindering potential cost savings and revenue generation
5. Social impacts	– Linear economy models can exacerbate social inequalities, particularly in regions where resource extraction occurs
	– Communities reliant on industries tied to the linear economy may face job instability, health hazards, and socio-economic challenges when these industries decline or face regulatory changes
6. Limited innovation	– The linear economy model often discourages innovation in sustainable practices and circular economy solutions
	Focusing on short-term profitability and linear supply chains may impede developing and adopting technologies and business models that promote resource efficiency and waste reduction
7. Vulnerability to shocks	– Dependency on linear supply chains makes economies vulnerable to disruptions like natural disasters, geopolitical conflicts, or supply chain bottlenecks
	– Lack of diversification and resilience in production and distribution networks increases the risk of economic shocks and supply shortages
8. Long-term unsustainability	– The linear economy model is inherently unsustainable, relying on finite resources and generating pollution and waste faster than the environment can absorb or mitigate
	– Continued adherence to linear economic practices jeopardises the Well-being of current and future generations by compromising environmental integrity and ecological balance

Fig. 32.2 Municipal solid waste: A resource mine for the circular economy. (Source: created by authors)

expanded and implemented in highly populated urban regions. Recycling has the potential to recover resources and reduce waste, but its effectiveness depends on the presence of efficient infrastructure for collecting, sorting, and processing. Despite the formidable obstacles, many municipalities have exhibited the capacity of inventive MSW management practices and policies. Implementing these measures, such as integrated waste management systems, source reduction, recycling incentives, and public awareness campaigns, has demonstrated their efficacy in diminishing waste production and fostering sustainable consumption practices. The achievement of this success should serve as a source of inspiration for policymakers and waste management professionals, encouraging them to investigate and adopt comparable strategies within their respective regions. Decentralised strategies for managing MSW, such as implementing community-based composting initiatives and establishing small-scale recycling enterprises, play a pivotal role in enabling local communities to assume responsibility for waste management. In addition to promoting social cohesion, this approach also involves the active participation of environmentalists and researchers. Furthermore, the progress made in waste-to-energy technologies, such as anaerobic digestion and gasification, exhibits the potential to convert organic waste into sustainable energy resources while mitigating adverse environmental consequences (Rada and Cioca 2017; Yadav et al. 2022; Derdera and Ogato 2023; Liu et al. 2023; Kurniawan et al. 2024; Lederer and Schuch 2024; Papamichael et al. 2024; Pomoni et al. 2024; Singh et al. 2024; Zhang et al. 2024).

Managing municipal solid waste poses intricate difficulties that necessitate comprehensive and multifaceted solutions that involve policy reforms, technological advancements, and community involvement. Municipalities have the potential to shift towards waste management systems that are more sustainable and resilient, thereby ensuring the protection of public health, the preservation of the environment, and the promotion of inclusive economic development. This can be achieved by emphasising waste prevention, resource recovery, and environmental stewardship. Nevertheless, attaining these objectives necessitates the cooperation of stakeholders, sustained financial investment, and a dedication to sustainable consumption and production practices. Managing MSW, which refers to the routine waste generated by residential, commercial, and municipal establishments, poses an increasingly complex issue. The amount of municipal solid waste produced varies considerably based on income level, consumption patterns, and urbanisation. Although high-income countries generally create a more significant amount of waste per person, there is a projected substantial increase in global MSW generation by 2050, necessitating the development of innovative solutions. Municipal solid waste is a diverse combination of organic and inorganic substances. Typically, 50% of municipal solid waste (SW) comprises organic materials such as food scraps and yard waste. An additional 17.5% of the waste comprises recyclable materials, including paper, plastic, glass, and metals. The remaining 31% of waste is classified as inert, encompassing construction debris and textiles. However, there can be significant variations in composition across different regions. Developing nations may exhibit a higher prevalence of organic waste due to reduced food processing and packaging levels. Comprehending this composition is imperative for the efficient management of waste. It aids in developing collection and segregation initiatives, identifying opportunities for waste reduction, reuse, and recycling, and planning suitable treatment and disposal facilities. Utilising waste for resource recovery mitigates environmental consequences, although obstacles can be overcome. The act of informal waste picking has the potential to modify the data gathered at landfill sites, and the composition of waste changes in response to shifting consumption patterns and packaging materials. Additional investigation is required to define waste in particular areas for customised waste management strategies and examine cutting-edge technologies for waste treatment and resource retrieval. The energy content (calorific value) of municipal solid waste (MSW) is crucial in waste-to-energy initiatives. In contrast, the moisture content significantly impacts waste management strategies and transportation expenses. By comprehending the generation and composition of MSW, it becomes possible to develop efficient methods to reduce waste, optimise resource recovery, and foster sustainable waste management practices. MSW presents a substantial contemporary challenge. With the ongoing expansion of global populations and the changing consumption patterns, there is a consistent rise in the quantity of MSW generated (Allevi et al. 2021; Bakalár et al. 2021; Bertanza et al. 2021; Munir et al. 2021; Mazzanti et al. 2022; Rebehy et al. 2023; Vines et al. 2023; Di Foggia and Beccarello 2023; Ersahin et al. 2023; Saravanakumar et al. 2024; Chenavaz and Dimitrov 2024; Gronba-Chyła et al. 2024; Pajura 2024). The efficient management of this waste stream is of utmost importance to ensure the protection of public health, the preservation of resources, and the preservation of the environment. Conventional techniques such as landfilling and incineration have historically served as the predominant means of municipal solid waste disposal, each possessing distinct merits and limitations. Globally, landfilling remains the predominant municipal solid waste disposal method. Sanitary landfills are purpose-built structures created to reduce the ecological consequences of waste disposal. Waste is accumulated in strata and compressed to decrease its size. The collection and treatment of leachate, a liquid byproduct resulting from the percolation of water through a landfill, serves the purpose of mitigating the risk of groundwater contamination. The capture and utilisation of landfill gas, produced through organic matter decomposition, can be employed for energy generation. Although landfills have disadvantages, they provide numerous notable benefits. First and foremost, they are comparatively cost-effective compared to certain alternative treatment methods. Furthermore, they can handle diverse waste materials, including those that pose challenges or are not recycling feasible. Landfills additionally serve as a protective measure for communities that have yet to establish comprehensive recycling or waste reduction initiatives, enabling them to efficiently handle their waste management needs. Nevertheless, it is crucial to acknowledge that landfilling lacks notable disadvantages (Ersahin et al. 2023; Rebehy et al. 2023; Chenavaz and Dimitrov 2024; Munir et al. 2024; Paes et al. 2024). The presence of landfills in densely populated regions can pose a significant concern due to the substantial land areas they occupy. Methane, a potent greenhouse gas, is produced by decomposing organic waste in landfills, contributing to climate change. Inadequate leachate management can lead to groundwater contamination, while ineffective capture and utilisation of landfill gas emissions can contribute to air pollution. Moreover, landfills possess a limited duration, and once they reach their maximum capacity, it becomes imperative to identify alternative disposal methods, underscoring the necessity for sustainable waste management solutions. Incineration is a conventional approach to handling municipal solid waste. Waste is incinerated in a regulated setting at elevated temperatures, substantially decreasing its volume. The thermal energy produced during incineration can be harnessed to generate electricity or steam, facilitating a waste-to-energy conversion mechanism. Incineration can efficiently eliminate specific hazardous substances found within the waste stream. Advocates of incineration highlight its capacity to significantly decrease waste volume, consequently prolonging the longevity of landfills. The energy derived from incineration has the potential to mitigate the reliance on fossil fuels, thereby presenting promising environmental advantages. Moreover, the process of incineration has the potential to effectively eliminate pathogens and other detrimental contaminants present in the waste. Nevertheless, incineration is not without its environmental implications. The combustion procedure initiates the emission of atmospheric contaminants, including nitrogen oxides, sulphur oxides, and particulate matter, which can potentially contribute to respiratory ailments and other health complications. In addition, incineration processes produce ash residue that may contain heavy metals and other hazardous substances, necessitating meticulous handling and proper disposal (Abad et al. 2019; Yadav et al. 2022; Gómez Montoya et al. 2024; Kurniawan et al. 2024;

Lederer and Schuch 2024; Papamichael et al. 2024; Pomoni et al. 2024; Poyyamozhi et al. 2024; Sikder et al. 2024; Singh et al. 2024). The variability in the efficiency of energy recovery from incineration is evident, and in certain instances, the environmental consequences of emissions may surpass the advantages associated with energy generation. The selection between landfilling and incineration as methods for managing municipal solid waste (MSW) is contingent upon a multitude of factors, encompassing the composition of the waste stream, the availability of resources, and the prevailing environmental regulations at the local level. Both methods are not flawless, and there is an increasing focus on waste reduction, reuse, and recycling strategies to minimise the overall amount of waste that needs to be disposed of. Although landfilling and incineration have been the primary methods of managing MSW for many years, there is an increasing acknowledgement of the necessity for more environmentally friendly approaches. Strategies to reduce waste, such as promoting product durability, encouraging reusable shopping bags, and implementing bans on single-use plastics, can substantially mitigate the amount of waste produced at its origin. Recyclable materials are diverted from the waste stream through the recycling programmes, which aim to reprocess them into new products. This practice serves to conserve resources and decrease dependence on landfills and incineration. Organic waste composting provides a natural method to transform food scraps and yard trimmings into a valuable soil amendment. Emerging as potential alternatives to conventional disposal methods are advanced waste treatment technologies. Anaerobic digestion is a process that decomposes organic waste in an environment with low levels of oxygen, producing biogas that can be utilised for energy generation. The mechanical biological treatment (MBT) process facilitates the segregation of waste into distinct fractions, thereby enabling the purification of specific materials and the transformation of organic substances into compost or fuel. Gasification entails the thermochemical conversion of waste materials into synthetic gas, which can be utilised for many applications. The future of MSW management is contingent upon adopting a comprehensive approach that emphasises waste reduction, reuse, and recycling. Conventional techniques such as landfilling and incineration are expected to remain relevant, but their utilisation should be reduced through effective waste diversion initiatives. The allocation of resources towards researching and developing novel waste treatment technologies exhibits potential for future advancements. To attain sustainable waste management, individuals, businesses, and governments must collaborate in adopting responsible consumption practices, implementing efficient waste reduction and diversion initiatives, and embracing innovative technologies that mitigate the environmental consequences associated with waste disposal. MSW is an increasingly problematic issue with substantial environmental, social, and economic ramifications. With the rapid growth of global populations and the escalating wastefulness of consumption patterns, there has been a concerning increase in the quantity of MSW generated. The efficient management of this continuously growing waste stream presents a multifaceted challenge for communities on a global scale (Tsai et al. 2020; Bakalár et al. 2021; Munir et al. 2021, 2024; Magazzino and Falcone 2022; Vines et al. 2023; Di Foggia and Beccarello 2023; Ersahin et al. 2023; Rebehy et al. 2023; Chenavaz and

Dimitrov 2024; Pajura 2024). Although historically prevalent, conventional approaches such as landfilling and incineration possess inherent limitations and are confronted with numerous challenges that can potentially worsen the MSW crisis if not proactively tackled. Environmental pollution is a significant and urgent challenge closely linked to the management MSW. Landfills, the predominant method of waste disposal worldwide, can pollute the atmosphere, water, and soil if not appropriately planned and managed. Landfill gas, produced through organic waste decomposition, is released into the atmosphere if not effectively captured and utilised. The gas in question comprises methane, a highly potent greenhouse gas with a considerably greater capacity for warming than carbon dioxide. If leachate, a liquid byproduct resulting from the percolation of rainwater through landfills, is not adequately collected and treated, it can infiltrate groundwater sources. Leachate contamination can make groundwater unsuitable for drinking or irrigation purposes, posing a significant risk to public health and the long-term viability of the environment. As a conventional disposal method, incineration also gives rise to environmental considerations. The process of waste combustion results in the emission of various air pollutants, such as nitrogen oxides, sulphur oxides, and particulate matter. These pollutants can potentially cause respiratory ailments, cardiovascular disorders, and other health complications, especially in communities near incineration facilities. The ash residue is a byproduct of incineration, potentially containing heavy metals and other toxic substances. The appropriate disposal of this ash is of utmost importance to mitigate the risk of soil and water resource contamination. Although certain incineration facilities utilise sophisticated emission control technologies, the efficacy of these systems can differ, and the ecological consequences of air emissions may surpass the advantages of using waste combustion for energy recovery. The management of municipal solid waste presents a notable challenge in terms of resource depletion. The prevailing worldwide economic framework, a linear economy, operates on a take-make-dispose paradigm. Virgin resources are obtained, transformed into goods, and disposed off as waste once they are no longer needed. The linear model exhibits inherent unsustainability due to the continuous depletion of finite resources, generating an expanding waste stream (Abad et al. 2019; Taleb and Al Farooque 2021; Yadav et al. 2022; Derdera and Ogato 2023; Kurniawan et al. 2024; Lederer and Schuch 2024; Papamichael et al. 2024; Poyyamozhi et al. 2024; Zhang et al. 2024). MSW management practices prioritising landfilling and incineration facilitate perpetuating this unsustainable cycle. In contrast, recycling and composting present prospects for retrieving valuable resources from waste materials and their subsequent reintegration into the production cycle. Emphasising waste reduction, reuse, and recycling can effectively preserve resources and facilitate shifting towards a circular economy, wherein materials are utilised for their maximum lifespan. The issue of landfill scarcity poses an increasingly significant challenge. The availability of appropriate land for new landfills is progressively diminishing, especially in areas with high population density. The lifespan of current landfills is limited, necessitating the identification of alternative disposal methods once they have reached their maximum capacity. The placement of landfills is frequently a subject of disagreement, as communities understandably oppose the creation of such facilities in their areas due to apprehensions regarding potential environmental and health consequences. The NIMBY (Not In My Back Yard) syndrome adds complexity to the siting process, making it difficult to find appropriate locations for new landfills. The escalating volume of MSW is primarily attributed to the substantial economic expansion observed in developing nations. With urbanisation and improving living standards in these nations, consumption patterns are emulating those of developed countries, resulting in a rise in waste production. Nevertheless, developing nations frequently require additional infrastructure and resources to establish efficient waste management systems. Insufficient waste collection services and a lack of public awareness and education regarding waste reduction and recycling worsen the problem in these regions. The practice of informal waste picking, which involves individuals scavenging for recyclable materials from waste dumps, has the potential to generate income for marginalised communities. However, it also presents significant health and safety hazards. The potential solution lies in integrating informal waste pickers into formal waste management systems, which can be facilitated through training, equipment, and social protection. The challenges of MSW management can be effectively addressed through technological advancements. Anaerobic digestion is a biological phenomenon characterised by the decomposition of organic waste in an environment lacking oxygen, producing biogas that can be utilised for energy generation. The mechanical biological treatment (MBT) process facilitates the segregation of waste into distinct fractions, thereby enabling the purification of specific materials and the transformation of organic substances into compost or fuel. Thermochemical gasification is a process that transforms waste into a synthetic gas, which can be utilised for a range of applications such as electricity production and transportation fuels. Further research and development are required to enhance the efficiency and cost-effectiveness of these advanced treatment technologies, which present potential alternatives to conventional disposal methods. Managing municipal solid waste (MSW) challenges go beyond environmental considerations. Insufficient waste management services can exert a substantial adverse influence on the well-being of the general population. Uncollected waste provides favourable conditions for the proliferation of disease-carrying vectors, including mosquitoes and rodents. Contact with leachate or impurities in the air (Table 32.2; Fig. 32.3).

32.4 Circular Economy Principles Applied to MSW Management

The escalating issue of managing MSW necessitates a fundamental change in perspective. The conventional linear waste management model, which heavily relies on landfills and incineration, is deemed unsustainable and has negative environmental consequences. A more viable strategy involves the implementation of circular economy principles in managing MSW. The primary objective of the circular economy is to dissociate economic growth from the depletion of resources and environmental pollution by maximising the lifespan of products and materials. Implementing these

Table 32.2 Closing the loop: managing resources in our communities

Aspect	Description
Municipal solid waste (MSW)	Households, commercial establishments, institutions, and industries in urban areas generate municipal solid waste. It includes many materials, such as paper, plastic, glass, metals, organic waste, and hazardous waste
Challenges in MSW management	
1. Rapid urbanisation	– Rapid population growth and urbanisation lead to increased municipal solid waste generation, placing strain on existing waste management infrastructure and services
	– Urbanisation also results in the concentration of waste generation in urban centres, exacerbating waste collection, transportation, and disposal challenges
2. Waste quantity and composition	– Increasing consumption patterns and changing lifestyles contribute to higher waste generation quantities, including a more significant proportion of non-biodegradable and hazardous materials
	– Varied waste composition requires tailored management approaches to handle different types of waste streams effectively
3. Inadequate infrastructure	– Many municipalities lack sufficient infrastructure for waste collection, transportation, treatment, and disposal, leading to inefficient waste management practices and environmental pollution
	– Insufficient funding, technical expertise, and politics will hinder waste management infrastructure development and maintenance
4. Environmental and health impacts	– Improper solid waste management practices contribute to environmental degradation, air and water pollution, soil contamination, and public health hazards
	– Open dumping and burning waste release greenhouse gases, toxic pollutants, and particulate matter, posing respiratory problems and other health risks to nearby communities
5. Land scarcity and landfill space	– The availability of suitable land for landfill sites is limited, particularly in densely populated urban areas, leading to challenges in finding and expanding existing sites
	– The shortage of landfill space necessitates exploring alternative waste management strategies, such as waste-to-energy facilities, recycling, and composting, to reduce reliance on landfills
Current practices in MSW management	
1. Waste segregation at source	– Encouraging households, businesses, and institutions to segregate waste at the source into recyclable, organic, and non-recyclable categories facilitates effective waste management downstream
	– Source segregation enables efficient recovery of recyclable materials and composting of organic waste, reducing the volume of waste sent to landfills or incinerators

(continued)

Table 32.2 (continued)

Aspect	Description
2. Collection and transportation systems	– Municipalities implement collection and transportation systems to collect segregated waste from households and commercial establishments
	– Various methods, such as door-to-door collection, community bins, and waste collection centres, are employed based on local needs and infrastructure availability
3. Recycling and resource recovery	– Recycling initiatives involve sorting, processing, and reusing materials such as paper, plastic, glass, and metals to conserve resources and reduce waste sent to landfills
	– Resource recovery facilities extract value from waste by composting organic matter, generating energy from waste-to-energy plants, and recovering recycling materials
4. Composting and organic waste management	– Composting facilities process organic waste into nutrient-rich compost, which can be used as a soil conditioner or fertiliser in agriculture and landscaping
	– Encouraging backyard and community composting initiatives reduces the volume of organic waste disposed of in landfills and mitigates methane emissions from decomposing organic matter
5. Waste-to-energy (WTE) facilities	– Waste-to-energy plants incinerate non-recyclable waste to generate electricity or heat, providing an alternative to landfilling while reducing the volume of waste and recovering energy from it
	– WTE facilities require proper emission controls and environmental safeguards to minimise air pollution and health risks associated with incineration
6. Public awareness and education	– Public awareness campaigns, educational programs, and community engagement initiatives raise awareness about waste management practices, promote waste reduction, and encourage citizen participation in recycling and source segregation efforts
	– Educational outreach programs target schools, workplaces, and community groups to foster a culture of responsible waste management and environmental stewardship
7. Regulatory framework and policy measures	– Governments enact regulations and policies to establish standards, guidelines, and incentives for effective solid waste management
	– Legislative measures may include waste management regulations, landfill taxes, extended producer responsibility (EPR) schemes, and incentives for recycling and waste reduction initiatives

principles in managing MSW makes it possible to convert waste from a burden into a valuable resource, promoting a more sustainable and resilient future. Waste reduction is the fundamental principle of the circular economy in MSW management. This entails implementing strategies to reduce the initial generation of waste. Product design is essential for minimising waste. The design of products should prioritise durability, repairability, and disassembly, thereby enabling prolonged lifespans and facilitating recycling processes upon reaching the end of their useful lifespans. Advocating for reusable alternatives to disposable items, such as shopping bags and water bottles, can greatly diminish waste production. Policy measures such as the implementation of bans on single-use plastics and the adoption of

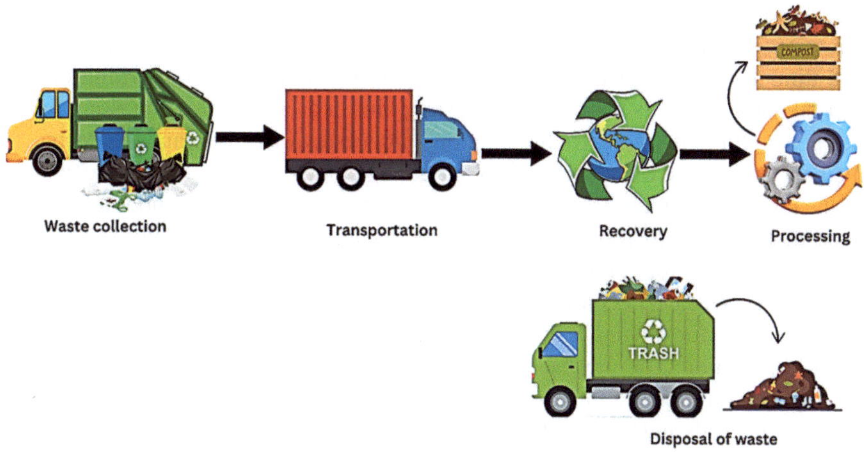

Fig. 32.3 Municipal solid waste management. (Source: created by authors)

extended producer responsibility schemes, which impose responsibility on manufacturers for the proper management of their products at the end of their life cycle, have the potential to provide incentives for the development of product redesign and waste reduction strategies (Allevi et al. 2021; Bakalár et al. 2021; Munir et al. 2021, 2024; Magazzino and Falcone 2022; Ambaye et al. 2023; Rebehy et al. 2023; Vines et al. 2023; Di Foggia and Beccarello 2023; Ersahin et al. 2023; Chenavaz and Dimitrov 2024; Gronba-Chyła et al. 2024). The principles of the circular economy prioritise the extension of product lifespans. Repair and refurbishment initiatives can potentially revitalise outdated products, prolonging their lifespan and mitigating the necessity for fresh manufacturing. Promoting the practice of repairing damaged items rather than disposing of them necessitates the provision of easily accessible and cost-effective repair services. The utilisation of product-as-a-service models, wherein consumers are charged for the utilisation of a product rather than its ownership, serves as a motivating factor for manufacturers to develop products that are both durable and repairable. Recycling is essential for achieving a closed-loop system in the circular economy of municipal solid waste management. Effective recycling programmes employ strategies to redirect valuable materials away from landfills and incineration, enabling their reintegration into the production cycle as secondary resources. Maximising recycling rates necessitates the investment in a comprehensive recycling infrastructure, encompassing collection facilities, sorting technologies, and reprocessing plants. The dissemination of knowledge to consumers regarding appropriate waste segregation practices and the cultivation of awareness regarding the advantages of recycling is imperative in fostering public engagement in such initiatives. Implementing policy measures such as compulsory recycling programmes and deposit-refund schemes can motivate individuals to engage in recycling activities. Composting can divert organic waste, a substantial part of MSW, from landfills. Composting is an inherent biological phenomenon that expedites the conversion of food scraps and yard trimmings into a precious soil amendment. Composting has been found to enhance soil fertility, increase water retention, and stimulate microbial activity, thereby facilitating optimal plant growth and diminishing the need for chemical fertilisers. Household composting can be carried out by utilising backyard composting bins or establishing composting facilities at a community scale. Redirecting organic waste away from landfills decreases methane emissions, thereby aiding in mitigating climate change. Within the context of the circular economy framework, waste streams possess the potential to serve as a valuable source of resources. Advanced waste treatment technologies provide novel approaches for harnessing the potential of waste materials. Anaerobic digestion is a process that decomposes organic waste in an environment with low levels of oxygen, producing biogas that can be utilised for generating electricity or as fuel for transportation. The mechanical biological treatment (MBT) process facilitates the segregation of waste into distinct fractions, thereby enabling the purification of specific materials and the transformation of organic substances into compost or fuel. Gasification is a thermochemical process that transforms waste materials into synthetic gas, which can be utilised for many applications. These technologies provide a method to convert waste into energy and resources, thereby decreasing dependence on landfills and natural resources. The effective implementation of a circular economy framework for MSW management necessitates the active participation and cooperation of all relevant parties involved. Governments' establishment of policy frameworks is crucial in incentivising waste reduction, reuse, and recycling. The imperative for driving change necessitates allocating resources towards infrastructure investment, advancing research and development in innovative technologies, and cultivating public awareness campaigns. It is the private sector's responsibility to create products that adhere to circularity principles, allocate resources to develop recycling infrastructure and investigate inventive business models that prolong the lifespan of products and reduce waste production (Rada and Cioca 2017; Abad et al. 2019; Taleb and Al Farooque 2021; Yadav et al. 2022; Derdera and Ogato 2023; Singh et al. 2024; Zhang et al. 2024; Gómez Montoya et al. 2024; Kurniawan et al. 2024; Lederer and Schuch 2024; Papamichael et al. 2024; Pomoni et al. 2024; Poyyamozhi et al. 2024; Sikder et al. 2024). The active involvement of consumers in embracing responsible consumption practices, engaging in recycling initiatives, and endorsing sustainable businesses is of utmost importance. Civil society organisations possess the capacity to enhance public consciousness, champion policy modifications, and advance community-driven waste management endeavours. Implementing a circular economy for MSW management presents various challenges. The economic feasibility of recycling programmes can be impeded by market failures, such as the frequently lower cost of virgin materials than recycled ones. Insufficient consumer awareness and a lack of behavioural change can impede the attainment of elevated recycling rates. Further development and investment are necessary to enhance advanced waste treatment technologies' technological maturity and cost-effectiveness. The potential advantages of implementing a circular economy in managing MSW are significant. Minimising dependence on landfills and incineration provides substantial environmental advantages, such as decreased greenhouse gas emissions, air and water pollution mitigation, and conservation of resources. A circular economy facilitates economic expansion by generating employment opportunities within the repair, refurbishment, and recycling industries. Moreover, it improves resource security by reducing reliance on the extraction of primary resources. Through the adoption of circular economy principles, it is possible to convert the management of MSW from a burdensome task into a promising opportunity, thereby fostering a future that is both sustainable and resilient for subsequent generations. A paradigm shift is necessary due to the increasing challenges of managing MSW. The circular economy presents a potentially advantageous framework for converting waste from a burden into a valuable asset. By emphasising waste reduction, prolonging the lifespan of products, optimising recycling and composting practices, and investigating novel waste-to-resource technologies, it is possible to establish a closed-loop system for managing MSW. Collaboration among governments, businesses, consumers, and civil society organisations is necessary for this approach. Despite obstacles, the potential environmental and economic advantages of implementing a circular economy for MSW management are indisputable. Ensuring a sustainable future for our planet and its inhabitants necessitates adopting this transformative approach. Efficiency and consumer participation can be enhanced by implementing standardised recycling symbols and collection procedures across various regions. The incorporation of informal waste pickers into formal waste management systems has the potential to augment social inclusivity and effectively mitigate safety concerns. Collaborations between the public and private sectors can utilise resources and specialised knowledge to create and execute efficient circular economy solutions. Maintaining ongoing research and development endeavours to enhance the efficacy and cost-efficiency of sophisticated waste treatment technologies is imperative (Yaashikaa et al. 2020; Awasthi et al. 2022; Yadav et al. 2022; Hoang et al. 2022; Marques and Teixeira 2022; Chen and Yu 2023; Zhang et al. 2024; Zhao et al. 2024; Kurniawan et al. 2024; Lederer and Schuch 2024; Munir et al. 2024; Papamichael et al. 2024; Pomoni et al. 2024; Taouahria 2024). By incorporating these supplementary factors in conjunction with the fundamental tenets of the circular economy, we can expedite the shift towards a more sustainable and accountable method of managing MSW, guaranteeing a healthier planet and a more prosperous future for everyone. Municipal solid waste is an escalating accumulation produced within residential, commercial, and institutional settings, presenting a substantial peril to the natural environment and human welfare. Conventional waste management methods, such as landfilling and incineration, exhibit numerous limitations, emphasising the pressing necessity for a more sustainable alternative. The fundamental principle of a circular economy for MSW management is waste reduction, which involves minimising the initial generation of waste. By implementing diverse strategies at the origin of waste generation, it is possible to substantially mitigate the strain on our environment and foster the development of a more sustainable future. The crux of waste reduction resides in the conceptualisation and production of goods. It is crucial to transition towards products designed with a focus on durability and longevity. This entails the utilisation of superior materials, resilient construction methodologies, and resilient designs that can endure the effects of stress and damage. Products with easily accessible spare parts and repair manuals must exhibit high repairability. Implementing modular designs that facilitate seamless upgrades or component replacements can potentially prolong the lifespan of products and mitigate the necessity for premature disposal. Governments can incentivise manufacturers to embrace these principles by implementing eco-design regulations and extended producer responsibility schemes. These measures aim to establish accountability among producers for properly managing their products at the end of their life cycle. One effective waste reduction strategy involves promoting the adoption of reusable alternatives to single-use products. Implementing basic measures such as reusable shopping bags, water bottles, and coffee mugs can effectively diminish our dependence on disposable commodities. Using reusable containers for food storage and takeaway meals can mitigate the production of plastic wrap and packaging materials. Investing in long-lasting dishware and cutlery eliminates the necessity for disposable plates and utensils. Policy measures such as the implementation of bans on single-use plastics, specifically plastic bags and straws, have the potential to incentivise consumers to embrace reusable alternatives. Businesses have the potential to contribute by providing incentives to customers who choose to utilise their reusable containers or bags. Packaging is a safeguard for products during transportation and storage (Allevi et al. 2021; Bakalár et al. 2021; Tong et al. 2021; Magazzino and Falcone 2022; Mazzanti et al. 2022; Vines et al. 2023; Ersahin et al. 2023; Chenavaz and Dimitrov 2024; Munir et al. 2024; Paes et al. 2024; Pajura 2024). However, abundant inefficient packaging greatly contributes to the MSW stream. Promoting packaging materials that are minimal, recyclable, and compostable is of utmost importance. Implementing refill and reuse systems, wherein consumers acquire products in large quantities and refill their reusable containers, can potentially substantially mitigate packaging waste. The implementation of standardised packaging designs has the potential to enhance recycling efficiency through the establishment of compatibility with pre-existing sorting systems. Governments possess the authority to implement regulatory measures aimed at curbing excessive packaging and promoting the adoption of sustainable packaging materials. Establishing partnerships among manufacturers, retailers, and consumers is imperative to foster the creation of inventive and environmentally friendly packaging solutions. The choices made by individual consumers significantly influence the amount of waste generated. It is essential to increase public consciousness regarding the environmental repercussions of excessive consumption and advocate for responsible consumption practices to decrease waste. The dissemination of knowledge regarding appropriate waste segregation and recycling practices to consumers enables them to implement waste management strategies actively. Promoting repairing damaged items rather than disposing of them, as well as favouring the purchase of pre-owned goods over new ones, has the potential to prolong the lifespan of products and mitigate the generation of waste. Marketing campaigns that advocate for sustainable consumption and emphasise the environmental advantages of waste reduction can influence consumers to make more conscientious decisions. Strategic planning and conscientious consumer behaviour can potentially mitigate household waste substantially. Meal planning and creating grocery lists based on specific needs can mitigate impulsive buying behaviour and minimise food waste. Gaining knowledge about food expiration dates and employing appropriate storage methods can effectively prolong the longevity of food products and mitigate the occurrence of spoilage. Food scraps and yard trimmings are redirected away from landfills and converted into valuable soil amendments by composting. Responsible resource management is promoted by endorsing businesses that reduce food waste, such as donating excess food to charitable organisations or utilising imperfect produce. The implementation of standardised labelling practices can significantly contribute to the reduction of waste. Providing unambiguous and uniform information regarding the duration of product lifespans, recycling details, and composting guidelines enables consumers to make well-informed decisions and appropriately manage waste materials. The efficiency of waste sorting and recycling programmes can be improved by implementing standardised recycling symbols and collection procedures across various regions. Collaborative efforts between governments and industry stakeholders can be employed to establish and execute unambiguous and uniform labelling standards for various products and packaging. Policy instruments are paramount in providing incentives for and promoting waste reduction behaviours (Abad et al. 2019; Taleb and Al Farooque 2021; Yadav et al. 2022; Derdera and Ogato 2023; Liu et al. 2023; Taouahria 2024; Gómez Montoya et al. 2024; He et al. 2024; Kumareswaran et al. 2024; Kurniawan et al. 2024; Poyyamozhi et al. 2024; Sikder et al. 2024; Singh et al. 2024). Extended producer responsibility (EPR) schemes, which impose responsibility on manufacturers for managing their products at the end of their life cycle, can incentivise them to develop products that are more amenable to repair, reuse, or recycling. Implementing prohibitions on disposable plastics and imposing restrictions on the disposal of recyclable or compostable materials in landfills can incentivise businesses and consumers to embrace more environmentally friendly practices. Implementing waste reduction targets and mandatory recycling programmes can serve as a structural framework for attaining waste reduction objectives at the national or regional level. The achievement of effective policy necessitates a collective endeavour involving governments, industry stakeholders, and environmental organisations to establish well-crafted regulations that can be enforced. Campaigns play a crucial role in cultivating a culture that prioritises waste reduction. Educational initiatives implemented within educational institutions can teach children about the ecological ramifications of waste, the significance of resource preservation, and the fundamental tenets of the circular economy. Community outreach programmes can enhance public awareness regarding waste reduction strategies, appropriate waste segregation practices, and the advantages of composting and reusing. Public awareness campaigns that employ diverse media channels, including television, radio, social media, and community events, have the potential to effectively engage a wide range of individuals and foster the adoption of responsible waste management practices. Collaborations among governmental entities, non-profit organisations, and waste management corporations have the potential to harness specialised knowledge and available resources to facilitate the creation of impactful educational and outreach initiatives. Technological advancements have the potential to provide novel solutions for the reduction of waste.

Product lifecycle management (PLM) software enables manufacturers to incorporate end-of-life considerations into their product design, streamlining the disassembly and recycling process. The utilisation of 3D printing technologies holds promise in waste reduction through the facilitation of on-demand production and the mitigation of surplus inventory and packaging requirements. Sharing economy platforms that facilitate the sharing of products and enable peer-to-peer rentals can potentially prolong the lifespan of products and diminish the need for new purchases. It is imperative to allocate resources towards research and the creation of cutting-edge technologies that promote waste reduction to cultivate a more sustainable future. Waste minimisation is the fundamental principle of a circular economy for MSW management (Tsai et al. 2020; Allevi et al. 2021; Bakalár et al. 2021; Munir et al. 2021, 2024; Magazzino and Falcone 2022; Vines et al. 2023; Di Foggia and Beccarello 2023; Ersahin et al. 2023; Rebehy et al. 2023; Chenavaz and Dimitrov 2024; Pajura 2024). By implementing a comprehensive strategy that includes product design, consumer behaviour modification, policy tools, education, and technological advancements, we can effectively decrease the quantity of waste produced at its origin. Implementing this approach alleviates the strain on landfills and incineration facilities, preserves valuable resources, mitigates greenhouse gas emissions, and promotes a more sustainable trajectory for the future. Adopting a waste-reduction mindset necessitates a collaborative endeavour involving governmental bodies, businesses, and individuals. By adopting waste reduction principles and promoting responsible consumption practices, it is possible to establish a global environment characterised by waste minimisation, resource conservation, and the flourishing of a circular economy. Municipal solid waste, the continuously increasing daily waste volume, presents a substantial environmental concern. Conventional waste disposal methods such as landfilling and incineration have disadvantages, emphasising the necessity for a more environmentally friendly approach. The circular economy framework emphasises the retrieval and utilisation of resources, and advocating for the reuse of products and materials is essential in redirecting waste away from landfills and diminishing our ecological impact. The conservation of resources, reduction of waste generation, and promotion of sustainability can be achieved by extending the lifespan of current products and materials. One crucial approach to fostering reuse entails advocating for restoring damaged items rather than their disposal. This approach requires repair services that are easily accessible and cost-effective. Manufacturers should prioritise repairability when designing products, incorporating readily replaceable components and ensuring repair manuals and spare parts are available. By endorsing autonomous repair establishments and enacting legislation that ensures consumers' access to repair information and parts, individuals are empowered to prolong the longevity of their possessions. The second-hand markets provide a platform for repurposing pre-owned items. Thrift stores, consignment shops, and online marketplaces allow consumers to buy high-quality items at reduced prices by facilitating the exchange of used goods. Advocating for these platforms and promoting the acquisition of pre-owned items instead of new ones can greatly diminish the need for new products and minimise waste production. Platforms for sharing products, such as carpooling services or clothing rental companies, enable individuals to temporarily access products, thereby diminishing the necessity for individual ownership and fostering resource efficiency. Refill and reuse systems substitute for disposable products that are used only once. Consumers acquire large quantities of goods and refill their reusable containers, reducing packaging waste. This methodology specifically pertains to detergents, cleaning agents, and refreshments. Business enterprises can incorporate refill stations within their retail establishments, motivating customers to engage in refill initiatives by offering reduced prices or loyalty incentives (Abad et al. 2019; Taleb and Al Farooque 2021; Liu et al. 2023; Gómez Montoya et al. 2024; Singh et al. 2024; Zhang et al. 2024; Kumareswaran et al. 2024; Kurniawan et al. 2024; Lederer and Schuch 2024; Papamichael et al. 2024; Pomoni et al. 2024; Poyyamozhi et al. 2024; Sikder et al. 2024). The establishment and execution of efficient refill and reuse systems necessitate the imperative collaboration among manufacturers, retailers, and consumers. The reusability of products can be greatly impacted by their design. Products should be constructed using resilient materials and engineered to endure numerous applications. Modular designs facilitating effortless upgrades or component replacements can prolong the product lifespan and encourage reuse. Implementing standardised product components, such as batteries or chargers, can enhance compatibility among various devices, thereby mitigating the necessity for frequent replacement transactions. Governments can encourage manufacturers to embrace these principles by implementing eco-design regulations that foster reusability and deter planned obsolescence. It is necessary to enhance public consciousness regarding its ecological advantages to foster a culture of reuse. Educational campaigns can potentially emphasise the environmental consequences associated with excessive consumption and underscore the significance of prolonging the lifespans of products. Considering repairs and second-hand alternatives to new purchases can shift consumption patterns towards a more sustainable paradigm. Social media campaigns and community workshops have the potential to demonstrate the ingenuity and adaptability linked to the practice of product reuse, thereby motivating individuals to incorporate these practices into their daily routines. The promotion of reuse can be significantly influenced by policy and infrastructure development. Providing tax incentives for repair services and financial rewards for businesses that adopt refill and reuse programmes can foster a conducive atmosphere for adopting these practices. Enhancing repair infrastructure through initiatives like technician training programmes and the creation of repair cafes can enhance the availability of cost-effective and dependable repair services. Implementing regulations limiting the use of disposable plastics and encouraging product durability can encourage businesses and consumers to make more environmentally friendly decisions. The process of transitioning towards a society that prioritises reuse poses specific challenges. The transformation of consumer attitudes away from the convenience associated with a “disposable” culture necessitates the implementation of ongoing awareness campaigns and educational initiatives. Promoting consumer confidence in reuse practices is contingent upon guaranteeing the hygiene and safety of second-hand goods, with a particular emphasis on clothing and food containers. Implementing uniformity across various brands and product categories can expedite repairs and enhance compatibility for refill systems. Notwithstanding these challenges, the prospects linked to the promotion of reuse are substantial. Prolonging the lifespans of products serves to preserve valuable resources, diminish energy consumption linked to manufacturing processes, and mitigate the release of greenhouse gas emissions. The practice of reusing utilised products contributes to the establishment of a circular economy, thereby diminishing our dependence on landfills and incineration. Moreover, the promotion of reuse can generate employment opportunities in the repair and second-hand goods industries, thereby fostering economic expansion and progress. Promoting product and material reuse is a fundamental aspect of a sustainable management strategy for MSW. To achieve a more circular economy, we can effectively decrease waste generation by promoting repair, embracing second-hand markets, endorsing refill and reuse systems, and advocating for product design that prioritises reusability (Tsai et al. 2020; Allevi et al. 2021; Munir et al. 2021, 2024; Tong et al. 2021; Bakalár et al. 2021; Bertanza et al. 2021; Magazzino and Falcone 2022; Mazzanti et al. 2022; Ambaye et al. 2023; Rebehy et al. 2023; Di Foggia and Beccarello 2023; Ersahin et al. 2023; Pajura 2024; Chenavaz and Dimitrov 2024; Gronba-Chyła et al. 2024). Establishing a society prioritising reuse requires a collective endeavour involving governmental bodies, businesses, and consumers. By allocating resources towards infrastructure development, raising awareness, and providing incentives for sustainable practices, it is possible to harness the vast potential of reuse and foster a future characterised by waste reduction, resource conservation, and the flourishing of a circular economy. Reuse principles have the potential to be implemented in diverse sectors to optimise resource recovery and mitigate waste generation. Promoting reclaimed construction materials, such as bricks, lumber, and doors, in fresh construction endeavours can substantially diminish the industry's environmental footprint. Implementing deconstruction techniques that prioritise meticulous disassembly and material retrieval instead of demolition can enhance the process of reusing materials. E-waste, also known as electronic waste, is an exponentially expanding type of waste. Advocating to repair and refurbish electronic devices can prolong their operational lifespan and mitigate the necessity for frequent replacements. Implementing take-back programmes, wherein manufacturers gather used electronics for refurbishment or responsible recycling, can effectively facilitate electronic waste management. Textile waste is produced in substantial quantities by the fashion industry. To offer consumers sustainable alternatives to buying new clothes, promoting clothing swaps, second-hand clothing stores, and clothing rental services is beneficial. Promoting the repair of damaged clothing and encouraging consumers to invest in high-quality, long-lasting garments can enhance the adoption of reuse practices in the textile industry. To reduce dependence on single-use disposable items, restaurants and cafes can adopt reusable dishware and cutlery systems. Implementing reusable take-out containers and promoting strategies aimed at reducing food waste within the food service industry have the potential to yield substantial reductions in waste generation. Technological advancements can further enhance the potential for reuse in MSW management. Online platforms that facilitate the connection between individuals needing repairs and proficient repair technicians can enhance the availability of repair services. The potential of 3D printing technologies to transform repair practices lies in their ability to facilitate the printing of replacement parts on demand, thereby diminishing the dependence on pre-manufactured components. Through the adoption of innovative practices and the cultivation of a reusing mindset, we can establish a more environmentally friendly future for the management of waste and the efficient use of resources. Advocating for the reutilisation of products and materials is an effective approach to redirecting waste away from landfills, preserving resources, and reducing our ecological footprint (Abad et al. 2019; Taleb and Al Farooque 2021; Yadav et al. 2022; Derdera and Ogato 2023; Liu et al. 2023; Singh et al. 2024; Zhang et al. 2024; Gómez Montoya et al. 2024; Kurniawan et al. 2024; Lederer and Schuch 2024; Papamichael et al. 2024; Pomoni et al. 2024; Poyyamozhi et al. 2024; Sikder et al. 2024). By implementing a comprehensive strategy that includes product design, consumer behaviour modification, policy tools, and technological progress, we can establish a society that places importance on reuse and prioritises the recovery of resources. Achieving a future focused on reuse necessitates the collaborative efforts of all parties involved. Through collaboration, we can harness the vast capacity of reusing and make a meaningful contribution to a circular economy, where waste is transformed into a valuable asset and sustainability is the standard. Conventional waste disposal methods such as landfilling and incineration have disadvantages, emphasising the necessity for a more environmentally friendly approach. Recycling, a fundamental aspect of the circular economy, presents an avenue for waste reduction and converting discarded materials into valuable resources. Improving recycling rates and infrastructure can establish a self-contained system for managing MSW, promoting environmental advantages, conserving resources, and stimulating economic development. Efficient recycling initiatives are essential for optimising the retrieval of resources. The accessibility, convenience, and user-friendliness of these programmes should be prioritised for residents. Implementing single-stream recycling, which allows for collecting all recyclable materials in a single bin, can enhance participation rates compared to multi-stream programmes that necessitate sorting based on material type. Implementing unambiguous and uniform labelling for recyclable materials and comprehensive public education initiatives enables residents to segregate their waste for proper recycling effectively. Implementing automated sorting technologies within recycling facilities can enhance operational efficiency and mitigate the risk of contamination from recyclable materials. An effective waste management practice relies on a strong recycling infrastructure encompassing collection facility, sorting plants, and reprocessing facilities. The implementation of contemporary recycling facilities equipped with advanced sorting technologies has the potential to enhance the overall quality of recycled materials, thereby increasing their appeal to manufacturers. Ensuring equitable access to recycling programmes and promoting inclusivity can be achieved by expanding collection services to underserved communities. Establishing effective transportation networks for storing and conveying recyclable materials to processing facilities is paramount in facilitating a streamlined recycling system. The economic sustainability of recycling programmes relies heavily on a robust market demand for recycled materials. Policy instruments such as extended producer responsibility (EPR) schemes, which impose responsibility on manufacturers for managing their products at the end of their life cycle, can encourage incorporating recycled content in producing new goods. Implementing government procurement policies that emphasise using recycled materials in public infrastructure projects can establish a consistent demand and foster the growth of the recycled materials market. Establishing strong markets for recycled materials necessitates collaboration among government agencies, manufacturers, and recycling companies. Technological progress presents auspicious prospects for augmenting the efficiency of recycling and the capabilities of processing. Implementing advanced sorting technologies, such as optical sorting and artificial intelligence, can potentially enhance the precision and effectiveness of material separation within the recycling process. Chemical recycling technologies can decompose intricate materials, like plastics, into their fundamental chemical components, resulting in the production of recycled materials of high quality. The potential for expanding the range of recyclable materials and enhancing the quality of recycled content is significant through investments in research and development of innovative recycling technologies. The behaviour of individual consumers heavily influences the success of recycling programmes. Public education campaigns that aim to raise awareness about the significance of recycling and the adoption of appropriate waste segregation practices have the potential to empower individuals to engage in resource recovery endeavours actively (Tsai et al. 2020; Allevi et al. 2021; Bakalár et al. 2021; Magazzino and Falcone 2022; Vines et al. 2023; Di Foggia and Beccarello 2023; Ersahin et al. 2023; Chenavaz and Dimitrov 2024; Gronba-Chyła et al. 2024; Munir et al. 2024; Pajura 2024). The cultivation of consumer consciousness regarding the ecological advantages of recycling and its contribution to the circular economy has the potential to cultivate a sense of accountability and promote conscientious waste management behaviours. Promoting environmentally friendly product design, prioritisation of recyclability, and the utilisation of recycled content can enhance consumer endeavours in waste reduction and contribute to the establishment of a circular economy. Implementing a strong recycling system poses specific difficulties. Fluctuations in the market price of virgin materials can occasionally diminish the competitiveness of recycled materials. The presence of non-recyclable materials in recyclables can substantially affect the quality and economic feasibility of the recycled materials. The lack of sufficient consumer awareness and knowledge about appropriate waste segregation can impede engagement in recycling initiatives. Notwithstanding these challenges, the potential benefits linked to improved recycling are significant. Recycling preserves precious resources, such as timber, metals, and plastics, by decreasing dependence on the extraction of new materials and minimising environmental harm. Recycling decreases the amount of greenhouse gas emissions linked to the production of new materials and aids in the efforts to mitigate climate change. A flourishing recycling sector generates employment opportunities in facilities dedicated to waste collection, sorting, and reprocessing, promoting economic expansion and advancement. By giving precedence to recycling and improving associated infrastructure, it is possible to transition the management of MSW from a linear "take-make-dispose" approach to a circular system emphasising resource recovery and reducing waste generation. This transition necessitates a comprehensive strategy that includes optimising programmes, developing infrastructure, creating markets, advancing technology, and changing consumer behaviour. Through collaboration, governmental bodies, enterprises, and individuals can harness the vast possibilities of recycling, thereby fostering a more sustainable trajectory for waste management and the efficient utilisation of resources. In municipal solid waste management, the circular economy framework encompasses more than waste reduction and resource recovery through recycling. Some waste cannot be easily recycled due to technological or economic constraints. This residual waste stream offers an avenue for additional resource extraction using advanced waste treatment technologies. The extraction of energy and valuable resources from this waste has the potential to reduce dependence on landfills and incineration, thereby promoting a future that is more sustainable and efficient in terms of resource utilisation. Waste-to-energy (WtE) technologies can harness energy from non-recyclable residual waste. Incineration is a crucial technology that involves burning waste in a regulated setting to produce heat, which can be utilised to generate electricity or steam for district heating. Advanced waste-to-energy (WtE) facilities employ advanced pollution control technologies to mitigate air emissions and reduce their environmental footprint. The prioritisation of waste reduction and recycling efforts is of utmost importance to minimise the quantity of waste directed towards incineration, despite its potential for energy recovery. Anaerobic digestion is a biological phenomenon characterised by decomposing organic waste in an environment lacking oxygen. The process yields biogas, a sustainable fuel source that can be utilised for generating electricity, fuelling transportation, or producing heat. Anaerobic digestion presents a significant avenue for redirecting organic waste away from landfills, where it undergoes anaerobic decomposition and produces methane, a highly potent greenhouse gas. Furthermore, the digestate, a residual product of the process, can enhance soil quality. Machine biological treatment (MBT) is a sequential procedure that effectively segregates waste materials into distinct fractions. The process of sorting and diverting recyclable materials for recycling is undertaken, whereas organic matter is subjected to composting or anaerobic digestion. The remaining residual fraction can be utilised for energy recovery via incineration or as refuse-derived fuel (RDF). MBT provides a flexible method for optimising the retrieval of resources from remaining waste streams. Gasification is a thermochemical procedure transforming waste materials into synthetic gas, commonly called syngas, within a regulated setting by subjecting them to elevated temperatures. Syngas possesses versatile applications, encompassing the generation of electricity, the production of transportation fuels, and the utilisation as a feedstock for chemical production. Plasma gasification employs elevated temperatures to decompose waste into pure syngas, presenting a potentially more environmentally friendly option than conventional incineration methods (Rada and Cioca 2017; Taleb and Al Farooque 2021; Yadav et al. 2022; Derdera and Ogato 2023; Liu et al. 2023; Kurniawan et al. 2024; Zhang et al. 2024; Lederer and Schuch 2024; Papamichael et al. 2024; Poyyamozhi et al. 2024; Sikder et al. 2024). Advanced thermal technologies can extract valuable resources and energy from complex waste streams. Determining the most effective strategy for extracting value from residual waste is contingent upon various factors, including the composition of the waste stream, the existing local infrastructure, and the energy requirements. The integration of various technologies has the potential to optimise resource recovery and mitigate environmental consequences. Thorough assessment and strategic deliberation are imperative in choosing the most suitable technologies for a given context. Although advanced waste treatment technologies show potential, some obstacles must be addressed. Implementing these technologies may be impeded by the substantial capital costs they entail. Transparent communication and stringent regulations must address public concerns about potential air emissions and environmental impacts. Moreover, the efficacy of these technologies is contingent upon the implementation of strong waste segregation practices at the point of origin to mitigate contamination and enhance the efficiency of treatment processes. Notwithstanding these challenges, the prospects linked to waste-to-resource recovery are substantial. The reduction of reliance on landfills and incineration, as well as the mitigation of environmental damage and greenhouse gas emissions, can be achieved through the extraction of energy and valuable resources from residual waste. Moreover, implementing these technologies has the potential to enhance energy security through diversifying renewable energy sources and mitigating reliance on fossil fuels. The extraction of energy and valuable resources from residual waste is paramount in managing municipal solid waste within a circular economy. Advanced waste treatment technologies provide novel approaches to reduce waste disposal activities and promote the efficient use of resources. Through meticulous selection and seamless integration of these technologies, in conjunction with comprehensive waste reduction and recycling initiatives, it is possible to establish a more sustainable trajectory for waste management and the efficient utilisation of resources. Effective waste-to-resource recovery strategies require collaboration among governments, waste management companies, and research institutions. This collaboration is crucial for transforming waste into valuable resources and promoting sustainability as the standard (Fig. 32.4).

32.5 Strategies and Initiatives for Optimising MSW Within the Circular Economy

Implementing a circular economy model for managing MSW requires a comprehensive strategy that includes policy revisions, technological progress, public education initiatives, and community involvement. Through the cultivation of cooperation among governments, businesses, and citizens, we can establish a framework that reduces the production of waste, enhances the retrieval of resources, and encourages the adoption of responsible waste management methods. Implementing efficient policy instruments and regulatory frameworks is pivotal in facilitating the shift towards a circular economy in managing MSW. EPR schemes impose responsibility on manufacturers to manage their products at the end of their life cycle. Manufacturers are motivated to develop products that possess enhanced

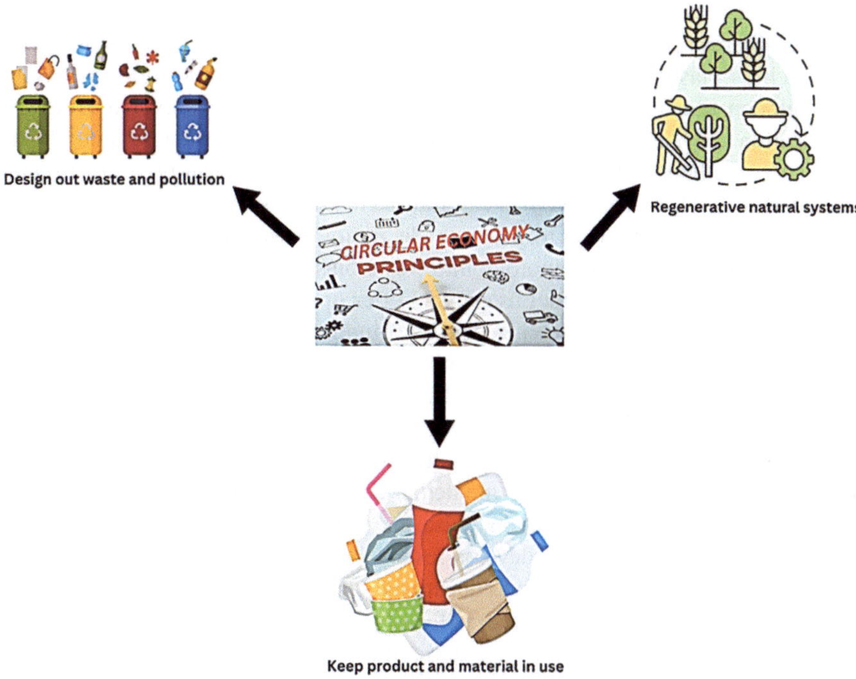

Fig. 32.4 Circular economy principles applied to MSW management. (Source: created by authors)

repairability, reusability, or recyclability, thereby fostering eco-design principles and mitigating waste generation at its origin. By establishing ambitious waste reduction objectives and enforcing prohibitions on the use of disposable plastics or the disposal of recyclable materials in landfills, a distinct trajectory for waste management practices can be established, encouraging businesses and consumers to adopt more sustainable alternatives. Implementing standardised recycling symbols and collection procedures across various regions can greatly enhance the effectiveness and engagement of recycling initiatives. Furthermore, providing explicit information regarding product lifespans, recycling details, and composting guidelines enables consumers to make well-informed decisions and appropriately manage their waste materials. Policy measures encompass a range of strategies, such as the provision of tax incentives for repair services, the allocation of subsidies to businesses that adopt refill and reuse programmes, and the imposition of financial penalties for failure to adhere to recycling regulations. These initiatives promote responsible waste management strategies and establish equitable conditions for sustainable enterprises. Technological progress is crucial in enhancing the efficiency and effectiveness of waste management across the entire value chain of the circular economy. Optical sorting employs lasers and cameras to identify and segregate various waste stream materials accurately. Artificial intelligence (AI) can potentially improve sorting efficiency through the analysis of data and the real-time adaptation of sorting processes. Traditional recycling primarily emphasises mechanical processing, whereas chemical recycling presents viable alternatives for intricate materials such as plastics. These technologies facilitate the decomposition of plastics into their constituent chemical elements, producing high-quality recycled materials. Gasification and plasma gasification are advanced thermal technologies that provide innovative methods for extracting valuable resources and energy from difficult waste streams. These technologies can transform waste into syngas, a sustainable fuel source that can be used to generate electricity or run vehicles. Incorporating sensors and cutting-edge technology into waste bins can enhance the efficiency of collection routes, minimise fuel usage, and offer up-to-date information regarding waste generation patterns (Allevi et al. 2021; Munir et al. 2021, 2024; Awasthi et al. 2022; Magazzino and Falcone 2022; Ambaye et al. 2023; Ersahin et al. 2023; Vines et al. 2023). The utilisation of this data has the potential to enhance waste management planning and optimise resource allocation. Promoting public awareness and community engagement is crucial for cultivating a cultural transformation towards a circular economy in managing municipal solid waste (MSW). From an early stage, incorporating waste reduction, reuse, and recycling education into school curricula can empower children to develop responsible consumer behaviour and waste management skills. Adult-focused educational campaigns can enhance awareness regarding the ecological consequences of waste and encourage adopting sustainable consumption practices. To cultivate a sense of ownership and promote engagement in responsible waste management practices, arranging community workshops, clean-up drives, and recycling competitions is advisable. Collaborating with community organisations can effectively utilise pre-existing networks and expand the scope of one's audience reach. By leveraging social media platforms, celebrity endorsements, and targeted messaging campaigns, it is possible to enhance public knowledge regarding circular economy principles and foster a shift towards sustainable behaviour. Emphasising the favourable environmental and economic consequences of responsible waste management has the potential to resonate with individuals and serve as a catalyst for their adoption of circular practices. Governments and waste management companies must communicate transparently with the public regarding waste management practices and infrastructure development. Public forums and citizen advisory committees facilitate community feedback and collaboration opportunities in shaping sustainable waste management solutions. The achievement of optimal management of municipal solid waste (MSW) within the context of the circular economy requires a comprehensive strategy that integrates efficient policy interventions, advanced technological solutions, and proactive involvement of the public. Through cultivating cooperation and encouraging originality, we can establish a forthcoming era in which waste is reduced, resources are retrieved and repurposed, and conscientious waste management practices become the standard. The concept of the circular economy presents a viable approach to achieving a more sustainable future in waste management. It aims to reduce environmental harm, preserve resources, and foster a prosperous circular economy.

32.6 Benefits of Integrating MSW into the Circular Economy

The conventional linear approach to municipal solid waste management, commonly called the "take-make-dispose" model, is unsustainable and approaching a critical threshold. Landfills are approaching their maximum capacity, and incineration, although it provides some energy recovery, has substantial environmental disadvantages. Incorporating MSW into a circular economy is not merely advantageous but imperative, as it enables the realisation of numerous advantages for our environment, economy, and resource security. An essential principle of the circular economy is to maximise the longevity of resources. The prioritisation of waste reduction, reuse, and recycling within the municipal solid waste management system can substantially decrease our dependence on virgin materials. The circular economy principles promote developing and producing durable products that can be repaired. This prolongs the lifespan of products, thereby decreasing the necessity for frequent product replacement and minimising the amount of waste produced. The process of recycling involves the conversion of waste materials into valuable resources that can be utilised in the production of new goods. Implementing a closed-loop system facilitates the continuous recycling of materials, thereby reducing the need to extract new resources. Composting is a viable method for diverting organic waste, constituting a substantial proportion of municipal solid waste. Applying compost enhances soil fertility, diminishing the need for chemical fertilisers and fostering the adoption of sustainable agricultural methods. However, it is essential to note that specific municipal solid waste (MSW) will invariably necessitate additional processing. The circular economy concept presents novel approaches to harnessing energy and retrieving valuable resources from the remaining waste stream. Incineration and gasification are technologies that can transform waste into either electricity or heat. Although not flawless, it provides a means for extracting energy from non-recyclable waste. The anaerobic digestion process involves the decomposition of organic waste, resulting in biogas production (Tsai et al. 2020; Allevi et al. 2021; Bakalár et al. 2021; Tong et al. 2021; Vines et al. 2023; Di Foggia and Beccarello 2023; Ersahin et al. 2023; Chenavaz and Dimitrov 2024; Munir et al. 2024; Paes et al. 2024; Saravanakumar et al. 2024). This renewable fuel source has utility in various applications such as electricity generation, transportation fuels, and heat production. Diverting organic waste from landfills contributes to a more environmentally friendly energy mix. Using waste as an energy source can reduce reliance on conventional fossil fuels, reduce greenhouse gas emissions, and promote a shift towards a more environmentally friendly future. The transition towards a circular economy in managing MSW offers promising economic prospects and the potential for job creation. The acquisition and application of novel recycling technologies, composting facilities, and waste-to-resource recovery plants necessitate the expertise of proficient individuals, thereby promoting the generation of employment opportunities within the green economy domain. The emphasis on product repair and reuse enhances the repair and second-hand goods industry, generating employment opportunities for repair technicians, product refurbishment specialists, and second-hand retail. Ongoing research and development in waste management technologies, material science, and product design are imperative for the successful transition towards a circular economy. This facilitates the cultivation of innovative ideas and presents prospects for enterprises and academic institutions to devise original approaches for managing sustainable waste. Incorporating municipal solid waste into the circular economy presents a holistic strategy for waste management, yielding advantages in terms of the environment, economy, and society (Table 32.3). To establish a more sustainable future for waste management and resource utilisation, it is crucial to prioritise resource conservation, investigate energy recovery alternatives, and promote the development of a green job market. The circular economy offers a mutually beneficial situation for our planet and its inhabitants, facilitating a future in which waste is transformed into a valuable resource, and sustainability is the standard (Fig. 32.5).

32.7 Challenges and Opportunities

The conventional linear approach to municipal solid waste management, commonly called the "take-make-dispose" model, is unsustainable and approaching a critical threshold. Landfills are approaching their maximum capacity, and incineration, although it provides some energy recovery, has substantial environmental disadvantages. Incorporating municipal solid waste into a circular economy is not merely advantageous but imperative, as it enables the realisation of numerous advantages for our environment, economy, and resource security. An essential principle of the circular economy is to maximise the longevity of resources. The prioritisation of waste reduction, reuse, and recycling within the municipal solid waste management system can substantially decrease our dependence on virgin materials. The circular economy principles promote developing and producing durable products that can be repaired. This prolongs the lifespan of products, thereby decreasing the necessity for frequent product replacement and minimising the amount of waste produced. The process of recycling involves the conversion of waste materials into valuable resources that can be utilised in the production of new goods. Implementing a closed-loop system facilitates the continuous recycling of materials, thereby reducing the need to extract new resources. Composting is a viable method for diverting organic waste, constituting a substantial proportion of MSW. Applying compost enhances soil fertility, diminishing the need for chemical fertilisers and fostering the adoption of sustainable agricultural methods. However, it is essential to note that specific MSW will invariably necessitate additional processing. The circular economy concept presents novel approaches to harnessing energy and retrieving valuable resources from the remaining waste stream. Incineration and gasification are technologies that can transform waste into either electricity or heat. Although not flawless, it provides a means for extracting energy from non-recyclable waste. The anaerobic digestion process involves the decomposition of organic waste, resulting in biogas production. This renewable fuel source has utility in various applications such as electricity generation, transportation fuels, and heat production. Diverting organic waste from landfills contributes to a more environmentally friendly energy

Table 32.3 Unlocking the value: benefits of including municipal solid waste in the circular economy

Strategy/initiative	Description
1. Waste prevention	– Emphasise waste prevention strategies such as reducing packaging, promoting reusable products, and minimising single-use items to prevent waste generation at the source
	– Encourage manufacturers and retailers to adopt eco-design principles, product life extension strategies, and sustainable consumption practices to minimise waste generation throughout the product's lifecycle
2. Source segregation	– Promote source segregation of waste at households, businesses, and institutions to separate recyclable, organic, and non-recyclable materials at the point of generation
	– Implement educational campaigns and provide infrastructure such as separate bins or collection points to facilitate convenient and effective source segregation practices
3. Circular design and product innovation	– Foster circular design principles in product development to promote material efficiency, durability, reparability, and recyclability
	– Encourage innovative product designs that enable easy disassembly, component reuse, and material recovery to facilitate circularity and reduce waste generation
4. Extended producer responsibility (EPR)	– Implement EPR schemes that hold producers responsible for managing the end-of-life of their products, encouraging product design for recyclability and incentivising collection and recycling infrastructure development
	– Require producers to take back and recycle their products, promote eco-design, and contribute to funding recycling programs and facilities
5. Recycling and material recovery	– Invest in recycling infrastructure and technology to increase the collection, sorting, and processing of recyclable materials such as paper, plastics, glass, and metals
	– Support the development of markets for recycled materials to stimulate demand and create economic incentives for recycling
6. Organic waste management	– Promote composting and anaerobic digestion facilities for waste management, converting organic matter into valuable compost or biogas for energy generation and soil enrichment
	– Encourage backyard composting, community composting programs, and decentralised organic waste processing facilities to divert organic waste from landfills and reduce methane emissions
7. Waste-to-energy (WTE)	– Invest in waste-to-energy facilities that use advanced technologies to convert non-recyclable waste into heat, electricity, or fuel, reducing the volume of waste sent to landfills and recovering energy from waste
	– Implement strict emission controls and environmental safeguards to minimise air pollution and ensure compliance with regulatory standards

(continued)

Table 32.3 (continued)

Strategy/initiative	Description
8. Product reuse and remanufacturing	– Promote product reuse and remanufacturing initiatives to extend the lifespan of products, reduce resource consumption, and minimise waste generation
	– Facilitate product refurbishment, repair services, and resale markets to encourage consumers to choose refurbished or pre-owned items instead of purchasing new products
9. Waste exchange platforms	– Establish waste exchange platforms or industrial symbiosis networks that connect businesses to exchange waste materials, by-products, or resources for reuse or recycling
	– Facilitate collaboration and partnerships among industries to identify opportunities for waste valorisation, resource sharing, and circular economy initiatives
10. Digital technologies for waste management	– Utilise digital technologies such as IoT sensors, smart bins, and data analytics to optimise waste collection routes, monitor bin fill levels, and improve operational efficiency in waste management processes
	– Implement waste management software and platforms for real-time waste stream tracking, reporting, and analysis to inform decision-making and resource allocation
11. Public awareness and behavioural change	– Conduct public awareness campaigns, educational programs, and community engagement initiatives to raise awareness about circular economy principles, waste reduction, and sustainable consumption practices
	– Promote behaviour change through incentives, rewards programs, and social marketing strategies that encourage individuals and businesses to adopt environmentally friendly habits and support circular economy initiatives
12. Policy support and regulatory framework	– Establish supportive policy frameworks, regulations, and incentives that promote circular economy principles, sustainable waste management practices, and resource efficiency
	– Implement waste management targets, landfill bans for recyclable materials, eco-labelling schemes, and tax incentives to encourage investment in circular economy solutions and foster a conducive environment for innovation and sustainability

mix. Using waste as an energy source can reduce reliance on conventional fossil fuels, reduce greenhouse gas emissions, and promote a shift towards a more environmentally friendly future. The transition towards a circular economy in managing MSW offers promising economic prospects and the potential for job creation. The acquisition and application of novel recycling technologies, composting facilities, and waste-to-resource recovery plants necessitate the expertise of proficient individuals, thereby promoting the generation of employment opportunities within the green economy domain. The emphasis on product repair and reuse enhances the repair and second-hand goods industry, generating employment opportunities for repair technicians, product refurbishment specialists, and second-hand retail. Ongoing research and development in waste management technologies, material science, and product design are imperative for the successful transition towards a

Fig. 32.5 Benefits of integrating MSW into the circular economy. (Source: created by authors)

circular economy. This facilitates the cultivation of innovative ideas and presents prospects for enterprises and academic institutions to devise original approaches for managing sustainable waste. Incorporating MSW into the circular economy presents a holistic strategy for waste management, yielding advantages in terms of the environment, economy, and society. To establish a more sustainable future for waste management and resource utilisation, it is crucial to prioritise resource conservation, investigate energy recovery alternatives, and promote the development of a green job market. The circular economy offers a mutually beneficial situation for our planet and its inhabitants, facilitating a future in which waste is transformed into a valuable resource and sustainability is the standard.

32.8 Conclusion

This investigation into implementing a circular economy in managing MSW has unveiled numerous advantages for the environment, economy, and resource security. To establish a closed-loop system that minimises waste generation, conserves resources, and encourages responsible waste management practices, we can focus on waste reduction, promote reuse and recycling, and implement innovative resource recovery technologies. The conventional perspective on municipal solid waste which regards it as a liability to be disposed of, is no longer viable. The circular economy presents a fundamental change in perspective, redefining MSW as a valuable asset that can be converted into novel products, produce sustainable energy, and contribute to a prosperous, environmentally friendly economy. The alteration in viewpoint holds significant importance in fostering innovation and advocating for conscientious waste management practices across society. Implementing a comprehensive waste management system necessitates a collective endeavour involving multiple stakeholders. Authorities can set ambitious goals for reducing waste, employ efficient policy tools such as extended producer responsibility, and allocate resources towards developing waste sorting, recycling, and resource recovery facilities. It is crucial to promote eco-design principles and encourage innovation in waste management. Businesses have the potential to significantly contribute by developing products that are durable, repairable, and recyclable. Demonstrating commitment to the circular economy can be achieved through investments in take-back programmes, refill and reuse systems, and sustainable packaging solutions. Another significant contribution is the provision of support for research and development in waste management technologies. Community involvement is crucial for cultivating a culture of waste minimisation and conscientious waste administration. The active engagement of individuals in the circular economy can be facilitated through implementing educational programmes, clean-up drives, and recycling initiatives. Promoting sustainable practices within local businesses conveys a compelling message. Each person can make a meaningful contribution to a circular economy. Engaging in informed decision-making regarding acquiring durable and repairable products, actively engaging in recycling initiatives, and adopting a more conscientious consumption approach are all pivotal measures in mitigating waste production. By implementing a cooperative strategy and acknowledging the significant capacity of MSW as a valuable asset, we can unleash the profound impact of the circular economy. It is conceivable to envision a prospective scenario wherein waste is reduced, resources are preserved, and a circular economy flourishes.

References

Abad V et al (2019) Promoting circular economy in the surroundings of an organic fraction of municipal solid waste anaerobic digestion treatment plant: biogas production impact and economic factors. Bioresour Technol 283:10–17. https://doi.org/10.1016/j.biortech.2019.03.064

Ahmad W et al (2024) Socio-economic benefits and policy implications of generating sustainable energy from municipal solid waste in Pakistan. Energy Clim Change 5:100124. https://doi.org/10.1016/j.egycc.2023.100124

Allevi E et al (2021) Municipal solid waste management in circular economy: a sequential optimization model. Energy Econ 100:105383. https://doi.org/10.1016/j.eneco.2021.105383

Ambaye TG et al (2023) Emerging technologies and sustainable strategies for municipal solid waste valorization: challenges of circular economy implementation. J Clean Prod 423:138708. https://doi.org/10.1016/j.jclepro.2023.138708

Awasthi SK et al (2022) Processing of municipal solid waste resources for a circular economy in China: an overview. Fuel 317:123478. https://doi.org/10.1016/j.fuel.2022.123478

Bakalár T et al (2021) Metal recovery from municipal solid waste incineration fly ash as a tool of circular economy. J Clean Prod 302:126977. https://doi.org/10.1016/j.jclepro.2021.126977

Bertanza G et al (2021) Implementation of circular economy in the management of municipal solid waste in an Italian medium-sized city: a 30-years lasting history. Waste Manag 126:821–831. https://doi.org/10.1016/j.wasman.2021.04.017

Chen C-Y, Yu T-T (2023) Towards a circular economy: recapturing battery, metal, and plastic from soil-size and gravel-size municipal solid waste incineration bottom ash using convolutional neural networks. J Clean Prod 432:139737. https://doi.org/10.1016/j.jclepro.2023.139737

Chenavaz RY, Dimitrov S (2024) From waste to wealth: policies to promote the circular economy. J Clean Prod 2024:141086. https://doi.org/10.1016/j.jclepro.2024.141086

Derdera SE, Ogato GS (2023) Towards integrated, and sustainable municipal solid waste management system in Shashemane city administration, Ethiopia. Heliyon 9(11):e21865. https://doi.org/10.1016/j.heliyon.2023.e21865

Di Foggia G, Beccarello M (2023) Designing circular economy-compliant municipal solid waste management charging schemes. Util Policy 81:101506. https://doi.org/10.1016/j.jup.2023.101506

Ersahin ME et al (2023) Nutrient recovery from municipal solid waste leachate in the scope of circular economy: recent developments and future perspectives. J Environ Manag 335:117518. https://doi.org/10.1016/j.jenvman.2023.117518

Gómez Montoya JP et al (2024) A thermo-economic analysis of a circular economy model for biomass in South America producing biofertilizers and biogas from municipal solid waste. Renew Energy 225:120254. https://doi.org/10.1016/j.renene.2024.120254

Gronba-Chyła A et al (2024) Municipal waste utilization as a substitute for natural aggregate in the light of the circular economy. J Clean Prod 440:140907. https://doi.org/10.1016/j.jclepro.2024.140907

He Y et al (2024) Driving sustainable circular economy in electronics: a comprehensive review on environmental life cycle assessment of e-waste recycling. Environ Pollut 342:123081. https://doi.org/10.1016/j.envpol.2023.123081

Hemali NA et al (2024) A method for determining the recycling value of unprocessed municipal solid waste in one cubic meter waste composition analysis technique. Methods X 12:102626. https://doi.org/10.1016/j.mex.2024.102626

Hoang AT et al (2022) Perspective review on municipal solid waste-to-energy route: characteristics, management strategy, and role in circular economy. J Clean Prod 359:131897. https://doi.org/10.1016/j.jclepro.2022.131897

Kumareswaran K et al (2024) Systematic review on liquid organic waste (LOW) characteristics, processing technologies, and their potential applications: towards circular economy and resource efficiency. J Clean Prod 447:141286. https://doi.org/10.1016/j.jclepro.2024.141286

Kurniawan TA et al (2024) Unlocking synergies between waste management and climate change mitigation to accelerate decarbonization through circular-economy digitalization in Indonesia. Sustainable Prod Consump 46:522–542. https://doi.org/10.1016/j.spc.2024.03.011

Lederer J, Schuch D (2024) The contribution of waste and bottom ash treatment to the circular economy of metal packaging: a case study from Austria. Resour Conserv Recycl 203:107461. https://doi.org/10.1016/j.resconrec.2024.107461

Liu Z et al (2023) Unlocking system transitions for municipal solid waste infrastructure: a model for mapping interdependencies in a local context. Resour Conserv Recycl 198:107180. https://doi.org/10.1016/j.resconrec.2023.107180

Magazzino C, Falcone PM (2022) Assessing the relationship among waste generation, wealth, and GHG emissions in Switzerland: some policy proposals for the optimization of the municipal solid waste in a circular economy perspective. J Clean Prod 351:131555. https://doi.org/10.1016/j.jclepro.2022.131555

Marques AC, Teixeira NM (2022) Assessment of municipal waste in a circular economy: do European Union countries share identical performance? Cleaner Waste Syst 3:100034. https://doi.org/10.1016/j.clwas.2022.100034

Mazzanti M, Mazzarano M, Zecca E (2022) A selection bias approach in the circular economy context: the case of organic municipal solid waste in Italy. J Clean Prod 348:131266. https://doi.org/10.1016/j.jclepro.2022.131266

Munir MT et al (2021) Municipal solid waste-to-energy processing for a circular economy in New Zealand. Renew Sust Energ Rev 145:111080. https://doi.org/10.1016/j.rser.2021.111080
Munir MT et al (2024) Green loops and clean skies: optimizing municipal solid waste management using data science for a circular economy. Environ Res 243:117786. https://doi.org/10.1016/j.envres.2023.117786
Nguyen THT et al (2024) Toward a circular economy for better municipal solid waste management in can Gio district, Ho Chi Minh city. Sustain Chem Pharm 38:101480. https://doi.org/10.1016/j.scp.2024.101480
Paes MX et al (2024) Waste management intervention to boost circular economy and mitigate climate change in cities of developing countries: the case of Brazil. Habitat Int 143:102990. https://doi.org/10.1016/j.habitatint.2023.102990
Pajura R (2024) Composting municipal solid waste and animal manure in response to the current fertilizer crisis - a recent review. Sci Total Environ 912:169221. https://doi.org/10.1016/j.scitotenv.2023.169221
Papamichael I et al (2024) Mobilisation of textile waste to recover high added value products and energy for the transition to circular economy. Environ Res 242:117716. https://doi.org/10.1016/j.envres.2023.117716
Pomoni DI et al (2024) Circular economy: a multilevel approach for natural resources and wastes under an Agri-food perspective. Water-Energy Nexus 7:103–123. https://doi.org/10.1016/j.wen.2023.12.003
Poyyamozhi M et al (2024) Elevating thermal comfort with eco-friendly concrete roof tiles crafted from municipal solid waste. J Build Eng 2024:109222. https://doi.org/10.1016/j.jobe.2024.109222
Rada EC, Cioca L (2017) Optimizing the methodology of characterization of municipal solid waste in EU under a circular economy perspective. Energy Procedia 119:72–85. https://doi.org/10.1016/j.egypro.2017.07.050
Rebehy PCPW et al (2023) Municipal solid waste management (MSWM) in Brazil: drivers and best practices towards to circular economy based on European Union and BSI. J Clean Prod 401:136591. https://doi.org/10.1016/j.jclepro.2023.136591
Rondón Toro E, Lobo A, Gallardo Izquierdo A (2022) Circularity indicator for municipal solid waste treatment plants. J Clean Prod 380:134807. https://doi.org/10.1016/j.jclepro.2022.134807
Saluja S et al (2024) Stabilized municipal solid waste as an alternative to natural sand in paver block construction. Process Saf Environ Prot 182:1035–1046. https://doi.org/10.1016/j.psep.2023.12.032
Sandhi A, Rosenlund J (2024) Municipal solid waste management in Scandinavia and key factors for improved waste segregation: a review. Cleaner Waste Syst 8:100144. https://doi.org/10.1016/j.clwas.2024.100144
Saravanakumar A et al (2024) Green circular economy of co-gasification with municipal solid waste and wood waste in a novel downdraft gasifier with rotating grate. Chem Eng J 479:147987. https://doi.org/10.1016/j.cej.2023.147987
Sikder M et al (2024) Green logistics and circular economy in alleviating CO2 emissions: does waste generation and GDP growth matter in EU countries? J Clean Prod 449:141708. https://doi.org/10.1016/j.jclepro.2024.141708
Singh M, Singh M, Singh SK (2024) Tackling municipal solid waste crisis in India: insights into cutting-edge technologies and risk assessment. Sci Total Environ 917:170453. https://doi.org/10.1016/j.scitotenv.2024.170453
Taleb MA; Al Farooque O (2021) Towards a circular economy for sustainable development: an application of full cost accounting to municipal waste recyclables. J Clean Prod 280:124047. https://doi.org/10.1016/j.jclepro.2020.124047
Taouahria B (2024) Predicting citizens municipal solid waste recycling intentions in Morocco: the role of community engagement. Waste Manage Bull 2(1):316–326. https://doi.org/10.1016/j.wmb.2024.02.008

Tong X, Yu H, Liu T (2021) Using weighted entropy to measure the recyclability of municipal solid waste in China: exploring the geographical disparity for circular economy. J Clean Prod 312:127719. https://doi.org/10.1016/j.jclepro.2021.127719

Tsai FM et al (2020) Municipal solid waste management in a circular economy: a data-driven bibliometric analysis. J Clean Prod 275:124132. https://doi.org/10.1016/j.jclepro.2020.124132

Vines V et al (2023) Understanding the trade-offs of national municipal solid waste estimation methods for circular economy policy. J Clean Prod 412:137349. https://doi.org/10.1016/j.jclepro.2023.137349

Yaashikaa PR et al (2020) Bioconversion of municipal solid waste into bio-based products: a review on valorisation and sustainable approach for circular bioeconomy. Sci Total Environ 748:141312. https://doi.org/10.1016/j.scitotenv.2020.141312

Yadav P et al (2022) Barriers in biogas production from the organic fraction of municipal solid waste: a circular bioeconomy perspective. Bioresour Technol 362:127671. https://doi.org/10.1016/j.biortech.2022.127671

Yong ZJ, Bashir MJK, Hassan MS (2021) Biogas and biofertilizer production from organic fraction municipal solid waste for sustainable circular economy and environmental protection in Malaysia. Sci Total Environ 776:145961. https://doi.org/10.1016/j.scitotenv.2021.145961

Zhang J et al (2024) Life cycle assessment of kerbside waste material for an open-looped and closed-loop production–towards circular economy designs. J Clean Prod 434:139991. https://doi.org/10.1016/j.jclepro.2023.139991

Zhao J et al (2024) Scenario analysis of the eco-efficiency for municipal solid waste management: a case study of 211 cities in western China. Sci Total Environ 2024:170536. https://doi.org/10.1016/j.scitotenv.2024.170536

Contribution of Agricultural Solid Waste (ASW) for Circular Bioeconomy

33

Anwesha Khanra

Abstract

Resolving global difficulties exacerbated by environmental pollution can be achieved through the implementation of a circular bio-economy. Producing bioenergy and biomaterials from waste can enhance the energy-environment relationship and create replacements for petroleum-based materials in feedstock production. All of these processes are able to develop a safe and low carbon footprint arena. The benefits of reutilization of agricultural solid waste (ASW) comprise abridged GHG emission and contribute towards the development of its bio-conversion to livestock. Circular bio-economy is considered as a noteworthy method for determining the worldwide problems raised by environmental pollution. The value-added products, those are derived by the conversion of ASW, remain in the supply chain and increase an entire presentation of agricultural waste management program.

Keywords

Agricultural solid waste · Circular economy · Biofuels · Biochar system and agriculture · Technological feasibility

33.1 Introduction

The development of biofuels and bioenergies has evolved to be less dependent of fossil fuel generation (Zhu et al. 2024; Filippov 2024). The manufacture of ASW has augmented owing to the rise of massive urbanization and industrialization

A. Khanra (✉)
Department of Biotechnology, Parul Institute of Technology, Parul University, Vadodara, Gujarat, India
e-mail: anwesha.khanra35689@paruluniversity.ac.in

A. Kumari et al. (eds.), *Valorization of Solid Wastes to Biofuels and Chemical Products for Sustainable World*, https://doi.org/10.1007/978-981-96-8594-3_33

(Voukkali et al. 2024; Ghosh et al. 2024). The waste derived from agricultural resources is framed as a renewable energy source which gains its popularity (Li et al. 2024; Liang et al. 2024). There are some areas that have shaped structures for actual nourishment management system meanwhile both industrialized and developing nations are harassed to achieve the enormous quantities of nutrition waste. The extreme manufacture of ASW and the robust need for a greener environment have an important influence on investigation exertions in the cohort of bioenergy production from ASW recently (Dubey et al. 2024; Singh et al. 2024). The residual portion or by-products fashioned by diverse farming activities are denoted as agricultural solid waste (ASW) which includes various crop residues, animal compost, food processing wastes, and other scraps (Ahmed et al. 2024; Wang et al. 2024). ASW management is critical owing to its financial value and impending potential to create the environment by using various techniques like composting, bioenergy production, and as raw materials for several industrial applications (Odejobi et al. 2024; Khan et al. 2024; Mehta et al. 2024). An efficient organization policy is indispensable to diminish ecological contamination and enhance the potential for changing wastes towards resource recovery and sustained agricultural strategies (Klotz et al. 2024). The agrarian segment is one of the areas that produce enormous ASW (Yang et al. 2024), and hence, there is a proper need to manage these kinds of wastes. By utilizing all the ASW, the circular bioeconomy as well as global health can be maintained (Jangde 2024). The reducing GHG emission and fossil fuel consumption are considered as the two assistances of reprocessing the ASW. Till now, people are dependent of commercial sources of fertilizers which include all the chemical pesticides, herbicides, and insecticides. An entire prohibition on the usage of insecticide estimates to decrease the worldwide food production by means of 42%. The atmosphere, environment, and human health system are extremely injured by indecorous treatment of ASW, thus everyone must work in an organized way to resolve this serious problem. In developing countries, agricultural solid waste is frequently disposed of irresponsibly in public spaces, resulting in dust, smoke, harmful gas emissions, soil contamination, and air pollution. These techniques may leave behind residue that contaminates water supplies and aquatic ecosystems.

33.2 Sources of ASW

Farm operations account for the majority of ASW, but ASW is also generated from manufacturing and other activities related to the food chain and agricultural practices. Any step or procedure in the agricultural-food system can produce ASW (Peng et al. 2023). Agricultural solid waste can be classified into several categories based on their origin, composition, and potential for reuse or disposal. Here are some common classifications:

33.2.1 Crop Residues

Several crop residues including leaves, maize, rice, wheat, oats, barley straws, and seed pods produced wastes in agricultural fields. The latest global crop residue output projection for 2024 is 2802 Mt/year. Very small portion of crop waste from rice, wheat, and maize is utilized globally to make animal feed or bioethanol. The rest is burned, which extremely troubles human health and the environment and makes agriculture less sustainable. Globally, 731 million tons of rice straw are generated annually. 667.6 Mt of straw is produced in Asia. By 2030, 221.8 Mt of rice straws are anticipated to be produced in India. An estimated 354.34 Mt of wheat straw are produced worldwide each year, with an average of 1–3 tons/acre (Kim and Dale 2004). Another significant crop that is frequently utilized to make lignocellulosic bioethanol is corn. Sarkar et al. and Kim and Dale estimate that the yearly output rate of maize stover (leaves, stalks, cobs, and husks) is 4.0 tons/acre, or 128.02 Mt/year.

33.2.2 Livestock Waste

Livestock waste refers to the excrement, urine, and other organic materials produced by animals raised in agricultural settings, such as farms, ranches, and smallholdings. This waste can include solid manure, liquid slurry, and bedding materials used in animal housing. Managing livestock waste is important due to its potential environmental impacts, including water pollution from nutrient runoff (like nitrogen and phosphorus), methane emissions (a potent greenhouse gas), and odor issues. Various methods are employed to handle livestock waste sustainably:

- Storage and treatment: Waste can be stored temporarily in lagoons or tanks to be treated or processed later. Treatment can involve composting, anaerobic digestion, or aerobic digestion to reduce pathogens and odor.
- Land application: Treated waste can be applied as fertilizer to agricultural fields, providing nutrients for crops. Proper application rates are crucial to prevent nutrient runoff into water bodies.
- Biogas production: Anaerobic digestion of livestock waste can generate biogas (methane), which can be used as a renewable energy for future work.
- Nutrient management: Careful management of nutrients in livestock waste is essential to prevent over-fertilization of soils and subsequent pollution of water bodies.

33.2.3 Agriculture Industry Waste

Agriculture industrial waste can be derived from the residues of the food processing sector. Defatted waste slurry or cake like cotton, mustard, rice bran, sesame, palm kernel, groundnut, soybean, rice husks, vegetables, fruit skin and pomades, eggshells, poultry, and meat from the corresponding industry are among them. This category including molasses and bagasse from the sugar industries. Sugarcane bagasse, a significant quantity of agro-industrial waste, builds up after sugarcane juice is recovered. From 85.84 million t of fresh fruit, edible oil firms worldwide create 35.19 Mt of palm cake and palm oil.

33.2.4 Aquaculture Waste

The demand for fish has increased together with the global population (de Jesús Gordillo Benavente & Hernández 2024). By 2030, it is anticipated that aquaculture, a fast expanding agricultural sector, will deliver around 2/3rd of the fish intended for worldwide consumption as a result of the agricultural intensification regime. Aquaculture, which encompasses the breeding, growing, and harvesting of freshwater fish, shellfish, and marine plants, as well as the preservation of endangered species, is one of the areas of the food production industry that is expanding at the fastest rate. Despite the recent pandemic's global expansion, aquaculture units increased globally in 2020, while there were regional and national variations. The majority of aquaculture's 122.6 Mt in live weight in 2020 came from 35.1 Mt of algae for food and other purpose, 700 t of shells and pearls for ornamental purposes, and 87.5 Mt of aquatic animal for human use. Compared to the 115.9 Mt in 2018, there has been a 6.7 Mt increase. Consequently, aiming for a "blue revolution," aquaculture is a creativity which aids in ensuring to fulfil the nutritious supplies of countless individuals. Nevertheless, as aquaculture raises and manufacture increases, issues like feed waste and water pollution become significant challenges that threaten the ecosystem and undermine the sustainability of aquaculture. The type of culture directly impacts the amount of feed required. Studies show that feed currently constitutes the leading source of waste in aquaculture. The primary contributors to solid waste in aquaculture include uneaten feed and fish waste. Solid waste significantly harms fish.

33.2.5 Food Processing By-Products

In order to reduce waste and increase resource efficiency, it is necessary to manage food processing by-products effectively (El Messaoudi et al. 2024). The following list of typical food processing by-products includes some potential applications for them. Food processing by-products are the remaining materials produced when raw agricultural commodities are processed into food items. These by-products may result from various food processing steps, such as cleaning, peeling, trimming, cooking, and extraction procedures.

- Fruit and vegetable peelings: The outer layers and peels that are thrown away after cleaning and peeling. These can be utilized for animal feed, composting, or as a source of bioactive substances for the cosmetics or pharmaceutical industries.
- Pulp and pomace: The solid residue that remains after fruit or vegetable juice is extracted. These can be composted, utilized as animal feed, or processed to derive pectin, nutritional fibers, or antioxidants.
- Oilseed meals: By-products of the extraction of oil, such as canola or soybean meal. These are frequently used as ingredients in livestock or aquaculture feed that are high in protein.

- Brewers' spent grain: Grain husks and malt leftover from making beer. It can be turned into value-added products like bioethanol or dietary fibers, or it can be utilized as compost or animal feed.
- Dairy processing by-products: Whey and lactose by-products from the production of cheese and yogurt, such as whey for protein supplements, lactose for food products, or biogas.
- Meat processing by-products: Trimmed fat, bones, and organs not used for primary meat products, which can be rendered for fats and oils, processed into pet food or animal feed, or used in pharmaceutical applications.
- Cereal processing by-products: Bran and germ separated during grain milling, which can be used in food products (like bran muffins), animal feed, or converted into dietary supplements.

33.2.6 Forestry Residues

The materials left over after forest operations like logging, thinning, pruning, and other forest management tasks are known as forestry residues. These residues may consist of:

- Smaller branches and tree tops that are unsuitable for timber or other principal forest products are referred to as branches and tops.
- Stumps and roots: After logging, the leftover portions of trees are either at the site or in the ground.
- Bark: A tree's exterior layer, which is frequently taken off during processing.

All of the aforementioned elements as well as other organic waste that remains on the forest floor following logging operations are referred to as "slash." Forestry residues can be used in various ways, including:

- Bioenergy production: Residues can be utilized as a raw material to create bioenergy, which can then be converted into biofuels, heat, or electricity.
- Mulching and composting: To enhance soil health, they can be chipped and utilized as mulch or compost.

33.2.7 Biochar System and Agriculture

Biochar are produced by the burning of agriculture waste under controlled environment, having potential uses in the improvement of climate change, sustainable plant growth, environmental control, and the creation of new materials (Gustafsson & Anderberg 2022). Biochar has a high capacity to absorb nutrients, water, and pollutants due to its porous structure and other properties. The large surface area gives soil microorganisms a large number of attachment sites. The stable carbon structure makes it a perfect material for long-term carbon storage. Since, it may work as a bio-fertilizer to improve soil quality (fertility), water-retaining capacity, and to stop land deterioration, biochar has emerged as a sustainable option. Numerous reactor systems and procedures are used in pyrolysis. The varieties of pyrolysis that produce distinct mixtures of charcoal, gas, and oil products are determined by temperature and heating rates. For quick pyrolysis, gasification, syngas production, biochar formation, and bio-oil, slow pyrolysis—which employs moderate temperatures and modest heating rates—is the recommended method. As the pyrolysis temperature rises, biochar's carbon content and aromatic condensation rise, but its yield decreases (Table 33.1).

Table 33.1 Operational conditions and product generation of pyrolysis

Parameters	Category	Details	Significances
Feedstock	Composition	Biomass: high cellulose, hemicellulose, lignin Plastics: polyethelene, polypropelene	Biomass with low moisture contents improve efficiency
Temperature	Process dependent	Low (<450 °C)—high biochar yield Moderate (450–600 °C)—optimal for bio-oil High (>650 °C)—maximizes syngas production	Fast pyrolysis occurs at approx. 500 °C
Heating rate	Reactor design	Low (<10 °C/min): slow pyrolysis, uniform heating Moderate: intermediate pyrolysis (10–100 °C/min) Elevated (>100 °C/min): rapid pyrolysis	High heating reduces biochar yield and enhances syngas production
Residence time	Dependent on feedstock	Fast pyrolysis: ~2 s; flash pyrolysis: <1 s; slow pyrolysis: 5–30 min	Controlled by different reactor type
Reaction environment	Inert gas atmosphere	N and Ar	Combustion is prevented in an oxygen-free atmosphere
Catalyst	Enhancement	Zeolites: bio-oil cracking Oxides of metals: boost the selectivity Metals from alkaline earths: Decrease oxygenates	Products-specific catalysts (e.g., hydrogen-rich syngas)
Product distribution	Temperature dependent	At 400 °C, biochar: about 30% About 50% bio-oil at 500 °C Over 600 °C, syngas: about 60%	Conditions and feedstock have an effect on product yield

33.3 Integrating ASW into the Circular Economy

The circular economy and agricultural waste have a profound and revolutionary link. Making the most of resources and reducing waste are the objectives of a circular economy. Crop residues, manure, and by-products of food processing are examples of agricultural waste that can be efficiently utilized in a number of ways to promote a circular economy. The following main ideas emphasize this relationship:

- Resource recovery: Bioenergy, biofertilizers, and bioplastics are just a few of the useful goods that may be made from agricultural waste. For instance, anaerobic digestion of crop leftovers can yield biogas, which can then be utilized to produce heat and power.
- Soil health and fertility: Soil health has been considered to measure the ability of the soil towards the growth of plant, animal, and human life. The compost as well as manure can be usually introduced into the soil for enhancing the soil fertility and structure. This phenomenon helps to lessen the use of artificial agriculture system.
- Reducing landfill waste: Utilizing the useful applications of ASW, landfill waste can be reduced which decreases the environmental effects of waste dispersion with the formation of leachate and GHG emission.
- Circular economy: The reutilization of biological resources opens a new avenue for developing a business model in urban and rural areas.

33.4 Technological Feasibility

In order to explore the technological feasibility of ASW, few things are essential to maintain like process technology, infrastructure development, and upscale variation (Lepe et al. 2024; Manna et al. 2021; Ni & Zhang 2024; Sánchez-García et al. 2024; Srivastava et al. 2020; Stürmer et al. 2021; Wang & Dong 2024). Due to the enormous amount of wastes generated by agricultural system, all the technologies used should be scalable. It should cover the mass operation as well as small scale operation for circularization. Apart from that, the infrastructure is highly mandatory for processing the ASW.

33.5 Economic Feasibility

The economic feasibility depends on the location of agricultural set up, relation of farmers, quality of farmers, and the investment of processing technologies. Apart from that, we should maintain the profitability index and the actual conversion of ASW into the valuable metabolites including bioenergy, biofertilizer, etc. Monetary support, subsidies, and inducements from the government bodies can be enhanced regarding the economic viability by using ASW.

33.6 Challenges and Considerations

The seasonality of ASW has consequence on the stable source of raw materials for processing facilities. In order to gather and conveyance ASW from grounds to dispensation, well-organized procedures should be in a proper place.

References

Ahmed RS, Abuarab ME, Baiomy MA, Ibrahim MM (2024) Heavy metals removal from industrial wastewater using bio-adsorbent materials based on agricultural solid wastes through batch and continuous flow mechanisms. J Water Process Eng 57:104665

de Jesús Gordillo Benavente L, Hernández CV (2024) The path to a circular economy through a green and digital transition. The paradigm shift in Mexican companies in the state of hidalgo. In: Entrepreneurship, technological change and circular economy for a green transition: research contributions for a more productive environment. Springer, Cham, pp 61–82

Dubey S, Parmar S, Upadhye VJ, Shrivastav A, Shah F, Mishra N, Bachani P, Mishra S (2024) Greener approach towards sustainable green plastics through eco-friendly upstream and downstream processing. In: Harnessing microbial potential for multifarious applications. Springer, Singapore, pp 253–269

El Messaoudi N, Miyah Y, Şenol ZM, Ciğeroğlu Z, Kazan-Kaya ES, Gubernat S, Georgin J, Franco DS (2024) Comprehensive analytical review of heavy metal removal efficiency using agricultural solid waste-based bionanocomposites. Nano-Struct Nano-Objects 38:101220

Filippov SP (2024) Transition to a carbon neutral economy: opportunities and limitations, current challenges. Therm Eng 71(1):18–35

Ghosh A, Kumar A, Biswas G (2024) Exponential population growth and global food security: challenges and alternatives. In: Bioremediation of emerging contaminants from soils. Elsevier, Amsterdam, pp 1–20

Gustafsson M, Anderberg S (2022) Biogas policies and production development in Europe: a comparative analysis of eight countries. Biofuels 13(8):931–944

Jangde V (2024) Sustainable circular economy: transforming fly ash into valuable compost. J Environ Nanotechnol 13(2):470–476

Khan MN, Sial TA, Ali A, Wahid F (2024) Impact of agricultural wastes on environment and possible management strategies. In: Frontier studies in soil science. Springer, Cham, pp 79–108

Klotz M, Oberschelp C, Salah C, Subal L, Hellweg S (2024) The role of chemical and solvent-based recycling within a sustainable circular economy for plastics. Sci Total Environ 906:167586

Lepe PT, Heredia KV, Namur EC, Fabián GC, García-Enriquez S (2024) Economics and commercialization of bioplastics. In: Bioplastics for sustainability. Elsevier, Amsterdam, pp 271–309

Li L, Tang Y, Bao Z, Tu W, Peng L, Zou L, Zhao C, Li Q (2024) When graphene meets circular agriculture: insights into agricultural sustainable development. Biosyst Eng 237:92–117

Liang W, Liu T, Bao H, Kumar V, Sun X, Sindhu R, Binod P, Zhang Z, Pandey A, Awasthi MK (2024) Novel animal waste management techniques. In: Engineering applications in livestock production. Academic, New York, pp 243–264

Manna MC, Rahman MM, Naidu R, Bari AF, Singh AB, Thakur JK, Ghosh A, Patra AK, Chaudhari SK, Subbarao A (2021) Organic farming: a prospect for food, environment and livelihood security in Indian agriculture. Adv Agron 170:101–153

Mehta A, Yadav A, Kumar A, Kumari K (2024) Waste utilization in horticulture: an overview. J Exp Agric Int 46(5):742–749

Ni Y, Zhang Z (2024) Comparison and selection of wet waste disposal modes for villages in agriculture-related towns taking Shanghai, China, as an example. Sustain For 16(12):4936

Odejobi OJ, Ajala OO, Osuolale FN (2024) Review on potential of using agricultural, municipal solid and industrial wastes as substrates for biogas production in Nigeria. Biomass Convers Biorefinery 14(2):1567–1579

Peng X, Jiang Y, Chen Z, Osman AI, Farghali M, Rooney DW, Yap PS (2023) Recycling municipal, agricultural and industrial waste into energy, fertilizers, food and construction materials, and economic feasibility: a review. Environ Chem Lett 21(2):765–801

Sánchez-García E, Martínez-Falcó J, Marco-Lajara B, Manresa-Marhuenda E (2024) Revolutionizing the circular economy through new technologies: a new era of sustainable progress. Environ Technol Innov 33:103509

Singh N, Gautam SB, Sawood GM, Yadav VL, Mishra G, Dixit S, Gupta SK (2024) Bioenergy from agro-waste: a sustainable solution for energy needs. In: Sustainable clean energy production using waste biomass: sustainable energy production and utilization. Springer, Singapore, pp 207–249

Srivastava P, Balhara M, Giri B (2020) Soil health in India: past history and future perspective. In: Soil health. Springer, Cham, pp 1–9

Stürmer B, Leiers D, Anspach V, Brügging E, Scharfy D, Wissel T (2021) Agricultural biogas production: a regional comparison of technical parameters. Renew Energy 164:171–182

Voukkali I, Papamichael I, Loizia P, Zorpas AA (2024) Urbanization and solid waste production: prospects and challenges. Environ Sci Pollut Res 31(12):17678–17689

Wang J, Dong H (2024) Preparation and application of multi-source solid wastes as clean aggregates: a comprehensive review. Constr Build Mater 418:135414

Wang N, He Y, Zhao K, Lin X, He X, Chen A, Wu G, Zhang J, Yan B, Luo L, Xu D (2024) Greenhouse gas emission characteristics and influencing factors of agricultural waste composting process: a review. J Environ Manag 354:120337

Yang Y, Zheng B, Luk C, Yuen KF, Chan A (2024) Towards a sustainable circular economy: understanding the environmental credits and loads of reusing modular building components from a multi-use cycle perspective. Sustainable Prod Consump 46:543–558

Zhu Y, Wu S, Li J, Jia Q, Zhang T, Zhang X, Han D, Tan Y (2024) Towards a carbon-neutral community: integrated renewable energy systems (IRES)–sources, storage, optimization, challenges, strategies and opportunities. J Energy Storage 83:110663

Industrial Solid Waste to Circular Economy

34

Feroz Shaik, Nayeemuddin Mohammed, Mohammed Ali Khasawneh, and Tasneem Sultana

Abstract

The management of industrial solid waste (ISW) presents a dual challenge, reducing environmental degradation while exploiting resources in a sustainable manner. Traditional linear ISW management processes, including landfilling and incineration, are environmentally destructive due to greenhouse gas emissions, land occupation, and the loss of valuable materials. Therefore, a shift towards circular economy is required. In this study, an integrated framework for converting ISW into high-value secondary products was developed. By employing different case-based industrial symbiosis models to identify the best waste valorization strategies such as closed-loop recycling, waste-to-energy conversion, and by-product industrial reuse. This approach is further substantiated by sectoral case studies that demonstrate savings of up to 65% raw materials and 48% in lifetime carbon emissions. The results indicate that the standardized implementation of CE marking not only reduces environmental impact, but also enhances industrial competitiveness, resource security and regulatory compli-

F. Shaik (✉)
Department of Mechanical Engineering, Prince Mohammad Bin Fahd University, Al Khobar, Kingdom of Saudi Arabia
e-mail: fshaik@pmu.edu.sa

N. Mohammed · M. A. Khasawneh
Department of Civil Engineering, Prince Mohammad Bin Fahd University, Al Khobar, Kingdom of Saudi Arabia
e-mail: mnayeemuddin@pmu.edu.sa; mkhasawneh@pmu.edu.sa

T. Sultana
Artificial Intelligence and Machine Learning, Godutai Engineering College for Women, Sharnbasva University, Kalaburagi, India

A. Kumari et al. (eds.), *Valorization of Solid Wastes to Biofuels and Chemical Products for Sustainable World*, https://doi.org/10.1007/978-981-96-8594-3_34

ance. It presents a useful reference for policy makers, industry stakeholders and researchers on connecting industrial waste management to SDGs.

Keywords

Circular economy · Industries · Solid waste · Sustainable · Reuse · Waste management

34.1 Circular Economy, Waste Management, and Sustainable Development

Circular economy (CE) methods in the industrial and service sectors have gained popularity in several nations in recent years. Medical waste management has to transition to a CE after relying on a linear economy for a long time. A shortage of raw resources combined with rising demand drives up the cost of goods, eventually impacting international markets. The CE distinguishes between technical and biological cycles and aims to maintain materials, goods, and parts at their maximum value and usefulness at all times. By its very nature, it is regenerative and healing. Using planning, resource allocation, procurement, production, and reprocessing as processes and outputs, the CE economic model seeks to optimize both ecological functioning and human well-being (Suárez-Eiroa et al. 2019). In the global production system, the CE symbolizes a paradigm change that encourages a move away from the linear model and towards maximizing the utilization of resources and value addition. The endeavor has caused a shift in the way that industrial waste management is seen, moving away from the statistically supported "expansion of recycling industry" and towards the use of "waste as resource" in order to achieve the highest possible resource recovery quality. The model's popularity is demonstrated by the increasing number of countries making CE their national policy. Through the implementation of the "new circular economy action plan," the European Commission seeks to promote sustainable job creation and reduce the demand on natural resources. The United Nations Economic Commission for Europe (UNECE) adopted the CE idea and launched new projects at each stage of the product life cycle to promote sustainable development. A number of methods have been used to assess the circular economy (CE), including life cycle assessment (LCA), material flow analysis (MFA), and environmentally extended input-output analysis (EEIOA). The circular economy is becoming increasingly significant.

In order to mitigate the detrimental effects of human activity on the environment, the concepts of "circular economy" (CE), "industrial waste management" (industrial WM), and "sustainable development" (SD) are connected. Because CE has a well-established track record of achieving SD, it serves as the foundation for many policy tools. SD stands for "development that meets the needs of the present without compromising the ability of future generations to meet their own necessities." With sustainability serving as the vision, SD is the policy's objective. The CE idea for industrial waste management has been embraced by nations worldwide in an effort to reduce the harmful effects of one of the biggest environmental concerns, industrial waste. Many criteria are used to describe the connection between the CE and SD. By placing a strong emphasis on pollution control and environmental preservation, CE is an economic model that aims to accomplish SD. CE is the combination of social, economic, and environmental sustainability. The CE has become known as a restorative and ecologically friendly economic system that seeks to minimize adverse effects from all angles and assists the sector in strategic sustainability decision-making while taking governance into account. Although there is not a single, widely recognized concept for the circular economy, a comparison of definitions revealed that attaining sustainability is a shared objective (Yang et al. 2023a, b).

The connections and distinctions between SD and CE, as well as the limits between the two concepts. According to them, the SD is described as a regenerative system that reduces energy losses, waste, emissions, and resource consumption by slowing and closing the cycles of energy and products through long-term planning, upkeep, repair, reuse, reconditioning, remanufacturing, or recycling of the system. For the benefit of current and future generations, the CE alludes to the harmonious integration of social inclusivity, environmental resilience, and economic performance (Negrete-Cardoso et al. 2022). CE might support SD, environmental management reform, and increased resource productivity and eco-efficiency. The idea of corporate ecology and its guiding principles of reduce, reuse, and recycle present difficulties because, despite its increasing significance, CE must enable the reconciliation of environmental care with commercial considerations (Sanguino et al. 2020). To build a hierarchy and make links between the main issues, a trial and evaluation laboratory technique for decision-making based on fuzzy theory might be used. (Kandasamy et al. 2022).

Population increase, industrialization, urbanization, and economic expansion are all contributing factors to the daily rise of solid waste. Accepting new sustainability rules and practices in India is mostly dependent on educating and raising people's understanding of efficient waste management techniques (Yaduvanshi et al. 2016). The CE's waste hierarchy aims to reduce waste and establish a zero-waste, sustainable environment. Better resource management and waste avoidance are made possible by the CE, which is established based on the planning and execution of effective waste management systems. Waste management (WM) strategies that promote the CE's objectives have received a lot of consideration. It is anticipated that novel companies will help maximize value production and minimize the utilization of resources. Employing the product leasing model, for instance, encourages manufacturers to apply circular design principles, which are advantageous for WM. Furthermore, buyers who care more about the environment are prepared to pay a premium for circular items, which encourages producers to include social and environmental features like sustainability certifications and green circular premiums into their cutting-edge products. Additionally, digital technology integration with CE has increased dramatically, which might increase productivity and sustainability. These technologies include radio frequency identity chips for garbage tracking, block chains for digital material passports, and artificial neural networks for expedited waste sorting. Research on the idea of the "circular economy rebound," which contends that systemic changes like improved efficiency and consumption may have a greater positive impact than the perceived advantages of CE strategies, have shown that this effect may lead to waste production and increased material use (Appolloni et al. 2022; Yang et al. 2023a, b).

Indonesia is dealing with several self-managed issues with the gathering, moving, processing (composting, recycling), and reliance on landfills for waste management. In industrialized nations, the application of Internet of Things (IoT) and information and communication technology (ICT) offers a new generation strategy to effectively and efficiently enhance the worldwide waste management system (Fatimah et al. 2020). The relationships between a number of the sustainable development goal objectives may be formed through CE activities. Additionally, the study highlights a number of possible trade-offs between the goals of decent labor, safe workplaces, and human health, and the CE's existing recycling procedures for wastewater, e-waste, and municipal garbage, while offering solutions for resolving these conflicts (Schroeder et al. 2019). Waste and emissions are reduced during building projects that use integrated sustainable construction concepts. Ogunmakinde et al. outlines many measures to minimize building waste, such as design, reuse, recycling, construction techniques, business models, waste and energy measuring instruments, and legislation. By incorporating these waste-reduction strategies into building projects, the overarching goals of sustainable development and construction might be met (Ogunmakinde et al. 2022). Metals, glass, and textiles have all been collected and used again in the past, demonstrating that waste can be a useful source of raw materials and energy. Increased knowledge of the necessity of waste separation led to the creation of innovative disposal methods and improved recycling capacities (Nelles et al. 2016).

The past few decades have seen substantial developments in waste management in many different countries. For example, Sweden has seen a decrease in the quantity of municipal solid waste disposed of in landfills, but an increase in material recycling and energy recovery. The Swedish Waste Management Environmental Assessment is one LCA instrument (SWEA). SWEA is intended to assess how different policy tools, under different future conditions, affect the environmental life cycle impacts of Swedish national waste management (Arushanyan et al. 2017). The inability of municipal organizations in emerging nations to manage the growing volumes of waste leads to the accumulation of uncollected waste on public roadways and spaces. Opening dumping sites or landfills are the most common and conventional methods of disposing of municipal waste management globally (Singh et al. 2011). Thailand's fast urbanization has led to a rise in the need for infrastructure and housing, which has increased construction waste production. The lack of landfill space and the growing uncontrolled dumping of building debris are major problems that are affecting many nations, including Thailand (Kofoworola and Gheewala 2009). The amount of electronic garbage, often known as waste electrical and electronic equipment (WEEE), is quickly garnering international attention. It is a relatively recent addition to the hazardous waste stream. An electrical or electronic item is considered to be a component of the e-waste stream after its useful life is over (Borthakur and Sinha 2013).

The wealthiest nations have successfully adopted the solid waste management (SWM) hierarchy and are now placing a high priority on recycling, reuse, and MSW reduction (Sharma and Jain 2020). Dasgupta et al. used the system dynamics approach (with STELLA software, version 8.0) to estimate the amount of e-waste that will be generated and its percentage distribution through various disposal pathways (landfill, second-hand market, and recycling) in order to improve management practices in India in the near future (Dasgupta et al. 2017). Conventional biodegradation methods have focused on degrading organic contaminants that are hazardous and have neglected to extract chemical energy from them, which is therefore lost as biomass. These superfluous biomass may impede mass transfer, clog bioreactors, and affect other bioprocesses (Senthilkumar and Naveen Kumar 2020). E-waste's enormous potential for usage in a variety of industries cannot be overlooked, however, due to its effects on society, the economy, and the atmosphere (Khandelwal et al. 2023).

34.2 Industrial Waste Generation Scenario

The industrial waste generation scenario refers to the situation or conditions surrounding the production and disposal of waste materials by industrial activities. Solid trash, liquid waste, and gaseous emissions are among the waste kinds that can be produced by industrial operations. If these waste items are not adequately handled, they may have negative impacts on both human health and the environment. The specific scenario of industrial waste generation can differ based on a number of variables, including the kind of industry, location, regulations, and waste management practices. However, there are some common trends and challenges associated with industrial waste generation. Quantity of waste is one of these challenges. Industrial activities often produce large volumes of waste due to the scale of their operations. This waste can include production scrap, packaging materials, chemicals, pollutants, and byproducts. Types of waste is also important factor. Industrial waste can be categorized into different types based on its physical and chemical properties. These include hazardous waste such as toxic chemicals and heavy metals, non-hazardous waste such as construction debris, and special waste like electronic waste. Pollution and environmental impact should also be taken into account. Improper disposal or inadequate treatment of industrial waste can lead to pollution of air, water, and soil. Contaminants from industrial waste can enter ecosystems, affecting biodiversity and posing risks to human health. Examples include chemical spills, air emissions, and improper disposal of hazardous substances.

Governments and international bodies have established regulations and guidelines to manage industrial waste as part of the regulatory framework mission. Controlling trash creation, encouraging recycling and reuse, and guaranteeing safe disposal procedures are the goals of these policies. The specific regulatory framework can vary from country to country. Moreover, waste management practices involve various strategies, including waste reduction at the source, recycling and reuse, treatment and neutralization, and safe disposal in designated facilities. The effectiveness of waste management practices might differ based on variables such infrastructure, technology, and financial resources. Finally, SD and CE is of crucial importance. There is an increasing emphasis on adopting sustainable practices and transitioning to a CE, where waste is minimized, and resources are reused. Industries are encouraged to implement cleaner production techniques, improve resource efficiency, and explore innovative waste management solutions. The fact that waste management has advanced and will continue to advance in several nations is noteworthy at this point. Two major reasons driving this change are the need for society to move towards a circular economy and more resource recovery in order to fulfill climate commitments. In order to minimize the negative environmental effects of industrial activity while supporting sustainable growth, Scenarios are helpful planning tools, evaluating potential upcoming advancements and regulations.

A complete life cycle assessment (LCA) model that encompasses the majority of a nation's waste streams and permits the inclusion of waste avoidance techniques was created in a 2017 research. It has been demonstrated that building such a model is feasible. The model is helpful for assessing potential waste situations in the future as well as the effects of policy tools that affect various waste categories and societal sectors. The whole model incorporates significant elements and possible outcomes of many policy instruments, but it is huge and complicated, with uncertainties arising from, among other things, data on environmental performances of systems and data gaps. The findings demonstrate how different future waste management scenarios in Sweden may have quite different environmental effects. Future possibilities that are more sustainable, waste management will still have a positive impact on the environment, but less so since there will be less pollution from nearby energy and transportation systems and less trash generated. According to valuation studies, the Swedish waste management system's most significant environmental effect categories are resource depletion, human toxicity, and climate change. Under these scenarios, waste incineration emissions of fossil CO_2 will remain a significant source of environmental consequences. Trash inclusion in a green power certification program, the resource tax, the incineration tax, and the weight-based collection fee are only a few of the policy instruments whose environmental effects are analyzed using this model. The majority of the time, the analyzed policy instruments have a limited impact when used alone, indicating that stronger policy tools and their combinations are required to achieve policy goals, such as those outlined in the European Union's (EU) action plan on circular economy (Arushanyan et al. 2017).

Solid waste is a sign of economic growth, but it also puts more strain on the environment. The main factor contributing to climate change and global warming is the global emissions of greenhouse gases, which are attributed to solid waste at a rate of 3%. The air quality in cities is deteriorated by massive solid waste creation and inefficient management, which eventually has a negative impact on human health. In megacities like Delhi, where there is a shortage of space for landfills, the issue of managing municipal solid waste is especially acute (Tahir et al. 2015). The public's active involvement and knowledge are crucial to successful municipal solid waste management; without them, the effort is unachievable. The general public does not separate garbage at the source, despite the urgent need to largely eradicate this "Urban Menace." Due to the size and scope of China's building sector, billions of tons of garbage from building and dismantling are produced, raising serious environmental issues that have not been well studied. Thus, the necessity for precise data, well-informed analysis, and further policy creation is crucial. Comprehensive surveys of construction and demolition sites, recyclers, and government agencies have been conducted in order to collect important data about demolition garbage from its generation to disposal. Case studies of Shenzhen, a rapidly growing flagship megacity in South China, have also been conducted.

Subsequently, sophisticated techniques have been developed to calculate and extrapolate the production, flows, and possible uses of demolition trash. The findings indicate that Shenzhen city produced almost 14 million tons of demolition garbage yearly between 2010 and 2015, and that this number is expected to rise in the near future. Based on available utilization possibilities, the recycling potential was estimated to be worth around 1.02 billion USD in 2015. If recycling rates are optimized, this amount might increase to 1.38 billion USD. The government would be responsible for the proper management of demolition waste, which would include landfill planning, rubbish collection, transportation, and alternatives for use. If this method were implemented in other Chinese megacities as well, it might provide insight into the measurement of demolition trash (Wu et al. 2016).

Environmental mishandling of plastic waste (MPW) is a growing worldwide problem. Targeting priority locations for the execution of mitigating measures requires an accurate understanding of waste creation. High-resolution distributions, long-term population and GDP projections, and country-level waste management information were used to predict global MPW generation at a 1000 m resolution from the current year to 2060. It is estimated that between 60 and 99 million metric tons (Mt) of MPW were generated worldwide in 2015. By 2060, this amount may quadruple to 155–265 Mt. under normal circumstances. Even in coming years, the continents of Asia and Africa will continue to have a disproportionately large MPW load. It has thus been shown that if global efforts are made to reduce the percentage of plastic in municipal solid waste and if developing countries invest heavily in waste management infrastructures as their GDP increases in the future, the amount of plastic trash generated may be reduced. Rivers are a significant source of the flow of plastic trash into the ocean, as evidenced by the fact that watersheds bigger than 100 km^2 carry the bulk of MPW (91%) (Lebreton and Andrady 2019).

The generation of municipal solid waste has increased noticeably on a global scale. Municipal solid waste output showed a clear trend and a positive correlation with worldwide economic growth in terms of the kilograms of solid rubbish produced daily per resident. The city's population would be represented by the total tons of solid rubbish generated daily during that specific year. In India, where trends in the production of solid waste and population growth are not always consistent, a city-specific relationship between population and solid trash output is not always appropriate. Over the course of 10 years (1981–1997), Mumbai's population rose at a pace of 49%, from 8.2 to 12.3 million. The increase in solid waste creation was mostly caused by the urbanization trend, which in India reached 27.8% in 2001 and is predicted to reach 50% by 2025. In 1995, it was discovered that the amount of solid trash generated in a few Indian cities was 0.64 kg in Kanpur, 0.52 kg in Lucknow, 0.4 kg in Varanasi, 0.59 kg in Ahmadabad, and 0.44 kg in Mumbai. India now generates over 42 million tons of solid trash yearly. The daily production of waste varies from 200 to 600 kg/capita, whereas the collection effectiveness varies from 50 to 90% (Rajput et al. 2009). India's enormous population is to blame for the country's enormous MSW output, which occurs at a rate of 400 g/person every day. It has been determined that the management of MSW produced in Indian cities is difficult and insufficient (Bhat et al. 2018). The body tasked with overseeing MSWM is facing an increasing amount of difficulties due to the shifting trends in MSW quantity and characteristics.

Additionally, the shifting waste composition trend highlights how crucial segregation is to the efficient running of waste management facilities. The authorities should support the collection, storage, and segregation of trash in a way that minimizes the creation of unsanitary and unhealthy circumstances in order to manage the MSW. According to their gross national income, 59 wealthy and developing countries' conditions for various MSW management technologies have been compared. A total of 19 selection parameters that affect the technologies' performance, operational suitability, and applicability have been investigated in order to help a city or town pick the optimal MSW management system or systems. Every danger and difficulty that may arise over the waste to energy (WtE) project's lifetime has also been covered. The most industrialized nations are currently aggressively emphasizing MSW reduction, reuse, and recycling after successfully implementing the solid waste management (SWM) hierarchy. However, as most of the MSW is dumped in an open manner and most of these nations rely on an inadequate waste infrastructure as well as the informal sector, solid waste management (SWM) has become a major issue in low-income and low-middle-income countries. Other significant obstacles are those pertaining to efficient waste management practices, funding accessibility, adequate technology selection, and a sufficient number of qualified personnel (Sharma and Jain 2020).

Resources that were intended for human consumption but were misplaced, contaminated, degraded, or thrown away are referred to as food waste. From collection to disposal, food waste is a growing issue that impacts all waste management sectors. Developing sustainable solutions includes not only retailers and end users but also the agricultural and industrial sectors, as well as other stakeholders in the food supply chains. To manage food waste properly, a number of strategies may be used, and they can be ranked similarly to a waste management hierarchy. Avoidance and donating edible portions to social services are the most sought-after remedies. Food waste is also utilized in industrial operations to produce biopolymers or biofuels. Subsequent actions entail composting in order to recover nutrients and fix carbon. Landfilling and incineration are the last and less ideal possibilities. Finding sustainable ways to manage food waste is one of society's largest challenges. In order to benefit the social, economic, and environmental domains, these solutions should be able to utilize the precious resources that food waste represents. Redistributing excess edible food to feed the underprivileged is a workable solution to the surplus production of food; but, governments must assist in order to enable food banks or social agencies to recover and redistribute the food. Food scraps may be utilized as feedstock to extract high-value components and produce biofuels and bioplastics. Ultimately, then, a crucial tactic for optimizing food waste utilization and increasing the potential revenue of the whole bioprocess chain is the integration of biotechnological processes in the co-production of biofuels and bioproducts (Girotto et al. 2015).

Globally, the construction business has been expanding quickly. In many developing nations, the growth has resulted in a significant issue with the creation of building trash. Construction wastes are divided into physical and non-physical waste categories, and each country's environment, economy, and society are more negatively impacted. It is crucial to comprehend the underlying causes of the generation before it can be effectively handled. 81 elements related to building operations have been identified using the triangulation approach as contributing to trash creation. Among these elements, 63 are linked with physical trash, and 73 with non-physical waste. The components were separated into seven categories: design, material handling, labor, management, site conditions, procurement, and external repercussions (Nagapan et al. 2012). The following guidelines are intended to limit construction waste output in any construction project:

1. Contractors and clients should communicate well.
2. A methodical approach to managing construction materials, equipment, and human resources should be used by construction actors.
3. Prior to beginning their employment, construction personnel must complete a training programmed on construction waste generation.
4. Workers in the construction industry should take up or modify any innovative methods for organizing and reducing the production of construction waste.
5. In advance of initiating any initiatives, site investigations must be carefully planned and executed to gather the appropriate kind and volume of information.
6. The procurement document on site has to be carefully developed and routinely checked.
7. Construction workers should be aware of climate change news before beginning work on a location.

China's hazardous waste generation rate is steadily growing. Approximately 11.62 million tons of industrial hazardous waste (IHW) were generated, making up 1.1% of the volume of industrial solid waste (ISW) in total. On average, 43.4% of IHW was recycled, 33.0% was stored, 23.0% was disposed of correctly, and 0.6% was released into the environment without any pollution control. For IHW management, there were 177 authorized treatment and disposal facilities. Only 416,000 tons of IHW were reported to have been disposed of at these centers, with 35% being burnt and 65% being landfilled. Among the IHW categories, waste acid and alkali were highest, making up 30.9%. 39.0% of IHW was provided by the industries producing chemical products and raw materials. One further significant source of hazardous waste is discarded household hazardous waste (HHW). Controlling hazardous waste contamination has emerged as a significant concern for Chinese environmental management (Duan et al. 2008).

Promoting clean production laws and lowering the amount of IHW from source control are the first two practical hazardous waste management strategies that have to be put into practice. In the second place, call for a much more specific definition of the phrase "hazardous waste"; develop workable standards and identification techniques, categorization guidelines, delisting and exclusion systems, and procedures for managing hazardous waste through national and local governance. Thirdly, establish the best possible solutions for recycling, storing, and disposing of hazardous waste in light of the existing circumstances in China by conducting research and developing the most advanced disposal technology and equipment in conjunction with the features of hazardous waste creation. Additionally, raise generator and local governors' understanding of hazardous waste management through targeted instruction, public relations, and training initiatives. Proteins, lipids, carbohydrates, and lignocelluloses are prevalent in the hazardous waste streams on terrestrial and aquatic ecosystems; thus, recycling, reuse, and value-adding provide an efficient method for reducing them while protecting the environment. National and international organizations establish policies aimed at decreasing trash output and enhancing the recyclable and reusable nature of the garbage produced. Implementing waste management policies that support the decrease in waste creation, reuse, recycling, and recovery can ultimately protect health of people and the environment while also supporting the circular economy idea (Gaur et al. 2020).

34.3 Industrial Waste to Energy

The World Bank projects that the amount of garbage produced per year would reach 2.01 billion metric tons; by 2050, this amount might rise to 3.40 billion metric tons. The majority of waste is buried in landfills, burnt, chemically processed, or discharged into streams. Forty percent of waste is just thrown in the trash without being cleaned up. The environment and human health are at risk due to improper waste disposal, which contaminates the air, water, and soil and encourages the spread of illness. Following new regulations and technological advancements that emphasize generating energy from garbage, people are becoming more conscious of the productivity of different waste products. Nevertheless, a few disadvantages lower waste management effectiveness and environmental sustainability. A substantial amount of the world's trash is made up of agro-industrial waste, which is created when agricultural products are processed industrially. Byproducts of liquid and solid processing, such as skins, kernels, pulp, and residual streams, are rich in biomass and contain many vital nutrients. The potential for these wastes to be used as inexpensive raw materials for the production of high-value goods including biofuels, enzymes, dyes, and bioactive compounds is encouraging.

New waste-reuse techniques that are more durable and economical than previous ones have been made possible by industrial biotechnology. Agro-industrial byproducts can be biotransformed via submerged and solid-state fermentation to create products that can be sold. Large-scale manufacturing of better products offers a solution to solve environmental issues and repurpose trash, which in turn promotes the green economy and ensures long-term sustainability. Circular economies (CEs) have been proposed as a workable way to move away from linear systems and towards more circular methods, addressing the planet's major problems with waste production and resource usage. In spite of their apparent benefits, completely circular systems are not necessarily self-sufficient. Based on the idea that moving to an economy based on renewable resources might lessen negative environmental consequences, the circular bioeconomy (CBE) concept was created. High-value bio-based products including food, feed, medications, biochemicals, and electricity are produced and transformed from sustainable natural resources to fuel the bio economy. Fertilizer, feed, biofuels for electricity and transportation, thermal energy, and building materials all depend on the boldness of organic matter. Nevertheless, the COVID-19 pandemic's effects on the three pillars of sustainability, especially those in the BE sector, have presented difficulties for all businesses recently. A thorough analysis of the procedures involved in recovery and recycling was conducted along with a CBE plan for food waste. According to reports, the main goals are to create jobs, minimize financial losses, manage resources more effectively, and have the power to change the behavior of stakeholders.

The utilization of microbial factories to transform agro-industry biowaste into biomaterials is a potential part of the circular economy. Current developments and the prospect of incorporating biowaste into the circular bioeconomy are also highlighted. A CE theory that takes this into account and stresses the value of inherent wastefulness for ecologically conscious crop management that makes use of renewable resources is put out. Using economically feasible techniques to decrease trash output and raise waste value, the agriculture sector has embraced a circular economy to better utilize the waste generated. Moreover, processing agricultural waste may provide a number of useful metabolites, energy, and minerals that may be sold to advance bioproduct technology. Waste is said to have undergone biological treatment if living organisms have controlled its transformation. Among these are biochemical conversion mechanisms as well. Biochemical processes are significantly slower than thermochemical changes, but they require a lot less energy to occur. Biofuels can be produced by conventional biological and chemical processes such photobiological processes, fermentation of alcohol, and anaerobic digestion. Figure 34.1 shows the various sources of biowaste and technologies implemented for the generation of energy from waste.

Biowaste primarily consists of biomass, which breaks down in both aerobic and anaerobic conditions. Encouraging living circumstances and environmental protection depend on effective biowaste management. Additionally, it has the ability to generate revenue while simultaneously resolving waste and energy management problems when paired with value addition. The third most significant pollutants are paper and pulp. The wood connections are broken by pulping. To turn wood into pulp, the paper industry uses both mechanical and chemical processes. Pulp quality and yield are impacted by the pulping process. Paper pulp used to make tissue and newspapers is generated mechanically. Paper, photographic film, and specialized rayon are all made using sulfite pulping. The process of making containerboard pulp is chemo-mechanical. Diterpenes, unsaturated fatty acids, juvabione, and

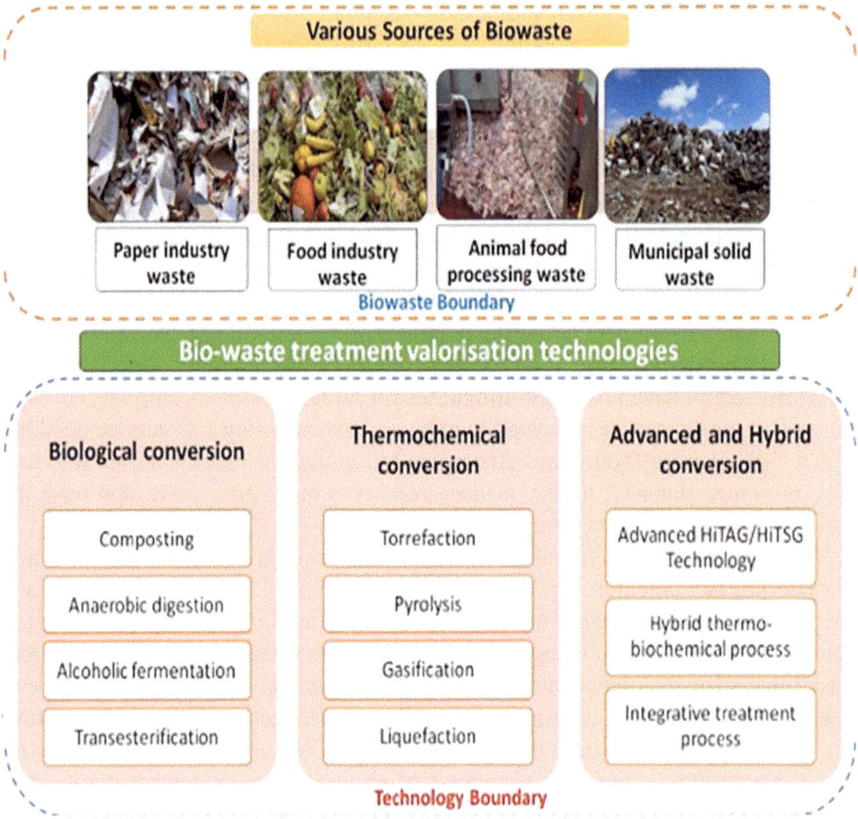

Fig. 34.1 Technologies for waste to energy generation (Mishra et al. 2023)

chlorinated resin acids are produced during the pulping process. Solid waste is produced by deinking, pulping, and wastewater treatment. For instance, 40–50 kg of sludge is produced from 1 ton of paper. If disposed of in the open, these industrial wastes rich in organic matter might damage ecosystems on land and in the ocean. These wastes can be put to immediate use in waste-to-energy systems such as anaerobic digestion.

Restaurants, grocery stores, and food manufacturing facilities have multiplied to keep up with the population growth. Food waste has accumulated as waste management techniques have advanced. Large volumes of garbage are produced by breweries, meat processing facilities, confectionery companies, and vegetable oil refineries. Filter sludge, pulp, fruit peels, and non-standard fruits are examples of solid wastes. On the other hand, starch, sugars, and organic solid residues make up the liquid wastes produced during the washing of fruits, vegetables, and meat. Cooking oils are thrown away after a few uses. It is anticipated that 18 million tons of used cooking oil would be produced annually by 2020. It is thrown into the environment straight since it does not dissolve in water, which is incredibly damaging to the ecology. Food waste is transformed into biofuels, enzymes, and nutraceuticals by biorefineries. Animal waste from dairy and poultry farms in developing nations is often produced as manure or leftovers of meat preparation. Meat treatment facilities dispose of animal wastes that include meat, bones, feathers, fur, and tallow remnants. These include items containing a large amount of organic materials, which can give off an odor and, if left untreated, encourage the growth of dangerous germs. Methane, a gas that is more dangerous than CO_2, is produced by this natural breakdown process. Furthermore, groundwater might get contaminated by animal waste tank discharge. Because of this, the process of turning animal waste into biomaterials has accelerated in recent years (Mishra et al. 2023).

Municipal solid waste (MSW) production in cities has increased due to rapidly expanding populations, making this issue increasingly important for both the day-to-day operations and long-term viability of cities. A person living in a poor or growing country generates 100–400 kg of MSW each year. Every 10 years, MSW production doubles, with 2.2 billion tons predicted by 2025 and 4.2 billion tons by 2050. There is evidence of MSW mistreatment in China, Bangladesh, India, and Thailand. Developing and developing nations' governments and waste management organizations face difficulties. Landfills, recycling, and biological and thermal treatments are the methods used to handle MSW. In order to recover energy, waste-to-energy combines landfill and waste combustion processes (Mishra et al. 2023; Tade et al. 2020). Leading Asian cities must create waste management strategies that mitigate climate change since it is a worldwide issue with growing urgency. Reducing health concerns, energy recovery, and greenhouse gas emissions would all be aided by the use of waste to energy (WTE) technologies that are tailored to the local environment. In order to achieve the co-benefits of the environment and economy, energy recovery efficiency is a critical factor (Menikpura et al. 2016). Planning for WTE heat and electricity is difficult, particularly because of heterogeneous waste. Over time, the qualities (lower heating value and composition) change. Furthermore, the external consumer receives live steam from the WTE plant (Touš et al. 2015).

Thermal treatment with energy recovery and conversion, incoming waste storage and pretreatment, wastewater treatment, flue gas treatment, ash and slag management, and so on can all be included in the basic linear framework of a WTE plant (Margallo et al. 2014). Growing populations and urbanization have resulted in more trash production. Waste production is rising more quickly than urbanization worldwide (Chen 2018). Rising technological breakthroughs, a big market potential, and favorable regulations imposed by national and municipal governments all give sufficient impetus for continuing expansion. However, the volume and quality of MSW, rising public opposition, and the sustainable development issue are the key obstacles to the waste-to-energy incineration industry's healthy and sustainable expansion in China (Cui et al. 2020). Numerous of these wastes have low concentrations of dissolved materials and suspended particles, which not only produce unpleasant odors and other moldering gases that are visually unpleasant, but also have negative environmental effects because they leach in landfills (Pham et al. 2015).

China's WTE incinerator facility, which employs incineration to turn municipal solid waste (MSW) into power, emerges as a viable option. In order to accomplish waste reduction, recycling, and harmlessness, it uses waste materials and converts them into power. This offers significant benefits for the environment and the economy and might meet the requirements of the circular economy. Most big and medium-sized cities employ fixed-site collection, which is the practice of homeowners placing bagged and sorted waste into designated rubbish bins outside their homes or by the side of the road (Xin-gang et al. 2016). Every process equipment designer's primary goal is to increase efficiency while reducing costs. Thus, it makes sense that apparatus with such a broad use as plate heat exchangers has garnered significant interest from scholars. Despite this, there are still fresh opportunities to get more performance out of them (Stehlík 2009).

Every process equipment designer's primary goal is to increase efficiency while reducing costs. Thus, it makes sense that apparatus with such a broad use as plate heat exchangers has garnered significant interest from scholars. Despite this, there are still fresh opportunities to get more performance out of them (Tabasová et al. 2012). If sufficient supervision is not provided for optimal resource utilization, there may be a combination of increased waste and pollution and economic loss. It may have a detrimental impact on the economy, as well as have an adverse effect on human health and cause significant environmental damage due to its hazardous emissions. As a result, it is crucial to continuously and thoroughly monitor resource recovery and utilization. On the other hand, the nation can guarantee both energy security and sustainable development if the waste's energy potential is effectively used (Bajic et al. 2015).

34.4 Recovery of Valuable Products from the Industrial Waste

Industrial wastes are not biodegradable, which makes their generation of solid and liquid wastes a significant problem. Hazardous and non-biodegradable trash make up between 10% and 15% of all industrial waste, and this waste category is predicted to expand at an annual rate. Fertilizer companies, chemical, pharmaceutical, and dye industries, iron and steel plants, fly ash, gypsum, glass products, plastics, and radioactive waste are all producers of silver foil. The harm these wastes do to ecosystems is still present. While biological, physical, and chemical methods may easily extract organic material from wastewater, the aforementioned approaches have trouble removing or separating inorganic metals because of their intrinsic characteristics, which include solubility and the propensity to form metal complexes (Krishnan et al. 2021). The production of biomaterials and the energy generation from biomass derived from agro-industrial waste have gained ground, showing a positive development trend. Agro-industrial wastes may be used to create a variety of goods and new investment possibilities that address environmental and economic problems (Freitas et al. 2021).

Chemical recycling turns waste polymers into feedstock for the production of chemicals, monomers, and fuels, whereas mechanical recycling turns waste polymeric materials into new polymer goods when the energy contained in plastics is released by burning. The impact of reaction parameters on products that are produced, agents and catalysts that are employed, and equipment that is used for certain chemical recycling techniques (Datta and Kopczyńska 2016). Advanced oxidation processes (AOPs) are widely employed for wastewater treatment because of their ability to degrade a wide spectrum of organic contaminants. Several variables can influence dye degradation/remediation, including pH, temperature, dye concentration, nutrition, salts, agitation, and oxygen. Optimization studies are required for the construction of industrial-scale treatment systems (Varjani et al. 2021). Alkaline leaching of vanadium, filtration, neutralization of sodium vanadate solution, precipitation of ammonium metavanadate, drying of the precipitate, and adsorption of residual vanadium from the filtrate on an anion are all processes required to recover vanadium from residue ashes after burning heavy oil fractions. Nickel is recovered from the residual ashes by acidic leaching, filtering, precipitating ammonium-nickelous sulphate, filtering, and drying (Chmielewski et al. 1997).

The anaerobic digestion of chrome-free tanning sludges (combined with pulverized shreds) can recover energy, with biogas yields equivalent to those achieved from home sludges. Composting may be used to treat digested tanning sludge and shreds, which are sometimes combined with pre-sorted residential garbage (Collivignarelli and Barducci 1984). Alum's high solids content in black liquor means that the amount of coagulant used has an impact on the solid process. Hydrochloric acid is used to change the pH of the solution and get rid of most of the particles from the black liquid. Silica may be extracted from other materials using sodium hydroxide since it generates sodium silicate (Masoud et al. 2017). Energy recovery from trash is crucial to the ecological management of waste materials in order to reduce waste-related issues and to promote the idea of interdisciplinary economic advantages (Ankush and Dubey 2020).

The chemical oxygen demand (COD) level of tannery wastewater can be decreased by projecting an appropriate treatment strategy. In order to generate reusable water, this method can involve sophisticated photochemical and biological treatment, followed by nanofiltration to separate and concentrate phosphate, chromium, and other impurities, and microfiltration to remove suspended particles. The economics of the treatment process can be improved by recovering value-added products from the concentrated wastewater treated by nanofiltration (Kannaujiya et al. 2019). The high alumina content of aluminum dose is used by modern methods to produce cement and concrete bricks. Aluminum dose reacts with water to produce hydrogen as well as other gases like methane and ammonia. Since very little gas is created, new techniques for using these gases must be developed, properly accounting for the solid residue that is formed following the reaction with water (Meshram and Singh 2018). Figure 34.2 displays the goods with additional value made from waste resources (Mishra et al. 2023).

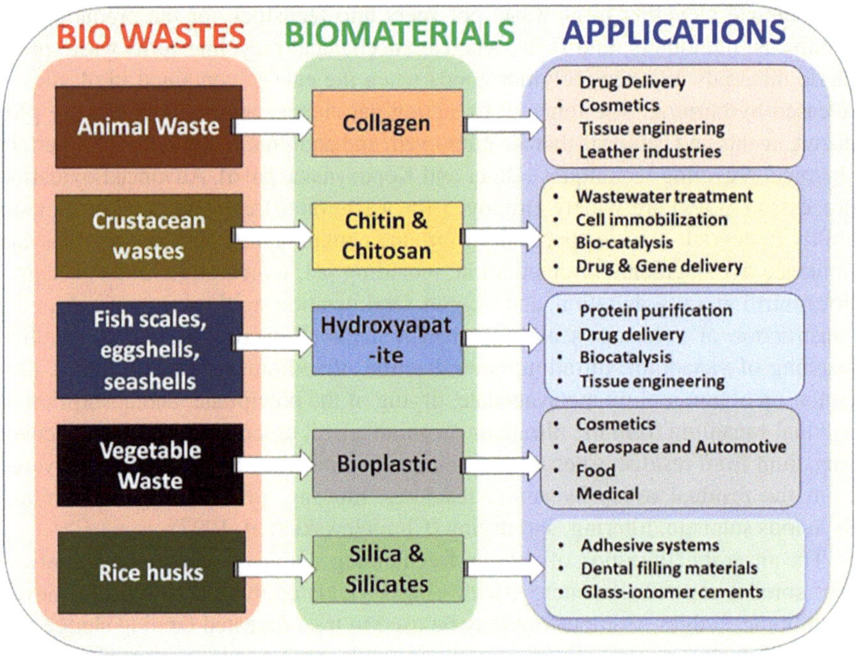

Fig. 34.2 Value added products from waste materials (Mishra et al. 2023)

34.5 Economic, Environmental, and Social Effects for Implementing Circular Economy from the Industrial Waste

Waste is a major contributor to the climate catastrophe, and governments need to solve this. One of these effective approaches to trash management is recycling. Since recycling is connected to the planet's atmosphere, it should consider the most recent recycling technology. There might be major advantages to improvements in waste collection, sorting, transportation, and final disposal. Linking the waste management process to community involvement and stakeholder involvement is crucial for increasing garbage sorting and recycling. It may spend more in garbage sorting facilities like collecting containers, transport vehicles, and transfer stations if it had additional financial support from public-private partnerships (Sasmoko et al. 2022). Given the centrality of consumers in a CE, social consciousness plays a critical role in the successful shift from a linear economy to one. The social awareness movement is still running strong, and in the last several years, there has been a notable growth in educational initiatives, public awareness campaigns, and seminars. Thus far, the public institutions have provided strong backing for this movement, which is also increasingly garnering support from the industry. Raising awareness necessitates a shift in perspective so that people stop thinking in terms of brand-new or used things and instead concentrate on the functionality and suitability of products (Lieder and Rashid 2016).

A company must adopt cleaner production methods in order to get CE; producers and customers must also be more accountable and conscious; if possible, renewable materials and technology must be utilized; thus it is necessary to implement suitable, dependable, and well-defined regulations and tools (Ghisellini et al. 2016). Redesigning a resource's life cycle to guarantee that all products are recycled and that no trash is transported to landfills, incinerators, or other conventional disposal techniques allows waste to be reused and repurposed. This is known as a zero-waste approach (Yang et al. 2023a, b). Using natural ecosystems as models, a novel approach to the industrial system aims to convert industrial operations into cyclical processes. It views the biosphere, of which the industrial system is a subsystem, as the source of its resources and services. As such, it emphasizes industrial system design ideas like dematerialization, energy efficiency, and the closing of material and energy loops (Bruel et al. 2019). The term "clean energy" also describes businesses that are moving towards renewable energy, which eliminates waste and hazardous chemicals. The benefits of this change are attributed to better systems, materials, products, and business models (Kyriakopoulos et al. 2019).

One of the biggest benefits of partially substituting discarded bricks for cement in powder form is that it can help decarbonize the construction industry. Because alkaline activators have detrimental impacts on human health, their use can significantly diminish the potential of alkali-activated materials (Fořt and Černý 2020). By focusing on reuse and recyclability, a CE may lengthen product life cycles and drastically change the present economic structure in the direction of sustainable growth. It also slows down and closes resource flows (Mies and Gold 2021). Fundamentally, clean energy affects access, cost, and the consumption of resources, energy, commodities, and services. It also has an effect on the employment it creates inside the "economic system." Among others, the European Commission praised the clean energy policy as a chance for social cohesiveness and integration (Vanhuyse et al. 2021). The idea of sustainable degrowth refers to a sustainable decrease in society's throughput. The author asserts that putting this concept into practice will inevitably improve the environment. In the past 10 years, there has been discussion over the distinction between steady-state economies and economic degrowth (Adami and Schiavon 2021).

References

Adami L, Schiavon M (2021) From circular economy to circular ecology: a review on the solution of environmental problems through circular waste management approaches. Sustain For 13(2):925. https://doi.org/10.3390/su13020925

Ankush K, Dubey KK (2020) Food industry waste biorefineries: future energy, valuable recovery, and waste treatment. In: Refining biomass residues for sustainable energy and bioproducts. Elsevier, Amsterdam, pp 391–406. https://doi.org/10.1016/B978-0-12-818996-2.00017-X

Appolloni A, Chiappetta Jabbour CJ, D'Adamo I, Gastaldi M, Settembre-Blundo D (2022) Green recovery in the mature manufacturing industry: the role of the green-circular premium and sustainability certification in innovative efforts. Ecol Econ 193:107311. https://doi.org/10.1016/j.ecolecon.2021.107311

Arushanyan Y, Björklund A, Eriksson O, Finnveden G, Ljunggren Söderman M, Sundqvist J-O, Stenmarck Å (2017) Environmental assessment of possible future waste management scenarios. Energies 10(2):2. https://doi.org/10.3390/en10020247

Bajic BZ, Dodić SN, Vucurovic DG, Dodic JM, Grahovac JA (2015) Waste-to-energy status in Serbia. Renew Sust Energ Rev 50:1437–1444. https://doi.org/10.1016/j.rser.2015.05.079

Bhat RA, Dar SA, Dar DA, Dar G (2018) Municipal solid waste generation and current scenario of its management in India. Int J Adv Res Sci Eng 7(2):419

Borthakur A, Sinha K (2013) Generation of electronic waste in India: current scenario, dilemmas and stakeholders. Afr J Environ Sci Technol 7(9):9

Bruel A, Kronenberg J, Troussier N, Guillaume B (2019) Linking industrial ecology and ecological economics: a theoretical and empirical foundation for the circular economy. J Ind Ecol 23(1):12–21. https://doi.org/10.1111/jiec.12745

Chen Y-C (2018) Evaluating greenhouse gas emissions and energy recovery from municipal and industrial solid waste using waste-to-energy technology. J Clean Prod 192:262–269. https://doi.org/10.1016/j.jclepro.2018.04.260

Chmielewski AG, Urbański TS, Migdał W (1997) Separation technologies for metals recovery from industrial wastes. Hydrometallurgy 45(3):333–344. https://doi.org/10.1016/S0304-386X(96)00090-4

Collivignarelli C, Barducci G (1984) Waste recovery from the tanning industry. Waste Manag Res 2(3):265–278. https://doi.org/10.1016/0734-242X(84)90031-4

Cui C, Liu Y, Xia B, Jiang X, Skitmore M (2020) Overview of public-private partnerships in the waste-to-energy incineration industry in China: status, opportunities, and challenges. Energ Strat Rev 32:100584. https://doi.org/10.1016/j.esr.2020.100584

Dasgupta D, Debsarkar A, Hazra T, Bala BK, Gangopadhyay A, Chatterjee D (2017) Scenario of future e-waste generation and recycle-reuse-landfill-based disposal pattern in India: a system dynamics approach. Environ Dev Sustain 19(4):1473–1487. https://doi.org/10.1007/s10668-016-9815-6

Datta J, Kopczyńska P (2016) From polymer waste to potential main industrial products: actual state of recycling and recovering. Crit Rev Environ Sci Technol 46(10):905–946. https://doi.org/10.1080/10643389.2016.1180227

Duan H, Huang Q, Wang Q, Zhou B, Li J (2008) Hazardous waste generation and management in China: a review. J Hazard Mater 158(2–3):221–227. https://doi.org/10.1016/j.jhazmat.2008.01.106

Fatimah YA, Govindan K, Murniningsih R, Setiawan A (2020) Industry 4.0 based sustainable circular economy approach for smart waste management system to achieve sustainable development goals: a case study of Indonesia. J Clean Prod 269:122263. https://doi.org/10.1016/j.jclepro.2020.122263

Fořt J, Černý R (2020) Transition to circular economy in the construction industry: environmental aspects of waste brick recycling scenarios. Waste Manag 118:510–520. https://doi.org/10.1016/j.wasman.2020.09.004

Freitas LC, Barbosa JR, Da Costa ALC, Bezerra FWF, Pinto RHH, Carvalho Junior RND (2021) From waste to sustainable industry: how can agro-industrial wastes help in the development of new products? Resour Conserv Recycl 169:105466. https://doi.org/10.1016/j.resconrec.2021.105466

Gaur VK, Sharma P, Sirohi R, Awasthi MK, Dussap C-G, Pandey A (2020) Assessing the impact of industrial waste on environment and mitigation strategies: a comprehensive review. J Hazard Mater 398:123019. https://doi.org/10.1016/j.jhazmat.2020.123019

Ghisellini P, Cialani C, Ulgiati S (2016) A review on circular economy: the expected transition to a balanced interplay of environmental and economic systems. J Clean Prod 114:11–32. https://doi.org/10.1016/j.jclepro.2015.09.007

Girotto F, Alibardi L, Cossu R (2015) Food waste generation and industrial uses: a review. Waste Manag 45:32–41. https://doi.org/10.1016/j.wasman.2015.06.008

Kandasamy J, Kinare YP, Pawar MT, Majumdar A, Agrawal R (2022) Circular economy adoption challenges in medical waste management for sustainable development: an empirical study. Sustain Dev 30(5):958–975. https://doi.org/10.1002/sd.2293

Kannaujiya MC, Mandal T, Mandal DD, Mondal MK (2019) Treatment of leather industry wastewater and recovery of valuable substances to solve waste management problem in environment. In: Bharagava RN (ed) Environmental contaminants: ecological implications and management, vol 14. Springer, Singapore, pp 311–340. https://doi.org/10.1007/978-981-13-7904-8_14

Khandelwal U, Bondre S, Jain MS, Singh KR (2023) Present scenario of E-waste generation, legislation, management and impacts: a review in Indian context. AIP Conf Proc 2721(1):070081. https://doi.org/10.1063/5.0153906

Kofoworola OF, Gheewala SH (2009) Estimation of construction waste generation and management in Thailand. Waste Manag 29(2):731–738. https://doi.org/10.1016/j.wasman.2008.07.004

Krishnan S, Zulkapli NS, Kamyab H, Taib SM, Din MFBM, Majid ZA, Chaiprapat S, Kenzo I, Ichikawa Y, Nasrullah M, Chelliapan S, Othman N (2021) Current technologies for recovery of metals from industrial wastes: an overview. Environ Technol Innov 22:101525. https://doi.org/10.1016/j.eti.2021.101525

Kyriakopoulos G, Kapsalis V, Aravossis K, Zamparas M, Mitsikas A (2019) Evaluating circular economy under a multi-parametric approach: a technological review. Sustain For 11(21):6139. https://doi.org/10.3390/su11216139

Lebreton L, Andrady A (2019) Future scenarios of global plastic waste generation and disposal. Palgrave Commun 5:6. https://doi.org/10.1057/s41599-018-0212-7

Lieder M, Rashid A (2016) Towards circular economy implementation: a comprehensive review in context of manufacturing industry. J Clean Prod 115:36–51. https://doi.org/10.1016/j.jclepro.2015.12.042

Margallo M, Dominguez-Ramos A, Aldaco R, Bala A, Fullana P, Irabien A (2014) Environmental sustainability assessment in the process industry: a case study of waste-to-energy plants in Spain. Resour Conserv Recycl 93:144–155. https://doi.org/10.1016/j.resconrec.2014.09.014

Masoud MS, Farag HA, Zaatout AA, Kamel HM, ElGreatly MH, Ahmed HM (2017) Recovery of some valuable components from industrial waste. Sep Sci Technol 52(3):512–519. https://doi.org/10.1080/01496395.2016.1251462

Menikpura SNM, Sang-Arun J, Bengtsson M (2016) Assessment of environmental and economic performance of waste-to-energy facilities in Thai cities. Renew Energy 86:576–584. https://doi.org/10.1016/j.renene.2015.08.054

Meshram A, Singh KK (2018) Recovery of valuable products from hazardous aluminum dross: a review. Resour Conserv Recycl 130:95–108. https://doi.org/10.1016/j.resconrec.2017.11.026

Mies A, Gold S (2021) Mapping the social dimension of the circular economy. J Clean Prod 321:128960. https://doi.org/10.1016/j.jclepro.2021.128960

Mishra B, Mohanta YK, Reddy CN, Reddy SDM, Mandal SK, Yadavalli R, Sarma H (2023) Valorization of agro-industrial biowaste to biomaterials: an innovative circular bioeconomy approach. Circ Econ 2(3):100050. https://doi.org/10.1016/j.cec.2023.100050

Nagapan S, Rahman IA, Asmi A (2012) Factors contributing to physical and non-physical waste generation in construction industry. Int J Adv Appl Sci 1(1):1. https://doi.org/10.11591/ijaas.v1.i1.pp1-10

Negrete-Cardoso M, Rosano-Ortega G, Álvarez-Aros EL, Tavera-Cortés ME, Vega-Lebrún CA, Sánchez-Ruíz FJ (2022) Circular economy strategy and waste management: a bibliometric analysis in its contribution to sustainable development, toward a post-COVID-19 era. Environ Sci Pollut Res 29(41):61729–61746. https://doi.org/10.1007/s11356-022-18703-3

Nelles M, Grünes J, Morscheck G (2016) Waste management in Germany—development to a sustainable circular economy? Procedia Environ Sci 35:6–14. https://doi.org/10.1016/j.proenv.2016.07.001

Ogunmakinde OE, Egbelakin T, Sher W (2022) Contributions of the circular economy to the UN sustainable development goals through sustainable construction. Resour Conserv Recycl 178:106023. https://doi.org/10.1016/j.resconrec.2021.106023

Pham TPT, Kaushik R, Parshetti GK, Mahmood R, Balasubramanian R (2015) Food waste-to-energy conversion technologies: current status and future directions. Waste Manag 38:399–408. https://doi.org/10.1016/j.wasman.2014.12.004

Rajput R, Prasad G, Chopra AK (2009) Scenario of solid waste management in present Indian context. Caspian J Environ Sci 7(1):45–53

Sanguino R, Barroso A, Fernández-Rodríguez S, Sánchez-Hernández MI (2020) Current trends in economy, sustainable development, and energy: a circular economy view. Environ Sci Pollut Res 27(1):1–7. https://doi.org/10.1007/s11356-019-07074-x

Sasmoko Z, Malik M, Awan U, Handayani W, Jabor MK, Asif M (2022) Environmental effects of bio-waste recycling on industrial circular economy and eco-sustainability. Recycling 7(4):60. https://doi.org/10.3390/recycling7040060

Schroeder P, Anggraeni K, Weber U (2019) The relevance of circular economy practices to the sustainable development goals. J Ind Ecol 23(1):77–95. https://doi.org/10.1111/jiec.12732

Senthilkumar K, Naveen Kumar M (2020) Generation of bioenergy from industrial waste using microbial fuel cell technology for the sustainable future. In: Kumar RP, Gnansounou E, Raman JK, Baskar G (eds) Refining biomass residues for sustainable energy and bioproducts. Academic, New York, pp 183–193. https://doi.org/10.1016/B978-0-12-818996-2.00008-9

Sharma KD, Jain S (2020) Municipal solid waste generation, composition, and management: the global scenario. Soc Responsibility J 16(6):917–948. https://doi.org/10.1108/SRJ-06-2019-0210

Singh RP, Tyagi VV, Allen T, Ibrahim MH, Kothari R (2011) An overview for exploring the possibilities of energy generation from municipal solid waste (MSW) in Indian scenario. Renew Sust Energ Rev 15(9):4797–4808. https://doi.org/10.1016/j.rser.2011.07.071

Stehlík P (2009) Contribution to advances in waste-to-energy technologies. J Clean Prod 17(10):919–931. https://doi.org/10.1016/j.jclepro.2009.02.011

Suárez-Eiroa B, Fernández E, Méndez-Martínez G, Soto-Oñate D (2019) Operational principles of circular economy for sustainable development: linking theory and practice. J Clean Prod 214:952–961. https://doi.org/10.1016/j.jclepro.2018.12.271

Tabasová A, Kropáč J, Kermes V, Nemet A, Stehlík P (2012) Waste-to-energy technologies: impact on environment. Energy 44(1):146–155. https://doi.org/10.1016/j.energy.2012.01.014

Tade RS, Nangare SN, Patil PO (2020) Agro-industrial waste-mediated green synthesis of silver nanoparticles and evaluation of its antibacterial activity. Nano Biomed Eng 12(1):57–66. https://doi.org/10.5101/nbe.v12i1.p57-66

Tahir DM, Hussain DT, Behaylu A (2015) Scenario of present and future of solid waste generation in India: a case study of Delhi mega City

Touš M, Pavlas M, Putna O, Stehlík P, Crha L (2015) Combined heat and power production planning in a waste-to-energy plant on a short-term basis. Energy 90:137–147. https://doi.org/10.1016/j.energy.2015.05.077

Vanhuyse F, Fejzić E, Ddiba D, Henrysson M (2021) The lack of social impact considerations in transitioning towards urban circular economies: a scoping review. Sustain Cities Soc 75:103394. https://doi.org/10.1016/j.scs.2021.103394

Varjani S, Rakholiya P, Shindhal T, Shah AV, Ngo HH (2021) Trends in dye industry effluent treatment and recovery of value added products. J Water Process Eng 39:101734. https://doi.org/10.1016/j.jwpe.2020.101734

Wu H, Duan H, Zheng L, Wang J, Niu Y, Zhang G (2016) Demolition waste generation and recycling potentials in a rapidly developing flagship megacity of South China: prospective scenarios and implications. Constr Build Mater 113:1007–1016. https://doi.org/10.1016/j.conbuildmat.2016.03.130

Xin-gang Z, Gui-wu J, Ang L, Yun L (2016) Technology, cost, a performance of waste-to-energy incineration industry in China. Renew Sust Energ Rev 55:115–130. https://doi.org/10.1016/j.rser.2015.10.137

Yaduvanshi NR, Krishnamurthy RM, S. (2016) Circular economy for sustainable development in India. Indian J Sci Technol 9(46):1–9. https://doi.org/10.17485/ijst/2016/v9i46/107325

Yang M, Chen L, Wang J, Msigwa G, Osman AI, Fawzy S, Rooney DW, Yap P-S (2023a) Circular economy strategies for combating climate change and other environmental issues. Environ Chem Lett 21(1):55–80. https://doi.org/10.1007/s10311-022-01499-6

Yang C-K, Ma H-W, Liu K-H, Yuan M-H (2023b) Measuring circular economy transition potential for industrial wastes. Sustainable Prod Consump 40:376–388. https://doi.org/10.1016/j.spc.2023.06.013

Correction to: Valorization of Solid Wastes to Biofuels and Chemical Products for Sustainable World

Abha Kumari, Monika Prakash Rai,
Ashokkumar Veeramuthu, and Abha Mishra

Correction to:
A. Kumari et al. (eds.), *Valorization of Solid Wastes to Biofuels and Chemical Products for Sustainable World*
https://doi.org/10.1007/978-981-96-8594-3

The original version of this book was inadvertently published without keywords in Chaps. 1, 5, 6, 7, 9, 13, 15, 19, and 33. These chapters have now been updated with keywords in the online version.

The updated version of this book can be found at
https://doi.org/10.1007/978-981-96-8594-3

A. Kumari et al. (eds.), *Valorization of Solid Wastes to Biofuels and Chemical Products for Sustainable World*, https://doi.org/10.1007/978-981-96-8594-3_35

MIX
Papier aus verantwortungsvollen Quellen
Paper from responsible sources
FSC® C105338

If you have any concerns about our products,
you can contact us on
ProductSafety@springernature.com

In case Publisher is established outside the EU,
the EU authorized representative is:
Springer Nature Customer Service Center GmbH
Europaplatz 3, 69115 Heidelberg, Germany

Printed by Libri Plureos GmbH
in Hamburg, Germany